Encyclopedia of
Plant Physiology

New Series   Volume 6

Editors

A. Pirson, Göttingen
M. H. Zimmermann, Harvard

Photosynthesis II

Photosynthetic Carbon Metabolism and Related Processes

Edited by

M. Gibbs and E. Latzko

Contributors

T. Akazawa L.E. Anderson J.A. Bassham E. Beck
S. Beer A. Ben-Amotz I.F. Bird C.C. Black P. Böger
B.B. Buchanan D.T. Canvin N.G. Carr E.A. Chapman
J. Coombs P. Dittrich G.E. Edwards E.F. Elstner
E.H. Evans M. Gibbs D. Graham M.G. Guerrero
H.W. Heldt S.C. Huber R.G. Jensen G.J. Kelly
A.J. Keys M. Kluge E. Latzko P.J. Lea C. Levi
R. McC. Lilley B.J. Miflin S. Miyachi E. Ohmann
R.P. Poincelot J. Preiss B. Quebedeaux O. Queiroz
T.B. Ray A. Schmidt A. Shomer-Ilan W.D.P. Stewart
N.E. Tolbert J.H. Troughton B. Vennesland
N.P. Voskresenskaya Y. Waisel D.A. Walker
C.P. Whittingham W. Wiessner I. Zelitch

Springer-Verlag Berlin Heidelberg New York 1979

Professor MARTIN GIBBS
Institute for Photobiology
of Cells and Organelles
Brandeis University
Waltham, MA 02254/USA

Professor Dr. ERWIN LATZKO
Westfälische Wilhelms-Universität
Botanisches Institut
Schloßgarten 3
4400 Münster/FRG

With 75 Figures

ISBN 3-540-09288-9 Springer-Verlag Berlin Heidelberg New York
ISBN 0-387-09288-9 Springer-Verlag New York Heidelberg Berlin

Library of Congress Cataloging in Publication Data. Main entry under title: Photosynthesis II. (Encyclopedia of plant physiology; new series, v.6) Bibliography: p. Includes index. 1. Photosynthesis. 2. Carbon-Metabolism. I. Gibbs, Martin, 1922 –. II. Latzko, E., 1924 –. III. Series. QK 711.2.E5 vol. 6 [QK 882] 581.1′08s [581.1′3342] 79-13944.

Typesetting, printing and bookbinding: Universitätsdruckerei H. Stürtz AG, Würzburg
2131/3130-543210

Contents

I. Introduction

II. CO_2 Assimilation

II A. The Reductive Pentose Phosphate Cycle

1. The Reductive Pentose Phosphate Cycle and Its Regulation
J.A. BASSHAM (With 1 Figure)

2. The Isolation of Intact Leaf Cells, Protoplasts and Chloroplasts
R.G. JENSEN

3. Studies with the Reconstituted Chloroplast System
R.McC. LILLEY and D.A. WALKER (With 1 Figure)

4. Autotrophic Carbon Dioxide Assimilation in Prokaryotic Microorganisms
E. OHMANN (With 1 Figure)

5. Light-Enhanced Dark CO_2 Fixation
S. MIYACHI (With 5 Figures)

II B. The C_4 and Crassulacean Acid Metabolism Pathways

6. The C_4 Pathway and Its Regulation
T.B. RAY and C.C. BLACK (With 4 Figures)

10. $\delta^{13}C$ as an Indicator of Carboxylation Reactions
J.H. TROUGHTON

II C. Factors Influencing CO_2 Assimilation

11. Interactions Between Photosynthesis and Respiration in Higher Plants
D. GRAHAM and E.A. CHAPMAN (With 2 Figures)

II D. Regulation and Properties of Enzymes of Photosynthetic Carbon Metabolism

II E. Metabolism of Primary Products of Photosynthesis

23. Metabolism of Starch in Leaves
J. PREISS and C. LEVI (With 2 Figures)

24. The Enzymology of Sucrose Synthesis in Leaves
C.P. WHITTINGHAM, A.J. KEYS, and I.F. BIRD

II F. Glycolic Acid and Photorespiration

25. Glycolate Synthesis
E. BECK

26. Glycolate Metabolism by Higher Plants and Algae
N.E. TOLBERT (With 1 Figure)

27. Photorespiration: Studies with Whole Tissues
I. ZELITCH (With 2 Figures)

28. Photorespiration: Comparison Between C_3 and C_4 Plants
D.T. CANVIN (With 12 Figures)

III. Ferredoxin-Linked Reactions

1. Transhydrogenase
P. Böger (With 1 Figure)

2. Oxygen Activation and Superoxide Dismutase in Chloroplasts
E.F. Elstner (With 2 Figures)

3. Ferredoxin-Linked Carbon Dioxide Fixation in Photosynthetic Bacteria
B.B. Buchanan (With 2 Figures)

8. Photosynthetic Assimilation of Sulfur Compounds
A. SCHMIDT (With 1 Figure)

9. Hydrogen Metabolism
A. BEN-AMOTZ (With 1 Figure)

List of Contributors

T. AKAZAWA
Research Institute for
Biochemical Regulation
School of Agriculture
Nagoya University
Chikusa, Nagoya 464/Japan

L.E. ANDERSON
Department of Biological Sciences
University of Illinois at Chicago Circle
Box 4348
Chicago, IL 60680/USA

J.A. BASSHAM
Laboratory of Chemical Biodynamics
Lawrence Berkeley Laboratory
University of California
Berkeley, CA 94720/USA

E. BECK
Lehrstuhl für Pflanzenphysiologie
Universität Bayreuth
8580 Bayreuth
Federal Republic of Germany

S. BEER
Department of Botany
Tel Aviv University
Tel Aviv/Israel

A. BEN-AMOTZ
Israel Oceanographic and
Limnological Research
Tel Shikmona
P.O. Box 8030
Haifa/Israel

I.F. BIRD
Rothamsted Experimental Station
Harpenden, Herts. AL5 2JQ
United Kingdom

C.C. BLACK, Jr.
Biochemistry Department
University of Georgia
Athens, GA 30602/USA

P. BÖGER
Lehrstuhl Physiologie und
Biochemie der Pflanzen
Universität Konstanz
7750 Konstanz
Federal Republic of Germany

B.B. BUCHANAN
Department of Cell Physiology
Hilgard Hall
University of California
Berkeley, CA 94720/USA

D.T. CANVIN
Department of Biology
Queen's University
Kingston, Ontario K7L 3NG/Canada

N.G. CARR
Department of Biochemistry
University of Liverpool
Liverpool, L69 3BX/United Kingdom

E.A. CHAPMAN
Plant Physiology Unit
CSIRO Division of Food Research and
School of Biological Sciences
Macquarie University
North Ryde, N.S.W. 2113/Australia

J. COOMBS
Group Research and Development
Tate and Lyle Ltd.
P.O. Box 68
Reading/United Kingdom

P. DITTRICH
Botanisches Institut der Universität
Menzingerstr. 67
8000 München 19
Federal Republic of Germany

G.E. EDWARDS
Horticulture Department
University of Wisconsin
Madison, WI 53706/USA

E.F. ELSTNER
Institut für Botanik und Mikrobiologie
Technische Universität München
Arcisstraße 21
8000 München 2
Federal Republic of Germany

E.H. EVANS
Biology Division
Preston Polytechnic
Corporation Street
Preston, Lancs./United Kingdom

M. GIBBS
Institute for Photobiology
of Cells and Organelles
Brandeis University
Waltham, MA 02254/USA

D. GRAHAM
Plant Physiology Unit
CSIRO Division of Food Research and
School of Biological Sciences
Macquarie University
North Ryde, N.S.W. 2113/Australia

M.G. GUERRERO
Departamento de Bioquímica
Facultad de Ciencias y C.S.I.C.
Universidad de Sevilla
Sevilla/Spain

H.W. HELDT
Lehrstuhl für Biochemie der Pflanzen
Untere Karspüle 2
3400 Göttingen
Federal Republic of Germany

S.C. HUBER
Crop Sciences Department
USDA-SEA
North Carolina State University
Raleigh, NC 27606/USA

R.G. JENSEN
Department of Biochemistry
University of Arizona
Tucson, AZ 85721/USA

G.J. KELLY
Botanisches Institut
der Universität Münster
Schloßgarten 3
4400 Münster
Federal Republic of Germany

A.J. KEYS
Rothamsted Experimental Station
Harpenden, Herts. AL5 2JQ
United Kingdom

M. KLUGE
Institut für Botanik, Technische Hochschule
Schnittspahnstraße 3–5
6100 Darmstadt
Federal Republic of Germany

E. LATZKO
Botanisches Institut
der Universität Münster
Schloßgarten 3
4400 Münster
Federal Republic of Germany

P.J. LEA
Biochemistry Department
Rothamsted Experimental Station
Harpenden, Herts. AL5 2JQ
United Kingdom

C. LEVI
Department of Biochemistry and
Biophysics
University of California
Davis, CA 95616/USA

R.McC. LILLEY
Department of Biology
The University of Wollongong
P.O. Box 1144
Wollongong
N.S.W. 2500/Australia

B.J. MIFLIN
Biochemistry Department
Rothamsted Experimental Station
Harpenden, Herts. AL5 2JQ
United Kingdom

S. MIYACHI
Institute of Applied Microbiology
University of Tokyo
Bunkyo-ku, Tokyo 113/Japan

E. OHMANN
Sektion Biowissenschaften der
Martin-Luther-Universität
Am Kirchtor 1
402 Halle/Saale
German Democratic Republic

R.P. POINCELOT
Department of Biology
Fairfield University
Fairfield, CT 06430/USA

J. PREISS
Department of Biochemistry
and Biophysics
University of California
Davis, CA 95616/USA

B. QUEBEDEAUX
Central Research and Development
Department
Experimental Station
E.I. du Pont de Nemours & Co
Wilmington, DE 19898/USA

O. QUEIROZ
Laboratoire du Phytotron
Centre National de la Recherche
Scientifique
91190 Gif-sur-Yvette/France

T.B. RAY
Central Research and Development
Department
Experimental Station
E.I. du Pont de Nemours & Co
Wilmington DE 19898/USA

A. SCHMIDT
Botanisches Institut
Universität München
Menzinger Str. 67
8000 München 19
Federal Republic of Germany

A. SHOMER-ILAN
Department of Botany
Tel Aviv University
Tel Aviv/Israel

W.D.P. STEWART
Department of Biological Sciences
University of Dundee
Dundee DD1 4HN/United Kingdom

N.E. TOLBERT
Department of Biochemistry
Michigan State University
East Lansing, MI 48824/USA

J.H. TROUGHTON
Research Division
Ministry of Agriculture and Fisheries
P.O. Box 2298
Wellington/New Zealand

B. VENNESLAND
Forschungsstelle Vennesland
der Max-Planck-Gesellschaft
Harnackstr. 23
1000 Berlin 33

N.P. VOSKRESENSKAYA
K.A. Timiriazev Institute of Plant
Physiology
Academy of Sciences of the USSR
Botanitcheskaya 35,
Moscow, 127106/USSR

Y. WAISEL
Department of Botany
Tel Aviv University
Tel Aviv/Israel

D.A. WALKER
Department of Botany
University of Sheffield
Western Bank, Sheffield S10 2TN/
United Kingdom

C.P. WHITTINGHAM
Rothamsted Experimental Station
Harpenden, Herts. AL5 2JQ
United Kingdom

W. WIESSNER
Pflanzenphysiologisches Institut
Abteilung Experimentelle Phykologie
Untere Karspüle 2
3400 Göttingen
Federal Republic of Germany

I. ZELITCH
Department of Biochemistry
The Connecticut Agricultural
Experiment Station
P.O. Box 1106
New Haven, CT 06504/USA

List of Abbreviations

ABA	abscisic acid
AMP, ADP, ATP	adenosine mono-, di-, triphosphate
ADPG	adenosine diphosphoglucose
APS	adenosine 5′-phosphosulfate
C_3 plants	plants with PGA as the primary product of photosynthetic CO_2 assimilation
C_4 plants	plants with OAA or a related 4C-acid as the first identifiable product of photosynthetic CO_2 assimilation
CAM	crassulacean acid metabolism
CCCP	m-chlorocarbonylcyanide phenylhydrazone
chl	chlorophyll
CMU	3-(p-chlorophenyl)-1,1-dimethylurea
CoA	coenzyme A
DBMIB	2,5-dibromo-3-methyl-6-isopropyl-p-benzoquinone
DCMU	3-(3′,4′-dichlorophenyl)-1,1-dimethylurea
DHAP	dihydroxyacetone phosphate
Diquat	1,1′-ethylene-2,2′-dipyridilium dibromide
DMO	dimethyloxazolidine
DSPD	disalicylidenepropanediamine
DTNB	5,5′-dithio-bis-(2-nitrobenzoic acid)
EDTA	ethylenediaminetetracetic acid
E4P	erythrose 4-phosphate
FAD	flavin adenine dinucleotide
FBP	fructose 1,6-bisphosphate
FCCP	carbonylcyanide p-trifluoromethoxyphenyl-hydrazone
Fd	ferredoxin
FMN	flavin mononucleotide
F6P	fructose 6-phosphate
GAP	glyceraldehyde 3-phosphate
(Glucosyl)$_n$	starch or α-glucan molecule
GMP, GDP, GTP	guanosine mono-, di-, triphosphate
Glu	glutamate

GOGAT	glutamine: α-ketoglutarate aminotransferase
G1P	glucose 1-phosphate
G6P	glucose 6-phosphate
GO	glycolate oxidase
HOQNO	2-n-heptyl-4-hydroxyquinoline-N-oxide
LED	light-enhanced dark CO_2 fixation
LEM	light effect mediator
NAD, NADH	oxidized, and reduced nicotinamide adenine dinucleotide
NADP, NADPH	oxidized, and reduced nicotinamide adenine dinucleotide phosphate
OAA	oxaloacetate
-P	-phosphate
Pi	inorganic orthophosphate
PAPS	3′-phosphoadenosine-5′-phosphosulfate
PEP	phosphoenolpyruvate
PGA	3-phosphoglycerate
pHMB	p-hydroxymercuribenzoate
PIB	post-illumination burst
PPGA	glycerate 1,3-bisphosphate
PPi	inorganic pyrophosphate
R5P	ribose 5-phosphate
RPP cycle	reductive pentose phosphate cycle
Ru5P	ribulose 5-phosphate
RuBP	ribulose 1,5-bisphosphate
SBP	sedoheptulose 1,7-bisphosphate
S7P	sedoheptulose 7-phosphate
SOD	superoxide dismutase
TCA cycle	tricarboxylic acid cycle
TDP	thymidine diphosphate
TN-NAD(P)	oxidized thionicotinamide NAD(P)
TPP	thiamine pyrophosphate
UMP, UDP, UTP	uridine mono-, di-, triphosphate
UDPG	uridine diphosphoglucose
Xu5P	xylulose 5-phosphate

I. Introduction

M. GIBBS and E. LATZKO

In the preface to his *Experiments upon Vegetables*, INGEN-HOUSZ wrote in 1779: "The discovery of Dr. PRIESTLEY that plants have a power of correcting bad air...shows...that the air, spoiled and rendered noxious to animals by their breathing in it, serves to plants as a kind of nourishment." INGEN-HOUSZ then described his own experiments in which he established that plants absorb this "nourishment" more actively in brighter sunlight. By the turn of the eighteenth century, the "nourishment" was recognized to be CO_2. Photosynthetic CO_2 assimilation, the major subject of this encyclopedia volume, had been discovered.

How plants assimilate the CO_2 was a question several successive generations of investigators were unable to answer; scientific endeavor is not a discipline in which it is easy to "put the cart before the horse". The horse, in this case, was the acquisition of radioactive isotopes of carbon, especially ^{14}C. The cart which followed contained the Calvin cycle, formulated by CALVIN, BENSON and BASSHAM in the early 1950's after (a) their detection of glycerate-3-P as the first stable product of CO_2 fixation, (b) their discovery, and that by HORECKER and RACKER, of the CO_2-fixing enzyme RuBP carboxylase, and (c) the reports by GIBBS and by ARNON of an enzyme (NADP-linked GAP dehydrogenase) capable of using the reducing power made available from sunlight (via photosynthetic electron transport) to reduce the glycerate-3-P to the level of sugars. INGEN-HOUSZ's observation of a connection between the plant's nourishment and sunlight could be finally described in cellular terms.

The subject of light-driven electron transport, coupled to phosphorylation and producing O_2, NADPH, and ATP, has been covered in the preceding Volume of this encyclopedia (*Photosynthesis I*, edited by A. TREBST and M. AVRON). This volume takes up where the other left off. The central theme is the Calvin cycle, which utilizes the above-mentioned NADPH and ATP to reduce CO_2 to carbohydrates which are then partitioned as needed for plant growth. Other closely connected assimilatory processes, particularly that of nitrogen, are included. The overall picture is quite extensive (Fig. 1) and clearly illustrates the importance of regulation for each photosynthetic cell to achieve its desired balance between the various metabolic pathways. Regulation is therefore a recurring topic in the pages of this book. The reader is also referred to the chapters by WALKER, HELDT, HATCH and OSMOND, and SCHNARRENBERGER and FOCK in Volume 3 of this encyclopedia (*Transport in Plants III*, edited by C.R. STOCKING and U. HEBER) for excellent outlines of subjects closely related to the present volume.

Many components of Figure 1 were largely unknown during the 1950's, hence it is appropriate to reflect on some of the advances made between the years during which the Calvin cycle was formulated and the present time. The mid-1960's is an appropriate period to base the discussion, since it was during these years that three significant developments, destined to influence research for the following decade, took place. These were (a) the isolation of chloroplasts capable of fixing CO_2 at rates equal to those achieved in vivo, (b) the detection of photorespiration and the discovery of peroxisomes, and (c) the discovery of the C_4 pathway.

Photosynthetically competent chloroplasts were first isolated from pea leaves by WALKER and from spinach leaves by JENSEN and BASSHAM. Experiments with such chloroplasts have provided invaluable clues as to how carbon flow during photosynthesis is controlled. Investigations by HELDT extended the concept of translocators to chloroplasts, so that the products of CO_2 fixation could no longer be imagined to diffuse aimlessly from the chloroplast, but rather only certain intermediates [most notably triose-P and Pi, and dicarboxylic acids (see Fig. 1)] could be visualized to move across the chloroplast envelope in a counter-exchange process. This revelation has had quite an impact on ideas concerning the formation and subsequent utilization of two end-products of photosynthesis, starch and sucrose. It has been known for over 100 years that starch synthesis is restricted to the chloroplast interior, but only recently has it been clarified, mainly by WHITTINGHAM and coworkers at the Rothamsted Agricultural Experiment Station, that the synthesis of sucrose does not take place here, but rather proceeds in the cytosol. At the same time, LEA and MIFLIN, also at Rothamsted, established that the amino acid pool can be rightly claimed as a third end-product of photosynthesis. These workers described a chloroplast enzyme which uses reduced ferredoxin for the reductive transfer of an amino group from glutamine to α-ketoglutarate, thus confirming that the chloroplast is capable of photosynthetically reducing nitrite to amino acids.

One last word about chloroplasts. Almost all the studies to date have been with spinach, and to a lesser extent, pea chloroplasts, and for good reason: these have been the only plants from which good chloroplasts could be consistently isolated using the original mechanical procedure. However, modern developments show that this restriction no longer applies. EDWARDS has introduced the technique of isolating chloroplasts from the fibrous leaves of plants such as wheat by gentle breakage of enzymatically isolated protoplasts. The time for comparative research on chloroplasts from different species has arrived.

A second event of the mid-1960's was the development of an appreciation for the extent and biochemical mechanism of photorespiration. The physiological extent of this process was largely established through gas exchange studies with whole leaves by KROTKOV, CANVIN and ZELITCH, while TOLBERT laid the foundations of a biochemical explanation (which included a light-dependent evolution of CO_2)

$\longrightarrow$

Fig. 1. Master scheme for photosynthetic carbon metabolism and related processes. Emphasis is placed on the situation in eukaryotic cells; outlines of a chloroplast, a microbody, and a mitochondrion are presented. Reactions shown outside these organelles occur in the cytosol. Of central importance is the Calvin cycle which supplies triose-P for starch synthesis in the chloroplast (*above, center*) or sucrose synthesis in the cytosol (*above, right*). Photosynthetic nitrogen assimilation is indicated inside the chloroplast (*left*) while photorespiration involving the glycolate pathway involves traffic between all three organelles (*below, right*). Cytosolic PEP carboxylation and subsequent release of the CO_2 prior to its incorporation by the Calvin cycle, a predominant feature of C_4 and CAM plants, is collected in the rectangular section labeled C_4 *metabolism* (*center, right*). ● and ■, translocators on the inner membrane of the chloroplast envelope. N_2 and H_2 assimilation, and the reductive tricarboxylic acid cycle (*dotted lines*) are in prokaryotic organisms only

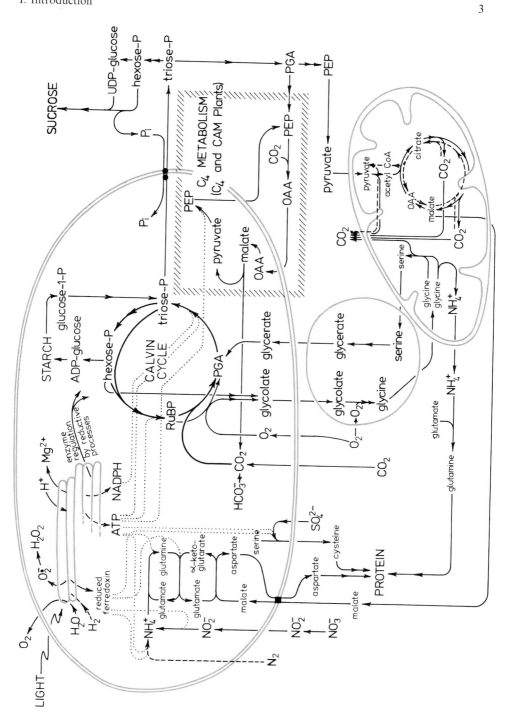

when he proposed the glycolate pathway, and several years later, discovered the peroxisome (see Fig. 1). At about the same time, BOWES, OGREN, and HAGEMAN obtained evidence that RuBP carboxylase could use O_2 in place of CO_2 as substrate, thereby effectively oxidizing RuBP, producing one molecule of glycerate-3-P and one of glycolate-2-P; the latter, after hydrolysis to glycolate, is believed to be the substrate for photorespiration. This mechanism of glycolate production has received great popularity, but it is not the only proposed mechanism (see chapter by BECK). In fact, there is much still to be learnt about photorespiration, particularly whether there is any need for this apparently wasteful process, and if not, whether it can be blocked so that plant productivity will correspondingly increase. This attractive goal is no doubt part of the reason for the recent proliferation of studies on RuBP carboxylase (chapter by AKAZAWA). One benefit of this renewed interest is that the old annoying problem of this enzyme's in vitro affinity for CO_2 being well below that required to support photosynthesis in vivo has been clarified.

Of the three developments which occurred during the mid-1960's, the C_4 pathway has no doubt proved the most controversial. Developed by KORTSCHAK and BURR at Honolulu, and by HATCH and SLACK at Indooroopilly from the results of experiments on photosynthesis by the leaves of maize and sugar cane, the C_4 pathway was originally viewed by some to be more or less an alternative to the Calvin cycle, but this outlook has proved unfounded since it does not constitute a mechanism for *net* CO_2 fixation which is required if the plant is to grow. Despite this, investigations on the C_4 pathway have led to a greatly broadened appreciation of the diversity of photosynthetic carbon metabolism in plants (now categorized as C_3 plants, C_4 plants, and CAM plants), to a re-evaluation of the value of PEP carboxylase detected first in plants by BANDURSKI (chapter by COOMBS), and to the discovery by HATCH and SLACK of the enzyme pyruvate Pi dikinase.

Reflecting on past successes, however, is not good for the soul of science unless some thought is also given to deficiencies in the state of knowledge achieved. The critical approach used by many authors in this volume fortunately provides this balance, and raises many questions. Will photosynthetic pathways not utilizing RuBP and PEP carboxylases be revealed in higher plants? What is the means of carbon communication between cells of plants characterized by the Kranz anatomy (chapters by RAY and BLACK, and EDWARDS and HUBER)? How do the many factors interact to influence the photosynthetic versatility of the CAM plant (chapters by KLUGE, QUEIROZ, and WAISEL). How is the flow of electrons through the ferredoxin molecule apportioned among the competing electron acceptors (chapters by VENNESLAND, SCHMIDT, BÖGER, and BEN-AMOTZ)? What advantage in crop productivity will result from the clear understanding of the relationship between biological nitrogen fixation and photosynthesis (chapters by STEWART and QUEBEDEAUX)? Why is a high activity of carbonic anhydrase present in photosynthetic cells (chapter by POINCELOT)? What is the mechanism(s) behind the light-mediated activation of several Calvin cycle enzymes (chapter by ANDERSON)? When will an activity of chloroplast SBPase sufficient to accommodate CO_2 fixation be demonstrated (chapter by LATZKO and KELLY)? What is the extent and purpose of mitochondrial respiration in illuminated photosynthetic cells (chapters by GRAHAM and CHAPMAN, and EVANS and CARR)? How is the effect of blue light on metabolism, outlined in the chapter by VOSKRESENSKAYA, mediated? Is the Calvin cycle truly absent from those photosynthetic bacteria which are reported to possess a reductive tricarboxylic acid cycle (chapters by BUCHANAN and OHMANN)? These and other problems are no doubt the reason why the viewpoints emphasized and opinions expressed by the contributors are often contradictory. For this we make no apology, but rather hope that they stimulate the curiosity of prospective investigators of photosynthetic carbon metabolism and related processes, so that future research will be as fascinating and fruitful as that of the past.

Acknowledgment. We record here our indebtedness to Dr. Grahame J. Kelly for his not inconsiderable input to each phase of this volume.

II. CO$_2$ Assimilation

II A. The Reductive Pentose Phosphate Cycle

1. The Reductive Pentose Phosphate Cycle and Its Regulation

J.A. BASSHAM

A. Introduction

The reductive pentose phosphate cycle (RPP cycle) is the basic biochemical pathway whereby carbon dioxide is converted to sugar phosphates during the process of photosynthesis (BASSHAM et al., 1954; BASSHAM and CALVIN, 1957). This pathway is apparently ubiquitous in all photoautotrophic green plants (NORRIS et al., 1955), as it has been found to occur in all eukaryotic photosynthetic cells and in all blue-green algae so far examined (Fig. 1). The RPP cycle also occurs in a variety of photosynthetic bacteria (STOPPANI et al., 1955), although another cycle has been reported for some photosynthetic bacteria (EVANS et al., 1966). In some higher plants, commonly designated C_4 plants, there is an additional pathway (C_4 cycle) via which carbon dioxide is first incorporated into four-carbon acids or amino acids and is later released to be refixed via the RPP cycle (KORTSCHAK et al., 1965; HATCH and SLACK, 1966, 1970).

The C_4 cycle is not an alternative to the RPP cycle. It does not result in any net reduction of CO_2. Its function appears to be to utilize ATP from the photochemical reactions to bring carbon into the chloroplasts where the RPP cycle is operating. By this process, C_4 plants maintain a higher level of CO_2 at the local environment of the carboxylation enzyme of the RPP cycle, ribulose 1,5-P_2 carboxylase (RuBPCase). The result is to diminish the oxygenase activity of this enzyme (BOWES et al., 1971; ANDREWS et al., 1973; LORIMER et al., 1973) which can convert the carboxylation substrate, ribulose 1,5-P_2 (RuBP) to glycolate, a substrate for the apparently wasteful process of photorespiration (see Chaps. II.25–28, this vol.). In addition, the C_4 pathway can recycle CO_2 that is produced within cells by photorespiration (TOLBERT, 1971).

The RPP cycle in eukaryotic cells occurs only inside the chloroplasts (ARNON et al., 1954), the subcellular organelles which are also the site of the primary photochemical reactions of photosynthesis in eukaryotic cells. The process of photosynthesis is complete within the chloroplasts, from the capture and conversion of light energy and the oxidation of water to molecular oxygen, to the uptake and reduction of carbon dioxide to starch and to triose phosphates which are exported from the chloroplasts to the cytoplasm (see Volumes 3 and 5 of this series).

The enzymes catalyzing steps of the RPP cycle are water-soluble and are located in the stroma region of the chloroplasts (ALLEN et al., 1957). Only three steps in the cycle out of a total thirteen actually use up cofactors (ATP and NADPH; BASSHAM et al., 1954; BASSHAM and CALVIN, 1957) which must be regenerated at the expense of cofactors formed by the light reactions in the thylakoids. ATP

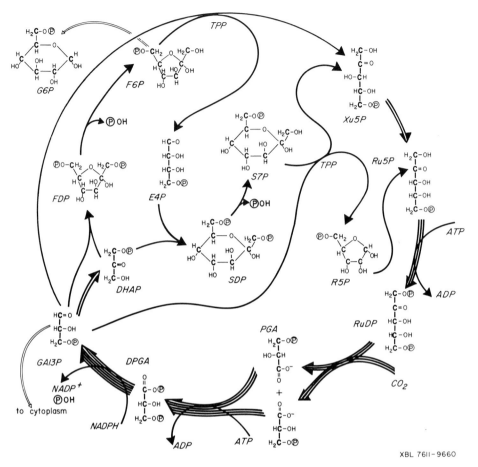

Fig. 1. The reductive pentose phosphate cycle. The *heavy lines* are for reactions of the RPP cycle, the *faint lines* indicate removal of intermediate compounds of the cycle for biosynthesis. The *number of heavy lines in each arrow* equals the number of times that each reaction occurs for one complete turn of the cycle, in which three molecules of CO_2 are converted to one molecule of GA13P. Abbreviations for this Fig. only: *RuDP*, ribulose 1,5-diphosphate; *DPGA*, 1,3 diphosphoglycerate; *GA13P*, 3-phosphoglyceraldehyde; *FDP*, fructose 1,6-diphosphate; *SDP*, sedoheptulose 1,7-diphosphate; ℗ represents the $-PO_3^{2-}$ and $-PO_3H^-$ radicals present at physiological pH, and ℗OH represents inorganic phosphate, HPO_3^{2-} (Pi^{2-})

is used directly, while NADPH is formed by reduction of $NADP^+$ at the expense of oxidation of two equivalents of ferredoxin.

$$2\,Fd_{red} + NADP^+ + H^+ \rightarrow 2\,Fd_{ox} + NADPH$$

There are no photochemical steps in the RPP cycle. There are, however, indirect effects of the light reactions on steps in the RPP cycle other than those requiring ATP or NADPH (see Chap. II.22, this vol.).

If the RPP cycle were the only metabolic pathway occurring in the stroma region inside the chloroplast envelope, and the only end product were starch, there might not be any need for metabolic regulation of the cycle. It would simply operate or not depending on the supply of NADPH and ATP from the light reactions and a supply of CO_2. Following the mapping of the RPP cycle, not much effort was made to investigate its metabolic regulation for several years. It later became apparent that the metabolism of the chloroplasts is not limited to the RPP cycle, nor is reduced carbon withdrawn from the cycle at a single point for a single product. Metabolic regulation of the cycle was clearly needed at rate-limiting steps to allow for shifts in metabolism in the chloroplasts (for example, between light and dark; BASSHAM and KIRK, 1968; BASSHAM, 1971b) and to keep the concentrations of intermediates of the RPP cycle within physiological ranges even as the demands on the cycle for various sugar phosphates as starting compounds for biosynthesis changed with the physiological needs of the plant (KANAZAWA et al., 1970; BASSHAM, 1971b).

Attention was focused on metabolic regulation of the RPP cycle by two kinds of experimental finding. One was the apparently anamolous transient changes in pool sizes of intermediate compounds in the chloroplasts observed when physiological conditions were suddenly changed. For example, reactions catalyzed by enzymes not requiring cofactors from the light were seen to change in rate between light and dark (PEDERSEN et al., 1966; BASSHAM and KIRK, 1968; BASSHAM, 1971b). The other kind of experimental result leading to a study of regulation was the apparent inadequacy of the activities of some of the isolated enzymes catalyzing steps in the cycle (STILLER, 1962; PETERKOFSKY and RACKER, 1961). The key carboxylation enzyme, ribulose 1,5-P_2 carboxylase (E.C. 4.1.1.39) appeared for years to have too high a K_mCO_2 value to accommodate the physiological levels of CO_2 normal for plants. The changes in enzyme activities and the apparent inadequacies of the activities under suboptimal conditions have by now been mostly explained as resulting from the needs of metabolic regulation (BASSHAM, 1971a; BASSHAM, 1973; BASSHAM, 1971b).

Information about the regulation of the RPP cycle has come from studies of the pool sizes of intermediate metabolites (often measured by using radioactively labeled substrates and measuring the labeling of the subsequently isolated metabolites (PEDERSEN et al., 1966; BASSHAM et al., 1968; KRAUSE and BASSHAM, 1969; BASSHAM and KRAUSE, 1969; KANAZAWA et al., 1972), and from measurement of O_2 evolution, phosphorylation, etc. in leaves, whole cells, isolated chloroplasts, and reconstituted chloroplasts (e.g., WALKER et al., 1967; HELDT and RAPLEY, 1970; SCHACTER et al., 1971; STOKES and WALKER, 1971; LILLEY et al., 1973; LILLEY and WALKER, 1975; WALKER, 1976; SLABAS and WALKER, 1976). Much additional knowledge has come from studies of the individual enzymes, both in crude extracts and in isolated form (Review, see KELLY et al., 1976). Many of these investigations are reviewed in more detail in subsequent chapters of this volume. The present article will describe the RPP cycle, its mapping, the requirements for regulation, and the kinetic evidence, based on measurement of labeled metabolites in cells, chloroplasts, etc., for regulation in vivo. Some examples, as appropriate, will be given for the detailed mechanism of regulation of the enzymes, as determined by studies of the properties of the isolated enzymes. For more complete review of these properties of isolated enzymes, and for various other specialized aspects, the reader should refer to the later articles.

B. The Reductive Pentose Phosphate Cycle

I. The Cyclic Path of Carbon Dioxide Fixation and Reduction

Fixation and reduction of carbon dioxide via the RPP cycle follows a cyclic pathway. In the initial step of this cycle, RuBP is carboxylated and hydrolytically split to give two molecules of 3-phosphoglycerate (PGA). This C_3 acid then is phosphorylated and reduced to glyceraldehyde-3-phosphate (GAP) via reactions using ATP and NADPH. The resulting triose phosphate then undergoes a series of isomerizations, condensations, and rearrangements which result in the conversion of five molecules of triose phosphate to three molecules of pentose phosphate, eventually in the form of ribulose-5-phosphate (Ru5P). This compound is then phosphorylated with ATP to regenerate the carboxylation substrate RuBP, thus completing the cycle.

A complete cycle, in which each reaction occurs at least once, is depicted in Figure 1. The number of arrows represent the number of products and reactants participating in one complete cycle. Three molecules of RuBP are carboxylated to give six molecules of PGA, and after reduction, six molecules of GAP. Only five of these GAP molecules (15 carbon atoms) are required to regenerate the three RuBP molecules. The sixth GAP molecule, equivalent in carbon to the three carbon dioxide molecules fixed, can either be converted by a reverse of glycolysis to glucose phosphate for starch synthesis, or can be exported from the chloroplast to the cytoplasm (BASSHAM et al., 1968; STOCKING and LARSON, 1969; HEBER et al., 1970; WERDEN and HELDT, 1971; BAMBERGER et al., 1975) for extrachloroplastic reactions. Other biosynthetic uses in the chloroplasts are also possible; for example, conversion of GAP to glycerol phosphate and eventually fats.

II. Individual Reactions of the RPP Cycle

1. The Carboxylation Reaction

In the initial step of the RPP cycle, the enzyme ribulose 1,5-bisphosphate carboxylase (RuBPCase) catalyzes the addition of CO_2 to C-2 of RuBP (QUAYLE et al., 1954; WEISSBACH and HORECKER, 1955; WEISSBACH et al., 1956; MÜLHOFFER and ROSE, 1965). It is believed that an ustable enzyme-bound six-carbon intermediate molecule results, and that this intermediate is hydrolytically split with a concurrent transfer of a pair of electrons from C-3 of the RuBP to C-2. The detailed mechanism of this complex reaction is discussed by AKAZAWA in Chapter II.17, this volume. The internal oxidation-reduction nature of this reaction led to the enzyme being called carboxydismutase for many years (CALVIN and BENSON, 1948).

This forward reaction has a negative Gibbs free energy change of nearly 10 Kcal (BASSHAM and KRAUSE, 1969). This conversion of chemical energy to heat provides a substantial part of the driving force for the cycle and also facilitates the functioning of this enzyme in important regulatory roles, as will be discussed later. It also means that this reaction is essentially irreversible.

2. Reduction of PGA to GAP

The reduction of PGA to GAP occurs in two steps. First, the PGA is converted to the acyl phosphate in a reaction using ATP and mediated by PGA kinase (E.C. 2.7.2.3).

With equal concentrations or activities of reactants this reaction is highly unfavorable ($\Delta G' \simeq 5$ Kcal) in the gluconeogenic direction, and it can only proceed as it does in photosynthesis by virtue of the relatively high physiological concentrations of PGA and such low concentrations of glycerate-1,3-bisphosphate (PPGA) that the latter compound is not normally detected in experiments in vivo or with whole or reconstituted chloroplasts using $^{14}CO_2$ and ^{32}P-labeled phosphate.

The PPGA is reduced with NADPH and triose phosphate dehydrogenase (NADPH specific; E.C. 5.3.1.1).

This reaction is also somewhat unfavorable energetically, but is helped by the fact that there are three products and only two reactants. The ratio NADPH/NADP$^+$ (free plus bound) is probably not more than $3:1$ (LENDZIAN and BASSHAM, 1976), and the concentration of GAP is not less than that of PPGA, but the production of P_i, when its concentration is about 1 mM contributes -4.1 Kcal to the reaction (BASSHAM and KRAUSE, 1969).

The overall reaction, whereby PGA is phosphorylated and reduced to GAP with ATP and NADPH, proceeds in the light under highly reversible conditions, and probably during photosynthesis is not subject to metabolic regulation. When the light is off, however, conversion of GAP to PGA may occur with a substantial negative ΔG^S, and it appears from reports described later that the oxidative sequence may well be regulated (ZIEGLER et al., 1968; MÜLLER, 1970; PUPILLO and GULIANI-PICCARI, 1975; ANDERSON and AVRON, 1976; WOLOSIUK and BUCHANAN, 1976) by mechanisms similar to those involved in other light–dark controlled steps. This sequence of reactions would thus be subject to light–dark regulation, but perhaps not to regulation during the period of active photosynthesis. This is in contrast to other regulated steps of the cycle that appear to be not only sites of light–dark regulation, but also to function as "fine-tuning" points of regulation in the light to keep in balance the concentrations of the various intermediate compounds of the RPP cycle.

3. Regeneration of Ru5P

A series of isomerizations and rearrangements are required for the conversion of five triose phosphate molecules to three pentose phosphate molecules. None of these reactions utilizes light-generated cofactors (ATP and NADPH), and most steps are highly reversible. Two steps which liberate P_i are rate-limiting and have substantial negative ΔG^S values. Both are sites of regulation.

Two molecules of GAP (per turn of the cycle) are converted with triose phosphate isomerase (E.C. 5.3.1.1) to dihydroxyacetone phosphate (DHAP). In the presence of aldolase (E.C. 4.1.2.13), the two triose phosphates condense to give fructose 1,6-bisphosphate (FBP) in a reversible step. FBP is then converted to fructose 6-phosphate (F6P) with fructose bisphosphatase (E.C. 3.1.3.11). This step has an estimated physiological $\Delta G^S = -7.0$ Kcal (Bassham and Krause, 1969).

The conversion of triose phosphates and hexose phosphates to pentose phosphates, requiring changes in chain length, is initiated with transketolase (E.C. 2.2.1.1) which transfers C-1 and C-2 (bound on the enzyme as the thiamine pyrophosphate adduct) from F6P to GAP, leaving the four carbon sugar phosphate, erythrose 4-phosphate (E4P), and forming xylulose 5-phosphate (Xu5P). This reaction is reversible.

The four-carbon aldose phosphate (E4P) can then condense with DHAP in a second reaction mediated by aldolase (E.C. 4.1.2.13) to give sedoheptulose, 1,7-bisphosphate (SBP). Like RuBP, this compound is unique to the RPP cycle and is not an intermediate compound in the oxidative pentose phosphate cycle (OPP cycle, also called the phosphogluconate pathway or the hexose monophosphate shunt.).

This seven-carbon ketose diphosphate is converted to sedoheptulose 7-phosphate and P_i with sedoheptulose 1,7-bisphosphatase. It now appears that this enzyme is distinct from the fructose 1,6-bisphosphatase. The reaction has a negative ΔG^S under physiological conditions of about -7 Kcal (Bassham and Krause, 1969), and is a regulated and rate-limiting step.

A second transketolase-mediated step follows in which C-1 and C-2 of S7P are transferred to GAP to give two pentose phosphates: Xu5P and ribose 5-phosphate (R5P). This completes the conversion of five GAP molecules to three pentose phosphate molecules.

Before they can be used to regenerate RuBP, the Xu5P and R5P must be converted to Ru5P (E.C. 5.3.1.6. and 5.1.3.4).

4. Formation of RuBP

The final step in the RPP cycle is the conversion of Ru5P to RuBP with ATP and phosphoribulokinase (E.C. 2.7.1.19). This reaction has an estimated ΔG^S of about -4 Kcal, so that it is intermediate between those reactions that are clearly reversible ($\Delta G^S = 0$ to -2 Kcal) and those that are almost completely irreversible ($\Delta G^S = -6$ to -10 or more Kcal). As discussed later and in subsequent articles there is much evidence that this step is metabolically regulated (Anderson and Avron, 1976; Wolosiuk and Buchanan, 1976; Wolosiuk et al., 1976).

III. Stoichiometry and Energetics

For each mol of CO_2 fixed by the cycle, one mol of ATP is required for the conversion of Ru5P to RuBP, the substrate for carboxylation. Two mol of ATP are required in the subsequent activation of the 2 mol of PGA formed, giving PPGA. The reduction of 2 mol of PPGA requires 2 mol of NADPH. For a complete turn of the RPP cycle, 3 mol of CO_2 are taken up, requiring the use of 9 mol of ATP and 6 mol of NADPH to make one mol of GAP. This overall result can be expressed as the sum of two equations representing the utilization of the cofactors from the light, and another equation for the conversion of CO_2, water, and P_i to GAP.

$6(NADPH + O_2 + H^+ \rightarrow NADP^+ + H_2O)$ $\qquad \Delta G' = -315.5$ Kcal

$9(ATP^{4-} + H_2O \rightarrow ADP^{3-} + P_i^{2-} + H^+)$ $\qquad \Delta G' = -\ 68.8$ Kcal

$\qquad\qquad\qquad$ Total $\qquad\qquad\qquad \Delta G' = -384.3$ Kcal

$3CO_2 + 3H_2O + P_i^{2-} \rightarrow GAP^{2-} + 3O_2$ $\qquad \Delta G' = +350.4$ Kcal

Subtracting the energy stored from the energy expended, we obtain $\Delta G' =$ only -33.9 Kcal as the driving energy for one turn of the cycle. The efficiency would be $350/384 = 91\%$. This calculation depends on the usual use of physiological free energy changes employed in biochemical energetics, but is somewhat misleading since actual physiological concentrations will always be much smaller than required to give activities of 1.0 – in fact, they are generally in the range of 10^{-5} to 10^{-2} M. A more realistic estimate for the energetics can be made by using measured or estimated concentrations of metabolites as approximate values for the activities of reactants and products, and correcting the $\Delta G'$ values to physiological G^S values (BASSHAM and KRAUSE, 1969). This is particularly important with respect to the energy of hydrolysis of ATP to ADP and P_i, where $\Delta G'$ (with 10 mM Mg^{2+}) is about -7.6 Kcal, but ΔG^S is in the range of -12.5 to -13.5 Kcal, depending on ATP/ADP ratios and P_i concentrations.

When metabolite concentrations, measured and estimated in photosynthesizing *Chlorella pyrenoidosa*, were used in this way to estimate ΔG^S values (BASSHAM and KRAUSE, 1969), the energy input for 6 mol of NADPH and 9 mol of ATP is -427.0 Kcal, and the energy stored in making GAP from CO_2, water, and P_i, with O_2 evolution is about $+353.6$ Kcal. The chemical free energy converted to heat to drive the cycle becomes 73.4 Kcal, and the efficiency of the cycle is 83%. Of the 73.4 Kcal expended, about 30 Kcal are used in the carboxylation reaction, 12 Kcal in the phosphorylation of RuBP, and 19 Kcal in the FBPase and SBPase reactions. These four metabolically regulated steps account for 83% of the total energy expenditure, while the remaining 17% is distributed among the remaining nine reactions, none of which is a control point during photosynthesis (although some become rate-limiting in the dark).

C. Utilization of the Products of the RPP Cycle

I. Starch Synthesis

Within the chloroplasts, the principal direct product of the cycle (in the absence of excessive photorespiration) is starch (GHOSH and PREISS, 1965; see also Chap. II.23, this vol.). The pathway to starch begins with conversion of GAP to F6P by reactions already discussed. There is no evidence to indicate that this conversion occurs at any different site or with different enzymes than those employed for the cycle. F6P is converted to glucose 6-phosphate (G6P) with hexose phosphate isomerase. Little or no free glucose is formed in photosynthesizing chloroplasts.

G6P is converted to G1P with phosphoglucomutase (E.C. 2.7.5.1). The reaction is somewhat "uphill" ($\Delta G' = +1.7$ Kcal) and is reversible, so that in vivo the ratio of G6P/G1P is about 20. The next step is the reaction of G1P with ATP, mediated by ADPglucose pyrophosphorylase and giving ADPglucose and inorganic pyrophosphate (PP_i). As first shown by GHOSH and PREISS (1966), the reaction catalyzed by ADPglucose pyrophosphorylase is an important regulatory point. The enzyme activity is stimulated by PGA and is strongly inhibited by high levels of P_i. The concentrations of both can change and are believed responsible for the regulation of starch biosynthesis. Also, along with the rapid drop in ATP in the dark, decreased activity of the enzyme resulting from increased level of P_i could account for the immediate drop in the level of ADPglucose when the light is turned off (BASSHAM, 1971a). Once formed, ADPglucose can transfer glucose to lengthen the amylose chain of a starch molecule.

$$(C_6H_{10}O_5)_n + ADPglucose \rightarrow (C_6H_{10}O_5)_{n+1} + ADP$$

II. Triose Phosphate Export

The triose phosphates, GAP and DHAP, were found to be the intermediate compounds of the chloroplasts that appeared to the largest extent in the medium of isolated spinach chloroplasts carrying out high rates of complete photosynthetic reduction of CO_2 (BASSHAM et al., 1968). This and other studies (STOCKING and LARSON, 1969; WERDEN and HELDT, 1971; HEBER and SANTARIUS, 1970) suggest that these compounds are a form of photosynthetic product exported to the cytoplasm. A specific phosphate translocator apparently exists whereby the transport out of the chloroplast of triose phosphate is balanced by the movement into the chloroplast of inorganic phosphate (WERDEN and HELDT, 1971; WALKER, 1976; HELDT and RAPLEY, 1970). Once triose phosphate appears in the cytoplasm it can provide, by its oxidation to PGA in the cytoplasm, both NADH and ATP (STOCKING and LARSON, 1969; HEBER and SANTARIUS, 1970). Besides the triose phosphates and P_i, it appears that PGA can move through the chloroplast envelope via the phosphate translocator mechanism.

Due to this exchange, it can be seen that an increase in P_i in the cytoplasm could stimulate the export of triose phosphates from the chloroplasts via the phosphate translocator, with P_i entering and triose phosphates emerging (for review, see WALKER, 1976a, b).

While it is not yet demonstrated that the cytoplasmic concentration of P_i is important in the in vivo regulation of photosynthetic metabolism, it seems entirely possible that such is the case. Many photosynthetic cells synthesize variable but substantial amounts of sucrose in the cytoplasm. This is true not only for higher plants, but also for the unicellular alga *Chlorella pyrenoidosa* (KANAZAWA et al., 1970). The formation of sucrose occurs via a mechanism similar to that in starch synthesis, but employing the reaction of G1P with UTP to produce UDPglucose which then reacts with F6P to form sucrose phosphate and finally sucrose (LELOIR, 1964). The rate of PP_i formation in the UDPglucose pyrophosphorylase reaction is equivalent to the rate of sucrose formation, which in turn can often account for 10% or so of CO_2 fixation in *Chlorella pyrenoidosa*, and presumably could be a much higher proportion in mature leaves of higher plants that are exporting most of the photosynthetic product to other parts of the plant. The rate of conversion of PP_i to P_i must equal the rate of formation of PP_i by the sucrose synthesis and other biosynthetic reactions (such as protein synthesis) that produce PP_i. The steady-state level of P_i could be sensitive to pyrophosphatase activity, which in turn could be regulated by such factors as Mg^{2+} level. This hypothesis illustrates one way that conditions in the cytoplasm might influence the RPP cycle which is inside the chloroplast. The rate of withdrawal of triose phosphates affects the degree in which regulated rate-limiting steps of the RPP cycle must be balanced in order to maintain workable levels of cycle intermediates in the chloroplast.

III. Glycolate Formation

Under most conditions of photosynthesis in chloroplasts, some glycolate is formed (SCHOU et al., 1950; BENSON et al., 1951). Under conditions which favor photorespiration, a very large part of the carbon fixed by the RPP cycle can be converted to glycolate (ZELITCH, 1966; ZELITCH, 1974). These conditions generally include high light intensity, low CO_2 pressure, more than a few percent O_2, and temperatures above normal (WILSON and CALVIN, 1955; BASSHAM and KIRK, 1962). Unicellular algae such as *Chlorella pyrenoidosa*, which do not evolve photorespiratory CO_2, form large amounts of glycolate under these conditions, and much of the glycolate is excreted into the medium. Even C_4 plants, which exhibit little or no photorespiration, form glycolate in amounts which increase with the conditions just listed. In the case of these plants, CO_2 formed by photorespiratory-type reactions is mostly recaptured in the leaves. Pathways of glycolate synthesis are discussed in detail by BECK (Chap. II.25, this vol.).

D. Mapping the RPP Cycle

I. Early History

Following the discovery of short-lived radiocarbon (^{11}C) which could be used to label the CO_2 taken up by the plants (RUBEN et al., 1940), Ruben and coworkers could prove that the fixation of CO_2 proceeds during the dark immediately following a period of illumination, and that the product of this fixation is an intermediate compound which can be subsequently transformed in the light (RUBEN and KAMEN, 1940b). With the discovery of the long-lived radioisotope ^{14}C (RUBEN and KAMEN, 1940a), the detailed pathway of photosynthetic carbon reduction could be mapped.

II. First Products of CO_2 Fixation

From 1946 to 1953, Calvin and coworkers used ^{14}C as a radioactive label to follow the path of CO_2 fixation and reduction in photosynthesis. The most useful separating and identification technique proved to be the resolution of the various acids and sugar phosphates by two-dimensional paper chromatography and radioautography (Benson et al., 1950). The paper chromotographic methods had been developed for the analysis of amino acids by Martin and Singe (1941). These methods plus radioautographic location of ^{14}C-labeled compounds on paper were applied to the analysis of ^{14}C-labeled amino acids formed during photosynthesis.

The most heavily labeled compound after very short periods of photosynthesis with $^{14}CO_2$ was PGA (Benson and Calvin, 1948). When this compound was chemically degraded, to permit measurement of the amount of ^{14}C in each carbon atom position, the predominant labeling at short times was in the carboxyl carbon (Schou et al., 1950). Such ^{14}C as was found in the remaining two carbon atoms was at all times equally distributed between those positions. Such a finding immediately suggested that a cyclic process was involved, in which the carboxylation product, PGA, was converted to a carboxylation substrate. During the cyclic process, the label had become equally distributed between the two carbon atoms supplied by this substrate.

The plants used for most of the studies were either unicellular algae, *Scenedesmus obliquus,* or young leaves of higher plants such as barley or soybean, which we now know to be C_3 plants. Nonetheless the C_4 acids, malic and succinic, were among the early products of ^{14}C fixation, and these acids for a short time were thought to play a role in the cyclic regeneration of CO_2 acceptor. It was soon found, however, that with malonate as an inhibitor, formation of these C_4 acids in algae could be suppressed without causing any decrease in either the photosynthetic uptake and reduction of CO_2 or in the appearance with time of ^{14}C label in the alpha and beta positions of PGA (Bassham et al., 1950). The C_4 acids were thus dismissed as intermediate compounds in photosynthetic CO_2 fixation and reduction, and were not to reappear in such a role until the discovery of the special metabolism in certain "tropical grasses" such as sugar cane and maize, and in a scattering of other species (Kortschak et al., 1965; Hatch and Slack, 1966; Hatch and Slack, 1970).

III. Sugar Phosphates

Application of the two-dimensional paper chromatographic method to analysis of products of photosynthesis with $^{14}CO_2$ soon revealed that a number of sugar monophosphates and diphosphates were formed during the first few seconds of exposure to the labeled $^{14}CO_2$. Among these were DHAP, FBP, and G6P (Benson et al., 1950). Chemical degradation of the hexose phosphates showed the ^{14}C first appearing in the C-3 and C-4 positions, after which the label appeared equally in C-1, C-2, and in smaller amounts in C-5 and C-6 (Bassham et al., 1954). Such kinetics and pattern of labeling strongly suggested that the PGA, once formed, was converted to hexose monophosphates by a portion of the gluconeogenesis pathway from phosphoenolpyruvate (PEP) to glucose. In turn, this proposition dictated that the cofactors required from the light reaction should be reduced pyridine nucleotides and ATP.

It was soon found that there were other sugar phosphates involved as early labeled intermediates. These were identified as SBP and S7P and as RuBP (BENSON et al., 1951; BENSON, 1951; CALVIN et al., 1951). Smaller amounts of Xu5P, Ru5P, and R5P were also found. Degradation of these compounds after short periods of photosynthesis with $^{14}CO_2$ revealed a pattern of labeling requiring the conversion of triose and hexose phosphates to pentose phosphates, a specific set of arrangements of carbon chain length, as appears in the final version of the cycle (BASSHAM et al., 1954; BASSHAM and CALVIN, 1957).

IV. Studies of Light–Dark and High–Low CO_2 Transients

Further information about the sequence of events in the cycle came from studies of the changes in levels of labeled compounds accompanying sudden changes in physiological conditions (CALVIN and MASSINI, 1952; WILSON and CALVIN, 1955). The unicellular alga, *Scenedesmus obliquus*, was allowed to photosynthesize with $^{14}CO_2$ in air for about 10 min. During this time samples of the algae were taken periodically, the algae in the sample were killed, and subsequently the labeled compounds were analyzed by two-dimensional paper chromatography and radioautography. After about 5 min photosynthesis with $^{14}CO_2$, the ^{14}C content of intermediate compounds of the cycle no longer increases, indicating that the compounds are fully labeled ("saturated"). After this time, the ^{14}C content may be taken as a measure of the actual concentrations of the compounds in the actively turning over pools in the cells.

After saturation with ^{14}C, the light was turned off, and more samples were taken in quick succession. Upon analysis, it was found that the concentration of PGA rises rapidly, indicating that cofactors generated by the light reaction are required for the subsequent conversion of PGA to sugar phosphates as expected (CALVIN and MASSINI, 1952). The rise in level of PGA for 20 s or longer also indicates that the carboxylation reaction itself proceeds in the dark for a time, suggesting that cofactors from the light are not required for the carboxylation reaction.

After about 30 s darkness, the PGA concentration falls. The carboxylation reaction by this time must have begun to deplete supply of carboxylation substrate, whereas the dark reactions utilizing PGA exported from the chloroplast continue. Although, as discussed earlier, carbon export from the chloroplasts during photosynthesis is in the form of triose phosphates, when the light is turned off, PGA can be exported.

When the light was turned off, the concentrations of several sugar phosphates declined, but the most significant drop was in the level of RuBP, which in many experiments fell below detectable limits. When this was first observed by CALVIN and MASSINI (1952), they concluded that the step involved in the regeneration of RuBP requires a light-produced cofactor, namely ATP, which is required for the conversion of Ru5P to RuBP with phosphoribulokinase.

In a similar experiment, Wilson and Calvin (1955) first established steady-state photosynthesis including ^{14}C-saturation with photosynthesizing *Scenedesmus*. Then they lowered the CO_2 level to nearly zero. In this case the carboxylation product, PGA, rapidly decreased in concentration, as expected, while the concentration of RuBP was the first among the sugar phosphates to rise. This provided direct in vivo evidence that RuBP is the carboxylation substrate in the RPP cycle. Since RuBP is a five-carbon compound, it was concluded that subsequent to the addition of CO_2 there must be a split to two three-carbon molecules. At least one of these products would have to be PGA, since PGA was shown to be the first product of CO_2 fixation. From a consideration of oxidation states of RuBP and CO_2, both three-carbon products must in fact be PGA, if there is no external oxidant or reductant supplied to the reaction. Later studies with isolated RuBP carboxylase have provided overwhelming evidence for the addition of CO_2 to the C-2 of the RuBP, and an internal oxidation-reduction of the six-carbon intermediate, with hydrolytic splitting to give two molecules of PGA (Quayle et al., 1954; Weissbach and Horecker, 1955; Weissbach et al., 1956; Mülhoffer and Rose, 1965).

V. Discovery of Enzymes of the RPP Cycle

During approximately the period when the RPP cycle was mapped through the use of labeled carbon, an oxidative pentose phosphate cycle (OPP cycle) was discovered by more classical biochemical methods in which the various enzymes required were isolated and characterized. Several of the reactions postulated for the OPP cycle appeared to be the reverse of reactions of the RPP cycle, and soon many of the required enzyme activities were isolated from green plants. For example, the transketolase enzyme, essential for both cycles, was purified from spinach by Horecker and coworkers (Horecker et al., 1953). Pentose phosphate isomerase was found in alfalfa (Axelrod and Bandurski, 1953). The finding of these and other enzyme activities of the OPP cycle and glycolysis provided much of the necessary supporting biochemical evidence for the RPP cycle.

There are, however, three enzyme activities unique to the RPP cycle. Of these, perhaps the most important in establishing the cycle is the RuBP carboxylase. The enzymic carboxylation of RuBP in vitro was first reported by Quayle and coworkers (Quayle et al., 1954) who demonstrated, in 1954, the formation of PGA, labeled with ^{14}C in the carboxyl group only, when RuBP and $^{14}CO_2$ were added to a cell-free extract obtained from *Chlorella*. The enzyme was purified and characterized by Weissbach and coworkers (Weissbach and Horecker, 1955; Weissbach et al., 1956) soon afterwards, and there has been a large amount of work on the properties of this enzyme since then (reviewed by Akazawa, Chap. II.17, this vol.).

A second enzyme unique to the RPP cycle is phosphoribulokinase, purified from spinach by Hurwitz and coworkers in 1956 (Hurwitz et al., 1956). The third unique enzyme is sedoheptulose 1,7-bisphosphatase (SBPase). For a long time it was thought that this enzyme might be identical to fructose 1,6-bisphosphatase, but recent work by Buchanan et al. (1976) suggests it is derived from FBPase when FBPase is converted from its dimer to its monomer form.

E. Metabolic Regulation of the RPP Cycle

I. In Vivo Kinetic Steady-State Studies with Labeled Substrates

The methods of kinetic analysis of measuring levels of labeled metabolites (see D. IV) have also proved useful in the identification of sites of metabolic regulation. The steady-state levels of radioactive intermediate compounds can be used to calculate the physiological free energy changes (ΔG^s) for a specified plant and set of physiological conditions. This information provides a direct measure of the reversibility of the reactions as they are occurring in vivo (BASSHAM and KRAUSE, 1969). It can be easily shown that the relation between ΔG^s and the reversibility of the reaction is given by:

$$\Delta G^s = -RT\ln(f/b)$$

where f is the forward reaction rate and b the back reaction rate.

In order for such measurements to be meaningful, accurate procedures for the maintenance of steady-state conditions and continuous measurement of CO_2, specific radioactivity, rapid sampling and killing, and quantitative analysis of radioactivity in each compound as a function of the amount of tissue sampled were developed (BASSHAM and KIRK, 1960). Initially, the steady-state kinetic method was used to demonstrate amino acid formation directly from photosynthate without the intermediacy of sucrose or starch in photosynthesizing *Chlorella pyrenoidosa* (SMITH et al., 1961). Some years later, the method was used to determine the ΔG^s values for reactions of the RPP cycle, as already discussed in B. II. and as summarized in Table 1 (BASSHAM and KRAUSE, 1969). The reactions shown to be rate-limiting in the light (during photosynthesis) were those mediated by RuBPCase, FBPase, SBPase, and phosphoribulokinase. As it happened, these were four of the five reactions for which there was already in vivo evidence for light–dark regulation. The three reactions with the highest negative free energy changes were catalyzed by enzymes which in their isolated state had been reported to have insufficient catalytic activity to accommodate the requirements of the RPP cycle (STILLER, 1962; PETERKOFSKY and RACKER, 1961).

II. Light–Dark Regulation

1. Respiratory Metabolism in Photosynthetic Cells and Chloroplasts

When the light is turned off with *Chlorella pyrenoidosa* that were previously photosynthesizing under steady-state conditions, a number of interesting transient changes in the pool sizes of intermediate compounds occur in addition to the changes in PGA and RuBP described in D. IV. Further changes were revealed by using ^{32}P-labeled phosphate in addition to $^{14}CO_2$, and by turning the light on again after allowing enough time to establish a new steady-state condition of repiratory metabolism in the dark (PEDERSEN et al., 1966; BASSHAM and KIRK, 1968; BASSHAM, 1971a; BASSHAM, 1971b; BASSHAM, 1973). There is an immediate

Table 1. Free energy changes of the RPP cycle. The standard physiological Gibbs free energy changes ($\Delta G'$) were calculated for unit activities, except $[H^+] = 10^{-7}$. The physiological free energy changes at steady state are for a 1% w/v suspension of *Chlorella pyrenoidosa* photosynthesizing with 0.4% $^{14}CO_2$ in air and with other conditions as described by Bassham and Krause (1969). The stroma concentrations were assumed to be four times the total cellular concentrations, and are used as approximations for activities

Reaction	$\Delta G'$	ΔG^s
$CO_2 + RuBP^{4-} + H_2O \rightarrow 2\ PGA^{3-} + 2\ H^+$	-8.4	-9.8
$H^+ + PGA^{3-} + ATP^{4-} + NADPH \rightarrow ADP^{3-} + GAP^{2-} + NADP^+ + P_i^{2-}$	$+4.3$	-1.6
$GAP^{2-} \rightarrow DHAP^{2-}$	-1.8	-0.2
$GAP^{2-} + DHAP^{2-} \rightarrow FBP^{4-}$	-5.2	-0.4
$FBP^{4-} + H_2O \rightarrow F6P^{2-} + P_i^{2-}$	-3.4	-6.5
$F6P^{2-} + GAP^{2-} \rightarrow E4P^{2-} + Xu5P^{2-}$	$+1.5$	-0.9
$E4P^{2-} + DHAP^{2-} \rightarrow SBP^{4-}$	-5.6	-0.2
$SBP^{4-} + H_2O \rightarrow S7P^{2-} + P_i^{2-}$	-3.4	-7.1
$S7P^{2-} + GAP^{2-} \rightarrow R5P^{2-} + Xu5P^{2-}$	$+0.1$	-1.4
$R5P^{2-} \rightarrow Ru5P^{2-}$	$+0.5$	-0.1
$Xu5P^{2-} \rightarrow Ru5P^{2-}$	$+0.2$	-0.1
$Ru5P^{2-} + ATP^{4-} \rightarrow RuBP^{4-} + ADP^{3-} + H^+$	-5.2	-3.8
$F6P^{-2} \rightarrow G6P^{2-}$	-0.5	-0.3

appearance of labeled 6-phosphogluconate in the dark, and an equally rapid disappearance in the light. This intermediate is unique to the OPP cycle.

With whole cells, the operation of the OPP cycle, indicated by the appearance of 6-phosphogluconate, could be occurring in the chloroplasts, the cytoplasm, or both. In fact, Heber et al. (1967) found that the unique enzymes of the OPP cycle, glucose-6-phosphate dehydrogenase and 6-phosphogluconate dehydrogenase, are present in both cytoplasm and chloroplasts of spinach and *Elodea*, with the larger amounts located in the cytoplasm.

Other studies of photosynthesis in isolated spinach chloroplasts in the presence of vitamin K_5, which is thought to divert electrons from the light reaction and thus mimic aspects of dark metabolism, demonstrated the formation of 6-phosphogluconate in chloroplasts (Krause and Bassham, 1969).

When both ^{14}C and ^{32}P labeled substrates were administered to photosynthesizing *Chlorella* and the light was turned off following over 11 min photosynthesis with $^{14}CO_2$, the curves diverged for the two isotopes in a given compound such as 6-phosphogluconate or any of the sugar phosphates, with the ^{32}P labeling becoming greater compared with ^{14}C labeling with time in the dark (Pedersen et al., 1966; Bassham and Kirk, 1968; Kanazawa et al., 1972). While intermediate pools were saturated with both labels in the light, and ^{32}P labeling continues to be saturated in the dark, it was clear that dark respiration utilized endogenous compounds that were not fully ^{14}C-labeled during the period of photosynthesis with $^{14}CO_2$. It is noteworthy that the level of ATP in the *Chlorella* cells, while dipping about 40% during the first minute in the dark, was reestablished at the same steady-state level as in the light after about 10 min of darkness.

Studies of the disappearance of labeled starch and sucrose in *Chlorella* in the dark following a long period of photosynthesis with $^{14}CO_2$ were particularly revealing with respect to the source of respiratory carbon in chloroplasts and cytoplasm. These studies were performed in the absence or presence of intracellular ammonium ion (KANAZAWA et al., 1972). The preliminary period of photosynthesis and the first part of the respiratory period were without added NH_4^+ in the suspending medium.

When the light was turned off, the level of labeled starch immediately began to decline, and continued to decline for the duration of the experiment at a very substantial but *constant* rate. Immediately after the light was turned off, the level of 6-phosphogluconate rose, and other changes in the sugar phosphate levels were indicative of operation of the OPP cycle. Since starch in *Chlorella* is in the chloroplasts, this OPP cycle activity probably occurs inside the chloroplasts.

When the light was turned off, labeled sucrose remained constant, and only began a steady decline (at about half the rate of starch utilization) when 1 mM NH_4^+ was added to the suspending medium (KANAZAWA et al., 1972). Coincident with this change was a second rise in 6-phosphogluconate, and temporary declines in the levels of FBP and SBP, followed by increases to new high levels. These and other changes suggested that the utilization of sucrose, located in the cytoplasm, was unaffected by darkness, but dependent on intracellular NH_4^+, with utilization probably occurring via the OPP cycle in the cytoplasm (KANAZAWA et al., 1972).

2. Light–Dark Regulation of FBPase and SBPase

Early examination of the activities of enzyme extracts from leaves showed limiting and possibly inadequate catalysis of the conversion of FBP and SBP to the respective monophosphates (PETERKOFSKY and RACKER, 1961; STILLER, 1962). It was therefore of interest when light–dark and dark–light transient changes in levels of these metabolites in *Chlorella*, previously labeled with ^{14}C and ^{32}P during photosynthesis, showed unexpected kinetics (BASSHAM and KIRK, 1968; PEDERSEN et al., 1966). When the light was turned off, the levels of both these compounds dropped rapidly, as would be expected with the cessation of the reduction of PGA to triose phosphates. Then, over a period of 10 min, the levels of these compounds rose again, a greater increase in ^{32}P labeling than in ^{14}C labeling. As discussed earlier, this indicates the respiratory breakdown of endogenous carbohydrates that were only partially labeled in the light.

When the light was turned on again, there was a very rapid buildup in the levels of FBP and SBP (as well as DHAP) for about 30 s, with the levels reaching higher than steady-state light levels. Then there was an equally rapid drop in these levels for another 30 s, followed by damped oscillations leading to a steady-state light level equal to that achieved in the previous light period.

The interpretation of these interesting kinetics is that when light is turned on, there is a rapid reduction of PGA to triosephosphates that are rapidly converted to FBP and SBP. The "overshoot" in the levels is attributed to the bisphosphatases having become inactive in the dark period, and requiring about 30 s in the light to become reactivated. During this period, the level of F6P and S7P also drops – further indication that the diphosphatases are inactive. With this and other blocks in the RPP cycle, regeneration of RuBP is limited, carboxylation does not reach steady-state rates, and the rate of formation of PGA and of its conversion to triose phosphates and FBP and ADP is limited. After 30 s, when the bisphosphatase becomes fully active, levels of these bisphosphates fall sharply, as these compounds are converted to sugar monophosphates and eventually to RuBP. With removal of the bisphosphatase blocks and other blocks, the RPP cycle reaches full velocity, and soon the rate of reduction of PGA to triose phosphates and consequent formation of FBP and SBP is sufficient to bring the levels of these compounds to their steady-state light values. Although such a control mechanism results in a surging of carbon around the RPP cycle, the oscillations are quickly damped (only about two are seen) due to the metabolite pools acting as capacitors and the rate-limiting steps acting as resistors.

The inactivation of the FBPase and SBPase activities in the dark is required to prevent the operation of futile cycles in the dark, when respiratory metabolism occurs in the chloroplasts. The buildup in the levels of FBP and SBP seen in the dark may be due to operation of the OPP cycle, of glycolytic conversion of G6P to FBP by hexose phosphate isomerase and phosphofructokinase, or both. As is well known in nonphotosynthetic cells, phosphofructokinase and FBPase activity, when both present in the same cellular compartment, must be so regulated that they are not both active at the same time, lest a futile cycle operate to hydrolyze ATP.

In the case of SBPase activity in the dark, a somewhat more extended futile cycle could operate, in which S7P and GAP would be converted to F6P and E4P by transaldolase (E.C. 2.2.1.2), which would be present although active only in the dark if there is an OPP cycle operating. This E4P plus a molecule of DHAP then would condense to give SBP. The F6P with ATP and phosphofructokinase would give FBP and hence GAP and DHAP, completing the futile cycle. Similar futile cycles would occur in the light, unless (as is likely) any phosphofructokinase activity present in the chloroplasts is inactivated in the light.

The prevention of such futile cycles requires that, in some cases, enzyme activites be capable of more variation than can be achieved with a single regulatory effector or inhibitor. FBPase and SBPase, like several other enzymes of the RPP cycle and the OPP cycle in chloroplasts, seem to be regulated by more than one factor that changes in concentration between light and dark. In common with some other regulated chloroplast enzymes, FBPase and SBPase respond to changes in pH and Mg^{2+}. Increased Mg^{2+} lowers the pH optima of these enzymes (PREISS et al., 1967; GARNIER and LATZKO, 1972). Since both Mg^{2+} (LIN and NOBEL, 1971; HIND et al., 1974; KRAUSE, 1974; BARBER et al., 1974; BARBER, 1976) and pH (SCHOU et al., 1950; HELDT et al., 1973) increase in the light in chloroplasts, the combined change has a substantial effect on enzyme activity.

A second major regulation of SBPase and FBPase depends on another important change between light and dark: the ratio of levels of reduced to oxidized cofactors (see Chap. II.22, this vol.).

3. Light–Dark Regulation of RuBP Carboxylase

When *Chlorella pyrenoidosa* are allowed to photosynthesize with $^{14}CO_2$ under steady-state conditions with a total CO_2 pressure comparable to that in air (0.03%), the steady-state level of RuBP is quite high: over 0.4 mM in the cells as a whole, and probably more than 2 mM in the stroma region of the chloroplasts (BASSHAM and KRAUSE, 1969). When the light is turned off, the level of RuBP declines rapidly for the first 2 min and then reaches a level about 0.05 of the light level from which it declines very slowly. Since the K_m for RuBP for the fully activated enzyme is about 0.035 mM (CHU and BASSHAM, 1975), and the $\Delta G'$ for the carboxylation reaction is -8.4 Kcal (BASSHAM and KRAUSE, 1969), this failure of the reaction to continue after two minutes of darkness means that the enzyme activity has greatly declined.

The light–dark inactivation of the RuBPcase is also evident with isolated spinach chloroplasts (BASSHAM and KIRK, 1968) where, following a period of photosynthesis with $^{14}CO_2$, the level of the RuBP in the dark declined to about one half the light value and then remained constant. When the light was again turned on, the level of RuBP rose very rapidly for 30 s and then declined to the light level. This behavior is analogous to that of the changes in FBP and SBP levels described above and attributed to dark inactivation of bisphosphatase activity, followed by light reactivation requiring 30 s.

When the drop in RuBP level in the isolated spinach chloroplasts was prevented by addition of ATP to the suspending medium just after the light was turned off, very little uptake of $^{14}CO_2$ occurred as long as the light was off (even though there was as much RuBP present in the chloroplasts as in the light). When the light was turned on again, high rates of $^{14}CO_2$ uptake resumed (JENSEN and BASSHAM, 1968). Although the rate of entry of ATP into whole chloroplasts may be low compared to the requirements of photosynthesis (HEBER and SANTARIUS, 1970; HELDT et al., 1971; STOKES and WALKER, 1971) this low rate is apparently sufficient to maintain the level of RuBP when it is not being consumed, once the RuBP carboxylase is inactivated.

Like the activities of FBPase and SBPase, RuBP carboxylase activity depends in part on pH, Mg^{2+} and reduced cofactors, but the mechanisms are different (see Chap. II.17, this vol.). The control of the carboxylase activity is complicated by the necessity for the plants to avoid, at least to some extent, the wasteful conversion of RuBP by oxygenase activity to phosphoglycolate and PGA (BOWES et al., 1971; ANDREWS et al., 1973; LORIMER et al., 1973). Molecular oxygen, O_2, can bind competitively at the CO_2 binding site of the carboxylase. This O_2 binding is thus favored by high O_2 and low CO_2 pressures. Low CO_2 pressures in the light can cause an increased level of RuBP, providing optimal conditions for the oxygenase reaction. To avoid this reaction, it is advantageous for the enzyme to be inactivated with respect to O_2 binding by a combination of high concentration of RuBP and low concentration of CO_2. Apparently, the binding of O_2 can only be decreased by a change in conformation of the enzyme which results in increased binding constants for both CO_2 and O_2.

Isolated RuBP carboxylase is activated by preincubation with CO_2 or bicarbonate and high levels of Mg^{2+} (e.g., 10 mM), *before* the enzyme is exposed to RuBP (PON et al., 1963; CHU and BASSHAM, 1973, 1974, 1975). Preincubation with physiological levels of RuBP in the absence of either bicarbonate or Mg^{2+} results in conversion of the enzyme to an inactive form with high K_m values for CO_2, and the enzyme does not recover its activity for many minutes upon subsequent exposure to physiological levels of bicarbonate and Mg^{2+} (CHU and

Bassham, 1973, 1974). Full activation of the isolated purified enzyme requires that the preincubation with CO_2 and Mg^{2+} also be carried out in the presence of either NADPH or 0.05 mM 6-phosphogluconate, each at physiological levels (Chu and Bassham, 1973, 1974, 1975).

With respect to light–dark regulation, it seems clear that the changes in Mg^{2+} levels and pH in the chloroplasts which affect FBPase and SBPase activities also result in changes in RuBP carboxylase activity, with the light-induced increases in pH and Mg^{2+} resulting in increased enzyme activity. The pH optimum of the isolated enzyme shifts towards the pH actually found in chloroplasts in the light (about 8) with increased Mg^{2+}, and the value of K_m for CO_2 is lower at pH 8 than at pH 7.2 (Sugiyama et al., 1968; Bassham et al., 1968; Lorimer et al., 1976).

The activation of the isolated enzyme by NADPH seems to be another part of the light–dark regulation, but the activation by 6-phosphogluconate is at first surprising, since this compound appears in the dark. Kinetic studies show that the 6-phosphogluconate is still present during the first 2 min of light after a dark period (Bassham and Kirk, 1968), and it may be that a useful activation occurs then, while the level of NADPH is still being built up (Chu and Bassham, 1973). In the dark, 6-phosphogluconate would not activate the carboxylase since the optimal conditions of pH and Mg^{2+} levels would not be met.

Although it appeared for many years that $KmCO_2$ for RuBP carboxylase is too low to support the RPP cycle, a number of laboratories have found evidence in recent years that the $KmCO_2$ is sufficiently low (see Chap. II.17, this vol.).

4. Light–Dark Regulation of Phosphoribulokinase

In vivo light–dark transient studies are not so revealing with respect to regulation of the conversion of Ru5P to RuBP since the level of the substrate ATP declines rapidly in the dark. In vivo evidence for the inactivation of the enzyme came from studies in which vitamin K_5 was added to photosynthesizing *Chlorella pyrenoidosa* (Krause and Bassham, 1969). The result was that electrons were diverted from the reduction of ferredoxin to the reduction of the oxidized form of vitamin K_5, but there was little effect on the level of ATP.

Consequently, those reactions needing reduced cofactors in the chloroplast were affected, but not via ATP levels.

Upon the addition of vitamin K_5 to the algae, there was an immediate increase in 6-phosphogluconate and in pentose monophosphates, but a rapid drop in the level of RuBP. It appears that the OPP cycle became activated, but that the conversion of Ru5P to RuBP ceased. This is in agreement with the known properties of the isolated enzyme (see Chap. II.19, this vol.).

5. Light–Dark Regulation of Glyceraldehyde 3-Phosphate Dehydrogenase

Since the reduction of phosphoryl 3-phosphoglycerate (PPGA) is the second step in the conversion of PGA to GAP, and the two steps require ATP and NADPH, light–dark transition studies reveal little about the possible regulation of the reactions. From steady-state kinetic tracer studies of photosynthesis of *Chlorella pyre-*

noidosa it seems clear that neither step is rate-limiting in the light (BASSHAM and KRAUSE, 1969), and it thus appears doubtful that regulation in the light is required. There is however, considerable enzymatic evidence that the enzymes catalyzing the oxidation of triose phosphates are subject to regulation (see Chap. II.22, this vol.).

In the light, during photosynthesis, the concentrations of PGA, NADPH, ATP and GAP are such that the overall reaction from PGA to GAP is highly reversible. In the dark, however, the drop in levels of both NADPH and ATP in the chloroplasts (LENDZIAN and BASSHAM, 1976) results in the oxidation reaction being highly favored, and limitation on the rate of oxidation may be required in order to maintain levels of GAP sufficient for operation of the OPP cycle in the chloroplasts.

6. Regulation of Glucose-6-Phosphate Dehydrogenase

Prevention of the OPP cycle in the chloroplasts during photosynthesis requires that the glucose-6-phosphate dehydrogenase be inactivated. The sudden appearance of 6-phosphogluconate in the dark and its disappearance in the light have already been mentioned. Not surprisingly, glucose-6-phosphate dehydrogenase is inactivated with increasing ratios of $NADPH/NADP^+$ (LENDZIAN and BASSHAM, 1975; WILDNER, 1975), and with changes in this ratio equal to those actually observed in chloroplasts between light and dark, there is a large change in the activity of this enzyme. The activity is further affected by RuBP and by pH in the directions expected to inactivate in the light (LENDZIAN and BASSHAM, 1975).

III. Regulation of the RPP Cycle During Photosynthesis

Besides the "on-off" kind of regulation required for transition from photosynthesis to respiration, a finer tuning of rate-limiting steps is required to keep in balance the concentrations of intermediate compounds as the physiological needs and rates of photosynthesis of the cells change (KANAZAWA et al., 1970). The possibilities for factors in the cytoplasm to influence the relative amounts of triose phosphates exported from the chloroplasts (see C. II) mean that the rates of formation and conversion of GAP via the cycle must be adjustable. The rate-limiting steps in the light are the formation and the carboxylation of RuBP and the conversions of FBP and SBP to F6P and S7P respectively. Thus the rate of the carboxylation reaction relative to the bisphosphatase reaction determines the rates of formation and utilization of triose phosphates. The reduction of PGA to triose phosphates and the conversion of triose phosphates to FBP and SBP are highly reversible, and therefore play no role in controlling triose phosphate concentrations during active, unimpaired photosynthesis.

Probably there is much more to be learned about the way in which this is accomplished, but one possible mechanism is in the sigmoidal dependence of FBPase activity on FBP concentration (PREISS et al., 1967). If the GAP level were to drop too much due to triose phosphate export, then the levels of DHAP and of FBP would also drop, the FBPase activity would decline, and triose phosphate concentration would build up. Undoubtedly, other mechanisms are required as well for controlling the levels of triose phosphates and also for the levels of hexose, heptose, and pentose monophosphates.

F. Concluding Remarks

Since the mapping of the RPP cycle 25 years ago, there has accumulated a wealth of information about the control of the cycle, the enzymes of the cycle and their regulation, the nature of export of photosynthate from the chloroplasts, and the physical and chemical variations in the chloroplasts with light, dark, and changing physiological conditions. The origin of glycolate, the substrate for photorespiration, remains in dispute, but much has been learned about glycolate formation by oxidation of sugar monophosphates and of RuBP. The C_4 cycle of certain plants was discovered, and we now know much about the way plants with this cycle minimize the adverse effects of photorespiration. Many of these topics have been mentioned very briefly in this article, but the chapters that follow will no doubt describe fully the progress in each. It is with this knowledge that the author has neglected much important work and concentrated to a great extent on work from his own laboratory in order to provide one perspective.

Acknowledgment. The preparation of this paper was sponsored by the Biomedical and Environmental Research Division of the U.S. Department of Energy under Contract No W-7405-ENG-48.

References

Allen, M.B., Whatley, F.R., Rosenberg, J.R., Capindale, J.B., Arnon, D.I.: In: Research in photosynthesis. Gaffron, H., Brown, A.H., French, C.C., Livingston, R., Rabinowitch, E.I., Strehler, B.C., Tolbert, N.E. (eds.), pp. 288–295. New York: Interscience Publ. Inc. 1957
Anderson, L.E., Avron, M.: Plant Physiol. *57*, 209–213 (1976)
Andrews, T.J., Lorimer, G.H., Tolbert, N.E.: Biochemistry *12*, 11–18 (1973)
Arnon, D.I., Allen, M.B., Whatley, F.R.: Nature (London) *174*, 394–398 (1954)
Axelrod, B., Bandurski, R.S.: J. Biol. Chem. *204*, 939–948 (1953)
Bamberger, E.S., Ehrlich, B.A., Gibbs, M.: Plant Physiol. *55*, 1023–1030 (1975)
Barber, J.: Trends Biochem. Sci. *1*, 33–36 (1976)
Barber, J., Mills, J., Nicholson, J.: FEBS Lett. *49*, 106–110 (1974)
Bassham, J.A.: In: Proc. 2nd. Int. Congr. Photosynthesis Res. Stresa, Forti, G., Avron, M., Melandri, A. (eds.), pp. 1723–1735. The Hague: Dr. W. Junk N.V. Publ. 1971a
Bassham, J.A.: Science *172*, 526–534 (1971b)
Bassham, J.A.: Symp. Soc. Exp. Biol. XXVII, Rate control of biological processes. pp. 461–483. Cambridge: Cambridge Univ. Press 1973
Bassham, J.A., Calvin, M.: In: The path of carbon in photosynthesis. pp. 1–107. Englewood Cliffs, N.J.: Prentice-Hall, Inc. 1957
Bassham, J.A., Kirk, M.R.: Biochim. Biophys. Acta *43*, 447–464 (1960)
Bassham, J.A., Kirk, M.R.: Biochem. Biophys. Res. Commun. *9*, 376–380 (1962)
Bassham, J.A., Kirk, M.: In: Comparative biochemistry and biophysics of photosynthesis. Shibata, K., Takamiya, A., Jagendorf, A.T., Fuller, R.C. (eds.), pp. 365–378. Univ. of Tokyo Press 1968

Bassham, J.A., Krause, G.H.: Biochim. Biophys. Acta *189*, 207–221 (1969)
Bassham, J.A., Benson, A.A., Calvin, M.: J. Biol. Chem. *185*, 781–787 (1950)
Bassham, J.A., Benson, A.A., Kay, L.D., Harris, A.Z., Wilson, A.T., Calvin, M.: J. Am. Chem. Soc. *76*, 1760–1770 (1954)
Bassham, J.A., Kirk, M., Jensen, R.G.: Biochim. Biophys. Acta *153*, 211–218 (1968a)
Bassham, J.A., Sharp, P., Morris, I.: Biochim. Biophys. Acta *153*, 901–902 (1968b)
Benson, A.A.: J. Am. Chem. Soc. *73*, 2971 (1951)
Benson, A.A., Bassham, J.A.: J. Am. Chem. Soc. *70*, 3939 (1948)
Benson, A.A., Calvin, M.: In: Cold Spring Harbor Symp. Quant. Biol. *13*, 6–10 (1948)
Benson, A.A., Bassham, J.A., Calvin, M., Goodale, T.C., Haas, V.A., Stepka, W.: J. Am. Chem. Soc. *72*, 1710–1718 (1950)
Benson, A.A., Bassham, J.A., Calvin, M.: J. Am. Chem. Soc. *73*, 2970 (1951)
Bowes, G., Ogren, W.L., Hageman, R.H.: Biochem. Biophys. Res. Commun. *45*, 716–722 (1971)
Buchanan, B.B., Schurmann, P., Wolosiuk, R.A.: Biochem. Biophys. Res. Commun. *69*, 970–978 (1976)
Calvin, M., Bassham, J.A., Benson, A.A., Lynch, V., Quellet, C., Schou, L., Stepka, W., Tolbert, N.E.: Soc. Exp. Biol. (G.B.) *5*, 284–305 (1951)
Calvin, M., Benson, A.A.: Science, *107*, 476–480 (1948)
Calvin, M., Massini, P.: Experientia *8*, 445–457 (1952)
Chu, D.K., Bassham, J.A.: Plant Physiol. *52*, 373–379 (1973)
Chu, D.K., Bassham, J.A.: Plant Physiol. *54*, 556–559 (1974)
Chu, D.K., Bassham, J.A.: Plant Physiol. *55*, 720–726 (1975)
Evans, M.C.W., Buchanan, B.B., Arnon, D.I.: Proc. Natl. Acad. Sci. USA *55*, 928–934 (1966)
Garnier, R.V., Latzko, E.: Proc. 2nd Int. Congr. Photosynthesis Res. Forti, G., Avron, M., Melandri, A. (eds.), pp. 1839–1845. The Hague: W. Junk 1972
Ghosh, H.P., Preiss, J.: J. Biol. Chem. *240*, 960–961 (1965)
Ghosh, H.P., Preiss, J.: J. Biol. Chem. *241*, 4491–4504 (1966)
Hatch, M.D., Slack, C.R.: Biochem. J. *101*, 103–111 (1966)
Hatch, M.D., Slack, C.R.: Annu. Rev. Plant. Physiol. *21*, 141–162 (1970)
Heber, U., Santarius, K.A.: Z. Naturforsch. *25b*, 718–728 (1970)
Heber, U., Hallier, U.W., Hudson, M.A.: Z. Naturforsch. *22b*, 1200–1215 (1967)
Heldt, H.W., Rapley, L.: FEBS Lett. *10*, 143–148 (1970)
Heldt, H.W., Sauer, F., Rapley, F.: In: Proc. 2nd Int. Congr. Photosynthesis Res. Stresa, Forti, G., Avron, M., Melandri, A. (eds.), *2*, 1345–1355 (1971)
Heldt, H.W., Werdan, K., Milovancev, M., Geller, G.: Biochim. Biophys. Acta *314*, 224–241 (1973)
Hind, G., Nakatani, H.Y., Izawa, S.: Proc. Natl. Acad. Sci. USA *71*, 1484–1488 (1974)
Horecker, B.L., Smyrniotis, P.Z., Klenow, H.: J. Biol. Chem. *205*, 661–682 (1953)
Hurwitz, J., Weissbach, A., Horecker, B.L., Smyrniotis, P.Z.: J. Biol. Chem. *218*, 769–783 (1956)
Jensen, R.G., Bassham, J.A.: Biochim. Biophys. Acta *153*, 227–234 (1968)
Kanazawa, T., Kanazawa, K., Kirk, M.R., Bassham, J.A.: Plant Cell Physiol. *11*, 149–160 (1970)
Kanazawa, T., Kanazawa, K., Kirk, M.R., Bassham, J.A.: Biochim. Biophys. Acta *256*, 656–669 (1972)
Kelly, G.J., Latzko, E., Gibbs, M.: Ann. Rev. Plant Physiol. *27*, 181–205 (1976)
Kortschak, H.P., Hartt, C.E., Burr, G.O.: Plant Physiol. *40*, 209–213 (1965)
Krause, G.H.: Biochim. Biophys. Acta *333*, 301–313 (1974)
Krause, G.H., Bassham, J.A.: Biochim. Biophys. Acta *172*, 553–565 (1969)
Leloir, L.F.: Biochem. J. *91*, 1–8 (1964)
Lendzian, K., Bassham, J.A.: Biochim. Biophys. Acta *396*, 260–275 (1975)
Lendzian, K., Bassham, J.A.: Biochim. Biophys. Acta *430*, 478–489 (1976)
Lilley, R. McC., Walker, D.: Plant Physiol. *55*, 1087–1092 (1975)
Lilley, R. McC., Schwenn, J.D., Walker, D.A.: Biochim. Biophys. Acta *325*, 596–604 (1973)
Lin, D.C., Nobel, P.S.: Arch. Biochem. Biophys., *145*, 622–632 (1971)
Lorimer, G.H., Andrews, T.J., Tolbert, N.E.: Biochemistry *12*, 18–23 (1973)
Lorimer, G.H., Badger, M.R., Andrews, T.J.: Biochemistry *15*, 529–536 (1976)

Martin, A.J.P., Singe, R.L.M.: Biochem. J. *35*, 1358–1368 (1941)

Müllhoffer, G., Rose, I.A.: J. Biol. Chem. *240*, 1341 (1965)

Müller, B.: Biochem. Biophys. Acta *205*, 102–109 (1970)

Norris, L., Norris, R.E., Calvin, M.: J. Exp. Bot. *6*, 64–74 (1955)

Pedersen, T.A., Kirk, M., Bassham, J.A.: Biochim. Biophys. Acta *112*, 189–203 (1966)

Peterkofsky, A., Racker, E.: Plant Physiol. *36*, 409–414 (1961)

Pon, N.G., Rabin, B.R., Calvin, M.: Biochem. Z. *338*, 7–9 (1963)

Preiss, J., Biggs, M., Greenberg, E.: J. Biol. Chem. *242*, 2292–2294 (1967)

Pupillo, P., Giuliani-Piccari, G.G.: Eur. J. Biochem. *51*, 475–482 (1975)

Quayle, J.R., Fuller, R.C., Benson, A.A., Calvin, M.: J. Am. Chem. Soc. *76*, 3610–3611 (1954)

Ruben, S., Kamen, M.D.: Phys. Rev. *57*, 549 (1940a)

Ruben, S., Kamen, M.D.: J. Am. Chem. Soc. *62*, 3451–3455 (1940b)

Ruben, S., Kamen, M.D., Hassid, W.A.: J. Am. Chem. Soc. *62*, 3443–3450 (1940)

Schacter, B., Eley, J.H., Gibbs, M.: Plant. Physiol. *48*, 707–711 (1971)

Schou, L., Benson, A.A., Bassham, J.A., Calvin, M.: Physiol. Plant. *3*, 487–495 (1950)

Slabas, R.R., Walker, D.A.: Biochim. Biophys. Acta *430*, 154–164 (1976)

Smith, D.C., Bassham, J.A., Kirk, M.R.: Biochim. Biophys. Acta *48*, 299–313 (1961)

Stiller, M.: Ann. Rev. Plant. Physiol. *13*, 151–170 (1962)

Stocking, C.R., Larson, S.: Biochem. Biophys. Res. Commun. *3*, 278–282 (1969)

Stokes, D.M., Walker, D.A.: In: Photosynthesis and respiration. Hatch, M.D., Osmond, C.B., Slatyer, R.O. (eds.), pp. 226–231. New York: John Wiley & Sons Inc. 1971

Stoppani, A.O.M., Fuller, R.C., Calvin, M.: J. Bacteriol. *69*, 491–501 (1955)

Sugiyama, T., Nakayama, N., Akazawa, T.: Biochem. Biophys. Res. Commun. *30*, 118–123 (1968)

Tolbert, N.E.: Annu. Rev. Plant. Physiol. *22*, 45–74 (1971)

Walker, D.A.: In: The intact chloroplast. Barber, J. (ed.), pp. 236–278. Netherlands: Elsevier/North-Holland Biomedical Press 1976a

Walker, D.A.: In: Encyclopedia of Plant Physiology, Vol. 3. Stocking, C.R., Heber, U. (eds.), pp. 85–136. Berlin, Heidelberg, New York: Springer 1976b

Walker, D.A., Cockburn, W., Baldry, C.W.: Nature (London) *216*, 597–599 (1967)

Weissbach, A., Horecker, B.L.: Fed. Proc. *14*, 302–303 (1955)

Weissbach, A., Horecker, B.L., Hurwitz, J.: J. Biol. Chem. *218*, 795–810 (1956)

Werden, K., Heldt, H.W.: In: Proc. 2nd Int. Congr. Photosynthesis Res. Stresa. Forti, G., Avron, M., Melandri, A. (eds.), *2*, 1337–1344 (1971)

Wildner, G.F.: Z. Naturforsch. *30c*, 756–760 (1975)

Wilson, A.T., Calvin, M.: J. Amer. Chem. Soc. *77*, 5948–5957 (1955)

Wolosiuk, R.A., Buchanan, B.B.: J. Biol. Chem. *251*, 6456–6461 (1976)

Wolosiuk, R.A., Schurmann, P., Breazeale, V.D., Buchanan, B.B.: Fed. Proc. *35*, 1932 (1976)

Zelitch, I.: Plant. Physiol. *41*, 1623–1631 (1966)

Zelitch, I.: Arch. Biochem. Biophys. *163*, 367–377 (1974)

Ziegler, H., Ziegler, I., Schmidt-Clausen, H.J.: Planta *81*, 181–192 (1968)

2. The Isolation of Intact Leaf Cells, Protoplasts and Chloroplasts

R.G. JENSEN

A. Introduction

The comprehension of plant metabolism requires more than the identification of pathways and enzymes involved; it requires an understanding of the contribution and function of various compartments, i.e., the organized cell and the subcellular organelles. Although metabolism during photosynthesis has been studied with whole leaves and leaf disks, such approaches are open to difficulties. One of the greatest problems is the control of stomatal aperature to allow CO_2 exchange. Also, intact leaves respond to hormonal control from other plant parts, translocation of products is regulated by sink demand, and leaf metabolism is affected by ion nutrition and water availability.

Marked improvements in isolation techniques have made separated leaf cells and protoplasts available for investigations in plant metabolism, genetic manipulation and plant virology (ZAITLIN and BEACHY, 1974; TAKEBE, 1975). For metabolic studies with photosynthetic tissues of higher plants, some distinct advantages are gained by using liquid suspensions of isolated leaf cells and protoplasts. Stomatal closure to CO_2 exchange is no longer a problem. Taking of uniform samples is convenient and it is much simpler to expose all the cells to equivalent quantities of added metabolites or herbicides and to remove them again. Separated leaf cells and protoplasts are being used to determine the function of various cell types in leaf photosynthesis. A major example is the knowledge of the metabolic compartmentation of C_4 photosynthesis, whose elucidation was made possible after separation of mesophyll protoplasts and bundle sheath cells. Questions concerning product translocation and phloem loading are being approached using separated cells and protoplasts.

Use of separated cells and protoplasts for photosynthesis and subsequent metabolic studies is not without problems. The cells are no longer connected by plasmodesmata and the plasmalemma membrane is only slightly permeable. The photosynthetic products mostly accumulate in the cells, causing potential metabolic feedback. With separated cotton leaf cells only 1% to 2% of the total carbon fixed per hour appeared in the suspending media (REHFELD and JENSEN, 1973). Likewise, the cells are quite impermeable to uptake of externally added compounds, so that cell permeability must be considered in designing experiments. With cotton leaf cells in 0.5 M inorganic phosphate, less than 2 mM phosphate was taken up over 1 h (JENSEN, unpublished). After isolation, the protoplasts, free of the protection of the cell wall, are quite fragile and prone to rupture upon handling.

The isolation of chloroplasts capable of high rates of photosynthetic CO_2 fixation has directly established that the chloroplast is the site of the complete process of photosynthesis in green plants. Studies with isolated chloroplasts are being used to demonstrate the operation of and interactions between the light-trapping

reactions and carbon metabolism. As opposed to separated leaf cells and protoplasts, chloroplasts release carbon compounds (mostly glycerate-3-P, triose-P and
glycolate) to the suspending medium and take up inorganic phosphate. Upon
isolation, this release of carbon to the suspending media can result in inhibition
of photosynthetic CO_2 fixation due to the depletion of Calvin cycle intermediates.

Selected aspects of the isolation and separation of cells, protoplasts, and chloroplasts, as developed for the retention and study of photosynthesis and carbon
assimilation, are presented and discussed. This commentary is not comprehensive;
but it does represent those developments which the author finds of greatest interest.

B. Isolation of Plant Leaf Cells and Protoplasts

I. Mechanical Methods

The initial methods developed for the separation and culturing of photosynthetically
active cells of seed plants were mechanical. The cells in the tissue were first plas-
molyzed in concentrated salt or sugar solutions. The high osmolarity caused water
loss from the living cells, resulting in retraction of the protoplast membrane from
the walls. The tissue was either teased apart or ground lightly in a mortar and
pestle to separate the cells (GNANAM and KULANDAIVELU, 1969). The tissue was
also sliced releasing protoplasts into the suspending medium (COCKING, 1972).
These methods usually yield only a relatively small number of intact cells or
protoplasts.

II. Enzymic Methods

In the last decade mechanical methods have been replaced by well-chosen crude
enzyme mixtures which separate the leaf cells by hydrolyzing the pectin of the
middle lamella (pectinase) and then free the protoplasts of the cellulosic wall
(cellulase). After plasmolysis, the procedure consists of two steps which can be
taken separately or combined. The tissue is macerated with a crude pectinase
(polygalacturonase) which, accompanied with some mechanical agitation, frees
the cells. Further treatment with a cellulase digests the walls and releases the
protoplasts (COCKING, 1972).

Many of the techniques used today are based on the work by TAKEBE et al.
(1968) who, by successfully adapting earlier procedures, described the rapid separation of large quantities of tobacco mesophyll cells. Leaf tissue was macerated
with a fungal pectinase in a hypertonic medium. These cells appeared morphologically intact and could synthesize tobacco mosaic virus. JENSEN et al. (1971) further
modified this approach for separation of mesophyll cells from tobacco leaves,
by using photosynthetic CO_2 fixation and metabolism as a guide to optimizing
the conditions of medium composition, pH, and luminance. These cell preparations
exhibited a sustained rate of CO_2 fixation for 20 to 25 h in the light with photosynthetic rates up to 58 μmol CO_2 fixed per mg chlorophyll per h. The cells were

also capable of taking up amino acids and nucleic acid precursors and incorporating them into proteins and nucleic acids (FRANCKI et al., 1971).

CATALDO and BERLYN (1974) also investigated conditions for release of large amounts of tobacco mesophyll cells capable of high rates (120 to 150 μmol CO_2 per mg chlorophyll per h) of photosynthesis as well as the release of minor veins. After 1.5 to 3.0 h of digestion with pectinase, the major portion of the cells were released by swirling for 30 s on a vortex mixer. This procedure supplied the necessary sheer force to separate the cells from the minor vein net and gave a high yield (80%–90%) of intact, functional mesophyll cells.

Palisade parenchyma cells and spongy parenchyma cells were isolated separately from *Vicia faba* L. leaflets by OUTLAW et al. (1976). The lower epidermis was first stripped away, macerated with pectinase (Macerase, Calbiochem), and preparations with 95% spongy mesophyll cells collected. After further maceration, the vascular nets were peeled away, leaving the upper epidermis attached to the palisade. Further maceration released the palisade cells.

Recently, PAUL and BASSHAM (1977) showed that the establishment and maintenance of high rates of photosynthetic CO_2 incorporation in mesophyll cells depended on a period of dark adaption immediately following isolation. In response to stress caused by the maceration, the cells increased in respiration and diminished in photosynthesis. If they were allowed a dark adaptive period of at least 12 h before subjecting the cell suspension to light, high rates of CO_2 fixation could be observed which were stable for over 80 h. Cells that were maintained at 8° C during the dark and at 22° C in the light recovered completely, while cells kept at a continuous temperature of 22° C did not recover as quickly, nor did they remain viable as long.

Using a combined pectinase-cell-stirring technique, SERVAITES and OGREN (1977) also offered some notable advances to the technology of cell separation. They were able to release between 50% to 70% of the mesophyll cells from soybean leaves within 15 min. The cells, isolated in physiological sorbitol medium (0.2–0.3 M) survived well the shock of isolation, as evidenced by the high rates of photosynthesis. Of note, the leaves from which the cells were separated had been shaded 1–2 days before removal from the plant.

Some tissues require more time to digest than tobacco or bean. AONO et al. (1974) reported that with spinach 5 h of maceration was required for 30% of the mesophyll cells to be released, based on leaf chlorophyll. NISHIMURA and AKAZAWA (1975) isolated spinach protoplasts from separated mesophyll cells, with the procedure taking over 6 h. In our hands, such long maceration times are quite detrimental to the subsequent photosynthetic activity of the separated mesophyll cells. Although spinach leaves are one of the slower tissues to release cells in the presence of pectinase, the process can be speeded up if the leaf epidermis is first brushed with a powdered abrasive (BEIER and BRUENING, 1975).

III. Cell and Protoplast Isolation from C_3 and C_4 Grasses

The mechanical approach of grinding with a mortar and pestle was modified to isolate mesophyll and bundle sheath cells from C_4 grasses. One of the earlier attempts was with *Digitaria sanguinalis* (crabgrass) leaves, where after gentle grinding the mesophyll cells were isolated by filtration on stainless steel sieves and nylon nets (Edwards and Black, 1971). More extensive grinding freed the bundle sheath strands, which upon careful use of a tissue homogenizer released the bundle sheath cells for collection on nylon nets. This approach with *D. sanguinalis* yielded 93% intact mesophyll cell preparations having about 15% of the leaf chlorophyll and contaminated with only 1% bundle sheath cells. Purified bundle sheath cell preparations could be obtained comprising up to 5% of the leaf chlorophyll, 60%–75% intact, with 5%–10% contamination with mesophyll cells. This mechanical procedure for separation of mesophyll and bundle sheath cells was improved by the addition of a digestion period with cellulase and pectinase (Chen et al., 1973). This further increased the yield and gave essentially homogeneous cell fractions.

Mechanical grinding without the assistance of macerating enzymes is a rather harsh procedure in which the mesophyll cells of most C_4 plants, with the exception of *D. sanguinalis* and *Kochia scoparia*, are usually broken (Huber et al., 1973). Huber and Edwards (1975) have recently optimized techniques and re-evaluated the various parameters necessary for cell and protoplast separation from C_3 and C_4 grasses capable of high rates of photosynthetic CO_2 fixation. Best results were obtained using small transverse leaf segments about 0.5 mm in width, incubated in the light at 30° C with 0.5 M sorbitol, 2% cellulase and 0.1% pectinase at pH 5.5. Again it was noted that leaf age and leaf segment size were important factors. Young tissue was more susceptible to enzymic digestion; with some species, as sorghum, no cell release was observed with plants over 21 days old. With wheat, highest activities were always observed with young tissue. The use of small uniform leaf segments eliminated the need for vacuum infiltration of the enzyme mixture into the leaf free space. Although shaking did provide higher yields of bundle sheath strands from *D. sanguinalis* and wheat, it reduced the yield and the activity for CO_2 fixation of the mesophyll protoplast. The pH of the medium has a dramatic effect on the yield and metabolic activity of the released cells or protoplasts. The pectinase and cellulase enzymes are optimally active for hydrolysis at pH 5.5–5.8. Also at this pH the most active cell preparations are obtained; typifying the observation that those factors which increase the rate of separation usually enhance CO_2 fixation of the released cells.

Temperature during isolation and storage is also important. Protoplasts and cells isolated at 30° C from *D. sanguinalis* were most active, while 22° C yielded the most active wheat mesophyll protoplasts. Storage was best at 4° C for wheat protoplasts and *D. sanguinalis* bundle sheath cells, while 25° C was best for storage of *D. sanguinalis* mesophyll protoplasts.

Kanai and Edwards (1973) have successfully used a polymer two-phase system for purification of mesophyll protoplasts and bundle sheath strands. The crude mesophyll protoplast preparation was added to a mixture containing 0.46 M sorbitol, 10 mM potassium phosphate, pH 7.5, 5.5% polyethylene glycol 6,000 and 10% dextran (MW 40,000). This mixture forms an aqueous two-phase system (Albertsson, 1971). Upon centrifugation at 300 *g* the intact protoplasts partition at the middle interface, while the chloroplasts and broken protoplasts disperse into the bottom phase. These and other methods as developed in Edwards' laboratory for isolation of photosynthetically active protoplasts were recently reviewed (Edwards et al., 1976 a).

C. Isolation of Intact Chloroplasts

The modern study of in vitro photosynthesis began with the quantitative experiments of HILL (1937). Hill discovered that with isolated plant chloroplasts photochemical O_2 evolution could be shown to occur in the absence of CO_2 assimilation. Subsequently, researchers have provided information upon which much of our present understanding of photosynthetic electron transport rests. Chloroplast isolation techniques have developed within the last decade, so that now it is routine with spinach or pea leaves to isolate functional intact chloroplasts capable of the entire act of photosynthesis of both O_2 evolution and CO_2 assimilation going to starch and sugar phosphates. Chloroplast photosynthesis requires only light, CO_2, H_2O, and Pi and with these, the rate of photosynthesis can be near that observed with intact plants. For details, the reader is referred to the publications of those investigators instrumental in developing these techniques (JENSEN and BASSHAM, 1966; KALBERER et al., 1967; WALKER, 1964, 1965. See WALKER, 1971, for a review).

For light-dependent CO_2 fixation, isolated chloroplasts must either have retained their outer envelopes intact with the stroma contents and lamellae inside or the stroma contents must be added back to the naked lamellae (reconstituted chloroplasts). HALL (1972) has proposed a useful nomenclature to be applied to isolated chloroplasts to define both their functional and morphological characteristics:

Type A Complete chloroplasts with intact outer envelopes exhibiting functioning translocases. They fix CO_2 at high rates (50–250 µmol per mg chl per h) without addition of substrates other than CO_2 and phosphate.

Type B Unbroken chloroplasts with surrounding envelopes and stroma but which fix CO_2 at rates less than 5 µmol per mg chl per h.

Type C Broken chloroplasts with intact lamellae systems slightly swollen, and washed free of the limiting outer envelopes and stroma. Little CO_2 fixation can be detected.

Type D Free-lamellae chloroplasts prepared from complete chloroplasts (Type A) by osmotic shock, but returned to an isotonic medium. The outer envelope and stroma are removed, but are still dispersed in the medium.

Type E Chloroplast fragments with swollen disks and irregularly arranged thylakoids, used often to study electron transport and photophosphorylation.

Type F Sub-chloroplast particles prepared by physical or chemical disruption of lamellae membranes.

I. Plant Material and Media

Experience has shown that two of the major factors required to prepare Type A chloroplasts capable of high rates of photosynthesis are the source and growth conditions of the leaf tissue. Leaf tissue from plants having high amounts of phenolic compounds and phenol oxidase usually yield chloroplast preparations having poor activity due to the interfering action of the polyphenols and quinones. With the addition of selective protective agents to the isolation medium, this problem may be reduced (LOOMIS, 1974), but there is no generally applicable standard method for all plant types. With sugar cane leaf extracts, BALDRY et al. (1970) found that the most protective compounds against o-diphenol:O_2 oxidoreductase activity were thioglycolate, β-mercaptoethanol, polyethylene glycol, and bovine serum albumin.

Chloroplasts isolated from spinach or pea have generally proven capable of the highest rates of light-dependent CO_2 fixation. They are also low in polyphenols. However, even if all isolation conditions are optimal, leaf tissue from stressed or senescing plants will yield chloroplasts with low rates of photosynthesis. With spinach, chloroplasts with the highest rates come from leaves of plants undergoing vigorous growth.

The compositon of the isolation media is also essential. Anxious to avoid potential metabolites in their pioneer work, Arnon and his co-workers used 0.35 M NaCl as an osmoticum rather than sucrose (KALBERER et al., 1967). These chloro-plasts exhibited only limited rates of photosynthesis (GIBBS and CALO, 1959). Subse-quently, WALKER (1964) introduced 0.33 M sorbitol, which is now used by most workers.

II. Isolation Methods

Isolated chloroplasts possess a somewhat fragile outer envelope, therefore it is important to prepare chloroplasts with rapid, but mild conditions. Grinding of the leaf tissue with a mortar and pestle is usually too harsh. Use of a Waring Blender or Sorvall Omni-mixer can be effective if one insures rapid cutting of tissue with a minimum of extra shear. We also attain good results using a dispersing homogenizer such as the Polytron (Kinematica GmbH, Lucerne, Switzerland), which within 1–2 s disrupts the tissue, giving good yields of intact chloroplasts.

After blending, the chloroplasts are filtered through two layers of Miracloth (Calbiochem). This removes the leaf debris, including whole cells larger in size than the chloroplasts. Centrifugation in 50 ml centrifuge tubes containing 20 ml of suspension is done for 80–100 s upon rapid acceleration to 600 g at 0°–4° C. Centrifugation time should be short, as chloroplasts often contain dense starch granules which can disrupt the outer envelope. After spinning, the supernatant is poured off and the loose chloroplasts on top, enriched with broken chloroplasts, are gently rinsed away from the harder-packed pellet by adding 2–3 ml of isolation medium carefully to the side of the tube and swirling slightly. The pellet is re-suspended in medium by slowly drawing in and out with a disposable capillary pipette, then recentrifuged. The chloroplasts store best as a pellet or thick suspen-sion cooled in ice.

Gross intactness of the chloroplasts can be easily observed by phase contrast microscopy (SPENCER and UNT, 1965; JENSEN and BASSHAM, 1966). Type A and Type B chloroplasts both appear bright and highly refractive without clear grana. A more sensitive and quantitative test to measure the portion of Type A plastids is to use ferricyanide reduction (Hill reaction). HEBER and SANTARIUS (1970) first showed that Type A chloroplasts were not capable of exhibiting ferricyanide reduc-tion as measured spectrophotometrically, and proposed that this property could be used to measure quantitatively the percentage of chloroplasts with intact envelopes. Walker modified this technique (LILLEY et al., 1975; LILLEY and WALKER, 1975) so that ferricyanide-dependent O_2 evolution was routinely measured as a criterion of chloroplast intactness. They note that although ferricyanide-dependent O_2 evolution appears to be suitable to estimate the proportion of chloroplasts with intact envelopes, it is not linearly related to the protein content or CO_2 fixation ability. This strongly suggested that during isolation some of the chloroplasts had resealed following rupture, and had lost some of the stromal contents.

Recently, a modified method for isolation of intact chloroplasts from peas and spinach was reported (NAKATANI and BARBER, 1977). This technique isolated the plastids in a cation-low medium and washed them in a cation-free medium with 0.33 M sorbitol. The authors reported that when the salt content of the medium was raised, a greater proportion of the pellet contained broken chloroplasts.

In order to evaluate these observations, we have compared our own isolation media (JENSEN, 1971) with that of NAKATANI and BARBER (1977) and of LILLEY and WALKER (1974). The maceration, filtration, and centrifugation procedures were the same in all preparations (as described above), only the composition of the maceration, wash, and suspension media was varied. Percent intactness was evaluated by ferricyanide-dependent O_2 evolution. All techniques gave preparations with greater than 85% intact chloroplasts. The results of one typical experiment are shown in Table 1. The biggest difference between the various isolation media was in the stability of the photosynthetic activity of the preparations during storage at 0° C. Preparations prepared and suspended in cation-free media (Media C, Table 1) were unstable and lost most of their capacity for photosynthetic CO_2-dependent O_2 evolution within 2 h, even though there was no increase in ferricyanide-dependent O_2 evolution over 4 h. If the chloroplasts, after the cation-free wash, were suspended in a cation-containing medium, stability was greatly enhanced (Media D). Of note, those chloroplasts prepared in the cation-free media did contain significantly more ribulose bisphosphate (RuBP) carboxylase activity, as measured after complete activation (BAHR and JENSEN, 1978). In our hands, chloroplasts isolated in a freshly prepared pyrophosphate medium (Media B) usually showed lower rates of photosynthesis, probably due to a deficiency of intermediates of the Calvin cycle as indicated by the lower amounts of RuBP. This problem might have been avoided if the maceration medium had been chilled to "a consistency resembling melting snow" (SCHWENN et al., 1973).

Other techniques for isolation of chloroplasts with retention of photosynthetic CO_2 fixation capability have been used. These techniques require considerably more effort and are not as rapid as those just outlined. It has generally been recognized that density gradient centrifugation can yield chloroplasts of greater purity, but most of these preparations have not retained significant photosynthetic capacity. Most likely the problem lies in the choice of the gradient material. At the density of chloroplasts, sucrose or sorbitol have too high osmotic potentials. Ficoll is extremely viscous. A purified silica sol, Ludox AM, has been used to form density gradients from which spinach chloroplasts were purified and photosynthesis retained (MORGENTHALER et al., 1974). LARSSON et al. (1971), have isolated spinach chloroplasts into three bands, essentially free from contamination, using countercurrent distribution in a dextran-polyethylene glycol two-phase system. One band contained broken chloroplasts, the second band had intact chloroplasts capable of good rates of CO_2 fixation, and the third band consisted of particles containing one to three chloroplasts surrounded by a membrane-bound cytoplasmic layer including mitochondria and microbodies (LARSSON and ALBERTSSON, 1974). This

Table 1. Properties of isolated spinach chloroplasts as prepared with different media

Isolation media[a]	Percent intactness	CO_2-dependent O_2 evolution (µmol per mg chl per h)			Maximal RuBP carboxylase[b] (µmol per mg chl per h)	nmol RuBP per mg chl[c]
		Time of storage				
		0 h	2 h	4 h		
A	90	266	195	127	350	25
B	90	180	105	26	337	13
C	89	216	6	8	407	28
D	89	296	155	99	398	ND

[a] Isolation Media A see JENSEN (1971), B see LILLEY and WALKER (1974), C see NAKATANI and BARBER (1977), D isolated and washed in media of NAKATANI and BARBER (1977), suspended in 0.33 M sorbitol, 50 mM HEPES-NaOH, pH 6.7, 2 mM EDTA, 1 mM $MgCl_2$ 1 mM $MnCl_2$, 0.5 mM K_2HPO_4, 2 mM Na isoascorbate.
[b] Assayed per BAHR and JENSEN (1978).
[c] Assayed enzymically per SICHER and JENSEN (in preparation).

last band was also capable of CO_2 fixation, but the product pattern included sucrose and more amino and organic acids, indicative of extrachloroplast metabolism.

The isolation techniques discussed here were developed to obtain cells, protoplasts, and chloroplasts which had retained significant rates of photosynthetic CO_2 fixation. With most preparations, researchers have not been concerned with contamination of extracellular or extrachloroplast components. A number of methods using marker enzymes have been used to estimate the extent of cytoplasmic contamination of chloroplast preparations. KELLY and GIBBS (1973) described a $NADP^+$-linked nonreversible D-glyceraldehyde-3-P dehydrogenase in photosynthetic tissues located outside the chloroplast. It has been used successfully to estimate cytoplasmic contamination of chloroplast preparations (KELLY and LATZKO, 1975). Pyruvate kinase is another suitable marker enzyme (HEBER, 1960). Contamination of mitochondria or peroxisomes has been estimated by measurement of cytochrome oxidase and catalase respectively (MIFLIN and BEEVERS, 1974). Measurement of the relative amounts of cytoplasmic ribosomal RNA compared to the ribosomal RNA of chloroplasts has also been used to test purity (BIRD et al., 1973).

III. Chloroplast Isolation from Other Plants

Although it is difficult to isolate functional chloroplasts capable of high rates of photosynthetic CO_2 fixation from plants other than spinach or pea, some recent approaches do offer potential solutions. When envelope-free thylakoid membranes are combined properly with a soluble chloroplast extract, substantial rates of photosynthetic CO_2 fixation can be shown (BASSHAM et al., 1974; LILLEY et al., 1974). This reconstructed chloroplast system offers a unique research tool whereby chloroplast metabolism can be studied without the limitations of the outer envelope. DELANEY and WALKER (1976) have recently used a reconstituted chloroplast system prepared from sunflower chloroplasts and measured good rates of photosynthesis, while the intact sunflower chloroplasts showed only very low rates. It appeared that the chloroplast outer envelope provided a protective barrier against damage by phenolic compounds during the isolation.

With the reconstituted system there are difficulties in obtaining an appreciable buildup of intermediates in the relatively large volume of the reaction mixture compared with the minute volume of the stromal compartment. This can be offset by adding excess stromal protein. BASSHAM et al. (1974) noted that nearly all the incorporated ^{14}C was found in intermediates of the Calvin cycle with very little starch or glycolate formed. Substantial amounts of ^{14}C were found in glycerate-3-P and hexose and heptulose monophosphates, while the levels of ^{14}C in dihydroxyacetone-P and fructose-1,6-P_2 were lower than the whole chloroplasts.

Another very promising technique for the isolation of functional chloroplasts is to first isolate leaf protoplasts. RATHNAM and EDWARDS (1976) showed that chloroplasts capable of high rates of photosynthesis could be obtained by careful breakage of protoplasts of several C_3 grasses (wheat, barley) and tobacco. These chloroplasts were almost totally intact and exhibited the same rate of CO_2 fixation as the protoplasts. The drawbacks that limit this approach are the same that limit the isolation of protoplasts. The inconveniences include the cost and availability of the pectinase and cellulase, and the several hours needed for protoplast release.

Although intact chloroplasts have been isolated from C_4 plants, the rates of light-dependent CO_2 fixation, especially in the absence of added substrates, such as pyruvate or phosphoenolpyruvate have been quite low. One exception was an early report that chloroplasts isolated from juvenile corn tissue could fix CO_2 at rates up to 45 µmol per mg chlorophyll per h (GIBBS et al., 1970). No attempt was made to separate mesophyll from bundle sheath chloroplasts. More recently it has been shown that corn mesophyll protoplasts do not contain ribulose bisphosphate carboxylase and only the bundle sheath chloroplasts contained the complete Calvin pathway (for review, see HATCH, 1976). It appears that complete operation of C_4 photosynthesis requires cooperation of both cell types, not just the operation of the mesophyll chloroplast as found in C_3 photosynthesis. As a representative of CAM plants, CO_2 fixation has been demonstrated in isolated *Kalanchoe* chloroplasts at rates up to 5.4 µmol per mg chlorophyll per h (LEVI and GIBBS, 1975).

By modification of approaches for direct leaf tissue disruption or gentle lysis of separated leaf protoplasts, the isolation and investigative use of chloroplasts of economically important plants will become more prevalent in the future. More advances in chloroplast isolation technology are now possible because the understanding of the requirements for photosynthetic CO_2 fixation in the plastid has been greatly expanded, i.e., measurement of intactness; factors involved in activation of the enzymic machinery, especially the RuBP carboxylase, and generation of RuBP in the light; the selective transport properties of the outer envelope; and, with C_4 plants, the unique functions of the mesophyll and bundle sheath chloroplasts.

References

Albertsson, P.A.: Partition of cell particles and macromolecules. New York-London-Sydney-Toronto. Wiley-Interscience, 1971
Aono, R., Kano, H., Hirata, H.: Plant Cell Physiol. *15*, 567–570 (1974)
Bahr, J.T., Jensen, R.G.: Arch. Biochem. Biophys. *185*, 39–48 (1978)
Baldry, C.W., Bucke, C., Coombs, J.: Planta *94*, 124–133 (1970)

Bassham, J.A., Levine, G., Folger, J. III.: Plant Sci. Lett. *2*, 15–21 (1974)
Beier, H., Bruening, G.: Virology, *64*, 272–276 (1975)
Bird, I.F., Cornelius, M.J., Dyer, T.A., Keys, A.J.: J. Exp. Bot. *24*, 211–215 (1973)
Cataldo, D.A., Berlyn, G.P.: Am. J. Bot. *61*, 957–963 (1974)
Chen, T.M., Campbell, W.H., Dittrich, P., Black, C.C.: Biochem. Biophys. Res. Commun. *51*, 461–467 (1973)
Cocking, E.C.: Ann. Rev. Plant Physiol. *23*, 29–50 (1972)
Delaney, M.E., Walker, D.A.: Plant Sci. Lett. *7*, 284–294 (1976)
Edwards, G.E., Black, Jr. C.C.: Plant Physiol. *47*, 149–156 (1971)
Edwards, G.E., Huber, S.C., Gutierrez, M.: In: Microbial and plant protoplasts. J.F. Peberdy, A.H. Rose, H.J. Rogers, E.C. Cocking (eds.), pp. 299–322. New York: Academic Press 1976
Francki, R.I.B., Zaitlin, M., Jensen, R.G.: Plant Physiol. *48*, 14–18 (1971)
Gibbs, M., Calo, N.: Plant Physiol. *34*, 318–323 (1959)
Gibbs, M., Latzko, E., O'Neal, D., Hew, C.S.: Biochem. Biophys. Res. Commun. *40*, 1356–1361 (1970)
Gnanam, A., Kulandaivelu, G.: Plant Physiol. *44*, 1451–1456 (1969)
Hall, D.O.: Nature (London) *235*, 125–126 (1972)
Hatch, M.D.: In: CO_2 metabolism and plant productivity. Burris, R.H., Black, C.C. (eds.), pp. 59–81. Baltimore: Univ. Park Press 1976
Heber, U.: Z. Naturforsch. *15b*, 100–109 (1960)
Heber, U., Santarius, K.A.: Z. Naturforsch. *25*, 718–728 (1970)
Hill, R.: Nature (London) *139*, 881–882 (1937)
Huber, S., Edwards, G.E.: Physiol. Plant. *35*, 203–209 (1975)
Huber, S.C., Kanai, R., Edwards, G.E.: Planta *113*, 53–66 (1973)
Jensen, R.G.: Biochim. Biophys. Acta *234*, 360–370 (1971)
Jensen, R.G., Bassham, J.A.: Proc. Natl. Acad. Sci. USA *56*, 1095–1101 (1966)
Jensen, R.G., Francki, R.I.B., Zaitlin, M.: Plant Physiol. *48*, 9–13 (1971)
Kalberer, P.P., Buchanan, B.B., Arnon, D.I.: Proc. Natl. Acad. Sci. USA *57*, 1542–1549 (1967)
Kanai, R., Edwards, G.E.: Plant Physiol. *52*, 484–490 (1973)
Kelly, G.J., Gibbs, M.: Plant Physiol. *52*, 111–118 (1973)
Kelly, G.J., Latzko, E.: Nature (London) *256*, 429–430 (1975)
Larsson, C., Albertsson, P.A.: Biochim. Biophys. Acta *357*, 412–419 (1974)
Larsson, C., Collin, C., Albertsson, P.A.: Biochim. Biophys. Acta. *245*, 425–438 (1971)
Levi, C., Gibbs, M.: Plant Physiol. *56*, 164–166 (1975)
Lilley, R.M., Fitzgerald, M.P., Rienits, K.G., Walker, D.A.: New Phytol. *75*, 1–10 (1975)
Lilley, R.M., Holborow, K., Walker, D.A.: New Phytol. *73*, 657–662 (1974)
Lilley, R.M., Walker, D.A.: Biochim. Biophys. Acta *368,* 269–278 (1974)
Lilley, R.M., Walker, D.A.: Plant Physiol. *55*, 1087–1092 (1975)
Loomis, W.D.: Methods Enzymol. *31*, 528–544 (1974)
Miflin, B.J., Beevers, H.: Plant Physiol. *53*, 870–874 (1974)
Morgenthaler, J.J., Price, C.A., Robinson, J.M., Gibbs, M.: Plant Physiol. *54*, 532–534 (1974)
Nakatani, H.Y., Barber, J.: Biochim. Biophys. Acta. *461*, 510–512 (1977)
Nishimura, M., Akazawa, T.: Plant Physiol. *55*, 713–716 (1975)
Outlaw, W.H., Jr., Schmuck, C.L., Tolbert, N.E.: Plant Physiol. *58*, 186–189 (1976)
Paul, J.S., Bassham, J.A.: Plant Physiol. *60*, 775–778 (1977)
Rathnam, C.K.M., Edwards, G.E.: Plant Cell Physiol. *17*, 177–186 (1976)
Rehfeld, D.W., Jensen, R.G.: Plant Physiol. *52*, 17–22 (1973)
Schwenn, J.D., Lilley, R.M., Walker, D.A.: Biochim. Biophys. Acta. *325*, 586–595 (1973)
Servaites, J.C., Ogren, W.L.: Plant Physiol. *59*, 587–590 (1977)
Spencer, D., Unt, H.: Aust. J. Biol. Sci. *18*, 197–210 (1965)
Takebe, I.: Annu. Rev. Phytopathol. *13*, 105–125 (1975)
Takebe, I., Otsuki, Y., Aoki, S.: Plant Cell Physiol. *9*, 115–124 (1968)
Walker, D.A.: Biochem. J. *92*, 22–23 (1964)
Walker, D.A.: Plant Physiol. *40*, 1157–1161 (1965)
Walker, D.A.: In: Methods in enzymology. San Pietro, A., (ed.), Vol. XXIII, pp. 211–220. New York: Academic Press 1971
Zaitlin, M., Beachy, R.N.: Adv. Virus Res. *19*, 1–35 (1974)

3. Studies with the Reconstituted Chloroplast System

R.McC. Lilley and D.A. Walker

A. Reconstituted Chloroplast Systems

I. Introduction

Photosynthetic carbon assimilation by isolated chloroplasts was first demonstrated by ARNON et al. in 1954. Work in the same laboratory also showed that the amount of CO_2 fixed was diminished by osmotic shock and that photosynthesis could be "reconstituted" by mixing osmotically shocked chloroplasts with "chloroplast extract". At first, only slow rates of CO_2 fixation by intact chloroplasts were reported, but improvements in technique eventually led to rates which equalled those of the parent tissue, (BUCKE et al., 1966; JENSEN and BASSHAM, 1966).

In retrospect the improvement has been attributed to two factors. The first was the introduction of new isolation procedures (WALKER, 1964) involving brief grinding, rapid centrifugation, and a return to Hill's practice of using a sugar (or sugar alcohol) as the osmoticum. It could be demonstrated (WALKER, 1965) that this gave preparations containing large numbers of chloroplasts with intact envelopes. The second involved incubating chloroplasts in mixtures which allowed them to display their full photosynthetic potential (JENSEN and BASSHAM, 1966). As before, osmotic shock virtually abolished CO_2 fixation.

In 1971 the question of the rapidity of movement of ATP and ADP across the spinach chloroplast envelope was still a matter for debate. The evidence against rapid movement was considerable and included the fact that (in isolated chloroplasts) exogenous ATP did not restore photosynthesis which had been inhibited by uncouplers. It seemed desirable, however, to see if the predicted recovery would occur in a system in which the chloroplast envelopes no longer offered a barrier to ATP, and accordingly a new attempt was made to "reconstitute" PGA-dependent O_2 evolution in envelope-free chloroplasts supplemented with a number of additives. In the event this was easily accomplished, the rates were equal to those displayed by the intact chloroplast, and the predicted restoration by ATP of the uncoupled reaction was demonstrated (STOKES and WALKER, 1971 a).

Following the first demonstration of PGA-dependent O_2 evolution, attempts were made to extend reconstitution so that it embraced more and more reactions of the pentose phosphate pathway (Fig. 1). When attention was paid to the Mg requirement of RuBP carboxylase it became possible to obtain CO_2-dependent O_2 evolution with R5P as precursor of the CO_2 acceptor (WALKER et al., 1971). Finally the circle was closed (Fig. 1) when BASSHAM et al. (1974) achieved CO_2 fixation with PGA as substrate. Similarly O_2 evolution has been recorded in mixtures containing CO_2 and triose phosphate as substrate (WALKER and SLABAS, 1976). In both laboratories evidence was obtained which implies that

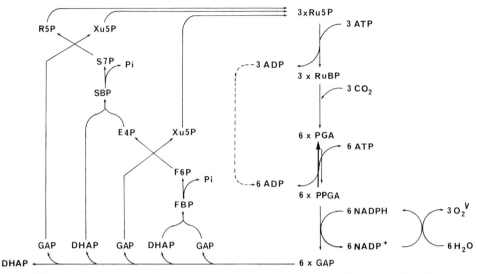

Fig. 1. The reductive pentose phosphate pathway. In the reconstituted system, CO_2 fixation with PGA (the product of CO_2 fixation) as substrate, and O_2 evolution with triose phosphate (the product of the only reductive step) as substrate, have both been observed (see Sect. B.III) indicating the operation of the full sequence of 13 partial reactions. In some circumstances evidence of several revolutions has been achieved

several revolutions of the cycle had occurred (Bassham et al., 1974; Walker and Lilley, 1974).

Definitive evidence of cycling in intact chloroplasts based on intramolecular ^{14}C distribution, demonstrated by Gibbs and Cynkin (1958) for intact chloroplasts, has not been attempted recently with reconstituted systems of high activity (but see Havir and Gibbs, 1963).

II. Definition

A reconstituted chloroplast system may be regarded as one in which the isolated photochemical apparatus of the chloroplast (thylakoid system) is reassociated with soluble components of the stroma in such a way that the mixture is capable of catalysing all or part of the reductive pentose phosphate pathway. According to this definition, a mixture of thylakoids and ferredoxin capable of reducing substrate concentrations of $NADP^+$ in the light would be an example of a Hill reaction, whereas an otherwise comparable mixture containing glyceric acid 1,3-bisphosphate (PPGA), triose phosphate dehydrogenase and only catalytic concentrations of $NADP^+$ would be a reconstituted system at its most simple.

A sequence involving the reactions between Ru5P and GAP approaches full photosynthesis to the extent that it involves simultaneous fixation of CO_2, evolution of O_2 and generates and utilises ATP and NADPH in much the same way as the entire cycle. Effective reconstitution of this type can be accomplished by adding back to osmotically shocked chloroplasts the soluble protein that is released, together with additional ferredoxin, NADP, ADP, and Mg. For some purposes additional stromal protein is used. Reconstitution using washed thylakoids and relatively pure protein fractions has recently been reported (SLABAS and WALKER, 1976a) and there seems to be no reason why rapid rates of photosynthetic carbon assimilation should not be eventually achieved in systems based on thylakoids plus fully purified enzymes and cofactors.

III. Methods of Preparation

In principle, intact chloroplasts are prepared by conventional methods (e.g., in 0.33 M sorbitol at pH 6.5) and subjected to osmotic shock (e.g., by resuspension in 1 part of medium + 24 parts of water). After a second centrifugation (which may be prolonged in order to remove any thylakoid fragments) the thylakoids are resuspended in full strength medium. The supernatant (stromal protein) is then added back to the chloroplasts to restore the original chlorophyll/protein ratio in a medium of normal osmolarity (i.e., containing the same concentrations of sorbitol, buffer etc.) used in the assay of intact chloroplasts. Alternatively excess stromal protein may be added, but this can usually be achieved only if the chloroplast extract is previously decreased in volume by concentration dialysis or similar procedure. Reaction mixtures normally contain additional ferredoxin, dithiothreitol, ascorbate, catalase, orthophosphate, Mg, and catalytic $NADP^+$ and ADP. For full details of normal typical procedures see e.g., LILLEY and WALKER (1974). Reconstituted systems can be prepared from more "difficult" species such as sunflower by including appropriate protective reagents (DELANEY and WALKER, 1976).

IV. Activities Achieved

The activity of whole (intact or Class A) chloroplasts is usually related to chlorophyll content. On this basis it is relatively easy to devise reconstituted systems which generate their own ATP and NADPH, and which have the same activity as the intact chloroplasts or the parent tissue from which they are derived. Indeed it is often easier to make a "good" reconstituted system than it is to prepare "good" chloroplasts. Most active systems so far employed did, however, contain additional stromal protein, although it is possible to get reasonable rates if the augmentation is restricted to the provision of additional ferredoxin. The requirement for extra ferredoxin is presumed to result from the fact that this is the only freely soluble protein component which interacts directly with the thylakoids (see Sect. B.I). In principle the rates achieved by each of the enzymes of the reductive pentose phosphate pathway should be the same whether they are spread evenly throughout a 1 ml reaction mixture or confined within chloroplasts which, for 100 µg chlorophyll, would occupy only 1 or 2 µl. This, however, presupposes both a lack of organisation and a lack of interaction (such as regulation of one step by the product of another), and clearly neither possibility can be lightly dismissed.

One aspect which has surprisingly failed to excite much comment is the rate of "endogenous" photophosphorylation maintained by the reconstituted system. PGA-dependent O_2 evolution requires two molecules of ATP per O_2 evolved and rates of 150–200 (i.e., 300–400 µmol ADP esterified per mg chl per h) are commonplace with this substrate (see e.g., Walker and Slabas, 1976). Although exceptionally high rates of photophosphorylation (1000–2000) have occasionally been attained under special conditions with phenazine methosulphate, or the related compound pyocyanine as cofactors, rates of ATP formation of 200–300 are much nearer the norm under conditions comparable to those used in the reconstituted system. It may therefore be inferred that, under these conditions, the reconstituted system can support non-cyclic photophosphorylation at rates which match or surpass those observed in the presence of non-physiological catalysts of cyclic photophosphorylation.

V. Advantages and Drawbacks

In a sense, the reconstituted system is like one large chloroplast in which the limiting envelope is replaced by the containing vessel. Accordingly, the permeability barrier imposed by the envelopes is removed and compounds which would normally pass this barrier slowly, if at all (see e.g., Heber, 1974; Walker, 1976), can be added at will. Similarly the proportions of the various fractions can be altered. For example, the soluble protein/chlorophyll ratio is easily changed and, if available, single components such as RuBP carboxylase can be boosted. Thylakoids from one species can be used with stromal protein from another (Delaney and Walker, 1976). Stromal protein is relatively stable and may be stored for several weeks. Since a high level of integrity is not needed in the initial stages of preparation, active reconstituted systems are more readily prepared than really active intact chloroplasts in which the activity declines strongly as a function of envelope damage.

The main drawback to date is that the stromal compartment of the reconstituted chloroplast may be larger in volume by a factor of 500 to 1000 (i.e., 100 µg chl is originally contained within a volume of about 1–2 µl and this is increased in the systems most frequently used to one ml or more). For this reason it is more difficult to build up cycle intermediates by the normal autocatalytic process (Sec. B.VI). At present the reconstituted system therefore lends itself more readily to the examination of straight chain sequences and control mechanisms (following the addition of an intermediate at substrate concentration) than to studies embracing the full cycle.

B. Factors Affecting the Activity of Partial Reaction Sequences in Reconstituted Chloroplast Systems

I. The Light Reactions

1. The Photoreduction of NADP$^+$

Ferredoxin, a soluble component of the stroma, may be regarded as having a special role in photoelectron transport since it transfers electrons from the insoluble photochemical apparatus in the thylakoid membranes to $NADP^+$ in the stroma of the chloroplast. In the first reported experiments with reconstituted chloroplasts, no extra ferredoxin was added (WHATLEY et al., 1956), and the system was dependent on endogenous ferrdoxin in the chloroplast extract. STOKES and WALKER (1971 b) found that the addition of ferredoxin to a preparation of osmotically shocked chloroplasts increased the rate of O_2 evolution with PGA as substrate; the degree of stimulation depended on the amount of chloroplast extract present (STOKES et al., 1972). Subsequent demonstrations of high activities of carbon fixation and of Calvin cycle turnover have been made with reconstituted chloroplasts containing about 10 µM of added ferredoxin (BASSHAM et al., 1974; WALKER and LILLEY, 1974).

A ferredoxin concentration of 10 µM should be adequate in reconstituted chloroplasts on the basis that 5–6 µM ferredoxin is sufficient to achieve the maximum rate of $NADP^+$-dependent O_2 evolution by envelope-free spinach chloroplasts in the absence of chloroplast extract (LILLEY and WALKER, unpublished). Using similar conditions but higher light intensities, ALLEN (1975) found that about 18 µM ferredoxin was required. Surprisingly, however, the actual concentration of ferredoxin in the stroma of intact chloroplasts is far higher than these values. Although early estimates put the ferredoxin content of spinach chloroplasts at one ferredoxin molecule per 400 molecules chl, or about 3 nmol ferredoxin per mg chl (TAGAWA and ARNON, 1962), much higher values have been measured recently, including 34 nmol ferredoxin per mg chl (MATSON and KIMURA, 1975) and 17 nmol ferredoxin per mg chl (LILLEY and WALKER, unpublished). Since the volume of the stroma compartment of spinach chloroplasts is about 23 µl per mg chl (HELDT et al., 1973), a ferredoxin content of 23 nmol per mg chl would be equivalent to a ferredoxin concentration of 1 mM in the stroma. The reason for this discrepancy, of two orders of magnitude, between the apparent ferredoxin requirement of reconstituted chloroplasts and the actual in vivo concentration has yet to be found.

The stimulation of CO_2-fixation in reconstituted chloroplasts by added pyridine nucleotides was first investigated by LOSADA et al. (1960). They found that 0.12 mM $NADP^+$ gave near-maximum stimulation, but added NAD^+ had little effect. While it is often possible to observe some activity due to endogenous NADP in the chloroplast extract (STOKES et al., 1972), added concentrations of 0.1 to 0.15 mM $NADP^+$ have usually been included in recently published experiments with reconstituted chloroplasts. LENDZIAN and BASSHAM (1976) estimated the concentrations of $NADP^+$ in illuminated intact and reconstituted chloroplasts and found that the $NADP^+$ concentrations in each were nearly identical, while the concentration of NADPH in intact chloroplasts was more than double that in reconstituted chloroplasts. However, these workers assumed a chloroplast internal volume of 0.2 ml per mg chl. If the volume of 23 µl per mg chl measured by HELDT et al. (1973) were used, the total concentration of $NADP^+$ plus NADPH would be 1.76 mM. Even if this higher figure is correct for the internal concentration of total NADP in intact chloroplasts, however, the disparity with the apparent requirement of reconstituted chloroplasts is not nearly as great with $NADP^+$ as with ferredoxin.

LENDZIAN and BASSHAM (1976) found that the NADPH/$NADP^+$ ratio reaches a maximum value of 1.1 in reconstituted chloroplasts photosynthesising under an atmosphere of nitrogen, but was reduced by about half in air. This contrasted with intact chloroplasts, in which higher $NADPH/NADP^+$ ratios (in excess of 2) were observed during photosynthesis, and which were similar under atmospheres of nitrogen and air.

2. Photophosphorylation

Most recently published experiments with reconstituted chloroplasts have included added concentrations of ADP of 0.1–0.2 mM. Alternatively, higher concentrations of added ATP of 1–2 mM have been used where it was desired to minimise effects such as the lag in O_2 evolution on addition of R5P (Walker et al., 1971; Slabas and Walker, 1976c). During photosynthesis by a reconstituted chloroplast system to which 0.1 mM ADP had initially been added, Lendzian and Bassham (1976) found that the ATP/ADP ratio varied between 5 and 2. The adenine nucleotide concentrations required by reconstituted chloroplasts are somewhat lower than those measured in intact chloroplasts. For example, Heber and Santarius (1970) found that chloroplasts, isolated by nonaqueous procedure from spinach, contained 60–100 nmol adenine nucleotides per mg chl. Assuming again a stromal volume of 23 μl per mg chl (Heldt et al., 1973), this is equivalent to an internal concentration of 2.6–4.3 mM.

The rate of formation of ATP by reconstituted chloroplasts during photosynthesis can be inferred from measured rates of carbon fixation, or of O_2 evolution with a particular added substrate. With 4 mM PGA as substrate, Walker and Slabas (1976) observed a rate of photophosphorylation of 342 μmol per mg chl per h, assuming the formation of 2 ATP molecules per O_2 evolved. The addition of substrate amounts of R5P to reconstituted chloroplasts causes a lag in O_2 evolution which continues until the bulk of this substrate is converted to RuBP (Slabas and Walker, 1976c). From the length of the initial lag caused by 1 mM R5P, Walker and Slabas (1976) estimated a photophosphorylation rate of approximately 240 μmol per mg chl per h, assuming complete conversion of R5P to RuBP. Since no O_2 evolution occurred, ATP production during this period was by cyclic and/or pseudo-cyclic photophosphorylation, presumably mediated by the transfer of electrons from reduced ferredoxin back to a component of the photoelectron transport chain, or on to molecular O_2, respectively (Arnon et al., 1967).

II. The Conversion of 3-Phosphoglycerate to Triose Phosphate

The two-step conversion of PGA to triose phosphate represents a short sequence of the Benson-Calvin cycle (Fig. 1). This sequence is of particular importance, because it not only includes the reaction in which photochemically generated reducing electrons enter the biochemical "dark" reactions of photosynthesis, but also contains an important site for the regulation of photosynthesis (see Sect. C.II.). The two enzymes involved, phosphoglycerate kinase and NADP-specific glyceraldehyde-3-phosphate dehydrogenase, are soluble components of the chloroplast stroma and are therefore present in chloroplast extract.

The conversion of added PGA to triose phosphate by reconstituted chloroplasts may be followed by measuring the rate of O_2 evolution (STOKES and WALKER, 1971 b). Alternatively, GBP may be added to follow specifically the second step of the reaction (SLABAS and WALKER, 1976 b). In reconstituted chloroplast systems exhibiting full Benson-Calvin cycle activity, the utilisation of added PGA may also be followed by measuring the consequent rate of carbon fixation (BASSHAM et al., 1974). The reaction can also be studied in chloroplast extract alone, without added thylakoids, by measuring the rate of oxidation of added NADPH substrate, with ATP (LOSADA et al., 1960), and for maximum activity, an ATP-regenerating system (LILLEY and WALKER, 1974). In this system 1 mol of NADPH is oxidised per mol PGA added, in the absence of Benson-Calvin cycle turnover, while in the full reconstituted chloroplast system, one mol of O_2 is evolved per two mol PGA added (LILLEY and WALKER, 1974).

Substantial rates of PGA-dependent O_2 evolution by reconstituted chloroplasts have been observed; for example WALKER and SLABAS (1976) reported a rate of 171 µmol O_2 per mg chlorophyll per h. However, even higher activities of PGA kinase and GAP dehydrogenase are present in chloroplast extract: LILLEY and WALKER (1974) measured rates of PGA-dependent NADPH oxidation of up to 700 µmol per mg chl per h. Thus, in reconstituted chloroplasts, the rates of regeneration of NADPH and ATP by photoelectron transport appear to be the limiting factors when PGA is the substrate, and in fact WALKER and SLABAS (1976) reported that O_2 evolution was not appreciably faster with $NADP^+$ than with PGA as substrate.

Both PGA-dependent O_2 evolution by reconstituted chloroplasts, and PGA-dependent NADPH oxidation by chloroplast extract are strongly inhibited by ADP, the effect being reversible by addition of an ATP-regenerating system (LILLEY and WALKER, 1974). SLABAS and WALKER (1976 b) assayed directly the activity of $NADP^+$-specific GAP dehydrogenase and found that ADP had no inhibiting effect. In a simplified reconstituted chloroplast system, containing purified PGA kinase and GAP dehydrogenase in place of chloroplast extract, these workers found that O_2 evolution with PGA as substrate was inhibited by ADP while that with PPGA was not. This showed that it was the PGA kinase reaction which is inhibited by ADP, presumably due to a mass action effect since this reaction is highly endergonic in the direction of PPGA formation. The activity of reconstituted chloroplasts is consequently inhibited by any reaction which increases the steady-state ADP concentration. Thus, the addition of uncoupler (LILLEY and WALKER, 1974), of R5P, or of glucose plus hexokinase (SLABAS and WALKER, 1976 c) all cause an inhibition of photosynthetic O_2 evolution which can be reversed by addition of an ATP-regenerating system. The high sensitivity of reconstituted chloroplasts to the photophosphorylation inhibitors arsenate and phloridzin (STOKES and WALKER, 1971 a) is probably due to the same effect.

LENDZIAN and BASSHAM (1976) measured the concentrations of certain metabolites in reconstituted chloroplasts, and calculated that the actual free energy change for the conversion of PGA to triose phosphate during photosynthesis in the presence of bicarbonate was between 0 and 0.5 $Kcal \cdot mol^{-1}$. They concluded that this reaction was unlikely to be modulated by allosteric inhibitors since the actual free energy change was so small. However, their results showed that PGA accumulated in reconstituted chloroplasts during photosynthesis in the presence of bicarbonate, even when the concentration of the precursor, RuBP, had fallen to a low value. The accumulation of PGA was matched by a complementary decrease in the level of dihydroxyacetone phosphate, which suggests that the rate of formation of triose phosphate from PGA was indeed rate-limiting due to the mass action effects discussed above.

III. The Conversion of Triose Phosphate to Pentose Phosphate

The conversion of triose phosphate to Ru5P may be regarded as the "regenerative" phase of the reductive pentose phosphate cycle and was first demonstrated in reconstituted chloroplasts by Bassham et al. (1974). This activity can be measured indirectly by assaying one reaction of the reductive pentose phosphate cycle after adding the product of that reaction as "primer" so that complete turnover of the cycle, including operation of the regenerative phase, must occur if activity is detected. Thus Bassham et al. (1974) measured rates of up to 120 µmol CO_2 fixed per mg chl per h with PGA as primer, and found that the amount of CO_2 fixed required far more RuBP than could be formed from the PGA initially present, and that some intermediates of the cycle accumulated, showing that regenerative activity and complete cycle turnover were occurring. A complementary demonstration of triose phosphate-dependent O_2 evolution was made by Walker and Slabas (1976).

The activity of the relatively complex regenerative pathway is determined by both the activity of each of the eight enzymes and the concentrations of the seven intermediate compounds involved. The rate of regeneration can be increased by adding larger amounts of chloroplast extract (thus increasing the activity of all enzymes), by the addition of extra FBPase alone, or by the addition of dithiothreitol, an FBPase activator (Walker et al., 1976). The regulatory role of this enzyme as a "gate" for the regenerative pathway (see also Sect. C.I) is further shown by the inhibitory effect of decreasing the Mg concentration from the optimum 20 mM or the pH from the optimum 7.9 (Walker et al., 1976); the activity of FBPase is very sensitive to Mg concentration and pH (Baier and Latzko, 1975). This sensitivity to pH was responsible for an earlier finding that utilisation of triose phosphate, and cycle turnover, was absent in reconstituted chloroplasts at pH 7.6 (Lilley and Walker, 1974).

The activity of the regenerative pathway is also affected greatly by the concentration of the intermediate F6P. Although regenerative activity in the presence of substrate quantities of this compound alone are low, if both F6P and GAP (provided from added FBP by aldolase activity) are present, very high regenerative activity occurs (Walker et al., 1976). Under these conditions the FBPase reaction is bypassed as F6P and GAP enter the transketolase reaction, yielding one molecule of pentose phosphate.

Regenerative activity in reconstituted chloroplasts is inhibited by DL-glyceraldehyde (Stokes and Walker, 1972), probably by affecting the transketolase reaction (Slabas and Walker, 1976a). Conversely, this activity is stimulated by antimycin A by an indirect effect, possibly resulting from an increase in the concentration of reduced ferredoxin (Walker et al., 1976).

IV. The Conversion of Ribulose-5-Phosphate to Ribulose-1, 5-Bisphosphate

The phosphoribulokinase reaction generates the carboxylation acceptor RuBP in one of the two ATP-utilising steps of the reductive pentose phosphate cycle. Reconstituted chloroplasts will fix CO_2 (SLABAS and WALKER, 1976c) and evolve O_2 (WALKER et al., 1971) at rates exceeding 100 µmol per mg chl per h when provided with substrate amounts of R5P. Ru5P and xylulose-5-phosphate can also be used as substrates (STOKES and WALKER, 1972).

When reconstituted chloroplasts are illuminated in the presence of R5P and bicarbonate, O_2 evolution commences only after a lag. This lag is caused, not by a delay in RuBP formation, nor in carboxylation, but by inhibition of the PGA kinase reaction by the ADP generated during the conversion of Ru5P to RuBP (SLABAS and WALKER, 1976c). The phosphoribulokinase reaction competes successfully for available ATP because its affinity (Km ATP = 0.42 mM) is higher than that of the PGA kinase (Km ATP = 2.0 mM; LAVERGNE et al., 1974). Also, as a highly exergonic reaction (BASSHAM and KRAUSE, 1969) it is relatively unaffected by the concentration of its product, ADP.

The rate of formation of RuBP is stimulated by dithiothreitol, an activator of phosphoribulokinase (SLABAS and WALKER, 1976c). Although this reaction is inhibited to some extent by DL-glyceraldehyde, it is less sensitive than the regenerative pathway (see Sect. B.III) to this inhibitor (STOKES and WALKER, 1972; BAMBERGER and AVRON, 1975; SLABAS and WALKER, 1976c).

V. The Fixation of Carbon Dioxide

The fixation of carbon dioxide by reconstituted chloroplasts was first reported by WHATLEY et al. (1956), who measured low rates of fixation of ^{14}C from bicarbonate into carbohydrate, which were stimulated severalfold in the presence of sugar phosphates. WALKER et al. (1971) obtained high rates of CO_2-dependent oxygen evolution (up to 175 µmol O_2 evolved per mg chl per h) in the presence of 1.5 mM R5P as substrate. These workers noted that the bicarbonate concentration required for maximum activity was reduced if the concentration of magnesium ions present was increased. BASSHAM et al. (1974) observed rates of carbon fixation of up to 120 µmol per mg chl per h with 1 mM PGA as substrate in a similar reconstituted system.

The effect of Mg concentration on photosynthesis in reconstituted chloroplasts was further investigated by LILLEY et al. (1974), who found that the rate of CO_2-dependent O_2 evolution with RuBP as substrate was very low in the presence of 0.5 mM Mg, but was restored to 70–90% of the maximum rate by increasing the Mg concentration to 2.5 mM. The maximum rate was obtained by either increasing the Mg concentration to 5 mM or the bicarbonate concentration from 1 to 40 mM. Since the rate of PGA-dependent O_2 evolution was relatively unaffected by Mg concentrations as low as 0.5 mM, the effect was attributed to the RuBP carboxylase reaction. It was also found that on addition of 5 mM $MgCl_2$ to reconstituted chloroplasts which had been preilluminated with 1 mM RuBP in the absence of Mg, there was a short lag before carbon fixation commenced, and the ultimate rate was much lower than when the reconstituted chloroplasts were preilluminated with Mg and carbon fixation initiated by addition of RuBP. Similar effects on purified RuBP carboxylase enzyme were first reported by Pon et al. (1963), and subsequently clarified by the demonstration that this enzyme is converted from an inactive to an active form by the reversible formation of an equilibrium complex with CO_2 and Mg^{2+} (Lorimer et al., 1976). The conversion of this enzyme from the inactive to the active form by Mg^{2+} and CO_2 is greatly slowed in the presence of RuBP (Chu and Bassham, 1975).

During photosynthesis in reconstituted chloroplasts under conditions where the optimal or near-optimal RuBP carboxylase activity is realised, the rate of CO_2 fixation can considerably exceed the rate of CO_2-dependent O_2 evolution (Lilley et al., 1974). Since very high activities of RuBP carboxylase (in excess of 500 μmol CO_2 fixed per mg chl per h) are observed in chloroplast extract (Lilley and Walker, 1975), PGA may be produced at a higher rate than it can be converted to triose phosphate, and must accumulate under these conditions.

The enzyme RuBP carboxylase also exhibits an intrinsic oxygenase activity which competes with the carboxylase activity and is believed to be responsible for photorespiration (Bowes et al., 1971; Lorimer and Andrews, 1973). The products of the oxygenase reaction are one molecule of PGA and one molecule of P-glycollate. Reconstituted chloroplasts should be a useful system for the study of photorespiratory phenomena such as P-glycollate production and its conversion to glycollate.

VI. Autocatalysis

The reductive pentose phosphate pathway must function as a breeder reaction producing more product than it consumes as substrate. If this were not so there would be no spare capacity for growth or response to improved environmental conditions. On paper, if all of the product is fed back into the cycle, the level of intermediates will double for every 15 molecules of CO_2 fixed. This is believed to be the basis of photosynthetic induction in which an initial lag on passing from prolonged dark to full illumination is followed by an autocatalytic acceleration to the steady-state rate allowed by other limiting factors such as light intensity. In intact chloroplasts this induction period can always be observed and may be prolonged by excess Pi which diverts product from recycling to export (Walker, 1976). In the reconstituted system, when substrate concentrations of intermediates are added to mixtures containing normal stromal protein/chlorophyll ratios, no lags are seen which cannot be attributed to causes other than autocatalysis. However, when the stromal protein/chlorophyll ratio is increased 12-fold, and in the absence of exogenous substrate, there is a clear acceleration in CO_2 fixation and O_2 evolution (Walker and Lilley, 1974). It is assumed that the additional protein offsets the dilution of products. Thus, within an intact chloroplast, the concentration of those products which cannot readily cross the envelope can build up rapidly, and even those which are exported in the presence of high Pi concentration are not lost in an uncontrolled fashion if Pi is present in physiological concentrations. Conversely, the product of the reconstituted system is not so restricted and, at chlorophyll concentrations normally employed, may diffuse freely through a volume

500 to 1000 times greater than that of the stromal compartment. Such are the respective affinities and activities of the stromal enzymes, however, that a 10–12-fold increase in enzyme concentration is sufficient to offset this substrate dilution and permit autocatalysis to be demonstrated.

C. Reconstituted Chloroplast Systems and the Regulation of Photosynthesis

I. The Role of Magnesium, pH and Reductants

Although rapid rates of reduction of $NADP^+$, and conversion of PGA to triose phosphate are observed in reconstituted chloroplasts at pH 7.6 in the presence of 0.5 mM Mg, for CO_2 fixation to occur the Mg concentration must be raised to at least 2.5 mM (LILLEY et al., 1974). However, substantial rates of conversion of triose phosphate to pentose phosphate are observed only after the Mg concentration is further increased to 5 mM (WALKER and LILLEY, 1974), or for optimum activity to 20 mM (WALKER et al., 1976), while the pH must also be raised to 7.9 (see Sect. B.III). Moreover, regenerative activity in reconstituted chloroplasts is further stimulated by the addition of dithiothreitol (WALKER et al., 1976). That most of the sensitivity of the regenerative reactions to Mg and pH resides in the FBPase reaction is illustrated by the findings that, during photosynthesis in reconstituted chloroplasts at pH 7.5 in 0.67 mM Mg, the concentration of FBP is relatively high (HAVIR and GIBBS, 1963) while at pH 8.0 and 30 mM Mg it is relatively low (BASSHAM et al., 1974).

These results strongly support the view that, in intact chloroplasts, the regulation of carbon fixation is mediated largely by changes in the pH and Mg concentration in the stroma, which result from the light-induced influx of protons and efflux of Mg^{2+} across the thylakoid membranes (WALKER, 1973). The pH in the stroma of intact spinach chloroplasts is now known to rise from about 7.4 in the dark to about 7.9 in the light (HELDT et al., 1973), while the Mg concentration increases by 1 to 3 mM on illumination (PORTIS and HELDT, 1976). Carbon fixation, but not PGA-dependent O_2 evolution by illuminated intact chloroplasts, is inhibited by treatments which decrease either the stromal pH (WERDAN et al., 1975) or the stromal Mg concentration (PORTIS and HELDT, 1976).

The kinetic properties of the enzymes RuBP carboxylase, FBPase, and sedoheptulose bisphosphatase are consistent with their apparent role in the regulation of the reductive pentose phosphate cycle (for reviews see KELLY et al., 1976). The activity of FBPase can be switched from zero to near maximum by a combination of increases in pH, Mg concentration, FBP concentration, and the addition of dithiothreitol (BAIER and LATZKO, 1975); the same variables affect FBP-dependent O_2 evolution in the reconstituted system in ways which are consistent with increased FBPase activity. For example, a mixture of FBP and F6P produces the same kinetic response as FBP together with increased pH and concentrations of Mg and dithiothreitol (WALKER et al., 1976). RuBP carboxylase appears to be less sensitive to changes in pH and Mg concentration than FBPase in reconstituted chloroplasts with 1 mM or higher bicarbonate concentrations. If however, the concentration of bicarbonate is lowered to reduce the level of dissolved CO_2 to that resulting from equilibrium with atmospheric CO_2, the activity of this enzyme exhibits a sensitivity to Mg and pH similar to that of FBPase (LORIMER et al., 1976). It therefore seems likely that the activity of RuBP carboxylase is also regulated by Mg and pH changes in vivo.

The precise role of light-generated reductant in the activation of the reductive pentose phosphate cycle is less clear (cf. Chap. II.22, this vol.). The addition of dithiothreitol to reconstituted chloroplasts results in a stimulation of the rate of photosynthesis, apparently by promoting FBPase activity (WALKER et al., 1976). Dithiothreitol is an activator not only of FBPase, but also of two other enzymes which are believed to be modulated by light, sedoheptulose bisphosphatase (ANDERSON, 1974) and phosphoribulokinase (AVRON and GIBBS, 1974). BUCHANAN and SCHÜRMANN (1973) have proposed that reduced ferredoxin plays a major role in activating several key enzymes of the reductive pentose phosphate cycle. The significance of this effect, and of the role of reduced thioredoxin (WOLOSIUK and BUCHANAN, 1977) deserves investigation in the reconstituted chloroplast system.

II. The Role of the ATP/ADP Ratio

Oxygen evolution in the reconstituted system can be markedly inhibited by ADP or by any reaction sequence which converts ATP to ADP (such as hexose plus hexokinase). Conversely, these inhibitions can be ameliorated or abolished by factors which enhance ATP formation, such as increased concentrations of ferredoxin or of creatine phosphate plus its kinase (LILLEY and WALKER, 1974; SLABAS and WALKER, 1976c; WALKER et al., 1976). The effect is seen at its best with R5P as a substrate because Ru5P acts as a sink for ATP. Thus CO_2 fixation starts virtually immediately but O_2 evolution only after a lag which is *prolonged* by increasing the R5P concentration. The lag can be abolished by creatine phosphate plus its kinase. Similarly steady-state O_2 evolution can be interrupted by ADP or by hexose plus hexokinase. The interruption is then a function of the amount of ADP or hexokinase added. The effect of ADP has been traced (SLABAS and WALKER, 1976b) to its displacement of the equilibrium between PGA and GBP (as catalysed by PGA kinase). The implications of regulation by ADP in this context have been discussed by WALKER (1976).

References

Allen, J.F.: Nature (London) *256*, 599–600 (1975)
Anderson, L.E.: Biochem. Biophys. Res. Commun. *59*, 907–913 (1974)
Arnon, D.I., Allen, M.B., Whatley, F.R.: Nature (London) *174*, 394–396 (1954)
Arnon, D.I., Tsujimoto, H.Y., McSwain, B.D.: Nature (London) *214*, 562–566 (1967)
Avron, M., Gibbs, M.: Plant Physiol. *53*, 136–139 (1974)
Baier, D., Latzko, E.: Biochim. Biophys. Acta *396*, 141–147 (1975)
Bamberger, E.S., Avron, M.: Plant Physiol. *56*, 481–485 (1975)

Bassham, J.A., Krause, G.H.: Biochim. Biophys. Acta *189*, 207–221 (1969)
Bassham, J.A., Levine, G., Forger, J.: Plant Sci. Lett. *2*, 15–21 (1974)
Bowes, G., Ogren, W.L., Hageman, R.H.: Biochem. Biophys. Res. Commun. *45*, 716–722 (1971)
Buchanan, B.B., Schürmann, P.: J. Biol. Chem. *248*, 4956–4964 (1973)
Bucke, C., Walker, D.A., Baldry, C.W.: Biochem. J. *101*, 636–641 (1966)
Chu, D.K., Bassham, J.A.: Plant Physiol. *55*, 720–726 (1975)
Delaney, M.E., Walker, D.A.: Plant Sci. Lett. *7*, 285–294 (1976)
Gibbs, M., Cynkin, M.A.: Nature (London) *182*, 1241–1242 (1958)
Havir, E.A., Gibbs, M.: J. Biol. Chem. *238*, 3183–3187 (1963)
Heber, U.: Ann. Rev. Plant Physiol. *25*, 393–421 (1974)
Heber, U., Santarius, K.A.: Z. Naturforsch. *25b*, 718–728 (1970)
Heldt, H.W., Werdan, K., Milovancev, M., Geller, G.: Biochim. Biophys. Acta *314*, 224–241 (1973)
Jensen, R.G., Bassham, J.A.: Proc. Natl. Acad. Sci. USA *56*, 1095–1101 (1966)
Kelly, G.J., Latzko, E., Gibbs, M.: Ann. Rev. Plant Physiol. *27*, 181–205 (1976)
Lavergne, D., Bismuth, E., Champigny, M.L.: Plant Sci. Lett. *3*, 391–397 (1974)
Lendzian, K., Bassham, J.A.: Biochim. Biophys. Acta *430*, 478–489 (1976)
Lilley, R.McC., Walker, D.A.: Biochim. Biophys. Acta *368*, 269–278 (1974)
Lilley, R.McC., Walker, D.A.: Plant Physiol. *55*, 1087–1092 (1975)
Lilley, R.McC., Holborow, K., Walker, D.A.: New Phytol. *73*, 657–662 (1974)
Lorimer, G.H., Andrews, T.J.: Nature (London) *243*, 359–360 (1973)
Lorimer, G.H., Badger, M.R., Andrews, T.J.: Biochem. *15*, 529–536 (1976)
Losada, M., Trebst, A.V., Arnon, D.I.: J. Biol. Chem. *235*, 832–839 (1960)
Matson, R.S., Kimura, T.: Biochim. Biophys. Acta *396*, 293–300 (1975)
Pon, N.G., Rabin, B.R., Calvin, M.: Biochem. Z. *338*, 7–19 (1963)
Portis, A.R., Heldt, H.W.: Biochim. Biophys. Acta *449*, 434–446 (1976)
Slabas, A., Walker, D.A.: Biochem. J. *153*, 613–619 (1976a)
Slabas, A., Walker, D.A.: Biochem. J. *154*, 185–192 (1976b)
Slabas, A., Walker, D.A.: Biochim. Biophys. Acta *430*, 154–164 (1976c)
Slabas, A., Walker, D.A.: Arch. Biochem. Biophys. *175*, 590–597 (1976d)
Stokes, D.M., Walker, D.A.: In: Photosynthesis and Photorespiration. Hatch, M.D., Osmond, C.B., Slatyer, R.O. (eds.), pp 226–231. New York: Wiley-Interscience 1971a
Stokes, D.M., Walker, D.A.: Plant Physiol. *48*, 163–165 (1971b)
Stokes, D.M., Walker, D.A.: Biochem. J. *128*, 1147–1157 (1972)
Stokes, D.M., Walker, D.A., McCormick, A.V.: In: Prog. Photosynthesis. Forti, G., Avron, M., Melandri, A. (eds.), Vol. 3., pp. 1779–1785. The Hague: W. Junk 1972
Tagawa, K., Arnon, D.I.: Nature (London) *195*, 537–543 (1962)
Walker, D.A.: Biochem. J. *92*, 22c–23c (1964)
Walker, D.A.: Plant Physiol. *40*, 1157–1161 (1965)
Walker, D.A.: New Phytol. *72*, 209–235 (1973)
Walker, D.A.: In: Current Topics in Cellular Regulation, Vol. 2., pp. 203–241. Horecker, B.L., Stadtman, E. (eds.). New York: Academic Press 1976
Walker, D.A., Lilley, R.McC.: Plant Physiol. *54*, 950–952 (1974)
Walker, D.A., McCormick, A.V., Stokes, D.M.: Nature (London) *233*, 346–347 (1971)
Walker, D.A., Slabas, A.: Plant Physiol. *57*, 203–208 (1976)
Walker, D.A., Slabas, A., Fitzgerald, M.P.: Biochim. Biophys. Acta *440*, 147–162 (1976)
Werdan, K., Heldt, H.W., Milovancev, M.: Biochim. Biophys. Acta *396*, 276–292 (1975)
Whatley, F.R., Allen, M.B., Rosenberg, L.L., Capindale, J.B., Arnon, D.I.: Biochim. Biophys. Acta *20*, 462–468 (1956)
Wolosiuk, R.A., Buchanan, B.B.: Nature (London) *266*, 565–567 (1977)

4. Autotrophic Carbon Dioxide Assimilation in Prokaryotic Microorganisms

E. OHMANN

A. Introduction

Besides higher plants and eukaryotic algae, several groups of prokaryotic microorganisms are capable of autotrophic carbon dioxide fixation. These are photosynthetic bacteria, blue-green algae, and chemolithotrophic bacteria.

On the basis of energy sources utilized microorganisms may be described in terms of the following nomenclature (LWOFF et al., 1946):

a) Photolithotrophy and photoorganotrophy – energy provided chiefly by photochemical reactions requiring exogenous inorganic hydrogen donors and organic hydrogen donors, respectively.

b) Chemolithotrophy and chemoorganotrophy – energy for growth in the dark is provided entirely by oxidation of exogenous inorganic and organic compounds, respectively.

In addition, the derivation of cell carbon may be designated by the prefixes "auto" and "hetero". Using the definition made by WOODS and LASCELLES (1954), autotrophy is based on the ability to use carbon dioxide as the main carbon source. It is of minor importance that several autotrophs are facultative and can occasionally grow with organic substances. All species of blue-green algae have the ability to grow in light with carbon dioxide as the principal carbon source. Blue-green algae therefore are often regarded as photolithoautotrophs par excellence. However, several strains are capable of photoheterotrophic growth and, in some cases, of true chemoheterotrophic growth in the dark (RIPPKA, 1972; PELROY et al., 1972; PELROY and BASSHAM, 1972, 1973a, b; STANIER, 1973; FOGG et al., 1973; WHITE and SHILO, 1975).

The photosynthetic bacteria display a remarkable diversity and versatility in their morphological and physiological properties. They comprise three major groups (VAN NIEL, 1944; PFENNIG, 1967; PFENNIG and TRÜPER, 1974). These are:

1. Green sulfur bacteria represented by the species *Chlorobium limicola* and *Chl. thiosulfatophilum* that are strictly anaerobic and obligately phototrophic, capable of photolithotrophic CO_2 assimilation with reduced sulfur compounds or molecular hydrogen as an electron donor. Acetate is the most widely used organic substrate that may be photoassimilated in the presence of both sulfide and CO_2.

2. Purple sulfur bacteria: Chromatiaceae (Thiorhodaceae), represented by the genus *Chromatium*. Most strains are strictly anaerobic and capable of photolithotrophic CO_2 assimilation in the presence of sulfide. Many strains are able to use molecular hydrogen as an electron donor. Some can use various organic compounds for photoassimilation or as electron donors. One species, *Thiocapsa roseopersicina*, is able to change its metabolism from phototrophy to chemolithoautotrophy (KONDRATIEVA et al., 1976).

3. Purple nonsulfur bacteria: Rhodospirillaceae (Athiorhodaceae), represented by *Rhodospirillum rubrum* and *Rhodopseudomonas spheroides*. They are photoorganotrophs, some strains being capable of photolithotrophic CO_2 assimilation with molecular hydrogen as an electron donor as well as of aerobic, chemoorganotrophic growth in the dark.

Chemolithoautotrophic microorganisms gain energy for CO_2 assimilation from the oxidation of inorganic compounds, such as reduced forms of sulfur, nitrogen, ferrous iron, or of hydrogen. Many of them are facultative, others are obligate chemolithoautotrophic. The biochemical background of obligate chemolithotrophy is still obscure, as it is for obligate photolithotrophy. For a more comprehensive treatment of chemolithotrophy the reviews of KELLY (1971) and SCHLEGEL (1975) are excellent. As shown in this brief survey there is no distinct borderline between the different types of metabolism, but rather there is a physiological overlapping with a series of gradual transitions. In both chemo- and phototrophs a continuous physiological spectrum is seen between obligate lithotrophy and complete heterotrophy. That special problems of regulation are expected is quite obvious.

B. Principles of Autotrophic Carbon Dioxide Assimilation in Prokaryotic Cells

In general the principles of autotrophic CO_2 assimilation in prokaryotic cells are the same as in eukaryotic cells. The main difference is the level of cellular compartmentation. Whereas photosynthesis, photorespiration, and dark respiration are separated in eukaryotic cells within specialized membrane-bound organelles (chloroplasts, peroxisomes, and mitochondria, respectively) no such segregation exists in prokaryotic cells. Other necessities of regulation therefore may be expected.

Two types of cyclic processes are proposed for the autotrophic assimilation of CO_2: the reductive pentose phosphate cycle (Calvin cycle) and the reductive carboxylic acid cycle (see Chap. III.3, this vol.). In both processes at the end of a series of reactions the acceptor for CO_2 is regenerated.

The justified conclusion that in a special group of organisms CO_2 is assimilated by a specific pathway should be based on different lines of evidence:

1. After short-time $^{14}CO_2$ fixation, the radioactivity is present in the intermediates of the cycle and the fraction of the total label present in the expected primary product decreases with increasing time of labeling.

2. The intermediates contain the labeled carbon in the expected positions.

3. The specific enzymes are present in cell-free extracts in an activity which is sufficient to satisfy the rate of CO_2 fixation by whole cells.

4. Mutants lacking one of the specific enzymes cannot grow autotrophically.

5. Isotope distribution is in accordance with the discrimination properties of the responsible carboxylation enzymes.

The use of only one approach in the absence of others can be very misleading, e.g., the presence and function of certain enzymes do not necessarily indicate a vital role for their action in vivo. Even more important than the nature of the primary product is the ability of the proposed photosynthetic reaction sequence to support growth. In recent reports WALKER (1974) and KELLY et al. (1976) draw attention to the demand that any CO_2 assimilation mechanism permitting growth must be an autocatalytic sequence able to generate more CO_2 acceptor than was present initially. Obviously this is much more important in fast-growing populations of microorganisms than in fully developed leaves of higher plants.

C. The Pathway of Carbon Dioxide Assimilation in Green Sulfur Bacteria

Conflicting views exist as to the mechanism of CO_2 fixation in green sulfur bacteria. On the basis of incorporation patterns during $^{14}CO_2$ assimilation and positive results concerning the presence of the enzymes necessary for the operation of the reductive pentose phosphate cycle it was previously assumed that CO_2 fixation in *Chlorobium* occurs via this cycle as in all other autotrophic organisms (Smillie et al., 1962; Fuller, 1969).

On the other hand CO_2 assimilation in green sulfur bacteria seems to be closely related to acetate metabolism. The utilization of acetate for cell growth in the light is dependent upon an exogenous supply of CO_2 and reducing power, e.g., sulfide. Working with intact cells of *Chlorobium limicola* and *Rhodospirillum rubrum*, respectively, Sadler and Stanier (1960), Hoare (1963) and Hoare and Gibson (1964) obtained a labeling pattern in the products from the photoassimilation of radioactive CO_2 and acetate that could not be explained exclusively on the basis of an operating reductive pentose phosphate cycle and normal citric acid cycle and that suggested the operation of other, or additional, fixation mechanisms.

Long before the sequences of the reductive pentose phosphate cycle had been discovered by Calvin and coworkers, van Niel (1949) had foreseen the possibility of a cyclic process for the reductive assimilation of CO_2, namely a kind of reversed citric acid cycle.

In 1966 Evans et al. proposed the reaction sequence of such a new carbon reduction cycle in *Chlorobium thiosulfatophilum*. This cycle in its "short" variant is illustrated in this volume (Chap. III.3). In one turn it incorporates two molecules of CO_2 and yields one molecule of acetyl-CoA that in the course of additional reactions may serve as a starting material for other biosynthetic reactions leading to amino acids, lipids or carbohydrates. For this purpose further reductions and carboxylation reactions are necessary. In its overall effect the reductive carboxylic acid cycle which generates acetyl-CoA from two molecules of CO_2 is a reversal of the Krebs citric acid cycle which degradates acetyl-CoA to two molecules of CO_2. The basic distinction is that the reversed cycle and the subsequent reaction steps now are endergonic. ATP and metabolic hydrogen must be invested, which may be provided by photosynthetic electron transport. The most important steps in this series of reactions are two ferredoxin-dependent carboxylation reactions catalyzed by pyruvate synthase and α-ketoglutarate synthase. Both make use of the strong reducing power of ferredoxin directly without the mediation of pyridine nucleotides to drive the energetically difficult carboxylations of acetyl-CoA intermediates leading to the formation of pyruvate and α-ketoglutarate:

$$\text{acetyl-CoA} + CO_2 + Fd_{red} \rightarrow \text{pyruvate} + CoA + Fd_{ox}$$

$$\text{succinyl-CoA} + CO_2 + Fd_{red} \rightarrow \alpha\text{-ketoglutarate} + CoA + Fd_{ox}$$

Another key enzyme of this cycle is the ATP: citrate lyase (or citrate lyase) which catalyzes the formation of acetyl-CoA (acetate) and oxaloacetate from citrate, a reaction which is essential for the operation of the reductive pathway described as a cyclic process.

In another variant of the cycle, the "long" variant, the sequence of reactions leading from acetyl-CoA to oxaloacetate is regarded as being integrated in the cyclic process itself (ARNON, 1969; EVANS et al., 1966). Pyruvate may be converted to oxaloacetate in a one-step reaction catalyzed by pyruvate carboxylase or with the help of PEP synthase and PEP carboxylase. One complete turn of the long reductive carboxylic acid cycle incorporates four molecules of CO_2 and results in the net synthesis of oxaloacetate. For CO_2 fixation in *Chlorobium* three possibilities are now discussed:

1. CO_2 assimilation occurs via the Calvin cycle and the reactions of the reductive carboxylic acid cycle are nonessential.

2. The two carbon cycles coexist, the Calvin cycle being mainly concerned with carbohydrate synthesis, whereas the reductive carboxylic acid cycle functions in the synthesis of amino acids and precursors of lipids and porphyrins.

3. CO_2 is exclusively fixed via the reductive carboxylic acid cycle and environmental conditions (CO_2 concentration) influence the products formed, favoring the synthesis of either carbohydrates or amino acids, respectively.

All enzymes of the reductive carboxylic acid pathway are apparently present in extracts of *Chlorobium* cells with the exception of citrate lyase. The evidence for this enzyme, obtained from experiments made by EVANS et al. (1966) is only indirect, because it is based on the formation of labeled aspartate from ^{14}C-isocitrate. BEUSCHER and GOTTSCHALK (1972), on the other hand, were not able to detect citrate cleavage in extracts of *Chl. thiosulfatophilum* and *Rsp. rubrum* examined under various conditions. Therefore the function of a reductive carboxylic acid cycle has been questioned.

Some, but not all, labeling data obtained after assimilation of $^{14}CO_2$ and other ^{14}C labeled compounds are consistent with the function of the reductive carboxylic acid cycle. The labeling pattern of glutamate observed earlier by HOARE and GIBSON (1964) is that expected on the basis of a functional reductive carboxylic acid cycle. More recently SIREVAG and ORMEROD (1970); SIREVAG (1974, 1975), and BUCHANAN et al. (1972) identified intermediates of the reductive carboxylic acid cycle as products of short-time CO_2 fixation in *Chl. thiosulfatophilum*. BUCHANAN et al. (1972) could show that the fixation products were profoundly influenced by the concentration of carbon dioxide. At a total concentration of 40 mM, most of the activity was fixed into glutamate and the intermediates of the reductive carboxylic acid cycle. The percentage of activity in these compounds showed a negative slope when plotted against time, as would be expected were these compounds early intermediates. In contrast, PGA was the principal labeled product after 5 s exposure to 0.8 mM HCO_3^- and the percentage of radioactivity incorporated into this compound showed a strong negative slope.

Contrary to the expectation for the function of a carboxylic acid cycle in a reductive direction are results obtained by BUCHANAN et al. (1972) after photoassimilation of ^{14}C-succinate and CO_2. In these experiments malate and fumarate were the earliest labeled products and appeared more rapidly than glutamate.

Nevertheless, the authors suggested that this bacterium fixes CO_2 exclusively via the reductive carboxylic acid cycle, and that so far there is no compelling evidence to support the existence of the reductive pentose phosphate cycle as a mechanism of CO_2 assimilation in *Chlorobium*. A strong line of argument is based on the fact that both BUCHANAN et al. (1972) and SIREVAG (1974) were unable to detect RuBP carboxylase, a key enzyme of the Calvin cycle, in *Ch. thiosulfatophilum*. By contrast TABITA et al. (1974) reported the existence of RuBP carboxylase in the same strain. They were able to purify it and described several characteristics including its stability, molecular weight, and quarternary structure. They concluded that there is no reason to rule out the Calvin cycle as the main path of CO_2 assimilation in *Chlorobium*. Still more recently BUCHANAN and SIREVAG (1976) failed to confirm the results of TABITA et al. (1974). In addition both SIREVAG (1974) and BUCHANAN and SIREVAG (1976) could not detect in the same extracts Ru5P kinase, the second enzyme which, like RuBP carboxylase, is a specific marker of the reductive pentose phosphate cycle. Arguments based on negative evidence are not very convincing. Therefore SIREVAG et al. (1977) used an approach which is based on different isotope discrimination properties of the key carboxylating enzymes. They found that *Chromatium* and *Rhodospirillum rubrum* had isotope discrimination similar to higher plants of the C_3 type, whereas *Chlorobium* showed properties substantially different.

To recapitulate: While the existence of several carboxylations which may contribute to the net assimilation of CO_2 in *Chlorobium* is well documented, the role of the reductive carboxylic acid cycle or the Calvin cycle as the major CO_2 fixation path is still open to different interpretations, each of which is supported by some experimental evidence. It seems, therefore, best to regard the entire subject as a still-open field of investigation. Final judgements must await future developments. It cannot be excluded that hitherto unknown mechanisms are involved (SIREVAG, 1976).

D. The Pathway of Carbon Dioxide Assimilation in Purple Bacteria

GLOVER et al. (1952), working with *Rsp. rubrum*, first established that PGA was the earliest product of photosynthetic $^{14}CO_2$ fixation, indicating that the reductive pentose phosphate cycle represents the major mechanism of carbon dioxide assimilation. Subsequent investigations done by other research groups confirmed these early results and gave more convincing evidence for the operation of the Calvin cycle in different genera and strains of both Chromatiaceae and Rhodospirillaceae. The kinetics of CO_2 incorporation, the elucidation of the labeling pattern in photosynthetic products, and the study of the specific activity of enzymes have been most effectively used. Although most work has been done with *Rsp. rubrum* (ANDERSON and FULLER, 1967a, b, c, 1969; FULLER, 1969; PORTER and MERRETT, 1972; SLATER and MORRIS, 1973a, b; TABITA and McFADDEN, 1974a, b), *Rps. capsulata* (STOPPANI et al., 1955), *Rps. palustris* (CHERNIADEV et al., 1969), *Chromatium* (FULLER et al., 1961; LOSADA et al., 1960; HURLBERT and LASCELLES, 1963; TRÜPER, 1964; LATZKO and GIBBS, 1966), and *Thiocapsa roseopersicina* (KONDRATIEVA et al., 1976) the results seem to be representative for purple bacteria in general.

For the reduction step of the Calvin cycle purple bacteria use NADH, as do other autotrophic bacteria.

Nevertheless, EVANS et al. (1966) and BUCHANAN et al. (1967) have speculated upon the contribution of the reductive carboxylic acid cycle in *Chromatium* and *Rsp. rubrum*. Incorporation of $^{14}CO_2$ in the intermediates of this cycle and in its expected products, mainly amino acids, constituted their evidence. But their interpretation was, however, without conclusive support since it did not take into consideration the influence of the mode of growth or the incubation conditions on the metabolic fate of CO_2. This was demonstrated repeatedly in Thiorhodaceae and Athiorhodaceae with respect to the presence of different carbon compounds or different light intensities (HURLBERT and LASCELLES, 1963; HOARE, 1963; ANDERSON and FULLER, 1967b; YOCH and LINDSTROM, 1967; FULLER, 1969; PORTER and MERRETT, 1972). The labeling pattern of glutamate after photoassimilation of ^{14}C-acetate in *Rsp. rubrum* is not in accord with the operation of a reductive carboxylic acid cycle because it would have resulted in the rapid incorporation of ^{14}C from carboxyl labeled acetate into carbon positions 3 and 4 of glutamate which did not occur (HOARE, 1963). In addition, BEUSCHER and GOTTSCHALK (1972) could not detect citrate cleavage in *Rsp. rubrum*. *Rps. spheroides* has no PEP carboxylase. Mutants of this organism, defective in pyruvate carboxylase, can grow phototrophically on acetate and CO_2 (PAYNE and MORRIS, 1969a, b). This means that they can use acetate by a route not involving carboxylations of pyruvate or PEP, and this route cannot be the reductive carboxylic acid cycle.

The data now available do indicate that in purple bacteria some auxiliary carboxylation reactions are involved in the process of CO_2 assimilation, at least under strong reducing conditions, but they do not question the dominant role of the Calvin cycle as the main route.

E. The Pathway of Carbon Dioxide Assimilation in Blue-Green Algae

The photosynthetic mechanism of blue-green algae differs distinctly from that in bacteria, both in the chemical nature of the pigments carrying out the photochemical reactions and the involvement of oxygen liberation from water. Here, in fact, we have present in prokaryotic cells the type of photosynthetic machinery characteristic of higher plants.

Investigations of the mechanism of CO_2 fixation in blue-green algae are still rather scarce. The data available at the moment are incomplete and derived from studies with very few species. The experimental approaches outlined in section B have never been applied in their entirety to a particular species of blue-green algae. Most information is derived from labeling experiments, only recently combined with enzyme investigations. In short-time experiments PGA was the most heavily labeled compound, followed by various sugar phosphates that serve as intermediate compounds of the reductive pentose phosphate cycle (NORRIS et al., 1955; KANDLER, 1961). This information is consistent with the operation of the Calvin cycle in these organisms. This view is supported by the finding that the distribution of ^{14}C within hexose molecules after a short period of photosynthesis in *Anacystis nidulans* is similar to that found in hexose from *Chlorella pyrenoidosa* (KINDEL and GIBBS, 1963). More recently PELROY and BASSHAM (1972) used four different strains of unicellular blue-green algae, including *Anacystis nidulans* and *Aphanocapsa*, and investigated carefully the kinetics of $^{14}CO_2$ incorporation in the light and during a subsequent dark period. The results obtained are in complete accord with the operation of the Calvin cycle as the dominant route of CO_2 assimilation in all strains investigated. However, still more evidence is essential before this pathway is firmly established as the sole route for the net assimilation of carbon dioxide in blue-green algae. In recent papers DÖHLER was able to show that pigment composition, growth and incubation conditions influence the pattern of $^{14}CO_2$ incorporation during photosynthesis in *Anacystis nidulans* (DÖHLER, 1974, 1976a, b). He concluded that depending on experimental conditions, carboxylation reactions of PEP or those of the reductive carboxylic acid cycle may provide supplementary possibilities for carbon dioxide assimilation. The presence of ferredoxin-dependent pyruvate synthase has been shown in some blue-green algae including *Anacystis* by LEACH and CARR (1971), BOTHE et al. (1974), and BOTHE and NOLTEERNSTING (1975). However, it was concluded that the physiological role of this reversible reaction is to generate reduced ferredoxin in the process of pyruvate cleavage.

All enzymes characteristic of the Calvin cycle have been detected in blue-green algae. Among them fructose bisphosphate aldolase is one which has excited most discussion, as it was suggested to be absent in blue-green algae. Later it was shown that blue-green algae contain a type II aldolase which is somewhat different from type I aldolase of chloroplasts (WILLARD et al., 1965). There is only one glyceraldehyde-3-phosphate dehydrogenase in blue-green algae which functions in glycolysis and in photosynthesis. It accepts either NADH or NADPH as hydrogen donor.

F. The Pathway of Carbon Dioxide Assimilation in Chemolithotrophic Bacteria

The chemolithotrophic bacteria fix CO_2 via the reductive pentose phosphate pathway. As in phototrophic bacteria, reducing power is required in the form of NADH rather than NADPH. The function of the Calvin cycle was first established for *Thiobacillus thioparus* by SANTER and VISHNIAC (1955) and for *Thiobacillus denitrificans* by MILHAUD et al. (1956), and TRUDINGER (1956). Intact cells fixed $^{14}CO_2$ most rapidly into the carboxylic group of PGA, and later intermediates of the cycle were labeled as expected.

Subsequent analogous studies using different methods of investigation, with both normal and mutant organisms, were done with numerous representatives of chemolithotrophs, e.g., *Hydrogenomonas* (BERGMANN et al., 1958; HIRSCH et al., 1963; McFADDEN, 1959; McFADDEN and TU, 1965; BOWIEN et al., 1976; PUROHIT and McFADDEN, 1977), *Thiobacillus thiooxydans* (SUZUKI and WERKMAN, 1958), *Thiobacillus ferrooxydans* (DIN et al., 1967; MACIAG and LUNDGREN, 1964), *Thioba-

cillus novellus (ALEEM and HUANG, 1965), *Thiobacillus intermedius* (PUROHIT et al., 1976a, b), *Nitrobacter* (ALEEM, 1965; MALAVOLTA et al., 1960), *Nitrosomonas* (RAO and NICHOLAS, 1966), and *Micrococcus denitrificans* (KORNBERG et al., 1960). According to their peculiar character as marker enzymes for autotrophic CO_2 fixation, Ru5P kinase and especially RuBP carboxylase attracted most interest. In all chemolithotrophic organisms so far investigated these enzymes have been detected. For more experimental evidence the review articles of PECK (1968), RITTENBERG (1969), KELLY (1971), and McFADDEN (1973) are recommendable.

Taken together, the experimental results justify the conclusion that the Calvin cycle is the major pathway of CO_2 assimilation in chemolithotrophic bacteria of different physiological types. Until now there seems to be no exception.

G. Regulatory Aspects of Carbon Dioxide Assimilation in Prokaryotic Microorganisms

As indicated in Section A, prokaryotic cells are faced with regulatory problems of CO_2 assimilation different from those of eukaryotic cells. The reason is their much lesser degree of cellular compartmentation. Very efficient mechanisms of regulation are expected especially for those organisms that are facultative heterotrophs, and that may switch their metabolism from photoautotrophic to dark heterotrophic growth. This must have consequences for the energy metabolism. As has been calculated, only 2.9 mol of ATP are required to convert one mol of carbon from carbohydrate to cell material, whereas 7.9 mol of ATP are needed to synthesize cell material from CO_2 (VISHNIAC, 1971). Special regulatory mechanisms ought to be involved to guarantee the energy harmony in the cell under different conditions. Interactions between the processes of carbohydrate degradation, tricarboxylic acid cycle, respiration, and CO_2 assimilation are not only possible but probably unavoidable. Although a detailed consideration of the mechanisms of the regulation of CO_2 assimilation is outside the scope of this article, it may be useful to mention briefly a few aspects peculiar to prokaryotic organisms. The enzymes involved in CO_2 fixation on the one hand and oxidative breakdown processes on the other are subject to induction and repression as well as to metabolic control of enzyme function. Because in many experiments the changing rates of CO_2 assimilation paralleled the amount of RuBP carboxylase, this enzyme is often used as a marker of Calvin cycle activity.

In Athiorhodaceae the activity of the Calvin cycle is high when they live as photolithoautotrophs and apply H_2 as reductant. Characteristically RuBP carboxylase and the capability for CO_2 fixation is reduced drastically in the presence of organic substrates, especially when the bacteria live aerobically by respiration in the dark (LASCELLES, 1966; ANDERSON and FULLER, 1967b; FULLER, 1969; KELLY, 1971; SLATER and MORRIS, 1973a, b; McFADDEN, 1973; TABITA and McFADDEN, 1974a). Also in Thiorhodaceae RuBP carboxylase is repressed to various degrees by the presence of organic substances in the medium (HURLBERT and LASCELLES, 1963). It is noteworthy that the extent of the RuBP carboxylase repression varied with the nature of the organic substrate (HURLBERT and LASCELLES,

1963; Tabita and McFadden, 1974). While under autotrophic conditions the function of the Calvin cycle is primarily to furnish the cell with carbon from CO_2, its role under photoorganotrophic conditions is assumed to provide a mechanism for removing excessive electrons originating during the assimilation of organic substrates which are more reduced than cell material. Added bicarbonate is required for growth on fatty acids (van Niel, 1944).

For facultative chemolithotrophic bacteria the general finding is that RuBP carboxylase is repressed during heterotrophic growth. The basal level of this enzyme as well as the extent of the repression varies, however, with the organism and with the growth substrate. For experimental details see the reviews of Rittenberg (1969), Kelly (1971), McFadden (1973) and Schlegel (1975). A particularly complex regulation has been found in several strains of *Hydrogenomonas*. Superimposed on the repression of enzyme synthesis are mechanisms of degradation or modification of the enzyme protein which cause a rapid decay of activity (Kuehn and McFadden, 1968).

It is commonly held that blue-green algae have only limited possibilities for transcriptional control and exhibit a rather strong inflexibility of their metabolic pattern compared with Athiorhodaceae (Pearce and Carr, 1967, 1969; Carr, 1973; Slater, 1975). *Chlorogloea fritschii* grown heterotrophically in the dark contained the same RuBP carboxylase activity as organisms grown autotrophically in the light (Carr, 1973). However, absence of mechanisms for enzyme repression and derepression is not invariable in blue-green algae, e.g., the G6P dehydrogenase content of *Aphanocapsa* varied considerably depending on the growth environment, principally the presence or absence of glucose (Pelroy et al., 1972). Nevertheless, control of metabolism in blue-green algae seems to be expected at the level of enzyme activity rather than in enzyme synthesis.

In all blue-green algae so far studied the oxidative pentose phosphate pathway appears to be the principal, if not the sole route of glucose oxidation (Cheung and Gibbs, 1965; Pearce and Carr, 1969; Rippka, 1972; Pelroy et al., 1972; Pelroy et al., 1976a, b). Experiments done by Pelroy et al. (1972) with *Aphanocapsa* showed that the conversion of glucose to CO_2 in the light is reduced to about 5% of the dark rate. Their results suggest that the operation of the oxidative pentose phosphate pathway might be subject to control by a product of light metabolism. Indeed, it was found that the key enzyme of this pathway, G6P dehydrogenase, is inhibited by RuBP and by high ratios of NADPH to NADP (Pelroy et al., 1972; Pelroy and Bassham, 1973b). It is noteworthy that RuBP is only detectable during photosynthesis (Pelroy and Bassham, 1973). Recent investigations of Pelroy et al. (1976a, b) supported evidence for the regulation of the Calvin cycle in blue-green algae. Photosynthetic CO_2 assimilation is blocked when cells of *Aphanocapsa* are placed in the dark. This loss of function does not result from a reduced supply of cellular energy, it is more likely the expression of a specific regulatory mechanism that is present when cells lose their capacity to carry out noncyclic photosynthetic electron transport (Pelroy et al., 1976a). Pelroy et al. (1976b) supported evidence that in *Aphanocapsa* the rate-limiting enzymes specific to CO_2 fixation are Ru5P kinase or RuBP carboxylase (or both) and FBPase plus SBPase and that the light–dark regulation of CO_2 assimilation is effected mainly in these reactions via a mechanism of "light activation" that is closely related to the production of reducing power in the course of photosynthetic electron transport. It has been demonstrated that a number of Calvin cycle enzymes of blue-green algae are stimulated in the light or by reducing agents (DUGGAN and ANDERSON, 1975; ANDERSON and DUGGAN, 1976). The "light inactivation" of G6P dehydrogenase observed by these authors seems to be an additional possibility to understand the restricted oxidative glucose breakdown in the light. Considerable information is now becoming available on the regulation of the Calvin cycle in purple bacteria and in chemolithotrophic bacteria. This regulation seems to be governed in part at least by the cellular energy metabolism and, in contrast to the oxygen-evolving photosynthetic organisms, does not involve "light modulation" of enzyme activity. JOHNSON and PECK (1965) discovered that AMP inhibited almost completely the ATP-dependent autotrophic CO_2 fixation in crude extracts of *Thiobacillus thioparus*. Inhibition of CO_2 fixation by AMP was subsequently observed with cell-free preparations of *Thiobacillus novellus* (ALEEM and HUANG, 1965) and *Hydrogenomonas facilis* (MCFADDEN and TU, 1965). Studies with extracts and with purified preparations of Ru5P kinase from several chemolithotrophic and phototrophic bacteria showed that the site of inhibition by AMP is this enzyme (GALE and BECK, 1966; JOHNSON, 1966; MCFADDEN and TU, 1967; MCELROY et al., 1968; RINDT and OHMANN, 1969). MCELROY et al. (1969) and RINDT and OHMANN (1969) detected that the activity of Ru5P kinase from *Hydrogenomonas facilis, Rps. spheroides*, and *Rsp. rubrum* depends on the allosteric activation caused by NADH. Later PEP was also reported to act as an inhibitor of Ru5P kinase in *Hydrogenomonas* (BALLARD et al., 1971; ABDELAL and SCHLEGEL, 1974), *Chromatium* (HART and GIBSON, 1971), and *Thiobacillus neapolitanus* (MCELROY et al., 1972).

Although only few autotrophs have been tested and a systematic investigation is still lacking, the experiments so far done show that the regulation of Ru5P kinase varies with the organism. Nevertheless, the most important effectors seem to be NADH and AMP, indicating that CO_2 fixation is under the control of energy charge and availability of reducing power. In many Athiorhodaceae, e.g., *Rps. spheroides* and *Rsp. rubrum*, which are capable of growing phototrophically in the light as well as heterotrophically in the dark, properties of the key enzymes of the main catabolic routes are also important for the regulation of the balance between autotrophic and heterotrophic aspects of their metabolism.

The schematic presentation of Fig. 1 summarizes the results obtained with *Rps. spheroides, Rps. capsulata*, and *Rsp. rubrum*. Because the regulatory properties of the enzymes may be different, generalizations are not allowed before other organisms have been studied in detail. The route of glucose breakdown in these organisms is the Entner-Doudoroff pathway and the tricarboxylic acid cycle with the key enzymes G6P dehydrogenase, pyruvate kinase, and citrate synthase. G6P dehydrogenase of *Rps. spheroides* (OHMANN et al., 1970) and of several other autotrophs including *Hydrogenomonas* (BLACKKOLB and SCHLEGEL, 1968) is allosterically inhibited by ATP. The enzyme of *Rps. spheroides* is also sensitive to RuBP (RIPPKA and STANIER, 1973). ATP is also a negative effector of pyruvate kinase from *Rps. spheroides* and *Rsp. rubrum* (KLEMME, 1973; SCHEDEL et al., 1975). BORRISS and OHMANN (1972) found that citrate synthase of *Rps. spheroides* is inhibited allosterically by NADH, and that this inhibition may be overcome by AMP and, to a lesser degree by ADP. The same situation has been found in the case of the citrate synthase of *Rps. capsulata* (EIDELS and PREIS, 1970).

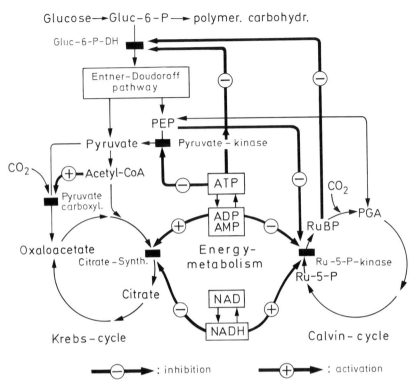

Fig. 1. Schematic presentation of the regulatory interaction of carbon dioxide assimilation and heterotrophic breakdown processes in some Athiorhodaceae

Under illumination the process of photosynthetic electron transport results in the production of NADH and ATP, the level of which would be expected to increase, while the intracellular AMP concentration would be expected to decrease (Jackson and Crofts, 1968; Ramirez and Smith, 1968). This may be received as a signal by Ru5P kinase and citrate synthase. As a consequence of the reciprocal effector sensitivities of the responsible key enzymes of the Krebs cycle and the Calvin cycle, the process of carbon dioxide assimilation is activated while oxidative breakdown via the Krebs cycle is depressed. In addition ATP, and ATP plus RuBP, may block carbon flow at the sites of pyruvate kinase and G6P dehydrogenase, respectively. At low light intensity the level of AMP as an indicator of energy depletion increases, while NADH and ATP concentrations decrease. This leads to deinhibition of citrate synthase, pyruvate kinase, and G6P dehydrogenase, resulting in high activity of degradative processes and a reduction of CO_2 assimilation via the Calvin cycle.

For several reasons it must be assumed that RuBP carboxylase is also important in regulating CO_2 assimilation in phototrophic and chemolithotrophic bacteria (cf. Chap. II.17, this vol.). Although a considerable amount of information exists on the regulation of CO_2 assimilation in prokaryotic microorganisms, it must be emphasized that only a very limited number of species have been studied and that most of the data describe the kinetic and regulatory behavior of enzymes in vitro; evidence for the importance of most of these properties in vivo is still lacking. Careful measurements of pool sizes of metabolic intermediates and other new approaches and ideas will certainly be needed for further critical investigations.

References

Abdelal, A.T.H., Schlegel, H.G.: Biochem. J. *139*, 481–489 (1974)
Aleem, M.J.H.: Biochim. Biophys. Acta *107*, 14–28 (1965)
Aleem, M.J.H., Huang, E.: Biochem. Biophys. Res. Commun. *20*, 515–520 (1965)
Anderson, L.E., Duggan, J.X.: In: Proc. 2nd Int. Symp. Photosynthetic Prokaryotes. Codd, G.A., Stewart, W.D.P. (eds.), pp. 179–181, Dundee 1976
Anderson, L.E., Fuller, R.C.: Plant Physiol. *42*, 487–490 (1967a)
Anderson, L.E., Fuller, R.C.: Plant Physiol. *42*, 491–496 (1967b)
Anderson, L.E., Fuller, R.C.: Plant Physiol. *42*, 497–502 (1967c)
Anderson, L.E., Fuller, R.C.: J. Biol. Chem. *244*, 3105–3109 (1969)
Arnon, D.I.: In: Progress in photosynthesis research. Metzner, H. (ed.), Vol. III, pp. 1444–1473. Tübingen: Laupp 1969
Ballard, R.W., McElroy, R.D.: Biochem. Biophys. Res. Commun. *44*, 614–618 (1971)
Bergmann, F.H., Towne, J.C., Burris, R.H.: J. Biol. Chem. *230*, 13–24 (1958)
Beuscher, N., Gottschalk, G.: Z. Naturforsch. *27b*, 967–973 (1972)
Blackkolb, F., Schlegel, H.G.: Arch. Mikrobiol. *63*, 177–196 (1968)
Bothe, H., Nolteernsting, U.: Arch. Microbiol. *102*, 53–57 (1975)
Bothe, H., Falkenberg, B., Nolteernsting, U.: Arch. Microbiol. *96*, 291–304 (1974)
Bowien, B., Mayer, F., Codd, G.A., Schlegel, H.G.: Arch. Microbiol. *110*, 157–166 (1976)
Buchanan, B.B., Evans, M.C.W.: Proc. Natl. Acad. Sci. USA *54*, 1212–1218 (1965)
Buchanan, B.B., Sirevag, R.: Arch. Microbiol. *109*, 15–20 (1976)
Buchanan, B.B., Bachofen, R., Arnon, D.I.: Proc. Natl. Acad. Sci. USA *52*, 839–847 (1964)
Buchanan, B.B., Evans, M.C.W., Arnon, D.I.: Arch. Mikrobiol. *59*, 32–40 (1967)
Buchanan, B.B., Schürmann, P., Shanmugam, K.T.: Biochim. Biophys. Acta *283*, 136–145 (1972)
Carr, N.G.: In: The biology of blue green algae. Carr, N.G., Whitton, B.A. (eds.), pp. 39–65. Oxford: Blackwell 1973
Cheung, W.Y., Gibbs, M.: Plant Physiol. *41*, 451–462 (1965)
Cherniadev, I.I., Kondratieva, E.N., Doman, N.G.: Izvestia Akad. Nauk SSSR *N 5*, 670–675 (1969)
Din, G.A., Suzuki, I., Lees, H.: Canad. J. Microbiol. *13*, 1413–1419 (1967)
Döhler, G.: Planta (Berl.) *118*, 259–269 (1974)
Döhler, G.: Planta (Berl.) *131*, 129–133 (1976a)
Döhler, G.: In: Proc. 2nd. Int. Symp. Photosynthetic Prokaryotes. Codd, G.A., Stewart, W.D.P. (eds.), pp. 176–178. Dundee 1976b
Duggan, J.X., Anderson, L.E.: Planta (Berl.) *122*, 293–297 (1975)
Eidels, L., Preiss, J.: J. Biol. Chem. *245*, 2937–2949 (1970)
Evans, M.C.W., Buchanan, B.B., Arnon, D.I.: Proc. Natl. Acad. Sci. USA *55*, 928–934 (1966)
Fogg, G.E., Stewart, W.D.P., Fay, P., Walsby, A.E.: The blue green algae, London, New York: Academic Press 1973
Fuller, R.C.: In: Prog. photosynthesis res. Metzner, H. (ed.), Vol. III, pp. 1579–1592. Tübingen: Laupp, 1969
Fuller, R.C., Smillie, R.M., Sisler, E.C., Kornberg, H.L.: J. Biol. Chem. *236*, 2140–2149 (1961)
Gale, N.L., Beck, J.V.: Biochem. Biophys. Res. Commun. *24*, 792–796 (1966)
Glover, J., Kamen, M.D., van Generen, H.: Arch. Biochem. Biophys. *35*, 384–408 (1952)

Hart, B.A., Gibson, J.: Arch. Biochem. Biophys. *144*, 308–321 (1971)
Hirsch, P., Georgiev, G., Schlegel, H.G.: Arch. Mikrobiol. *46*, 79–95 (1963)
Hoare, D.S.: Biochem. J. *87*, 284–301 (1963)
Hoare, D.S., Gibson, J.: Biochem. J. *91*, 546–559 (1964)
Hoare, D.S., Hoare, S.C., Moore, R.B.: J. Gen. Microbiol. *49*, 351–370 (1967)
Hurlbert, R.E., Lascelles, J.: J. Gen. Microbiol. *33*, 445–458 (1963)
Jackson, J.B., Crofts, A.R.: Biochem. Biophys. Res. Commun. *32*, 908–915 (1968)
Johnson, E.J.: Arch. Biochem. Biophys. *114*, 178–183 (1966)
Johnson, E.J., Peck, H.D.: J. Bacteriol. *89*, 1041–1050 (1965)
Kandler, O.: Naturwissenschaften *48*, 604 (1961)
Kelly, D.P.: Ann. Rev. Microbiol. *25*, 177–210 (1971)
Kelly, G.J., Latzko, E., Gibbs, M.: Ann. Rev. Plant Physiol. *27*, 181–205 (1976)
Kindel, P., Gibbs, M.: Nature (London) *200*, 260–261 (1963)
Klemme, H.J.: Arch. Mikrobiol. *90*, 305–322 (1973)
Kondratieva, E.N., Zhukov, V.G., Ivanovsky, R.N., Petushkova, Y.P., Monosov, E.Z.: Arch. Microbiol. *108*, 287–292 (1976)
Kornberg, H.L., Collins, J.F., Bigley, D.: Biochim. Biophys. Acta *39*, 9–24 (1960)
Kuehn, G.D., McFadden, B.A.: J. Bacteriol. *95*, 937–946 (1968)
Lascelles, J.: J. Gen. Microbiol. *23*, 499–509 (1966)
Latzko, E., Gibbs, M.: Plant Physiol. *41*, IV (1966)
Leach, C.K., Carr, N.G.: Biochim. Biophys. Acta *245*, 165–174 (1971)
Losada, M., Trebst, A., Ogata, S., Arnon, D.I.: Nature (London) *186*, 753–760 (1960)
Lwoff, A., van Niel, C.B., Ryan, F.J., Tatum, E.L.: Cold Spring Harbor Symp. Quant. Biol. *11*, 302–303 (1946)
Maciag, W.J., Lundgren, D.G.: Biochem. Biophys. Res. Commun. *17*, 603–607 (1964)
Malavolta, E., Delviche, C.C., Borge, W.D.: Biochem. Biophys. Res. Commun. *2*, 245–249 (1960)
McElroy, R.D., Johnson, E.J., Johnson, M.K.: Biochem. Biophys. Res. Commun. *30*, 678–682 (1968)
McElroy, R.D., Johnson, E.J., Johnson, M.K.: Arch. Biochem. Biophys. *131*, 272–275 (1969)
McElroy, R.D., Mack, H.M., Johnson, E.M.: J. Bacteriol. *112*, 532–538 (1972)
McFadden, B.A.: J. Bacteriol. *77*, 339–343 (1959)
McFadden, B.A.: Bacteriol. Rev. *37*, 289–319 (1973)
McFadden, B.A., Tu, C.L.: Biochem. Biophys. Res. Commun. *19*, 728–733 (1965)
Milhaud, G., Aubert, J.P., Millet, J.: C.R. Acad. Sci. *243*, 102–105 (1956)
Niel, C.B. van: In: Photosynthesis in plants. Franck, J., Loomis, W.E. (eds.), p. 473. Ames, Iowa: Iowa State College Press 1949
Niel, C.B. van: Bacteriol. Rev. *8*, 1–118 (1944)
Niel, C.B. van: Adv. Enzymol. *1*, 263–328 (1941)
Norris, L., Norris, R.E., Calvin, M.: J. Exp. Bot. *6*, 64–74 (1955)
Ohmann, E., Borriss, R., Rindt, K.P.: Z. Allg. Mikrobiol. *10*, 37–53 (1970)
Payne, J., Morris, J.G.: FEBS Lett. *4*, 52–54 (1969a)
Payne, J., Morris, J.G.: J. Gen. Microbiol. *59*, 97–101 (1969b)
Pearce, J., Carr, N.G.: J. Gen. Microbiol. *49*, 301–313 (1967)
Pearce, J., Carr, N.G.: J. Gen. Microbiol. *54*, 451–462 (1969)
Peck, H.D.: Ann. Rev. Microbiol. *22*, 489–518 (1968)
Pelroy, R.A., Bassham, J.A.: Arch. Mikrobiol. *86*, 25–38 (1972)
Pelroy, R.A., Bassham, J.A.: J. Bacteriol. *115*, 937–942 (1973a)
Pelroy, R.A., Bassham, J.A.: J. Bacteriol. *115*, 943–948 (1973b)
Pelroy, R.A., Kirk, M.R., Bassham, J.A.: J. Bacteriol. *128*, 623–632 (1976a)
Pelroy, R.A., Levine, G.A., Bassham, J.A.: J. Bacteriol. *128*, 633–643 (1976b)
Pelroy, R.A., Rippka, R., Stanier, R.Y.: Arch. Mikrobiol. *87*, 303–322 (1972)
Pfennig, N.: Ann. Rev. Microbiol. *21*, 285–324 (1967)
Pfennig, N., Trüper, H.: In: Bergey's manual of determinative bacteriology. 8th ed. Buchanan, R.E., Gibbons, W.E. (eds.), pp. 24–64. Baltimore: Williams and Wilkins, 1974
Porter, J., Merrett, M.J.: Plant Physiol. *50*, 252–255 (1972)
Purohit, K., McFadden, B.A.: J. Bacteriol. *129*, 415–421 (1977)
Purohit, K., McFadden, B.A., Cohen, A.L.: J. Bacteriol. *127*, 505–515 (1976a)

Purohit, K., McFadden, B.A., Shaykh, M.M.: J. Bacteriol. *127*, 516–522 (1976b)
Ramirez, J., Smith, L.: Biochim. Biophys. Acta *153*, 466–478 (1968)
Rao, P.S., Nicholas, D.J.D.: Biochim. Biophys. Acta *124*, 221–232 (1966)
Rindt, K.P., Ohmann, E.: Biochem. Biophys. Res. Commun. *36*, 357–364 (1969)
Rippka, R.: Arch. Mikrobiol. *87*, 93–98 (1972)
Rippka, R., Stanier, R.Y.: In: Abstr. Symp. Prokaryotic Photosynthetic Organisms, pp. 135–138. Freiburg, 1973
Rittenberg, S.C.: Adv. Microb. Physiol. *3*, 151–196 (1969)
Sadler, W.R., Stanier, R.Y.: Proc. Natl. Acad. Sci. USA *46*, 1328–1334 (1960)
Santer, M., Vishniac, W.: Biochim. Biophys. Acta *18*, 157–158 (1955)
Schedel, M., Klemme, J.H., Schlegel, H.G.: Arch. Microbiol. *103*, 237–245 (1975)
Schlegel, H.G.: In: Marine ecology. Kinne, O. (ed.), Vol. II, Part I, pp. 9–60. London: John Wiley and Sons 1975
Sirevag, R.: Arch. Microbiol. *98*, 3–18 (1974)
Sirevag, R.: Arch. Microbiol. *104*, 105–111 (1975)
Sirevag, R., Ormerod, J.G.: Science *169*, 186–188 (1970)
Sirevag, R., Buchanan, B.B., Berry, J.A., Throughton, J.H.: Arch. Microbiol. *112*, 35–38 (1977)
Slater, J.H.: Arch. Microbiol. *103*, 45–49 (1975)
Slater, J.H., Morris, J.: Arch. Mikrobiol. *88*, 213–223 (1973a)
Slater, J.H., Morris, J.: Arch. Mikrobiol. *92*, 235–244 (1973b)
Smillie, R.M., Rigopoulos, N., Kelly, H.: Biochim. Biophys. Acta *56*, 612–614 (1962)
Stanier, R.Y.: In: The biology of blue green algae. Carr, N.G., Whitton, B.A. (eds.), pp. 501–518. Oxford: Blackwell 1973
Stoppani, A.D.M., Fuller, R.C., Calvin, M.: J. Bacteriol. *69*, 491–501 (1955)
Suzuki, J., Werkman, C.H.: Arch. Biochem. Biophys. *76*, 103–111 (1958)
Tabita, F.R., McFadden, B.A.: J. Biol. Chem. *249*, 3453–3458 (1974a)
Tabita, F.R., McFadden, B.A.: J. Biol. Chem. *249*, 3459–3464 (1974b)
Tabita, F.R., McFadden, B.A., Pfennig, N.: Biochim. Biophys. Acta *341*, 187–194 (1974)
Trudinger, P.A.: Biochem. J. *64*, 274–286 (1956)
Trüper, H.G.: Arch. Mikrobiol. *49*, 23–50 (1964)
Vishniac, W.: Symp. Soc. Gen. Microbiol. *21*, 355–366 (1971)
Walker, D.A.: In: Plant carbohydrate biochemistry. Proc. 10th Symp. Phytochem. Soc. Pridham, J.B. (ed.), pp. 7–26. London, New York: Academic Press 1974
White, A.W., Shilo, M.: Arch. Microbiol. *102*, 123–127 (1975)
Willard, J.M., Schulman, M., Gibbs, M.: Nature (London) *206*, 195 (1965)
Woods, D.D., Lascelles, J.: Symp. Soc. Gen. Microbiol. *4*, 1–27 (1954)
Yoch, D.C., Lindstrom, E.S.: Biochem. Biophys. Res. Commun. *28*, 65–69 (1967)

5. Light-Enhanced Dark CO$_2$ Fixation

S. MIYACHI

A. Light-Enhanced Dark CO$_2$ Fixation in C$_3$ Plants

I. Introduction

In 1939, McALISTER discovered that wheat leaves can "pick up" CO$_2$ in darkness immediately following a high rate of photosynthesis. BENSON and CALVIN (1947, 1948) found that dark ^{14}CO$_2$ fixation in green algae (*Chlorella* and *Scenedesmus*) was greatly enhanced by prior illumination in the absence of CO$_2$. The prior illumination in the absence of CO$_2$ and the following enhanced dark CO$_2$ fixation are termed "preillumination" and "light-enhanced dark CO$_2$ fixation (LED)", respectively. Analysis of ^{14}CO$_2$ fixation products revealed that the radioactivity was incorporated into phosphoglyceric acid (PGA), alanine and also sugars during LED. Degradation of ^{14}C-glucose obtained after LED in *Chlorella* showed that 75%–79% of the total radioactivity was in the 3 and 4 positions. BENSON and CALVIN therefore assumed that a certain amount of reducing power accumulated in green algae during preillumination and could later be used for CO$_2$ reduction. On the other hand, GAFFRON et al. (1951) found that 96% of the total ^{14}C fixed during LED (60 s) was in PGA and pyruvic acid. After 10 s of photosynthetic ^{14}CO$_2$ fixation, only 62% of the total ^{14}C was incorporated into these compounds, although the total radioactivity incorporated was practically equal to that observed after 60 s of LED. Therefore they concluded that after preillumination, the dark reaction consists of only one step, namely, the fixation of the added CO$_2$ into the carboxyl groups of PGA and pyruvic acid. There the reaction stops and reduction beyond the stage of PGA occurs in light only.

II. Capacity for Light-Enhanced Dark CO$_2$ Fixation

With *Chlorella ellipsoidea*, MIYACHI et al. (1955b) showed that at 25° C, LED came to cessation within 30 s after ^{14}C-bicarbonate addition. When the algal cells were preilluminated, the capacity of LED measured by 30-s dark ^{14}CO$_2$ fixation increased rapidly at first then more slowly to attain a stationary value (after 20–40 min preillumination at saturating light intensity) of 5–7 μmol ^{14}CO$_2$ fixed per g d.w. (TAMIYA et al., 1957). When the preillumination was discontinued, the capacity for LED decreased in the manner of a first-order reaction (half-life, 2–3 min; MIYACHI et al., 1955a). When *Chlorella* cells in the "lollipop" (thickness, 2 cm; cell density, 6 ml packed cell volume/liter of 5 mM phosphate buffer, pH 7.0) which had been illuminated from both sides for 30 min (intensity, 2 × 15,000 lux), were provided with various concentrations of NaH^{14}CO$_3$ in the following darkness, the capacity for LED increased in parallel with the bicarbonate concentration and attained the maximum level at 1.5 mM (MIYACHI, unpublished). MIYACHI (unpublished) also found that the following relationship holds between the capacity for LED and the light intensity [(L)] during preillumination:

$$\text{Capacity for LED} = (L)^3/[\Phi^3 + (L)^3]$$

where Φ is a constant corresponding to the light intensity which causes 50% of the maximum LED (2×850 lux). LED practically attained its maximum level with preillumination at an intensity of 2×2500 lux. Thus, the light intensity required for saturating LED is far lower than that required for saturating photosynthesis and the response of LED to the light intensity is quite different from that of photosynthesis. Changes in the capacity for LED during the course of the algal life cycle have also been reported (NIHEI et al., 1954).

Observations that the stationary level of LED was lowered and the rate of decrease in the capacity for LED after cessation of preillumination was enhanced by the introduction of various oxidants such as O_2 (MIYACHI et al., 1955 b), p-benzoquinone (MIYACHI et al., 1955 a), 2,6-dichlorophenolindophenol (DPIP; KATOH et al., 1958) and H_2O_2 (HIROKAWA et al., 1958) are in accord with the inference that the reducing substance produced during preillumination is used for the reduction of CO_2 in the following darkness. The finding that the dark reduction of DPIP was greatly enhanced by preillumination of *Chlorella* cells (KATOH et al., 1958) also supported this inference. TAMIYA et al. (1957) found that the dark decay of CO_2-fixing capacity which had been produced by preillumination was greatly accelerated by cyanide. Based on these observations, they assumed that during illumination, a reducing substance (R) is formed by the action of a reductant produced via photolysis of water. They further proposed the modes of reactions between R and various oxidants including PGA by which all of the above observations could be coherently explained.

III. Products

KANDLER and HABERER-LIESENKÖTTER (1963) showed that more than 80% of the total ^{14}C fixed after 6 s of LED in *Chlorella* was in PGA, while the radioactivity incorporated into sugar phosphates was only 7% of the total. TOGASAKI and GIBBS, who extended LED to the blue-green alga *Anacystis nidulans* (TOGASAKI and GIBBS, 1963), found that $^{14}CO_2$ was mostly incorporated into the carboxyl group of PGA during LED under He (TOGASAKI and GIBBS, 1967). They also showed that glucose 6-phosphate (G6P) obtained after LED was predominantly labeled in carbon atom 4 (70%–80%) and had very little isotope in the C-5 and C-6 positions. The very uneven distribution of tracer in the G6P may resemble the tracer distribution of the compound formed in an early period of photosynthesis. With *C. ellipsoidea*, HOGETSU and MIYACHI (1970) also showed that the main initial $^{14}CO_2$ fixation product in LED was PGA. Under aerobic conditions (21% O_2), the amount of radioactivity in PGA was about half that found under anaerobic conditions. The percent incorporation of radioactivity into PGA rapidly decreased during the rest of the dark period. Irrespective of the presence or absence of O_2 in the atmosphere, the radioactivity transferred to sugar phosphates accounted for only 10% of the total ^{14}C. Recently, OKABE (1975) extended the studies on the effect of O_2 on LED to other green algae (*Chlorella vulgaris* 11 h and *Chlorella protothecoides*), red alga (*Porphyridium cruentum*) and blue-green algae (*Nostoc muscorum* and *A.nidulans*). In all the species tested, the highest $^{14}CO_2$ incorporation was found in PGA during the initial phase of LED. The radioactivity in PGA was transferred to other compounds during the later phase. In all the species, and irrespective of the presence or absence of O_2, its transfer to sugar phosphates accounted for only 10%–15% of the total ^{14}C fixed. Thus, it is now clear that during the initial period of LED, CO_2 is mostly incorporated into PGA and less than 15% of the PGA is reduced to sugar phosphates during the subsequent period.

IV. RuBP, NADPH, and ATP Levels

The above-mentioned levels of ^{14}C incorporation into PGA under aerobic and anaerobic conditions indicate that the formation of ribulose 1,5-bisphosphate (RuBP) during preillumination is regulated by O_2. MUTO and MIYACHI (unpublished) showed by an enzymatic method that the RuBP which accumulated in *C. ellipsoidea* during preillumination (μmol per g d.w.) in the absence or presence of O_2 (21%) amounts to ca. 1.6–2.4 and 0.8–1.2, respectively. The same values obtained with uniformly ^{32}P-labeled *C.ellipsoidea* (KANAI and MIYACHI, unpublished) were 4.0 and 2.4, respectively. The addition of CO_2 simultaneous with the cessation of preillumination greatly enhanced the decrease of RuBP.

OH-HAMA and MIYACHI (1959) showed that when *C. ellipsoidea* was preilluminated, NADPH increased following a time course similar to that of the rise in the capacity for LED. The very small increment of NADPH (μmol per g d.w.; 0.12 under N_2 and 0.03 under CO_2-free air) as compared to the capacity for LED is in accord with a limited operation of the reductive pentose phosphate cycle.

KANAI and MIYACHI showed that the ATP level in *C.ellipsoidea* decreased rapidly after cessation of preillumination (Fig. 1). Thus, the limited reduction of PGA in LED may also be caused by a shortage of ATP required for phosphorylating PGA.

When DCMU and NaH$^{14}CO_3$ were added to *C.ellipsoidea* under continued illumination, initial rapid $^{14}CO_2$ fixation followed by slow incorporation was observed. The time course of total $^{14}CO_2$ fixation, as well as the radioactivities incorporated into various products, were similar to those observed in LED, the maximum level of radioactivity incorporated into sugar phosphates being only 7% of the total ^{14}C incorporated (Fig. 2). This effect of DCMU suggests that the reduction of PGA immediately stops when the light-induced supply of NADPH and/or ATP is discontinued, although CO_2 fixation (via RuBP carboxylase) proceeds to some extent. TOGASAKI and BOTOS (1972) found that LED was almost completely eliminated when *Chlamydomonas reinhardi* was preilluminated in N_2 in the presence of DCMU. In accord with their observation, MUTO and MIYACHI (unpublished) found that the RuBP level in *C.ellipsoidea* which had been preilluminated in the presence of 0.1 mM CMU was as low as 0.08–0.16 μmol per g d.w., and the level did not change when CO_2 was added after cessation of preillumi-

Fig. 1. Decrease in the ATP level in *Chlorella ellipsoidea* in the dark following preillumination. Uniformly ^{32}P-labeled cells were preilluminated for 21 min under N$_2$, then the light was switched off with (*closed circle*) or without (*open circle*) simultaneous change of the bubbling gas to N$_2$ containing 2% CO$_2$. Samples were taken at the indicated time intervals and ATP was separated by two-dimensional paper chromatography (KANAI and MIYACHI, original)

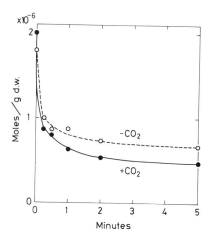

Fig. 2. Effect of DCMU on time course of ^{14}CO$_2$ incorporation into products in *Chlorella ellipsoidea* in light. DCMU (1 μM) was added to the algal cells (which had been preilluminated for 30 min) then after 15 s NaH^{14}CO$_3$ (1 mM) was added under continued illumination and aeration with CO$_2$-free air. *Ala*, alanine; *ASP*, aspartate; *Glu*, glutamate; *Mal*, malate (OSHIO and MIYACHI, original)

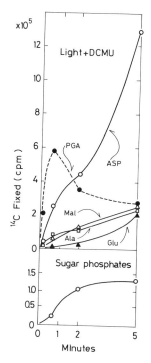

nation. TOGASAKI and BOTOS (1972) further showed that LED was restored when the preillumination in the presence of DCMU was carried out under H$_2$.

These results indicate that in LED, CO$_2$ reacts with RuBP which had accumulated during the period of preillumination, and due to the limited supply of NADPH and/or ATP, only a part of PGA thus formed is reduced to triose phosphate. However, two observations are not consistent with this concept.

1. The effect of various inhibitors (MIYACHI, 1959) showed that p-chloromercuribenzoic acid markedly accelerated the dark decay of the CO$_2$-fixing capacity pro-

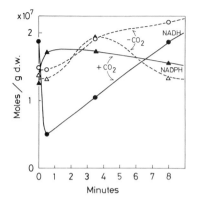

Fig. 3. Effect of CO_2 on the levels of NADH and NADPH in *Chlorella ellipsoidea* in the dark following preillumination. Cells were preilluminated for 50 min, then the light was switched off with (*closed symbol*) or without (*open symbol*) simultaneous addition of $NaHCO_3$ (1 mM). The algal suspensions were bubbled with N_2 throughout the experiment (Oh-hama and Miyachi, original)

duced by preillumination, while the same compound did not show a significant effect on the rate of $^{14}CO_2$ fixation during LED in *Chlorella* (Miyachi, 1960).

2. When CO_2 was added immediately after preillumination of *Chlorella* under aerobic conditions, both NADPH and NADH increased sharply (Miyachi et al., 1960). Under anaerobic conditions, addition of CO_2 after preillumination caused a sharp temporary decrease in the NADH level in the dark, while no such change was observed in the NADPH level (Fig. 3). One of the important conclusions drawn from these experiments was that NAD is converted into NADP by light.

V. Fate of PGA

Hogetsu and Miyachi (1970) found that the fate of PGA formed during LED in *Chlorella* was strongly dependent on the presence or absence of O_2 in the atmosphere. Under anaerobic conditions, the radioactivity in PGA was eventually transferred to alanine, while the main end product under aerobic conditions was aspartate.

Okabe (1975) showed that in *C.protothecoides* and *C.vulgaris*, the modes of radioactivity transfer from PGA were essentially the same as those observed in *C.ellipsoidea*. In *P.cruentum*, the radioactivity in PGA was mostly transferred to alanine under anaerobic conditions, but no enhancement of ^{14}C transfer to aspartate was observed under aerobic conditions. In blue-green algae, ^{14}C incorporation into alanine was very slow. Remarkable ^{14}C incorporation into aspartate was observed irrespective of the presence or absence of O_2 in the atmosphere. Most interesting was the finding that, unlike all the other species tested, O_2 enhanced LED in *A.nidulans* (Fig. 4).

Miyachi and Hogetsu (1970) studied the effects of preillumination with monochromatic red (679 nm) and blue (453 nm) light on subsequent dark $^{14}CO_2$ fixation in *C.ellipsoidea*. Under aerobic conditions, the rate of transfer of radioactivity from PGA to aspartate observed after preillumination with blue light was two to three times higher than that after preillumination with red light. Under anaerobic conditions, the transfer of radioactivity into alanine and lactate was more pronounced during LED caused by preillumination with red than with blue light.

Fig. 4. Oxygen enhancement of LED in *Anacystis nidulans*. *Solid lines*, LED; *dotted lines*, dark $^{14}CO_2$ fixation without preillumination; *closed circles*, $^{14}CO_2$ fixation under N_2; *open circles*, $^{14}CO_2$ fixation under aeration with CO_2-free air. NaH$^{14}CO_3$ was added at 0 min

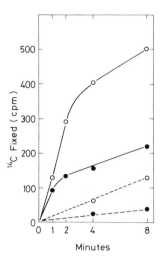

VI. Higher Plants

LED was also observed in leaves of higher plants such as spinach, soybean (MIYACHI and HOGETSU, 1970) and tobacco (POSKUTA, 1972). POSKUTA (1972) reported that O_2 greatly lowered the magnitude of LED, although such an effect was not observed by MIYACHI and HOGETSU (1970). POSKUTA also showed that in the presence of 3.5% O_2, about 10% of the $^{14}CO_2$ was incorporated into the sugar fraction, while practically no radioactivity was detected in this fraction at O_2 concentrations higher than 10%. Although MIYACHI and HOGETSU (1970) found LED in chloroplasts isolated from spinach leaves, the dark "pick-up" of CO_2 after photosynthetic CO_2 fixation has not been observed in isolated chloroplasts (WALKER et al., 1968; LATZKO and GIBBS, 1969). In contrast to whole cells, the magnitude of LED and its decay in the dark in isolated chloroplasts were not influenced by the presence of O_2.

B. Light-Enhanced Dark CO$_2$ Fixation in C$_4$ Plants

POSKUTA (1969) reported that maize leaves can pick up CO_2 in the dark following photosynthesis. This was followed by the discovery of LED in maize by MIYACHI and HOGETSU (1970). SAMEJIMA and MIYACHI (1971) showed that $^{14}CO_2$ is mostly incorporated into malate and aspartate during LED. Only about 10% of the total radioactivity incorporated appeared in phosphate esters. The radioactivity in the C$_4$ acids was quickly transferred to phosphate esters when the light was switched on again, indicating that the transfer of carboxyl carbon of C$_4$ acids to form PGA is light-dependent. The same authors (SAMEJIMA and MIYACHI, 1978) further reported that similar to that observed in *Chlorella*, a C$_3$ plant (MIYACHI et al., 1955b), the capacity of LED in maize leaves decayed following time course

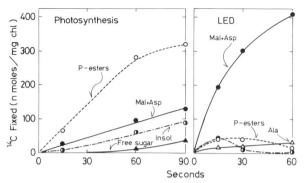

Fig. 5. Time course of ^{14}C incorporation into products during photosynthetic fixation (*left*) and light-enhanced dark fixation (*right*) of ^{14}C-bicarbonate infiltrated through vascular bundles of maize leaf. Maize leaf was cut horizontally across the veins to obtain a rectangular segment. The lower cut edge of the segment was immersed in NaH^{14}CO$_3$ solution (50 mM) and the upper edge was connected to a vacuum pump. The vacuum infiltration of NaH^{14}CO$_3$ in the dark was completed within 2 s. *Left*, light turned on immediately after vacuum infiltration; *right*, NaH^{14}CO$_3$ vacuum-infiltrated immediately after preillumination

characteristics of a first-order reaction. Since phosphoenolpyruvate (PEP)-level under CO$_2$-free air did not decrease in the dark subsequent to preillumination, they concluded that the dark decay of the LED-capacity is caused by a decrease of PEP carboxylase activity. In sharp contrast to their result, however, Laber et al. (1974) reported that PEP-level increased sharply after cessation of preillumination. Samejima and Miyachi (1978) also studied the photosynthetic ^{14}CO$_2$ fixation products of maize leaf segments after vacuum-infiltrating NaH^{14}CO$_3$ through the vascular bundles. When ^{14}CO$_2$ was fed directly to bundle sheath cells by this technique, the main initial ^{14}CO$_2$-fixation products were found in phosphate esters (55% of the total ^{14}C fixed was in PGA after 15 s), indicating that bundle sheath cells can fix CO$_2$ via the reductive pentose phosphate cycle. On the other hand, the main initial ^{14}CO$_2$-fixation products were malate and aspartate even when ^{14}CO$_2$ was provided through vascular tissues in the dark immediately following preillumination (Fig. 5). This indicates that the amount of RuBP accumulated in maize leaves during preillumination is far smaller than that of PEP, or RuBP carboxylase activity decays very rapidly after cessation of illumination.

Samejima and Miyachi (unpublished) studied the effect of light intensity of the second illumination on the transfer of carbon from C$_4$ acids in maize leaves. The transfer was observed at intensities higher than 1 k lux with almost maximum transfer at ca. 3 k lux. At 500 lux, the radioactivity in malate decreased and that in aspartate increased but little or no radioactivity was transferred to phosphate esters.

Detailed studies on changes in the pool size of various intermediates of the reductive pentose phosphate cycle as well as of C$_4$ acids during preillumination and a subsequent CO$_2$ fixation have been carried out by Farineau (1971). The results of similar experiments by Latzko et al. (1971) were consistent with the view that RuBP serves as a primary CO$_2$ acceptor in maize leaves. Supporting evidence for the above inference was obtained from degradation of the malate formed during the light-enhanced dark ^{14}CO$_2$ fixation which showed that about

25% of the label was in the C-1 carbon atom (LABER et al., 1972). In contrast, USUDA et al. (1973) found that most of the radioactivity in malic acid was in C-4, supporting the view that PEP serves as a primary acceptor of ^{14}CO$_2$ in maize.

STAMIESZKIN et al. (1972) reported that the capacity for LED in maize was greatly inhibited by O$_2$, although the pattern of ^{14}C incorporation was not affected. On the other hand, no O$_2$ effect was observed by MIYACHI and HOGETSU (1970). POSKUTA and FRANKIEWICZ-JOSKO (1975) also reported that dark pick-up of CO$_2$ subsequent to photosynthesis with blue light illumination was higher than that after red light illumination. Note, however, that none of the above studies examined the effect of O$_2$ on the stomatal aperture which controls CO$_2$ fixation.

References

Benson, A.A., Calvin, M.: Science *105*, 648 (1947)
Calvin, M., Benson, A.A.: Science *107*, 476–480 (1948)
Farineau, J.: In: Photosynthesis and photorespiration. Hatch, M.D. et al. (ed.), pp. 202–210. New York: Wiley-Interscience 1971
Gaffron, H., Fager, E.W., Rosenberg, J.L.: In: Symp. Soc. Exp. Biol. No. 5, pp. 262–283. Cambridge: Univ. Press 1951
Hirokawa, T., Miyachi, S., Tamiya, H.: J. Biochem. *45*, 1005–1010 (1958)
Hogetsu, D., Miyachi, S.: Plant Physiol. *45*, 178–182 (1970)
Kandler, O., Haberer-Liesenkötter, I.: Z. Naturforsch. *18b*, 718–730 (1963)
Katoh, T., Hirokawa, T., Miyachi, S.: J. Biochem. *45*, 907–912 (1958)
Laber, L.J., Latzko, E., Levi, C., Gibbs, M.: In: Proc. IInd Int. Congr. Photosynthesis Res. Forti, G. et al. (ed.), pp. 1737–1744. The Hague: Dr.W. Junk N.V. Publishers 1972
Laber, L.J., Latzko, E., Gibbs, M.: J. Biol. Chem. *249*, 3436–3441 (1974)
Latzko, E., Gibbs, M.: Plant Physiol. *44*, 396–402 (1969)
Latzko, E., Laber, L., Gibbs, M.: In: Photosynthesis and photorespiration. Hatch, M.D. et al. (ed.), pp. 196–201. New York: Wiley-Interscience 1971
McAlister, E.D.: J. Gen. Physiol. *22*, 613–636 (1939)
Miyachi, S.: Plant Cell Physiol. *1*, 1–15 (1959)
Miyachi, S.: Plant Cell Physiol. *1*, 117–130 (1960)
Miyachi, S., Hirokawa, T., Tamiya, H.: J. Biochem. *42*, 737–743 (1955a)
Miyachi, S., Hogetsu, D.: Can. J. Bot. *48*, 1203–1207 (1970)
Miyachi, S., Hogetsu, D.: Plant Cell Physiol. *11*, 927–936 (1970)
Miyachi, S., Izawa, S., Tamiya, H.: J. Biochem. *42*, 221–244 (1955b)
Miyachi, S., Oh-hama, T., Tamiya, H.: Plant Cell Physiol. *1*, 151–153 (1960)
Nihei, T., Sasa, T., Miyachi, S., Suzuki, K., Tamiya, H.: Arch. Mikrobiol. *21*, 156–166 (1954)
Oh-hama, T., Miyachi, S.: Biochim. Biophys. Acta *34*, 202–210 (1959)
Okabe, K.: Ph. D. Thesis, Faculty of Science, Univ. Tokyo (1975)
Poskuta, J.: Physiol. Plant. *22*, 76–85 (1969)
Poskuta, J.: In: Proc. IInd Int. Congr. Photosynthesis Res. Forti, G. et al. (ed.), pp. 2121–2127. The Hague: Dr. W. Junk 1972
Poskuta, J., Frankiewicz-Josko, A.: In: Environmental and biological control of photosynthesis. 1911, pp. 89–105. 1975
Samejima, M., Miyachi, S.: In: Photosynthesis and photorespiration. Hatch, M.D. et al. (ed.), pp. 211–217. New York: Wiley-Interscience 1971

Samejima, M., Miyachi, S.: Plant Cell Physiol. *19*, 907–916 (1978)
Stamieszkin, I., Maleszewski, S., Poskuta, J.: Z. Pflanzenphysiol. *67*, 180–182 (1972)
Tamiya, H., Miyachi, S., Hirokawa, T.: In: Research in photosynthesis. Gaffron, H. et al. (ed.), pp. 213–223. New York: Interscience Publ. Inc. 1957
Togasaki, R.K., Botos, C.R.: In: Proc. IInd Int. Congr. Photosynthesis. Forti, G. et al. (ed.), pp. 1759–1772. The Hague: Dr. W. Junk 1972
Togasaki, R.K., Gibbs, M.: In: Special issue of plant cell physiol. pp. 505–511. Tokyo: Univ. Tokyo Press 1963
Togasaki, R.K., Gibbs, M.: Plant Physiol. *42*, 991–996 (1967)
Usuda, H., Samejima, M., Miyachi, S.: Plant Cell Physiol. *14*, 423–426 (1973)
Walker, D.A., Baldry, C.W., Cockburn, W.: Plant Physiol. *43*, 1419–1422 (1968)

II B. The C_4 and Crassulacean Acid Metabolism Pathways

6. The C_4 Pathway and Its Regulation

T.B. RAY and C.C. BLACK

A. Discovery of C_4 Photosynthesis

Nearly a century ago botanists first examined and described an arrangement of green leaf cells known today as Kranz leaf anatomy with its characteristic green mesophyll and bundle sheath cells. In evaluating this pioneering work HABERLANDT (1914) states:

> "—the (bundle) sheath— contains numerous chloroplasts, which are moreover often large and very brightly colored. It is uncertain whether this green inner sheath merely represents an unimportant addition to the chlorophyll-apparatus of the plant, or whether there exists some as yet undiscovered division of labour between the chloroplasts in the sheath and those in the girdle (mesophyll) cells."

Research within the last decade has revealed this "undiscovered division of labor" between mesophyll and bundle sheath cells and chloroplasts. The physical localization of chloroplasts in these adjacent leaf cells initially made direct studies difficult and contributed to some controversy about this division of labor. However to understand C_4 photosynthesis one must have a clear picture of Kranz leaf anatomy because two green cell types are present in all plants in which C_4 photosynthesis has been well documented.

The discovery of the division of labor between cells during C_4 photosynthesis resulted from continuing the biochemical studies of KORTSCHAK, BURR and co-workers (1957, 1965) who found that the first ^{14}C-labeled products of sugarcane leaf photosynthesis were mostly 4-carbon organic acids, malic and aspartic; rather than 3-phosphoglyceric acid (PGA), the traditional first stable product of photosynthesis in $^{14}CO_2$. The aim of this contribution is to consider the regulation of C_4 photosynthesis in leaves under natural conditions. The literature on Kranz leaf anatomy will be summarized and the range of environmental pressures which contribute to the regulation of C_4 photosynthesis will be considered. Then the biochemical features of C_4 photosynthesis will be presented with emphasis upon regulation via cellular compartmentation and upon the efficiency of the C_4 pathway in CO_2 assimilation. Finally, we will discuss the recent discovery of C_3–C_4 intermediate plants plus the possibility of other types of plant photosynthesis and outline the criteria needed to establish the presence of C_4 photosynthesis in a given plant. Each of these topics, leaf anatomy, environment, biochemistry, and cellular compartmentation, will be unified in conclusions on the regulation of C_4 photosynthesis in intact leaves.

B. Kranz Leaf Anatomy

In modern times the realization that a division of labor might be involved in C_4 leaf photosynthesis occurred simultaneously in at least four laboratories late in 1967 (DOWNTON and TREGUNNA, 1968; LAETSCH, 1968; DOWNES and HESKETH, 1968; JOHNSON and HATCH, 1968). These and other workers soon realized that all known plants with C_4 photosynthesis possessed Kranz leaf anatomy. However, even earlier HABERLANDT (1914) realized that different metabolic activities occurred in specific green leaf cells. He states that:

> "A leaf of sugarcane picked just after sunrise, contained no starch, either in the (green mesophyll) cells or in the (green) bundle sheaths; in other words, the entire synthetic products (of the previous day) had been removed during the night. A leaf picked about 3 P.M., after a sunny morning, also contained no starch in the (mesophyll) cells; however, the chloroplasts of the bundle sheaths enclosed large numbers of minute starch-grains."

By observing a sequential depletion and synthesis of starch in leaf cell types, he clearly thought in terms of a division of functions and metabolic activities in adjacent green cells. Today it is known that sugarcane is a C_4 plant.

Kranz leaf anatomy as seen in a light microscope is illustrated in Figure 1. This corn leaf in cross-section is a typical example of a C_4 leaf with its large green bundle sheath cells flanked by a layer of green mesophyll cells. Such an ordered cell arrangement is easy to see in a monocot leaf with their regularly spaced vascular tissues, but a similar arrangement is present in leaves of C_4 dicots. Electron microscopy has been extensively employed to investigate further the structural differences on both a cellular and organelle level which may exist in leaves of specific C_4 plants. Crabgrass (*Digitaria sanguinalis*; Fig. 2) has mesophyll chloroplasts which are strikingly different in ultrastructure from bundle sheath chloroplasts. However such a striking ultrastructural difference is not necessary for C_4 photosynthesis to operate (BLACK and MOLLENHAUER, 1971). Indeed one cannot safely generalize about chloroplast and cell, size or structure, regarding C_4 photosynthesis except to state that two green cell types are involved in C_4 photosynthesis.

From comparative biochemical studies on C_4 photosynthesis, today we recognize three major types of C_4 plants as illustrated in Table 1. It is possible to make some general correlations of chloroplast ultrastructure with these three groups. The $NADP^+$-malic enzyme plants have reduced grana development in bundle sheath chloroplasts (Fig. 2); the NAD^+-malic enzyme plants have good grana development of their chloroplasts, which are located centripetally within bundle sheath cells; and the PEP carboxykinase plants have good grana development of their chloroplasts, which tend to be located randomly or centrifugally within the bundle sheath cells. While these general correlations hold within grasses, dicots and sedges do not follow clear patterns. Even knowing these correlations it is not clear what relationships exist between the regulation of C_4 photosynthesis and chloroplast ultrastructure or the location of chloroplasts within the bundle sheath cells. Other consistent cytological/structural characteristics are known for C_4 photosynthesis. These are the dense concentration of chloroplasts, mitochondria, and peroxisomes in bundle sheath cells compared to mesophyll cells, the presence of numerous plasmodesmata between mesophyll and bundle sheath cells, and the thick-walled nature of grass bundle sheath cells. Each of these characteristics has

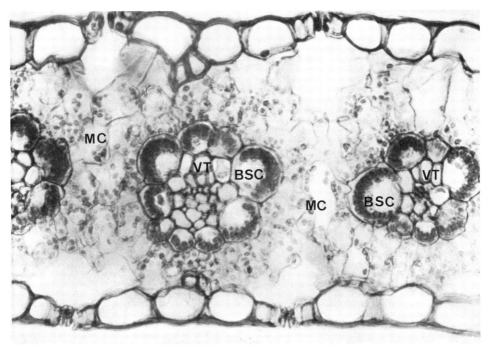

Fig. 1. Light photomicrograph of a maize (*Zea mays*) leaf in cross-section showing Kranz leaf anatomy. *BSC*, bundle sheath cells; *MC*, mesophyll cells; *VT*, vascular tissues. ($\times 185$)

been studied regarding the operation of the C_4 pathway and are discussed in other articles in this volume.

One of the most informative ways to investigate the relationship of leaf anatomy to CO_2 exchange capabilities is to evaluate three-dimensional leaf cell arrangements via scanning electron microscopy. In studies on the internal cell arrangements in C_4 leaves (EDWARDS and BLACK, 1971; BLACK et al., 1973; CHEN et al., 1974), it has been noted that open air spaces extend from the atmosphere to all photosynthetic cells of crabgrass; this should facilitate efficient CO_2 uptake. In C_4 grass leaves the green bundle sheath cells are tightly packed in a double spiral around the vascular tissues, whereas the mesophyll cells are very loosely arranged in spirals around the bundle sheath cells. Generally, there are two layers of green mesophyll cells between adjacent vascular strands in C_4 plants (CROOKSTON and MOSS, 1974). In contrast, C_3 leaves have many (9–15) mesophyll cells between the vascular tissues and numerous other morphological characteristics such as bundle sheath cells which lack chloroplasts in most C_3 type plants.

I. Variations in Leaf Anatomy

Numerous variations in C_4 Kranz-type leaf anatomy in specific plants have been reported. These include such features as two circular layers of green cells around

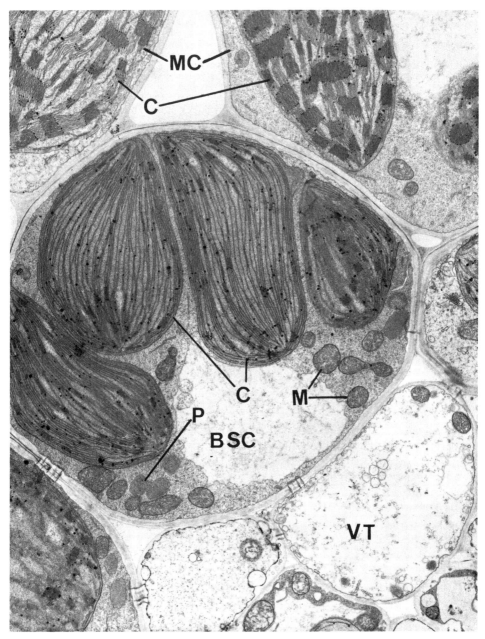

Fig. 2. Electron photomicrograph of a crabgrass (*Digitaria sanguinalis*) leaf in cross-section. *BSC*, bundle sheath cell; *C*, chloroplast; *M*, mitochondrion; *MC*, mesophyll cells; *P*, peroxisome; *VT*, vascular tissues. ($\times 16,500$)

Table 1. Variations in the biochemistry of C_4 photosynthesis found in specific C_4 plants

Major BSC[a] decarboxylase	Energetics of decarboxylation in BSC	Major substrate moving from:		
		MC to BSC	BSC to MC[b]	Representative plants
$NADP^+$ malic enzyme	Production of 1 $NADPH/CO_2$	Malate	Pyruvate	*Zea mays* *Digitaria sanguinalis*
NAD^+ malic enzyme	Production of 1 $NADH/CO_2$	Aspartate	Alanine/ pyruvate	*Atriplex spongiosa* *Portulaca oleracea*
PEP carboxy-kinase	Consumption of 1 ATP/CO_2	Aspartate	PEP[c]	*Panicum maximum* *Sporobolus poiretii*

[a] MC = mesophyll cell; BSC = bundle sheath cell.
[b] 3-PGA also moves into the mesophyll cells to support the synthesis of hexoses and starch in all three groups of plants.
[c] Nitrogen balance also must be maintained via an aminotransferase-type shuttle likely involving alanine. PEP also may be converted to pyruvate with the formation of ATP in BSC.

the circumference of a leaf, nongreen central bundle sheath cells, and green bundle sheath cells without a vascular system. Also included are double layers of green bundle sheath cells in *Astridia,* and double and single layers of green bundle sheath cells separated from the green mesophyll cells by a nongreen cell layer in Cyperaceae (JOHNSON, 1964; BLACK and MOLLENHAUER, 1971; CROOKSTON and MOSS, 1973; OLESEN, 1974; SHOMER-ILAN, BEER and WAISEL, 1975. Other specialized anatomical modifications such as colorless bulliform cells between vascular bundles and bundle sheath cell layers which do not totally surround the vascular tissues occur as in *Uniola* (MAYNE et al., 1977). However, it must be noted that in all cases of documented C_4 photosynthesis (Sect. H) two green cell types are present!

C. Environmental Regulation of C_4 Photosynthesis

Plants with C_4 photosynthesis not only possess two green cell types, they also exhibit characteristic responses to the environment. The aim of this section is to summarize the range of photosynthetic responses to various environmental pressures and then to consider the site or mechanism within the C_4 pathway whereby the environment exerts a regulatory influence. However, it must be noted that each environmental pressure probably elicits several responses and environmental changes in nature do not occur separately. Since our current knowledge on environmental regulation is so limited, it is difficult to decide on the key regulatory mechanisms associated with each of these environmental parameters.

I. Light Intensity

Soon after the first presentation of a biochemical scheme for the C_4 pathway (HATCH and SLACK, 1966) and with the realization that C_4 photosynthesis occurs in a variety of higher plants (HATCH et al., 1967; SLACK and HATCH, 1967; JOHNSON and HATCH, 1968), it was realized that there are many physiological features common to these plants including a hyperbolic response of leaf photosynthesis to increasing light intensity approaching values near those observed in full sunlight. Previous physiological work on leaf photosynthesis had clearly established that some plants, now known to be C_4, exhibit nonsaturating-type light curves (HESKETH and MOSS, 1963; HESKETH, 1963; HARTT, 1965; BROWN et al., 1966). This characteristic response of the C_4 pathway has been widely confirmed. More recently a plant, *Tidestromia oblongifolia*, has been found that exhibits a linear response of leaf photosynthesis even at the high light intensities found in Death Valley, California (BJÖRKMAN et al., 1972). Thus C_4 plants exhibit a nonsaturating response curve at the light intensities found in nature. Of course, it is widely recognized that light intensity during growth influences the development of chloroplasts, leaf thickness, cell size, and other features of the photosynthetic apparatus, but it is unknown how such long-term regulation occurs. Indeed, a detailed study on the influence of growth light intensity on C_3 photosynthesis (BJÖRKMAN et al., 1972) demonstrated that structural, photochemical, and biochemical components all changed. However no single component of photosynthesis can be pinpointed as the most sensitive site where light intensity influences C_4 photosynthesis.

During C_4 photosynthesis it has been calculated that 5 molecules of ATP and 2 molecules of NADPH are used per molecule of CO_2 reduced (CHEN et al., 1969). Thus C_4 plants must require more energy from their photochemical reactions than C_3 plants, hence a greater dependence of C_4 photosynthesis on photochemical-type reactions. However when one isolates mesophyll or bundle sheath cells or chloroplasts from C_4 plants and studies their light responses, these isolated systems tend to saturate near $1/3$ to $1/4$ of full sunlight. There are at least two light activated enzymes in C_4 photosynthesis, pyruvate Pi dikinase and $NADP^+$ malate dehydrogenase (SLACK, 1968; JOHNSON and HATCH, 1970), but they are rapidly activated and cannot explain the long-term leaf responses to high light intensities. The reasons for the hyperbolic or linear light responses of intact C_4 leaf photosynthesis remain unknown.

II. CO_2 Concentration

Early physiological work indicated that C_4 plants tend to require lower levels of CO_2 in the atmosphere than C_3 plants (HESKETH, 1963). Today it is generally accepted that air CO_2 concentrations (0.03%) are close to (within ~ 75–85%) saturation for C_4 photosynthesis while C_3 plants in air respond linearly to increases in CO_2. Thus C_4 photosynthesis seems to be an adaptation to utilize efficiently the current levels of CO_2 in our atmosphere. The reasons for the efficient CO_2 assimilation by C_4 plants can largely be attributed to the efficient trapping of CO_2 by PEP carboxylase which will be discussed later.

III. O_2 Concentration

Again, physiological studies established that lowering or raising O_2 levels in the atmosphere had little immediate effect on leaf photosynthesis in C_4 plants (FORRESTER et al., 1966b; HESKETH, 1967; DOWNES and HESKETH, 1968) in contrast to C_3 photosynthesis which is stimulated by lowering O_2 and inhibited by raising O_2 relative to air (FORRESTER et al., 1966a; BJÖRKMAN, 1966; DOWNES and HESKETH, 1968). The insensitivity of net C_4 photosynthesis in short time exposures to various O_2 levels is somewhat surprising considering RuBP carboxylase/oxygenase is present in the bundle sheath cells of C_4 plants. However, a lack of influence of O_2 on net leaf photosynthesis in short term studies is not an indication that the biochemistry is insensitive. Recent work on products of C_4 photosynthesis has demonstrated that O_2 does shift the flow of carbon from organic acids into the C_3 cycle (FOSTER and BLACK, 1977). How this shift occurs is unknown but C_4 plants do make biochemical adjustments to O_2 levels in the air which are not reflected in net CO_2 uptake rates.

IV. Temperature

C_4 photosynthesis in general has a high (30 °–45 °C) temperature optimum (MURATA and IYAMA, 1963; COOPER and TAINTON, 1968). In specific C_4 plants growing in particular environments, optima ranging from near 30 °C in *Spartina townsendii* to 47 °C in *Tidestromia oblongifolia* have been reported (BJÖRKMAN et al., 1972; LONG et al., 1975). In addition a characteristic C_4 feature is a rapid drop in photosynthesis as the temperature is lowered. The cold sensitivity of C_4 photosynthesis has been related to the cold sensitivity of a key enzyme, pyruvate Pi dikinase, which generates PEP from pyruvate (HATCH and SLACK, 1968) to form the initial CO_2 acceptor. This is a unique enzyme in C_4 photosynthesis and its sensitivity to low temperature would stop photosynthesis. The underlying regulatory factors at high temperatures are not known although it is known that several C_4 enzymes have high temperature optima for maximum activity. The clearest indication of a sensitive site at high temperatures is the sensitivity of Photosystem II in C_4 plants (BJÖRKMAN et al., 1976). Photosystem II even in C_3 plants is very sensitive to heat (VERNON and ZUAGG, 1960), but the sensitive component is unknown.

V. Water

The efficient utilization of water by specific plants which we now know as C_4, was reported early in this century (SHANTZ and PIEMEISEL, 1927), and recently confirmed by DOWNES (1969). Water use efficiency is tightly coupled to CO_2 uptake since both water vapor and CO_2 pass simultaneously through the stomatal pores. C_4 plants are more efficient at taking CO_2 from air than C_3 plants, so it is reasonable for C_4 plants to lose less water per CO_2 fixed. Therefore, the primary biochemical basis for their efficient water usage is the efficient carboxylation of CO_2 which we will attribute later to PEP carboxylase. Efficient water usage by C_4 plants also has been evoked as a reason for their adaptation and evolution in some xeric environments.

VI. Salinity

In their natural habitat many C_4 plants are adapted to growing in saline environments. In addition sodium is a required nutrient for the growth of C_4 plants (Brownell, 1965; Brownell and Crossland, 1972), but the role of sodium in C_4 plants is unknown. It also has been reported that NaCl can shift a balance from C_4 to C_3 photosynthesis in *Aeluropus litoralis* leaves (Shomer-Ilan and Waisel, 1973). Downton and Törökfalvy (1975) were unable to reproduce the work with *Aeluropus* but this latter study has been questioned (Shomer-Ilan and Waisel, 1975). At this time there appears to be a role for sodium in C_4 photosynthesis but a regulatory role if any for salt is very uncertain (see also Chap. II.15, this vol.).

VII. Nitrogen Supply

Recently the hypothesis has developed that C_4 plants utilize nitrogen more efficiently than C_3 plants (Brown, 1977). The supporting data show ~ two fold greater dry matter production per unit of leaf nitrogen and nearly a two fold greater rate of photosynthesis with C_4 plants than with C_3 plants. In addition, C_4 plants only invest about 10%–25% of their leaf nitrogen supply in RuBP carboxylase in contrast to C_3 plants which invest 40%–60%. Thus C_4 plants have regulated the amount of nitrogen invested in RuBP carboxylase in addition to sequestering the enzyme in their bundle sheath cells. Again the presence of the efficient CO_2 trap, PEP carboxylase, allows C_4 plants to use less RuBP carboxylase. How this regulation of RuBP carboxylase level occurs is unknown but it results in a very efficient utilization of available soil nitrogen. Therefore, C_4 plants can grow on nitrogen-poor soils, and low soil nitrogen also may be a selection pressure which led to the evolution of C_4 photosynthesis (Brown, 1977).

D. Biochemical Schemes for the C_4 Pathway

Despite the availability in the literature of these anatomical and environmental data, the discovery of the biochemical aspects of C_4 photosynthesis was the catalysis for the rapid expansion and unifying of knowledge on C_4 photosynthesis which we know today. Data showing that sugarcane (Kortschak et al., 1957, 1965) and corn (Tarchevskii and Karpilov, 1963) fix CO_2 into 4-carbon acids appeared sporadically in the literature prior to Hatch and Slack's now classical paper in 1966. Hatch and Slack confirmed the data on sugarcane and built on the biochemical ideas of Kortschak and co-workers (1965). More important, however, they presented a testable biochemical scheme for the C_4 pathway! In this remarkably

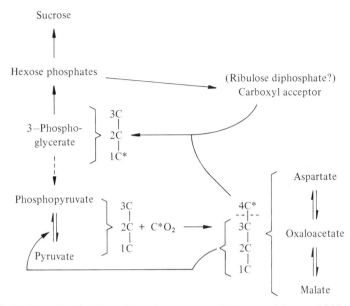

Fig. 3. Original biochemical scheme for the C_4 pathway in sugarcane (HATCH and SLACK, 1966)

incisive manuscript they presented the scheme shown in Figure 3 (HATCH and SLACK, 1966).

If one studies Figure 3 and then studies a current model for the C_4 pathway (Fig. 4), the similarities are striking. A decade of research has modified HATCH and SLACK's 1966 scheme somewhat, but that first scheme was basically correct in outline. A major difference in the two schemes is the compartmentation of portions of their scheme (Fig. 3) into mesophyll cells and bundle sheath cells. Other major changes include the substitution of a decarboxylation reaction for a proposed transcarboxylation (HATCH and SLACK, 1966) and the direct conversion of pyruvate to PEP via pyruvate Pi dikinase. Thus today much of the work on the C_4 pathway agrees with the schemes shown in both Figures 3 and 4.

Other workers have proposed alternate schemes such as that of COOMBS and BALDRY (1972) in which the C_3 cycle is presented as being in the mesophyll cells and the bundle sheath chloroplasts act simply as amyloplasts. Variations on the idea of bundle sheath cell chloroplasts carrying out the C_3 cycle by either directly fixing CO_2 from the air (LABER et al., 1974) and/or using the C_4 acids as an alternate pathway to supply their CO_2, have been put forward (POINÇELOT, 1972). The data to support these alternative schemes comes from experiments which have serious errors or limitations. In the cell extraction procedures of COOMBS and BALDRY, and POINCELOT, the cross contamination of mesophyll and bundle sheath cells introduces an error into the data which these workers did not take into account (BLACK, 1973). LABER and co-workers (1974) compared total leaf pool turnover for PEP, RuBP and PGA in corn and spinach leaves preilluminated in the absence of CO_2. Similar responses were observed for these metabolite pools in both leaves upon the addition of CO_2 either in the light or in the dark. They concluded that RuBP can serve as a primary CO_2 acceptor in both corn and spinach leaves. An intractable limitation of such data on total metabolite pools in

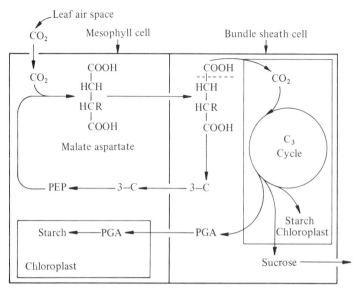

Fig. 4. Current scheme for the C_4 pathway showing the cooperation between mesophyll and bundle sheath cells in assimilating CO_2. 3-C is the 3-carbon unit remaining after C_4 acid decarboxylation in a bundle sheath cell (Table 1)

C_4 leaves is the presence of two distinct green cell types. The cellular location of the metabolites must be known before the results can be interpreted.

Our strong support for the scheme shown in Figure 4 comes from its consistency with whole leaf short-time $^{14}CO_2$ fixation studies both in turnover of ^{14}C-labeled intermediates and in the labeling patterns of specific carbon atoms, the strict localization of enzyme activities in mesophyll and/or bundle sheath cells, and $^{14}CO_2$ fixation and O_2 evolution studies with isolated mesophyll cells, protoplasts, and organelles. This supporting data will be presented next and it is consistent with the overall operation of the C_4 pathway as shown (Fig. 4) even though there are variations in specific C_4 plants particularly in leaf anatomy and in the leaf decarboxylase (Table 1). In addition, uncertainties remain on explanations for the efficiency of CO_2 fixation, on the possibility of direct fixation of CO_2 in bundle sheath cells via the C_3 cycle, and on the movement of metabolites between mesophyll and bundle sheath cells in quantities sufficient to support the rapid rate of C_4 leaf photosynthesis. These uncertainties, however, do not detract from the acceptance of the overall scheme shown in Figure 4.

E. Regulation via Enzymatic and Metabolic Compartmentation into Leaf Cell Types

A division of labor through cellular compartmentation of enzymes in C_4 leaves has proven to be an integral part of C_4 photosynthesis models (BURRIS and BLACK, 1976). The most widely accepted models unify Kranz leaf structure with the demonstrated metabolic functions of C_4 mesophyll and bundle sheath cells.

SLACK (1969) and SLACK et al. (1969) provided some of the first qualitative evidence that mesophyll cells and bundle sheath cells of C$_4$ leaves may have different enzyme complements. By nonaqueous methods they isolated chloroplasts from mesophyll and bundle sheath cells and found indications that PEP carboxylase activity was enriched in mesophyll cell chloroplasts and RuBP carboxylase activity was enriched in bundle sheath cells. The quantitative evidence of cellular compartmentation of C$_4$ photosynthesis came from the work of EDWARDS et al. (1970) when they developed techniques for separating mesophyll and bundle sheath cells from *Digitaria sanguinalis* (crabgrass). The technique allowed the isolation of large quantities of cells into homogeneous populations of mesophyll and bundle sheath cells (EDWARDS and BLACK, 1971a) permitting the quantitative determination of the enzyme complement of mesophyll cells and bundle sheath cells of C$_4$ plants. PEP carboxylase and NADP$^+$-malic dehydrogenase were localized ($>98\%$) in the mesophyll cells while RuBP carboxylase, ribose-5-phosphate isomerase, ribulose-5-phosphate kinase, and malic enzyme were localized ($>98\%$) in bundle sheath cells. Since then a large amount of quantitative data using isolated cells or protoplasts has accumulated on the compartmentation of leaf enzymes in a variety of C$_4$ plants (KANAI and BLACK, 1972; CHEN et al., 1973; KANAI and EDWARDS, 1973; CHEN et al., 1974; HATCH and KAGAWA, 1976 and BURRIS and BLACK, 1976; Chap. II.7, this vol.). All of these reports are consistent with PEP carboxylase being localized in the mesophyll cells and RuBP carboxylase in bundle sheath cells.

As a result of this strict intercellular enzyme compartmentation in C$_4$ plant leaves, extensive transport of metabolites between mesophyll and bundle sheath cells must occur during C$_4$ photosynthesis (Fig. 4). There is a great deal of uncertainty about this aspect of C$_4$ photosynthesis. This topic has been reviewed in an earlier series volume (HATCH and OSMOND, 1976). The limited data available on transport demonstrate that it occurs very rapidly. OSMOND (1971), in localizing the initial products of C$_4$ photosynthesis in maize leaves, found that after 2 s of photosynthesis ^{14}C-labeled compounds were present in both mesophyll and bundle sheath cells. The presence of plasmodesmata connecting the two cell types (LAETSCH, 1971a, b) led to the speculation that photosynthetic intermediates are transported through plasmodesmata. There is little direct data at present to support this speculation. The rapid labeling of photosynthetic carbon intermediates and the lack of reliable quantitative techniques for localizing water soluble compounds in intact leaves have made it difficult to design in vivo experiments relating to the transport of intermediates between mesophyll and bundle sheath cells during C$_4$ photosynthesis.

In addition to questions about how compounds are transferred there are questions about what is transported between mesophyll and bundle sheath cells (Table 1). An initial approach taken to answer this question was to observe the effects of various photosynthetic intermediates on photosynthesis in mesophyll and bundle sheath cells (EDWARDS et al., 1970). By feeding isolated intact mesophyll and bundle sheath cells from leaves of C$_4$ plants various photosynthetic intermediates and observing the effects these intermediates have on photosynthesis and on labeling patterns of photosynthetic intermediates, it was possible to determine which portions of the C$_4$ pathway occur in mesophyll and bundle sheath cells (EDWARDS and BLACK, 1971b; KANAI and BLACK, 1972; SALIN et al., 1973; SALIN and BLACK,

1974; Huber and Edwards, 1975a, 1975b; Hatch and Kagawa, 1976). It has been demonstrated that, depending upon the plant species, either malate or aspartate, or in some species both, will serve as a source of CO_2 in isolated bundle sheath cells (Hatch and Kagawa, 1976; Burris and Black, 1976) of C_4 plants. Thus malate and aspartate probably are the major compounds transported from mesophyll to bundle sheath cells (Table 1).

Based on similar cell feeding experiments, alanine, pyruvate, and PGA have been proposed as the major metabolites moving back to mesophyll cells from bundle sheath cells (Burris and Black, 1976). Of course, with the movement of amino acids and phosphorylated intermediates between cells, the appropriate nitrogen and phosphorous balances must be maintained. With the transport of numerous compounds between cell types the prospect of regulation of C_4 photosynthesis by regulating metabolite transport is a distinct possibility, however, there are no data available which demonstrate this type of control.

The regulation of C_4 photosynthesis on an enzymatic level is another area about which little is known. The separation of the pathway into two cell types makes this problem more complex. At the same time, one of the bases for regulating the efficiency of C_4 plants, is the compartmentation of enzymes. There have been several reports on the regulation of enzymes in the C_4 pathway. $NADP^+$-malic dehydrogenase and pyruvate Pi dikinase undergo light mediated activation and the latter is also sensitive to cold temperatures and requires a second protein for activation (see Sect. C; Sugiyama, 1974; Sugiyama and Boku, 1976). In addition to these enzymes, there are reports that PEP carboxylase is a regulated enzyme (see Chap. II.20, this vol.).

With mesophyll cell protoplasts, Huber and Edwards (1975) demonstrated that depending upon the species of C_4 plant (Table 1) either malate or aspartate inhibited pyruvate induced $^{14}CO_2$ fixation. The degree of inhibition was greater with malate in NAD^+-malic enzyme species (75% at 2 mM) than with aspartate. In the $NADP^+$-malic enzyme species aspartate was the more effective inhibitor. PEP carboxykinase species were not affected by either acid. The site of these inhibitions was concluded to be the β-carboxylation step. This conclusion was supported by data showing inhibition of PEP carboxylase by malate and aspartate in crude leaf extracts of crabgrass. Several other investigators, however, have found either small or no inhibitory effects of malate and aspartate on PEP carboxylase (Lowe and Slack, 1971; Nishikido and Takanashi, 1973). Ting and Osmond (1973) were able to detect a 19% inhibition of purified PEP carboxylase from *Atriplex spongiasa* with 1 mM malate and only a 7% inhibition with 1 mM aspartate. Fifty percent inhibition was achieved with 5–10 mM malate. These inhibiting levels are substantially higher than those reported by Huber and Edwards (1975). The immediate product of PEP carboxylase, oxaloacetic acid, also has been reported to inhibit PEP carboxylase (Lowe and Slack, 1971). Fifty percent inhibition was achieved with 0.25 mM OAA. The physiological significance of this inhibition is questionable since the levels of OAA required for inhibition probably are well above the OAA level in leaves of C_4 plants (Hatch and Slack, 1966; Kanai and Black, 1972; Huber and Edwards, 1975). In contrast, the levels of malate and aspartate causing inhibition of pyruvate induced $^{14}CO_2$ fixation in mesophyll cells are within the range of concentrations that these compounds are found in leaves of C_4 plants (Hatch, 1971a, b; Huber and Edwards, 1975). From this inconsistent and limited amount of information, it is difficult to conclude that the levels of malate and aspartate in C_4 leaves control the activity of PEP carboxylase.

In the course of studying the C_4 pathway of photosynthesis it has been found that the green C_4 mesophyll cell is a unique oxygen evolving cell. The C_4 mesophyll cell lacks RuBP carboxylase and ribulose-5-phosphate kinase, two required enzymes for the C_3 cycle. However carbon reduction via glyceraldehyde 3-phosphate dehydrogenase is quite active. CO_2 fixation is via β-carboxylation to form oxaloacetate which is reduced to form malate. Both of these carbon reduction reactions are tightly coupled to O_2 evolution (SALIN et al., 1973; SALIN and BLACK, 1974; BURRIS and BLACK, 1976). Unquestionably this unique photosynthetic metabolism is partially responsible for the efficiency of C_4 plants.

F. Efficiency of C_4 Leaf Photosynthesis

The intercellular compartmentation of enzymes has profound effects on the net CO_2 assimilation in C_4 plants. Perhaps the two most important effects are that C_4 plants do not exhibit photorespiration and generally have high rates of photosynthesis. Two reasons have been proposed to explain these phenomena. The first was that C_4 plants have a "CO_2 trap" which refixes photorespiratory CO_2 before it can leave the leaf (FORRESTER et al., 1966 b). Secondly, C_4 plants may have in effect a "CO_2 pump" which maintains high CO_2 levels in bundle sheath cells (HATCH, 1971 b). Both hypotheses are supported by the localization of PEP carboxylase in mesophyll cells and organic acid decarboxylation plus RuBP carboxylase/oxygenase activity in the bundle sheath cells.

I. CO_2 Pools

The flux of organic acids into the bundle sheath cell and their decarboxylation provide a high enough CO_2 concentration to effectively saturate RuBP carboxylase according to HATCH's (1971 b) calculations. The CO_2 pool in leaves of maize and *Amaranthus* during steady state photosynthesis are between 0.8 and 1.9 mM CO_2 (HATCH, 1971 b). These relatively high levels of CO_2 are far above the KmCO$_2$ for RuBP carboxylase and probably are saturating concentrations for the enzyme (JENSEN and BAHR, 1976). In addition to allowing high rates of CO_2 fixation by RuBP carboxylase, the high CO_2 levels in bundle sheath cells would also inhibit RuBP oxygenase activity which would result in lower rates of photorespiration. This hypothesis assumes that the leaf CO_2 pool is a single pool and that RuBP carboxylase is at the same pool site within the leaf, namely, the bundle sheath cell chloroplast.

II. CO_2 Trapping

The hypothesis on the refixation of photorespiratory CO_2 was proposed by FORRESTER et al. (1966 b) when they were unable to detect an O_2 inhibition of photosynthesis in maize in short-term exposures to O_2. They suggested that maize can recycle CO_2 internally. With the discovery of the C_4 pathway it became apparent that due to enzyme compartmentation and Kranz anatomy, bundle sheath cells are essentially surrounded by PEP carboxylase. Thus any CO_2 released through photorespiration has to pass out of, through, or around mesophyll cells before it can move out of the leaf. It could then be subject to refixation by PEP carboxylase. It has been demonstrated that bundle sheath cells of C_4 plants possess the enzymes of the photorespiratory pathway (LIU and BLACK, 1972; BURRIS and BLACK, 1976). In addition it has also been demonstrated that CO_2 fixation in isolated bundle sheath cells is inhibited by O_2 and that isolated bundle sheath possess RuBP oxygenase activity and produce glycollate (CHOLLET, 1976). Thus isolated bundle sheath cells have the potential for carrying out photorespiration. Refixation of photorespiratory CO_2 would require additional ATP and NADPH produced photo-synthetically. Thus one would expect the quantum efficiency for photosynthesis to drop under conditions which promote high photorespiration. EHLERINGER and BJÖRKMAN (1977), however, found that the quantum efficience for C_4 photosyn-thesis was unaffected by O_2. It may be that the CO_2 concentrations in bundle sheath cells can be maintained at high enough levels even under high O_2 levels to prevent photorespiration. It follows that by lowering ambient CO_2 levels which could lower the CO_2 concentration in the bundle sheath cells and at the same time raising the O_2 levels, some effect on the efficiency of C_4 photosynthesis should be observed.

In conclusion, due to the strict compartmentation of enzymes in mesophyll cells (particularly PEP carboxylase) and in bundle sheath cells, no photorespiration is apparent in intact C_4 plant leaves. In relation to this, by severely limiting photorespiration particularly in regard to net CO_2 loss, C_4 plants have evolved a method for regulating photorespiration.

III. CO_2 Fixation by Bundle Sheath Cells

Another aspect of C_4 photosynthesis which arises from the compartmentation of leaf enzymes concerns the ability of the C_3 cycle in bundle sheath cells to fix CO_2 directly from the atmosphere. There is still some question if atmospheric CO_2 from the intercellular spaces can move directly into bundle sheath cells for fixation. In the original work of KORTSCHAK et al. (1965), HATCH and SLACK (1966), KANAI and BLACK (1972), and BLACK et al. (1973), numerous photosynthetic pulse-chase labeling experiments demonstrated that ^{14}C from $^{14}CO_2$ could enter PGA at very early time points. For example when leaves of sugarcane were exposed to $^{14}CO_2$ for 4 s 15% of the incorporated label was found in PGA and 7% in hexose monophosphates (HATCH and SLACK, 1966). KORTSCHAK et al. (1965) reported that after 0.6 s in $^{14}CO_2$ $\sim 12\%$ of the $^{14}CO_2$ was incorporated in PGA plus glycerate. Thus even though PGA does not represent a very large portion of $^{14}CO_2$ initially fixed, these experiments suggest that PGA is a primary product of CO_2 fixation in C_4 plants. Anatomical data also support the direct fixation of CO_2 by the C_3 cycle in bundle sheath cells of C_4 plants. Portions of bundle sheath cells are exposed to the intercellular air spaces in leaves of C_4 plants (EDWARDS and BLACK, 1971; BLACK and MOLLENHAUER, 1971). Thus CO_2 could diffuse directly into these cells. In addition isolated bundle sheath cells will fix CO_2 particularly in the presence of C_3 cycle intermediates (EDWARDS et al., 1970; BURRIS and BLACK, 1976). Finally, it has been established that organic acids are decarboxylated in the bundle sheath cells and the resulting CO_2 is refixed by the C_3 cycle (BURRIS and BLACK, 1976; HATCH and KAGAWA, 1976).

It has been suggested that a CO_2-impermeable barrier exists around bundle sheath cells (TROUGHTON, 1971; LAETSCH, 1971a, b). LAETSCH has published electron photomicrographs which show an electron opaque layer between bundle sheath and mesophyll cells. He speculates that this layer may be impermeable to CO_2, however, there is no evidence to support this speculation and the layer is absent in some C_4 plants. HATCH has presented data which support the hypothesis that all the CO_2 fixed by the C_3 cycle during C_4 photosynthesis arises from the decarboxylation of organic acids and no CO_2 is fixed directly (HATCH, 1969; JOHNSON and HATCH, 1969). The kinetics of labeling of the CO_2 pool in leaves of maize and *Amaranthus* were such that it would be derived from malate and aspartate (HATCH, 1971b). The CO_2 pool was saturated with ^{14}C after malate but before PGA. Likewise during pulse-chase experiments the label was lost first from malate then CO_2 and then PGA. JOHNSON and HATCH (1969) calculated the rate constants for the loss of ^{14}C from the C-4 position of malate and aspartate during pulse-chase experiments and found that the calculated value for such a rate agreed very closely with the rates of ^{14}C accumulation in the C-1 position of PGA. They concluded that all of the label entering the C-1 position of PGA is derived from the C-4 position of malate and aspartate. HATCH (1976) recently demonstrated that extrapolation of curves showing the percent of ^{14}C fixed into PGA during short term labeling experiments with C_4 leaves places the line through the origin at zero time. From these data and the kinetics of labeling of the malate, aspartate, CO_2 and PGA pools, HATCH concludes that the C_3 cycle does not fix CO_2 directly but rather all the CO_2 fixed in bundle sheath cells is derived from the decarboxylation of C_4 acids.

If leaves of certain C_4 plants carrying out steady-state photosynthesis are suddenly darkened, large amounts of CO_2 will be rapidly evolved from the leaves for 2 or 3 m (DOWNTON, 1970; BROWN and GRACEN, 1972). This post-illumination CO_2 burst (PIB) is thought to arise from the decarboxylation of C_4 acids in the bundle sheath of C_4 plants (WYNN et al., 1973). If CO_2 from the bundle sheath cells can rapidly leave the leaf, CO_2 also should rapidly enter the bundle sheath cells.

Recent work of RAY and BLACK (1976, 1977) approached this problem using a specific inhibitor of decarboxylation, 3-mercaptopicolinic acid (3-MPA). 3-MPA inhibits the enzyme phosphoenolpyruvate carboxykinase which acts to decarboxylate oxaloacetate in bundle sheath cells of some C_4 plants (Table 1). By inhibiting PEP carboxykinase, 3-MPA should remove the main source of CO_2 cycle in bundle sheath cells. When leaves of *Panicum maximum* were treated with 3-MPA, the turnover rates of malate and aspartate were severely reduced as expected if C_4 acid metabolism is blocked (RAY and BLACK, 1977). However, 3-MPA also severely reduced the rates of formation of PGA and C_3 cycle products. This suggests that CO_2 fixation by the C_3 cycle in bundle sheath cells of C_4 plants is almost entirely dependent on the decarboxylation of C_4 acids as proposed by HATCH. EDWARDS and BLACK (1971a, b) have estimated that up to 15% of the CO_2 fixed during C_4 photosynthesis is fixed directly by the C_3 cycle. If this were the case then treatment of leaves with 3-MPA would remove at least 85% of the CO_2 normally present in bundle sheath cells. Under these conditions the C_3 cycle may not fix CO_2 fast enough to maintain the proper levels of photosynthetic intermediates. Thus the observed effects of 3-MPA on CO_2 fixation with *P. maximum* leaves may be the result of the C_3 cycle running down.

The concepts discussed in this section have emphasized the intercellular compartmentalization of enzymes and other metabolic activities in C_4 leaves and, as a result, the spatial separation of various portions of the pathway (Fig. 4). This spatial arrangement within intact leaves apparently is essential in maintaining an efficient mode of CO_2 fixation during C_4 photosynthesis.

G. C_3–C_4 Intermediate Plants

In addition to the specific spatial features of C_4 plants some of the most often cited differences between C_3 and C_4 plants are that C_4 plants have higher rates of photosynthesis and no apparent photorespiration (DOWNES and HESKETH, 1968; BLACK et al., 1969). Overall C_4 plants are considered to be more efficient at fixing CO_2 than C_3 plants (BLACK et al., 1969). It should be noted that optimum environmental conditions for photosynthesis, growth, and productivity are not the same for C_3 and C_4 species. Therefore, direct comparisons are difficult to make. In general C_4 plants have higher rates of photosynthesis and lower rates of photorespiration than C_3 plants.

Since photosynthesis is a primary process in plant productivity, increasing the efficiency of photosynthesis has long been a goal of plant research. Recently one approach to this problem has been to search for, or to try to develop through breeding programs, plant species which are intermediate to C_3 and C_4 plants. The possibility that C_3–C_4 intermediates exist or can be produced is strengthened by the knowledge that there are 18 genera of plants which contain both C_3 and C_4 species (BURRIS and BLACK, 1976). The close genetic relationship between many of these plants theoretically could allow crosses between C_3 and C_4 species. BJÖRKMAN and co-workers have carried out extensive investigations of crosses between C_3 and C_4 species of *Atriplex* (NOBS et al., 1971). Of the several crosses made only that between the C_4 *Atriplex rosea* and the C_3 *Atriplex triangularis* (*A. patula* ssp. *A. hastata* Hull) has been carried past the F_1 generation (NOBS, 1976). Morphologically the F_1 was intermediate to the parents including an intermediate Kranz anatomy (BOYNTEN et al., 1971). On a biochemical level the F_1 contained intermediate levels of PEP carboxylase and RuBP carboxylase (PEARCY and BJÖRKMAN, 1971). In studies using $^{14}CO_2$ to determine intermediates of photosynthesis, C_4 acids (malate and aspartate) contained the same percentage of ^{14}C after 6 s of photosynthesis as C_3 cycle intermediates (PGA + sugar · P). Pulse labeling experiments indicated, however, that the C_4 acids did not turn over and that PGA behaved as the primary product of photosynthesis (PEARCY and BJÖRKMAN, 1971). Apparently the F_1 hybrid in addition to possessing the C_3 cycle contains a β-carboxylation system but lacks the enzymes required for the metabolism of C_4 acids. Rates of net photosynthesis in the F_1 hybrid were lower than those of either parent (BJÖRKMAN et al., 1971). In addition photosynthesis in the F$_1$ hybrid was inhibited by O$_2$. Clearly, although CO$_2$ could be fixed by PEP carboxylase, the resulting C$_4$ acids did not play a major role in net photosynthesis.

Other attempts to cross C$_3$ and C$_4$ species have not met with success. Moss reported that crosses between related species of *Panicum* or *Euphorbia* were not successful (Moss et al., 1969; CROOKSTON and MOSS, 1971; MOSS, 1971). Other approaches taken were to induce mutations with ionizing radiation and then screen for C$_4$ photosynthesis by looking for a low CO$_2$ compensation point (MENZ et al., 1969; CANNELL et al., 1969; MOSS, 1971). After screening thousands of oat, wheat and soybean seedlings none were found which contained a low CO$_2$ compensation point. Thus inducing C$_4$ characteristics in C$_3$ plants by mutagenic agents or by breeding for C$_4$ characteristics through genetic crosses have not been successful in developing plants with C$_4$ characteristics.

To date two naturally occurring species have been described as C$_3$–C$_4$ intermediate type plants. KENNEDY and LAETSCH (1974) reported that *Mollugo verticillata* has both C$_3$ and C$_4$ characteristics. Unlike the Atriplex hybrids, *M. verticillata* did not have an intermediate leaf anatomy but did have numerous chloroplasts in the bundle sheath cells. In fact, the gross leaf anatomy was very similar to a C$_3$ dicot with two types of mesophyll cells, spongy and pallisade. The rate of net photosynthesis in *M. verticillata* was the same as that found in *M. cerviana* a C$_4$ type. Likewise photorespiration in *M. verticillata* was only slightly higher than that in *M. cerviana* and far below rates found in *Hordeum vulgare*, a C$_3$ species. When leaves of *M. verticillata* were exposed to ^{14}CO$_2$ for 5 s, about 20% of the ^{14}C incorporated was found in malate and aspartate and 18% was in PGA and sugar phosphates. Interestingly nearly 54% of the ^{14}C was in alanine. Thus equal amounts of C$_3$ and C$_4$ primary products were found at early times during ^{14}C labeling. The important question of whether the C$_4$ acids turn over releasing CO$_2$ which can be incorporated into the C$_3$ cycle was not answered. Conflicting data concerning the intermediate nature of *M. verticillata* have been reported by BJÖRKMAN (1976), who found this species to have C$_3$ type photorespiration and a C$_3$ compensation point. Furthermore, *M. verticillata* had a typical C$_3$ δ^{13}C value (TROUGHTON et al., 1974; BROWN and BROWN, 1975). This latter parameter, however, may not be a valid measure of the intermediate nature of a plant since its value would depend upon the contribution of β-carboxylation to overall photosynthesis (this topic is discussed in Chap. II.10, this vol.). Thus whether or not *M. verticillata* is a C$_3$–C$_4$ intermediate species remains to be established.

The best known naturally occurring C$_3$–C$_4$ intermediate is *Panicum milioides* (BROWN and BROWN, 1975; BROWN, 1976). A thorough investigation of leaf photosynthesis and photorespiration in *P. milioides* was made first by BROWN and BROWN (1975) and the results were confirmed by KECK and OGREN (1976). *P. milioides* photosynthesis has an intermediate response to O$_2$ levels, an intermediate rate of CO$_2$ release into CO$_2$ free air, and an intermediate CO$_2$ compensation point. *P. milioides* leaf anatomy also is intermediate to C$_3$ and C$_4$ anatomy in that chloroplasts are in the bundle sheath cells and the bundle sheath cells have slightly thicker cell walls than normally found in C$_3$ species (BROWN and BROWN, 1975; KANAI and KASHIWAGA, 1975). In addition the interveinal distance in *P. milioides* leaves is intermediate to C$_3$ and C$_4$ species. The biochemical characteristics of photosynthesis in *P. milioides* has been investigated by several groups (KANAI and KASHIWAGI, 1975; KESTLER et al., 1975; GOLDSTEIN et al., 1976; KU et al., 1976; KECK and OGREN, 1976). All report that the levels of PEP carboxylase activity in *P. milioides* is intermediate to C_3 and C_4 species. The activities of the decarboxylating enzymes in *P. milioides* although slightly higher than the activities found in C_3 plants are far lower than C_4 species. The compartmentation of enzymes in bundle sheath and mesophyll cells characteristic of C_4 species was not apparent in *P. milioides* since RuBP carboxylase and PEP carboxylase activity was found in both cell types (KU et al., 1976). Pulse labeling experiments demonstrated that PGA is the primary product of photosynthesis in *P. milioides* (KANAI and KASHIWAGI, 1975; KESTLER et al., 1975) and that ^{14}C is transferred from PGA into sugar phosphates and sucrose. KANAI and KASHIWAGI (1975) reported, however, that about 5% of the ^{14}C fixed by *P. milioides* leaves after a short exposure to $^{14}CO_2$ is incorporated into malate and aspartate but that these compounds do not feed label into C_4 intermediates. In similar experiments GOLDSTEIN et al. (1976) could not detect label in C_4 acids. Thus during steady state photosynthesis in air *P. milioides* fixes CO_2 by the C_3 cycle.

Using *P. milioides, P. maximum,* and *H. vulgare,* KESTLER et al. (1975) performed pulse labeling experiments near the CO_2 compensation point. From data obtained in these experiments a hypothesis was proposed concerning the biochemical components of CO_2 compensation point (KESTLER et al., 1975). Essentially the CO_2 compensation point is governed by the relative amounts of carbon moving through photosynthesis and photorespiration. The enzymes responsible for producing the primary products of these two competing pathways at the compensation point are PEP carboxylase for photosynthesis and RuBP oxygenase for photorespiration. At the low CO_2 concentration of the compensation point there would be very little RuBP carboxylase activity. The ratio of the activities of PEP carboxylase and RuBP oxygenase thus determines the CO_2 compensation point. A striking correlation is found between the ratios of PEP carboxylase and RuBP oxygenase, the rates of photorespiration, and the CO_2 compensation point (KESTLER et al., 1975). For example, *P. maximum* has a low ratio of RuBP oxygenase to PEP carboxylase, a low rates of photorespiration and a low compensation point. *H. vulgare* has a high enzyme ratio, high rates of photorespiration and a high compensation point. *P. milioides* is intermediate in all these respects. This in air *P. milioides* behave like a C_3 type plant. Under conditions which would favor relatively high rates of photorespiration over photosynthesis *P. milioides* becomes more like a C_4 type plant.

From the C_3–C_4 intermediates which have been described it is apparent that even though features of C_4 photosynthesis can be incorporated into C_3 plants there is no guarantee that the plant will be more efficient at fixing CO_2 than the C_3 parent or precursor. It may be that C_4 pathway can only be efficient if all components of the system are present. This is suggested by the work of BJÖRKMAN and co-workers with *Atriplex.* If this is true, then to increase CO_2 uptake, the problem is to convert C_3 plants to C_4 types. The results of work with *P. milioides* counters this line of reasoning. *P. milioides* contains only a few features of C_4 plants and yet has an intermediate level of photorespiration. This is encouraging in that it may be possible to select a few key characteristics of C_4 photosynthesis and incorporate them into C_3 species.

In addition to C_3–C_4 intermediate plants there are numerous plants about which questions remain concerning CO_2 fixation and photorespiration. For example, C_3 plants are known which have rates of CO_2 fixation like those found in C_4 plants such as *Typha latifolia* (McNaughton and Fullem, 1970), *Camissania claviformis* (Mooney et al., 1976) and *Helianthus annuus* (sunflower) (Hesketh, 1963). This latter plant has a response to light intensity and CO_2 concentration which is intermediate to C_3 and C_4 species. Benedict and Scott (1976) found a C_4 pattern of photosynthesis in the marine grass *Thalissia testodinum* but Kranz leaf anatomy is not present. Closer examination of plants with anomalous patterns of photosynthesis, photorespiration, and leaf anatomy could result in the discovery or development of new types of C_3–C_4 intermediate plants.

H. Criteria for the Presence of C_4 Photosynthesis

Since the discovery of C_4 photosynthesis the pathway has been found in a wide range of plants (Downton, 1975) and, as just discussed, even C_3–C_4 intermediate plants are now known (Brown and Brown, 1975). However using various characteristics of C_4 photosynthesis as criteria, several investigators have claimed the discovery of C_4 photosynthesis in such diverse photosynthetic tissues as blue–green algae, a marine grass, and an undifferentiated plant tissue culture (Döhler, 1974; Benedict and Scott, 1976; Kennedy, 1976). Many of the characteristics used for distinguishing C_4 photosynthesis can be equivocal if used as sole proof for the presence of C_4 photosynthesis. Since some confusion is arising in the literature, it would be useful to review some of the criteria used to establish C_4 photosynthesis originally.

Perhaps the best single criterion for establishing the presence of C_4 photosynthesis is the synthesis of malate and aspartate as early (< 10 s) products of photosynthesis and the subsequent turnover of these compounds into C_3 cycle intermediates. These were the features of the C_4 pathway which first led to its discovery (Kortschak et al., 1965; Hatch and Slack, 1966). During C_4 photosynthesis $^{14}CO_2$ is incorporated first into the C-4 position of malate or aspartate. The carbon then is transferred to the C-1 position of PGA through a decarboxylation reaction and subsequent refixation of the CO_2 by RuBP carboxylase. It is important to emphasize that the carbon from the C-4 position is rapidly and quantitatively transferred to the C-1 position of PGA (Johnson and Hatch, 1969).

Some of these features of C_4 photosynthesis have been used to claim that the C_4 pathway occurs in a particular plant. It was reported that the blue–green alga *Anacystis nidulans* carries out C_4 photosynthesis (Döhler, 1974; Colman et al., 1976). Earlier work demonstrated that this alga produces substantial amounts of aspartate during photosynthesis (Janz and MacLean, 1973). The conclusion that C_4 photosynthesis is present in *A. nidulans* was based on data showing aspartate as a primary product of photosynthesis. In addition, the alga was found to have PEP carboxylase activity. However, in feeding experiments with ^{14}C-aspartate, the aspartate was converted directly to other amino acids rather than serving as a precursor for C_3 cycle intermediates (Döhler, 1974). Thus as concluded by

JANZ and MACLEAN (1973), β-carboxylation and the subsequent formation of malate and aspartate only makes a minor contribution to photosynthesis in *A. nidulans*. The synthesis of amino acids from malate and aspartate in C_3 plants was suggested by BASSHAM and CALVIN (1957). In 1970 TAMÀS and co-workers demonstrated that $^{14}CO_2$ initially was incorporated by greening barley leaves into malate and aspartate. This labeling probably was accomplished through the action of PEP carboxylase. Rather than finding ^{14}C in sucrose and starch as would be expected if C_4 photosynthesis were occurring, the label was found in protein. As the greening process continued, increasing amounts of ^{14}C was found initially in PGA and ultimately in sucrose while the incorporation into malate and aspartate remained constant. Eventually CO_2 fixation into sucrose via the C_3 cycle became the dominant path of CO_2 fixation with only a small percentage of ^{14}C being incorporated into C_4 acids. So C_3 plants can fix CO_2 by β-carboxylation reactions into C-4 acids, but the contributions of this pathway to photosynthesis is minor. Neither the presence of PEP carboxylase nor the appearance of label in organic acids in a plant tissue is a good sole criterion of C_4 photosynthesis.

Another criterion for the presence of C_4 photosynthesis is the presence of an active 4-carbon acid decarboxylase (Table 1). A decarboxylase is required for the C_4 pathway model shown in Figure 4. However, this is not an unambiguous criterion since CAM plants also require an active decarboxylase (BURRIS and BLACK, 1976).

Kranz leaf anatomy often has been used as a criterion for the presence of C_4 photosynthesis (DOWNTON, 1971). Caution also must be used when defining photosynthesis by leaf anatomy since there are variations in Kranz-type anatomy in C_4 type plants (Sect. B.I) as well as in C_3–C_4 intermediate plants such as *P. milioides*, which was classified as a C_4 plant by DOWNTON in 1971 based on Kranz leaf anatomy. However in all cases of careful C_4 documentation two green cell types are present.

A useful criterion for C_4 photosynthesis is the carbon isotope discrimination ratio (SMITH and EPSTEIN, 1971; TROUGHTON et al., 1974). C_4 plants have been found to have $\delta^{13}C$ values closer to that of the CO_2 in air than do C_3 plants. The postulated reason for the difference is that RuBP carboxylase discriminates against ^{13}C, while PEP carboxylase does not (WHELAN et al., 1973). C_4 plants have $\delta^{13}C$ values between $-10^o/_{oo}$ to $-18^o/_{oo}$ while C_3 plants have values from $-23^o/_{oo}$ to $-34^o/_{oo}$ (BENDER, 1968; SMITH and EPSTEIN, 1971). This difference has proved almost unequivocal for C_3 and C_4 plants. There are, however, exceptions which must be considered when using isotope discrimination for classifying C_3 and C_4 plants. First, plants which carry out CAM can have either a C_3 or C_4 type $\delta^{13}C$ value, depending upon whether or not the plants are fixing CO_2 by β-carboxylation or via the carboxylation of RuBP (BENDER et al., 1973; CREWS et al., 1976). Corrections also must be allowed in comparing marine plants with terrestrial plants since the $\delta^{13}C$ values of sea water and air are different (BLACK and BENDER, 1976). Finally, the $\delta^{13}C$ value may not be useful as a criterion for C_3–C_4 intermediates. Both of the postulated naturally occurring intermediates, *P. milioides* and *Mollugo verticillata*, have C_3 values for $\delta^{13}C$ (TROUGHTON et al., 1974; BROWN and BROWN, 1975; BROWN, 1976). Although it is a useful criterion for studying carbon assimilation, the carbon isotope ratio of plants also can be ambiguous.

Of the criteria discussed, no single one should be used to unequivocally establish the presence of C_4 photosynthesis in a species. Several though, taken together where possible, should be sufficient. These could include the kinetics of ^{14}C-labeling of malate, aspartate, and PGA; two types of green cells; the leaf $\delta^{13}C$ value; plus high activity of PEP carboxylase, a decarboxylase (Table 1), and other unique C_4 enzymes such as pyruvate pi dikinase and NADP$^+$-malic dehydrogenase. In addition, a variety of whole leaf or whole plant physiological responses such as the CO_2 compensation point, the response of photosynthesis to O_2, CO_2, light intensity, and temperature (reviewed in BLACK, 1973; BURRIS and BLACK, 1976) can be used to document C_4 photosynthesis.

I. Conclusions in the Regulation of C_4 Photosynthesis in Leaves

C_4 plants have been known for sometime to respond to various environmental pressures in a different manner than C_3 species. Photosynthesis in C_4 species tends to saturate near full sunlight, has a higher temperature optimum than C_3 photosynthesis, saturates near air concentrations of CO_2, and is little affected by O_2. In addition, C_4 species have greater water and nitrogen use efficiencies than C_3 species. Exactly how each of these environmental parameters affect the daily regulation of C_4 photosynthesis is not known.

A key component in the regulation of C_4 photosynthesis is the presence of two distinct green cell types in all known C_4 species. A strict compartmentation of different enzymes into these two cell types results in a unique biochemistry in each cell and in a strong cooperation between the two cells. The green mesophyll cell lacks the carboxylating portion of the C_3 cycle and possesses a very active β-carboxylation and organic acid metabolism system. The green bundle sheath cell has an active 4-carbon organic acid decarboxylase and the complete C_3 cycle.

In explaining the efficiency of C_4 photosynthesis, both the CO_2 trap and the CO_2 pump hypothesis are based on the localization of PEP carboxylase in mesophyll cells plus a decarboxylase and RuBP carboxylase in bundle sheath cells. These characteristics and hypotheses have been used to explain the efficient fixation of CO_2 and the efficient use of water and nitrogen by C_4 species.

The hypothesis for the regulation of photorespiration in C_4 plants by sequestering RuBP carboxylase/oxygenase in the bundle sheath cells also arises from the compartmentation of PEP carboxylase in mesophyll cells such that it surrounds the bundle sheath cells. The proposed mechanism for regulating photorespiration in C_4 species is based on the assumption that the CO_2 concentration in bundle sheath cells is maintained at levels high enough to saturate RuBP carboxylase and that photorespiratory CO_2 released from the bundle sheath cells is refixed by the PEP carboxylase in the mesophyll cells.

Although a biochemical scheme for C_4 photosynthesis involving two cell types has been widely accepted (Fig. 4), the sites of biochemical regulation of the C_4 pathway are not known. NADP$^+$ malic dehydrogenase and pyruvate Pi dikinase are enzymes activated by light mediated processes. The latter also is cold-inactivated and requires a separate protein for activation. Yet the possible regulatory functions of these enzymes on leaf photosynthesis have not been fully explored. PEP carboxylase may be another regulatory enzyme but regulation such as feedback inhibition by C-4 organic acids has not been demonstrated consistently in vitro nor in vivo. Since two cell types with specific enzyme contents are necessary for the C_4 pathway it is particularly difficult to extrapolate from data on regulation of an enzyme in vitro and from data on total leaf pool sizes to the intact leaf situation. The in vivo regulation may have no relationship to in vitro enzyme regulation studies. Other sites of regulation may be the transport of metabolites between bundle sheath and mesophyll cells. However, the mechanism of transport has yet to be elucidated so no data are available on regulation of metabolic intermediate movements between cell types.

An approach to the regulation of C_4 photosynthesis could be through the study of C_3–C_4 intermediate plants. The naturally occurring intermediate *P. milioides* has only a few C_4 characteristics but it has lower rates of photorespiration than C_3 species. Apparently several critical components are required for the regulation of photorespiration in C_4 plants and an understanding of how these components regulate photorespiration may assist in understanding why C_4 plants fix CO_2 more efficiently than C_3 plants.

Finally, there are some definite characteristics which can be used in combination to define a C_4 plant such as the photosynthetic kinetics of ^{14}C-labeling of C-4 organic acids and PGA; two green cell types; the $\delta^{13}C$ value; and high activities of PEP carboxylase and an organic acid decarboxylase in green cells. Definite identification of a plant as possessing the C_4 pathway demands data on a variety of these characteristics.

References

Bassham, J.A., Calvin, M.: The path of carbon in photosynthesis, p. 73. Englewood Cliffs, N.J.: Prentice-Hall 1957

Bender, M.M.: Radiocarbon *10*, 468–472 (1968)

Bender, M.M., Rouhani, I., Vines, H.M., Black, C.C.: Plant Physiol. *52*, 427–430 (1973)

Benedict, C.R., Scott, J.R.: Plant Physiol. *57*, 876–880 (1976)

Björkman, O.: Physiol. Plant. *19*, 618–633 (1966)

Björkman, O.: In: CO_2 metabolism and plant productivity. Burris, R.H., Black, C.C. (eds.), pp. 287–310. Baltimore-London-Tokyo: University Park Press 1976

Björkman, O., Pearcy, R.W., Nobs, M.A.: Carnegie Inst. Washington. Yearb. *69*, 640–648 (1971)

Björkman, O., Boardman, N.K., Anderson, J.M., Thorne, S.W., Goodchild, D.J., Pyliotis, N.A.: Carnegie Inst. Washington Yearb. *71*, 115–135 (1972a)

Björkman, O., Pearcy, R.W., Harrison, T.A., Mooney, H.: Science *175*, 786–789 (1972b)

Björkman, O., Boynton, J., Berry, J.: Carnegie Inst. Washington Yearb. *75*, 400–407 (1976)

Black, C.C.: Ann. Rev. Plant Physiol. *24*, 253–286 (1973)

Black, C.C., Bender, M.M.: Aust. J. Plant Physiol. *3*, 25–32 (1976)

Black, C.C., Mollenhauer, H.H.: Plant Physiol. *47*, 15–23 (1971)

Black, C.C., Chen, T.M., Brown, R.H.: Weed Sci. *17*, 338–343 (1969)

Black, C.C., Edwards, G.E., Kanai, R., Mollenhauer, H.H.: In: II Int. Congr. Photosynthesis. 1745–1757, 1971

Black, C.C., Campbell, W.H., Chen, T.M., Dittrich, P.: Quart. Rev. Biol. *48*, 299–313 (1973)

Boynten, J.E., Nobs, M.A., Björkman, O., Pearcy, R.W.: Carnegie Inst. Washington Yearb. *69*, 629–632 (1971)

Brown, R.H.: In: CO_2 metabolism and plant productivity. Burris, R.H., Black, C.C. (eds.), pp. 311–325. Baltimore-London-Tokyo: University Park Press 1976

Brown, R.H.: Crop Sci. In press (1977)

Brown, R.H., Brown, W.V.: Crop Sci. *15*, 681–685 (1975)

Brown, R.H., Gracen, V.E.: Crop Sci. *12*, 30–33 (1972)

Brown, R.H., Blaser, R.E., Dunton, H.L.: Proc. Int. Grassland Congr. *10*, 108–113 (1966)

Brown, W.V.: Am. J. Bot. *62*, 395–402 (1975)

Brownell, P.F.: Plant Physiol. *40*, 460–468 (1965)

Brownell, P.F., Crossland, C.J.: Plant Physiol. *49*, 794–797 (1972)

Burr, G.O., Hartt, C.E., Brodie, H.W., Tanimoto, T., Kortschak, H.P., Takahashi, D., Ashton, F.M., Coleman, R.E.: Ann. Rev. Plant Physiol. *8*, 275–308 (1957)

Burris, R.H., Black, C.C. (Eds.): CO_2 metabolism and plant productivity, p. 431. Baltimore, London, Tokyo: University Park Press 1976

Cannell, R.Q., Brun, W.A., Moss, D.N.: Crop Sci. *9*, 840–841 (1969)

Chen, T.M., Brown, R.H., Black, C.C.: Plant Physiol. *44*, 649–654 (1969)

Chen, T.M., Campbell, W.H., Dittrich, P., Black, C.C.: Biochem. Biophys. Res. Commun. *51*, 461–467 (1973)

Chen, T.M., Dittrich, P., Campbell, W., Black, C.C.: Arch. Biochem. Biophys. *163*, 246–262 (1974)

Chollet, R.: In: CO_2 metabolism and plant productivity. Burris, R.H., Clack, C.C. (eds.), pp. 327–341. Baltimore, London, Tokyo: Univ. Park Press 1976

Colman, B., Cheng, K.H., Ingle, R.K.: Plant Sci. Let. *6*, 123–127 (1976)

Coombs, J., Baldry, C.W.: Nature (London) *238*, 268–270 (1972)

Cooper, J.P., Tainton, M.N.: Herbage Abstr. *38*, 167–176 (1968)

Crews, C.E., Williams, S.L., Vines, H.M., Black, C.C.: In: CO_2 metabolism and plant productivity. Burris, R.H., Black, C.C. (eds.), pp. 235–250. Baltimore, London, Tokyo: Univ. Park Press 1976

Crookston, R.K., Moss, D.N.: Plant Physiol. *52*, 397–402 (1973)

Crookston, R.H., Moss, D.N.: Crop Sci. *14*, 123–125 (1974)

Döhler, G.: Planta *118*, 259–269 (1974)

Downes, R.W.: Planta *88*, 261–273 (1969)

Downes, R.W., Hesketh, J.D.: Planta *78*, 79–84 (1968)

Downton, W.J.S.: Can. J. Bot. *48*, 1795–1800 (1970)

Downton, W.J.S.: In: Photosynthesis and photorespiration. Hatch, M.D., Osmond, C.B., Slayter, R.O. (eds.), pp. 554–558. New York, London, Sydney, Toronto: John Wiley and Sons 1971

Downton, W.J.S.: Photosynthetica *9*, 96–105 (1975)

Downton, W.J.S., Törökfalvy, E.: Z. Pflanzenphysiol. *75*, 143–150 (1975)

Downton, W.J.S., Tregunna, E.B.: Can. J. Bot. *46*, 207–215 (1968)

Edwards, G.E., Black, C.C.: In: Photosynthesis and photorespiration. Hatch, M.D., Osmond, C.B., Slayter, R.O. (eds.), pp. 153–168. New York, London, Sydney, Toronto: Wiley Interscience 1971a

Edwards, G.E., Black, C.C.: Plant Physiol. *47*, 149–156 (1971b)

Edwards, G.E., Lee, S.S., Chen, T.M., Black, C.C.: Biochem. Biophys. Res. Commun. *39*, 389–395 (1970)

Edwards, G.E., Kanai, R., Black, C.C.: Biochem. Biophys. Res. Commun. *45*, 278–285 (1971)

Ehleringer, J., Björkman, O.: Plant Physiol. *59*, 86–90 (1977)

Forrester, M.L., Krotkov, G., Nelson, C.D.: Plant Physiol. *41*, 422–427 (1966a)

Forrester, M.L., Krotkov, G., Nelson, C.D.: Plant Physiol. *41*, 428–431 (1966b)

Foster, F.A., Black, C.C.: Photosynthetic organelles. Special issue of plant and cell physiol. Miyachi, S., Katoh, S., Fujita, Y., Shibata, K. (eds.), pp. 325–340. Japan, Jpn. Soc. Plant Physiol. (1977)

Goldstein, L.D., Ray, T.B., Kestler, D.P., Mayne, B.C., Brown, R.H., Black, C.C.: Plant Sci. Let. *6*, 85–96 (1976)

Haberlandt, G.: Physiological plant anatomy. Translation of 4th Edition, p. 777. London: Macmillian and Co. 1914

Hartt, C.E.: Plant Physiol. *40*, 718–724 (1965)

Hatch, M.D.: In: Photosynthesis and photorespiration. Hatch, M.D., Osmond, C.B., Slayter, R.O. (eds.), pp. 139–152. New York, London, Sydney, Toronto: Wiley Interscience 1971a

Hatch, M.D.: Biochem. J. *125*, 425–432 (1971b)

Hatch, M.D.: In: CO_2 metabolism and plant productivity. Burris, R.H., Black, C.C. (eds.), pp. 59–82. Baltimore, London, Tokyo: Univ. Park Press 1976

Hatch, M.D., Kagawa, T.: Arch. Biochem. Biophys. *175*, 39–53 (1976)

Hatch, M.D., Osmond, B.: In: Encyclopedia of plant physiology. Heber, U., Stocking, C.R. (eds.), Berlin, Heidelberg, New York: Springer (1976)

Hatch, M.D., Slack, C.R.: Biochem. J. *101*, 103–111 (1966)

Hatch, M.D., Slack, C.R.: Biochem. J. *106*, 141–146 (1968)

Hatch, M.D., Slack, C.R., Johnson, H.S.: Biochem. J. *102*, 417–422 (1967)

Hesketh, J.D.: Crop Sci. *3*, 493–496 (1963)

Hesketh, J.D.: Planta *76*, 371–374 (1967)

Hesketh, J.D., Moss, D.N.: Crop Sci. *3*, 107–110 (1963)

Huber, S.G., Edwards, G.E.: Plant Physiol. *55*, 835–844 (1975a)

Huber, S.C., Edwards, G.E.: Plant Physiol. *56*, 324–331 (1975b)

Jansz, E.R., MacLean, F.I.: Can. J. Microbiol. *19*, 497–504 (1973)

Jensen, R., Bahr, J.T.: In: CO_2 metabolism and plant productivity. Burris, R.H., Black, C.C. (eds.), pp. 3–18. Baltimore, London, Tokyo: Univ. Park Press 1976

Johnson, H.S., Hatch, M.D.: Phytochemistry *7*, 375–380 (1968)

Johnson, H.S., Hatch, M.D.: Biochem. J. *114*, 127–134 (1969)

Johnson, H.S., Hatch, M.D.: Biochem. J. *119*, 273–280 (1970)

Johnson, M.C. Sr.: Ph. D. Thesis, Univ. of Texas (1964)

Kanai, R., Black, C.C.: In: Net carbon dioxide assimilation in higher plants. Black, C.C. (ed.), pp. 75–93. Raleigh, N.C.: Cotton Inc., 1972

Kanai, R., Edwards, G.E.: Plant Physiol. *51*, 1133–1137 (1973)

Kanai, R., Kashiwagi, M.: Plant Cell Physiol. *16*, 669–679 (1975)

Keck, R.W., Ogren, W.L.: Plant Physiol. *58*, 552–555 (1976)

Kennedy, R.A.: Plant Physiol. *58*, 573–575 (1976)

Kennedy, R.A., Laetsch, W.M.: Science *184*, 1087–1089 (1974)

Kestler, D.P., Mayne, B.C., Ray, T.B., Goldstein, L.D., Brown, R.H., Black, C.C.: Biochem. Biophys. Res. Commun. *66*, 1439–1445 (1975)

Kortschak, H.P., Hartt, C.E., Burr, G.O.: Proc. Hawaii. Acad. Sci. *21* (1957)

Kortschak, H.P., Hartt, C.E., Burr, G.O.: Plant Physiol. *40*, 209–213 (1965)

Ku, S.B., Edwards, G.E., Kanai, R.: Plant Cell Physiol. *17*, 615–620 (1976)

Laber, L.J., Latzko, E., Gibbs, M.: J. Biol. Chem. *249*, 3436–3441 (1974)

Laetsch, W.M.: Am. J. Bot. *55*, 875–883 (1968)

Laetsch, W.M.: In: Photosynthesis and photorespiration. Hatch, M.D., Osmond, C.B., Slayter, R.E. (eds.), pp. 323–349. New York, London, Sydney, Toronto: John Wiley & Sons 1971a

Laetsch, W.M.: Sci. Prog. Oxf. *57*, 323–351 (1971b

Liu, A., Black, C.C.: Arch. Biochem. and Biophys. *149*, 269–280 (1972)

Long, S.P., Incoll, L.D., Woolhouse, H.W.: Nature (London) *257*, 622–624 (1975)

Lowe, J., Slack, C.R.: Biochim. Biophys. Acta *235*, 207–209 (1971)

Mayne, B.C., Ray, T.B., Black, C.C.: Proc. Natl. Acad. Sci. USA. In press (1977)

McNaughton, S.J., Fullem, L.W.: Plant Physiol. *45*, 703–707 (1970)

Menz, K.M., Moss, D.N., Cannell, R.Q., Brun, W.A.: Crop Sci. *9*, 692–694 (1969)

Mooney, H.A., Ehleringer, J., Berry, J.A.: Science *194*, 322–323 (1976)

Moss, D.N.: In: Photosynthesis and photorespiration. Hatch, M.D., Osmond, C.B., Slayter, R.O. (eds.), pp. 120–123. New York, London, Sydney, Toronto: John Wiley & Sons 1971

Moss, D.N., Krenzer, E.G., Brun, W.A.: Science *164*, 187–188 (1969)

Murata, Y., Iyama, J.: Proc. Crop. Sci. Soc. Japan *31*, 315–322 (1963)

Nishikido, T., Takanashi, H.: Biochem. Biophys. Res. Comm. *53*, 126–133 (1973)

Nobs, M.A.: Carnegie Inst. Washington Yearb. *75*, 421–423 (1976)

Nobs, M.A., Björkman, O., Pearcy, R.W.: Carnegie Inst. Washington Yearb. *69*, 625–629 (1971)

Olesen, P.: Bot. Notiser *127*, 352–363 (1974)
Osmond, C.B.: Aust. J. Plant Physiol. *24*, 159–163 (1971)
Pearcy, R.W., Björkman, O.: Carnegie Inst. Washington Yearb. *69*, 632–640 (1971)
Poincelot, R.P.: Plant Physiol. *50*, 336–340 (1972)
Ray, T.B., Black, C.C.: J. Biol. Chem. *251*, 5824–5826 (1976)
Ray, T.B., Black, C.C.: Plant Physiol. In press (1977)
Salin, M., Black, C.C.: Plant Sci. Let. *2*, 303–308 (1974)
Salin, M., Campbell, W.H., Black, C.C.: Proc. Natl. Acad. Sci. USA *70*, 3730–3734 (1973)
Shantz, H.L., Piemeisel, L.N.: J. Agrie. Res. *34*, 1093–1189 (1927)
Shomer-Ilan, A., Waisel, Y.: Physiol. Plant. *29*, 190–193 (1973)
Shomer-Ilan, A., Waisel, Y.: Z. Pflanzenphysiol. *77*, 272–273 (1975)
Shomer-Ilan, A.S., Beer, S., Waisel, Y.: Plant Physiol. *56*, 676–679 (1975)
Slack, C.R.: Biochem. Biophys. Res. Commun. *30*, 483–488 (1968)
Slack, C.R.: Phytochem. *8*, 1387–1391 (1969)
Slack, C.R., Hatch, M.D.: Biochem. J. *103*, 660–665 (1967)
Slack, C.R., Hatch, M.D., Goodchild, D.J.: Biochem. J. *114*, 489–498 (1969)
Smith, B.N., Epstein, S.: Plant Physiol. *47*, 380–384 (1971)
Sugiyama, T.: Plant Cell Physiol. *15*, 723–726 (1974)
Sugiyama, T., Boku, K.: Plant Cell Physiol. *17*, 851–854 (1976)
Tamàs, I.A., Yemm, E.E., Bidwell, R.G.S.: Can. J. Bot. *48*, 2313–2317 (1970)
Tarchevskii, I.A., Karpilov, Y.S.: Fiziol. Rast. *10*, 229–231 (1963)
Ting, I.P., Osmond, C.B.: Plant Physiol. *51*, 439–447 (1973)
Troughton, J.H.: Planta *100*, 87–92 (1971)
Troughton, J.H., Card, K.A., Hendy, C.H.: Carnegie Inst. Yearb. *73*, 768–780 (1974)
Vernon, L.P., Zaugg, W.S.: J. Biol. Chem. *235*, 2728–2733 (1960)
Whelan, T., Sackett, W.M., Benedict, C.R.: Plant Physiol. *51*, 1051–1054 (1973)
Wynn, T., Brown, R.H., Campbell, W.H., Black, C.C.: Plant Physiol. *52*, 288–291 (1973)

7. C$_4$ Metabolism in Isolated Cells and Protoplasts

G.E. EDWARDS and S.C. HUBER

A. Introduction

General methods for studying photosynthetic metabolism in C$_4$ or CAM plants might ideally consist of studies at the following levels of complexity: (1) whole leaf, (2) isolated cells or protoplasts and (3) chloroplasts. Fairly extensive studies both with whole leaves (BLACK, 1973; HATCH and OSMOND, 1976), and isolated photosynthetic cells and protoplasts (EDWARDS et al., 1976a, b) have been made with C$_4$ plants although isolation of functional chloroplasts from C$_4$ species has been limited (O'NEAL et al., 1972; HATCH and KAGAWA, 1973; HUBER and ED-WARDS, 1975a, b; RATHNAM and EDWARDS, 1977a).

HATCH and OSMOND (1976) have reviewed evidence to support a general concept of C$_4$ photosynthesis and some variation in metabolism among different species while regulatory aspects of photosynthesis in C$_4$ plants are reviewed by RAY and BLACK (Chap. II.6, this vol.). The purpose of this section is to discuss recent applications of isolated cells and protoplasts in the study of C$_4$ metabolism, including some of the limitations of these studies and future areas of interest. Some of the advantages of cells or protoplasts for studying photosynthesis in C$_4$ plants may also apply to C$_3$ plants (Chap. II. 2, this vol.; EDWARDS et al., 1976b) and CAM plants although there has been little success in isolating functional cells and protoplasts (KANAI and EDWARDS, 1973a; ROUHANI et al., 1973) or chloroplasts (LEVI and GIBBS, 1975) from CAM plants.

B. Three Groups of C$_4$ Plants

C$_4$ plants have been divided into three groups on the basis of the C$_4$ acid decarboxylating mechanisms found in the bundle sheath: NADP-malic enzyme, NAD-malic enzyme, and PEP (phosphoenolpyruvate) carboxykinase species (GUTIERREZ et al., 1974a; HATCH et al., 1975; GUTIERREZ et al., 1976). PEP carboxykinase and NAD-malic enzyme species have a mesotome sheath inside the bundle sheath cells while NADP-malic enzyme species lack a mesotome sheath (HATTERSLEY and WATSON, 1976; BROWN, W.V., 1977).

C. Localization of Enzymes of C$_4$ Metabolism in C$_4$ Plants

I. Intercellular Localization

There is considerable evidence with isolated mesophyll cells and protoplasts and bundle sheath cells of C$_4$ plants for: (1) the localization of PEP carboxylase and

other enzymes of the carboxylation phase of the C$_4$ pathway in mesophyll preparations, (2) the decarboxylation enzymes of the decarboxylation phase of the C$_4$
pathway in bundle sheath cells, and (3) RuBP carboxylase, phosphoribulokinase,
and ribose-5-phosphate isomerase of the Calvin-pathway in bundle sheath cells
(Ku and EDWARDS, 1975; HATCH and OSMOND, 1976; EDWARDS et al., 1976a).
A key question concerning the mechanism of photosynthesis in C$_4$ plants has
been the intercellular localization of RuBP carboxylase. Besides a lack of RuBP
carboxylase activity in isolated C$_4$ mesophyll protoplasts or cells, mesophyll protoplasts from C$_4$ plants lack the native protein molecule and subunits of fraction I
protein (RuBP carboxylase) when chloroplast soluble and insoluble proteins are
analyzed by gel electrophoresis (HUBER et al., 1976). This strongly suggests that
RuBP carboxylase is not present in C$_4$ mesophyll chloroplasts in either an active
or inactive form. Additional compelling evidence has come recently from HATTERS-
LEY et al., 1977. Using a fluorescent antibody to RuBP carboxylase they found
strong fluorescence from bundle sheath cells and no fluorescence from mesophyll
cells in cross-sections of leaves from several species. These results leave little doubt
that this protein is localized exclusively in bundle sheath cells of C$_4$ plants. *Panicum
milioides* and *Panicum hians*, species having C$_3$ photosynthesis but apparent Kranz
anatomy, have RuBP carboxylase in both mesophyll and bundle sheath cells (Ku
et al., 1976; HATTERSLEY et al., 1977).

II. Intracellular Localization

C$_4$ mesophyll protoplasts are useful for studying intracellular localization of enzymes as they provide a convenient source of intact organelles and cytoplasmic
enzymes. Mesophyll chloroplasts (90%–95% intact) can be obtained from protoplasts by gently rupturing the plasma membranes (GUTIERREZ et al., 1975; HUBER
and EDWARDS, 1975b). Extracts obtained from pure preparations of mesophyll
protoplasts and bundle sheath cells of C$_4$ plants have extended and strengthened
earlier studies of enzyme localization performed using nonaqueous techniques and
leaf homogenates (see HATCH and OSMOND, 1976 for earlier studies). Using both
differential centrifugation and separation of organelles on sucrose gradients, intracellular localization of enzymes of mesophyll and bundle sheath cells have been
studied in all three groups of C$_4$ plants (GUTIERREZ et al., 1975; KAGAWA and
HATCH, 1975; RATHNAM and EDWARDS, 1975). Among enzymes of the carboxylation phase of the C$_4$ pathway, PEP carboxylase is in the cytosol, while pyruvate
Pi dikinase and NADP-malate dehydrogenase are in the chloroplast. In NAD-malic
enzyme and PEP carboxykinase species relatively high levels of aspartate and
alanine aminotransferases are found in the cytosol of the mesophyll cells (GUTIER-
REZ et al., 1975). Homogenates of bundle sheath cells contain the enzymes of the
decarboxylation phase of the C$_4$ pathway with the following localization of decarbox-
ylases: NADP-malic enzyme and PEP carboxykinase are in the bundle sheath
chloroplasts (RATHNAM and EDWARDS, 1975) while NAD-malic enzyme is in the
bundle sheath mitochondria (KAGAWA and HATCH, 1975; RATHNAM and ED-
WARDS, 1975).

D. Criteria for Intactness of Cellular Preparations

Successful isolation of photosynthetically functional protoplasts and cells from various species has proceeded in some ways as slowly as those for isolating functional chloroplasts, with success depending on species and plant age which may explain some of the variability in reports among experimenters. Both mechanical methods and enzymatic digestion with cellulase and pectinase have been used for isolating mesophyll and bundle sheath preparations. Both mesophyll and bundle sheath cells were initially isolated mechanically with *Digitaria sanguinalis* (EDWARDS et al., 1970). Mesophyll protoplasts and bundle sheath strands were initially isolated enzymatically from maize by KANAI and EDWARDS, 1973 (Fig. 1). Mesophyll protoplasts and bundle sheath strands can now be separated enzymatically from a large number of species, though suitable techniques are not available for C_4 dicots.

The following criteria indicate functional preparations have been isolated.

I. Mesophyll Preparations

1. Exclusion of Evans Blue stain by protoplasts or cells (KANAI and EDWARDS, 1973a) indicates maintenance of semipermeability of the plasma membrane.

2. High rates of light-dependent CO_2 fixation in the presence of pyruvate or alanine with mesophyll protoplast extracts with malate or aspartate as major products, respectively. This indicates a photochemically and biochemically functional carboxylation phase of the C_4 pathway (HUBER and EDWARDS, 1975a, b). High rates of PEP-induced fixation with mechanically isolated mesophyll cells in the light or dark, with oxaloacetate as the major product, and low light dependent fixation with pyruvate suggests the lack of a completely functional carboxylation phase of the C_4 pathway (EDWARDS et al., 1970; SALIN et al., 1973; CHEN et al., 1974).

C_4 mesophyll chloroplasts of high integrity both in terms of enzyme retention and function have been obtained from protoplasts of *D. sanguinalis*, *Urochloa panicoides* and *Eleusine indica* (GUTIERREZ et al., 1975; HUBER and EDWARDS, 1975a, b). Although maize and sorghum are of interest for agronomic reasons and mesophyll protoplasts and bundle sheath cells are readily isolated, there are unresolved problems in obtaining mesophyll preparations that are stable and functionally equivalent to that of *D. sanguinalis* (KAGAWA and HATCH, 1974b; HUBER and EDWARDS, unpublished).

II. Bundle Sheath Preparations

1. Exclusion of Evans Blue stain by isolated bundle sheath cells (GUTIERREZ et al. 1974b) indicates maintenance of semipermeability of the cell.

2. High rates of light dependent CO_2 fixation without addition of exogenous substrates such as ribose-5-phosphate, adenine nucleotides, or NADP indicates maintenance of semipermeability properties and photosynthetic capacity i.e., *Panicum capillare* (GUTIERREZ et al., 1974b), *Atriplex spongiosa* (KAGAWA and

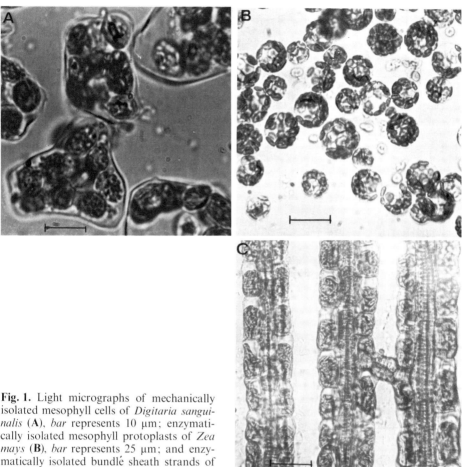

Fig. 1. Light micrographs of mechanically isolated mesophyll cells of *Digitaria sanguinalis* (**A**), *bar* represents 10 μm; enzymatically isolated mesophyll protoplasts of *Zea mays* (**B**), *bar* represents 25 μm; and enzymatically isolated bundle sheath strands of *Zea mays* (**C**), *bar* represents 40 μm

HATCH, 1974a), and *Eriochloa borumensis* (RATHNAM and EDWARDS, 1977a). A large stimulation of CO$_2$ fixation by addition of ribose-5-phosphate and adenine nucleotides (EDWARDS et al., 1970; CHOLLET and OGREN, 1973; HUBER et al., 1973; CHEN et al., 1974; FARINEAU, 1975; RAY and BLACK, 1976) suggests the loss of semipermeability properties of the cells and chloroplasts since C$_3$ chloroplasts have not been found to readily transport these metabolites (WALKER, 1974).

3. High rates of light dependent O$_2$ evolution with C$_4$ acids suggests the maintenance of both the photosynthetic capacity of the Calvin–Benson pathway and carboxyl donation (HATCH and KAGAWA, 1976; RATHNAM and EDWARDS, 1977a, b).

Success in isolating functional bundle sheath cells with various techniques (mechanically or enzymatically with cellulase and pectinase) depends in part on the species. Using the above criteria relatively active bundle sheath cells have been obtained by mechanical isolation from *A. spongiosa* (KAGAWA and HATCH, 1974a;

Hatch and Kagawa, 1976; and *E. borumensis* (Rathnam and Edwards 1977a). However, the latter are exceptions as relatively inactive preparations result with a number of other species including *D. sanguinalis* and maize (Edwards et al., 1970; Chollet and Ogren, 1973; Huber et al., 1973; Farineau, 1975; Hatch and Kagawa, 1976). By enzymatic digestion relatively active bundle sheath cells have been obtained from *P. capillare* (Gutierrez et al., 1974b), *D. sanguinalis* (Huber and Edwards, 1975c), and *E. borumensis* (Rathnam and Edwards, 1977a); relatively inactive preparations have been obtained from a number of other species e.g., *Panicum maximum* and *Eleusine indica* (Gutierrez et al., 1974b; Ray and Black, 1976). Combined enzymatic–mechanical procedures gave inactive preparations (Chen et al., 1974; Hatch and Kagawa, 1976).

Some loss of integrity of bundle sheath preparations by enzymatic digestion may be due to prolonged incubation (5–6 h), while some species such as *P. maximum* and *E. indica* seem inherently difficult. The inclusion of bovine serum albumin (0.05%) and sodium ascorbate (50 mM) may give more active bundle sheath preparations (Gutierrez et al., 1974b; Rathnam and Edwards, 1977a, b) but the degree of benefit with various species needs further evaluation. Active preparations of bundle sheath cells from enzymatic digestion are obtained with a minimum of digestion time (2–3 h) with tissue preparation and temperature of incubation being important factors (Huber and Edwards, 1975c). In NADP-malic enzyme species such as maize and sugarcane, CO_2 fixation in bundle sheath cells is apparently limited by the capacity of photosystem II rather than a loss of integrity during isolation (Ku et al., 1974; Mayne et al., 1975; Edwards et al., 1976a).

E. Variations in C_4 Metabolism

I. Mesophyll Cells of C_4 Plants

Among the three groups of C_4 plants that have been identified there is considerable quantitative variation in the levels of certain enzymes (Ku and Edwards, 1975; Hatch and Osmond, 1976) and in the photochemical characteristics of the cell types (Ku et al., 1974; Mayne et al., 1975; Edwards et al., 1976a). However, mesophyll protoplasts of all three groups are somewhat similar as they all contain the carboxylation phase of the C_4 pathway. All three groups of C_4 plants have high levels of pyruvate Pi dikinase of the C_4 pathway (Hatch et al., 1975) and high rates of light and pyruvate-dependent CO_2 fixation can be demonstrated with mesophyll preparations (Huber and Edwards, 1975b). With pyruvate as a precursor the principal products are oxaloacetate and malate while alanine serves as a precursor for light-dependent fixation of CO_2 to form oxaloacetate and aspartate. Both in vivo and in vitro the NADP-malic enzyme species *D. sanguinalis* favors malate formation while PEP carboxykinase and NAD-malic enzyme species favor aspartate formation (Huber and Edwards, 1975b). NAD-malic enzyme and PEP carboxykinase species have higher levels of aspartate and alanine aminotransferases in mesophyll cells than do NADP-malic enzyme species while the reverse is true for NADP-malate dehydrogenase (GUTIERREZ et al., 1975; KU and EDWARDS, 1975; HATCH and OSMOND, 1976).

II. Bundle Sheath Cells of C$_4$ Plants

There are variations in the pathway of C$_4$ acid decarboxylation in bundle sheath cells depending on the primary decarboxylase. While NAD-malic enzyme appears to be the primary decarboxylase in NAD-malic enzyme species (cf. HATCH and OSMOND, 1976), it is also a secondary decarboxylase in the bundle sheath of NADP-malic enzyme and PEP-carboxykinase species (GUTIERREZ et al., 1974a; RATHNAM and EDWARDS, 1975, 1977a). This complexity was not included in most of the original schemes for the transport of C$_4$ pathway metabolites between mesophyll and bundle sheath cells. A malate-pyruvate exchange was suggested with NADP-malic enzyme species and aspartate-alanine exchange with PEP carboxykinase and NAD-malic enzyme species. However, this now appears to be an oversimplification as both of these exchange systems are thought to be operating in a given species (RATHNAM and EDWARDS, 1975, 1977b). Based on both enzyme localization and C$_4$ acid· decarboxylation by bundle sheath cells, chloroplasts and mitochondria of all three groups, RATHNAM and EDWARDS (1975, 1977a, b) proposed aspartate as a transport metabolite from mesophyll cells to bundle sheath mitochondria and malate as transport metabolite from mesophyll cells to bundle sheath chloroplasts.

In NADP-malic enzyme species malate transport to bundle sheath chloroplasts would result in decarboxylation through NADP-malic enzyme (the primary decarboxylase), while aspartate would be transported to bundle sheath mitochondria for decarboxylation through NAD-malic enzyme. In NAD-malic enzyme species, malate would be transported to bundle sheath chloroplasts for conversion to OAA (oxaloacetate) through NAD-malate dehydrogenase; the OAA formed would be subsequently metabolized by the bundle sheath mitochondria through the mitochondrial malate dehydrogenase and NAD-malic enzyme (RATHNAM and EDWARDS, 1975, 1977b). There is no known chloroplastic decarboxylase in NAD-malic enzyme species although there is some evidence to suggest that malate oxidation by the bundle sheath cells is linked to the reduction of 3-phosphoglycerate (RATHNAM and EDWARDS, 1977b). However aspartate transport to the bundle sheath mitochondria in NAD-malic enzyme species would be the primary transport metabolite from mesophyll cells for decarboxylation through NAD-malic enzyme (KAGAWA and HATCH, 1975; HATCH and OSMOND, 1976).

In PEP carboxykinase species malate transported from the mesophyll cell to bundle sheath chloroplasts would be oxidized to OAA and decarboxylated through PEP carboxykinase. In this case there may be a malate + Pi-PEP exchange between mesophyll and bundle sheath rather than a malate-pyruvate exchange (RATHNAM and EDWARDS, 1977a). Aspartate, transported from the mesophyll cell to the bundle sheath mitochondria, would be converted to OAA; part would be decarboxylated through NAD-malic enzyme in the mitochondria, and part would be transported to the chloroplast and decarboxylated through PEP carboxykinase.

There has been limited success in isolating intact organelles from bundle sheath cells (KAGAWA and HATCH, 1975; RATHNAM and EDWARDS, 1975, 1977a, b) and further evaluation of the suggested metabolism in bundle sheath cells is dependent on the isolation and study with functional chloroplasts and mitochondria.

F. Energetics in C_4 Metabolism

As previously discussed, one of the major roles of the C_4 mesophyll chloroplast is to convert pyruvate to PEP (Sect. E. I). The enzyme involved, pyruvate Pi dikinase, appears to operate together with adenylate kinase and pyrophosphatase, such that the inhibitory end products AMP and PPi are continually removed. This greatly favors enzyme action in the direction of PEP formation and also increases the energy requirement to 2 ATP/pyruvate (one ATP is required by pyruvate Pi dikinase and one by adenylate kinase). This proposed stoichiometry (see HATCH and OSMOND, 1976) was verified in vitro with intact mesophyll chloroplasts of *D. sanguinalis* (HUBER and EDWARDS, 1976a).

Three sources of ATP for the conversion of pyruvate to PEP have been demonstrated: noncyclic, pseudocyclic and cyclic photophosphorylation (EDWARDS et al., 1976a). Which source predominates in vitro depends on the species and the experimental conditions. Noncyclic electron flow is only possible if the pyridine nucleotide pool is turning over rapidly. Considering the enzymology of the C_4 pathway, this could be accomplished by the reduction of OAA to malate in the chloroplast. Mesophyll chloroplasts of *D. sanguinalis* (and NADP-malic enzyme species, "primary malate former") contain high levels of NADP-malate dehydrogenase, such that the potential for noncyclic photophosphorylation is high. Experimentally, noncyclic electron flow can be induced by adding exogenous OAA. The potential for OAA dependent noncyclic photophosphorylation is highest in mesophyll chloroplasts of NADP-malic enzyme species (EDWARDS et al., 1976a).

Pseudocyclic electron flow involves the flow of electrons from H_2O (through both photosystems) to O_2. O_2 and NADP have been shown to compete for electrons from H_2O (EDWARDS et al., 1976a). If the concentration of organic oxidants (OAA or 3-phosphoglycerate) is suboptimal for keeping the pyridine nucleotide pool oxidized, then O_2 will act as an electron acceptor. With OAA as the oxidant, O_2 will act as an electron acceptor with chloroplasts of *D. sanguinalis*, until the concentration of OAA is 0.5 mM or higher. This is also the concentration of OAA which experimentally was found to induce maximum rates of noncyclic electron flow. Without NADP as a competing electron acceptor, the rate of pseudocyclic electron flow continues to increase up to 100% O_2; half-maximal stimulation by O_2 occurs at roughly 30% O_2 (EDWARDS et al., 1976a). In mesophyll chloroplasts of C_4 plants representing the PEP carboxykinase and NAD-malic enzyme species, where the principal products are oxaloacetate and aspartate, pseudocyclic and/or cyclic electron flow are probably more important than noncyclic electron flow as a source of ATP since reductive power is not required by the carboxylation phase from alanine to aspartate. With the different species tested, the rate of pseudocyclic photophosphorylation determined indirectly from pyruvate dependent CO_2 fixation ranges from 40–60 µmol ATP mg chl^{-1} h^{-1} under 20% O_2 (EDWARDS et al., 1976a).

Cyclic electron flow is the third source of ATP that has been identified in C_4 mesophyll chloroplasts. Cyclic flow appears to operate in vitro at a rate of

60–80 µmol ATP mg chl^{-1} h^{-1} with various species, providing energy for conversion of pyruvate to PEP independent of the operation of noncyclic or pseudocyclic electron flow. The cyclic contribution to total photophosphorylation can be readily assessed by the use of the classic inhibitor antimycin A, which has been found to be a specific and potent inhibitor of cyclic electron flow in mesophyll chloroplasts of C$_4$ plants (EDWARDS et al., 1976a; HUBER and EDWARDS, 1976b).

In intact C$_4$ mesophyll chloroplasts in vitro, cyclic electron flow appears to contribute a constant amount of ATP to the total photophosphorylation of the chloroplast. The contribution of noncyclic electron flow will, by definition, depend on the presence and concentration of organic oxidants. Pseudocyclic electron flow can be envisioned as a regulatory process. Where the demand for ATP relative to NADPH is greater than that provided by noncyclic and cyclic electron flow, pseudocyclic electron flow will provide the additional ATP. The sources of ATP identified with intact chloroplasts in vitro (in the absence of any added cofactors) are probably potential sources in vivo as well.

Cyclic, noncyclic, and pseudocyclic electron flow are also potential sources of ATP in the bundle sheath chloroplast, although these have not been sufficiently studied in vitro (theoretical energy requirements for bundle sheath cells of three groups of C$_4$ plants have been previously formulated, EDWARDS et al., 1976a). What is known to date, however, is that NADP-malic enzyme species have a partial photosystem II deficiency in bundle sheath chloroplasts. The degree of deficiency is species dependent (EDWARDS et al., 1976a) and is compensated for by NADP-malic enzyme, which donates both CO$_2$ and NADPH to the Calvin cycle (EDWARDS et al., 1976a; HATCH and OSMOND, 1976). In these species, cyclic photophosphorylation is a logical source of ATP for the Calvin cycle, although this remains to be demonstrated.

G. Future Studies on C$_4$ Metabolism with Cells and Protoplasts

I. Transport Studies

Schemes for C$_4$ photosynthesis require massive intercellular transport of metabolites which is, as of yet, only speculative. HATCH and OSMOND (1976), on the basis of calculated flux rates, have suggested that concentration gradients could drive metabolite transport by diffusion, though ATP dependent transport is not excluded. If transport is through the desmotubules of the plasmodesmata, then there may be transport across the endoplasmic reticulum as well, although electron microscopy evidence on these structural components is very limited. Future in vitro studies aimed at clarifying this transport should include purification (from both cell types) and examination of the transport properties of the plasma membrane and microsomal fractions. With mesophyll protoplasts of *D. sanguinalis*, the plasma membrane may represent a barrier to the uptake of pyruvate, as mesophyll protoplasts have a lower maximum velocity for pyruvate-dependent CO$_2$ fixation than do mesophyll protoplast extracts (HUBER and EDWARDS, 1975a). Alternatively, this could be explained on the basis of end product inhibition of PEP carboxylase. If the synthesized C_4 acids reach a relatively high concentration in the cytoplasm of the protoplast, CO_2 fixation would be reduced, as malate and aspartate are potent inhibitors of PEP carboxylase from C_4 plants (HUBER and EDWARDS, 1975d).

In addition, direct transport studies of the flux of metabolites across the chloroplast envelope are required, as current schemes envision the transport of compounds (e.g., pyruvate and PEP across the mesophyll chloroplast envelope) which are not thought to be readily transported metabolites in spinach (C_3) chloroplasts. With C_4 mesophyll chloroplasts recent studies indicate that a phosphate translocator catalyzes the uptake of Pi in exchange for PEP and a pyruvate translocator catalyzes pyruvate uptake (HUBER and EDWARDS, 1977a, b).

II. Screening for Inhibitors of C_4 Photosynthesis

Three functional assay systems have been developed with isolated protoplasts and cells of C_4 plants which can serve as a means of screening compounds for specific blocks of certain phases of C_4 photosynthesis. These are:

a) Light-dependent $^{14}CO_2$ fixation with pyruvate using C_4 mesophyll protoplast extracts which measure the carboxylation phase of the C_4 pathway (HUBER and EDWARDS, 1975a, b). This provides an in vitro experimental system apparently equivalent to the carboxylation phase in vivo which requires enzymatic reactions of carbon assimilation, exchange of metabolites across chloroplast membranes, and photochemical production of energy. Thus potential inhibitors might block any one of a number of steps required in this phase. Effective inhibitors would need to be analyzed for their specificity by looking at partial reactions of the pathway.

b) The decarboxylation phase of the C_4 pathway in bundle sheath cells. Three decarboxylation mechanisms have been identified which can be assayed with bundle sheath cells of the three C_4 groups by following the decarboxylation of 4-^{14}C malate or aspartate; or carboxyl donation from C_4-dicarboxylic acids to the Calvin–Benson pathway by monitoring C_4 acid-dependent O_2 evolution (KAGAWA and HATCH, 1974a; HATCH and KAGAWA, 1976; RATHNAM and EDWARDS, 1977a, b).

c) Direct fixation of $^{14}CO_2$ via the Calvin–Benson pathway in bundle sheath cells. Isolated bundle sheath cells capable of reasonably high rates of light-dependent fixation of $^{14}CO_2$ without addition of exogenous substrates can be used to test potential inhibitors of C_4 photosynthesis on the Calvin–Benson pathway.

In these various assays it is important to use completely functional systems (Sect. D) in order to establish whether a compound differentially inhibits certain pathways of C_4 photosynthesis. Table 1 summarizes some recent measurements of differential effects of a few inhibitors on these pathways. Oxalic acid at 0.1 mM inhibits the decarboxylation phase through NADP-malic enzyme with relatively little effect on other phases. 3-Mercaptopicolinic acid at 0.1 mM, an inhibitor of PEP carboxykinase in C_4 plants (RAY and BLACK, 1976) and animals (KOSTOS et al., 1975) inhibits the decarboxylation through PEP-carboxykinase but not $^{14}CO_2$ fixation by the bundle sheath, or the carboxylation phase through the C_4 pathway. DL-Glyceraldehyde strongly inhibits $^{14}CO_2$ fixation by bundle sheath cells with less effect on the carboxylation phase of the C_4 pathway. Loss of $^{14}CO_2$ through decarboxylation of 4-^{14}C dicarboxylic acids by bundle sheath cells is stimulated by DL-glyceraldehyde apparently by preventing fixation of the $^{14}CO_2$ by the Calvin–

Table 1. Differential inhibition of photosynthetic metabolism in mesophyll and bundle sheath preparations of some C$_4$ species

Inhibitor	Concentration mM	C$_4$ carboxylation mesophyll protoplast extract[a] μmol CO$_2$ fixed mg chl^{-1} h^{-1}	^{14}CO$_2$ fixation Calvin–Benson pathway bundle sheath cells		C$_4$ decarboxylation bundle sheath cells μmol ^{14}CO$_2$ evolved mg chl^{-1} h^{-1}		
			NADP[a] type	PEP[b] type	NADP[a] type	NAD[c] type	PEP[b] type
A. Control	0	80	40	203	294	342	256
Oxalic acid	0.1	74 (92)	42 (105)	195 (96)	40 (14)	362 (106)	249 (97)
Phenylpyruvate	5	38 (47)	40 (100)	–	–	–	–
DL-Glyceraldehyde	75	60 (75)	–	41 (20)	–	–	–
B. Control	0	80	31	241	176	353	281
3-Mercaptopicolinic acid	0.1	76 (95)	31 (100)	239 (99)	173 (98)	345 (98)	65 (23)

Carboxylation in C$_4$ pathway was in the presence of 1 mM pyruvate, 6 mM NaH^{14}CO$_3$. ^{14}CO$_2$ fixation with bundle sheath cells was in the presence of 5 mM NaH^{14}CO$_3$. Decarboxylation through NADP-malic enzyme included 10 mM 4-^{14}C malate, 6 mM 3-phosphoglycerate and 25 mM DL-glyceraldehyde. Decarboxylation through NAD-malic enzyme and PEP-carboxykinase included 4-^{14}C aspartate, 10 mM 2-α-ketoglutarate, 25 mM DL-glyceraldehyde (RATHNAM and EDWARDS, 1977a, b, c for other details). A and B represent separate experiments. Figures in parentheses give activity as percentage of control rate.
[a] Species used was *Digitaria sanguinalis.*
[b] Species used was *Eriochloa borumensis.*
[c] Species used was *Panicum miliaceum.*

Benson pathway (RATHNAM and EDWARDS, 1977a, b). Phenylpyruvate is a competitive inhibitor (Ki = 5.6 mM) of the carboxylation phase of the C$_4$ pathway but does not affect ^{14}CO$_2$ fixation through the Calvin–Benson pathway. Although some inhibitors may be relatively specific for blocking at one site (i.e., 3-mercaptopicolinic acid inhibition of PEP carboxykinase (RAY and BLACK, 1976), partial reactions of each phase will need to be analyzed further to localize the observed effects to enzymes of the pathway, transport, or photochemical reactions. Meaningful studies at the cellular level with various species will continue to be dependent on improved techniques for isolation and separation of functional chloroplast and cell types.

Acknowledgment. Certain of the authors' studies included in this review were supported in part by research grants (BMS-74-09611, PCM = 77-09384) from the National Science Foundation.

References

Black, C.C.: Ann. Rev. Plant Physiol. *24*, 253–286 (1973)
Brown, W.V.: Memoirs of the Torrey botanical club, Kiger, R.W. (ed.), Vol. 23, pp. 1–97. Durham, N.C.: Fisher-Harrison Corporation 1977

Chen, T.M., Dittrich, P., Campbell, W.H., Black, C.C.: Arch. Biochem. Biophys. *163*, 246–262 (1974)
Chollet, R., Ogren, W.L.: Plant Physiol. *51*, 787–792 (1973)
Edwards, G.E., Lee, S.S., Chen, T.M., Black, C.C.: Biochem. Biophys. Res. Commun. *39*, 389–395 (1970)
Edwards, G.E., Huber, S.C., Ku, S.B., Gutierrez, M., Rathnam, C.K., Mayne, B.C.: In: CO$_2$ metabolism and productivity of plants. Burris, R.H., Black, C.C. (eds.), pp. 83–112. Baltimore, Md.: Univ. Park Press 1976a
Edwards, G.E., Huber, S.C., Gutierrez, M.: In: Microbial and plant protoplasts. Peberdy, J.F., Rose, A.H., Rogers, H.J., Cocking, E.C. (eds.), pp. 299–322. New York: Academic Press 1976b
Farineau, J.: Physiol. Plant. *33*, 300–309 (1975)
Gutierrez, M., Gracen, V.E., Edwards, G.E.: Planta *119*, 279–300 (1974a)
Gutierrez, M., Kanai, R., Huber, S.C., Ku, S.B., Edwards, G.E.: Z. Pflanzenphysiol. *72*, 305–319 (1974b)
Gutierrez, M., Huber, S.C., Ku, S.B., Kanai, R., Edwards, G.E.: In: III. Int. Congr. Photosynthesis Res. Avron, M. (ed.), pp. 1219–1230. Amsterdam, Oxford, New York: Elsevier Sci. Publ. Co. 1975
Gutierrez, M., Edwards, G.E., Brown, W.V.: Biochem. Syst. Ecol. *4*, 47–49 (1976)
Hatch, M.D., Kagawa, T.: Arch. Biochem. Biophys. *159*, 842–853 (1973)
Hatch, M.D., Kagawa, T.: Arch. Biochem. Biophys. *175*, 39–53 (1976)
Hatch, M.D., Osmond, C.B.: In: Encyclopedia of plant physiology. New series, Pirson, A., Zimmermann, M.H. (eds.), Vol. III, pp. 144–184. Berlin, Heidelberg, New York: Springer 1976
Hatch, M.D., Kagawa, T., Craig, S.: Aust. J. Plant Physiol. *2*, 111–128 (1975)
Hattersley, P.W., Watson, L.: Aust. J. Bot. *24*, 297–308 (1976)
Hattersley, P.W., Watson, L., Osmond, C.B.: Aust. J. Plant Physiol. *4*, 523–539 (1977)
Huber, S.C., Edwards, G.E.: Plant Physiol. *55*, 835–844 (1975a)
Huber, S.C., Edwards, G.E.: Plant Physiol. *56*, 324–331 (1975b)
Huber, S.C., Edwards, G.E.: Physiol. Plant. *35*, 203–209 (1975c)
Huber, S.C., Edwards, G.E.: Can. J. Bot. *53*, 1925–1933 (1975d)
Huber, S.C., Edwards, G.E.: Biochim. Biophys. Acta. *440*, 675–687 (1976a)
Huber, S.C., Edwards, G.E.: Biochim. Biophys. Acta. *449*, 420–433 (1976b)
Huber, S.C., Edwards, G.E.: Biochim. Biophys. Acta. *462*, 603–612 (1977a)
Huber, S.C., Edwards, G.E.: Biochim. Biophys. Acta. *462*, 583–602 (1977b)
Huber, S.C., Kanai, R., Edwards, G.E.: Planta *113*, 53–56 (1973)
Huber, S.C., Hall, T.C., Edwards, G.E.: Plant Physiol. *57*, 730–733 (1976)
Kagawa, T., Hatch, M.D.: Biochem. Biophys. Res. Commun. *59*, 1326–1332 (1974a)
Kagawa, T., Hatch, M.D.: Aust. J. Plant Physiol. *1*, 51–64 (1974b)
Kagawa, T., Hatch, M.D.: Arch. Biochem. Biophys. *167*, 687–696 (1975)
Kanai, R., Edwards, G.E.: Plant Physiol. *52*, 484–490 (1973a)
Kanai, R., Edwards, G.E.: Plant Physiol. *51*, 1133–1137 (1973b)
Kostos, V., Di Tullio, N.W., Rush, J., Cieslinski, L., Saunders, H.L.: Arch. Biochem. Biophys. *171*, 459–465 (1975)
Ku, S.B., Edwards, G.E.: Z. Pflanzenphysiol. *77*, 19–32 (1975)
Ku, S.B., Gutierrez, M., Kanai, R., Edwards, G.E.: Z. Pflanzenphysiol. *72*, 320–337 (1974)
Ku, S.B., Edwards, G.E., Kanai, R.: Plant Cell Physiol. *17*, 615–620 (1976)
Levi, C., Gibbs, M.: Plant Physiol. *56*, 164–166 (1975)
Mayne, B.C., Dee, A.M., Edwards, G.E.: Z. Pflanzenphysiol. *74*, 275–291 (1975)
O'Neal, D., Hew, C.S., Latzko, E., Gibbs, M.: Plant Physiol. *49*, 607–614 (1972)
Rathnam, C.K.M., Edwards, G.E.: Arch. Biochem. Biophys. *171*, 214–225 (1975)
Rathnam, C.K.M., Edwards, G.E.: Planta *133*, 135–144 (1977a)
Rathnam, C.K.M., Edwards, G.E.: Arch. Biochem. Biophys. *182*, 1–13 (1977b)
Rathnam, C.K.M., Edwards, G.E.: Plant Cell Physiol. *18*, 963–968 (1977c)
Ray, T.B., Black, C.C.: J. Biol. Chem. *251*, 5824–5826 (1976)
Rouhani, I., Vines, H.M., Black, C.C.: Plant Physiol. *51*, 97–103 (1973)
Salin, M.L., Campbell, W.H., Black, C.C.: Proc. Natl. Acad. Sci. USA *70*, 3730–3734 (1973)
Walker, D.A.: In: Medical and technical publications. Int. Rev. Sci. Biochem. Ser. I, Northcote, D.H. (ed.), Vol. 11, pp. 1–49. London: Butterworths 1974

8. The Flow of Carbon in Crassulacean Acid Metabolism (CAM)

M. KLUGE

A. Introduction

It is now generally accepted that the Crassulacean Acid Metabolism (CAM) may be interpreted as a variant of the photosynthetic CO_2 fixation. CAM is performed mainly but not exclusively by succulents of the Crassulacean family (BLACK and WILLIAMS, 1976; SZAREK and TING, 1977; KLUGE and TING, 1978). There is increasing evidence that CAM is of significance as an adaptational mechanism facilitating photosynthesis in arid environments (cf. KLUGE, 1976; KLUGE and TING, 1978).

The phenomenon CAM has been repeatedly reviewed during the last two decades. The first extended analysis of the metabolic pathways involved in CAM were made by WOLF (1960) and RANSON and THOMAS (1960). Metabolic aspects of CAM have been considered by BEEVERS et al. (1966); TING (1971); TING et al. (1972); BLACK (1973); OSMOND (1975, 1976); KLUGE (1976, 1977). Recently, OSMOND (1978) considered a variety of different aspects of CAM, and the most inclusive analysis of the CAM phenomenon has been provided by the monograph of KLUGE and TING (1978).

The goal of this chapter is to outline the basic principles of carbon flow characteristic for CAM. For a detailed discussion of the typical gas exchange phenomena linked to CAM, and of ecological aspects of CAM see OSMOND, 1978; KLUGE and TING, 1978.

B. Basic Phenomena of CAM

The most striking feature of CAM is a diurnal variation in the malic acid content of the photosynthesizing cells, with accumulation of malic acid at night and its disappearance during the day (Fig. 1). This diurnal rhythm of malic acid is accompanied by an inverse diurnal rhythm of polyglucan (mainly starch). Acids other than malic acid are not normally involved in the diurnal acid cycle, with the exception of citric acid, which may show oscillations parallel to those of malic acid, but with lower amplitude. Isocitric acid which in the leaves of certain Crassulaceae is the quantitatively dominating acid at the end of the day (i.e., when the malic acid content is low), lacks substantial rhythmic alterations (e.g., see detailed investigations of VICKERY, 1953).

The diurnal malic acid fluctuation is in the order of 100–200 µeq per g fresh weight, sufficient to be recognized by taste, a peculiarity which led to the discovery of CAM by HEYNES in 1815 (WOLF, 1960).

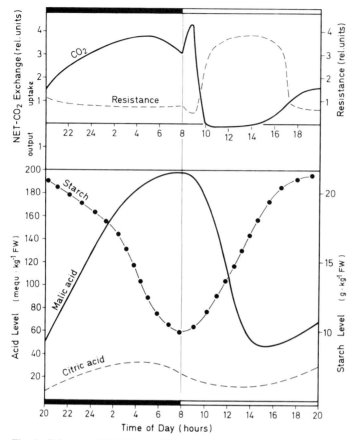

Fig. 1. Scheme of CAM activities as performed during a diurnal cycle by a typical CAM plant (e.g., *Kalanchoe daigremontiana*) under laboratory conditions. *Upper part:* CO_2 exchange and stomatal resistance. *Lower part:* Levels of organic acid and starch. (Kluge, Proc. 4th Int. Cong. Photosynth. pp. 335, 1977)

CAM plants normally perform nocturnal net fixation of CO_2 from the atmosphere ("De Saussure Effect"; Wolf, 1949; Thomas, 1949). During the day, uptake of external CO_2 is largely depressed in CAM plants (Fig. 1). The behavior of CO_2 exchange by the mesophyll cells causes a pattern of stomata movements inverse to that of normal plants, i.e., CAM plants open their stomata at night and close them during the day (Fig. 2). It is this stomatal behavior to which CAM owes its ecological advantage (Kluge and Fischer, 1967; Osmond, 1978; Kluge and Ting, 1978).

C. The Metabolic Sequences of CAM

CAM is a metabolic sequence where two groups of reactions are connected in series, although separated in time, i.e., certain reactions are dominant during the

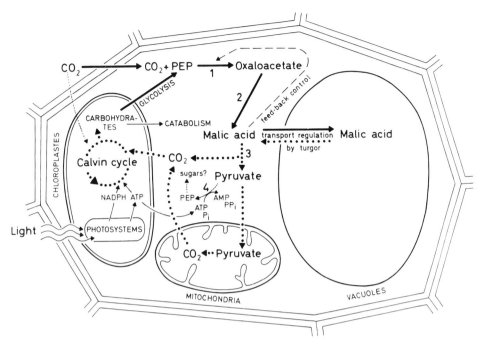

Fig. 2. Simplified model of total carbon flow in CAM. *1*, PEP-carboxylase reaction; *2*, malate dehydrogenase reaction; *3*, decarboxylation of malate; *4*, pyruvate-P_i-dikinase reaction. Reactions dominating during night (———); reactions dominating during day (......); regulatory properties (------) (KLUGE, 1976)

night while others being strictly "light-dependent" (Sect. C. II. 5). Hence, it is reasonable to discuss the metabolic reactions of the dark and the light phases separately, and then consider how the events of dark and light metabolism are linked together (Fig. 2.).

I. The Flow of Carbon During the Night

1. The Initial Step: Dark CO_2 Fixation

There is now unequivocal agreement that nocturnal malic acid synthesis in CAM is initiated by dark CO_2 fixation, involving the β-carboxylation of phosphoenolpyruvate (PEP) to malate.

This reaction is catalyzed by PEP carboxylase which was first shown to occur in plants by BANDURSKI and GREINER (1953) and in particular in CAM by WALKER (1956, 1957). Since the reaction equilibrium greatly favors oxaloacetate formation, and the affinity of PEP carboxylase for HCO_3^- is high (about 0.02 mM; TING and OSMOND, 1973a, b), the value of the PEP carboxylase reaction as a CO_2 harvesting step is obvious. Oxaloacetate derived from the PEP carboxylase reaction is subsequently reduced to malate. This reaction is catalyzed by NAD-linked malate dehydrogenase which has been shown to be present in CAM plants with high activity (KLUGE and OSMOND, 1972).

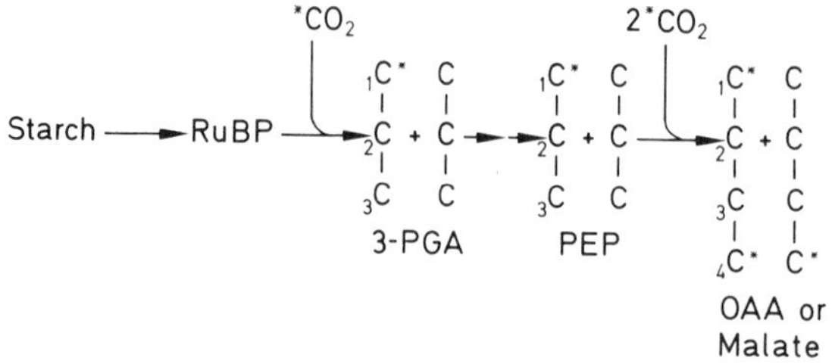

```
        *CO₂              2*CO₂                    Fig. 3. Scheme of the double CO₂
         |                 |                       fixation in nocturnal malate synthe-
         |  ₁C*  C         |  ₁C*  C      ₁C*  C   sis as proposed by BRADBEER et al.
         |   |   |         |   |   |       |   |   (1958)
Starch → RuBP → ₂C + C → ₂C + C → ₂C + C
             |   |   |         |   |       |   |
            ₃C   C         ₃C   C      ₃C   C
         3-PGA             PEP          |   |
                                       ₄C*  C*

                                      OAA or
                                      Malate
```

Fig. 3. Scheme of the double CO_2 fixation in nocturnal malate synthesis as proposed by BRADBEER et al. (1958)

If nocturnal CO_2 fixation occurs by β-carboxylation according to the above scheme, all the label (from $^{14}CO_2$) found in ^{14}C-oxaloacetate and in ^{14}C-malate derived from this oxaloacetate should occur in the C-4 atom. This is in fact the case when ^{14}C-malate is synthetized in vitro using PEP carboxylase and malate dehydrogenase (SUTTON and OSMOND, 1972). ^{14}C-Malic acid isolated from CAM plants which had fixed $^{14}CO_2$ in the dark often shows an asymmetric label distribution, with one-third of the label being located in C-1 and two-thirds in C-4 (VARNER and BURELL, 1950; BRADBEER et al., 1958; STILLER, 1959; COCKBURN, 1965). Such a distribution of label is inconsistent with the PEP carboxylase reaction being the initial step of dark CO_2 fixation, but is consistent with two carboxylation steps (Fig. 3) connected in series, i.e., carboxylation of RuBP (catalyzed by RuBP carboxylase), proceeding a β-carboxylation of PEP (BRADBEER et al., 1958). Hence, the observed asymmetric labeling of malate caused this double carboxylation scheme to become a dogma in CAM research for nearly two decades. Recent observations argue against this mechanism. Firstly SUTTON and OSMOND (1972) showed with improved methods that ^{14}C-malate isolated from CAM plants within seconds after dark CO_2 fixation had more than 90% of the label located in C-4. They blamed artifacts produced by the technique of malate degradation used by BRADBEER et al. (1958) for the earlier observed label in C-1. This conclusion was further supported by WINTER and KANDLER (1976). Secondly KLUGE et al. (1974, 1975) found no alteration in the distribution of label in ^{14}C-malate if unlabeled PEP was fed to tissue slices of CAM plants fixing $^{14}CO_2$. Thirdly BJÖRKMAN and OSMOND (1974) showed that the rate of dark CO_2 fixation in contrast to CO_2 fixation in the light, was not lowered by increased oxygen concentrations. This suggested that RuBP carboxylase (which is inhibited by oxygen) is not involved in the nocturnal harvesting of CO_2. Finally, COCKBURN and McAULAY (1975) showed unequivocally by mass spectrometric detection and quantification of the isotope species of malic acid synthetized from $^{13}CO_2$ during dark CO_2 fixation that no double-labeled malate is synthesized.

Present theories therefore no longer consider double carboxylation as a principal feature of dark CO_2 fixation in CAM, although it is true that it may occur during certain times of the day (see below).

Despite these findings, there is no doubt that the distribution of ^{14}C in malate synthesized in the dark from $^{14}CO_2$ dark fixation may approach a 1:2 rate between C-1 and C-4 (KLUGE et al., 1974, 1975). However, it has been suggested by TING (1971), by COCKBURN and McAULAY (1975), and demonstrated by DITTRICH (1976), that this effect might be due to randomization of label, originally located in C-4 by the action of fumarase.

2. The Origin of the CO_2 Acceptor in the Dark CO_2 Fixation

As shown in Figure 1, there is an inverse relationship between the pool sizes of malic acid and of storage carbohydrates such as starch and other glucans (WOLF, 1960; RANSON and THOMAS, 1960; SUTTON, 1975a, b). Hence, it is reasonable to postulate that the CO_2 acceptor (PEP) is derived from a reserve glucan. This postulation is also supported by $^{14}CO_2$ pulse chase experiments showing that ^{14}C carbon from starch previously labeled in the light is transferred to malic acid in the following dark period (KLUGE et al., 1975). Finally, dark consumption of glucan is clearly inhibited when the demand for CO_2 acceptor molecules is kept low by application of CO_2 free air in the dark, or by inhibition of dark CO_2 fixation by other methods (KLUGE, 1969a).

The pathway of glucan conversion to PEP is still being investigated. Recent work on enzymes of the glycolytic pathway in CAM plants (SUTTON, 1975a, b) confirms that glycolysis could play a major role in the production of PEP.

The recently disproved assumption of a double CO_2 fixation being involved in the nocturnal harvesting of exogenous CO_2, with RuBP reaction as primary carboxylating step, anticipated a pathway capable of providing Ru5P as precursor of RuBP. The work of BRADBEER (1954), STILLER (1959) and KHAN (1969) suggest that the oxidative pentose phosphate pathway may operate in CAM plants in order to generate pentose phosphates. Since the hypothesis of a double carboxylation in darkness has been disproved, the role of the oxidative pentose phosphate pathway in nocturnal malate synthesis of CAM remains to be evaluated.

There is some controversy concerning the initial step of starch breakdown. SUTTON (1975a, b) showed phosphorylase but not α-amylase to be active in leaves of *Kalanchoe daigremontiana*, suggesting that starch breakdown proceeds mainly via phosphorolysis rather than hydrolysis. However, VIEWEG and FEKETE (1977) found that α-amylase activity in *K. daigremontiana* is high enough to permit substantial breakdown. At the moment it can only be assumed that both α-amylase and phosphorylase participate in starch breakdown.

3. Depletion of Malic Acid in the Dark, and Secondary Products of Nocturnal CO_2 Fixation

In darkened leaves of CAM plants, malic acid accumulation is followed by a period where the malic acid level remains constant and finally declines if the plants are held in prolonged darkness (WOLF, 1960). This loss of malic acid is accompanied by a net output of CO_2, both processes being increased by increasing temperature (WOLF, 1960). There is little doubt that at least part of the CO_2 produced during prolonged darkness originates from malic acid. Even during a normal dark period some CO_2 is generated from malic acid (KLUGE, 1968) although admittedly at a low rate when compared with rates during the light period or prolonged darkness. It remains to be clarified whether nocturnal and day-time CO_2 production from malic acid proceed via the same pathways (Sect. C. II). Operation of the Krebs cycle in the dark is indicated from the labeling of a variety of Krebs cycle intermediates including citric, isocitric, succinic and fumaric acid (cf. KLUGE and TING, 1979), after $^{14}CO_2$ dark fixation. The Krebs cycle may be stimulated if leaves of CAM plants are sectioned (KINRAIDE and BE-

HAN, 1975); under these (artificial) conditions the respiratory sequences may successfully compete with malic acid accumulation.

A variety of amino acids can be found among the secondary products of dark CO_2 fixation. Aspartate and glutamate may contain about 10%, and alanine and glycine in the order of 5% or less of the label incorporated from $^{14}CO_2$ in darkness. Labeling of aspartate is expected because it is formed by amination of oxaloacetate (the primary product of dark CO_2 fixation). Labeling of other amino acids could result from operation of the Krebs cycle.

4. Problems of Malic Acid Storage and the Compartmentation of Metabolic Sequences

Malic acid may accumulate during the night up to concentrations of 100 mM in cells performing CAM. The cytoplasm could not tolerate acid concentrations of this order, and it is clear that most of this malic acid is stored in the vacuoles (KLUGE and OSMOND, 1972; KLUGE and HEININGER, 1973; BUSER and MATILE, 1977) which are particularly voluminous in CAM plant cells.

An early suggestion was that the storage behavior of the vacuoles might be part of a mechanism controlling CAM (WOLF, 1960; KLUGE and HEININGER, 1973). Malic acid transport at the tonoplast of CAM cells has been studied by KLUGE and HEININGER (1973), V. WILLERT and KLUGE (1973), LÜTTGE and BALL (1974), and LÜTTGE et al. (1975). From these investigations it was concluded that the transport of malic acid from the cytoplasm into the vacuoles must be an active process, while export, however, occurs passively (DENIUS and HOMANN, 1972). Furthermore, from results obtained with leaf slices of *Kalanchoe daigremontiana* suspended in solutions of various osmotic potential, LÜTTGE et al. (1975) postulated that the net export of malic acid from the vacuole into the cytoplasm might be triggered by turgor pressure. It remains to be clarified whether such a turgor mechanism also acts in vivo (LÜTTGE and BALL, 1977).

It has also been suggested but not yet proved, that light (KLUGE, 1971), particularly far red light (NALBORCZYK et al., 1975) may also be able to trigger the release of malic acid from the vacuole and thus initiate malic acid metabolism.

The location of enzymes involved in nocturnal CO_2 fixation and malic acid synthesis has not been systematically investigated. There is evidence that the bulk of the PEP carboxylase is cytoplasmic (TING and OSMOND, 1973a, b). However, PEP carboxylase activity has also been found to be associated with chloroplasts (GARNIER-DARDART, 1965; MUKERJI and TING, 1968). It should be noted that enzyme extraction from CAM plants is difficult and could be the cause of certain discrepancies in the results from different laboratories.

II. The Flow of Carbon During the Day

1. A General Scheme of Carbon Flow

It can be shown from tracer studies that carbon fixed into malic acid during the night is transferred into phosphorylated compounds and sugars during a following period of illumination (KUNITAKE and SALTMAN, 1958). There is no doubt that photosynthesis is the central mechanism involved. This can be concluded from numerous observations that CAM plants produce O_2 in the light even when no external CO_2 is being taken up. Hence, the CO_2 derived from malic acid is made available to photosynthesis and is refixed by the Calvin cycle. LEVI and GIBBS (1974, 1975) have isolated chloroplasts from *Kalanchoe* leaves capable of $^{14}CO_2$ fixation and concluded that only this C_3 pathway of CO_2 fixation operates in CAM plant chloroplasts. RuBP carboxylase has been shown to be active in CAM plants with rates of 0.5–15.8 µmol per min per mg chl (OSMOND, 1976).

The carbon flow from malic acid to the end products of photosynthesis (Fig. 2) can be thought of in terms of the nocturnally synthetized and accumulated malic acid as reservoir of CO_2 in the CAM-performing leaves. This reservoir is filled up during the night and emptied during the day. This capability to store CO_2 permits the ecological advantage that CO_2 is taken up at night when water loss by transpiration through open stomata is minimal; during the day the stomata can remain closed (KLUGE, 1976; KLUGE and TING, 1979).

2. The Breakdown of Malic Acid

Malic acid, released from the vacuole into the cytoplasm, is consumed in the light by decarboxylation yielding free CO_2 and a 3-C residue. Groups of CAM plants may be distinguished according to their decarboxylation mechanism (OS-MOND, 1976, 1978). One group utilizes an oxidative decarboxylation of malic acid by NADP- or NAD-linked (GARNIER-DARDART, 1965, DITTRICH, 1975) malic enzyme. In the other group of CAM plants, PEP carboxykinase is present at high activity (DITTRICH et al., 1973). In this latter group of CAM plants malic acid is first oxidized to oxaloacetate which is then decarboxylated yielding CO_2 and PEP. The in vivo site of malic acid decarboxylation is unclear. There is evidence that NADP-malic enzyme is bound to mitochondria (BRANDON, 1967), and there are reports that isoenzymes of NADP-malic enzyme can be found in chloroplasts, mitochondria and cytoplasm (GARNIER-DARDART, 1965; MUKERIJ and TING, 1968). Substantial activity of PEP carboxykinase is pelleted along with mitochondria and chloroplasts (DITTRICH et al., 1973).

3. The Fate of CO_2 Derived from Malic Acid

CO_2 produced by malic acid decarboxylation is normally retained in the leaves for refixation in the Calvin cycle, but under low light intensity, and in particular when the temperature is high, a substantial amount of this CO_2 can be lost to the atmosphere (KLUGE et al., 1973; NOSE et al., 1977).

4. The Fate of the 3-C Residue

In the "malic enzyme" group of CAM plants, pyruvate remains as the 3-C residue. The question arises whether this 3-C skeleton is incorporated directly into carbohydrates or degraded to CO_2 prior to subsequent reassimilation. CHAM-PIGNY (1960) favored the latter interpretation. However, HAIDRI (1955), by infiltrating labeled pyruvate at a known position into deacidifying leaves of *Kalanchoe*, suggested total incorporation of the pyruvate into hexoses by reversed glycolysis. This would require conversion of pyruvate to PEP, and indeed, KLUGE and OSMOND

(1971) and Sugiyama and Laetsch (1975) have shown pyruvate-P_i-dikinase to be active in CAM plants. In the "PEP carboxykinase group" of CAM plants, PEP is a direct product from decarboxylation of the C-4 skeleton. In general, metabolism of the 3-C residue during the light period still remains a gap in the overall picture of carbon flow in CAM.

5. Fixation of External CO_2 in the Light

It can be seen from the CO_2 exchange of CAM (Fig. 1) that in the light three phases of gas exchange occur (Osmond, 1976; Fig. 4; cf. Neales, 1975 and Kluge and Ting, 1978).

Phase A: an initial burst of CO_2 uptake;

Phase B: a depression of CO_2 uptake where the curve approaches the compensation
    point or even net CO_2 output may occur dependent on external factors, and

Phase C: a phase in the late light period where net CO_2 uptake again occurs
    and where a more or less steady state of gas exchange may exist.

This classical 3-phase pattern can show pronounced variations either between different CAM plant species or in a given individual under varying environmental conditions (Neales, 1975; Kluge, 1976). Figure 1 is for a CAM plant (e.g., *Kalanchoe daigremontiana*) performing the standard pattern of CAM-type gas exchange.

A detailed analysis of the CO_2 metabolism occurring during each phase of the light period has been provided by Osmond (1976). Phase A (Fig. 4): during the initial burst of CO_2 uptake, a lag in the consumption of malic acid can be observed (Kluge, 1968a), the acid level may even increase during the first hour of light (Kluge, unpublished). During this time exogenously supplied $^{14}CO_2$ is incorporated mainly into malic acid, although phosphorylated compounds and sugars are also labeled (Osmond and Allaway, 1974). This indicates that both PEP carboxylase and RuBP carboxylase are operating simultaneously with dominance of the first. The proposed carbon flow pattern (Osmond, 1976) during the initial burst is shown in Fig. 4a. Phase B (Fig. 4b): the depression of CO_2 uptake following the initial burst reflects the reassimilation by the Calvin cycle of CO_2 derived from malic acid (compare Kluge, 1968b). Exogenous CO_2 is obviously excluded (although it may be fixed at a low rate if the epidermis is removed; label from $^{14}CO_2$ fed in this way appears mainly in Calvin cycle products rather than in malic acid, suggesting that PEP carboxylase must be somehow inhibited during phase B (Sect. D). Phase C (Fig. 4c): CO_2 uptake occurs in the late light period after all the malic acid stored during the previous night has been consumed. Both products of the Calvin cycle and malic acid are labeled from exogenously supplied $^{14}CO_2$, indicating that PEP carboxylase is again active. Under low light intensity (about 7000 lx), malate may even represent the major labeled product (Kluge, 1969, 1971). There is evidence that the carbon appearing in this malate does not enter directly via PEP carboxylase, but rather via the double carboxylation pathway (Fig. 3; Osmond and Allaway, 1974; Kluge et al., 1975; Osmond and Björkman, 1975).

An increase of the malic acid level becomes clearly measurable if the light period is prolonged, or if the plants are transferred to continuous illumination

Fig. 4. Scheme of carbon flow in CAM during distinct phases of the light period. *Phase A*, initial burst of CO_2 uptake after onset of light. *Phase B*, period of deacidification where endogenous CO_2 is used and external CO_2 is excluded. *Phase C*, late hours of the light period where exogenous CO_2 is fixed (modified after OSMOND, 1976)

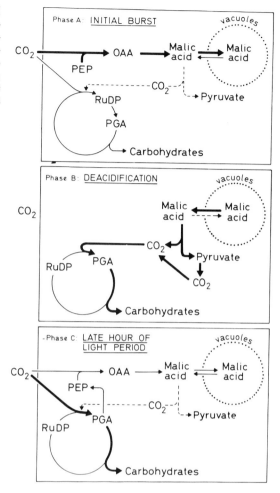

(KLUGE, 1969; LÜTTGE and BALL, 1977). This indicates that during the late afternoon the vacuole again acts as a sink for malic acid (KLUGE, 1977; Sect. D).

6. Photorespiration in CAM Plants

Measurements of photorespiratory CO_2 production is difficult in CAM plants since CO_2 is also produced from malic acid. Nevertheless, it is almost certain that photorespiration (i.e., CO_2 production via the glycolate pathway) proceeds in CAM plants (OSMOND, 1976). RuBP carboxylase from *Kalanchoe daigremontiana* has an oxygenase activity, and thus must produce P-glycolate (BADGER et al., 1975). Glycine and serine (both intermediates of photorespiration) collect the label after $^{14}CO_2$ fixation in the light (KLUGE, 1969), although carbon flow through the glycine/serine pool seems to be relatively slow (OSMOND, 1976).

Further evidence in favor of the occurrence of photorespiration in CAM plants comes from CO_2 exchange experiments. As shown by Björkmann and Osmond (1974) and Osmond and Björkmann (1975), CO_2 uptake occurring in the late light phase (Phase C, Fig. 1) is reversibly inhibited by O_2 concentrations higher than 2%, and an O_2 sensitive compensation point of 50 µl/l can be measured (Allaway et al., 1974), comparable to that of C_3 plants which show photorespiration. However, inhibition of CO_2 fixation by oxygen is not affected by changes in temperature between 10° C and 30° C, and inhibition of CO_2 by 21% oxygen could not be completely reversed by high tensions of CO_2. Both findings led to the conclusion that photorespiration in CAM plants might be "similiar to that in C_3 plants, but however, not identical" (Osmond, 1976).

In CAM plants, during malic acid decarboxylation the CO_2 concentration in the subepidermal spaces should increase and thus would suppress the oxygenase reaction of RuBP carboxylase (Osmond, 1976) and hence, photorespiration. Indeed, findings of Kluge (1969) show that glycine/serine collects less label after $^{14}CO_2$ fixation during malic acid consumption (Phase B) than during the late phase of the light period (Phase C), when production of CO_2 from stored malic acid is finished.

D. Regulation of Carbon Flow in CAM

Apart from ecological aspects, metabolic regulation of CAM is currently the most intensively investigated field of research and has received several reviews (Queiroz, 1974; Kluge, 1976; Kluge, 1977).

The main problem of regulation, as far as the diurnal cycle of CAM is considered, concerns the metabolism proceeding in the light, when free CO_2 is apparently available to both PEP carboxylase and RuBP carboxylase, yet is nearly quantitatively incorporated into the Calvin cycle rather than being at least partially recycled by PEP carboxylase into malic acid.

Considering the higher activity of PEP carboxylase compared with RuBP carboxylase in CAM plants (PEP carboxylase in the order of 500 µmol per mg chl per min, RuBP carboxylase in the order of 150 µmol per mg chl per min; cf. Osmond, 1978), and the high affinity of PEP carboxylase for CO_2 ($Km_{total\ CO_2}$ 0.2 mM; Ting and Osmond, 1973a, b) the preference for RuBP carboxylase becomes difficult to understand.

Queiroz (see Chap. II. 9, this vol.) interprets CAM regulation in terms of endogenously oscillating enzyme activities ("biological clock" model) with PEP carboxylase activity being high during the night and low during the day. However, the basic oscillator controlling the enzyme activities remain speculative.

Osmond and Allaway (1974) and Osmond (1976) argue in favor of the regulation by competition between the two carboxylating pathways for 3-C skeletons. During the day the main source of PEP is considered to be 3-PGA derived from the Calvin cycle, but this net 3-PGA production is thought to be insufficient for supplying PEP for the PEP carboxylase reaction. This model cannot account for the fact that there is considerable CO_2 fixation via PEP carboxylase when plants are illuminated during the regular dark period (KLUGE, 1969)

An alternative explanation (KLUGE, 1969; KLUGE and OSMOND, 1972) is that PEP carboxylase might be inhibited during the light period by accumulated malate. Evidence in favor of this model has been provided from in vivo experiments with *Kalanchoe tubiflora* phyllodia or *Sedum praealtum* leaves (KLUGE, 1969, 1971; AVADHANI et al., 1971) and from in vitro studies with PEP carboxylase from CAM plants which showed that malate is a strong inhibitor of PEP carboxylase (KLUGE and OSMOND, 1971; TING and OSMOND, 1973a, b). A more recent version of this "feedback-model" includes the storage behavior of the vacuole for malic acid (KLUGE and HEININGER, 1973; LÜTTGE et al., 1975). There are reasonable arguments for a mechanism where malic acid flux across the tonoplast might be controlled in CAM plants by turgor (LÜTTGE and BALL, 1974; LÜTTGE et al., 1975). This hypothesis needs further evidence (LÜTTGE and BALL, 1977).

From the foregoing discussion it is clear that a general picture of CAM regulation is far from complete. It can be hoped that investigations on those plant species which can "switch" from C_3-photosynthesis to CAM depending on environmental conditions (eg., in *Mesembryanthemum crystallinum*; WINTER and LÜTTGE, 1976), will advance our knowledge of CAM regulation considerably.

E. Conclusions

In spite of the remarkable ability of CAM plants to cover their demand for CO_2 by nocturnal net CO_2 fixation, and even to exist for long periods without net CO_2 uptake in the light (KLUGE and FISCHER, 1967; SZAREK et al., 1973; KLUGE and TING, 1979), CAM does not provide an alternative to the reductive pentose phosphate cycle for carbon assimilation in plants. Rather, as true in all other photoautotrophic plants, the Calvin cycle represents the central event in carbon flow. All metabolic sequences apart from the Calvin cycle and which contribute to CAM may be interpreted as auxiliary mechanisms developed during evolution in order to facilitate photosynthesis in arid habitats.

In nature, CAM plants are confronted with the problem of how to guarantee maximum possible carbon gain with a minimum loss of water (OSMOND, 1975; KLUGE, 1976; TING, 1976, KLUGE and TING, 1978). This problem is solved by temporal separation of primary CO_2 fixation via PEP carboxylase (i.e., CO_2 uptake from the atmosphere through open stomata at night) from light driven carbon reduction (i.e., recycling of CO_2 endogenously supplied by malate decarboxylation behind closed stomata during the day). In comparison, in C_4 plants these two steps are spatially rather than temporally separated. The key points of the CAM mechanism are the reversible storage of CO_2 in the form of malic acid and the storage of light energy which is trapped via photosynthesis into glucan (starch). This glucan provides, the following night, the source of carbon and energy for generation of the CO_2 acceptor PEP. This allows CO_2 fixation to proceed at night in a "down hill" reaction without a direct requirement for light energy.

References

Allaway, W.G., Austin, B., Slatyer, R.O.: Aust. J. Plant Physiol. *1*, 397–405 (1974a)
Avadhani, P.N., Osmond , C.B., Tan, K.K.: In: Photosynthesis and photorespiration. Hatch, M.D., Osmond, C.B., Slatyer, R.O. (eds.), pp. 288–293. New York: Wiley-Interscience 1971
Badger, M.R., Andrews, T.J., Osmond, C.B.: In: Proc. 3rd Int. Congr. Photosynthesis. Avron, M. (ed.), pp. 1421–1429. Amsterdam: Elsevier 1975
Bandurski, R.S., Greiner, C.M.: J. Biol. Chem. *204*, 781–786 (1953)
Beevers, H., Stiller, M.L., Butt, V.S.: In: Plant physiol. Stewart, F.C. (ed.), Vol. 4, pp. 119–262. New York: Academic Press 1966
Björkman, O., Osmond, C.B.: Carnegie Inst. Yearb. *73*, 852–858 (1974)
Black, C.C.: Ann. Rev. Plant Physiol. *24*, 253–286 (1973)
Black, C.C., Williams, S.: In: CO_2 metabolism and plant productivity. Burris, R.H., Black, C.C. (eds.), pp. 407–424. Baltimore: Univ. Park Press 1976
Bradbeer, J.W.: Doctoral Thesis, Durham Univ. Durham, England (Newcastle Division) (1954)
Bradbeer, J.W., Ranson, S.L., Stiller, M.: Plant Physiol. *33*, 66–70 (1958)
Brandon, P.C.: Plant Physiol. *42*, 977–984 (1967)
Buser, Ch., Matile, Ph.: Z. Pflanzenphysiol. *82*, 462–466 (1977)
Champigny, M.L.: Ph.D. Thesis, Univ. Paris (1960)
Cockburn, W.: Ph.D. Thesis, Univ. Newcastle on Tyne, England (1965)
Cockburn, W., McAulay, A.: Plant Physiol. *55*, 87–89 (1975)
Denius, H.R., Homann, P.: Plant Physiol. *49*, 873–880 (1972)
Dittrich, P.: Plant Physiol. *57*, 310–314 (1975)
Dittrich, P.: Plant Physiol. *58*, 288–291 (1976)
Dittrich, P., Campbell, W.H., Black, C.C. Jr.: Plant Physiol. *52*, 357–361 (1973)
Garnier-Dardat, J.: Physiol. Veg. *3*, 215–227 (1965)
Haidri, D.: Plant Physiol. *30*, 4 (1955)
Khan, A.A.: Ph.D. Thesis, Locknow Univ. Lucknow (1969)
Kinraide, T.B., Behan, M.J.: Plant Physiol. *56*, 830–835 (1975)
Kluge, M.: Planta *80*, 255–263 (1968a)
Kluge, M.: Planta *80*, 359–377 (1968b)
Kluge, M.: Planta *86*, 142–150 (1969a)
Kluge, M.: Planta *88*, 113–129 (1969b)
Kluge, M.: In: Photosynthesis and photorespiration. Hatch, M.D., Osmond, C.B., Slatyer, R.O. (eds.), pp. 283–287. New York: Wiley-Interscience (1971a)
Kluge, M.: Planta *98*, 20–30 (1971b)
Kluge, M.: In: Water and plant life. Problems and modern approaches. Lange, O.L., Kappen, L., Schulze, E.D. (eds.), pp. 331–320. Ecol. Stud. *19*, 313–323 (1976a)
Kluge, M.: In: CO_2 metabolism and plant productivity. Burris, R.H., Black, C.C. (eds.), pp. 205–216. Baltimore, London, Tokyo: Univ. Park Press 1976b
Kluge, M.: In: Integration of activity in the higher plant. Symp. Soc. Exp. Biol. Jennings, D.H. (ed.), Vol. XXXI, pp. 155–175. Cambridge, New York, Melbourne: Cambridge Univ. Press 1977
Kluge, M., Fischer, K.: Planta *77*, 212–223 (1967)
Kluge, M., Heininger, B.: Planta *113*, 333–343 (1973)
Kluge, M., Ting, I.P.: Ecological Studies; Vol. 30. Berlin, Heidelberg, New York, Springer-Verlag 1978
Kluge, M., Ting, I.P.: Berlin, Heidelberg, New York: Springer (in press) 1979
Kluge, M., Lange, O.L., Eichmann, M. von, Schmid, R.: Planta *112*, 357–372 (1973)
Kluge, M., Kriebitsch, Ch., v. Willert, D.: Z. Pflanzenphysiol. *72*, 460–467 (1974)
Kluge, M., Bley, L., Schmid, R.: In: Environmental and biological control of photosynthesis. Marcelle, R. (ed.), pp. 281–288. The Hague: Dr. W. Junk Publishers 1975
Kunitake, G., Saltmann, P.: Plant Physiol. *83*, 400–403 (1958).
Levi, C., Gibbs, M.: Plant Physiol. *53* Suppl., 36 (1974)
Levi, C., Gibbs, M.: Plant Physiol. *56*, 164–166 (1975)
Lüttge, U., Ball, E.: Z. Pflanzenphysiol. *73*, 326–338 (1974)

Lüttge, U., Ball, E.: Oecologia *31*, 85–94 (1977)
Lüttge, U., Kluge, M., Ball, E.: Plant Physiol. *56*, 613–616 (1975)
Mukerji, S.K., Ting, I.P.: Phytochemistry *7*, 903–911 (1968)
Nalborczyk, E., LaCroix, L.J., Hill, R.D.: Can. J. Bot. *53*, 1132–1138 (1975)
Neales, T.F.: In: Environmental and biological control of photosynthesis. Marcelle, R. (ed.), pp. 299–310. The Hague: Dr. W. Junk Publishers 1975
Nose, A., Shiroma, M., Miyomatsu, K., Murayama, S.: Jpn. J. Crop Sci. *45*, 579–588 (1977)
Osmond, C.B.: In: CO_2 metabolism and plant productivity. Burris, R.H., Black, C.C. (eds.), pp. 217–233. Baltimore, London, Tokyo: Univ. Park Press 1976
Osmond, C.B.: Ann. Rev. Plant Physiol. *29*, 379–414 (1978)
Osmond, C.B., Allaway, W.G.: Aust. J. Plant Physiol. *1*, 503–512 (1974)
Osmond, C.B., Björkmann, O.: Aust. J. Plant Physiol. *2*, 155–162 (1975)
Queiroz, O.: Ann. Rev. Plant Physiol. *24*, 115–134 (1974)
Ranson, S.L., Thomas, M.: Ann. Rev. Plant Physiol. *11*, 81–110 (1960)
Stiller, M.L.: Ph.D. Thesis, Purdue Univ., Lafayette, Ind. (1959)
Sugiyama, T., Laetsch, W.M.: Plant Physiol. *56*, 605–607 (1975)
Sutton, B.G.: Aust. J. Plant Physiol. *2*, 377–388 (1975a)
Sutton, B.G.: Aust. J. Plant Physiol. *2*, 389–402 (1975b)
Sutton, B.G., Osmond, C.B.: Plant Physiol. *50*, 360–365 (1972)
Szarek, S.R., Johnson, H.B., Ting, I.P.: Plant Physiol. *52*, 539–541 (1973)
Szarek, S.R., Ting, I.P.: Photosynthetica *11*, 330–342 (1977)
Thomas, M.: New Phytol. *48*, 390–420 (1949)
Ting, I.P.: In: Photosynthesis and photorespiration. Hatch, M.D., Osmond, C.B., Slatyer, R.O. (eds.), pp. 169–185. New York: Wiley-Interscience 1971
Ting, I.P., Johnson, H.B., Szarek, S.R.: In: Net carbon dioxide assimilation in higher plants. Black, C.C. (ed.), pp. 26–53. Proc. Symp. S. Sect. Raleigh: Am. Soc. Plant Physiol. Cotton Inc. 1972
Ting, I.P., Osmond, C.B.: Plant Physiol. *51*, 439–447 (1973a)
Ting, I.P., Osmond, C.B.: Plant Physiol. *51*, 448–453 (1973b)
Varner, J.E., Burrell, M.: Arch. Biochem. *25*, 280–287 (1950)
Vickery, H.B.: J. Biol. Chem. *205*, 369–381 (1953)
Vieweg, G.H., Fekete, de M.A.R.: Z. Pflanzenphysiol. *81*, 74–79 (1977)
Walker, D.A.: Nature (London) *178*, 593–594 (1956b)
Walker, D.A.: Biochem. J. *67*, 73–79 (1957)
Willert, von D.J., Kluge, M.: Plant Sci. Lett. *1*, 391–397 (1973)
Winter, J., Kandler, O.: Z. Pflanzenphysiol. *78*, 103–112 (1976)
Winter, K., Lüttge, U.: In: Water and plant life. Lange, O.L., Kappen, L., Schulze, E.D. (eds.), pp. 323–332. Ecol. Studies. *19* (1976)
Wolf, J.: Planta *37*, 510–534 (1949)
Wolf, J.: In: Encyclopedia of plant physiology. Ruhland, W. (ed.), Vol. XII, pp. 809–889. Berlin, Heidelberg, New York: Springer 1960

9. CAM: Rhythms of Enzyme Capacity and Activity as Adaptive Mechanisms

O. QUEIROZ

A. Introduction

There are some explanations for the fact that circadian endogenous rhythmicity, which is a basic general property of the physiology of eukaryotes, appears as a rather disturbing phenomenon for the physiologist, both from conceptual and experimental points of view:

a) The existence of circadian rhythms discloses the annoying situation that very little is known about the inner processes which, in eukaryotic organisms, confer an adaptive ability for identifying local time and measuring the progress of a season. Nevertheless, substantial advances have been achieved in recent years on understanding the chemical and genetic mechanisms of this biological clock.

b) The ubiquity of circadian rhythms results in metabolic networks shifting continuously between repeated periodic patterns (even in the absence of external periodisms). Hence iterative measurements over 24 h are necessary if one intends to obtain a realistic description of a complex physiological function, particularly when two or more pathways are likely to be involved. In CAM, the "double carboxylation" mechanism for malate synthesis, the source of phosphoenolpyruvate (PEP), the fate of pyruvate, and the variability of $\delta^{13}C$ are probably good examples of the mixing of different pathways in different ratios according to the time of day. Attempting to integrate the effects of endogenous rhythmicity obviously makes interpretation of data and elaboration of comprehensive models more difficult.

B. Endogenous Versus Nonendogenous Rhythms: A Necessary Distinction

Eukaryotes have evolved a special class of rhythms, the so-called endogenous rhythms, which exhibit a specific, well-defined set of properties (PITTENDRIGH, 1972, 1976; BÜNNING, 1973; HASTINGS et al., 1976). It is not within the scope of this paper to analyze these properties, nevertheless some of them can be recalled to assist the following discussion on CAM rhythms:

a) endogenous rhythms persist spontaneously in the absence of external signals;

b) in contrast to "forced" light-sensitive oscillations (which can respond to any light/dark schedule) endogenous rhythms will synchronize only to a defined light/ dark range; outside this range the rhythm will keep its own period irrespective of the external signals: for instance in *Kalanchoë* the rhythm of petal movement synchronizes to external periods ranging from 18 h to 28 h, but not to 16 h nor to 30 h periods (JOHNSSON et al., 1973);

c) night-breaks by red light flashes maintain low levels of PEP carboxylase only when the flashes are given during a well-defined part of the night (Fig. 8);

d) two discrete light pulses given h hours apart produce the same effect as that obtained by a continuous light period of h hours; this result suggests special roles for dusk and dawn signals (see HILLMAN, 1976; PITTENDRIGH, 1976, for reviews), as evidenced in the control of CAM (Sect. E.II.3.).

These properties of endogenous rhythms show that (a) they depend on control mechanisms different from those controlling forced oscillations; (b) the control mechanisms involved in endogenous rhythms imply an ability to measure time; (c) endogenous rhythms could have homeostatic and adaptive roles (PITTENDRIGH and CALDAROLA, 1973; QUEIROZ, 1974).

A high degree of complexity is necessarily involved, and ultimately one is led to consider the role of biological clocks, i.e., a class of processes capable of measuring the time elapsed between two signals. It has been shown (HASTINGS, 1960; PITTENDRIGH, 1976) that this clock is separable from the overt rhythms it drives. In the case of CAM, endogenous time-measurement is particularly involved in (a) measuring the season (photoperiod) by distinguishing the time interval between dusk and dawn, and (b) programming the appropriate sequences of metabolic events.

C. Enzyme Capacity and Enzyme Activity: Two Distinct Levels of Control

Variations in the total amount of an enzyme present in a tissue (referred to as enzyme capacity, or sometimes enzyme concentration) reflect changes in the enzymic potentiality of that tissue (e.g. ROWSELL, 1962; SCRUTTON and UTTER, 1968; QUEIROZ and MOREL, 1974; QUEIROZ, 1974, 1976). Enzyme capacity should not be confused with enzyme activity (i.e., that fraction of enzyme capacity which actually operates at a given moment and which can be measured by metabolic flow), since their relaxation times (GOODWIN, 1963) are different and they involve different classes or regulatory processes (STADTMAN, 1970); for instance in CAM feedback inhibition by malate, which can actually modify the activity of PEP carboxylase, should not be invoked to explain the observed rhythms of capacity of this enzyme (see Sect. E.II.2); in contrast, these rhythms could result from rhythmical variation in the balance between synthesis and degradation of the enzyme.

Variations of the excess of capacity over activity probably have an important role in metabolism (ATKINSON, 1969; SCHIMKE, 1969; QUEIROZ, 1974); particularly in the case of CAM we assume that (a) they define the homeostatic adaptability of the tissue, and (b) their relationships to external periodic factors are governed by the circadian clock (see Sect. E).

D. Rhythms Connected with CAM

I. Components of the Malate Rhythm

a) Two distinct sets of rhythmical functions contribute to the rhythm of malate content in the leaves:

i) The rhythms of CO_2 uptake, of PEP carboxylase capacity, of PEP carboxylase activity, of malate dehydrogenase capacity and of malate dehydrogenase activity, and their specific biophysical and biochemical control mechanisms, are the components of the event "malate synthesis" represented by the functional box I in the block diagram of Fig. 1.

ii) The rhythms of CO_2 emission, of malic enzyme capacity and of malic enzyme activity, and their specific control systems, are the components of the event "malate utilization" (box III in Fig. 1).

b) The storage step (box II in Fig. 1), which is a functional intermediate between the two sets of rhythms, operates as a "delay block" (JOHNSSON and KARLSSON, 1972). Its role is fundamental in CAM timing, and its parameters control the input to box III, so that the "malate utilization" event (box III) becomes independent of the kinetics of operation of the "malate synthesis" event (box I; see comments in the next section).

Earlier results (QUEIROZ, 1966) showed that the drop in malate content during the day is linearly related to the amount of malate pooled in the leaves at the end of the night. Another important feature of this storage step is that with each cycle, under CAM-promoting conditions, the amount of malate synthesized during the night is not totally depleted during the following day: hence the leaf enlarges progressively its permanent pool of malate. When a threshold (which depends on the leaf age and growing conditions) is reached in the size of this permanent pool, changes in control mechanisms for CAM can be expected (see Sect. E.II.2 and Fig. 2).

II. CO$_2$ Uptake and CO$_2$ Output

The enzymes responsible for the peak of CO_2 uptake during the night and for the peak of CO_2 production during the morning (PEP carboxylase and malic enzyme respectively) were identified in the fifties (see for example MOYSE and JOLCHINE, 1957; WALKER, 1957; WALKER, 1960) but without mention of rhythmical properties. Control of CO_2 exchange by photoperiod with involvement of phytochrome were reported in 1954 (GREGORY et al., 1954). The rhythm of CO_2 output in response to light and temperature has been established by WILKINS since 1959 (WILKINS, 1959, 1960, 1965, 1973; HARRIS and WILKINS, 1976); these studies, which utilized CO_2-free air, provided evidence for the presence of a clock mechanism underlying this step of CAM. The endogenous rhythm of CO_2 uptake was also established (NUERNBERGK, 1961; QUEIROZ, 1970), utilizing normal air. Circadian rhythmicity of the CO_2 compensation point has been reported more recently (JONES and MANSFIELD, 1972; JONES, 1973) and appears to correlate with enzymic data. Other approaches to the regulation of CO_2 exchange, particularly with respect

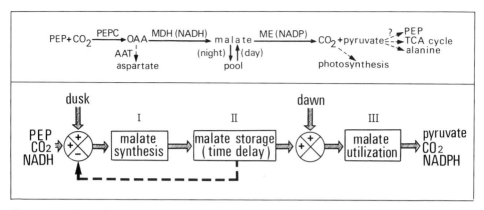

Fig. 1. *Top,* The main CAM pathway (*solid arrows*) and its associated reactions (*dashed arrows*). *Bottom,* Block diagram of the component steps of the overall night/day rhythm of malate concentration. The rhythms of blocks *I* and *III* and function of block *II* are detailed in the text, see Section D.I. Positions where dusk and dawn signals enter the model are shown (see Sect. E.II.3); *dashed arrow* indicates negative feedback (see Sect. E.II.2). *OAA,* oxaloacetate: *PEP,* phosphoenol pyruvate; *AAT,* aspartate aminotransferase; *ME,* malic enzyme; *MDH,* malate dehydrogenase; *PEPC,* PEP carboxylase; *TCA,* tricarboxylic acid cycle

to the involvement of phytochrome, are currently being studied in the laboratories of Wilkins (WILKINS and HARRIS, 1976), LaCROIX (NALBORCZYK et al., 1975), as well as in Gif-sur-Yvette.

Recent claims (WILKINS, 1973; WILKINSON and SMITH, 1976) that the rhythm of CO_2 output could be attributable to the rhythm of PEP carboxylase might be questioned since the existence of the malate storage step rules out direct kinetic connections between the nocturnal PEP carboxylase step and the production of CO_2 from the pooled malate by malic enzyme during the following day (or, alternatively, in continuous darkness). Formal correlation between the rhythm of PEP carboxylase (capacity) and the kinetics of CO_2 emission by no means implies a direct biochemical linkage: it only shows that the whole pathway is time-coherent (Sect. E.II).

III. PEP Carboxylase

1. Detecting the Rhythm of Enzyme Capacity

This is an important point attending to some contradictions found in the literature. Our experience is that a rhythm of PEP carboxylase capacity can be found in all CAM species examined so far, on condition that the following minimal experimental precautions are fulfilled:

a) Photoperiod conductive to CAM (and also, incidentally, to flower induction) favors the detection of the rhythm: maximum CAM and amplitude of the PEP carboxylase rhythm are obtained in the case of "short-day plants" (e.g., *Kalanchoe blossfeldiana*) by a sufficient, specific number of short days: in the case of "long-short-day plants" (e.g., *Bryophyllum daigremontianum*) by an appropriate combination of long days+short days; and in the case of "long-day plants" (*Sedum* sp.) by a sufficient number of long days.

b) The rhythm can also be clearly detected under unfavorable photoperiods (although at a low level and with small amplitude) if the semi-specific inhibitors of the phenolic type synthesized under these photoperiods (Sect. E.I.3) are completely eliminated during the extraction. The composition of the extraction medium depends on the leaf age and photoperiodic history. Absence of the enzyme rhythm reported in some instances may have been due to such inhibitors.

c) The objection that, in crude extracts, variations in the malate content according to time of day could result in artifactual "rhythms" of enzyme capacity is ruled out by the fact that the rhythm can also be observed after malate removal by Sephadex filtration of the extracts, and by the results obtained after mixing extracts in different proportions (QUEIROZ and MOREL, 1974a). It can also be noted that the increase of malate content in leaves, as a function of the number of CAM-promotive photocycles, corresponds to an increase (and not decrease) of the enzyme capacity measured in total extracts (QUEIROZ and MOREL, 1974a).

2. Rhythmical Patterns

a) Both synchronism and nonsynchronism between maximum capacity and maximum activity can be observed; the physiological meaning of these different patterns will be discussed in a later section (see E.II.2 and Fig. 2).

b) Dependence on photoperiod and on thermoperiod of level (mean value), amplitude, and phase of the rhythm of enzyme capacity has been established by studies with *K.blossfeldiana* (Fig. 3; GREGORY et al., 1954; QUEIROZ, 1968; BRULFERT et al., 1975) and other CAM plants: the short-day plant *K.velutina*

→

Figs. 2–9. Enzyme capacity measured in mmoles/g dry wt/h; enzyme activity and label incorporation in c.p.m./g dry wt $\times 10^{-7}$; short days=09.00 to 18.00 h, long days=09.00 to 01.00 h. Data from *Kalanchoe blossfeldiana*

Fig. 2. Synchronized (**A**) and desynchronized (**B**) rhythms of PEP carboxylase capacity and activity in leaves with small (**A**) and large (**B**) permanent pools of malate (see Sect. E.II.2)
Fig. 3. Rhythms of PEP carboxylase capacity under continuous light (*LL*), long days (*LD*) and after 26 and 40 short days (*SD*) (see Sect. E.II.1)
Fig. 4. Rhythm of malic enzyme (*ME*) capacity compared to the rhythm of malic enzyme activity (production of labeled pyruvate). *Inset:* Detailed data between 08.00 and 11.00 h (see Sect. E.II.3). Compared with label incorporation into isocitrate (see Sect. E.I.2c)
Fig. 5. Rhythms of capacity of phosphofructokinase (*PFK*) and enolase after 40 short days (see Sect. D.VII)
Fig. 6. Rhythms of incorporation of labeled acetate into citrate, succinate+α−ketoglutarate (αKG) and glutamate (see Sect. D.VIII)
Fig. 7. Profile of separation of PEP carboxylase forms by acrylamide gel electrophoresis from extracts of leaves under long days (*LD*) and short days (*SD*); (see Sect. E.I.1)
Fig. 8. Changes in the level of PEP carboxylase capacity obtained by short days (P_6) or by different positions of 10 h darkness defined by two night breaks (30 min red light; P_1 to P_5). Effects expressed as a ratio to the level of capacity in long days (P_L). Main light periods and night breaks were followed by 15 min far-red light in order to eliminate the P-730 form of phytochrome (see Sect. E.II.3)
Fig. 9. Block diagram of the path of information flow proposed for the control of CAM by dusk and dawn signals (see Sects. E.II.2 and E.II.3 for discussion)

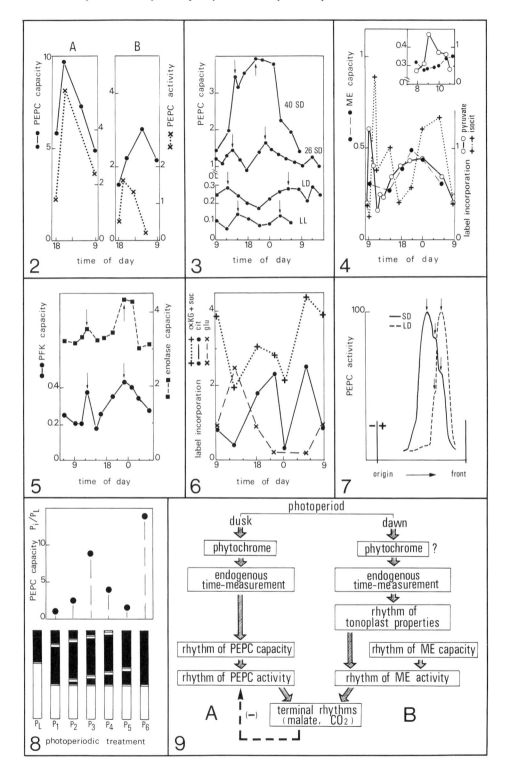

shows a similar behavior; in the case of the long-short-day plant *B. daigremontianum* after completion of the juvenile phase (10–12 weeks under long days) a further treatment of about 35 short days (thermoperiod 22° C/12° C) brings the rhythm to its highest level and amplitude; in contrast the rhythm is hardly detectable in continuous short days if a previous long-day treatment of the proper duration has not been given. *B. tubiflorum*, another long-short-day plant, shows a similar behavior. In *Sedum telephium*, which has strict long-day requirement for flowering, the rhythm of PEP carboxylase capacity is two to six times higher in long days than in short days.

c) Phytochrome mediates photoperiodic control of CAM (GREGORY et al., 1954; QUEIROZ, 1968; BRULFERT et al., 1975; Fig. 8).

d) An after-effect of the CAM promotive short-day treatment was shown in *K. blossfeldiana* when the plants were returned to noninductive photoperiods (BRULFERT et al., 1975; BRULFERT et al., 1976).

e) In full-CAM operation (NEALES, 1975) in *K. blossfeldiana* the peak of the rhythm of enzyme capacity occurs during the night. Recent more detailed analysis (12–14 measurements per cycle) shows the presence of a more or less distinct shoulder, or small peak occurring during the day (Fig. 3). In long days, as in continuous light, two fairly equivalent peaks can be detected per 24-h cycle (Fig. 3; see Sect. E.II.1 for discussion).

f) Leaf age modifies the mean level and the amplitude of the rhythm of enzyme capacity, but not the phase: under entraining (photoperiodic) conditions the rhythm is synchronized for the whole plant.

IV. Malate Dehydrogenase

On the assumption that the very high capacity of malate dehydrogenase makes it unsuitable for fine temporal regulation of the CAM pathway, studies of its rhythm have received little attention. Nevertheless, a better knowledge of the regulatory role of balances between enzyme capacities may lead to a modification of this point of view.

V. Malic Enzyme (NADP)

Malic enzyme (NADP-dependent) is responsible for the kinetics of malate decarboxylation in a broad category of CAM plants (DITTRICH et al., 1973).

a) The rhythm of capacity shifts much faster than the rhythm of PEP carboxylase capacity (QUEIROZ et al., 1972; QUEIROZ and MOREL, 1974a): this suggests that different "endogenous oscillators" could be involved (see Sect. E.II.3).

b) The rhythm of activity, measured by pyruvate production from labeled malate fed to leaves, has a complex profile (Fig. 4; MOREL, 1976): the first main peak is always locked to the "lights-on" signal in day/night conditions (this massive pyruvate production during the first hours of light corresponds to the well-known peak of CO_2 production in CAM); the second main peak is synchronized to the peak of enzyme capacity and shifts with it (see Sect. E.II.3).

VI. Aspartate Aminotransferase

Notwithstanding that aspartate is one of the first metabolites derived from CO_2 fixation by PEP carboxylase, the oxaloacetate → aspartate branch is often neglected by authors on the basis that the aminotransferase activity involved is much lower than that of malate dehydrogenase. Data obtained with *K. blossfeldiana* show that (1) the capacity of aspartate aminotransferase is always much higher than PEP carboxylase capacity (2) both increase exponentially under short days showing an allometric linkage, and (3) aspartate aminotransferase capacity exhibits a 24 h-period rhythm, the phase of which is correlated to that of the rhythm of PEP carboxylase capacity (MOREL et al., 1972; QUEIROZ et al., 1972; QUEIROZ and MOREL, 1974 a). These facts obviously bear physiological significance. Even if the actual carbon flow through aspartate is low relative to that through malate, the possibility that it plays a role in amino acid or in nucleotide synthesis connected to CAM cannot be discarded.

VII. Enzymes of the Glycolytic Pathway

Coordination between temporal changes in CAM and in glycolysis is currently being investigated in this laboratory at the levels of enzyme capacity and activity. Light/dark variations of capacity of phosphofructokinase, aldolase, 3-P-glyceralde-hyde dehydrogenase, P-glycerate mutase and enolase have been followed during long-day and short-day treatments in the leaves of *K. blossfeldiana* (Fig. 5). Except for aldolase, strict allometry was shown between the increase in capacity of these enzymes and that of PEP carboxylase under short days (PIERRE and QUEIROZ, 1972; PIERRE and QUEIROZ, 1979). Recent results established that the mechanisms controlling glycolysis are different depending on the level of CAM (PIERRE and QUEIROZ, 1979).

VIII. Tricarboxylic Acid Cycle

1. ^{14}C-Acetate or ^{14}C-pyruvate are incorporated into intermediates of the TCA cycle in leaves of *K. blossfeldiana* according to a bi-modal 24-h rhythm (Fig. 6; MOREL, 1976). Shifting of the rhythm suggests looseness or absence of entrainment by photoperiod. Interestingly, labeling of glutamate presents a single peak per 24-h cycle (Fig. 6).

2. ^{14}C-Malate is incorporated into the TCA cycle in a day/night pattern which strictly conforms to the pattern of labeled pyruvate production by malic enzyme (Fig. 4; MOREL, 1976).

E. Working Hypothesis and Models

I. Seasonal Adaptation

1. Isofunctional Forms of PEP Carboxylase and Photoperiodism

A significant finding by BRULFERT (unpublished) was that transferring *K. blossfeldiana* from long days to short days modifies the PEP carboxylase isozymic pattern obtained from DEAE cellulose chromatography, electrofocusing, and acrylamide

gel electrophoresis (Fig. 7). This problem deserves further research on the characteristics and localization of the enzymic forms which respond to photoperiodic
control.

2. Metabolic Flexibility

a) Optional patterns of CO_2 fixation are a basic aspect of CAM flexibility (BLACK,
1973; ALLAWAY et al., 1974; OSMOND and ALLAWAY, 1974; OSMOND and BJÖRKMAN,
1975). These results, taken together with evidence for diversity in CAM enzymic
patterns (EDWARDS et al., 1971; DITTRICH et al., 1973; DITTRICH, 1976), imply
that metabolic events can follow flexible programs, to which rhythmicity is able
to provide the necessary time-measurement frame. Isotope discrimination analysis,
repeated at different appropriate moments in combination with different entraining
conditions during growth (SUTTON and OSMOND, 1972; BENDER et al., 1973; LERMAN
and QUEIROZ, 1974; OSMOND et al., 1976), can provide information on the history
of the plant. A deeper insight into the timing of the contributions from different
CO_2-donors could then be achieved, as suggested by current investigations in
MOYSE's laboratory (LERMAN et al., 1974; DELEENS and GARNIER-DARDART, 1977).

b) Resistance to drought is considered to be a basic feature of CAM (see
for example reviews by TING, 1971 and TING et al., 1972), both in the case of
cold desert plants as reported by MEDINA and DELGADO (1976), and in the case
of hot desert as shown in *Opuntia* (SZAREK and TING, 1974, 1975). Results by
SZAREK and TING (1974, 1975) suggest that in this plant the external annual variation in water potential could be the primary entraining agent for the annual variation of CAM activity, which is high during the winter (short-day) season. The
question arises whether there is in this material a photoperiodically controlled
process (e.g., variation in PEP carboxylase capacity) providing the organism with
a basis for measuring the advance of seasons: if it is the case, one could speculate
on a selectively evolved and seasonally coherent multi-level control (water status,
enzymic capacity) affording maximum metabolic efficiency under favorable environmental conditions.

c) Incorporation of carbon chains into the TCA cycle, resulting from pyruvate
production by malic enzyme (Fig. 4) shows that:
i) the activity of malic enzyme is a timer for the activity of the TCA cycle during
full-CAM operation (MOREL, 1976).
ii) CAM has an anaplerotic role towards the TCA cycle; the significance of this
role of CAM is obvious during the seasonal strong morphogenetic activity of
the plant (flowering) which coincides with full-CAM activity in all species examined
so far.

3. Balance Between CAM and Synthesis of Phenols: A Regulatory Process?

Substances of the tannin type extracted from leaves of *K. blossfeldiana* under long
days can inhibit PEP carboxylase, malate dehydrogenase, and aspartate aminotransferase, but not malic enzyme or several other non-CAM enzymes (BRULFERT et
al., 1973). The content of these substances varies diurnally. Transfer of plants
to short days results in a rapid decrease of their levels. According to the kinetics

of these processes, taken together with the inhibitory properties, the hypothesis that photoperiod might regulate PEP distribution between the CAM pathway and the shikimate pathway is favored.

II. Timing CAM

1. Entrained and Nonentrained Rhythms of PEP Carboxylase

Different profiles of PEP carboxylase rhythms can be observed in *K. blossfeldiana* under either continuous light, long days or short days (Fig. 3): the rhythm of enzyme capacity presents two peaks per 24 h under constant conditions and under long days, but is changed into a circadian rhythm after transfer to short days. Rhythms with apparent periods of about 12–15 h have been reported, particularly in *Chenopodium* (FROSCH and WAGNER, 1973; DEITZER et al., 1974; WAGNER, 1976), for the capacity of several enzymes under constant (nonentraining) conditions. In the case shown in Figure 3 the two peaks of PEP carboxylase clearly respond differently to entrainment by photoperiod: in long days the rhythm behaves as in constant light, but after transfer to short days one of the peaks increases markedly (the other peak finally appears only as a more or less distinct shoulder), and this increase coincides with that of the dark CO_2 fixation. In addition, the interval between the two peaks is not constant. In connection with these results, it is important to recall the development in short days of a particular form of the enzyme (Fig. 7; Sect. E.I.1).

In conclusion, the results of entraining the rhythm by different photoperiods and the corresponding isozymic patterns suggest the existence of two different rhythmic populations of PEP carboxylase, both circadian but differently phased, and exhibiting different behavior in respect to photoperiod. One of the forms could be specific to CAM.

2. Feedback Control

a) Significance and Limitations

In earlier explanations for the rhythmicity of PEP carboxylase (QUEIROZ, 1967, 1968) it was assumed that a feedback inhibition by accumulated malate caused a drop in PEP carboxylase activity until the malate was utilized by malic enzyme the following morning (or, in prolonged darkness, during an endogenous cyclic increase in malic enzyme activity). However, observed rhythms of total PEP carboxylase were erroneously confused with actual enzyme activity in vivo, an error which is still present in discussions on CAM: data on rhythms of enzyme capacity are often referred to as if they represent rhythms of activity. In future, a distinction between the two levels of organization should be clearly made.

The author is still firmly convinced that feedback inhibition of PEP carboxylase by malate is an important component of CAM regulation, at least under well-defined ranges of age, temperature, and photoperiod. But the author cannot agree with claims that control of PEP carboxylase by malate can by itself explain the day/night rhythm of CAM. For instance, the feedback hypothesis cannot explain the following observations: (1) the rhythms of PEP carboxylase capacity; (2) the rhythms of capacity of enzymes other than PEP carboxylase, their coherent phase behavior and their allometric variations (Queiroz and Morel, 1974a); (3) the exponential increase in the level of PEP carboxylase in short days, and the quantitative negation of this effect by appropriate night-break schedules (see Gregory et al., 1954; Queiroz, 1968; Fig. 8); (4) the fact that the minimum malate content (towards the middle of the day) corresponds to low levels of activity (and capacity) of PEP carboxylase.

Points (2) and (3) are typical of systems controlled by the so-called clock mechanism.

b) Prospects

A clear understanding of the physiological role of feedback inhibition by malate as a timer for the PEP carboxylase step implies answering the following questions: (1) is feedback inhibition a necessary component of CAM cycling or does it intervene only in a limited range of cases or conditions (age, environment)? (2) how can its actual intervention be experimentally detected? (3) how might feedback control be combined with other control mechanisms operating during various times of the 24-h cycle? Morel and Queiroz (1974) and Queiroz and Morel (1974b) have contributed to answering questions (1) and (2) by showing that the rhythms of capacity and activity of PEP carboxylase can be either synchronized or desynchronized. The rationale is that synchronism (and direct proportionality) can be expected between the variations in enzyme capacity and the variations in enzyme activity as long as no specific external or internal control influences the latter. Such a direct relationship appears to be the case for the data shown in Figure 2A: increase and decrease in PEP carboxylase activity appear to be governed by the rhythm of enzyme capacity. In contrast, the curves in Figure 2B are clearly desynchronized; data in this case comes from leaves which had built up a large permanent pool of malate: it can be supposed that in such leaves feedback inhibition begins early in the night so that enzyme activity drops, although the capacity keeps on increasing.

An interesting approach to the role of malate in the control of enzyme activity (Queiroz, 1967; Kluge and Osmond, 1972; Garnier-Dardart and Queiroz, 1974), as developed by the Darmstadt group (Lüttge and Ball, 1974; Lüttge et al., 1975; Kluge, 1976), is to consider changes in membrane conformation during the dark/light transition. The involvement of membranes in the biological clocks is under extensive investigation; but even in the most elaborated membrane hypothesis (Njus et al., 1974) one is ultimately led to consider enzyme rhythms. The assumption that membrane-mediated changes in malate availability are responsible for the peak of malic enzyme activity is ruled out by Morel's feeding experiments (Fig. 4): labeled malate supplied at different times before and after lights-on, or during continuous darkness, was shown to reach the locality of the malic enzyme at all times, but nevertheless the curve for its decarboxylation to pyruvate is similar to that for CO_2 evolution by nontreated plants. It can be concluded that a factor other than malate availability is responsible for the usual pattern of malic enzyme operation.

3. Dusk and Dawn Signals

a) The lights-off signal (Fig. 1) appears to affect the rhythm of PEP carboxylase capacity, as evidenced with different CAM species by experiments in which shifting solely this signal (changing from long days to short days) results in changes in the level (Figs. 3 and 8) and phase of the rhythm of capacity (Fig. 3; QUEIROZ et al., 1972; QUEIROZ, 1968, 1976; BRULFERT et al., 1975). The fact that the time at which a night-break (30 min red light) is given determines its depressive effect on the capacity of the enzyme (Fig. 8) establishes that an endogenous rhythm of sensitivity to light, i.e., a system in which the response depends on the coincidence between the light input and a specific phase of the clock, (see PITTENDRIGH, 1972; BÜNNING, 1973) controls the capacity of PEP carboxylase. In conclusion, the path of information flow between the external agent (photoperiod) and the PEP carboxylase activity appears to be mediated by components as schematized in Figure 9A.

b) The lights-on signal (Fig. 1) clearly entrains malic enzyme activity, but not its capacity (Fig. 4). A first peak of activity is locked to the dawn signal irrespective of the time of peak capacity; the fact that this peak of activity also occurs under continuous darkness, together with properties established for the corresponding rhythm of CO_2 output (Sect. D.II), indicate that the underlying mechanism is of the clock type and not merely a "forced" effect of light. In conclusion, the path of information flow differs from that proposed for the dusk signal and PEP carboxylase: as schematized in Figure 9B, the dawn signal appears to influence the level of malic enzyme activity (without changing its capacity), probably through effects resulting from changes in clock-controlled rhythms of tonoplast properties (see Sect. E.II.2).

c) CAM therefore appears to be entrained by the combined effect of two signals (dusk and dawn) driving two different clock-controlled oscillators (the PEP carboxylase step and the malic enzyme step respectively):

— day/night entrainment of CAM (i.e., synchronization of the metabolic program to the 24-h cycle) would be obtained by coupling the rhythm of malic enzyme activity to the dawn signal;

— measure of the season (i.e., sensitivity to photoperiod) would be obtained by coupling the rhythm of PEP carboxylase capacity to the dusk signal, affording a reliable basis for adaptive behavior: depending on the phase of the endogenous clock at which dusk occurs, the capacity and (at least in *K. blossfeldiana*) the isozymic pattern of PEP carboxylase are changed.

References

Allaway, W.G., Osmond, C.B., Troughton, J.H.: In: Mechanisms of regulation of plant growth. Bieleski, R.L., Ferguson, A.R., Cresswell, M.M. (eds.), pp. 195–202. Wellington: Bull. 12, R. Soc. N.Z. 1974

Atkinson, D.E.: In: Current topics in cellular regulation. Horecker, B.L., Stadtman, E.R. (eds.), Vol. 1, pp. 29–43. New York-London: Academic Press, 1969

Bender, M.M., Rouhani, I., Vines, H.M., Black, C.C.: Plant Physiol. *52*, 427–430 (1973)

Black, C.C.: Ann. Rev. Plant Physiol. *24*, 253–286 (1973)

Brulfert, J., Guerrier, D., Queiroz, O.: Plant Physiol. *51*, 220–222 (1973)
Brulfert, J., Guerrier, D., Queiroz, O.: Planta *125*, 33–44 (1975)
Brulfert, J., Imhoff, C., Fontaine, D.: In: Etudes de biologie végétale. Hommage au Professeur P. Chouard. Jacques, R. (ed.), pp. 443–455. Paris 1976
Bünning, E.: The physiological clock. Berlin, Heidelberg, New York: Springer 1973
Deitzer, G.F., Kempf, O., Fischer, S., Wagner, E.: Planta *117*, 29–41 (1974)
Deleens, E., Garnier-Dardart, J.: Planta *135*, 241–248 (1977)
Dittrich, P.: Plant Physiol. *57*, 310–314 (1976)
Dittrich, P., Campbell, W.H., Black, C.C.: Plant Physiol. *52*, 357–361 (1973)
Edwards, G.E., Kanai, R., Black, C.C.: Biochem. Biophys. Res. Commun. *45*, 278–285 (1971)
Frosch, S., Wagner, E.: Can. J. Bot. *51*, 1521–1528 (1973)
Garnier-Dardart, J., Queiroz, O.: Phytochemistry *13*, 1695–1702 (1974)
Goodwin, B.C.: Temporal organization in cells. p. 10, 11, 13, 94–96. London, New York: Academic Press 1963
Gregory, F.G., Spear, I., Thimann, K.V.: Plant Physiol. *29*, 220–229 (1954)
Harris, P.J.C., Wilkins, M.B.: Planta *129*, 253–258 (1976)
Hastings, J.W.: Cold Spring Harbor Symp. Quant. Biol. *25*, 131–143 (1960)
Hastings, J.W., Aschoff, J.W.L., Bünning, E., Edmunds, L.N., Hoffmann, K., Pittendrigh, C.S., Winfree, A.T.: In: The molecular basis of circadian rhythms. Hastings, J.W., Schweiger, H.G. (eds.), pp. 49–62. Berlin: Dahlem Konferenzen 1976
Hillman, W.S.: Ann. Rev. Plant Physiol. *27*, 159–179 (1976)
Johnsson, A., Karlsson, H.G.: J. Theor. Biol. *36*, 153–174 (1972)
Johnsson, A., Karlsson, H.G., Engelmann, W.: Physiol. Plant. *28*, 134–142 (1973)
Jones, M.B.: Ann. Bot. *37*, 1027–1034 (1973)
Jones, M.B., Mansfield, T.A.: Planta *103*, 134–146 (1972)
Kluge, M.: In: CO_2 metabolism and productivity in plants. Proc. 54th Steenbock Symp. Burris, R.H., Black, C.C. (eds.), pp. 205–216, 1976
Kluge, M., Osmond, C.B.: Z. Pflanzenphysiol. *66*, 97–105 (1972)
Lerman, J.C., Deleens, E., Nato, A., Moyse, A.: Plant Physiol. *53*, 581–584 (1974)
Lerman, J.C., Queiroz, O.: Science *183*, 1207–1209 (1974)
Lüttge, U., Ball, E.: Z. Pflanzenphysiol. *73*, 326–338 (1974)
Lüttge, U., Kluge, M., Ball, E.: Plant Physiol. *56*, 613–616 (1975)
Medina, E., Delgado, M.: Photosynthetica *10*, 155–163 (1976)
Morel, C.: In: Etudes de biologie végétale. Hommage au Professeur P. Chouard. Jacques, R. (ed.), pp. 457–466. Paris 1976
Morel, C., Celati, C., Queiroz, O.: Physiol. Vég. *10*, 743–763 (1972)
Morel, C., Queiroz, O.: J. Interdiscip. Cycle Res. *5*, 206–217 (1974)
Moyse, A., Jolchine, G.: Bull. Soc. Chim. Biol. *39*, 725–745 (1957)
Nalborczyk, E., LaCroix, L.J., Hill, R.D.: Can. J. Bot. *53*, 1132–1138 (1975)
Neales, T.F.: In: Environmental and biological control of photosynthesis. Marcelle, R. (ed.), pp. 299–310. The Hague: Dr. W. Junk 1975
Njus, D., Sulzman, F.M., Hastings, J.W.: Nature (London) *248*, 116–120 (1974)
Nuernbergk, E.L.: Planta *56*, 28–70 (1961)
Osmond, C.B., Allaway, W.G.: Aust. J. Plant Physiol. *1*, 503–511 (1974)
Osmond, C.B., Bender, M.M., Burris, R.H.: Aust. J. Plant Physiol. *3*, 787–799 (1976)
Osmond, C.B., Björkman, O.: Aust. J. Plant Physiol. *2*, 155–162 (1975)
Pierre, J.N., Queiroz, O.: C.R. Acad. Sci (Paris) *275*, 2881–2884 (1972)
Pierre, J.N., Queiroz, O.: Planta *144*, 143–151 (1979)
Pittendrigh, C.S.: Proc. Natl. Acad. Sci. (USA) *69*, 2734–2737 (1972)
Pittendrigh, C.S.: In: The molecular basis of circadian rhythms. Hastings, J.W., Schweiger, H.G. (eds.), pp. 11–48. Berlin: Dahlem Konferenzen 1976
Pittendrigh, C.S., Caldarola, P.C.: Proc. Natl. Acad. Sci. USA *70*, 2697–2701 (1973)
Queiroz, O.: Physiol. Vég. *4*, 323–339 (1966)
Queiroz, O.: C.R. Acad. Sci (Paris) *265*, 1928–1931 (1967)
Queiroz, O.: Physiol. Vég. *6*, 117–136 (1968)
Queiroz, O.: Physiol. Vég. *8*, 75–110 (1970)
Queiroz, O.: Ann. Rev. Plant Physiol. *25*, 115–134 (1974)
Queiroz, O.: Physiol. Vég. *14*, 629–639 (1976)

Queiroz, O., Celati, C., Morel, C.: Physiol. Vég. *10*, 765–781 (1972)
Queiroz, O., Morel, C.: Plant Physiol. *53*, 596–602 (1974a)
Queiroz, O., Morel, C.: J. interdiscipl. Cycle Res. *5*, 217–222 (1974b)
Rowsell, E.V.: In: Methods in enzymol. *5*, 685–697 (1962)
Schimke, R.T.: In: Current topics in cellular regulation. Horecker, B.L., Stadtman, E.R. (eds.), Vol. 1, pp. 77–124. New York, London: Academic Press 1969
Scrutton, M.C., Utter, M.F.: Ann. Rev. Biochem. *37*, 249–302 (1968)
Stadtman, E.R.: In: The enzymes. Boyer, P.D. (ed.), Vol. 1, pp. 397–459. New York, London: Academic Press 1970
Sutton, B.G., Osmond, C.B.: Plant Physiol. *50*, 360–365 (1972)
Szarek, S.R., Ting, I.P.: Plant Physiol. *54*, 76–81 (1974)
Szarek, S.R., Ting, I.P.: Am. J. Bot. *62*, 602–609 (1975)
Ting, I.P.: In: Photosynthesis and photorespiration. Hatch, M.D., Osmond, C.B., Slatyer, R.O. (eds.), pp. 169–185. New York: Wiley Interscience 1971
Ting, I.P., Johnson, H.B., Szarek, S.R.: In: Net carbon dioxide assimilation in higher plants. Black, C.C. (ed.), pp. 26–53. Proc. Symp. S. Sect. Am. Soc. Plant Physiol., Cotton Inc. Raleigh, N.C. 1972
Wagner, E.: J. Interdiscip. Cycle Res. *7*, 313–332 (1976)
Walker, D.A.: Biochem. J. *67*, 73–79 (1957)
Walker, D.A.: Biochem. J. *74*, 216–223 (1960)
Wilkins, M.B.: J. Exp. Bot. *10*, 377–390 (1959)
Wilkins, M.B.: J. Exp. Bot. *11*, 269–288 (1960)
Wilkins, M.B.: In: Circadian clocks, Aschoff, J. (ed.), pp. 145–163. Amsterdam: North Holland 1965
Wilkins, M.B.: J. Exp. Bot. *24*, 488–496 (1973)
Wilkins, M.B., Harris, P.J.C.: In: Light and plant development. Smith, H. (ed.), pp. 399–417. London: Butterworths 1976
Wilkinson, M.J., Smith, H.: Plant Sci. Lett. *6*, 319–324 (1976)

10. $\delta^{13}C$ as an Indicator of Carboxylation Reactions

J.H. TROUGHTON

A. Introduction

The nomenclature $\delta^{13}C$ describes quantitatively the phenomenon of fractionation of carbon isotopes. Discrimination between isotopes emphasizes the differences in thermodynamic and kinetic properties of the isotopic species and generally assumes the chemical behavior of the species is similar. UREY (1947) advanced the theory of isotope fractionation in a paper on the thermodynamic properties of isotopic substances, and this work predicted that fractionation would occur between the two stable isotopes of carbon, ^{12}C and ^{13}C. Atmospheric CO_2 is a mixture of the stable isotopes $^{12}C^{16}O^{16}O$, $^{13}C^{16}O^{16}O$ and $^{12}C^{18}O^{18}O$ and fractionation between the stable carbon isotopes can be expected during the assimilation of CO_2 in photosynthesis.

As reviewed in this chapter, the carboxylation reactions of photosynthesis are the major sites of carbon isotope fractionation in higher plants. This phenomenon can be used to identify particular carboxylation reactions in plants and to study heavy-atom isotope effects associated with enzymes. The results of these studies have now been extended to disciplines as diverse as paleoecology, animal physiology, agronomy and medicine.

B. Carbon Isotope Fractionation and Its Measurement

Carbon in atmospheric CO_2 is 1% ^{13}C. The difficulty in detecting small differences in isotope abundance in samples is overcome by expressing the isotope ratio as the difference in isotope ratio between the sample and a standard (CRAIG, 1957). The standard is calcite in a specimen of *Belemnitella americana* found in the Peedee Formation in South Carolina, United States of America, abbreviated to PDB.

Precise determination of the carbon isotope ratio became possible with the development of special mass spectrometric techniques (NIER, 1947; MCKINNEY et al., 1950). The nomenclature is now standardised and expressed as the delta value (δ) in parts per thousand ($^0/_{00}$).
Thus:

$$\delta^{13}C\ (^0/_{00}) = \left[\frac{^{13}C/^{12}C\ \text{sample} - {}^{13}C/^{12}C\ \text{standard}}{^{13}C/^{12}C\ \text{standard}} \right] \times 1000$$

δ is the per millilitre deviation of the sample from the PDB standard. Most plant $\delta^{13}C$ values are negative, indicating that with respect to the standard, the material is depleted in the heavier of the two isotopes.

C. Variation in $\delta^{13}C$ Values Between Plants

Atmospheric CO_2 has a $\delta^{13}C$ value between -6.4 and $-7.0^0/_{00}$ (CRAIG, 1953; CRAIG, 1954; KEELING, 1958) unless there is restricted air circulation as in glasshouses, forests, caves, the soil or in crops with intense photosynthetic or respiratory activity (LERMAN and TROUGHTON, 1975). In aquatic environments the temperature and pH dependence of the equilibrium reaction between CO_2 and bicarbonate in solution complicates the $\delta^{13}C$ value of the carbon source (DEGENS, 1969). CO_2 for photosynthesis is derived from the atmosphere, therefore the usefulness of the isotope technique depends on a relatively stable $\delta^{13}C$ value of the atmosphere throughout time. Fluctuations may occur in the isotope ratio, but they should be small compared with the generation of the biological entities (TROUGHTON, 1971b). Almost all the plant samples that have been measured have been depleted in ^{13}C compared with the atmosphere (CRAIG, 1953; BENDER, 1968; TROUGHTON, 1971a; TROUGHTON et al., 1971; BENDER, 1971; TROUGHTON et al., 1974). There are numerous possible causes for variation in carbon isotope ratios among plants and many of these are discussed in the following sections.

I. Discrimination Caused by the Photosynthetic Pathway

Extensive measurements of $\delta^{13}C$ values from an extremely wide range of plant species growing in diverse habitats indicate two major categories of plants, C_3 plants with values from -22 to $-40^0/_{00}$ and C_4 plants with values from -9 to $-19^0/_{00}$ (TROUGHTON, 1971b; BENDER, 1971; SMITH and EPSTEIN, 1971). The mean value for C_3 plants was $-27.8 \pm 2.75^0/_{00}$ and for C_4 plants $-13.56 \pm 1.62^0/_{00}$ (TROUGHTON et al., 1974). These results suggest the major source of variation in $\delta^{13}C$ values is associated with the primary carboxylation events, and this conclusion is confirmed by results from CAM plants which can vary in their isotope ratio depending on which carboxylation enzyme is involved.

II. Variation in $\delta^{13}C$ Values Between Plant Varieties and Species

TROUGHTON (1972) compared wheat lines from the primitive diploid wild *Triticum boeoticum* to the modern hexaploid *Triticum spelta* and *Triticum aestivum* (var. Gabo). Seedlings had similar $\delta^{13}C$ values, but in older plants there was some variation apparently associated with the vegetative or reproductive state of the plants. Five varieties of each of six commercial crop plants were also measured and the $\delta^{13}C$ value only varied by about $2^0/_{00}$.

There now exists extensive documentation of $\delta^{13}C$ values for numerous plants which can assist plant taxonomists to evaluate the distribution of C_3, C_4 or CAM pathways among different plant genera. This technique has been used in phytogenetic studies in families such as Agrinonae, Cyperaceae, Asteraceae, Bromeliaceae and Euphorbiaceae (Lerman and Raynal, 1972; Smith and Brown, 1973; Medina and Troughton, 1974; Smith and Turner, 1975). The $\delta^{13}C$ technique has also been used to distinguish between C_3 and C_4 plants within genera which have both plant types, e.g., *Atriplex, Cyperus, Bassia, Kochia, Euphorbia* and *Panicum* (Troughton et al., 1971; Hatch et al., 1972; Smith and Robbins, 1974; Pearcy and Troughton, 1974).

III. Variation in $\delta^{13}C$ Values Within a Plant

Small differences of up to $3^0/_{00}$ have been measured in tissue from different plant parts (Park and Epstein, 1961; Troughton, 1972). Differences in $\delta^{13}C$ values of up to $4^0/_{00}$ have been measured between rings in the same tree growing under natural conditions (Craig, 1954; Jansen, 1963; Farmer and Baxter, 1974; Freyer and Wiesberg, 1974; Pearman, 1976). Variation in the chemical composition of the ring may explain some of the differences, as the $\delta^{13}C$ value of cellulose and lignin can be as much as $3^0/_{00}$ different (Freyer and Wiesberg, 1974). Climatic variation between seasons or years may also be involved which will be of significance to paleoecological studies.

Large variations in $\delta^{13}C$ values between chemical constituents within the same plant have been measured (Abelson and Hoering, 1961; Park and Epstein, 1961; Whelan et al., 1970). Differences of up to $10^0/_{00}$ between lipids and cellulose occur in the same plant. These differences must be considered if differential preservation of cellular components is likely to occur, as in fossil specimens.

IV. Fractionation Associated with Carboxylation Enzymes

In vitro measurements of carbon isotope fractionation confirm the importance of the enzymes associated with the primary carboxylation events as being a major site of fractionation. For RuBP carboxylase Christeller et al. (1976) measured a value of $-28.3^0/_{00}$ which confirmed a corrected earlier estimate of Park and Epstein (1960). This value is close to the mean value for whole C_3 plants.

The fractionation caused by enzyme phosphoenolpyruvate (PEP) carboxylase in vitro is about $-9^0/_{00}$ compared with a mean value for C_4 plants of $-13.56^0/_{00}$ (Whelan et al., 1973; Troughton et al., 1974). The closeness of the two values indicates the importance of the enzyme in determining the value in vivo. Whelan et al. (1973) suggest the discrepancy could be due to CO_2 directly entering the bundle sheath cells, and thereby by-passing PEP carboxylase in the mesophyll cells.

The carbon isotope fractionation technique has provided evidence for a new carboxylation pathway in *Chlorobium* (Sirevag et al., 1977).

V. Compartmental Organisation and Isotope Fractionation

The presence, absence or proportion of the particular enzymes, RuBP or PEP carboxylase does not in itself determine the carbon isotope ratio as it is affected by the organisation of the enzymes and their role in the plant. It is the primary recipient of CO_2 from the atmosphere that is the major cause of the difference in $\delta^{13}C$ value between the atmosphere and C_3, C_4 and CAM plants.

The ^{13}C discriminated against during the carboxylation event must be able to escape the site of the reaction if fractionation is to occur. The reaction must therefore occur in an "open" compartment. This explains fractionation in C_3 plants, but in C_4 plants the RuBP carboxylase apparently does not cause an additional fractionation after the initial one caused by PEP carboxylation. If, as expected, RuBP carboxylase is fixing CO_2 in the bundle sheath cells, then this must be a closed compartment (BERRY and TROUGHTON, 1974). Another explanation involves an effect of high CO_2 concentration in the bundle sheath cells on fractionation, but it is not due to CO_2 concentration as an in vitro analysis indicated that fractionation by RuBP carboxylase is unaffected by CO_2 concentration (CHRISTELLER et al., 1976).

Isotope fractionation by CAM plants also supports the compartmental hypothesis. CO_2 is fixed in the C_4 acids by PEP carboxylase in the dark and stored as acid. In the light CO_2 is released from the acid, fixed by RuBP carboxylase and subsequently enters the normal carbon cycle (LERMAN et al., 1974). Plants fixing CO_2 in the dark have a C_4-like $\delta^{13}C$ value, while those fixing CO_2 in the light have a C_3-like value (ALLAWAY et al., 1974). The decarboxylation of dark-fixed CO_2 must occur in a closed compartment if there is no evidence of discrimination due to RuBP carboxylase.

Artificial tests of the compartment theory support it as a concept to explain isotope fractionation under some conditions. The difference in $\delta^{13}C$ value between C_3 and C_4 plants in an open system was $13^{0}/_{00}$, whereas there was no difference between these plants in a closed system (BERRY and TROUGHTON, 1974). NALBORCZYK and co-workers (1975) grew *Kalanchoë* with CO_2 supplied either solely in the dark or light, or in both dark and light. When CO_2 was applied solely in the dark the $\delta^{13}C$ reflected PEP carboxylase fractionation, whereas when CO_2 was provided in the light it reflected the $\delta^{13}C$ value of RuBP carboxylase.

The $\delta^{13}C$ value primarily reflects the proportion of atmospheric CO_2 fixed by either RuBP or PEP carboxylase. The validity of this hypothesis has major consequences in the interpretation of $\delta^{13}C$ values in both higher plants and throughout the carbon cycle in nature. Three examples of the application of these results are discussed. HATCH et al. (1972) found no variation among plants with the different isoenzymes of PEP carboxylase. Numerous hybrids between C_3 and C_4 species in the *Atriplex* genus have now been analysed for their $\delta^{13}C$ value, and in all cases the results indicate the C_3 pathway is operating, even though in some plants there are high levels of PEP carboxylase (BJÖRKMAN et al., 1973). Presumably the structural organisation necessary for effective co-operation between PEP and RuBP carboxylase based systems had broken down. It has been suggested that *Mollugo verticillata* is a natural hybrid between C_3 and C_4 plants (KENNEDY and LAETSCH, 1974) but $\delta^{13}C$ measurements indicate RuBP carboxylase is a major species involved in carboxylation (TROUGHTON et al., 1974).

VI. Respiration

Carbon dioxide evolved from cells in the vicinity of photosynthetic reactions will influence the $\delta^{13}C$ value of the plant material if there is the possibility that it is recycled or if there is fractionation during the decarboxylation process. The $\delta^{13}C$ value of CO_2 evolved in the dark is generally similar to the plant tissue (BAERTSCHI, 1953; PARK and EPSTEIN, 1960; SMITH, 1972; TROUGHTON, 1972). It is experimentally extremely difficult to measure unequivocally the $\delta^{13}C$ value of CO_2 evolved in the light under normal physiological conditions in higher plants. TROUGHTON (1972) and HSU and SMITH (1972) have suggested the CO_2 evolved from a leaf or shoot tissue in the light in both monocotyledon and dicotyledon C_3 plants can be up to $12^0/_{00}$ enriched in ^{13}C compared with the bulk of the plant tissue. It is even more difficult to collect sufficient CO_2 from C_4 plants in the light, but the same trend to enrichment of ^{13}C in the respired CO_2 was confirmed by TROUGHTON (1974). Reassimilation of some of the evolved CO_2 would explain the enrichment in ^{13}C of the respired CO_2 in both plant types.

D. Environmental Effects on the $\delta^{13}C$ Value of Plants

I. Temperature

In C_3 and C_4 plants TROUGHTON (1972), SMITH et al. (1973), and TROUGHTON and CARD (1975a) found a small tendency to more negative $\delta^{13}C$ values with increasing temperature over the range 15–35° C. The results did not support a major effect as predicted theoretically by LIBBY (1972). There may be a more significant, yet small, trend in CAM plants, possibly caused by an effect of temperature on the proportion of CO_2 fixed in the light or dark (TROUGHTON and CARD, 1975a; NEALES, 1975). Temperature has a marked effect on the $\delta^{13}C$ value of phytoplankton due to the temperature (and pH) dependence of the CO_2-bicarbonate equilibrium (DEGENS, 1969).

WHELAN et al. (1973) measured major temperature effects on carbon isotope discrimination by RuBP carboxylase in vitro, but CHRISTELLER et al. (1976) could only detect a change in $\delta^{13}C$ value under the same conditions from $-26.6^0/_{00}$ at 15° C to $-29.3^0/_{00}$ at 35° C.

II. Carbon Dioxide Concentration

PARK and EPSTEIN (1960) grew tomato plants at 0.05% and 1.5% CO_2 concentration and plants at the higher level were 2% more negative in $\delta^{13}C$ value than plants at the lower concentration. This result was of interest because of its possible relevance to understanding the operation of RuBP carboxylase in vivo. Recent evidence suggests that this enzyme can exist in two states, depending on the substrate CO_2 concentration. One interpretation of the PARK and EPSTEIN results could suggest that the two enzyme states discriminate differently against ^{13}C. To test this idea, the effect of bicarbonate concentration on fractionation by RuBP carboxylase was measured in vitro, but there was no evidence of any CO_2 concentration dependent effect (CHRISTELLER et al., 1976). If the result had been positive it would have supported the suggestion of two enzyme states, but this conclusion cannot be separated from an alternative hypothesis that the enzyme has an active and inactive form.

III. Oxygen Concentration Effects on Discrimination

The influence of oxygen concentration on CO_2 exchange in plants is often used to distinguish between C_3 and C_4 plants. The $\delta^{13}C$ value of C_4 plants has been shown to be unaffected by O_2 concentration, but reducing the concentration from 21% to 4% caused a change from $-20^0/_{00}$ to $-17^0/_{00}$ in C_3 plants (BERRY et al., 1972).

IV. Light Level

PARK and EPSTEIN (1960) failed to show any difference in $\delta^{13}C$ value of tissue from plants grown in 1000 foot-candles or in full sunlight.

V. Availability of Water

The influence of plant water status on carbon isotope fractionation in C_3 and C_4 is not known, but it is a feature of explaining the ecological behaviour of CAM plants and the resultant $\delta^{13}C$ value. CAM plants have the facility to adapt the photosynthetic pathway from light to dark CO_2 fixation, depending on environmental conditions and particularly the degree of aridity.

In the absence of water stress, CO_2 fixation in the light in CAM plants will result in C_3-like $\delta^{13}C$ values. Water stress induces daytime stomatal closure and apparently enhances the ability of the plant to assimilate CO_2 in the dark. In CAM plants the $\delta^{13}C$ value can be expected to reflect the proportion of CO_2 assimilated in the light or dark (TROUGHTON, 1972). This postulate was confirmed in laboratory investigations (OSMOND et al., 1973; OSMOND et al., 1976).

The ability of water stress to cause a shift in metabolism from C_3 to the C_4 type has received strong support from two completely independent experimental studies, one by BLOOM in California and the other by WINTER in Israel (BLOOM and TROUGHTON, 1977; WINTER and TROUGHTON, 1978). *Mesembryanthemum* plants were grown and water-stressed to a varying degree. Water stress induced dark CO_2 fixation and under these natural conditions there was a high correlation between water stress, titratable acidity, and $\delta^{13}C$ value. Other CAM plants such as the stem-succulent Opuntias show less propensity to switch from dark to light carbon assimilation (SZAREK and TROUGHTON, 1976; SUTTON et al., 1976).

The natural variation in $\delta^{13}C$ value among CAM plants has been investigated in a diverse range of ecological sites in many parts of the globe (MEDINA and TROUGHTON, 1974; OSMOND and ZIEGLER, 1975; MOONEY et al., 1977a; TROUGHTON et al., 1977). Considerable variation in isotope content occurred, and although water status was a significant parameter in causing the variation, genetic factors, life form, stage of growth and other environmental factors were also involved.

VI. Salinity and Carbon Isotope Fractionation

The carbon isotope ratio of a variety of monocotyledon and dicotyledon C_3 and C_4 plants has been shown to be unaffected by salinity over a wide range (MAHALL et al., 1974).

E. Implications of Variation in $\delta^{13}C$ Values Among Plant Species

All parts of the carbon cycle in nature will be "labelled" as a result of fractionation associated with the carboxylation reaction and subsequent events. The implications of this are numerous and range from the necessity to correct C-14 dates because of the difference in fractionation between C_3, C_4 and CAM plants, to the possibility of using the effect to trace carbon throughout the ecosystem.

I. Natural Products

Plant products are "labelled" according to the plant of its origin, therefore the $\delta^{13}C$ value may be of use to detect adulteration of food. Sugar from sugarcane is $-11^0/_{00}$ whereas sugar from sugarbeet is $-25^0/_{00}$ (TROUGHTON et al., 1974). Natural vanilla has a value of $-20^0/_{00}$ whereas synthetic vanilla from lignin measures $-28^0/_{00}$. Variations of this kind may also be useful in tracing the origins of other products, such as alcohol, oil, fats, paper, etc.

II. Paleoecology

TROUGHTON (1971) reported relatively constant $\delta^{13}C$ values for coal spanning time back to the Permian period. This suggests the $\delta^{13}C$ content of the atmosphere has been relatively constant over that period and/or there has been little change in RuBP carboxylase, at least as it affects carbon isotope fractionation. Carbon isotope measurements of organic material up to 3.2 billion years ago gave similar $\delta^{13}C$ values to present-day material (BARGHOORN et al., 1974). These and other applications are fully discussed elsewhere (STOUT et al., 1975; LERMAN and TROUGHTON, 1975; TROUGHTON and CARD, 1975b; TROUGHTON et al., 1974).

III. Physiology – Plant, Animal, and Human

The differential labelling of plants with ^{13}C provides a natural tracer of carbon for plant, animal and human systems. Large amounts of material can be labelled, the isotope is safe and it can be monitored over long periods of time. Its value in plant physiology is limited because of the availability of ^{14}C and ^{13}C is mainly used to indicate the carbon pathway. In animal studies, however, it can trace carbon metabolism in all tissues over long periods of time. The carbon isotope ratio of animals grazing C_3 grass pastures (e.g., South Australia) are naturally isotopically different from cows grazing C_4 grass (e.g., Queensland) as illustrated by $\delta^{13}C$ values of hair or milk (MINSON et al., 1975). The whole animal is labelled in a manner reflecting the plant species grazed and switching the diet from plants with one pathway to the other causes a gradual change in isotope label throughout the whole animal. It therefore allows the calculation of kinetic parameters to describe the flux of carbon compounds throughout the body (JONES et al., 1978).

In agronomy, the proportion of C_3 and C_4 species in a pasture can be relatively simply obtained from a $\delta^{13}C$ measurement of a bulked sample of the pasture. A measurement before and after animal grazing of the pasture would indicate any preferential selection by the animal of either C_3 or C_4 species (LUDLOW et al., 1976). It is also possible to make estimates of the proportion of roots that are C_3 or C_4 (JONES et al., 1978).

Carbon is the basis of human metabolism and the non-radioactive character of ^{13}C makes it an ideal tracer in human studies. There have been few applications yet, but it is used as a diagnostic tool in medical studies either in humans, or in animals to illustrate its potential use in humans in relation to such problems as diabetes (LACROIX et al., 1973; LEFEBVRE et al., 1975; MOSORA et al., 1974).

F. Conclusions

In the short time since the variation in $\delta^{13}C$ values between C_3 and C_4 plants was interpreted, there has been an explosion in the use of the technique in an extremely wide range of disciplines. The use of the differently labelled plant materials as a tracer does not require an understanding of the causes of variation in $\delta^{13}C$ values among plants. Currently the most serious deficiency in the subject is the absence of a detailed biophysical description of the mechanism of carbon isotope discrimination at the enzyme level.

References

Abelson, P.H., Hoering, T.C.: Proc. Natl. Acad. Sci. USA 47, 623–632 (1961)
Allaway, W.G., Osmond, C.B., Troughton, J.H.: R. Soc. N. Z. Bull. 12, 195–202 (1974)
Baertschi, P.: Helv. Chim. Acta 36, 773–781 (1953)

Barghoorn, E.S., Troughton, J.H., Margulis, L.: Am. Sci. *62*, 389 (1974)
Bender, M.M.: Radiocarbon *10*, 468–472 (1968)
Bender, M.M.: Phytochemistry *10*, 1239–1244 (1971)
Berry, J.A., Troughton, J.H.: Carnegie Inst. Wash. Yearb. *73*, 785–790 (1974)
Berry, J.A., Troughton, J.H., Björkman, O.: Carnegie Inst. Wash. Yearb. *71*, 158–161 (1972)
Björkman, O., Troughton, J.H., Nobs, M.A.: Brookhaven Symp. Biol. *25*, 206–276 (1973)
Bloom, A.J., Troughton, J.H.: Oecologia (submitted) (1977)
Christeller, J.T., Laing, W.A., Troughton, J.H.: Plant Physiol. *57*, 580–582 (1976)
Craig, H.: Geochim. Cosmochim. Acta *3*, 53–92 (1953)
Craig, H.: J. Geology *62*, 115–149 (1954)
Craig, H.: Geochim. Cosmochim. Acta *12*, 133–149 (1957)
Degens, E.T.: In: Organic geochemistry. Eglinton, G., Murphy, M.T.J. (eds.), pp. 302–329. Berlin, Heidelberg, New York: Springer 1969
Farmer, J.G., Baxter, M.S.: Nature (London) *247*, 273–275 (1974)
Freyer, H.D., Wiesberg, L.: Nature (London) *252*, 757 (1974)
Hatch, M.B., Osmond, C.B., Troughton, J.H., Björkman, O.: Carnegie Inst. Wash. Yearb. *71*, 35–41 (1972)
Hsu, J.C., Smith, B.N.: Plant Cell Physiol. *13*, 689–694 (1972)
Jansen, H.S.: Nature (London) *196*, 84–85 (1963)
Jones, R.L., Ludlow, M.M., Troughton, J.H., Blunt, C.G.: J. Agric. Sci. Camb. *89* (submitted) (1978)
Keeling, C.D.: Geochim. Cosmochim. Acta *13*, 322–334 (1958)
Kennedy, R.A., Laetsch, W.M.: Science *184*, 1087 (1974)
Lacroix, M., Mosora, F., Pontus, M., Lefebvre, P., Luyckx, A., Lopez-Habib, G.: Science *181*, 445–446 (1973)
Lefebvre, P., Mosora, P., Lacroix, M., Luyckx, A., Lopez-Habib, G., Duchesue, J.: J. Am. Diabetes Assoc. 24, 185–189 (1975)
Lerman, J.C., Raynal, J.: C.R. Acad. Sci. Paris Ser. *D 275*, 1391–1394 (1972)
Lerman, J.C., Troughton, J.H.: Proc. 2nd Int. Conf. Stable Isotopes (in press) (1975)
Lerman, J.C., Deleens, E., Nato, A., Moyse, A.: Plant Physiol. *53*, 581–584 (1974)
Libby, L.M.: J. Geophys. Res. *77*, 4310–4317 (1972)
Ludlow, M.M., Troughton, J.H., Jones, R.L.: J. Agric. Sci., Camb. *87*, 625–632 (1976)
Mahall, B., Card, K.A., Troughton, J.H.: Carnegie Inst. Wash. Yearb. *73*, 784–785 (1974)
McKinney, C.R., McCrea, J.M., Epstein, S., Allen, H.A., Urey, H.C.: Rev. Sci. Instrum. *21*, 724–730 (1950)
Medina, E., Troughton, J.H.: Plant Sci. Lett. *2*, 357–362 (1974)
Minson, D.J., Ludlow, M.M., Troughton, J.H.: Nature (London) *256*, 602 (1975)
Mooney, H.A., Troughton, J.H., Berry, J.A.: Oecologia *30*, 295–306 (1977)
Mosora, F., Duchesne, J., Lacroix, M.: C.R. Acad. Sci. (Paris) *278*, 1119–1122 (1974)
Nalborczyk, E., Croix, L.J., Hill, R.D.: Can. J. Bot. *53*, 1132–1138 (1975)
Neales, T.F.: In: Environmental and biological control of photosynthesis. Marcelle, R. (ed.), pp. 369–371. The Hague: W. Junk 1975
Nier, A.O.: Rev. Sci. Instrum. *18*, 398–411 (1947)
Osmond, C.B., Ziegler, H.: Oecologia *18*, 209–217 (1975)
Osmond, C.B., Allaway, W.G., Sutton, B.G., Queiroz, O., Luttge, U., Winter, K.: Nature (London) *246*, 41–42 (1973)
Osmond, C.B., Bender, M.M., Burris, R.H.: Aust. J. Plant Physiol. *3*, 787–799 (1976)
Park, R., Epstein, S.: Geochim. Cosmochim. Acta *21*, 110–126 (1960)
Park, R., Epstein, S.: Plant Physiol. *36*, 133–138 (1961)
Pearcy, R.W., Troughton, J.H.: Plant Physiol. *55*, 1054–1056 (1975)
Pearman, G.I., Francey, R.J., Fraser, P.J.B.: Nature (London) *260*, 771–772 (1976)
Sirevåg, R., Buchanan, B.B., Berry, J.A., Troughton, J.H.: Arch. Microbiol. *112*, 35–38 (1977)
Smith, B.N.: Bioscience 22, 226–231 (1972)
Smith, B.N., Brown, W.V.: Am. J. Bot. *60*, 505–513 (1973)
Smith, B.N., Epstein, S.: Plant Physiol. *47*, 380–384 (1971)
Smith, B.N., Robbins, M.J.: Proc. 3rd Int. Congr. Photosynthesis, 1579–1587 (1974)
Smith, B.N., Turner, B.L.: Am. J. Bot. *62*, 541–545 (1975)
Smith, B.N., Herath, H.M.W., Chase, J.B.: Plant Cell Physiol. *14*, 177–182 (1973)

Stout, J.D., Rafter, T.A., Troughton, J.H.: In: Quarternary studies. Suggate, R.P., Cresswell, M.M. (eds.), pp. 279–286. Wellington: The R. Soc. N. Z. 1975

Sutton, B.G., Ting, I.P., Troughton, J.H.: Nature (London) *261*, 42–43 (1976)

Szarek, S.R., Troughton, J.H.: Plant Physiol. *58*, 367 (1976)

Troughton, J.H.: Proc. Radiocarbon User's Conf. pp. 37–46 (1971a)

Troughton, J.H.: In: Photosynthesis and photorespiration. Hatch, M.D., Osmond, C.B., Slatyer, R.O. (eds.), pp. 124–127. New York: Wiley-Interscience 1971b

Troughton, J.H.: Proc. Int. Conf. Radiocarbon Dating 8th, Lower Hutt, 421–438 (1972)

Troughton, J.H., Card, K.A.: Planta *123*, 185–190 (1975a)

Troughton, J.H., Card, K.A.: What's New Plant. Physiol. *7*, 1–5 (1975b)

Troughton, J.H., Hendy, C.H., Card, K.A.: Z. Pflanzenphysiol. *65*, 461–464 (1971)

Troughton, J.H., Hendy, C.H., Card, K.A.: Carnegie Inst. Wash. *73*, 768–780 (1974a)

Troughton, J.H., Wells, P.V., Mooney, H.A.: Science *185*, 610–612 (1974b)

Troughton, J.H., Mooney, H.A., Berry, J.A., Verity, D.: Oecologia *30*, 307–312 (1977)

Urey, H.C.: J. Chem. Soc. 562–581 (1947)

Whelan, T., Sackett, W.M., Benedict, C.R.: Plant Physiol. *51*, 1051–1054 (1973)

Whelan, T., Sackett, W.M., Benedict, C.R.: Biochem. Biophys. Res. Commun. *41*, 1205–1210 (1975)

Winter, K., Troughton, J.H.: Z. Pflanzenphysiol. *88*, 153–162 (1978)

II C. Factors Influencing CO_2 Assimilation

11. Interactions Between Photosynthesis and Respiration in Higher Plants

D. GRAHAM and E.A. CHAPMAN

A. Introduction

I. The Relevance of Photosynthetic and Respiratory Interactions

Interaction between photosynthesis and respiration is an enigma. Ever since photosynthetic gas exchange was first measured the question has been posed whether dark respiration continues in the light, and, if so, at what rate (RABINOWITCH, 1945; HEATH, 1969). The question has some importance for estimation of the growth of plants and of photosynthesis since apparent photosynthetic CO_2 fixation (including both photosynthetic CO_2 and light-enhanced dark CO_2 fixation (see Chap. II.5, this vol.) must be corrected for loss of CO_2 by respiratory activity. It is now known that such respiratory activity comprises at least two components, photorespiration which is a light-dependent process, and dark respiration which may continue during illumination. Photorespiration (see Chap. 5.II. 25–28, this vol.) is a light-induced evolution of CO_2, and concomitant O_2 uptake, involving oxidation of C_2-substrates via a complex series of reactions in the chloroplast, peroxisome, and mitochondrion. Dark respiration, on the other hand, involves oxidation of sugars via glycolysis and the tricarboxylic acid (TCA) cycle and the oxidative pentose phosphate (OPP) pathway. The main subject of this chapter is the interaction between photosynthesis and dark respiration during illumination, although the interactions between dark respiration and photorespiration cannot be ignored.

One way of approaching the complexities of photosynthetic and respiratory interactions is to consider whether the green higher-plant cell requires dark respiratory activity in the light. While dark respiration may not be necessary in the light to supply the energy requirements of the cell (ATP and NADPH, NADH) it will be necessary to supply carbon skeletons derived from the TCA cycle and required in biosynthesis, e.g., succinate for porphyrins and α-ketoglutarate for the glutamate family of amino acids. Such synthetic reactions for production of chlorophylls and protein are likely to be of greatest magnitude in young, growing cells, but turnover of these components continues in mature cells (BROWN, 1972; HUFFAKER and PETERSON, 1974) so that continued activity of the TCA cycle in such cells is to be expected, though possibly at lower rates.

B. Physiological Observations

I. Plants with C_3-Type Photosynthesis

1. Photorespiration

Photorespiration, usually measured as CO_2 release, (see CANVIN and FOCK, 1972; Chap. II.28 by CANVIN, this vol. for appraisal of methods) is estimated to exceed dark respiration by about 3- (LUDWIG and CANVIN, 1971) to 8.5-fold (ZELITCH, 1975). Should dark respiration continue in the light then it could account for a significant proportion of the CO_2 evolved (RAVEN, 1972 a, b).

The complexities of interpreting physiological gas exchange measurements have been formalised in schemata by several authors (JACKSON and VOLK, 1970; LAKE, 1967; SAMISH and KOLLER, 1968; SAMISH et al., 1972). From these it is apparent that in the light the contributions of dark respiration and refixation of evolved CO_2 are major stumbling blocks to the quantitative determination of photorespiration.

2. Distinction Between Photorespiration and Dark Respiration

Photorespiration and dark respiration respond differently to oxygen concentration, since RuBP oxygenase and the oxidative terminal oxidases are differentially sensitive to oxygen. Photorespiration is stimulated by increasing oxygen concentration in the range 1%–100%, whereas dark respiration is unaffected (FORRESTER et al., 1966; TREGUNNA et al., 1966; POSKUTA, 1968). There is also substantial evidence to show that the substrates of photorespiration are early products of photosynthesis and are distinct from those of dark respiration (GOLDSWORTHY, 1966; LUDWIG and CANVIN, 1971; D'AOUST and CANVIN, 1972; MANGAT et al., 1974; CANVIN et al., 1976).

3. Rate of Dark Respiration in the Light

There is a variety of physiological evidence which suggests, but does not prove, that dark respiration is inhibited in the light to a greater or lesser degree.

a) Light Intensity

When CO_2 efflux of *Rumex acetosa* leaves is measured as a function of light intensity, a minimum in the curve can be observed at about $2 \cdot 10^3$ erg cm^{-2} s^{-1} at low CO_2 concentration (1 to 7 µl/l; HOLMGREN and JARVIS, 1967). This was interpreted to indicate an inhibition of dark respiratory CO_2 efflux at very low light intensities, an effect which is over-ridden by increasing photosynthesis and photorespiration at higher light intensities. This effect is analogous to that shown in algae in which O_2 influx was measured (HOCH et al., 1963) also at very low CO_2 concentration. CO_2 evolution in sunflower leaves at low light intensity (5000 lux) was also found to be slightly below that of dark respiration using a method involving extrapolation to zero CO_2 in the atmosphere which respresents the minimum value for CO_2 evolution (HEW et al., 1969).

b) Oxygen Effect on CO_2 Compensation Point

Measurement of the CO_2 compensation point as a function of oxygen concentration (0%–100%) for leaves gives a linear relationship which usually extrapolates through the origin (Tregunna et al., 1966; Forrester et al., 1966a; Björkman et al., 1970) although this is not always the case (Poskuta, 1968). Dark respiration was not affected by oxygen concentration. The former results have been adduced to support the view that both dark respiration and photorespiration are totally inhibited at low oxygen ($<1\%$) in the light. A positive value at zero oxygen would be expected if dark respiration occurred in the light.

c) Specific Activity of $^{14}CO_2$ Evolved in Light

The Canvin group has shown that there is a close similarity in specific activities of evolved CO_2 and compounds of the glycolate pathway (see above) and concludes this "leaves no room for dilution of the specific activity of the evolved CO_2 by CO_2 from dark respiratory sources which have low specific activity" (Canvin et al., 1976). While this argument is persuasive it cannot be considered conclusive, since the relative specific activities are rarely 100%. In short-term $^{14}CO_2$ feeding experiments in the light the TCA cycle will be inhibited (see below) while in longer-term feeding ^{14}C will enter the TCA cycle and be evolved as $^{14}CO_2$ which would contribute to the CO_2 release in the light.

4. Reassimilation of Evolved CO_2

Most of the methods of estimating photorespiration are prone to the criticism that they do not take into account refixation of CO_2. Estimates of re-assimilation of evolved CO_2 are impossible to determine directly with present methods of gas exchange measurement. Moss (1966), however, has proposed that the post-illumination burst of CO_2 evolution, a characteristic of photorespiration, is a reasonable minimum measure of total CO_2 production in the light. The difference between that value and CO_2 evolution in the light would represent re-fixation. Ludwig and Canvin (1971) using this method estimate 34% of total CO_2 production in the light is refixed. Bravdo (1968) has estimated refixation at 21% to 25% of net photosynthesis for several species, while Raven (1972b) estimates a value of up to 17% of total CO_2 fixation.

5. Effect of Light Quality on Respiration

Blue light has been shown to stimulate respiration in higher plants (Voskresenskaya, 1972) possibly through the mediation of flavin cofactors. Recently Hillman (1977) has shown that respiratory metabolism in *Lemna* is subject to phytochrome control which sets in train a photoperiodic rhythm in dark respiratory activity. The effects of circadian periodicity and phytochrome-actuation on respiratory gas exchange (Hillman, 1976a, b) is a field which has received very little attention so far in considering light effects on respiratory (both photorespiration and dark respiration) activity.

6. Studies in the Absence of Photosynthesis or Photorespiration

Various studies have been made in which photosynthesis was absent either because of a lack of chlorophyll in etiolated or albino leaves (ROSENSTOCK and RIED, 1960; HEW and KROTKOV, 1968), or because of inhibition by DCMU (EL-SHARKAWY et al., 1967; POSKUTA et al., 1967; DOWNTON and TREGUNNA, 1968; CHAPMAN and GRAHAM, 1974a). The general conclusion is that respiratory gas exchange in light and dark are similar when photosynthesis and photorespiration are absent. Since the effect of light on respiration is most likely mediated through photosynthetic or photorespiratory metabolism it is perhaps not surprising that dark respiration continues unaffected in these circumstances.

II. Plants with C_4-type Photosynthesis

Plants having the C_4-type photosynthetic pathway (see Chap. II.6, this vol.) are characterized, among other properties, by having very low (<10 ppm) to zero CO_2 compensation points under most conditions and a carboxylation mechanism with a high affinity for CO_2. This situation may result from a limited activity of the glycolate pathway of photorespiration or a high capacity for refixation of photorespiratory CO_2 produced, or a combination of both (JACKSON and VOLK, 1970; FORRESTER et al., 1966b). The absence of CO_2 evolution in the light in plants such as maize, sorghum, and *Amaranthus* has been attributed mainly to refixation of CO_2 produced by photorespiration and dark respiration (EL-SHARKAWY et al., 1967; MEIDNER, 1970a, b) but DOWNTON and TREGUNNA (1968), POSKUTA (1969), and BULL (1969) have concluded that photorespiration and presumably also dark respiration were inhibited in the light and efficient endogenous recycling of CO_2 was not a factor. Recent work with isolated bundle sheath cells of C_4 plants (reviewed by BLACK, 1973) indicates, however, that the glycolate pathway is present, possibly in a modified form (MAHON et al., 1974).

These observations for C_4 plants make it even more difficult than is the case in C_3 plants to determine the effects of light on respiratory activity and the question must be left open at present.

C. Biochemical Observations

I. Plants with C_3-Type Photosynthesis

1. Location of Respiratory Metabolism in the Cell

It is generally accepted that glycolysis and the oxidative pentose phosphate pathway are principally located in the cytosol and the mitochondrion is the site of the TCA cycle (DAVIES et al., 1964). However, there was initially much confusion about the presence of respiratory enzymes in the chloroplast, due mainly to the use of inadequate techniques for the separation of other cellular components from the chloroplast, especially before the characterization of the peroxisome. The application of non-aqueous techniques and other improved organelle isolation methods has resulted in considerable clarification to the extent that the fully operational TCA cycle is now believed to be confined to the mitochondrion but that various enzymes especially malate and isocitrate dehydrogenases have isoenzymes in the cytosol, peroxisomes and chloroplasts (TOLBERT, 1971). It is apparent that the chloroplast does not contain a functional TCA cycle (ELIAS and GIVAN, 1977). All of the enzymes of the OPP pathway are also present in the chloroplast (HEBER et al., 1967).

2. Carbon Metabolism

As in algae (see Chap. II.12, this vol.), photosynthetically incorporated $^{14}CO_2$ does not appear readily to enter the organic acids or related amino acids (glutamate and aspartate) of the TCA cycle in leaves of barley (BENSON and CALVIN, 1950), rice (NISHIDA, 1962), mung bean (GRAHAM and WALKER, 1962) and spinach (HEBER and WILLENBRINK, 1964). Extinguishing the light leads to an immediate influx of ^{14}C into compounds of the TCA cycle. Considering the ready export from chloroplasts of triose phosphates and PGA, (HEBER, 1974; WALKER, 1976) which would be ^{14}C-labeled in such experiments, it has been concluded (JACKSON and VOLK, 1970; HEBER, 1974) that dark respiration is inhibited in the light. Similar conclusions have been derived from comparable experiments with algae (see Chap. II.12, this vol.). A comparison of the distribution of ^{14}C in the various C atoms of sucrose, alanine (as a measure of pyruvate) and malate (an intermediate of the TCA cycle) in sunflower leaves after photosynthesis of $^{14}CO_2$ in weak (~ 700 lux) or strong illumination (~ 5000 lux) led GIBBS (1953) to conclude that strong illumination inhibited entry of pyruvate to the TCA cycle. BIDWELL et al. (1955) came to a similar conclusion after feeding glucose-U-^{14}C to wheat leaves in the light. Compared with results in the dark, little ^{14}C-label entered the TCA cycle intermediates or alanine, which suggested to them that the blockage occurred before pyruvate. However, when ^{14}C-labelled organic acids of the TCA cycle are fed to mung bean leaves in the light they are interconverted to all components of the cycle (GRAHAM and WALKER, 1962; GRAHAM and COOPER, 1967; CHAPMAN and GRAHAM, 1974a, b) suggesting that turnover of the cycle can occur in the light. Similarly from experiments in which glutamine-^{14}C was fed to wheat leaves (BIDWELL et al., 1955), glutamic acid-^{14}C to barley leaves (NAYLOR and TOLBERT, 1956) or aspartic acid-^{14}C to a wide variety of higher plant leaves (NAYLOR et al., 1958) it was concluded that TCA cycle activity in the light was responsible for redistribution of label into intermediates of the cycle.

The presence of inorganic nitrogen in the form of ammonium nitrate stimulates the entry of ^{14}C to the TCA cycle intermediates and related amino acids from $^{14}CO_2$ or glucose-^{14}C, fed in the light to leaves of wheat (BIDWELL et al., 1955) and rice (NISHIDA, 1962). Ammonium ion particularly stimulates entry to the TCA cycle as shown in *Lemna* (ERISMANN and KIRK, 1969) and mung bean (GRAHAM, unpublished). This effect is also observed in algae (see Chap. II.12, this vol.; HILLER, 1970). The interpretation of these observations may now be complicated by the recent finding of organic nitrogen-synthesising systems in isolated chloroplasts. ELIAS and GIVAN (1977) have shown that it is possible to generate α-keto-glutarate by isocitrate dehydrogenase in pea chloroplasts. The location of GOGAT and glutamine synthetase in the chloroplast (MITCHELL and STOCKING, 1975; MIFLIN and LEA, 1977) would permit synthesis of glutamate and gluatamine from α-ketoglutarate although the source of isocitrate is uncertain (ELIAS and GIVAN, 1977). Thus in experiments with whole leaves it may be uncertain whether mitochondrial or chloroplastic systems are responsible for the observations of the effects of inorganic nitrogen.

Long-term changes in organic acids were compared in cotyledons and leaves of *Pisum sativum* L. (BARTHOVÁ and LEBLOVÁ, 1969). They showed accumulation of succinate and lactate in light-grown plants which they concluded was due to blockage of the TCA cycle at succinate dehydrogenase and presumably diversion of pyruvate to lactate. The accumulation of lactate was quite small and since the methods involved gross analysis of the tissue, much of the accumulation of succinate was presumably in the cell vacuole and would not necessarily involve inhibition of the TCA cycle.

There have been few studies of the effect of light on higher plant glycolysis. SANTARIUS and HEBER (1965) found a decrease in pyruvate on illumination of leaves of several species which they concluded was caused by light-induced inhibition of glycolysis, mediated by ATP/ADP ratios, presumably at 3-phosphoglycerate kinase although other possible sites would be phosphofructokinase and pyruvic kinase. Their experiments refer to a few minutes of illumination after darkness and would therefore agree with our suggestion (see below) that there is an initial inhibition of respiratory activity which, at least in the case of the TCA cycle, is followed by resumption of a rate similar to that in the dark. Such activity will require acetyl-CoA units, derived either from pyruvate via glycolysis or from fatty acid oxidation. Synthesis of acetyl CoA, presumably for use in fatty acid synthesis, is possible in the chloroplast and it has been suggested that this compound may be derived from photosynthetically produced PGA (GIVAN and HARWOOD, 1976).

The conflicting evidence presented above for or against the operation of the TCA cycle and glycolysis in the light can be resolved on two grounds. The first is that most of the experiments with labeled compounds have been carried out in the initial few minutes of illumination following darkness. It was first suggested by GRAHAM and WALKER (1962) and subsequently confirmed (GRAHAM and COOPER, 1967; CHAPMAN and GRAHAM, 1974a, b) that rapid readjustments in metabolic pools occur during the first minutes of illumination (Fig. 1) which manifest themselves as an apparent inhibition of the TCA cycle (Fig. 2). Subsequently the cycle appears to establish new steady-state levels of intermediates (Fig. 1) and operates at a rate similar to that in the dark (CHAPMAN and GRAHAM, 1974a, b). Therefore, conclusions based on short-term experiments are likely to be misleading because respiration is inhibited during initial periods in the light.

The second concerns the synthetic status of the leaf as related to age. It has been stated above that supply of inorganic nitrogen can cause a stimulation of carbon flow into the TCA cycle in the light. This phenomenon is related apparently to the synthesis of amino acids, particularly of the glutamate family, which probably are used in protein synthesis. It is interesting to note, therefore, that NISHIDA (1962) showed that a much greater proportion of ^{14}C entered TCA cycle intermediates from $^{14}CO_2$ in the light in young growing rice leaves than in mature leaves. Younger leaves are likely to require higher synthetic capabilities than older leaves.

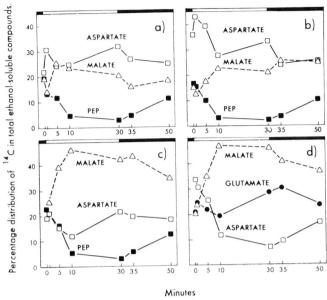

Fig. 1a–d. Transient changes in some ^{14}C-labeled intermediates during dark-light-dark transitions. Each sample of five leaves was fed acetate-2-^{14}C (**a**), citrate-1,5-^{14}C (**b**), fumarate-2,3-^{14}C (**c**), or succinate-2,3-^{14}C (**d**), in the dark for 2 h, then transferred to water in an air draught for 15 min before the experiments. *Solid bar* represents darkness and *open bar* represents illumination (Chapman and Graham, 1974b)

3. Quantitative Role of the TCA Cycle in the Light

Very little work has been done on this aspect of respiratory metabolism in the light. The difficulty is, of course, the extreme complexity of the situation in the illuminated green cell, with several compartments (chloroplast, peroxisome, mitochondria, cytosol, and vacuole) with multiple sites of certain respiratory enzymes, particularly the dehydrogenases, and the possibility of interaction of photosynthesis and respiration at substrate and coenzyme levels. In mung bean leaves a comparison in light and dark of the effects of two inhibitors of the TCA cycle, malonate and fluoroacetate, on carbon traffic in the cycle led to the conclusion that in the transition from dark to light there was an initial inhibition of the TCA cycle which subsequently in the light established a new steady state at a rate comparable with that in the dark (Chapman and Graham, 1974a, b). Similar conclusions that the TCA cycle continues in the light have been proposed by Stepanova and Baranova (1972) from experiments in which ^{14}C-succinate or ^{14}C-acetate were fed in light or dark for 30 min to rhubarb or tobacco leaves. They attributed lowered ^{14}CO$_2$ evolution from these substrates in light to reassimilation, a conclusion which was supported by the appearance of ^{14}C in carbohydrates, especially in the light. Both Zelitch (1971; 1973) and Raven (1972a, b) have reviewed the evidence and conclude that dark respiration continues in the light.

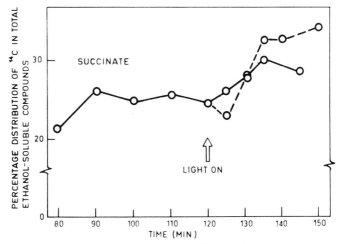

Fig. 2. Effects of illumination on the percentage distribution of ^{14}C in ^{14}C-succinate in the presence of 0.01 M malonate at pH 4.0. $^{14}CO_2$ was fed in the dark to samples of five leaves for 2 h and the leaves were transferred to the inhibitor in the dark. Sampling in the dark was begun after 80 min and a series of leaf samples was illuminated from 120 min. Dark, *solid line;* light, *broken line* (CHAPMAN and GRAHAM, 1974a)

4. Control by Adenine Nucleotides, Nicotinamide Adenine Dinucleotides and Metabolite Translocators

The most likely mediators of the interactions between photosynthesis and respiration are the adenine nucleotides, ATP, ADP, AMP, and inorganic phosphate (P_i) and the nicotinamide adenine dinucleotides, NADP, NADPH and NAD, NADH. These cofactors are known to control many respiratory enzymes (TURNER and TURNER, 1975). Light effects remarkable changes in redox levels, in general, increasing both the reduced/oxidized ratio of nicotinamide adenine dinucleotides (HEBER and SANTARIUS, 1965) and energy charge levels (SANTARIUS and HEBER, 1965). Evidence obtained mainly with spinach leaf chloroplasts indicates that they are impermeable to nicotinamide adenine dinucleotides and to adenylates (WALKER, 1976) and it is now considered that transfer from chloroplast to cytosol of reducing potential is effected by a dicarboxylate (oxalacetate/malate) shuttle. Transfer of energy in the form of ATP is effected not by an adenylate translocator (HELDT et al., 1972), but by a triosephophate/PGA shuttle which also transfers (NADH) reducing potential (STOCKING and LARSON, 1969; KRAUSE and HEBER, 1976). Plant mitochondria are able to oxidize exogenous NADH and the mitochondrial adenine nucleotide translocator allows entry of ADP (WISKICH, 1977). NADPH may be oxidised in the cytosol by a variety of synthetic systems. Thus a complex system of interchange of cofactors exists whereby the effects of light can be transduced to respiratory metabolism. The major points of such light action on respiration are likely to be phosphofructokinase, the triosephosphate dehydrogenase system and pyruvate kinase of glycolysis and the dehydrogenases of the TCA cycle, particularly isocitrate and malate dehydrogenases. Some interaction with the initial dehydrogenase steps of the OPP pathway would also be anticipated. All these enzymes are subject to control by either adenylates or the nicotinamide adenine dinucleotides or both.

The interaction of the components of the photosynthetic phosphate translocator (DHAP, PGA and Pi; Werdan and Heldt, 1972) with their cytoplasmic counterparts must also be considered a likely influence on glycolytic metabolism and may contribute to both synthesis of sucrose in the cytosol and provision of 3C units for continuance of glycolysis in the oxidative direction.

Complex allosteric control (Kelly and Latzko, 1977) and light-induced inactivation (Kachru and Anderson, 1975) of phosphofructokinase is possible in plants, although for the other enzymes mentioned a variety of more straightforward controls are mediated mainly by the adenylates and ratios of oxidised/reduced nicotinamide adenine dinucleotides (for details see Turner and Turner, 1975; Heber, 1974).

Relatively little information is available on specific light-induced controls in green leaves, although more is available for algae (see Chap. II.12, this vol.). Control by oxidised/reduced pyridine nucleotide has been proposed to account for the increase in the malate/aspartate ratio on transition from dark to light in mung bean leaves (Graham and Walker, 1962; Graham and Cooper, 1967; Chapman and Graham, 1974b). During the initial period of illumination increase in NADH/NAD, mainly through a massive conversion of NAD to NADP in the light (Graham and Cooper, 1967), is suggested to inhibit the malate dehydrogenase and also isocitrate dehydrogenase steps of the TCA cycle.

5. Synthetic Function of the TCA Cycle in the Light

In order to function as a synthetic, anaplerotic mechanism it is necessary for the TCA cycle to be continuously replenished with C4 organic acids. While this could occur for short periods from storage reserves in the cell vacuole, the major route is almost certainly via CO_2 fixation by PEP carboxylase to give oxaloacetate and so malate by reduction or aspartate by transamination (Walker, 1962). Several authors have shown active $^{14}CO_2$ fixation into TCA cycle intermediates in the dark (see for example Benson and Calvin, 1950; Graham and Walker, 1962; Chapman and Graham, 1974a, b), but as shown above only a small proportion of $^{14}CO_2$ fixed in the light appears in TCA cycle intermediates. However, malate is usually the most prominently labeled of those compounds (see also Chap. II.5, this vol.).

6. The Oxidative Pentose Phosphate Pathway in the Light

Raven (1972a) has carefully reviewed this topic and concluded that the OPP pathway occurs in green cells in the light but most of the evidence has been obtained with algae.

Following our approach of establishing a need for the operation of a pathway in the light it is clear that the OPP pathway is not likely to be essential during illumination since the reductive pentose phosphate (RPP) pathway is operational in the chloroplasts for the production of carbon skeletons, e.g., (a) erythrose-4-phosphate required in shikimic acid synthesis and hence the related syntheses of phenolics and aromatic amino acids and (b) pentose monophosphates for nucleotide synthesis. However, the inability to export most phosphorylated substrates from the chloroplast (HELDT, 1976) to the cytoplasm where syntheses of, for example, aromatic compounds may occur (STAFFORD, 1974) could justify operation of the OPP pathway in the light. The other major function of the OPP pathway is the synthesis of NADPH but, since NADPH can be transferred from the chloroplast to the cytoplasm by the dihydroxyacetone phosphate/PGA shuttle (STOCKING and LARSON, 1969; KRAUSE and HEBER, 1976) using the non-reversible NADP-triosephosphate dehydrogenase (KELLY and GIBBS, 1973), it is unlikely that this particular function of the OPP pathway is essential in the light.

Apart from the wide distribution of the two enzymes unique to the OPP pathway, glucose-6-P and 6-phosphogluconate dehydrogenases, the evidence for the operation of this pathway in the light in higher plants rests on the finding of C1/C6 ratios greater than 1.0 from specifically labeled glucose fed to leaves of tobacco (MacLACHLAN and PORTER, 1959), grape berries (LEFÈBVRE and RIBÉREAU-GAYON, 1970), and leaves of *Gingo biloba* (L.), *Helianthus annuus* (L.) and *Iris pseudacorus* (L.) (RAVEN, J.A. unpublished results reported in RAVEN, 1972a).

The OPP pathway is operative in isolated spinach chloroplasts as shown by the labelling of 6-phosphogluconate from $^{14}CO_2$ in the light in the presence of vitamin K_5 (KRAUSE and BASSHAM, 1969). All the enzymes of the OPP pathway have been found in isolated chloroplasts (HEBER et al., 1967).

7. Interaction of Photorespiration and Dark Respiration

In view of the presence of the respiratory enzymes malate and isocitrate dehydrogenases in peroxisomes (TOLBERT, 1971) and glycine decarboxylation in mitochondria (WOO and OSMOND, 1976; 1977) it may be supposed that interaction between the carbon metabolism of photorespiration and dark respiration occurs. If, as we have supposed above, dark respiration can continue in the light, there must be means for separating the two activities especially in the mitochondrion. It is not clear at present how this is achieved since if glycine decarboxylation is a major source of CO_2 in photorespiration the redox status of the mitochondrion needs to be oxidised to maintain a flow of carbon through the glycine decarboxylase step. There would thus be competition for NAD^+ between the photorespiratory reaction and the dehydrogenases of the TCA cycle assuming location inside the inner membrane of the mitochondrion for the glycine decarboxylation and the malate-oxaloacetate shuttle as proposed by WOO and OSMOND (1977).

II. Plants with C_4-Type Photosynthesis

The information available on dark respiratory metabolism in the light in C_4 plants is meagre. ZELITCH (1966) has shown that acetate-1-^{14}C fed to maize leaf disks in the light produced amounts of $^{14}CO_2$ similar to that produced by tobacco leaf disks and the ratios of $^{14}CO_2$ released/^{14}C assimilated were similar, suggesting that the C_4 plant maize has TCA cycle activity similar to the C_3 plant tobacco in the light. CHAPMAN and OSMOND (1974) on the basis of experiments feeding ^{14}C-labeled TCA cycle intermediates to leaves of the C_4 plants *Atriplex spongiosa* and *Sorghum bicolor* and the use of TCA cycle inhibitors malonate or fluoroacetate concluded that the TCA cycle continued to operate in the light in C_4 plants.

They also emphasised the role of the mitochondrial TCA cycle in the transfer of malate and aspartate for photosynthetic metabolism in bundle sheath cells.

The ADP or oxaloacetate-dependent glycine decarboxylation systems in mitochondria of the C_3 plant spinach (Woo and Osmond, 1976) have also been found inside the inner membrane of mitochondria of the C_4 plants, *A. spongiosa* and *Panicum miliaceum* although not in maize (Woo and Osmond, 1977). Interaction between photorespiration and dark respiration in C_4 plants is therefore likely as in C_3 plants.

D. General Conclusions

On the basis of physiological evidence many investigators favour the view that dark respiration is inhibited in the light either partially or totally. There has, however, been a tendency by some to ignore the contribution of re-assimilation of CO_2 produced by photorespiration and dark respiration in the light. Nevertheless where the relative specific activity of evolved $^{14}CO_2$ approaches 100% it seems likely that the contribution from endogenous reserves of low specific activity must be small. Such high relative specific activities are not found commonly. Therefore, in many circumstances some dark respiratory activity may continue in the light.

The above conclusion is amplified by the somewhat more definitive biochemical experiments which indicate that initially in the light the TCA cycle is inhibited, but subsequently it can operate at a rate comparable with that in the dark. This would imply supply of acetyl-CoA via glycolysis or fatty acid oxidation also at a rate comparable with that in the dark. It is possible that such a situation occurs in younger, growing tissues while in mature tissues the TCA cycle may function at a lower rate and be subject to some inhibition in the light. This view would accord with the essential function of the TCA cycle as a source of carbon skeletons for synthetic reactions which are of greater magnitude in young than in mature tissues.

Acknowledgements. We are grateful to many colleagues who sent us reprints and preprints prior to publication and for helpful comments, particularly R.G.S. Bidwell, D.T. Canvin, C.B. Osmond and D.A. Walker. For help in the preparation of the manuscript we are grateful to Mrs. Val Petro.

References

D'Aoust, A.L., Canvin, D.T.: Photosynthetica *6,* 150–157 (1972)
Barthová, J., Leblová, S.: Biol. Plant. *11,* 97–109 (1969)
Benson, A.A., Calvin, M.: J. Exp. Bot. *1,* 63–68 (1950)
Bidwell, R.G.S., Krotkov, G., Reed, G.B.: Can. J. Bot. *33,* 189–196 (1955)
Björkman, O., Gauhl, E., Nobs, M.A.: Carnegie Inst. Wash. Yearb. *68,* 620–633 (1970)
Black, C.C.: Ann. Rev. Plant Physiol. *24,* 253–286 (1973)
Bravdo, B.: Plant Physiol. *43,* 479–483 (1968)

Brown, J.S.: Ann. Rev. Plant Physiol. *23*, 73–86 (1972)

Bull, T.A.: Crop Sci. *9*, 726–729 (1969)

Canvin, D.T., Fock, H.: Methods enzymol. *24*, 246–260 (1972)

Canvin, D.T., Lloyd, N.D.H., Fock, H., Przybylla, K.: In: CO_2 metabolism and plant productivity. Burris, R.H., Black, C.C. (eds.), pp. 161–176. Baltimore: Univ. Park Press 1976

Chapman, E.A., Graham, D.: Plant Physiol. *53*, 879–885 (1974a)

Chapman, E.A., Graham, D.: Plant Physiol. *53*, 886–892 (1974b)

Chapman, E.A., Osmond, C.B.: Plant Physiol. *53*, 893–898 (1974)

Davies, D.D., Giovanelli, J., ap Rees, T.: Plant biochemistry. Oxford: Blackwell Scientific Publications 1964

Downton, W.J., Tregunna, E.B.: Plant Physiol. *43*, 923–929 (1968)

Elias, B.A., Givan, C.V.: Plant Physiol. *59*, 738–740 (1977)

El-Sharkawy, M.A., Loomis, R.S., Williams, W.A.: Physiol. Plant. *20*, 171–186 (1967)

Erismann, K.H., Kirk, M.R.: In: Prog. Photosynthetic Res. Metzner, H. (ed.), Vol. III, pp. 1538–1545. Tübingen: Int. Union Biol. Sci. 1969

Forrester, M.L., Krotkov, G., Nelson, C.D.: Plant Physiol. *41*, 422–427 (1966a)

Forrester, M.L., Krotkov, G., Nelson, C.D.: Plant Physiol. *41*, 428–431 (1966b)

Gibbs, M.: Archiv. Biochem. Biophys. *45*, 156–160 (1953)

Givan, C.V., Harwood, J.L.: Biol. Rev. *51*, 365–406 (1976)

Goldsworthy, A.: Phytochem. *5*, 1013–1019 (1966)

Graham, D., Cooper, J.E.: Aust. J. Biol. Sci. *20*, 319–327 (1967)

Graham, D., Walker, D.A.: Biochem. J. *82*, 554–560 (1962)

Heath, O.V.S.: The physiological aspects of photosynthesis. London: Heinemann Educational Books Ltd. 1969

Heber, U.: Ann. Rev. Plant Physiol. *25*, 393–421 (1974)

Heber, U., Santarius, K.A.: Biochim. Biophys. Acta *109*, 390–408 (1965)

Heber, U., Willenbrink, J.: Biochim. Biophys. Acta *82*, 313–324 (1964)

Heber, U., Hallier, U.W., Hudson, M.A.: Z. Naturforsch. *22b*, 1200–1215 (1967)

Heldt, H.W.: In: The intact chloroplast. Barber, J., (ed.), pp. 215–234. Amsterdam: Elsevier/North-Holland Biomedical Press 1976

Heldt, H.W., Sauer, F., Rapley, L.: In: Prog. Photosynthesis. Forti, G., Avron, M., Melandri, A., (eds.), pp. 1345–1355. Proc. 2nd Int. Congr. Photosynth. Res. The Hague: Junk N.V. 1972

Hew, C-S., Krotkov, G.: Plant Physiol. *43*, 464–466 (1968)

Hew, C-S., Krotkov, G., Canvin, D.T.: Plant Physiol. *44*, 662–670 (1969)

Hiller, R.G.: J. Exp. Bot. *21*, 628–638 (1970)

Hillman, W.S.: Ann. Rev. Plant Physiol. *27*, 159–179 (1976a)

Hillman, W.S.: Proc. Natl. Acad. Sci. USA *73*, 501–504 (1976b)

Hillman, W.S.: Nature (London) *266*, 833–835 (1977)

Hoch, G., Owens, O.v.H., Kok, B.: Arch. Biochem. Biophys. *101*, 171–180 (1963)

Holmgren, P., Jarvis, P.G.: Physiol. Plant. *20*, 1045–1051 (1967)

Huffaker, R.C., Peterson, L.W.: Ann. Rev. Plant Physiol. *25*, 363–392 (1974)

Jackson, W.A., Volk, R.J.: Ann. Rev. Plant Physiol. *21*, 385–432 (1970)

Kachru, R.B., Anderson, L.E.: Plant Physiol. *55*, 199–202 (1975)

Kelly, G.J., Gibbs, M.: Plant Physiol. *52*, 674–676 (1973)

Kelly, G.J., Latzko, E.: Plant Physiol. *60*, 295–299 (1977)

Krause, G.H., Bassham, J.A.: Biochim. Biophys. Acta *172*, 553–565 (1969)

Krause, G.H., Heber, U.: In: Topics in photosynthesis. The intact chloroplast. Barber, J. (ed.), Vol. 1, pp. 171–214. Amsterdam: Elsevier 1976

Lake, J.V.: Aust. J. Biol. Sci. *20*, 487–493 (1967)

Lefèbvre, A., Ribéreau-Gayon, P.: C.Rr. hebd. Seanc. Acad. Sci. Paris *270D*, 1727–1729 (1970)

Ludwig, L.J., Canvin, D.T.: Plant Physiol. *48*, 712–719 (1971)

MacLachlan, G.A., Porter, H.K.: Proc. R. Soc. *B 150*, 460–473 (1959)

Mahon, J.D., Fock, H., Höhler, T., Canvin, D.T.: Planta (Berl.) *120*, 113–123 (1974)

Mangat, B.S., Levin, W.B., Bidwell, R.G.S.: Can. J. Bot. *52*, 673–681 (1974)

Meidner, H.: J. Exp. Bot. *21*, 1067–1075 (1970a)

Meidner, H.: Nature (London) *228*, 1349 (1970b)

Miflin, B.J., Lea, P.J.: Ann. Rev. Plant Physiol. *28*, 299–329 (1977)

Mitchell, C.A., Stocking, C.R.: Plant Physiol. *55*, 59–63 (1975)

Moss, D.M.: Crop Sci. *6*, 351–354 (1966)

Naylor, A.W., Rabson, R., Tolbert, N.E.: Physiol. Plant. *11*, 537–547 (1958)

Naylor, A.W., Tolbert, N.E.: Physiol. Plant. *9*, 220–229 (1956)

Nishida, K.: Physiol. Plant. *15*, 47–58, (1962)

Poskuta, J.: Physiol. Plant. *21*, 1129–1136 (1968)

Poskuta, J.: Physiol. Plant. *22*, 76–85 (1969)

Poskuta, J., Nelson, C.D., Krotkov, G.: Plant Physiol. *42*, 1187–1190 (1967)

Rabinowitch, E.I.: Photosynthesis and related processes. Vol. I. New York: Interscience Publishers Inc. 1945

Raven, J.A.: New Phytol. *71*, 227–247 (1972a)

Raven, J.A.: New Phytol. *71*, 995–1014 (1972b)

Rosenstock, G., Ried, A.: In: Handbuch der Pflanzenphysiologie. Ruhland, M. (ed.), Vol. XII, pp. 259–333. Berlin, Heidelberg, New York: Springer 1960

Samish, Y., Koller, D.: Plant Physiol. *43*, 1129–1132 (1968)

Samish, Y.B., Pallas, J.E., Jr., Dornhoff, D.M., Shibles, R.M.E.: Plant Physiol. *50*, 28–30 (1972)

Santarius, K.A., Heber, U.: Biochim. Biophys. Acta *102*, 39–54 (1965)

Stafford, H.: Ann. Rev. Plant Physiol. *25*, 459–486 (1974)

Stepanova, A.M., Baranova, A.A.: Biokhimiya *37*, 520–526 (1972)

Stocking, C.R., Larson, S.: Biochem. Biophys. Res. Commun. *37*, 278–282 (1969)

Tolbert, N.E.: Ann. Rev. Plant Physiol. *22*, 45–74 (1971)

Tregunna, E.B., Krotkov, G., Nelson, C.D.: Physiol. Plant *19*, 723–733 (1966)

Turner, J.F., Turner, D.H.: Ann. Rev. Plant Physiol. *26*, 159–186 (1975)

Voskresenskaya, N.P.: Ann. Rev. Plant Physiol. *23*, 219–234 (1972)

Walker, D.A.: Biol. Rev. *37*, 215–256 (1962)

Walker, D.A.: In: Encyclopedia of plant physiology. New Series. Pirson, A., Zimmerman, M. (eds.), Vol. I, pp. 85–136. Berlin, Heidelberg, New York: Springer 1976

Werdan, K., Heldt, H.W.: In: Proc. 2nd Int. Congr. Photosynthesis Res. Forti, G., Avron, M., Melandri, A. (eds.), pp. 1337–1344. The Hague: J.V. Junk 1972

Wiskich, J.T.: Ann. Rev. Plant Physiol. *28*, 45–69 (1977)

Woo, K.C., Osmond, C.B.: Aust. J. Plant Physiol. *3*, 771–785 (1976)

Woo, K.C., Osmond, C.B.: Plant Cell Physiol. Spec. Issue 315–323 (1977)

Zelitch, I.: Plant Physiol. *41*, 1623–1631 (1966)

Zelitch, I.: Photosynthesis, photorespiration and plant productivity. New York: Academic Press 1971

Zelitch, I.: Proc. Natl. Acad. Sci. USA *70*, 579–584 (1973)

Zelitch, I.: Ann. Rev. Biochem. *44*, 123–145 (1975)

12. The Interaction of Respiration and Photosynthesis in Microalgae

E.H. Evans and N.G. Carr

A. Introduction

For the purposes of this article we shall consider that group of micro-organisms that are capable of oxygenic photosynthesis, thus encompassing examples of euka-ryotic red and green algae as well as prokaryotic blue-green algae. The relationship between respiration and photosynthesis in these organisms can be considered to be the interaction between light and dark growth and is therefore closely related to the potential for photoautotrophy, photoheterotrophy and heterotrophy exhib-ited by each alga. The difference in cellular arrangement of respiratory and photo-synthetic apparatus between prokaryotic and eukaryotic cells is central to under-standing the interaction of these processes in different organisms. The separation, in all eukaryotes, of the two processes into distinct organelles (mitochondria and chloroplasts) is without parallel in prokaryotes. These do not possess specialised organelles, and photosynthetic and respiratory electron transport must occur in close juxtaposition to each other, possibly on the same membrane. The best charac-terised prokaryotes in this respect are the Athiorhodaceae, which photosynthesise using hydrogen or organic substrates as reductant.

The inhibition of respiration in light in photosynthetic bacteria was observed and documented 40 years ago, (NAKAMURA, 1937; VAN NIEL, 1941) and shown to occur in the bacterial chromatophore (pigment-containing membrane vesicles; KATOH, 1961). HORIO and KAMEN (1962) examined the competition between respira-tion and photosynthesis in *Rhodospirillum rubrum* in some detail. *Oscillatoria limnetica* is one of several blue-green algae that can switch from oxygenic two-photosystem photosynthesis to anoxygenic one-photosystem photosynthesis which uses sulphide as an electron donor (COHEN et al., 1975a, 1975b). The fact that the same microorganism may carry out bacterial and green plant photosynthesis according to the environmental conditions prevailing encourages the speculation that the interaction between respiration and photosynthesis in blue-green algae may be similar to that in Athiorhodaceae.

B. The Kok Effect

Interest in the interaction of respiration and photosynthesis in algae stemmed from the observation by KOK (1949) that the net oxygen exchange of an illuminated algal suspension was not a linear function of light intensity. This non-linearity has become known as the "Kok Effect" and was subsequently demonstrated in a number of unicellular algae, including *Haematococcus* sp. (Kok, 1949), *Oscilla-toria* sp. and *Symploca* sp. (Šetlik, 1957), *Anacystis nidulans* and *Anabaena variabilis* (Hoch et al., 1963), *Fragillaria sublinearis* Bunt, 1969), and *Chlorella fusca* (Ga-brielson and Vejlby, 1959). These organisms provide a reasonable cross section of algae, and only a few organisms, for example *C. fusca*, do not show a significant Kok effect (Jones and Myers, 1963). Kok (1949) suggested that the effect was due to a light-induced depression of respiration, and it is noteworthy that the blue-green alga *A. variabilis* required higher light intensities to produce the Kok effect than did *A. nidulans* (Jones and Myers, 1963) since the former organism also has a higher endogenous respiration rate (Kratz and Myers, 1955). Hoch et al. (1963) used an improved mass spectrometer inlet system to investigate $^{18}O_2$ uptake in the light and showed that two separate systems were functioning. Dark respiration was inhibited by light, but at higher light intensities an oxygen uptake dependant on light intensity was resolved. This has become known as photorespira-tion. Unfortunately, Hoch et al. (1963) were only able to use low oxygen concentra-tions in order to conserve the tracer, so the rates of photorespiration measured were low.

Earlier work on the relationship between respiration and photosynthesis based on the Kok effect has been reviewed by Gibbs (1962) and the remainder of this article will deal with later results and related problems.

C. Electron Transport Mechanisms for the Kok Effect

I. General Considerations

The characteristics of the Kok effect have been investigated by several workers. The light inhibition of respiration was increased by light activation of photosystem I only (Hoch et al., 1963; Jones and Myers, 1963), and was DCMU-insensitive. In *Scenedesmus* sp., the Kok effect was seen only in the presence of DCMU (Hoch et al., 1963) and anaerobiosis inhibited the effect (Healey and Myers, 1971). Following a short period of illumination, transients are seen in the rate of oxygen uptake, and these may also be related to light control of respiration (Ried, 1968). The components c and d of Ried, an oscillation in oxygen uptake in *Chlorella fusca* following the light period, were concluded to be due to dark respiration. The characteristics of this oscillation were affected by light, in particular far red light, exciting photosystem I. The uptake was stimulated by 2,4 dinitrophe-nol and inhibited by CCCP and antimycin A. The inhibitor effects led Ried to suggest that the regulation of the light effect on respiration was mediated by the energy charge ratio of the cytoplasm, and that the effect required a high ATP concentration. However, uncouplers have been shown to inhibit neither the Kok effect nor the light-off oxygen transients exhibited by *Chlamydomonas reinhar-dii* (Healey and Myers, 1971). These workers concluded, using their own results, and those of Ried (1968) that both the light-off transients and the Kok effect are manifestations of photodepression of oxygen uptake dependent on photosys-tem I.

RIED and co-workers (RIED and ŠETLIK, 1970; ŠETLIK et al., 1973) have analysed the light-off transients in considerable detail in seven algal species. They found differences in the damping of oscillations in these algae, but found that the least amount of damping was exhibited by *A. nidulans*, but on the whole, the same characteristics were exhibited by both prokaryotic blue-green algae and eukaryotic green algae.

ŠETLIK et al. (1973) further analysed the oxygen transients in terms of four possible variables.

a) Photosynthetic evolution of oxygen.
b) Photosynthetic inhibition of respiration.
c) Photosynthetic stimulation of respiration.
d) Photosynthetic uptake of oxygen.

Conclusive analysis of each curve in these terms was not possible, but it seemed that reproducible kinetics were given by cells in the same stage of growth.

HOCH et al. (1963) proposed that cyclic photophosphorylation mediated by photosystem I caused the ATP:ADP ratio in the cytoplasm to rise, either due to direct movement of ATP or the involvement of a shuttle system as discussed later (Sect. E) and this depressed respiration in the mitochondria (of eukaryotic algae). This theory would be supported by the apparent inhibition of glycolysis under these conditions (HEBER et al., 1964) and the failure of recently fixed ^{14}C to enter the tricarboxylic acid cycle in the light (BASSHAM et al., 1956). HEALEY and MYERS (1971) prefer the theory that, in eukaryotic algae, reductant from the mitochondria is moved, in the light, to the chloroplasts, and oxidised via photosystem I. That this is possible has been demonstrated by investigations of hydrogen production by algae.

STUART and KALTWASSER (1970), using *Scenedesmus* sp., showed that light-induced hydrogen production exhibited an action spectrum of photosystem I, and was not inhibited by DCMU, but was inhibited by CCCP. A mutant lacking cytochrome f showed no light-induced hydrogen production, whereas one lacking photosystem II did so. Similar results were found using *Chlamydomonas moewusii* (HEALEY, 1970). In addition, HEALEY demonstrated that acetate stimulated light-induced hydrogen production and that monofluoracetic acid was an inhibitor, thereby confirming that the origin of the chloroplast reducing power was oxidative carbon metabolism.

RIED and ŠETLIK (1970) speculated that photophosphorylation controls the flux through the glycolytic pathway by allosteric inhibition of phosphofructokinase by ATP. This hypothesis would relate the oscillations in oxygen transients to the glycolytic oscillations in yeast observed by CHANCE et al. (1964). Reductant other than that generated by photosynthesis can be used for carbon dioxide fixation by the blue-green alga, *Chlorogloea fritschii* (EVANS et al., 1977). Following dark heterotrophic growth, CO_2 fixation recovery is immediate, and only inhibited to 70% of the control rate by DCMU or DBMIB. Photosystem I-dependent oxidation of organic acids in lamellar fragments of photoautotrophically grown *A. variabilis* has been demonstrated (MURAI and KATOH, 1975) and *Oscillatoria limnetica* will grow by oxidizing sulfide via photosystem I (COHEN et al., 1975a, b). Thus it seems possible, in blue-green algae, that, under certain conditions, photosystem II can be bypassed. RIPPKA (1972) has characterised this capacity for facultative photoheterotrophy in unicellular blue-green algae, as determined by their ability to grow in the light in the presence of DCMU.

Controversy has existed for several years about a relationship between the respiratory and photosynthetic electron transport chains in algae. At the present time, it seems possible that a connection between the two chains may exist in blue-green algae, but no evidence can be presented to suggest this for eukaryotes.

II. Prokaryotes

The respiratory activities of blue-green algae are well documented in whole cells, however, the low rates of oxygen uptake – Q_{O_2} values of 1–10 being common – have made cell free analysis difficult (CARR, 1973). LEACH and CARR (1968, 1969, 1970) observed both NADH and NADPH oxidation in darkness in homogenates of *Anabaena variabilis,* and found this oxidation to be linked to phosphorylation (LEACH and CARR, 1970). These workers, and BIGGINS (1969) using *Anacystis nidulans,* found that NADPH oxidase activity was higher than NADH oxidase, and that it was localised in the supernatant of a high speed centrifugation of an algal homogenate. It might be possible, therefore, that this activity was contaminated by an active NADP-ferredoxin reductase in the supernatant, which has been shown, when isolated from spinach chloroplasts, to be able to give NADPH-cytochrome f reductase activity (FORTI and STURANI, 1968). However, studies with intact *A. nidulans* (BIGGINS, 1969) have shown fluctuation of $NADP^+$ pools, and not NAD^+, on alteration in the aeration supply, providing further evidence that NADPH is the major source of respiratory electrons in this organism. HORTON (1968) was able to demonstrate NADH oxidase activity in membrane preparations of *A. nidulans,* which was sensitive to the inhibitors cyanide, HOQNO, rotenone and amytal. EVANS and GRIFFITHS have been able to confirm this using membrane preparations also giving measurable photosystem I and II activity (unpublished data). Endogenous respiration of *A. nidulans* in vivo has been shown to be sensitive to cyanide, although not completely inhibited (BIGGINS, 1969). In *Anabaena variabilis* no substantial difference is found between the ATP levels in the dark and the light (IMAFAKU and KATOH, 1976), and this is also true of *Chlorella fusca* (LEWENSTEIN and BACHOFEN, 1972). However, using *Anacystis nidulans,* BORNEFELD and SIMONIS (1974) demonstrated a lower level of ATP in dark-incubated cells. In all photosynthetic cells, under anaerobic dark conditions, the ATP levels fall. The lower dark ATP level in *A. nidulans* may be a reflection of its low respiratory rate.

In view of the contradictory data on respiratory measurements using cell-free preparations of blue-green algae, it is not surprising that definitive results on the relationship between the photosynthetic and respiratory chains are unavailable. BISALPUTRA et al. (1969) showed that reduced tellurite was deposited on the photosynthetic lamellae of *Nostoc sphaericum* both in the light, and in the dark in the presence of succinate. This led them to suggest that the photosynthetic lamellae also function as mitochondrial equivalents. HOLTON and MYERS (1967) have proposed that cytochrome c (549) identified by BIGGINS (1967) could be a link between the respiratory and photosynthetic chains of *Phormidium luridum.* They point out, that although the cytochrome c-549-carbon monoxide adduct of *A. nidulans* had a similar spectrum to the cytochrome c-carbon monoxide adduct of colourless blue-green algae (WEBSTER and HACKETT, 1966), certain features implied that cytochrome c-549 was not the oxidase of *A. nidulans*. PULICH (1977) has isolated a c-type, carbon monoxide-binding cytochrome from an endophytic *Nostoc* sp. This cytochrome c-548 was proposed to participate in the respiratory chain, being reduced by NADPH in the presence of a heat labile component present in cell free extracts. PULICH speculates that this component might be ferredoxin-NADP reductase functioning as NADP-cytochrome reductase. This idea has already been mentioned in connection with *A. nidulans*; whether such an interaction exists in the intact algae is untested. Kinetic measurements of cytochromes c-549 (*Phormidium luridum*) and cytochrome c-548 (*Nostoc* sp.) in light in the presence and absence of oxygen might resolve their function.

III. Eukaryotes

It was suggested by HIJAMA et al., (1969) that cytochrome b-563 (cytochrome b_6) could perform a similar function in the pale green mutant (ATCC 18302) of *Chlamydomonas reinhardii* as cytochrome c-549 of *Phormidium luridum*. Because of the interaction of the cytochrome with oxygen and NAD(P)H, the scheme shown in Fig. 1 was proposed, involving a proposed respiratory chain, either within or without the chloroplast. However, subsequent kinetic data on the role of cytochrome b_6, reviewed recently by CRAMER and WHITMARSH (1977), places cytochrome b_6 on the pathway of cyclic photophosphorylation associated with photosystem I, but does not confirm any respiratory link. Cytochrome b_6 is isolated either in a digitonin-produced photosystem I particle or as complex with cytochrome f, where the limiting stoichiometry is 2 cytochrome b_6:1 cytochrome f (NELSON and RACKER, 1972). Cytochrome b_6 appears to be unstable to isolation, and can also appear to be labile during experiments in situ. Thus the nature of the interaction between cytochrome b_6 and oxygen, proposed by HIJAMA et al. (1969) is difficult to assess.

JAMES and LEECH (1964) also found that cytochrome b_6 was auto-oxidisable, but BOHME and CRAMER (1972), using anaerobic chloroplasts, showed that DBMIB, an inhibitor of the electron transport chain acting at plastoquinone, inhibited photosystem I-mediated cytochrome b_6 photooxidation. It has also been suggested that cytochrome b_6 is oxidised by cytochrome f (DOLAN and HIND, 1974). DOLAN and HIND considered that the reoxidation kinetics of cytochrome b_6 corresponded to the re-reduction kinetics of cytochrome f in the light and the dark. Thus it may be that cytochrome b_6 is oxidised via plastoquinone and/or cytochrome f. Any of these three components then might be oxidised by oxygen.

D. The Interaction of Oxygen with the Photosynthetic Electron Transport Chain

Inhibition of photosynthesis by oxygen was first observed by WARBURG (1920), and the nature of the interaction of oxygen with the photosynthetic chain has

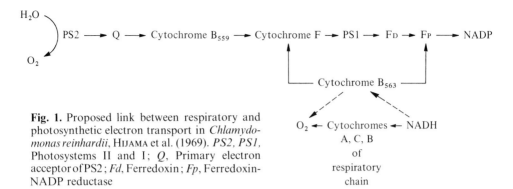

Fig. 1. Proposed link between respiratory and photosynthetic electron transport in *Chlamydomonas reinhardii*, HIJAMA et al. (1969). *PS2, PS1*, Photosystems II and I; *Q*, Primary electron acceptor of PS2; *Fd*, Ferredoxin; *Fp*, Ferredoxin-NADP reductase

been investigated by many authors (eg., JAMES and LEECH, 1964; HIJAMA et al., 1969; ŠETLIK et al., 1973; DOLAN and HIND, 1974; CRAMER and WHITMARSH, 1977).

HEBER and FRENCH (1968) considered that many effects could be explained by the interaction of oxygen with the primary reductant of photosystem I. Photosystem II fluorescence was quenched by oxygen. This quenching was inhibited by DCMU and in mutants lacking photosystem I. However, these results would also be observed if cytochrome b_6 were reduced by photosystem I and reoxidised by oxygen, as suggested by HIJAMA et al. (1969). ŠETLIK et al. (1973) found that photosynthetic oxygen uptake was stimulated by light exciting photosystem II, and diminished considerably in the presence of light exciting photosystem I. They suggested that oxygen uptake was due to reoxidation of an electron transport carrier, which, in a reduced form, caused inhibition of electron transport. Increased light-induced oxygen uptake was associated with a state of increased demand for ATP in the cell. Thus oxygen released the constraint on noncyclic electron transport; and photosystem I-mediated cyclic electron transport, by raising the ATP level, inhibited light-induced oxygen uptake.

Prolonged dark incubation resulted in a marked decrease in respiration in *Chlorella pyrenoidosa, Synechococcus* sp. (BROWN, 1953; GOEDHEER, 1963) and *Plectonema boryanum* (PADAN et al., 1971). This has been explained as interaction between the photosynthetic and respiratory chains. This effect was more pronounced in *Synechococcus* sp. than *C. pyrenoidosa*, and respiration was stimulated by subsequent illumination (GOEDHEER, 1963).

Presumably, this stimulation of oxygen uptake following illumination could involve reoxidation of photosynthetic electron transport mediators distinct from energy-yielding dark oxidation of substrates via the respiratory chain. GOEDHEER (1963) suggests that oxygen may play a role in poising the redox potential of the photosynthetic chain for maximum efficiency. Such an effect has been suggested for photosynthetic bacteria (NEWTON and KAMEN, 1957; CUSANOVICH et al., 1968; EVANS and CROFTS, 1974; DUTTON et al., 1977). EVANS and CROFTS (1974) have shown that photosynthetic electron transport in *Rhodopseudomonas capsulata* increases over a discrete redox potential range, a finding confirmed by DUTTON et al. (1977) using *Rhodopseudomonas spheroides*. GOEDHEER (1963) investigated delayed fluorescence, in particular, induction effects of *C. pyrenoidosa* and *Synechococcus* sp. following periods of illumination. He explained the weak induction peak after 20 min darkness as a consequence of the reoxidation of intermediate carriers between the two photosystems via a respiratory chain. This effect was most pronounced with *Synechecoccus* sp. and *Anacystis nidulans* leading him to suggest that this oxidative capacity was 10 times greater in these blue-green algae than in *C. pyrenoidosa*.

However, the dark respiratory rate of *A. nidulans* (1.6 µl O_2 per gm dry wt. per h) is similar to that of *C. pyrenoidosa* (1.0–1.4 µl O_2 per gm dry wt. per h). After phototrophic growth, PADAN et al. (1971) suggested that the increase in respiratory rate following illumination was a consequence of photoassimilation of CO_2, and light-induced protein synthesis, in *Plectonema boryanum*. They felt that the respiratory rate in this blue-green alga was governed solely by the physiological state of the cells, and that the requirement for carbon dioxide assimilation was explained by a need for oxidative substrate. EVANS et al. (1977) have observed a decrease in the respiratory rate of *Chlorogloea fritschii* grown heterotrophically in darkness, following introduction into the light, which may be the consequence of a diversion of substrate from the respiratory chain to carbon dioxide fixation.

There may, therefore, be three possible interactions between photosynthesis and respiration; between electron transport mediators; indirectly via endogenous redox potential, and via availability of substrate to the two systems, a consequence of compartmentation. All these possibilities may exist in both prokaryotes and eukaryotes, but, at the present time, it seems likely that interaction between electron transport chains does not occur in eukaryotes. (See CRAMER and WHITMARSH, 1977).

E. Metabolically Mediated Control of Oxygen Uptake

The relative permeability of chloroplasts isolated from higher plants, and mitochondria to NAD(P)H and ATP may contribute to the relationship between respiration and photosynthesis.

The energetics of intact chloroplasts have recently been reviewed by KRAUSE and HEBER (1976). That reducing power from the cytoplasm may be made available inside the chloroplast for hydrogen production has already been mentioned, (STUART and KALTWASSER, 1970; HEALEY, 1970). However, it is probable that the $NADP^+$: NADPH ratio in the chloroplast is controlled internally. The oxidative reactions catalysed by glucose 6-phosphate dehydrogenase and 6-phosphogluconate dehydrogenase produce NADPH, and both have been detected in chloroplasts (HEBER et al., 1967). The former enzyme, but not the latter, is inactivated by a light-mediated process (LENDZIAN and ZIEGLER, 1972). Thus NADPH may be produced in the chloroplast in darkness. Several lines of evidence indicate that the chloroplast envelope does not permit penetration of pyridine nucleotides HEBER and SANTARIUS, 1965; HEBER et al., 1967; MATHIEU, 1967; ROBINSON and STOCKING, 1968; HELDT and SAUER, 1971), and similarly, the rates of penetration of adenylates are very low (HEBER et al., 1967; HEBER and SANTARIUS, 1970). The direct transfer of ATP, ADP and AMP resulting from diffusion and specific translocation is

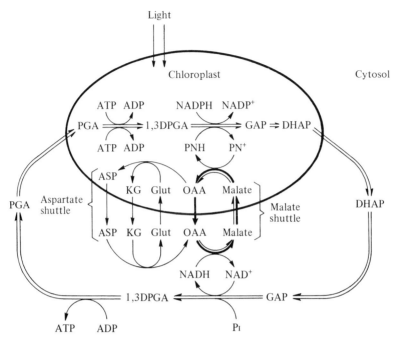

Fig. 2. Schematic representation of metabolite transfer between chloroplast and cytosol designed to export ATP from chloroplasts in the light. Shuttle transfer of dihydroxyacetone phosphate (DHAP) and phosphoglycerate (PGA) mediates indirect transport of NADH and ATP. Back transfer of NADH is possible by cyclic transfer of malate and oxaloacetate (OAA) or of malate, glutamate (Glut), α-ketoglutarate (KG) and aspartate (ASP). KRAUSE and HEBER (1976). *DPGA,* diphosphoglyceric acid; *GAP,* glyceraldehyde-3-phosphate

probably limited to rates lower than 5 µmol per mg chl per h (KRAUSE and HEBER, 1976). This is insufficient for observed changes in cytoplasmic ATP levels and for ATP utilisation of chloroplasts in darkness. It must, therefore, be concluded that both NAD(P)H and adenylates are transported by indirect methods in and out of the chloroplast. These shuttle systems have been reviewed (HEBER, 1974; KRAUSE and HEBER, 1976), and, briefly summarising, ATP may be transported from chloroplasts in the light by a shuttle transfer of dihydroxyacetone phosphate and phosphoglycerate. Transfer of NADH is possible by cyclic transfer of malate and oxaloacetate, or of malate, glutamate, α-ketoglutarate and aspartate. These shuttle systems are summarised in Fig. 2. It must be emphasised, however, that this information has been derived from the spinach chloroplast, and that algal chloroplasts may well differ in some respects.

In darkness, the energy demands of an alga must be met by glycolysis and respiration. In animal cells the respiratory rate is controlled by the phosphorylation potential of the mitochondria in relationship to that of the cytoplasm (CHANCE and WILLIAMS, 1956). Thus it seems reasonable to suppose that, in the light, ATP production by the chloroplast suppresses respiration, and, in darkness, this control is lifted (HOCH et al., 1963; KRAUSE and HEBER, 1976).

However, it has been suggested by PALMER (1976) that a non-phosphorylating branch of the respiratory electron transport chain exists, and that the major role of plant mitochondria may not be ATP production, but rather the production of carbon skeletons for biosynthetic purposes via the tricarboxylic acid cycle. Thus, a non-phosphorylating electron transport chain removing NADH, may allow this role to be fulfilled in the presence of high levels of ATP, produced by the chloroplast.

F. Synopsis

Many algae are capable of both heterotrophic growth in the dark and photoheterotrophic and/or photoautotrophic growth in the light. This could be considered a measure of the degree of co-operation between respiration and photosynthesis at cellular level.

Certain strains of *Chlorella* have a higher capacity for oxidative phosphorylation when grown in the dark compared with the light, despite a higher growth rate under photolithotrophic conditions (SOROKIN and KRAUSS, 1962; PULICH and WARD, 1973; ENDO et al., 1974). The lower capacity for oxidative phosphorylation in photolithotrophic cells is correlated with the number of mitochondria per cell, which may comprise almost 20% of the cell volume in the apochlorotic alga *Chlamydomonas* sp. and *Chlorella pyrenoidosa* (SCHOTZ et al., 1972; ATKINSON et al., 1974). In the illuminated algal cell, evidence suggests that the tricarboxylic acid cycle is operational, and functioning as a carbon skeleton synthesis pathway (RAMBECK and BASSHAM, 1973; RAVEN, 1972). The supply of reductant for processes other than carbon fixation is provided by the photosynthetic electron transport chain (see discussion in RAVEN, 1976). Insufficient energy is produced by respiration to be a significant contribution to the energy requirements of photolithotrophic growth, (RAVEN, 1972).

Blue-green algae are restricted in their generation of respiratory reductant by their possession of an interrupted tricarboxylic acid cycle (see CARR, 1973). The absence of α-ketoglutarate dehydrogenase limits the oxidation of two-carbon units, and is perhaps the ultimate expression of the tricarboxylic acid "cycle" being used for biosynthetic rather than oxidative purposes. However, phosphorylation-linked respiratory electron flow does occur and clearly can contribute to the maintenance, if not the growth, of all blue-green algae. The conservation of this mode of ATP generation in the presence of light is accomplished by the light-inhibition of respiration. This interaction may be via a direct connection between the two processes in blue-green algae, but is more likely, in eukaryotic microalgae, to be an indirect control exerted by variation of metabolite, ATP and NAD(P)H pools in chloroplast, mitochondrion and cytoplasm.

References

Atkinson, Jr., A.W., John, P.C.L., Gunning, B.E.S.: Protoplasma *81*, 77–110 (1974)
Bassham, J.A., Shibata, K., Steenberg, K., Bourden, J., Calvin, M.: J. Am. Chem. Soc. *78*, 4120–4124 (1956)

Biggins, J.: Plant Physiol. *42*, 1442–1447 (1967)

Biggins, J.: J. Bacteriol. *99*, 570–575 (1969)

Bisalputra, T., Brown, D.L., Weier, T.E.: J. Ultrastruct. Res. *27*, 182–197 (1969)

Böhme, H., Cramer, W.A.: Biochim. Biophys. Acta. *283*, 302–315 (1972)

Bornefeld, T., Simonis, W.: Planta *115*, 309–318 (1974)

Brown, A.H.: Am. J. Bot. *40*, 719–729 (1953)

Bunt, J.: Nature (London), *207*, 1373–1375 (1969)

Carr, N.G.: In: The biology of blue-green alga. Carr, N.G., Whitton, B.A. (eds.), pp. 39–65. Oxford: Blackwells Scientific Publications Ltd. 1973

Chance, B., Williams, G.R.: Adv. Enzymol. *17*, 65–98 (1956)

Chance, B., Estabrook, R.W., Ghosh, A.: Proc. Natl. Acad. Sci. USA *51*, 1244–1251 (1964)

Cohen, Y., Padan, E., Shilo, M.: J. Bacteriol. *123*, 855–861 (1975a)

Cohen, Y., Jørgensen, B.B., Padan, E., Shilo, M.: Nature (London) *257*, 489–492 (1975b)

Cramer, W.A., Whitmarsh, J.: Ann. Rev. Plant Physiol. *28*, 133–172 (1977)

Cusanovich, A.M., Bartsch, R.G., Kamen, M.D.: Biochim. Biophys. Acta. *153*, 397 (1968)

Dolan, E., Hind, G.: Biochim. Biophys. Acta. *357*, 380–385 (1974)

Dutton, P.L., Prince, R.C., van der Berg, W.H., Takamiya, K.: Proc. 4th Int. Cong. Photosynthesis, Reading, Abstract. p. 100 (1977)

Endo, H., Nakajima, K., Chino, R., Shirata, M.: Agric. Biol. Chem. *38*, 9–18 (1974)

Evans, E.H., Crofts, A.R.: Biochim. Biophys. Acta *357*, 78–88 (1974)

Evans, E.H., Carr, N.G., Evans, M.C.W.: Biochim. Biophys. Acta *501*, 165–173 (1977)

Forti, G., Sturani, E.: Eur. J. Biochem. *3*, 461 (1968)

Gabrielson, E.K., Vejlby, K.: Physiol. Plant. *12*, 425–440 (1959)

Gibbs, M.: In: Physiology and biochemistry of algae. Lewin, R.A. (ed.), pp. 61–90. London, New York: Academic Press Inc. 1962

Goedheer, J.C.: Biochim. Biophys. Acta *66*, 61–71 (1963)

Healey, F.P.: Plant Physiol. *45*, 153–159 (1970)

Healey, F.P., Myers, J.: Plant Physiol. *47*, 373–379 (1971)

Heber, U.: In: Proc. 3rd Int. Congr. Photosynthesis. Avron, M. (ed.), Vol. II, pp. 1335–1348. Amsterdam: Elsevier/North Holland

Heber, U., French, C.S.: Planta. (Berl.) *79*, 99–112 (1968)

Heber, U., Santarius, K.A.: Biochim. Biophys. Acta. *109*, 390–408 (1965)

Heber, U., Santarius, K.A.: Z. Naturforsch. *25b*, 718–728 (1970)

Heber, U., Santarius, K.A., Urbach, W., Ullrich, W.: Z. Naturforsch. *19b*, 576–587 (1964)

Heber, U., Hallier, U.W., Hudson, M.A.: Z. Naturforsch. *22b*, 1200–1215 (1967)

Heldt, H.W., Sauer, F.: Biochim. Biophys. Acta *234*, 83–91 (1971)

Hijama, T., Nishimura, M., Chance, S.: Plant. Physiol. *40*, 1220–1227 (1969)

Hoch, G.O., Owens, vH.O., Kok, B.: Arch. Biochem. Biophys. *101*, 171–180 (1963)

Holton, R.W., Myers, J.: Biochim. Biophys. Acta. *131*, 362–374 (1967)

Horio, T., Kamen, M.D.: Biochemistry *6*, 1141, 1157 (1962)

Horton, A.A.: Biochem. Biophys. Res. Commun. *32*, 839–845 (1968)

Imafaku, H., Katoh, T.: Plant Cell Physiol. *17*, 515–524 (1976)

James, W.O., Leech, R.M.: Proc. Soc. Lond. *160B*, 13–24 (1964)

Jones, L.W., Myers, J.: Nature (London) *199*, 670–672 (1963)

Katoh, S.: J. Biochem. *49*, 126–132 (1961)

Kok, B.: Biochim. Biophys. Acta. *3*, 625–631 (1949)

Kratz, W.A., Myers, J.: Plant Physiol. *30*, 275 (1955)

Krause, G.H., Heber, U.: In: The intact chloroplast. Barber, J. (ed.) Elsevier/North Holland: Biomedical Press 1976

Leach, C.K., Carr, N.G.: Biochem. J. *109*, 4–5 (1968)

Leach, C.K., Carr., N.G.: Biochem. J. *112*, 125–126 (1969)

Leach, C.K., Carr, N.G.: J. Gen. Microbiol. *64*, 55–70 (1970)

Lendzian, K., Ziegler, H.: In: Proc. 2nd Inst. Cong. Photosynthesis. Forti, G., Avron, M., Melandri, A. (ed.), Vol. III, pp. 1831–1838. The Hague: Dr. W. Junk 1972

Lewenstein, A., Bachofen, R.: Biochim. Biophys. Acta *267*, 80–85 (1972)

Mathieu, Y.: Photosynthetica *1*, 57–63 (1967)

Murai, I., Katoh, T.: Plant Cell Physiol. *16*, 789–797 (1975)

Nakamura, H.: Acta Phytochem. *9*, 189–234 (1937)

Nelson, N., Racker, E.: J. Biol. Chem. *247*, 3848–3853 (1972)
Newton, J.W., Kamen, M.D.: Biochim. Biophys. Acta. *25*, 462 (1957)
Padan, E., Raboy, B., Shilo, M.: J. Bacteriol. *106*, 45–50 (1971)
Palmer, J.M.: Ann. Rev. Plant Physiol. *27*, 133–157 (1976)
Passam, H.C., Coleman, J.O.D.: J. Exp. Bot. *26*, 536–543 (1975)
Pulich, W.H.: J. Phycol. *13*, 40–45 (1977)
Pulich, W.H., Ward, C.H.: Plant Physiol. *51*, 337–344 (1973)
Rambeck, W.A., Bassham, J.A.: Biochim. Biophys. Acta *304*, 725–735 (1973)
Raven, J.A.: New Phytol. *71*, 227–247 (1972)
Raven, J.A.: In: The intact chloroplast, Barber, J. (ed.), pp. 403–443. Elsevier/North Holland: Biomedical Press 1976
Ried, A.: Biochim. Biophys. Acta *153*, 653–663 (1968)
Ried, A., Šetlik, I.: Czech. Acad. Sci. Inst. Microbiol. Annu. Rep. Lab. Algol. 1969, Trebon (1970)
Rippka, R.: Arch. Mikrobiol. *76*, 341–348 (1972)
Robinson, J.M., Stocking, C.R.: Plant Physiol. *42*, 1597–1604 (1968)
Schotz, F., Bathelt, H., Arnold, C.G., Schimmer, O.: Protoplasma *75*, 229–254 (1972)
Šetlik, I.: Cesk. Biol. *6*, 424–429 (1957)
Šetlik, I., Ried, A., Berkova, E.: Czech. Acad. Sci. Inst. Microbiol. Annu. Rep. Lab. Algol. 1970. Trebon (1973)
Sorokin, C., Krauss, R.W.: Plant Physiol. *37*, 37–42 (1962)
Stuart, T.S., Kaltwasser, H.: Planta (Berl.) *91*, 302–313 (1970)
van Niel, C.B.: Adv. Enzymol. *1*, 263–328 (1941)
Warburg, O.: Biochem. Z. *103*, 188–217 (1920)
Webster, D.A., Hackett, D.P.: Plant Physiol. *41*, 599 (1966)

13. Effect of Light Quality on Carbon Metabolism

N.P. Voskresenskaya

A. Introduction

The first experimental evidence for the possibility of controlling photosynthetic carbon metabolism by light quality was obtained in the middle of our century. By the time of this writing, the total effects of blue (BL) and red (RL) light on photosynthetic carbon metabolism have come to be well known. BL activates an accumulation of protein in newly synthesized organic material and RL activates accumulation of carbohydrates. Moreover, the phenomenon seems to be widely distributed both in higher plants and in algae. Further, it has been found that many vital biosyntheses can be regulated by simple alteration of light quality. Since many reviews exist (Pirson and Kowallik, 1964; Krotkov, 1964; Kowallik, 1970; Voskresenskaya, 1972), this article is largely limited to a characterization of our present knowledge of the problem.

B. Principal Effects of Blue and Red Light on Carbon Metabolism

When green plants grown under white light are illuminated for a few minutes or even seconds by BL or RL (under similar conditions for photosynthesis) significant differences in the rates of formation of some photosynthetic intermediates are seen. In a broad range of CO_2 concentrations and light intensities BL (λ max = 460–470 nm) activates the incorporation of labeled carbon into amino acids (alanine, aspartate, glutamate, and sometimes serine and glycine), and organic acids (malate and citrate). Additionally, the enhanced ^{14}C incorporation into the lipid fraction in the presence of blue light as compared with red light alone was observed by Miyachi (1977). When the exposures are more prolonged, activation of protein biosynthesis is also observed under BL. Such an effect of BL does not disappear in the presence of RL. It is saturated at low light intensity (Krotkov, 1964; Miyachi et al., 1977; 1978) and has an after-action. RL (λ max = 670–680 nm) when comparing with BL activates the incorporation of label into sugar phosphates, sucrose, and starch (Kowallik, 1970; Voskresenskaya, 1972; Miyachi et al., 1978). Recently, similar effects of BL and RL on photosynthetic carbon metabolism have been also observed in experiments with the red alga *Porphyridium caldazium* (Döhler et al., 1976). Such regulatory effects of BL have been also obtained with nongreen *Chlorella* mutants (Georgi, 1974; Kamiya and Miyachi, 1974) and with a green *Chlorella* whose photosynthesis was inhibited by DCMU (Andersag and Pirson,

1976). In all these organisms BL activated the formation of noncarbohydrate compounds and the biosynthesis of protein at the expense of endogenous carbohydrates or exogenous glucose. RL as compared with darkness does not induce any metabolic alterations in nongreen *Chlorella*. In green *Chlorella* (in the presence of DCMU), RL activates the conversion of exogenous glucose into various carbohydrates, especially into starch. It is supposed that the effect of RL is associated with cyclic photophosphorylation. On transferring these *Chlorella* cells from RL to BL protein formation increases and the accumulation of carbohydrate decreases (ANDERSAG and PIRSON, 1976). Thus the responses of carbon metabolism to BL action are similar both in green and in nongreen plants. In green plants, however, this nonphotosynthetic reaction induced by BL interferes with photosynthetic carbon metabolism, and as a result the pattern of carbon metabolism except for the Calvin cycle is found to differ from that seen with RL.

C. Specific Features of Blue Light Action on Carbon Metabolism

I. In the Absence of Photosynthesis

Acceleration of the glycolytic conversion of carbohydrates may be the cause of observed BL effects on carbon metabolism. Such a possibility is indicated by a coincidence of action spectra of glucose utilization and protein biosynthesis activation (GEORGI, 1974). In addition only BL is found to activate phosphoenol pyruvate (PEP) carboxylase (KAMIYA and MIYACHI, 1975) and pyruvate kinase (PK); these activations are due to extra enzyme biosynthesis in comparison with their synthesis in darkness (KOWALLIK and RAYTERS, 1976). The action of BL on glycolytic enzymes, however, is mostly mediated by substrates. Further, the activation of PK biosynthesis has been observed also in darkness after an addition of glucose into the medium. It is suggested, therefore, that BL enhances the efflux of glucose from certain cell compartments. There are also some other considerations concerning the activation of sugar transport across the cell membrane by BL (LAUDENBACH and PIRSON, 1969; KOWALLIK and KIRST, 1975; ANDERSAG and PIRSON, 1976). It has been shown that respiration is activated simultaneously with an alteration of metabolism induced by BL. The phenomenon has been detected both in nongreen and in green higher plants and algae, and it has been thoroughly studied by KOWALLIK (1970). The respiration activated by BL differs from dark respiration by its temperature optimum (KOWALLIK and KIRST, 1975). It is unclear, however, whether it is the cause or the consequence of metabolic changes. Effects of BL on respiration and metabolism have similar action spectra and equally low light saturation (~ 800 erg cm^{-2} s^{-1}). Based on the action spectra, the effects of BL are more likely linked with an excitation of flavins (KAMIYA and MIYACHI, 1974; GEORGI, 1974). In certain cases where BL acted on plastid rudiments of a yellow Chlorella mutant (KOWALLIK and RAYTERS, 1976) and on the chloroplast envelope of green plants, the participation of carotenoids cannot be excluded (JEFFREY et al., 1974).

II. In the Presence of Photosynthesis

In C_4 plants more malate and aspartate accumulate under BL than under RL (Grishina et al., 1974). This fact is interpreted to reflex a rapid activation of PEP carboxylase (Poskuta et al., 1975). In C_3 plants the enforcement of synthesis of noncarbohydrate products proceeds with the participation of Calvin cycle intermediates (Bassham, 1973). So it is not excluded that BL enhances the efflux of PGA out of the cycle. This is in accordance with the fast inhibition of NADP-linked glyceraldehyde 3-P dehydrogenase (GAP dehydrogenase) by BL (Ziegler et al., 1972). It is also possible that the products of the Calvin cycle are formed in equal proportions under BL and RL, but that there are higher chances under BL for an efflux of PGA and triose phosphates out of the chloroplasts. In this case, accumulation of various sugars of the Calvin cycle and starch inside the chloroplast under RL would be observed. Finally, the presence of glycolytic enzymes in chloroplasts (Kelly et al., 1976) and the possibility of their activation by BL allows the secondary formation of triose phosphates, PGA and PEP from hexose phosphates of the Calvin cycle and from starch. All these possibilities have not yet been tested experimentally.

Various metabolic changes of terminal and intermediate products of the Calvin cycle (proceeding outside the cycle) can be linked with the oxygen uptake which appears in green plants under illumination, and which increases in a similar manner as photosynthesis with increasing light intensity and proceeds more actively under BL than under RL (Voskresenskaya, 1965; 1975). The presence of oxygen is vitally necessary for all metabolic effects of BL. Elimination of oxygen from the medium leads to the disappearance of all the usual metabolic effects of BL both at low and at saturating light intensities (Voskresenskaya et al., 1972).

D. Direct Regulation of Certain Enzymes by Blue Light in Vitro and Its Possible Realization in Vivo

Those enzymes which contain flavins (FMN or FAD) can be directly affected by BL. This is assumed to be the result of an intramolecular interaction of the triplet of flavin with the apoenzyme. According to Schmid (1970) the activity of glycolate oxidase (GO) in vitro in the presence of FMN is inhibited by weak BL, but does not change under RL of any intensity. Inhibitory effects of BL on malate dehydrogenase and transketolase were later observed (Codd, 1972a; 1972b). In the cases of GO and malate dehydrogenase their apoenzymes interact directly with flavin. However, oxygen also participated in the modification of transketolase apoenzyme by BL and FMN. In vivo one can observe a quenching of an excited state of FMN by various cell metabolites and by oxygen itself (Schmid, 1970). Although there is no specific spectral dependency for ribulose bisphosphate (RuBP) carboxylase activity in the absence of FMN (Wildner et al., 1972), in the presence of FMN BL inhibits the enzyme activity in vitro; simultaneously, the oxygenase properties of the enzyme are activated (STEWART and CODD, 1976). It remains unknown, however, whether the possible interaction of RuBP carboxylase with FMN occurs in vivo.

At least the glycolate pathway and photorespiration are not activated by BL in most cases (VOSKRESENSKAYA, 1972). On the contrary an inhibition of the former pathway is sometimes observed (KEERBERG, 1975). This inhibition more likely results from a decrease in possibilities for glycolate formation (ROBINSON and GIBBS, 1974) than from an inhibition of glycolate oxidase activity under the influence of BL. For instance, this effect may be due to BL activation of stomata opening (KEERBERG et al., 1971; VOSKRESENSKAYA and POLYAKOV, 1975). There may in addition be cases in algae when the formation of glycolate is equal under RL and BL, but where the glycolate is excreted into the medium only under RL. As a consequence, glycolate metabolism under BL is accompanied by CO_2 evolution (DÖHLER and KOCH, 1972). Under saturating intensities of BL an activation of photorespiration and an enhancement of glycolate, glycine, and serine formation have been also observed (GALMICHE, 1972; VOSKRESENSKAYA, 1972).

Recently (APARICIO et al., 1976) a fast reactivation of nitrate reductase previously inhibited by a high NADPH content was observed under BL in the presence of flavins. It was suggested that FAD, the enzyme cofactor, is a modulator of enzymatic activity in vivo. However the initial state of mineral nitrogen (oxidized or reduced) does not obviously play any significant role in the acceleration of biosynthesis of nitrogenous compounds by BL (GEORGI, 1974).

Reversible photoreduction of cytochromes b and c has been discovered and investigated in vitro and in vivo in some lower fungi (POFF and BUTLER, 1974; SCHMIDT and BUTLER, 1976) and corn seedlings (BRIGGS, 1976). Action spectra of cytochrome photoconversion indicate that flavins are the pigments that excite the reaction. Therefore the possibility exists that BL may primarily act at the level of electron-transport chain reactions.

E. Long-Term Effects of Light Quality on Biosyntheses and Chloroplast Organization

Preferential accumulation of carbohydrates (particularly starch) is a characteristic metabolic feature of plants photosynthesizing under RL during a long period. In plants photosynthesizing under BL the carbon assimilated is preferentially utilized for biosynthesis of nucleic acids, proteins (PIRSON and KOWALLIK, 1964; KOWALLIK, 1970) and phospho-organic compounds including ATP (STEUP and PIRSON, 1974). These peculiarities in metabolism provoked by light quality are accompanied by alterations in chloroplast organization. Under RL as compared with BL less active chloroplasts are formed. The chloroplast membrane system has a very unusual structure; it is freely disrupted and is easily aged (VOSKRESENSKAYA, 1972; VETTERMANN, 1973). The plants have limited CO_2 gas exchange, weak photophosphorylation, low activity of electron transport from water (VOSKRESENSKAYA, 1972;

Drozdova et al., 1976a, 1976b; Harnischfeger, 1974; Voskresenskaya et al., 1977), and equally low activity of enzymes taking part in carbon metabolism inside and outside the chloroplasts (Stabenau, 1972; Poyarkova et al., 1973; Harnischfeger et al., 1974; Feierabend, 1975). BL activates the biosynthesis of chlorophyll and carotenoids, alters the composition of the lipids and lipoquinones of thylakoids (Grumbach and Lichtenthaler, 1975; Senger, 1976; Tevini, 1977), and accelerates the formation of active chloroplasts whose fine structure, enzymatic activities, and activity of electron-transport chain reactions are similar to those of plants grown under white light.

The enhancement of all these characteristics in plants grown unter BL is accompanied by the appearance of some distinctions both in electron-transport chain organization (Drozdova et al., 1976a, 1976b) and in carbon metabolism. The principal feature of the latter is that the activities of RuBP and PEP carboxylases (in C_3 and C_4 plants, respectively) become significantly higher in plants grown under BL. The GAP dehydrogenase in the BL plants is activated insignificantly (Poyarkova et al., 1973). Such distinctions make it possible that leakage of PGA from the Calvin cycle is more significant under BL than under RL. The ratio of the activities of glycolate oxidase and RuBP carboxylase differs insignificantly in plants grown under RL or BL (Voskresenskaya, 1975). Detectable effects of BL on various biosyntheses (Harnischfeger, 1974; Feierabend, 1975; Steup, 1975; Voskresenskaya, 1975) and on the organization of the photosynthetic apparatus (Vlasova et al., 1971; Voskresenskaya, 1972; Senger, 1976; Tevini, 1977) have low light saturation and are also observed in plants grown under white light.

Thus, for biogenesis of active chloroplasts not only the excitation of phytochrome, but also that of pigments which absorb BL (that is, of flavins or carotenoids) is necessary (Feierabend, 1976; Voskresenskaya, 1975; Tevini, 1977). The activity and structure of chloroplasts in plants progressively aged under RL can be restored by transferring the plants to BL. In this case an enhancement of the photosynthetic activity of chloroplasts is linked with the depletion of starch, the enforcement of carbon incorporation into noncarbohydrate products, and the induction of an additional biosynthesis of nucleic acids and enzymes involved in carbohydrate metabolism (Voskresenskaya, 1972; 1975; Vettermann, 1973; Poyarkova et al., 1973; Andersag and Pirson, 1975). It is not excluded that the links between photosynthetic carbon metabolism and general carbon metabolism peculiar to plants grown under BL involve an interaction between the protein-synthesizing system of chloroplasts and that of the cell. Thus, in fern gametophytes, enhancement in the contents of RNA and protein induced by BL is linked, first of all, with an increase in the label incorporated into nuclear RNA. All manifestations of BL action at the level of cytoplasmic and chloroplast ribosomes may therefore represent a secondary phenomena (Raghavan, 1968). An increase of cooperation between the protein-synthesizing systems of chloroplasts and the cell under BL was shown also in experiments with *Chlorella* (Steup, 1975). In this case, a faster incorporation of label into rRNA under BL as compared with RL was observed. In addition, under BL the label preferentially incorporated into cytoplasmic rRNA, while under RL it mainly appeared in chloroplastic rRNA. The primary causes of activation of protein biosynthesis by BL remain to be solved.

F. Conclusion

Photosynthetic carbon metabolism is modulated by means of photoreactions triggered by a pigment which absorbs in ultraviolet and short-wave visible wavelength ranges. Continuous excitation of the pigment under white light provides the active course both for photosynthesis and for the various essential biosyntheses. Prolonged absence (under RL) of the BL-dependent photoreaction leads to the functional failure and eventual degradation of chloroplasts. Since RL excites chlorophyll and phytochrome to an extent similar to that of BL, but nevertheless does not ensure optimal conditions for photosynthetic carbon metabolism, it is necessary to consider the pigment excited by BL and the associated system of photoregulation as an indispensable constituent of plants.

The most favorable illumination conditions for the proper growth of plants will be those which ensure an interaction of all photoreactions (white light). Growing plants under light of different quality may, however, become a means for the effective control of various biosyntheses, the integration of the various essential activities of plants, and the realization of their potential for productivity.

References

Andersag, R., Pirson, A.: Biochem. Physiol. Pflanz. *169*, 71–85 (1976)
Aparicio, P.J., Roland, J.M., Calero, F.: Biochem. Biophys. Res. Commun. *70*, 1071–1077 (1976)
Bassham, J.A.: In: Symp. Soc. Exp. Bot. Vol. XXVII, pp. 461–483. Cambridge Univ. Press 1973
Briggs, W.R.: In: Symp. Abstr. VII Int. Congr. Photobiol. p. 54. Rome, Italy 1976
Codd, G.A.: FEBS Lett. *20*, 211–214 (1972a)
Codd, G.A.: Z. Naturforsch. *27b*, 701–704 (1972b)
Döhler, G., Bürstell, H., Jilg-Winter, G.: Biochem. Physiol. Pflanz. *170*, 103–110 (1976)
Döhler, G., Koch, R.: Planta (Berl.) *105*, 352–359 (1972)
Drozdova, I.S., Grishina, G.S., Voskresenskaya, N.P.: Dokl. Acad. Sci. USSR *226*, 221–224 (1976a)
Drozdova, I.S., Krendeleva, T.E., Verkhoturov, V.N., Timofeev, K.N., Matorin, D.N., Voskresenskaya, N.P.: Fiziol. Rast. (USSR) *23*, 861–868 (1976b)
Feierabend, J.: Planta *123*, 63–77 (1975)
Galmiche, J.M.: In: Proc. II Int. Congr. Photosynthesis Res. Forti, G., Avron, M., Melandri, A. (eds.), Vol. III, pp. 1875–1882. The Hague: Junk Publishers 1972
Georgi, M.: Dissertation, Köln (1974)
Grishina, G.S., Maleszewski, S., Frankiewicz, A., Voskresenskaya, N.P., Poskuta, J.: Z. Pflanzenphysiol. *73*, 189–197 (1974)
Grumbach, K.H., Lichtenthaler, H.K.: Z. Naturforsch. *30c*, 337–341 (1975)
Harnischfeger, G., Treharne, K., Feierabend, J.: Plant Sci. Lett. *3*, 61–66 (1974)
Jeffrey, S.W., Douce, R., Benson, A.A.: Proc. Natl. Acad. Sci. USA *71*, 807–809 (1974)
Kamiya, A., Miyachi, S.: Plant Cell Physiol. *15*, 927–937 (1974)
Kamiya, A., Miyachi, S.: Plant Cell Physiol. *16*, 729–736 (1975)
Keerberg, O.F.: In: Photoregulation of plant metabolism and morphogenesis. Kursanov, A.L., Voskresenskaya, N.P. (eds.) pp. 158–170. Moscow: Nauka 1975
Keerberg, H., Keerberg, O., Pärnik, T., Viil, J., Värk, E.: Photosynthetica *5*, 99–106 (1971)
Kelly, G.J., Latzko, E., Gibbs, M.: Ann. Rev. Plant Physiol. *27*, 181–205 (1976)

Kowallik, W.: In: Photobiology of microorganisms. Halldal, P. (ed.), pp. 165–185. London, New York, Sydney: Wiley 1970
Kowallik, W., Kirst, R.: Planta *124*, 261–266 (1975)
Kowallik, W., Rayters, G.: Planta *128*, 11–14 (1976)
Krotkov, G.: Trans. R. Soc. Can. *2*, 205–215 (1964)
Laudenbach, B., Pirson, A.: Arch. Microbiol. *67*, 226–242 (1969)
Miyachi, S., Kamiya, A., Miyachi Shiziko: In: Biological solar energy conversion. Mitsui, A., Miyachi, S., San-Pietro, A., Tamura, S. (eds.), pp. 167–182. Acad. Press Inc. 1977
Miyachi, S., Miyachi Shiziko, Kamiya, A.: Plant Cell Physiol. *19*, 244–256 (1978)
Pirson, A., Kowallik, W.: Photochem. Photobiol. *4*, 489–497 (1964)
Poff, K.L., Butler, W.: Nature (London) *248*, 799–801 (1974)
Poskuta, J., Frankiewicz-Jozko, A.: In: Environmental and biological control of photosynthesis. Marcelleth (ed.), pp. 89–105. The Hague: Junk 1975
Poyarkova, N.M., Drozdova, I.S., Voskresenskaya, N.P.: Photosynthetica *7*, 58–66 (1973)
Raghavan, V.: Planta (Berl.) *81*, 38–48 (1968)
Robinson, J.M., Gibbs, M.: Plant Physiol. *53*, 790–797 (1974)
Schmid, G.H.: Ber. Dtsch. Bot. Ges. *83*, 399–415 (1970)
Schmidt, W., Butler, W.L.: Photochem. Photobiol. *24*, 77–80 (1976)
Senger, H.: In: Symp. Abst. VII. Int. Congr. Photobiology. p. 255. Rome, Italy 1976
Stabenau, H.: Z. Pflanzenphysiol. *67*, 105–112 (1972)
Steup, M.: Arch. Microbiol. *105*, 143–151 (1975)
Steup, M., Pirson, A.: Biochem. Physiol. Pflanz. *166*, 447–459 (1974)
Stewart, R., Codd, G.A.: Plant Physiol. *57* (Suppl.) 6 (1976)
Tevini, M.: In: Lipids and lipid polymers in higher plants. Tevini, M., Lichtenthaler, H.K. (eds.), pp. 121–145. Berlin, Heidelberg, New York: Springer 1977
Vettermann, W.: Protoplasma *76*, 261–278 (1973)
Vlasova, M.P., Drozdova, I.S., Voskresenskaya, N.P.: Fiziol. Rast. USSR *4*, 5–11 (1971)
Voskresenskaya, N.P.: Fotosintez i spektralnii sostav sveta. Moskwa: Nauka 1965
Voskresenskaya, N.P.: Ann. Rev. Plant Physiol. *23*, 219–234 (1972)
Voskresenskaya, N.P.: In: Photoregulation of plant metabolism and morphogenesis. Kursanov, A.L., Voskresenskaya, N.P. (eds.) pp. 16–36. Moscow: Nauka 1975
Voskresenskaya, N.P., Polyakov, M.A.: Plant Sci. Lett. *5*, 333–338 (1975)
Voskresenskaya, N.P., Viil, J.A., Grishina, G.S., Pärnik, T.K.: In: Proc. IInd Int. Congr. Photosynthesis Res. Forti, G., Avron, M., Melandri, A. (eds.), pp. 2129–2136. The Hague: Junk 1972
Voskresenskaya, N.P., Drozdova, I.S., Krendeleva, T.E.: Plant Physiol. *59*, 151–154 (1977)
Wildner, G.F., Zilg, H., Criddle, R.S.: Proc. IInd Int. Congr. Photosynthesis Res. Forti, G., Avron, M., Melandri, A. (eds.), pp. 1825–1830. The Hague: Junk 1972
Ziegler, I., Schmidt-Clausen, N.D., de Beauclair, B.: In: Teoreticheskie osnovii fotosinteticheskoi produktivnosti. Nichoporovich, A. (ed.), pp. 225–229. Moskwa: Nauka 1972

14. Photoassimilation of Organic Compounds

W. WIESSNER

A. Introduction

Photoassimilation of organic compounds is generally understood as the light-dependent or light-promoted utilization of organic substrates as carbon sources for the growth of chlorophyll- or bacteriochlorophyll-containing organisms. This process has been unequivocally established primarily for photosynthetic bacteria, procaryotic cyanophyta and eukaryotic algae. However, photometabolism of organic substrates by photosynthetically active bacteria and by blue-green algae differs significantly from this process in eukaryotic algae and will be treated in the contribution of OHMANN (Chap. II.4, this vol.). Therefore, this chapter will concentrate on the latter group of organisms. This is reasonably justified, since photosynthetic bacteria, blue-green cyanophyta and eukaryotic algae differ from each other with respect to composition and organization of their photosynthetic apparatus. Some other relevant topics closely connected to photometabolism, such as the mechanism of the uptake of organic compounds into cells and their organelles, and general aspects of dark fermentation and respiration, are also omitted.

Useful summaries on the growth of algae with organic compounds as carbon sources and their metabolism, including detailed references to older literature, have been presented by PETERSEN (1935), PARKER (1961), DANFORTH (1962), GIBBS (1962a, 1962b), WIESSNER (1970), ZAJIC and CHIU (1970), NEILSON et al. (1973) and more recently by NEILSON and LEWIN (1974).

B. Definitions

Certain terms are used by different authors to describe the several nutritional possibilities with respect to growth on organic substrates. The following terminology will be used in the chapter:

a) Phototrophy: The energy required for nutritional processes is obtained by the absorption of light energy and associated photosynthetic energy conservation, i.e., formation of ATP and reduction of $NADP^+$. Thus the prefix photo- in this sense describes the dependence of the carbon assimilation process on light as the energy source.

b) Photoautotrophy: CO_2 is incorporated into cell carbon compounds by the mechanism of photosynthetic CO_2 fixation for which water serves as the source of the required reduction equivalents. A synonym is the short term autotrophy, which commonly describes the photosynthetic nutrition of algae and higher plants. Photolithotrophy, often used in the sense of photoautotrophy is, however, a more general term for the use of any inorganic substrate (such as H_2S, H_2, H_2O) as source of the reduction equivalents for CO_2 fixation.

c) *Photoheterotrophy:* An organic substrate is the main, if not exclusive source of cellular carbon in the course of the light-dependent metabolism. This term does not refer to any special mechanism by which carbon from this organic compound is incorporated into cell constituents.

d) *Photomixotrophy:* CO_2 and an organic carbon compound are supplied together in the growth medium. They are both assimilated simultaneously by the organism in question. The use of this term does not imply that either the inorganic or organic carbon source is absolutely required for growth of the organism or for the operation of the carbon metabolic pathways.

Sometimes this terminology is used to classify algal species into different nutritional types according to their favored growth conditions. This, however, is avoided in this article.

C. Pathways and Products of Photometabolism

The light-dependent metabolism of organic compounds by algae has so far been studied intensively only in a limited number of species. The main reason is the necessity to cultivate the organisms in mass culture under axenic conditions. This has been achieved for very few species. Nevertheless, no unique metabolic pathways have been demonstrated. Acetate, glucose, and fructose are used by most algae. Pyruvate, several organic acids of the tricarboxylic acid cycle (citrate, malate, succinate, fumarate), mannose, sucrose, glycerol, and ethanol can also serve as carbon sources for light-dependent growth.

Metabolic pathways involved in the photometabolism of the various organic compounds differ according to the chemical nature of the substrate and to the systematic position of the organism in question. In spite of this, several aspects are common to all substrates and all algae. (1) Photometabolism requires the conservation of light energy into ATP either by cyclic or by open chain (noncyclic) light-dependent electron transport and photophosphorylation (see Part III, Vol. Photosynthesis I this series). In cyclic electron transport through photosystem I, ATP is the only product of the light reaction. If this process of energy conservation dominates, metabolic pathways which require ATP only for the metabolism of the organic compound are favored. In noncyclic electron transport $NADP^+$ generally functions as a terminal electron acceptor for electrons removed from water at photosystem II and NADPH is furnished in addition to ATP. The NADPH must be reoxidized in order to facilitate continuous photophosphorylation. This can be achieved by reduction of intermediates derived from the metabolized organic substrate. Thus noncyclic electron transport can favor transformation of the organic substrate into more reduced compounds. For example, higher fatty acids can be formed from acetate, or glyoxylate formed during acetate metabolism can be converted to glycollate by the NADPH-requiring glyoxylate reductase. In vivo, O_2

can replace $NADP^+$ as terminal electron acceptor in a Mehler type reaction (MEHLER 1951). Under these conditions ATP is the only product of photosynthetic electron transport. In organisms capable of this type of photophosphorylation, products of photometabolism are generally the same as in those in which cyclic photophosphorylation dominates. Thus the type of photophosphorylation predominating is a significant determinant for the nature of the products formed during photometabolism. (2) The synthesis of amino acids in the presence of external nitrogen sources or high internal nitrogen pools can reduce the transformation of photoassimilated carbon compounds into carbohydrate storage products. (3) The presence of CO_2 can lead to a competition between the photoassimilation of organic carbon and photosynthetic CO_2 fixation for the limited amount of ATP and/or NADPH, thus decreasing the photometabolic process or vice versa. (4) The simultaneous occurrence of photosynthesis and photometabolism can lead to interactions between metabolic pathways at the level of allosteric regulation of enzyme reactions. (5) The observation of a light-dependent incorporation of carbon from an organic carbon source into cell material does not necessarily imply that the organic molecule is incorporated in toto into cell constituents or that ATP and/or NADPH are required only for metabolism of the organic molecule in question. For example, acetate carbon can be transformed into polysaccharide storage products either by (a) direct photoassimilation of acetate; this requires ATP for activation of the acetate to acetyl-coenzyme A, which then passes via the glyoxylic acid cycle into gluconeogenesis, utilizing more ATP in this process. In this case half the acetate molecules are incorporated into glucose without separation of their carbon bonds. Or, (b), the activated acetate molecule enters the tricarboxylic acid cycle and is oxidized to CO_2 which is subsequently refixed during normal photosynthesis. In this case the light requirement is twofold: first, ATP is needed for the acetate activation, and second, ATP and NADPH are utilized for photosynthesis. Such a process is called indirect assimilation of carbon from the organic carbon compound. A typical example is Lewin's *Chlamydomonas dysosmos* mutant D 2075 (LEWIN, 1954) which completely oxidizes acetate under photoheterotrophic growth conditions, and finally assimilates the CO_2 photosynthetically.

It has been claimed occasionally that eukaryotic algae can grow photoheterotrophically under anaerobic conditions. Careful reinvestigation of this problem, however, revealed anaerobic growth only when photosynthesis supplied internally required oxygen (WIESSNER and GAFFRON, 1964; NÜHRENBERG et al., 1968).

D. Photoassimilation of Acetate

In the following the main features and rules of photometabolism will be described using the examples of acetate and glucose metabolism. The metabolism of other organic compounds obeys, in general, the same rules, and is brought about by similar dissimilatory pathways.

Acetate can be assimilated in the light by a wide range of eukaryotic algae. Some are obligate phototrophs incapable of heterotrophic growth in the dark, e.g., strains of *Chlamydomonas dysosmos* (LEWIN, 1954), *Chlamydomonas mundana* (EPPLEY et al., 1963), *Chlamydomonas pulsatilla* (DROOP, 1971), *Chlamydobotrys stellata* (PRINGSHEIM and WIESSNER, 1960), certain strains of *Pandorina morum* (PALMER and STARR, 1971; PALMER and TOGASAKI, 1971), *Phaeodactylum tricornutum* (COOKSEY, 1972, 1974), *Cocconeis sp.* (BUNT, 1969), *Cocconeis diminuta* (COOKSEY, 1971, 1972), *Diplostauron elegans* (LYNN and STARR, 1970), and *Euglena gracilis* L strain (COOK, 1967). Light-dependent acetate assimilation begins with an ATP-dependent activation by either acetyl-coenzyme A synthetase, as in *E. gracilis* (ABRAHAM and BACHHAWAT, 1962; OHMANN, 1964a) and *C. stellata,* or acetate kinase as in *Chlorella fusca, Stichococcus bacillaris* and *Scenedesmus* sp. (OHMANN, 1964b). Following activation, the acetate can be incorporated predominantly into lipids, as has been shown with *Chlorella* (SYRETT et al., 1964; GOULDING and MERRETT, 1966; KYDAR and DOMAN, 1968) and *Cocconeis diminuta* (COOKSEY, 1971). However, in *Chlamydobotrys stellata* (PRINGSHEIM and WIESSNER, 1961) and *Chlamydomonas mundana* (EPPLEY et al., 1963) the acetate preferentially enters into carbohydrates and proteins. The reasons for these preferences have not been elucidated completely, although the following may play a significant role: The type of photophosphorylation responsible for ATP formation, the presence of the glyoxylate pathway, and the presence of CO_2 during the assimilatory process.

Of the algae, which predominantly incorporate acetate into carbohydrates and amino acids, *C. stellata* and *C. mundana* have been studied most intensively. In both members of the volvocean family, acetate metabolism in the light depends strictly on photophosphorylation. This has been shown in experiments using 2,4-dinitrophenol which prevents the dark assimilation of acetate but is ineffective in the light. In addition, photoassimilation of acetate is almost unaffected by DCMU, (an inhibitor of photosynthetic electron transport). It proceeds with highest quantum efficiency at wavelengths longer than 680 nm (EPPLEY and MACIASR, 1962a, b; EPPLEY et al., 1963; GAFFRON et al., 1963; WIESSNER, 1963, 1965, 1966a, b; WIESSNER and GAFFRON, 1964a, b). Both facts indicate that in these two species cyclic photophosphorylation is the preferred mechanism for ATP synthesis during acetate photometabolism. The incorporation of acetate carbon into cell constituents proceeds with the participation of the tricarboxylic acid cycle, the glyoxylate cycle and gluconeogenesis to hexosephosphates from phosphoenolpyruvate. This has been demonstrated by examining the fate of methyl- and carboxyl-labeled carbon-[14]-acetate. After its application to the algae, most of the tricarboxylic acid cycle intermediates are labeled within 10–60 s, prior to carbohydrates and amino acids. Up to 90% of the CO_2 evolved during acetate assimilation is derived from the acetate carboxyl, whereas the methyl carbon is almost completely incorporated into carbohydrates. The characteristic enzymes of the glyoxylate pathway, isocitrate lyase and malate synthetase, have been demonstrated in crude extracts from both species (WIESSNER, 1962; WIESSNER and KUHL, 1962; EPPLEY et al., 1963; GOULDING and MERRETT, 1967a). The tricarboxylic acid cycle seems to be incomplete in *Chlamydobotrys stellata.* MERRETT (1976) could detect only very low activities of α-ketoglutarate dehydrogenase and succinate thiokinase. Thus in *Chlamydobotrys stellata* and *Chlamydomonas mundana* the presence of true direct photoassimilation of acetate is well established, requiring only cyclic photophosphorylation as an energy yielding process.

In contrast, the light dependent carbon assimilation from acetate in almost all other algae so far tested (*Chlorella, Scenedesmus, Ulva lactuca, Chlorogonium elongatum, Cocconeis diminuta, Phaeodactylum tricornutum, Euglena gracilis* L strain) behaves quite differently. It is sensitive to DCMU at concentrations which inhibit photosynthetic electron transport. The action spectrum is the same as for photosynthetic CO_2-fixation. Lipids and amino acids are the main assimilatory products. DCMU induces a pattern of acetate assimilation in the light similar to that in the dark: it decreases the carbon incorporation into lipids as well as the formation of carbohydrates. At the same time, CO_2 evolution increases (CALVIN et al., 1951; SCHLEGEL, 1956, 1959; COOK, 1967; GOULDING and MERRETT, 1967b; MERRETT and GOULDING, 1967a, b; WIESSNER, 1968; COOKSEY, 1971, 1972, 1974; PISKUNKOVA and PIMENOVA, 1971, 1973; GEMMILL and GALLOWAY, 1974). These results favor operation of the normal open chain photosynthetic electron transport during acetate photoassimilation, yielding ATP and NADPH which both favor the observed synthesis of lipids. In *Chlorella* glycollate production from acetate is also increased in the light, and in this case the reaction requiring NADPH would be the glyoxylate reductase (MERRETT and GOULDING, 1967b). The decrease in carbohydrate formation by DCMU can be explained differently. First, CO_2 liberated during acetate oxidation could be reutilized via photosynthesis. In *Gonium* species, for example, light-dependent carbon assimilation from acetate can be lowered experimentally by absorbing the CO_2 formed from acetate with a solution of KOH (HENRY-HISS, 1977). Second, the inhibition of oxygen-dependent photophosphorylation would lower the direct conversion of acetate into carbohydrates. The participation of this type of photophosphorylation, however, has not yet been definitely confirmed.

Regulation of acetate photometabolism can take place either at the enzymatic level or at the level of photosynthetic electron transport and energy conversion. A key enzyme of the glyoxylate pathway, isocitrate lyase, is subject to catabolic repression. It exists with very low activity in autotrophic algae and in the presence of glucose. Exceptions are *Chlamydobotrys stellata* and *Chlamydomonas mundana*, whose crude extracts show high activity even from autotrophically cultivated cells. Enzyme activity in all other algae increases significantly in the presence of acetate (REEVES et al., 1962; WIESSNER and KUHL, 1962; HARROP and KORNBERG, 1963; SYRETT et al., 1963, 1964; HAIGH and BEEVERS, 1964a, b; COOK and CARVER, 1966; GOULDING and MERRETT, 1966; SYRETT, 1966; JOHN et al., 1970; McCULLOUGH and JOHN, 1972). One exception is an obligate phototrophic strain of *Euglena gracilis,* where acetate does not induce operation of the glyoxylate by-pass (COOK, 1967). As a consequence of high glyoxylate cycle activity, fat formation from acetate is reduced in favor of carbohydrate and protein synthesis (SYRETT et al., 1964; FISCHER and WIESSNER, 1968). Light during growth decreases the glyoxylate pathway activity (SYRETT et al., 1963; COOK and CARVER, 1966), even though photophosphorylation promotes synthesis of isocitrate lyase, e.g., in *Chlorella fusca* (SYRETT, 1966). In *Chlamydobotrys stellata* and *Chlamydomonas mundana* light dependent growth on acetate markedly reduces the activity of the Calvin cycle enzyme, ribulose-1,5-bisphosphate carboxylase, without affecting the chlorophyll content of the algae (MERRETT, 1967a; WIESSNER and FRENCH, 1970). In other algae, such as *Euglena gracilis,* growth on acetate in the light represses chlorophyll synthesis together with the activity of the Calvin cycle (APP and JAGENDORF, 1963; STABENAU, 1969).

The influence of acetate photometabolism on the mechanism of photosynthetic energy conversion and its consequence for the regulation of photoassimilation has so far been studied mainly in *C. stellata*. The most striking consequences of a change from autotrophic to photoheterotrophic nutrition are: (a) The amount of a special chlorophyll-a-protein, P-695, is increased even though the total chlorophyll a content remains unchanged. The amount of chlorophyll b is lowered to about 50% of its value in autotrophic cells (WIESSNER, 1969a, b; WIESSNER and FRENCH, 1970). Grana stacks disappear almost completely from the chloroplasts (WIESSNER and AMELUNXEN, 1969a, b; THIEDE, 1976). (b) The dark reduction of cytochrome f after its light dependent oxidation proceeds with only about 10% of its normal velocity and cannot be influenced by DCMU (WIESSNER and FORK, 1971). The photosystem II dependent dichlorophenolindophenol Hill-reaction is reduced by about 70% (MENDE et al., 1978). The slow 515 nm change is not present (WIESSNER and FORK, 1971). (c) The amount of variable fluorescence emitted at 683 nm is reduced (WIESSNER and FRENCH, 1970; MENDE and WIESSNER, 1977; WIESSNER et al., 1977). It remains to be answered whether or not the decrease in photosystem II activity and in photosynthetic electron transport between the two photosystems is due to a reduced capacity of the oxygen evolving system of photosynthesis. MENDE and WIESSNER (1975) in their studies on the nutritional dependence of the delayed light emission payed attention to the possibility of a special electron carrier system in photoheterotrophically cultivated *Chlamydobotrys stellata*. This system might operate on the acceptor side of photosystem II and donate electrons either to oxygen or in a cyclic pathway to the donor side of photosystem II (MENDE et al., 1978). The observation of a correlation in the half time of dark decay of a light-induced absorbance change at 560 nm and a 60 µs component of delayed light emission has led them to speculate about the possible participation of cytochrome b-559 in this regulatory system. The physiological explanation of this mechanism for inactivation of the photosystem II mediated electron flow might be the requirement for a very efficient photosystem I electron transport for photophosphorylation which proceeds in chloroplast preparations at high rates only under conditions of reduced electron flow from photosystem II (TAGAWA et al., 1963; Trebst et al., 1963; HEBER, 1969). Additional proof for this hypothesis comes from the observation of a light minus dark absorbance change at 563 nm. It signals the redox reactions of the photosystem I operated cytochrome b-563. In photoheterotrophic *C. stellata* the turn over rate is ten times higher than in autotrophic cells (MENDE et al., 1978).

E. Photoassimilation of Glucose

Glucose and other hexoses can be photometabolized by algae under aerobic as well as under anaerobic conditions. Such metabolism has been observed with *Chlorella* (MYERS, 1947; KANDLER, 1954, 1955; BERGMANN, 1955; BUTT and PEEL, 1963; WHITTINGHAM et al., 1963; MOSES et al., 1959; DVOŘÁKOVÁ-HLADKA, 1966), *Ankistrodesmus* and *Scenedesmus* (SIMONIS, 1956; TAYLOR, 1960; BISHOP, 1961; DVOŘÁKOVÁ-HLADKA, 1966), *Chlamydomonas humicola* (LUKSCH, 1933) and *Nitella translucens* (SMITH, 1967).

In *Chlorella vulgaris* under aerobic conditions, as in most other green algae, light only slightly increases glucose assimilation above the dark rate. The high capacity for dark metabolism of glucose is evidence for the ability of oxidative phosphorylation to supply ATP for all the energy requiring reactions of glucose assimilation (e.g., the hexokinase reaction). The enzymatic sequences involved in glucose metabolism involve both the EMBDEN-MEYERHOF pathway and the oxidative pentose phosphate cycle. To some extent light favors the formation of glycollic acid under aerobiosis (WHITTINGHAM et al., 1963; MARKER and WHITTINGHAM, 1966). Anaerobically, more than 85% of the assimilated glucose enters into oligo- and polysaccharides (KANDLER and TANNER, 1966). The significance of cyclic and noncyclic photophosphorylation for anaerobic glucose assimilation by algae has been studied most carefully in Kandler's laboratory. Strongest support for its dependence on cyclic photophosphorylation comes from results on quantum requirements, which were 6.0 Einstein at 658 nm and 4.1 Einstein at 712 nm (TANNER et al., 1965, 1966, 1968). The process can be inhibited up to 70% by antimycin A, a potent inhibitor of cytochrome b-dependent photophosphorylation. Elegant proof comes from Bishop's studies on *Scenedesmus* mutants. Mutant 11, which has an incomplete photosystem I, does not show any significant glucose assimilation (WEAVER and BISHOP, 1963), whereas mutant 8, which has a defective light reaction II exhibits very active glucose assimilation (TANNER et al., 1967). Furthermore, photoreduction is drastically reduced by glucose assimilation because of the competition for ATP (BISHOP, 1961). DCMU diminishes anaerobic glucose photoassimilation to about 45%. As pointed out in Section D, DCMU action can be explained differently: (a) indirect carbon assimilation from glucose, under aerobic and even under anaerobic conditions, when fermentation yields CO_2; (b) participation of oxygen-dependent photophosphorylation; (c) inhibition of cyclic photophosphorylation at high DCMU concentrations (ASAHI and JAGENDORF, 1963). Final proof for any of these possibilities has not been obtained. The first seems likely, because aerobic glucose uptake does not decrease the capacity for photosynthetic CO_2 fixation of light-grown cells except at very low concentrations of inorganic phosphate (SIMONIS, 1956; MOSES et al., 1959; LATZKO and GIBBS, 1969).

References

Abraham, A., Bachhawat, B.K.: Biochim. Biophys. Acta *62*, 376–384 (1962)
App, A.A., Jagendorf, A.T.: J. Protozool. *10*, 340–343 (1963)
Asahi, T., Jagendorf, A.T.: Arch. Biochem. Biophys. *100*, 531–541 (1963)
Bergmann, L.: Flora *142*, 493–539 (1955)
Bishop, N.I.: Biochim. Biophys. Acta *51*, 323–332 (1961)
Bunt, J.S.: J. Phycol. *5*, 37–42 (1969)
Butt, V.S., Peel, M.: Biochem. J. *88*, 31p (1963)
Calvin, M., Bassham, J.A., Benson, A.A., Lynch, V.H., Ouellet, C., Schou, L., Stepka, W., Tolbert, N.E.: Symp. Soc. Exp. Biol. *5*, 284–305 (1951)
Cook, J.R.: J. Protozool. *14*, 382–384 (1967)
Cook, J.R., Carver, M.: Plant Cell Physiol. *7*, 377–383 (1966)
Cooksey, K.E.: J. Gen. Microbiol. *66*, V (1971)
Cooksey, K.E.: Plant Physiol. *50*, 1–6 (1972)
Cooksey, K.E.: J. Phycol. *10*, 253–257 (1974)
Danfort, W.F.: In: Physiology and biochemistry of algae. Lewin, R.A. (ed.), pp. 99–123. New York: Academic Press 1962

Droop, M.R.: Rev. Algol. *4*, 247–248 (1971)

Dvořáková-Hladká, J.: Biol. Plantarum *8*, 354–360 (1966)

Eppley, R.W., Maciasr, F.M.: Physiol. Plant. *15*, 72–79 (1962a)

Eppley, R.W., Maciasr, F.M.: Am. J. Bot. *49*, 671 (1962b)

Eppley, R.W., Maciasr, F.M.: Limnol. Oceanogr. *8*, 411–418 (1963)

Eppley, R.W., Gee, R., Saltman, P.: Physiol. Plant. *16*, 777–792 (1963)

Fischer, E., Wiessner, W.: Ber. Dtsch. Bot. Ges. *81*, 347–348 (1968)

Gaffron, H., Wiessner, W., Homann, P.: Photosynthetic mechanism of green plants. Publ. 1145 Natl. Acad. Sci. pp. 436–440. Wash. D.C.: Natl. Res. Council 1963

Gemmill, E.R., Galloway, R.A.: J. Phycol. *10*, 359–366 (1974)

Gibbs, M.: In: Physiology and biochemistry of algae. Lewin, R.A. (ed.), pp. 61–90. New York: Academic Press 1962a

Gibbs, M.: In: Physiology and biochemistry of algae. Lewin, R.A. (ed.), pp. 91–97. New York: Academic Press 1962b

Goulding, K.H., Merrett, M.J.: J. Exp. Bot. *17*, 678–689 (1966)

Goulding, K.H., Merrett, M.J.: J. Gen. Microbiol. *48*, 127–136 (1967a)

Goulding, K.H., Merrett, M.J.: J. Exp. Bot. *18*, 620–630 (1967b)

Haigh, W.G., Beevers, H.: Arch. Biochem. Biophys. *107*, 147–151 (1964a)

Haigh, W.G., Beevers, H.: Arch. Biochem. Biophys. *107*, 152–157 (1964b)

Harrop, L.C., Kornberg, H.L.: Biochem. J. *88*, 42p (1963)

Heber, U.: In: Prog. Photosynthesis Res. Metzner, H. (ed.), Vol. II, pp. 1082–1090. Tübingen: H. Haupt Jr. 1969

Henry-Hiss, Y.: Diss. Univ. Paris XI/Orsay 1977

John, P.C.L., Thurston, C.F., Syrett, P.J.: Biochem. J. *119*, 913–919 (1970)

Kandler, O.: Z. Naturforsch. *9b*, 625–644 (1954)

Kandler, O.: Z. Naturforsch. *10b*, 38–46 (1955)

Kandler, O., Tanner, W.: Ber. Dtsch. Bot. Ges. Generalversammlungsht. *79*, 48–57 (1966)

Kydar, M.M., Doman, N.G.: Priklad. Biokhim. Mikrobiol. *4*, 727–730 (1968)

Latzko, E., Gibbs, M.: Plant Physiol. *44*, 295–300 (1969)

Lewin, R.A.: J. Gen. Microbiol. *11*, 459–471 (1954)

Luksch, I.: Beih. Bot. Centralbl. Erste Abt. *50*, 64–69 (1933)

Lynn, R.I., Starr, R.C.: Arch. Protistenk. *112*, 283–302 (1970)

Marker, A.F.H., Whittingham, C.P.: Proc. R. Soc. *B65*, 473–478 (1966)

McCullough, W., John, P.C.L.: Nature (London) *239*, 402–405 (1972)

Mehler, A.H.: Arch. Biochem. Biophys. *33*, 65–77 (1951)

Mende, D., Niemeyer, H., Hecker, S., Wiessner, W.: Photosynthetica *12*, 440–448 (1978)

Mende, D., Wiessner, W.: In: Proc. 3rd Int. Congr. Photosynthesis. Avron, M. (ed.), Vol. I, pp. 505–513. Amsterdam: Elsevier Scientific Publ. Comp. 1975

Mende, D., Wiessner, W.: Abstr. 4th Int. Congr. Photosynthesis p. 248. 1977

Merrett, M.J.: Plant Physiol. *58*, 179–181 (1976)

Merrett, M.J., Goulding, K.H.: J. Exp. Bot. *18*, 128–139 (1967a)

Merrett, M.J., Goulding, K.H.: Planta (Berl.) *75*, 275–278 (1967b)

Moses, V., Holm-Hansen, O., Bassham, J.A., Calvin, M.: J. Mol. Biol. *1*, 21–29 (1959)

Myers, J.: J. Gen. Physiol. *30*, 217–226 (1947)

Neilson, A.H., Lewin, R.A.: Phycologia *13*, 227–264 (1974)

Neilson, A.H., Blankley, W.F., Lewin, R.A.: In: Handbook of phycological methods. Stein, J.R. (ed.), pp. 275–285. London: Cambridge Univ. Press 1973

Nührenberg, B., Lesemann, D., Pirson, A.: Planta (Berl.) *79*, 162–180 (1968)

Ohmann, E.: Biochim. Biophys. Acta *82*, 325–335 (1964a)

Ohmann, E.: Biochim. Biophys. Acta *90*, 249–259 (1964b)

Palmer, E.G., Starr, R.: J. Phycol. *7*, 85–89 (1971)

Palmer, E.G., Togasaki, R.G.: J. Protozool. *18*, 640–644 (1971)

Parker, B.C.: Ecology *42*, 381–386 (1961)

Petersen, J.B.: Dansk. Bot. Ark. *8*, 1–183 (1935)

Piskunkova, N.F., Pimenova, N.N.: Mikrobiologiya *40*, 962–965 (1971)

Piskunkova, N.F., Pimenova, N.N.: Mikrobiologiya *42*, 9–13 (1973)

Pringsheim, E.G., Wiessner, W.: Nature (Lond.) *188*, 919–921 (1960)

Pringsheim, E.G., Wiessner, W.: Arch. Mikrobiol. *40*, 231–246 (1961)

Reeves, H.C., Kadis, S., Ajl, S.J.: Biochim. Biophys. Acta *57*, 403–404 (1962)

Schlegel, H.G.: Planta (Berl.) *47*, 510–526 (1956)

Schlegel: H.G.: Z. Naturforsch. *14b*, 246–253 (1959)

Simonis, W.: Z. Naturforsch. *11b*, 354–363 (1956)

Smith, F.A.: J. Exp. Bot. *18*, 348–358 (1967)

Stabenau, H.: Diss. Göttingen (1969)

Syrett, P.J.: J. Exp. Bot. *17*, 641–654 (1966)

Syrett, P.J., Merrett, M.J., Bocks, S.M.: J. Exp. Bot. *14*, 249–264 (1963)

Syrett, P.J., Bocks, S., Merrett, M.J.: J. Exp. Bot. *15*, 35–47 (1964)

Tagawa, K., Tsujimoto, H.Y., Arnon, D.I.: Proc. Natl. Acad. Sci. (USA) *49*, 567–572 (1963)

Tanner, W., Dächsel, L., Kandler, O.: Plant Physiol. *40*, 1151–1156 (1965)

Tanner, W., Loos, E., Kandler, O.: In: Currents in photosynthesis. Thomas, J.B., Goedheer, J.C. (ed.), pp. 243–251. Rotterdam: Ad. Donker Publ. 1966

Tanner, W., Zinecker, U., Kandler, O.: Z. Naturforsch. *22b*, 358–359 (1967)

Tanner, W., Loos, E., Klob, W., Kandler, O.: Z. Pflanzenphysiol. *59*, 301–303 (1968)

Taylor, F.J.: Proc. R. Soc. Lond. *B151*, 483–490 (1960)

Thiede, B.: Protoplasma *87*, 361–385 (1976)

Trebst, A., Eck, H., Wagner, S.: In: Photosynthetic mechanisms of green plants. Publ. 1145. Natl. Acad. Sci. pp. 174–194. Wash. D.C.: Natl. Res. Council 1963

Weaver, E.C., Bishop, N.I.: Science *140*, 1095–1097 (1963)

Wiessner, W.: Arch. Microbiol. *43*, 402–411 (1962)

Wiessner, W.: Plant Physiol. *38*, XXVIII (1963)

Wiessner, W.: Nature (London) *205*, 56–57 (1965)

Wiessner, W.: Nature (London) *212*, 403–404 (1966a)

Wiessner, W.: Ber. Dtsch. Bot. Ges. *79*, 1. Generalversammlungsht. 58–62 (1966b)

Wiessner, W.: Planta (Berl.) *79*, 92–98 (1968)

Wiessner, W.: In: Prog. Photosynthesis Res. Metzner, H. (ed.), Vol. I, pp. 442–449. Tübingen H. Haupt Jr. 1969a

Wiessner, W.: Photosynthetica *3*, 225–232 (1969b)

Wiessner, W.: In: Photobiology of microorganisms. Halldal, P. (ed.), pp. 95–133. London, New York, Sydney, Toronto: Wiley-Interscience 1970

Wiessner, W., Amelunxen, F.: Arch. Mikrobiol. *66*, 14–24 (1969a)

Wiessner, W., Amelunxen, F.: Arch. Mikrobiol. *67*, 357–369 (1969b)

Wiessner, W., Fork, D.C.: In: Carnegie Inst. Wash. Yearb. 695–699 (1971)

Wiessner, W., French, C.S.: Planta (Berl.) *94*, 78–90 (1970)

Wiessner, W., Gaffron, H.: Fed. Proc. *23*, 226 (1964b)

Wiessner, W., Gaffron, H.: Nature (London) *201*, 725–726 (1964a)

Wiessner, W., Kuhl, A.: In: Beiträge zur Physiologie und Morphologie der Algen. pp. 102–108. Stuttgart: Gustav Fischer Verlag 1962

Wiessner, W., Dubertret, G., Mende, D.: Abstr. 4th Int. Congr. Photosynthesis p. 410. 1977

Whittingham, C.P., Bermingham, M., Hiller, R.G.: Z. Naturforsch. *18b*, 701–706 (1963)

Zajic, J.E., Chiu, Y.S.: In: Properties and products of algae. Zajic, J.E. (ed.), pp. 1–47. New York: Plenum Publishing Corp. 1970

15. Biochemical Basis of Ecological Adaptation

A. Shomer-Ilan, S. Beer and Y. Waisel

A. Introduction

Processes of plant adaptation to various environments are to be found at all levels of plant function and organization. However, while many adaptational responses to certain environmental conditions have been described at the levels of plant structure and organization, the biochemical basis of such adaptations is often not fully understood. This can be exemplified by the subject of this chapter: the biochemical basis of adaptation of photosynthetic systems to extreme environments.

It is known that photosynthesis involves a long chain of mechanisms, enzymes, and intermediate products, which are regulated by a large number of internal and external parameters. Environmental stresses such as drought, salinity, low and high temperatures, light intensities, etc., may affect the overall process of carbon fixation at different points. Such effects may change the rate and direction of the fixation process even by modifying one single enzyme. Such an event would be most critical if a major reaction is affected and/or if a key enzyme is involved.

A major obstacle to advances in the clarification of the biochemical level of plant adaptation is the fact that experiments concerning biochemical processes often must be carried out in vitro with isolated enzymes. Analysis of plant behavior at that molecular level is distant and intermittent from the level and site at which responses are actually taking place in the intact plant. Deductions of biochemical data to the overall ecological responses found at higher levels of plant organization are consequently difficult to obtain and frequently unsatisfactory.

Evolutionary as well as direct responses to environmental changes and prevention of metabolic damage (cf. SCULTHORPE, 1967; LEVITT, 1972; WAISEL, 1972) under such conditions may occur via one or a combination of the following systems:

1. Production of structural or chemical protection agents e.g., accumulation of sugars, polyols, proline, betaine, reducing systems, etc.

2. Induction of changes in the organization, composition and structure of cell membranes, thereby improving control of water and solute permeability.

3. Changes in metabolic pathways from those which are affected to existing resistant alternatives.

4. Induction of new metabolic systems which are better suited to the new ecological situation.

Almost all known systems involve the production and activation of new enzymes, or of enzymes with modified properties.

In the following we shall focus on adaptational mechanisms concerning primary carbon metabolism through the three carbon fixation pathways, i.e., C_3, C_4, and CAM, and on possible interreactions between them.

B. Biochemical Variations in C_3 Plants

In spite of the large mass of information on photosynthesis, knowledge of the biochemical changes which contribute to ecological adaptation in C_3 plants is insufficient and is largely based on speculations. Ribulose bisphosphate carboxylase (RuBP carboxylase) has a key role in the C_3 carbon reduction pathway. Therefore, changes in this enzyme may reflect adaptation to environmental stresses. Comparative studies on the structure of RuBP carboxylase from different plant sources have revealed its molecular diversity (ANDERSON et al., 1968; KAWASHIMA and WILDMAN, 1970; TAKABE and AKAZAWA, 1975). RuBP carboxylases of either higher plants or green algae are composed of 8 large subunits (type A) and 8 small subunits (type B); some photosynthetic bacteria also contain RuBP carboxylase with the two subunits (A and B), whereas others like *Rhodospirillum rubrum* and *Chlorobium thiosulfatophilum* are composed of A subunits only (TABITA et al., 1974). (See also Chap. II.17, this vol.).

The amino acid composition of the large catalytic subunits (type A) have a low level of heterogeneity. The fact that they are quite similar in different plant sources reflects a possible single ancestral origin (TAKABE and AKAZAWA, 1975). In contrast, the amino acid composition of the smaller regulatory subunits B differs in various plant groups (TAKABE and AKAZAWA, 1975). The variation in the composition of the subunits may explain the wide range of environmental conditions that the enzyme can tolerate and under which it can be active. Differences were observed even in the same plant species: Two types of the large catalytic subunit A of RuBP carboxylase were found in *Beta vulgaris* L. (KUNG et al., 1975), as well as in the photosynthetic bacterium *Hydrogenemonas eutropha* (PUROHIT and McFADDEN, 1976). In *Brassica oleracea* L., two types of subunit A and two of subunit B were found after cold-hardening (SHOMER-ILAN and WAISEL, 1975).

Several studies have been made on the thermal responses of RuBP carboxylase, as well as of PEP carboxylase, from various plants adapted to extreme temperatures. Plants showed optimal growth and photosynthesis when grown in their optimal native environmental temperature. The apparent Km for RuBP carboxylase was not temperature-dependent, whereas the activation energy of this enzyme reflects the adaptation of plants to extreme environments. RuBP carboxylase from desert plants has a higher activation energy as compared to the enzyme extracted from a tropical variant of tomato (WEBER et al., 1977). Furthermore, the activation energy of RuBP carboxylase, extracted from *Caltha intraloba* plants which had adapted to low temperatures, was lower than that of species from warmer environments (PHILLIPS and McWILLIAM, 1971). These differences can be explained by the involvement of hydrophobic interactions at higher temperatures (BIGELOW, 1967; LEVITT, 1972; SHOMER-ILAN and WAISEL, 1975). Proteins usually exhibit a reversible denaturation between 10° C and 0° C. Such a change results from a weakening of the hydrophobic bonds, which are responsible for the tertiary structure (TANFORD, 1962). Due to various internal and external influences, such denaturation may change into an irreversible aggregation, causing damage to the tissues (BRANDTS, 1967; LEVITT, 1972). A decrease in the hydrophobicity of an enzyme protein from adapted or hardened plants lowers the chances of denaturation and aggregation. This is a basic process leading to adaptation to cold. For example, a lower average hydrophobicity (according to BIGELOW, 1967) of Fraction 1 protein was found in cold-hardened young cabbage leaves when compared with the protein extracted from nonhardened leaves (SHOMER-ILAN and WAISEL, 1975). Furthermore, more hydrophobic bonds would stabilize proteins structure when exposed to high temperatures. Indeed, proteins of thermophilic micro-organisms showed higher average hydrophobicity than proteins of mesophilic species (BIGELOW, 1967).

Mineral nutrition and age were also found to affect the average hydrophobicity of RuBP carboxylase in cabbage leaves, i.e., cause isozymic substitution. Such substitution is another mechanism which contributes to the adaptation of plants to environmental stress (ROBERTS, 1969; SHOMER-ILAN and WAISEL, 1975).

Similar responses were observed for phosphoenolpyruvate carboxylase (PEP carboxylase), where an increase in the activation energy was shown to occur in the cold-sensitive C_4 plants *Zea mays* and *Pennisetum typhoides* after exposure to chilling. Such changes suggest a configurational alteration of the enzyme. On the other hand, no changes in the activation energy of PEP carboxylase were observed in the chilling-resistant plant *Sorghum bipinata* var. *halepense* (PHILLIPS and McWILLIAM, 1971).

C. Biochemical Adaptation of CAM and C_4 Plants

There is no question today as to the universality of the C_3-carbon fixation pathway (Calvin cycle) for net CO_2 incorporation by all plants. The two other pathways, i.e., the C_4 and CAM pathways, are consequently looked upon not as alternatives, but as additional systems supporting this basic C_3–pathway (cf. KELLY et al., 1976). C_3 photosynthesis is, therefore, more basic and ancestral to the C_4 and CAM systems (SMITH and ROBBINS, 1974). It must be pointed out that even though these two additional systems certainly bear importance with respect to plant adaptation to extreme environments, variations in the Calvin cycle and its key enzymes are also relevant to the ecological adaptation of C_4 and CAM plants.

The enzymatic system involved in the C_4 and CAM pathways are essentially alike. PEP carboxylase is a pace-limiting enzyme in both pathways (BLACK, 1973; KELLY et al., 1976). Carbon, probably in the form of HCO_3^- (COOPER and WOOD, 1971; COOMBS et al., 1975), is fixed by PEP carboxylase in the cytoplasm of CAM and C_4 plants (GIBBS et al., 1970; KAGAWA and HATCH, 1974) to form oxaloacetate and thence the first stable products malate and/or aspartate. These products are subsequently transported to another site of the cell (CAM plants) or the leaf (most C_4 plants), decarboxylated, and the CO_2 is reassimilated in chloroplasts having the Calvin cycle (HATCH and SLACK, 1970; BLACK, 1973). The PEP carboxylase system provides CAM plants with malate which is stored during the dark period until it can be refixed in the light by the Calvin cycle (BLACK, 1973; QUEIROZ, 1976).

The C_4-system involves the activity of several specific enzymes. Nevertheless, not all of them were absent before the evolution of the C_4 and the CAM pathways. For example, the carboxylating and decarboxylating enzymes do exist in C_3 plants

(DOWNTON, 1971). What was really limiting activity at high rates was the supply of PEP. Thus, presumably, just one mutation which would enable the formation of a good supplier of PEP, like pyruvate-Pi-dikinase, which had evolved in C_4 plants, would also enable a high activity of PEP carboxylase. This would complete an additional system for an efficient capture of carbon.

Dark fixation by PEP carboxylase is known to exist in all living cells. The enzyme is active at different rates in various plants, various organs, and even in various cells of the same organ. PEP carboxylase extracted from various higher plants (C_3, C_4 and CAM) showed a wide variation in their Km and Vmax values (TING and OSMOND, 1973a; BLACK, 1973; COOMBS et al., 1973) and substrate regulation properties (COOMBS and BALDRY, 1975; GOATLY and SMITH, 1974; NISHI-KIDO and TAKANASHI, 1973; MIZIORKO et al., 1974). Three isozymes of PEP carboxylase differing in properties were isolated from cotton leaves (MUKERJI and TING, 1971). It is assumed that such variations in enzyme properties enable plant adaptation to a wide range of environmental factors (cf. JOHNSON, 1974).

Some plant species have developed the ability to exhibit CAM under certain conditions. The induction of CAM in these potential CAM plants involves not only the carbon-fixing enzyme, but also the activation of transaminating and decarboxylating enzymes. CAM is induced, in part, by short diurnal photoperiods (QUEIROZ and MOREL, 1974). Low night temperatures seem to be necessary for the accumulation of malate during the long nights. The reason for this depends partly on the decarboxylating NADP-linked malic enzyme which is involved in this process. At low temperatures, this enzyme changes its conformation to one with a lower affinity for malate, thus enabling the accumulation of malate in the cells (BRANDON and VAN BOEKEL-MOL, 1973). Malate which is accumulated during cool nights may regulate PEP carboxylase activity by feedback inhibition (TING and OSMOND, 1973b).

The possibility of a rapid switch to β-carboxylation was shown to occur in *Chlorella* following a change in CO_2 concentration (GRAHAM and WHITTINGHAM, 1968). C_4 metabolism was found in stem callus cultures of one type of cells of the C_4 plants *Froelichia gracilis* (LAETSCH and KORTSCHAK, 1972) and *Portulaca oleracea* (KENNEDY et al., 1977), and in some unicellular algae (COLMAN et al., 1976). Thus, C_4 metabolism does not necessarily involve different types of cells and can operate even in single cells. C_4 metabolism was shown also in *Psilotum nudum* (BLACK, 1974) and in some marine algae (KAREKAR and JOSHI, 1973) which in certain conditions possess C_4 metabolism, although their chlorenchyma is not differentiated. Nevertheless, cells of most plants with the complete C_4 pathway (C_4 plants) do undergo biochemical specialization. For example, the mesophyll cells specialize in β carboxylation and nitrate assimilation. They are rich in PEP carboxylase (cf. BLACK, 1973, 1974) and nitrate-assimilating enzymes (RATHNAM and EDWARDS, 1976; HAREL et al., 1977). The ability of their chloroplasts to fix carbon by RuBP carboxylase had been lost due to lack of the enzyme (HUBER et al., 1976; WALBOT, 1977). Similar results were obtained for the outer chlorenchymatous layer in *Suaeda monoica*, a C_4 plant lacking the Kranz anatomy (SHOMER-ILAN et al., 1978). Other groups of cells which are specialized for the reduction of carbon by the Calvin cycle exhibit high activities of RuBP carboxylase and decarboxylating enzymes (cf. BLACK, 1973, 1974). The biochemical specialization is often thought to be closely correlated with a special leaf anatomy (Kranz anatomy) found in most C_4 plants (HATCH and SLACK, 1970; BLACK, 1973; LAETSCH, 1974; BROWN, 1975). However, as mentioned above, not all plants exhibiting the C_4 pathway of photosynthesis have this type of anatomy. Thus, the interdependence of these two features does not seem to be absolute (SHOMER-ILAN et al., 1975; JOSHI, 1975; BENEDICT and SCOTT, 1976; BOURDU, 1976; WINTER et al., 1977).

I. Adaptive Value of C_4 Metabolism

One of the questions which arises in this connection is: what might be the adaptive value of a biochemical structural and physiological development such as that found in most plants possessing the C_4 metabolism? As indicated by the distribution of C_4 and CAM plants, the β carboxylation system of carbon fixation seems to have evolved in plants exposed to extreme environments (LAETSCH, 1968; SMITH and ROBBINS, 1974; PEARCY and TROUGHTON, 1975). Such environments are characterized by high temperatures, high light intensities, salinity, and water stress conditions, all of which prevail in arid regions.

Even though the photosynthetic potential of C_3 and C_4 plants may differ, the productivities of these plant groups are similar when compared in their respective natural environments (GIFFORD, 1974). However, under stress conditions most C_4 plants seem to perform better than C_3 plants with respect to net photosynthetic rates. This can be exemplified by the relative independence of the quantum yield (mol CO_2 fixed per absorbed einstein) of C_4 plants, found over a wide temperature range, as compared to that of C_3 plants. In the latter, quantum yields decrease to values below those of C_4 plants at temperatures above $30°$ C (EHLERINGER and BJÖRKMAN, 1977). On the other hand, in temperate cool (cf. DOWNTON, 1971; BJÖRKMAN, 1971) and nonsaline conditions (WAISEL, 1972; WAISEL et al., 1974; BEER, 1975), many C_4 plants perform poorly.

The question of what was the precise factor that triggered the development of the C_4 system also remains unanswered. We know that high abscisic acid (ABA) levels in plants favor a high PEP carboxylase activity (SANKHLA and HUBER, 1975). As ABA rapidly appears in stressed plants (cf. MEYER, 1974), it is tempting to assume that this hormone was also involved in the appearance of the C_4 and CAM pathways.

1. Affinity for CO_2

The exact biochemical advantage of the β carboxylation system in the above-mentioned extreme environments is difficult to evaluate. It was thought that the low Km (HCO_3^-) of PEP carboxylase should make the C_4-system more efficient than the C_3-system in carbon fixation. Such a system would be of prime importance under conditions where high temperature and water stress would limit CO_2 diffusion into the leaf by closing the plants' stomates. Indeed, one of the adaptational advantages of C_4 and CAM plants in arid areas is their high water use efficiency, i.e., high ratio of CO_2 fixed per unit of water transpired (cf. BLACK, 1973).

The affinities of RuPB carboxylase and PEP carboxylase have recently been reevaluated. RuPB carboxylase extracted from C_3 as well as from C_4 plants was shown to yield Km (CO_2) values of approximately 15 µM (BAHR and JENSEN,

1974a, 1974b). An estimated Km (HCO_3^-) value for *Zea mays* PEP carboxylase was in the same range (UEDAN and SUGIYAMA, 1976). Even if the low Km (CO_2) value for RuBP carboxylase would be somewhat underestimated, its high V_{max} value makes it competitive with PEP carboxylase in respect to carbon assimilation (LILLEY and WALKER, 1975).

CO_2 assimilation is usually limited by low CO_2 supply combined with high diffusion resistances of the mesophyll. An increase in CO_2 concentration near those cells possessing the Calvin cycle would consequently raise their net CO_2 fixation rates. Indeed, it is accepted by many investigators that the high activity of PEP carboxylase which is found in the mesophyll cells of C_4-plants, followed by malate translocation and decarboxylation, may eventually concentrate CO_2 in the locality of the C_3 system (HATCH, 1971; KELLY et al., 1976). The C_4-system might have another advantage based on local pH changes and consequently on the carbon species available in the mesophyll. If the pH of the mesophyll of C_4-plants tends to increase, for example under saline conditions due to accumulation of alkali ions (WAISEL and ESHEL, 1971), or due to high OH^- fluxes into the cytoplasm, carbon at those sites will be available rather in the form of HCO_3^- than of CO_2. Such a situation would be optimal for the development of an HCO_3^--utilizing system such as the PEP carboxylase system found in C_4 plants. In fact, certain halophytes tend to have higher pH values of leaf extracts and higher pH optima for PEP carboxylase activity when grown under saline than when grown under nonsaline conditions (WAISEL et al., 1974; BEER, 1975).

2. Decreased Photorespiration

A higher intracellular CO_2 concentration would also partly revoke the performance of RuBP carboxylase as an oxygenase (BADGER and ANDREWS, 1974; BOWES et al., 1975; LAING et al., 1974). This effect would be more pronounced in C_4 plants of the NADP-linked malic enzyme type where bundle sheath chloroplasts often lack significant O_2 production from photosystem II (cf. DOWNTON, 1971).

Another explanation for the advantage of the C_4-system is the lack of apparent photorespiration, i.e., light- and oxygen-dependent CO_2 release derived from glycolate metabolism (cf. ZELITCH, 1971). Photorespiration may lead to a considerable loss of carbon in plants, especially under high light and high temperature conditions (BADGER and ANDREWS, 1974; LAING et al., 1974) and in alkaline environments (KROPF and JACOBY, 1975). Since the enzymatic potential for photorespiration is present also in C_4 plants, it was suggested that one of the advantages of the C_4-system is its effective capability of refixing internally released CO_2. Carbon dioxide which is released by photorespiration in the bundle sheath cells is recaptured by PEP carboxylase in the cytoplasm of the outer mesophyll cells (cf. KELLY et al., 1976).

3. Effects of Light and Temperature

C_4-plants have high light-saturation and high temperature optima, but their efficiency under low light and low temperature conditions is low (LAETSCH, 1968; DOWNTON, 1971). Two enzymes probably play a role in limiting photosynthesis under low temperature conditions. Both enzymes, NADP-dependent malic dehydrogenase and pyruvate-Pi-dikinase, are located in mesophyll-cell chloroplasts, and are light-activated (Hatch et al., 1975). Their activities are extremely reduced below 12° C. The high activation energy of pyruvate-Pi-dikinase (E.C. 2.7.9.1) under low temperature would hinder plant growth under such conditions (cf. Taylor et al., 1974). Far-reaching conclusions have even blamed this enzyme for the limitation of corn distribution in the corn belt of North America. However, low temperature sensitivity is not equal in different plants or varieties; pyruvate-Pi-dikinase extracted from cold-resistant cultivars of *Zea mays* showed a longer half-life than the isoenzymes extracted from cold-sensitive varieties under low-temperature treatments (Sugiyama and Boku, 1976). This is further proof of the important role played by adapted isozymes in the process of plant adaptation.

D. Induced Variations in Carbon Fixation Pathways

The fixation of atmospheric CO_2 into PGA, malate, or aspartate depends on the relative activities and/or compartmentation of PEP carboxylase and RuBP carboxylase. Although the genetic potential for β-carboxylation is present in most plant groups (cf. Downton, 1971 and others), the potential biochemical specialization and/or coordination between the two carbon fixation pathways have actually evolved in C_4 and CAM plants. The relative contributions of the various carbon fixation pathways may be determined by the plants' developmental stage (Shomer-Ilan and Waisel, 1973; Kennedy and Laetch, 1973; Khanna and Sinha, 1973) as well as by environmental factors (Shomer-Ilan and Waisel, 1973; Tew et al., 1974; Kennedy, 1977).

Expression of CAM in potential CAM plants can be induced by various environmental conditions. Carbon in such plants is fixed via the C_3 system as long as the plants are kept under long photoperiods and high-temperature conditions. When plants are moved to short photoperiod conditions and to night temperatures below 18° C, a shift to CAM can be observed (cf. Black, 1973; Queiroz, 1976). Such a shift is due to induction of the carboxylating (Waisel et al., 1974; Treichel et al., 1974) and decarboxylating enzymes (Black, 1976), disappearance of inhibitors (Brulfert et al., 1973) and lower activity of the NADP-linked malic enzymes during cool nights (Brandon and von Boekel-Mol, 1973). CAM can also be induced by subjecting plants to salinity or water stress conditions (Waisel et al., 1974; Winter, 1974a, b; von Willert et al., 1976).

I. Effects of Age, CO_2-Concentration, and Nitrogen Nutrition

Age has an apparent effect on carbon fixation systems. Chloroplasts from the leaves of young maize plants apparently fix CO_2 via the C_3 and not via the C_4 system. They lack the decarboxylating enzymes at that stage of development and therefore cannot utilize C_4 acids (Gibbs et al., 1970). Similar results were reported for young and sodium-depleted *Aeluropus litoralis*. However, in *Aeluropus*, the young leaves showed minimal activity of PEP carboxylase. Addition of NaCl to the growth medium of such plants induced the apparent activity of PEP carboxylase and shifted the carbon fixation from the C_3 towards the C_4 pathway (SHOMER-ILAN and WAISEL, 1973). In the above-mentioned cases leaves of older age performed the C_4 pathway, and the plants are then regarded as C_4 plants.

Opposite trends were reported for other plant species. Young leaves of *Sorghum bicolor* and *Portulaca oleracea* (KHANNA and SINHA, 1973; KENNEDY and LAETCH, 1973) possess the C_4-pathway, whereas older leaves of these plants gradually change towards the C_3-pathway. It was shown for *Portulaca* plants that RuBP carboxylase activity was less reduced in old leaves than was PEP carboxylase activity (KENNEDY, 1976).

The immature green barley pericarp fixes atmospheric CO_2 primarily by PEP-carboxylase, although the plant is a C_3 plant (DUFFUS and ROSIE, 1973). The same was found in developing seeds of *Pisum sativum,* and it was suggested that high activities of PEP carboxylase appear in those stages of plant development that exhibit high respiration/photosynthesis ratios, i.e., contain high internal CO_2 concentrations (HEDLEY et al., 1975).

In contrast, *Chlorella pyrenoidosa* grown under low CO_2 (0.03%) or high CO_2 (5%) concentrations showed typical C_3 characteristics. However, when plants were moved from a high to a low CO_2 medium, a shift towards C_4-metabolism could be observed (GRAHAM and WHITTINGHAM, 1968).

The relative significance of C_3 versus C_4 carboxylation pathways can be altered also by supplying plants with various forms of nutrients. *Themeda triandra,* a plant with a typical Kranz anatomy, shows the C_4-pathway when grown on nitrate as its source of nitrogen. However, when such plants are given ammonium instead of nitrate, C_3 characteristics become more apparent (TEW et al., 1974).

Mineral nutrition and chilling may affect the carbon metabolism in plants in other ways. By changing the source of nitrogen (TEW et al., 1974) or by lowering the temperature (BROOKING and TAYLOR, 1973) a change could be induced in the balance between malate and aspartate. Such a change probably results from alterations in the rate of transamination, but its adaptive value remains obscure.

Induction of changes in carbon fixation pathways could be controlled by ABA. The content of this hormone in plants is known to increase due to water stress, salinity, temperature shocks, oxygen deficiency (MIZRAHI and RICHMOND, 1972; MEYER, 1974), and senescence (EVEN-CHEN and ITAI, 1975). Indeed, an increase in PEP-carboxylase activity following ABA treatment was actually found by SANKHLA and HUBER (1975). However, the question of whether these two variables do have a cause-and-effect relationship still needs investigation.

II. Effect of NaCl

It is well known that both CAM and C_4 plants are frequently halophytes or of halophytic origin (cf. LAETSCH, 1974; LIPHSCHITZ and WAISEL, 1974). Further-more, many nonhalophytic C_4 plants seem to require sodium as a micronutrient (BROWNELL and CROSSLAND, 1972). Such a requirement for sodium was also shown in *Bryophyllum tubiflorum* when performing CAM, but not when photosynthesizing via the C_3 system (BROWNELL and CROSSLAND, 1974). Even though many halophytes are not C_4 plants, a trend towards C_4 carboxylation seems to be induced by salt in such plants.

Various halophytes investigated exhibited higher rates of PEP carboxylase activity when exposed to high salinity levels (SHOMER-ILAN and WAISEL, 1973; TREICHEL et al., 1974; WAISEL et al., 1974; BEER, 1975; KENNEDY, 1977). However, there is no rule without exceptions, and in this case no increase in PEP carboxylase activity could be found in the C_4 halophyte *Atriplex spongiosa* (OSMOND and GREENWAY, 1972). In some CAM plants the salt treatment increased the ratio of PEP carboxylase protein: total extractable protein (WAISEL et al., 1974; BEER, 1975).

Another enzyme which is closely related to the C_4 metabolism of the malate type, malate dehydrogenase, was also found at higher activities in the leaves of high-salt halophytes (VON WILLERT, 1974).

Enzymes are generally sensitive to the presence of inorganic ions in vitro, their activities being inhibited even at low salt concentrations. This is true of RuBP carboxylase and PEP carboxylase extracted from most higher plants. PEP carboxylase extracted from a salt-tolerant *Atriplex* was shown to be more salt-sensitive than the enzyme taken from a salt-sensitive species (OSMOND and GREENWAY, 1972). Furthermore, PEP carboxylase from some halophytic species was found to be more sensitive to NaCl when extracted from high-salt grown plants (TREICHEL et al., 1974). Such negative effects of NaCl could in part be overcome by high PEP concentrations (TING and OSMOND, 1973).

In some cases it seems that salt effects can be of a competitive type or can occur at the regulatory site of the enzyme, perhaps sometimes altering the enzyme configuration.

Certain exceptions to the negative in vitro salt effects were also found. PEP carboxylase from the coastal C_3 halophyte *Cakile maritima* (BEER et al., 1975) and from two seagrass species was found to be stimulated by various chloride salts at possible physiological concentrations. Under the very same conditions the activity of RuBP carboxylase of the two seagrasses was strongly inhibited (BEER, unpublished). A similar in vitro stimulating effect of salt was found for malate dehydrogenase of the salt marsh halophyte *Borrichia frutescens* (CAVALIERI and HUANG, 1977). Exceptions were also found in some salt-tolerant halophytic bacteria, in which not only salt-resistant but also salt-activated enzyme systems have evolved (BAXTER and GIBBONS, 1954; MEVARECH and NEUMAN, 1977).

The precise effects of salt on the in situ activity of photosynthetic enzymes are little known, and much more information is needed. The seemingly contradictory effects of salts upon the activity of PEP carboxylase in vitro and in vivo may not be a disparity. Explanations should be sought in intra-cellular compartmentation of the ions, which may be excluded from the sites of enzyme activity or formation. Such salt compartmentation would raise the need for balancing the cytoplasmic and vacuolar water potentials. Proline (STEWART and LEE, 1974; TREICHEL, 1975), betaine (WYN JONES et al., 1976) and organic acids (OSMOND, 1965) seem to be the major osmoregulators in higher plants in response to salinity (WAISEL, 1972; FLOWERS et al., 1977). As proline levels up to 600 mM (STEWART and LEE, 1974) and betaine levels up to 1 M (WYN JONES et al., 1976) do not inhibit enzyme activity in vitro, they may also have an important role as cytoplasmatic osmoregulators. Organic acids are toxic at high concentrations and therefore seem to be mostly restricted to the vacuoles (OSMOND, 1965). There, they may balance excess cation levels (WAISEL, 1972; FLOWERS et al., 1977). Sucrose and reducing sugars are also known to be osmoregulators in cold and dessication tolerance (LEVITT, 1972). However, they seem to play a minor role in adjustment of higher plants to salinity (FLOWERS et al., 1977). On the other hand, many microalgae and fungi adjust very rapidly to changes in external osmotic potential. This is done by glycerol production in the halophytic alga *Dunaliella* (BEN AMOTZ and AVRON, 1973), by isofloridoside in *Ochromonas* (KAUSS, 1967) and by mannitol in *Platymonas* (HELLEBUST, 1976).

E. Concluding Remark

Most people would regard biochemistry and ecology as two distinct scientific disciplines. As evident in the present article, such a distinction is unjustified. We should come to understand that what we really speak of when discussing biochemical aspects of ecology are the interactions between the inseparable and interdependent two ends of one system. Nothing could better phrase this issue than the words of Herbert Spencer: "Life is a continuous adjustment of the internal conditions to the external environment."

References

Anderson, L.E., Price, G.B., Fuller, R.C.: Science *161*, 482–486 (1968)
Badger, M.R., Andrews, T.J.: Biochem. Biophys. Res. Commun. *60*, 204–210 (1974)
Bahr, J.T., Jensen, R.G.: Plant Physiol. *53*, 39–44 (1974a)
Bahr, J.T., Jensen, R.G.: Biochem. Biophys. Res. Commun. *57*, 1180–1185 (1974b)
Baxter, R.M., Gibbons, N.E.: Can. J. Biochem. Physiol. *32*, 206–217 (1954)
Beer, S.: M. Sc. Thesis. Tel Aviv Univ. Dept. Bot. (1975)
Beer, S., Shomer-Ilan, A., Waisel, Y.: Physiol. Plant *34*, 293–295 (1975)
Ben-Amoz, A., Avron, M.: Plant Physiol. *51*, 875–878 (1973)
Benedict, C.R., Scott, J.R.: Plant Physiol. *57*, 876–880 (1976)
Bigelow, C.C.: J. Theor. Biol. *16*, 187–211 (1967)
Björkman, O.: In: Photosynthesis and photorespiration. Hatch, M.D., Osmond, C.B., Slatyer, R.O. (eds.), pp. 18–32. New York: Wiley-Interscience 1974
Black, C.C.: Ann. Rev. Plant Physiol. *24*, 253–286 (1973)
Black, C.C.: In: Proc. 3rd Int. Congr. Photosynthesis. Avron, M. (ed.), pp. 1201–1208. Amsterdam: Elsevier (1974)
Black, C.C.: In: Symp. S. Sect. A.S.P.P. Benedict, C.R. (ed.), pp. 51–73 (1976)
Bourdu, R.: Physiol. Veg. *14*, 551–561 (1976)
Bowes, G., Ogren, W.L., Hageman, R.H.: Plant Physiol. *56*, 630–633 (1975)
Brandon, P.C., Van Boekel-Mol, T.N.: Eur. J. Biochem. *35*, 62–69 (1973)
Brandts, J.F.: In: Thermobiology. Rose, A.H. (ed.), pp. 25–70. New York: Academic Press (1967)
Brooking, J.R., Taylor, A.O.: Plant Physiol. *52*, 180–182 (1973)
Brown, W.V.: Am. J. Bot. *62*, 395–402 (1975)
Brownell, P.F., Crossland, C.J.: Plant Physiol. *49*, 794–797 (1972)
Brownell, P.F., Crossland, C.J.: Plant Physiol. *54*, 416–417 (1974)
Brulfert, J., Guerrier, D., Queiroz, O.: Plant Physiol. *51*, 220–222 (1973)

Cavalieri, A.J., Huang, A.C.H.: Physiol. Plant. *41*, 78–84 (1977)
Colman, B., Cheng, K.H., Ingle, R.K.: Plant. Sci. Lett. *6*, 123–127 (1976)
Coombs, J., Baldry, C.W.: Planta *124*, 153–158 (1975)
Coombs, J., Baldry, C.W., Bucke, C.: Planta (Berl.) *110*, 95–107 (1973)
Coombs, J., Maw, S.L., Baldry, C.W.: Plant. Sci. Lett. *4*, 97–102 (1975)
Cooper, T.G., Wood, H.: J. Biol. Chem. *246*, 5488–5490 (1971)
Downton, W.J.S.: In: Photosynthesis and photorespiration. Hatch, M.D., Osmond, C.B.,
    Slatyer, R.O. (eds.), pp. 3–17. New York: Wiley-Interscience (1971)
Duffus, W.M., Rosie, R.: Planta (Berl.) *114*, 219–226 (1973)
Ehleringer, J., Björkman, O.: Plant Physiol. *59*, 86–90 (1977)
Even-Chen, Z., Itai, C.: Physiol. Plant. *34*, 97–100 (1975)
Flowers, T.J., Troke, P.F., Yeo, A.R.: Annu. Rev. Plant Physiol. *28*, 89–121 (1977)
Gibbs, M., Latzko, E., O'Neal, D., Hew, C.S.: Biochem. Biophys. Res. Commun. *40*, 1356–1361
    (1970)
Gifford, R.M.: Aust. J. Plant Physiol. *1*, 107–117 (1974)
Goatly, M.B., Smith, H.: Planta (Berl.) *117*, 67–73 (1974)
Graham, D.G., Whittingham, C.P.: Z. Pflanzenphysiol. *58*, 418–427 (1968)
Harel, E., Lea, P.J., Miflin, B.J.: Planta (Berl.) *134*, 195–200 (1977)
Hatch, M.D.: In: Photosynthesis and photorespiration. Hatch, M.D., Osmond, C.B., Slatyer,
    R.O. (eds.), pp. 139–152. New York: Wiley-Interscience (1971)
Hatch, M.D., Kagawa, T.: Aust. J. Plant Physiol. *2*, 111–128 (1975)
Hatch, M.D., Slack, C.R.: Annu. Rev. Plant Physiol. *21*, 141–162 (1970)
Hedley, C.L., Harvey, D.M., Keely, R.J.: Nature (London) *258*, 352–354 (1975)
Hellebust, J.A.: Can. J. Bot. *54*, 1735–1741 (1976)
Huber, S.C., Hall, T.C., Edwards, G.E.: Plant Physiol. *57*, 730–733 (1976)
Johnson, B.: Science *184*, 28–37 (1974)
Joshi, G.V.: In: Studies in photosynthesis under saline conditions. Kolhapur, India: Shivaji
    University Press 195 pp (1975)
Kagawa, T., Hatch, M.D.: Aust. J. Plant Physiol. *1*, 57–64 (1974)
Karekar, M.D., Joshi, G.V.: Bot. Mar. *16*, 216–220 (1974)
Kawashima, N., Wildman, S.G.: Annu. Rev. Plant Physiol. *21*, 325–358 (1970)
Kauss, H.: Z. Pflanzenphysiol. *56*, 453 (1967)
Kelly, G.J., Latzko, E., Gibbs, M.: Annu. Rev. Plant Physiol. *27*, 181–205 (1976)
Kennedy, R.A.: Planta (Berl.) *128*, 149–154 (1976)
Kennedy, R.A.: Z. Pflanzenphysiol. *83*, 11–24 (1977)
Kennedy, R.A., Laetsch, W.M.: Planta (Berl.) *115*, 113–124 (1973)
Kennedy, R.A., Barnes, J.E., Laetsch, W.M.: Plant Physiol. *59*, 600–603 (1977)
Khanna, R., Sinha, S.K.: Biochem. Biophys. Res. Commun. *52*, 121–124 (1973)
Kropf, G., Jacobi, G.: Plant. Sci. Lett. *5*, 67–71 (1975)
Kung, S.D., Gray, J.C., Wildman, S.G., Carlson, P.S.: Science *187*, 353–355 (1975)
Laetsch, W.M.: Am. J. Bot. *55*, 875–883 (1968)
Laetsch, W.M.: Annu. Rev. Plant Physiol. *25*, 27–52 (1974)
Laetsch, W.M., Kortschak, H.P.: Plant Physiol. *49*, 1021–1023 (1972)
Laing, W.A., Ogren, W.L., Hageman, R.H.: Plant Physiol. *54*, 678–685 (1974)
Levitt, J.: Responses of plants to environmental stresses. New York: Academic Press 732 pp.
    (1972)
Lilley, R. McC., Walker, D.A.: Plant Physiol. *55*, 1087–1092 (1975)
Liphschitz, N., Waisel, Y.: New Phytol. *73*, 507–513 (1974)
Mevarech, M., Neuman, E.: Biochemistry *16*, 3786–3791 (1977)
Meyer, F.H.: Annu. Rev. Plant Physiol. *25*, 259–307 (1974)
Miziorko, H.M., Nowak, T., Mildvan, A.S.: Arch. Biochem. Biophys. *163*, 378–389 (1974)
Mizrahi, Y., Richmond, E.E.: Plant Physiol. *50*, 667–670 (1972)
Mukerji, S.K., Ting, I.P.: Arch. Biochem. Biophys. *143*, 297–317 (1971)
Nishikido, T., Takanashi, H.: Biochem. Biophys. Res. Commun. *53*, 126–133 (1973)
Nishimura, M., Akazawa, T.: Biochemistry *13*, 2277–2281 (1974)
Osmond, C.B.: Nature (London) *198*, 503–504 (1965)
Osmond, C.B., Greenway, H.: Plant Physiol. *49*, 260–263 (1972)
Pearcy, R.W., Troughton, J.: Plant Physiol. *55*, 1054–1056 (1975)

Phillips, P.J., McWilliam, J.R.: In: Photosynthesis and photorespiration. Hatch, M.D., Osmond, C.B., Slatyer, R.O. (eds.), pp. 97–104. New York: Wiley-Interscience (1971)
Purohit, K., McFadden, B.A.: Biochem. Biophys. Res. Commun. *71*, 1220–1222 (1976)
Queiroz, O.: Physiol. Veg. *14*, 629–639 (1976)
Queiroz, O., Morel, C.: Plant Physiol. *53*, 596–602 (1974)
Rathman, C.K.M., Edwards, G.E.: Plant Physiol. *57*, 881–885 (1976)
Roberts, D.W.: Int. Rev. Cytol. *26*, 303–328 (1969)
Sankhla, N., Huber, E.: Z. Pflanzenphysiol. *74*, 267–271 (1975)
Sculthorpe, C.D.: The biology of aquatic vascular plants. London: E. Arnold Publ. 610 pp. 1967
Shomer-Ilan, A., Waisel, Y.: Physiol. Plant. *29*, 190–193 (1973)
Shomer-Ilan, A., Waisel, Y.: Physiol. Plant. *34*, 90–96 (1975)
Shomer-Ilan, A., Beer, S., Waisel, Y.: Plant Physiol. *56*, 676–679 (1975)
Shomer-Ilan, A., Neumann-Ganmore, R., Waisel, Y.: Proc. FESPP meeting, Organised by Society of Experimental Biology University of Edinburgh 481–482 1978
Smith, B.N., Robbins, M.J.: In: Proc. 3rd Int. Congr. Photosynthesis. Avron, M. (ed.), pp. 1579–1588. Amsterdam: Elsevier (1974)
Stewart, G.R., Lee, J.A.: Planta (Berl.) *120*, 279–289 (1974)
Sugiyama, T., Boku, K.: Plant Cell Physiol. *17*, 851–854 (1976)
Tabita, F.R., McFadden, B.A., Pfenning, N.: Biochem. Biophys. Acta *341*, 187–194 (1974)
Takabe, T., Akazawa, T.: Plant Cell Physiol. *16*, 1049–1060 (1975)
Tanford, C.: J. Am. Chem. Soc. *84*, 4240–4247 (1962)
Taylor, A.D., Slack, C.R., McPherson, H.G.: Plant Physiol. *54*, 696–701 (1974)
Tew, J., Cresswell, C.F., Fair, P.: In: Proc. 3rd Int. Congr. Photosynthesis. Avron, M. (ed.), pp. 1249–1266. Amsterdam: Elsevier (1974)
Ting, I.P., Osmond, C.B.: Plant Physiol. *51*, 448–453 (1973a)
Ting, I.P., Osmond, C.B.: Plant Science Lett. *1*, 123–128 (1973b)
Treichel, S.P.: Z. Pflanzenphysiol. *76*, 56–68 (1975)
Treichel, S.P., Kirst, G.O., von Willert, D.J.: Z. Pflanzenphysiol. *71*, 437–449 (1974)
Uedan, K., Sugiyama, T.: Plant Physiol. *57*, 906–910 (1976)
von Willert, D.J.: Oecologia *14*, 127–137 (1974)
von Willert, D.J., Kirst, G.O., Treichel, S., von Willert, K.: Plant. Sci. Lett. *7*, 341–346 (1976)
Waisel, Y.: Biology of halophytes. New York: Academic Press 395 pp. (1972)
Waisel, Y., Eshel, A.: Experientia *27*, 231–232 (1971)
Waisel, Y., Beer, S., Shomer-Ilan, A.: Dtsch. Bot. Ges. Tagung in Würzburg. pp. 114 (1974)
Walbot, V.: Plant Physiol. *60*, 102–108 (1977)
Weber, D.J., Andersen, W.R., Hess, S., Hansen, B.J., Gunasekaren, M.: Plant Cell Physiol. *18*, 693–699 (1977)
Winter, K.: Oecologia *15*, 383–390 (1974a)
Winter, K.: Plant. Sci. Lett. *3*, 279–281 (1974b)
Winter, K., Kramer, D., Troughton, J.H., Card, K.A., Fischer, K.: Z. Pflanzenphysiol. *81*, 341–346 (1977)
Wyn Jones, R.G., Storey, R., Pollard, A.: Int. Workshop transmembrane ionic exchanges in plants, Rouen (1976)
Zelitch, I.: Photosynthesis, photorespiration and plant productivity. New York: Academic Press 347 pp. 1971

II D. Regulation and Properties of Enzymes of Photosynthetic Carbon Metabolism

16. Light-Dependent Changes of Stromal H^+ and Mg^{2+} Concentrations Controlling CO_2 Fixation

H.W. HELDT

A. Background

Illumination of chloroplasts causes a transport of protons across the thylakoid membrane (NEUMANN and JAGENDORF, 1964) and leads to a pH decrease in the thylakoid space (DEAMER et al., 1967; RUMBERG and SIGGEL, 1969; HAGER, 1969; GAENSSLEN and McCARTY, 1971; ROTTENBERG et al., 1972). Furthermore, a light-dependent release of Mg^{2+} from the thylakoids has been observed (DILLEY and VERNON, 1965; HIND et al., 1974). These earlier studies with chloroplasts which had lost their envelope suggested that light-dependent changes of the H^+ and Mg^{2+} concentrations may also occur in the chloroplast stroma. There has been a great deal of speculation that CO_2 fixation may be regulated by the stromal concentrations of these cations.

B. Measurement of the pH in the Stroma and the Thylakoid Space of Intact Spinach Chloroplasts

When radioactively labeled methylamine, a weak base, and dimethyloxazolidine (DMO), a weak acid, are added to illuminated intact chloroplasts, measurements by silicone layer filtering centrifugation reveal that both substances are accumulated in the chloroplasts (HELDT et al., 1973). This shows that illuminated chloroplasts contain two compartments. The compartment accumulating the DMO anion is more alkaline than the medium, and the other accumulating the methylammonium ions is more acidic than the medium; these two compartments corresponding to the stroma and the thylakoid space. Taking into account the size of these compartments, the pH in both compartments can be calculated from the uptake of methylamine and DMO. With isolated spinach chloroplasts kept in a medium of pH 7.6 in the dark, the stromal pH is found about 7.0 (Table 1). This pH difference is also found when the envelope has been rendered permeable to protons by the addition of m-chlorocarbonylcyanide phenylhydrazone (CCCP) and seems to be due to a Donnan equilibrium (GIMMLER et al., 1975).

When the chloroplasts are illuminated, the stromal pH is shifted from pH 7.0 to 8.0. This strong alkalization of the stroma is mainly due to proton transport into the thylakoid space. However, it may be also in part due to light-dependent H^+ transport from the stroma to the external medium, as observed with illuminated isolated chloroplasts kept in a slightly buffered medium (GIMMLER et al., 1975). The nature of this proton pump across the envelope, probably requiring ATP,

Table 1. pH Measurement in intact spinach chloroplasts. Medium pH 7.6, 20 °C. HELDT et al. (1974)

	light	dark
pH stroma	8.01	6.95
pH thylakoid space	5.34	6.77
Δ pH	2.67	0.18

is not known. It could be identical with Mg^{2+}-dependent ATP-ase of the envelope
(DOUCE et al., 1973) and may have an important function in maintaining a pH
gradient across the envelope.

The light-induced alkalization of the stroma can be decreased by low concentrations of CCCP (0.3 μM) which do not fully uncouple photophosphorylation (WER-
DAN et al., 1975) and also by nitrite (PURCZELD et al., 1978). The effect of nitrite
is probably caused by a proton shuttle across the envelope due to the permeation
of the dissociated and undissociated nitrous acid (HEBER and PURCZELD, 1977).
A decrease of the proton gradient across the envelope is also observed with high
concentrations (20 mM) of acetate or bicarbonate (WERDAN et al., 1975).

The pH measurements have been carried out with spinach chloroplasts. Also
in studies with the prokaryotic alga *Anacystis nidulans* a light-dependent proton
gradient across the thylakoid membrane and a concomitant alkalization of the
cytoplasm has been observed. This indicates that light-dependent pH changes in
the metabolic compartment of the reductive CO_2 fixation cycle are a general
phenomenon (FALKNER et al., 1976).

C. pH Dependence of CO_2-Fixation

CO_2 fixation has a very strong dependence on the stromal pH, as shown in Fig. 1.
The pH optimum of CO_2 fixation is generally found to be about 8.0, whereas
the rate of CO_2 fixation at pH 7.0 is about zero. Obviously, those pH changes
occurring in the stroma during a dark–light transient are sufficient to switch CO_2
fixation from zero to maximal activity. These findings strongly suggest that CO_2
fixation is regulated by the pH in the stroma.

The question arises which enzymatic steps of the reductive CO_2 fixation cycle
are being controlled by pH changes. The reduction of 3-phosphoglycerate, a partial
reaction of the cycle requiring ATP and NADPH, is almost independent of changes
of the stroma pH between 7.0 and 8.5 (WERDAN et al., 1975). This allows the
conclusion that the pH-dependent step controlling CO_2 fixation is neither the
generation of ATP nor of NADPH.

In order to identify the pH-sensitive step of CO_2 fixation, the stromal levels
of the intermediates of CO_2 fixation were analyzed. The inhibition of CO_2 fixation
by adding nitrite which lowers the stromal pH from 7.8 to 7.2 caused about
a tenfold increase of the levels of fructose-1,6-bisphosphate and sedoheptulose-1,7-
bisphosphate, whereas the levels of the hexose- and heptose- monophosphates

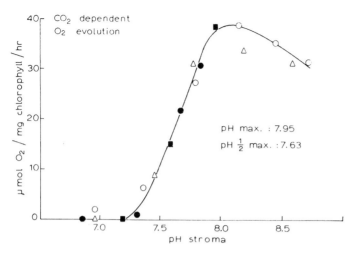

Fig. 1. pH dependence of CO_2-dependent O_2 evolution; ○,0; ●,10; △,20; ■,40 mM acetate in the medium. WERDAN et al. (1975)

were reduced to about 25% of the normal values (PURCZELD et al., 1978). These data indicated that in these experiments with isolated chloroplasts the inhibition of CO_2 fixation caused by a decrease of the stromal pH was ultimately due to an inhibition of the fructose- and sedoheptulose bisphosphatases.

D. Measurement of the Stromal Mg^{2+} Concentration in Intact Spinach Chloroplasts

Since the chloroplast envelope appears to be virtually impermeable to Mg^{2+} (GIMMLER et al., 1975; PFLÜGER, 1973) changes of the stromal Mg^{2+} concentration cannot be detected in the medium, unless the envelope has been rendered permeable for Mg^{2+}. There have been indications from the measurement of chlorophyll fluorescence in broken and intact chloroplasts that the stromal Mg^{2+} concentration may be increased by illumination (KRAUSE, 1974; BARBER et al., 1974). For the quantitative determination of the stromal Mg^{2+} concentration, three different experimental approaches have been employed. Firstly the chloroplasts were kept under defined metabolic conditions, and subjected to brief osmotic shock in order to rupture the envelope and thus release the stromal Mg^{2+} to the medium. In one experiment 379 nmol Mg^{2+}/mg chl were released from chloroplasts kept in the dark, and 405 from illuminated chloroplasts (PORTIS and HELDT, 1976). Assuming that the released Mg^{2+} originated from the stroma, and taking into account an average stromal space of 25 µl/mg chl, these data indicated that illumination caused an increase of the stromal Mg^{2+} concentration of about 1 mM. Secondly the chloroplasts were incubated with the ionophore A23187, known to render membranes permeable to Mg^{2+} (PFEIFFER et al., 1974). After 30 s treatment with this substance, Mg^{2+} equivalent to a stromal concentration of 3 mM was released from chloroplasts kept in the dark and 5 mM from illuminated chloroplasts (PORTIS and HELDT, 1976). Finally the stromal Mg^{2+} concentration was measured with the metallochromic indicator Eriochrome Blue S E. From the change of the absorption signal, a light-dependent increase in the concentration of free Mg^{2+} in the stroma of about 2 mM was calculated (KRAUSE, 1977). This method did not enable the measurement of the stromal Mg^{2+} concentration in the dark.

These measurements obtained with entirely different methods demonstrate that illumination increases the stromal Mg^{2+} concentration by about 2 mM. It appears that this increase of the stromal Mg^{2+} is due to a transport across the thylakoid membrane. Since the light dependent H$^+$ transport across this membrane in intact chloroplasts amounts to about 400 nmol per mg chl (PURCZELD et al., 1978), about 25% of the H$^+$ transport seems to be compensated by a countertransport of Mg^{2+}. It may be noted, that very large amounts of Mg^{2+} appear to be present in the stroma even in the dark.

E. Mg^{2+} Dependence of CO$_2$ Fixation

On the addition of ionophore A2 3187 to intact chloroplasts, CO$_2$ fixation was inhibited within 30 s, and this inhibition was mostly reversed by the subsequent addition of Mg^{2+} to the medium (PORTIS and HELDT, 1976). The data allowed the conclusion that the inhibition of CO$_2$ fixation was mainly due to a loss of Mg^{2+} from the stroma. It has been calculated from these experiments that 0.6–1.2 mM stromal Mg^{2+} were sufficient for half maximal restoration of CO$_2$ fixation, which is in the range of the concentration changes observed in illuminated chloroplasts.

These findings support the possibility that CO$_2$ fixation might be regulated by light-dependent changes of the Mg^{2+} concentration in the stroma, but the question still remains whether this feasible regulatory mechanism is of physiological significance. A regulation of CO$_2$ fixation by Mg^{2+} would imply that, during darkness, the concentration of free Mg^{2+} in the stroma is below 1 mM. This would mean that the relatively large amount of Mg^{2+}, which is also found in the stroma during darkness, has to be in a bound state. Whether this is really the case cannot be decided from the available experimental data. The fact, however, that such a binding of larger amounts of cations is principally possible may be illustrated from the case of Ca^{2+}, a potent inhibitor of fructose-1,6-bisphosphatase (BAIER, 1976) which is found in actively photosynthesizing chloroplasts in similar amounts as Mg^{2+}. According to MIGINIAC-MASLOW and HOARAU (1977) Ca^{2+} of the chloroplast is not free.

For the elucidation of the Mg^{2+}-sensitive steps of the CO$_2$ fixation cycle it was of interest that those concentrations of ionophore A23187 causing a strong inhibition of CO$_2$ fixation had almost no effect on phosphoglycerate reduction (PORTIS and HELDT, 1976). This led to the conclusion that the Mg^{2+}-sensitive step was not the generation of ATP or NADPH but a step in the CO$_2$ fixation cycle between the triosephosphates and 3-phosphoglycerate. Subsequent measurements of the stromal metabolite levels showed that the inhibition of CO$_2$ fixation by Mg^{2+} deficiency in the stroma was accompanied by a three to four fold increase of the levels of fructose- and sedoheptulose bisphosphates and a five fold decrease of the levels of hexosemonophosphates (Portis et al., 1977). These data clearly indicated that the inhibition of CO_2 fixation in the isolated chloroplasts was due to the inhibition of fructose- and sedoheptulosebisphosphatases.

F. Concluding Remarks

There are considerable changes of the H^+ and Mg^{2+} concentrations observed in the stroma of spinach chloroplasts. Each of these changes alone might be sufficient to switch CO_2 fixation from zero to maximal activity. Furthermore, evidence has been presented that CO_2 fixation may be also regulated by reducing equivalents. The addition of these three different mechanisms for a regulation of CO_2 fixation may be important in obtaining a very rigid light control of CO_2 fixation. The measurement of the stromal metabolite levels showed that the changes of the stromal Mg^{2+} and H^+ concentrations and also a change of reducing equivalents affect the fructose- and sedoheptulose bisphosphatases (Heldt et al., 1977), which concurs with the known properties of the isolated enzymes (Latzko and Kelly, Chap. II. 19, this vol.). These findings identify the two phosphatases as important regulation sites of the reductive CO_2 fixation cycle.

References

Baier, D.: Doktor Dissertation Fakultät für Landwirtschaft und Gartenbau Weihenstephan der Technischen Universität München (1976)
Barber, J., Mills, J., Nicolson, J.: FEBS Lett. 49, 106–110 (1974)
Deamer, D.W., Crofts, A.R., Packer, L.: Biochim. Biophys. Acta 131, 81–96 (1967)
Dilley, R.A., Vernon, L.P.: Arch. Biochem. Biophys. 111, 365–375 (1965)
Douce, R., Holtz, R.B., Benson, A.A.: J. Biol. Chem. 248, 7215–7222 (1973)
Falkner, G., Horner, F., Werdan, K., Heldt, H.W.: Plant Physiol. 58, 717–718 (1976)
Gaensslen, R.E., McCarty, R.E.: Arch. Biochem. Biophys. 147, 55–65 (1971)
Gimmler, H., Schäfer, G., Heber, U.: Proc. 3rd Int. Congr. Photosynthesis. Avron, M. (ed.), pp 1381–1392. Amsterdam: Elsevier 1975
Hager, A.: Planta (Berl.) 89, 224–243 (1969)
Heber, U., Purczeld, P.: In: Proc. 4th Int. Congr. Photosynthesis. Hall, D.O., Coombs, J., Goodwin, T.W. (eds.), pp 107–118. London: The Biochemical Soc. 1977
Heldt, H.W., Werdan, K., Milovancev, M., Geller, G.: Biochim. Biophys. Acta 314, 224–241 (1973)
Heldt, H.W., Fliege, R., Lehner, K., Milovancev, M., Werdan, K.: In: Proc. 3rd Int. Congr. Photosynthesis. Avron, M. (ed.), pp 1369–1379. Amsterdam: Elsevier 1974
Heldt, H.W., Chon, C.J., Lilley, R. McC., Portis, A.R.: In: Proc. 4th Int. Congr. Photosynthesis. Hall, D.O., Cooms, J., Goodwin T.W. (eds.), pp 469–478. London: The Biochemical Soc. 1977
Hind, G., Nakatani, H.Y., Izawa, S.: Proc. Natl. Acad. Sci USA 71, 1484–1488 (1974)
Krause, G.H.: Biochim. Biophys. Acta 333, 301–313 (1974)
Krause, G.H.: Biochim. Biophys. Acta 460, 500–510 (1977)

Miginiac-Maslow, M., Hoarau, A.: Plant Sci. Lett. 9, 7–15 (1977)

Neumann, J., Jagendorf, A.T.: Arch. Biochem. Biophys. 107, 109–119 (1964)

Pfeiffer, D.R., Reed, P.W., Lardy, H.A.: Biochemistry 13, 4007–4014 (1974)

Pflüger, R.: Z. Naturforsch. 28c, 779–780 (1973)

Portis, A.R., Heldt, H.W.: Biochim. Biophys. Acta 449, 434–446 (1976)

Portis, A.R., Chon, J.C., Mosbach, A., Heldt, H.W.: Biochim. Biophys. Acta 461, 313–325 (1977)

Purczeld, P., Chon, C.J., Portis, A.R., Heldt, H.W., Heber, U.: Biochim. Biophys. Acta 501, 488–498 (1978)

Rottenberg, H., Grunwald, T., Avron, M.: Eur. J. Biochem. 25, 54–63 (1972)

Rumberg, B., Siggel, U.: Naturwissenschaften 56, 130–132 (1969)

Werdan, K., Heldt, H.W., Milovancev, M.: Biochim. Biophys. Acta 396, 276–292 (1975)

17. Ribulose-1,5-Bisphosphate Carboxylase

T. Akazawa

A. Fraction-1-Protein and RuBP Carboxylase

Biochemical studies on leaf proteins carried out by Wildman and Bonner (1947) revealed the presence of a major protein component having a large molecular size (18 s) which was designated as fraction-1-protein. The ubiquitous distribution of this protein in green plant leaves and green algae, as determined by analytical ultracentrifugation and immunological precipitation methods, stimulated later studies on its enzymic nature (Dorner et al., 1958). The independent investigation of the path of carbon in photosynthetic CO_2 fixation, together with that on the enzymic machinery of the reductive pentose phosphate cycle, led to the discovery of ribulose-1,5-bisphosphate (RuBP) carboxylase (E.C. 4.1.1.39; carboxydismutase) catalyzing the following key reaction [Eq. (1); Quayle et al., 1954; Weissbach et al., 1954, 1956; Chap. II.1, this vol.]

$$RuBP + CO_2 \xrightarrow{Mg^{2+}} 2\,(PGA) \tag{1}$$

The presence of the enzyme in various photoautotrophic organisms was demonstrated by Fuller and Gibbs (1959). It was subsequently established that the enzymic entity of fraction 1 protein is indeed RuBP carboxylase, and currently it is agreed that these two proteins are synonymous (for review, Akazawa, 1970; Kawashima and Wildman, 1970; Ellis, 1973; Kelly et al., 1976; Kung, 1977; Jensen and Bahr, 1977).

A possible unknown role (structural or functional) of fraction 1 protein in situ cannot be excluded, and the isolation of polyhedral inclusion bodies (carboxysomes) containing RuBP carboxylase activity from microbial sources has been reported (Shively et al., 1973; Codd and Stewart, 1976; Purohit et al., 1976b).

The specific localization of RuBP carboxylase in chloroplasts has been reported by several investigators, using aqueous or nonaqueous isolation techniques (Park and Pon, 1961; Heber et al., 1963; Smillie, 1963). The isolation of intact chloroplasts from mechanically ruptured spinach leaf protoplasts has given additional proof for the enzyme being exclusively associated with chloroplasts ($d = 1.21$; Fig. 1; Nishimura et al., 1976). In germinating castor bean endosperm, RuBP carboxylase was found to be specifically localized in plastid fractions, although its metabolic role in this nonphotosynthetic organ remains to be elucidated (Osmond et al., 1975; Nishimura and Beevers, 1978).

In the initial stage of research on C_4-photosynthesis conducted by Slack and Hatch (1967), it was reported that the lower specific activity of RuBP carboxylase in leaf extracts from C_4-plants such as sugarcane, maize, and sorghum as compared to the C_3-plants paralleled the much lower content of fraction 1 protein discernible

Fig. 1. Specific localization of RuBP carboxylase in chloroplasts separated from mechanically ruptured spinach leaf protoplasts by sucrose density gradient centrifugation. *Arrow* indicates the major band of RuBP carboxylase in the polyacrylamide gel electrophoresis of d = 1.21 fractions. (After NISHIMURA et al., 1976)

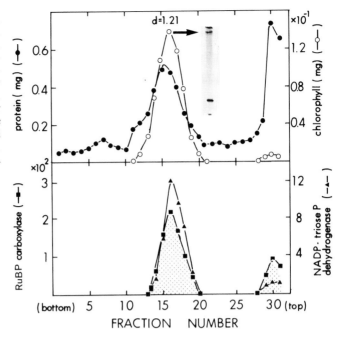

by analytical ultracentrifugation. However, with the discovery of dimorphic leaf structure of C_4-plants (Kranz anatomy) and compartmentation of the two different types of CO_2-fixation mechanism (see Chap. II.8, this vol.), it is now well established that the enzyme is localized in chloroplasts of the bundle sheath cells in C_4-plants (SLACK et al., 1969; MATSUMOTO et al., 1977). The early experimental finding by BJÖRKMAN and GAUHL (1969) reporting high recovery of RuBP carboxylase activity from repetitively ground C_4-leaf tissue (*Atriplex rosea*), is an instructive example of the caution required in interpreting apparently negative results.

In the third important class of plants, called CAM (crassulacean acid metabolism) plants, the primary CO_2 fixation reaction in the dark is into C_4 acids, and the incorporation of CO_2 derived from the deacidification of these acids into carbohydrate occurs in the light (Chap. II.8, this vol.). RuBP carboxylase participates in this light-phase reaction mechanism, and enzyme compartmentation and regulation of changes in enzyme activities in dark/light phases is the subject of intensive investigations (Chap. II.8, this vol.; OSMOND, 1978).

Great interest has been aroused in the assessment of the contribution of primary CO_2 fixation reactions in photosynthetic systems, and the carbon isotope discrimination ratio (expressed as a $\delta^{13}C$ value) has been determined for a wide variety of plant species (C_3, C_4 and CAM; Chap. II.10, this vol.; TROUGHTON et al., 1974; LERMAN, 1975). It is now accepted that the more negative $\delta^{13}C$ values in C_3-plants reflect the enzymic isotope discrimination effect exerted by RuBP carboxylase in vitro (CHRISTELLER et al., 1976).

Experimental results obtained by HATTERSLEY et al. (1977) on immunofluorescence labeling of thin leaf transsections of forty C_3 and C_4 plants and one CAM plant using the fluorescence-conjugated rabbit antisera raised against RuBP carboxylase and its subunit also lend support to the presence of the enzyme in chloroplasts. In particular, almost exclusive localization of the enzyme in the bundle sheath cells of C_4-plants has been demonstrated by this technique.

Recent experiments employing different techniques have shown that fraction 1 protein comprises the peripheral molecules associated with chloroplast thylakoid membranes (KANNANGARA et al., 1970; STROTMANN et al., 1973; HENRIQUES and PARK, 1976; STAEHLIN, 1976). Clarification of the nature of membrane-attached RuBP carboxylase molecules is of particular importance from the view of enzyme biosynthesis, as well as the regulatory mechanism of the enzyme reaction (AKAZAWA, 1978).

B. Molecular Structure of RuBP Carboxylase

In comparison to many other plant enzymes, the isolation and purification of RuBP carboxylase can be done relatively easily. WEISSBACH et al. (1956) isolated a nearly homogeneous enzyme preparation from spinach (reported MW, 3×10^5) by employing only repeated $(NH_4)_2SO_4$ precipitation. However, advancement of research techniques for purifying enzyme proteins, e.g., ion-exchange column chromatography, Sephadex or Sepharose gel filtration, and isoelectric focusing among many others, has made possible the isolation of extremely pure RuBP carboxylase preparations from various leaf tissues by simple steps. A particularly useful technique is the one-step sucrose density gradient centrifugation developed by GOLD-THWAITE and BOGORAD (1971), a method which has proven applicable to plant as well as microbial carboxylases (TABITA and McFADDEN, 1974b; McFADDEN et al., 1975; OSMOND et al., 1975). The Danish group (STRØBAEK and GIBBONS, 1976) has reported another simple method for obtaining homogeneous enzyme preparations from barley (*Hordeum vulgare*) and tobacco (*Nicotiana tabacum*) leaves, employing Sephadex G-25 and Sepharose 6B gel filtrations followed by ultrafiltration.

Numerous investigations have been carried out to determine the physicochemical properties, e.g., molecular weights, amino acid compositions, and subunit structures, of purified RuBP carboxylase from various origins, such as photosynthetic bacteria, algae, and higher plants. Some representative data on the molecular weight and subunit composition of enzymes from published data are compiled in Table 1. The enzyme from eukaryotic organisms invariably has a large molecular weight (MW, about 5.5×10^5). Since the first report by RUTNER and LANE (1967) showing the presence of two different subunits in spinach leaf RuBP carboxylase as demonstrated by Na-dodecyl sulfate polyacrylamide gel electrophoresis, this technique has been widely applied to the enzymes from other sources. It is now generally agreed that the plant-type enzyme contains two different polypeptide chains, large subunit (A); (MW about 5×10^4) and small subunit (B); (MW about 1.2×10^4). It has been reported that some higher plant and microbial enzymes contain multiple polypeptides in each constituent subunit (KUNG et al., 1974; KUNG, 1976; PUROHIT and McFADDEN, 1977). Although their chemical nature has not been well characterized, STRØBAEK et al. (1976) have discussed the idea that the polymorphism of subunit B in RuBP carboxylase from *Nicotiana tabacum*

Table 1. Quaternary structure of RuBP carboxylase (partial list)

Source	M_r	Subunit A	Subunit B	Structural model[a]	Authors
Bacteriophyta					
Rhodospirillum rubrum	1.14×10^5	+	−	A_2	TABITA and McFADDEN (1974a)
Chlorobium thiosulfatophilum	3.6×10^5	+	−	A_6	TABITA et al. (1974a)[b]
Thiobacillus intermedius	4.6×10^5	+	−	A_8	PUROHIT et al. (1976b)
Chromatium vinosum	5.2×10^5	+	−	A_8B_8	TAKABE and AKAZAWA (1975b)
Alcaligenes eutrophus or	5.05×10^5	+	+	A_8B_8	BOWIEN et al. (1976)
Hydrogenomonas eutropha	4.9×10^5	+	+	$A_5A_3B_8$	PUROHIT and McFADDEN (1977)
Cyanophyta					
Agmenellum quadruplicatum	4.5×10^5	+	−	A_8	TABITA et al. (1974b)[c]
Anabaena cylindrica	4.5×10^5	+	−	A_8	TABITA et al. (1974b)[c]
	5×10^5	+	+	A_8B_8	TAKABE (1977)[c]
Plectonema boryanum	5×10^5	+	+	A_8B_8	TAKABE et al. (1976)
Anabaena variabilis	5×10^5	+	+	A_8B_8	TAKABE et al. (1976)
Cyanophora paradoxa	5.25×10^5	+	+	A_8B_8	CODD and STEWART (1977a)[c]
Aphanocapsa	5.25×10^5	+	+	A_8B_8	CODD and STEWART (1977b)
Microcystis aeruginosa	5.18×10^5	+	+	A_8B_8	STEWART et al. (1977)
Euglenophyta					
Euglena gracilis	5.25×10^5	+	+	A_8B_8	McFADDEN et al. (1975)
Chlorophyta					
Chlorella ellipsoidea	5×10^5	+	+	A_8B_8	SUGIYAMA et al. (1971)
Chlamydomonas reinhardtii	$5-5.6 \times 10^5$	+	+	A_8B_8	GIVAN and CRIDDLE (1972)
Tracheophyta (Angiosperms)					
Spinacia oleracea	5.15×10^5	+	+	A_8B_8	NISHIMURA et al. (1973)
Nicotiana tabacum	$5-5.5 \times 10^5$	+	+	A_8B_8	BAKER et al. (1975; 1977a, b)
Beta vulgaris	5.85×10^5	+	+	A_8B_8	STRØBAEK and GIBBONS (1976)
Hordeum vulgare	5.1×10^5	+	+	A_8B_8	STRØBAEK and GIBBONS (1976)

[a] Except only a few cases proposal of the quaternary structural models have not been based on the firm chemical or physical data. They are mostly derived from the molecular weight determinations of the two constituent subunits.

[b] BUCHANAN and SIREVAG (1976) and TAKABE and AKAZAWA (1977) could not detect RuBP carboxylase in this bacterium.

[c] Removal of subunit B during the enzyme purification step employing the acid treatment can be envisaged from these papers.

is derived from two allelic genes of parent species, *Nicotiana sylvestris* and *Nicotiana tomentosiformis*.

Based on the experimental results showing that there are 8 binding sites for RuBP as well as for the structural analog of the hypothetical C_6-intermediate [CRBP; see Sect C, Eq. (4)] per mol spinach enzyme, a model structure for the basic protomer containing eight pairs each of large and small subunits was proposed (WISHNICK et al., 1970; SIEGEL et al., 1972). Studies dealing with molecular weight determinations and amino acid analyses of constituent subunits from plant enzymes are compatible with this structural model (SUGIYAMA and AKAZAWA, 1970; SUGIYAMA et al., 1971; NISHIMURA et al., 1973). Additional support for the symmetrical structure came from an investigation by BAKER et al. (1975; 1977a, b) who reported that X-ray diffraction, electron microscopy and optical diffraction analyses of the crystalline tobacco leaf RuBP carboxylase agree with a molecular structure of the composition A_8B_8. Figure 2 shows the molecular packing diagram of the enzyme, which may be identical to crystalline inclusion bodies in chloroplasts. Recent experimental data obtained by ROY et al. (1978) dealing with the molecular arrangement of subunit B in the pea leaf RuBP carboxylase using the cross-linking reagents support this structural model.

Enzymes from some photosynthetic bacteria (prokaryotes) such as *Chromatium vinosum* and *Hydrogenomonas eutropha* (*Alcaligenes eutrophus*) have a structural make-up analogous to the plant enzyme (TAKABE and AKAZAWA, 1975a, b; BOWIEN et al., 1976; PUROHIT and McFADDEN, 1977). However, it has been reported that the enzyme from *Rhodospirillum rubrum* contains only a large subunit dimer (A_2; TABITA and McFADDEN, 1974a) and that of *Thiobacillus intermedius* is an octamer of subunit A (A_8; PUROHIT et al., 1976a). As listed in Table 1, contradictory observations were reported for RuBP carboxylases from some prokaryotic organisms, e.g., bacteria and blue-green algae. Further investigations are needed to establish whether unusual structural combinations of the two subunits are indeed present in certain prokaryotes.

The comparison of the structural organization of RuBP carboxylases between primitive prokaryotes and eukaryotic green plants is of great interest from the standpoint of molecular evolution of autotrophy (McFADDEN, 1973; McFADDEN and TABITA, 1974; TAKABE and AKAZAWA, 1975c). The existence of the plant-type RuBP carboxylase in blue-green algae appears to be consistent with the current concept of the endosymbiotic origin of chloroplasts (MARGULIS, 1971; TAKABE, 1978). Naturally the final answer for the molecular mechanism of the phylogenetic development of RuBP carboxylase must await the determination of the amino acid sequences of enzyme molecules, and structural studies have been initiated for determining the primary sequence of the small subunit from several higher plant carboxylases (HASLETT et al., 1976; POULSEN et al., 1976). It is this writer's view that the use of immunological microcomplement assay procedures of WILSON and coworkers (KING and WILSON, 1975) may also provide fruitful information on this subject.

Using the statistical treatment of amino acid compositions of the two constituent subunits of the enzymes from various species, SΔQ values have been calculated following the method of MARCHALONIS and WELTMAN (1971), and indicated that the structurally homologous large subunit (A) among enzymes of divergent organisms has probably evolved from a common ancestral gene (TAKABE and AKAZAWA,

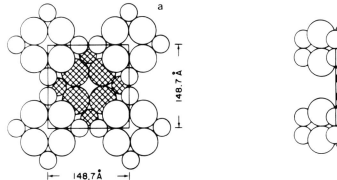
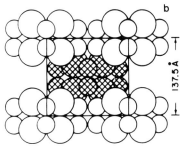

Fig. 2a and b. Molecular packing diagram of Form III RuBP carboxylase crystals from tobacco leaf. (After BAKER et al., 1977b)

1975c). This notion is consistent with the fact that subunit A carries the catalytic site of the enzyme molecule (see below). One can assume that the catalytically important core portion is genetically conserved during the evolutionary history of the enzyme. In this context experiments to determine the sequence homology around the catalytic lysyl residues of spinach and *Rhodospirillum rubrum* enzymes will be particularly rewarding (SCHLOSS and HARTMAN, 1977a, b). On the other hand, the structure of subunit B appears to be species-specific, and the rate of amino acid substitutions in the molecule has been calculated to be significantly greater than that for subunit A (TAKABE, 1978).

Among several criteria for characterizing subunit A of the enzyme molecule as the catalytic entity, the following four are judged to be most convincing.

1. Spinach RuBP carboxylase (A_8B_8) dissociates into the constituent subunits (A_8, octamer of subunit A, and B_1, monomer of subunit B, etc.) upon treatment with p-mercuribenzoate (PMB) and alkaline pH [Eq. (2); NISHIMURA and AKAZAWA, 1973; NISHIMURA et al., 1973].

$$A_8B_8 \xrightleftharpoons[\text{DTT, pH 7.0}]{\text{PMB, pH 9.0}} [A_8+A_7+A_6 \ldots]^* + [B_1+B_2]^* \qquad (2)$$

* PMB-modified form

Similarly, the enzyme from *Chromatium vinosum* (A_8B_8) dissociates into two subunit components upon brief exposure to alkali [pH 9.0; Eq. (3); TAKABE and AKAZAWA, 1973a).

$$A_8B_8 \xrightleftharpoons[\text{pH 7.0}]{\text{pH 9.0}} [A_8] + [B_1] \qquad (3)$$

In both cases, the dissociation process is reversible and the original enzyme molecule can be reformed upon neutralization; in the case of the spinach enzyme the addition of excess dithiothreitol (DTT) is essential (NISHIMURA and AKAZAWA, 1974b, c). More important, however, is the demonstration that the octameric form of the large subunit, A_8, retains partial carboxylase activity in the absence of the small subunit (specific activity, about 20%). The argument as to whether this residual enzyme activity is ascribed to the presence of the undissociated native enzyme in the separated A_8 preparations should be examined against the observation that Na-dodecyl-sulfate gel electrophoresis can readily resolve 15% contaminants[1]. Addition of subunit B to A_8 causes an enhancement of enzyme activity, accompanied by a partial reformation of the original 18s enzyme molecule. But the magnitude of reconstitution and the activity recovery were not great, indicating that a structural constraint prevents the complete reassembly of the molecule, which is conformationally identical to the original native enzyme, from separated units.

2. The specific localization of functionally important SH-residues in subunit A was substantiated with both spinach and *Chromatium* enzymes (SUGIYAMA and AKAZAWA, 1970; TAKABE and AKAZAWA, 1975a). This finding has been further reinforced by the identification of two catalytically essential lysine residues in the large subunit by means of affinity labeling of the spinach enzyme with 3-bromo-1,4-dihydroxy-2-butanone-1,4-bisphosphate or N-bromoacetyl ethanolamine phosphate (HARTMAN et al., 1973; NORTON et al., 1975; SCHLOSS and HARTMAN, 1977b; STRINGER and HARTMAN, 1978; SCHLOSS et al., 1978).

Additional proof for the presence of lysyl residues in the active site region of the enzyme molecule has been provided by chemical modification studies on the bacterial and spinach enzymes using pyridoxal-5'-P (PAECH et al., 1977), as well as the use of cyanate on the crystalline tobacco leaf enzyme (CHOLLET and ANDERSON, 1978). The latter studies show the preferential carbamylation of lysyl residue located in subunit A by cyanate, which is largely protected by the presence of the substrate, RuBP.

3. The rabbit antiserum developed against subunit A of the spinach enzyme [anti-(A)] specifically inhibits the carboxylase activity, whereas the one against subunit B [anti-(B)] is totally ineffective (NISHIMURA and AKAZAWA, 1974a; TAKABE and AKAZAWA, 1975b; see also Sect. D, RuBP oxygenase). The inhibitory effect displayed by anti-(A) on the carboxylase activities from other enzyme sources indicates that there exist interspecific structural homologies as far as the large subunit is concerned. There is now further evidence showing that subunit B contributes to the regulatory function of the enzyme, relating to the Mg^{2+} effect which will be discussed in Section C.

4. The finding that some bacterial enzymes contain only the large subunit, A_2 for *Rhodospirillum rubrum* and A_8 for *Thiobacillus intermedius*, respectively, supports the idea that the catalytic site exists in subunit A (see Table 1).

C. Reaction Mechanism and Regulation

Earlier investigations have shown that the enzyme catalyzes the formation of 2 mol of PGA from RuBP [Eq. (1)], and the reaction sequence depicted in Eq. (4) was proposed by CALVIN (1956)

---

[1] It has not been determined, however, if an oligomeric form such as A_8B_2 can retain a partial enzyme activity, which would have escaped detection by this method.

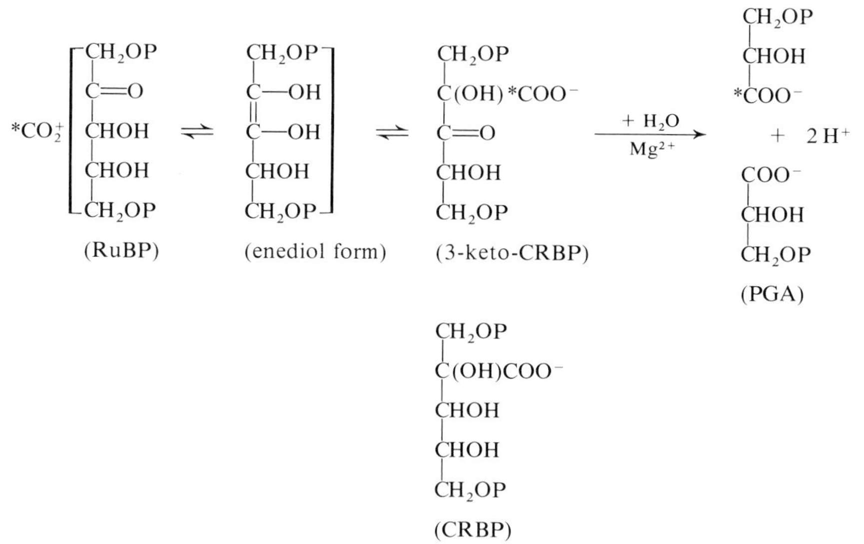

$$
*CO_2^+ \begin{bmatrix} CH_2OP \\ | \\ C=O \\ | \\ CHOH \\ | \\ CHOH \\ | \\ CH_2OP \\ (RuBP) \end{bmatrix} \rightleftharpoons \begin{bmatrix} CH_2OP \\ | \\ C-OH \\ \| \\ C-OH \\ | \\ CHOH \\ | \\ CH_2OP \\ (\text{enediol form}) \end{bmatrix} \rightleftharpoons \begin{matrix} CH_2OP \\ | \\ C(OH)*COO^- \\ | \\ C=O \\ | \\ CHOH \\ | \\ CH_2OP \\ (3\text{-keto-CRBP}) \end{matrix} \xrightarrow[Mg^{2+}]{+ H_2O} \begin{matrix} CH_2OP \\ | \\ CHOH \\ | \\ *COO^- \\ \\ + \\ \\ COO^- \\ | \\ CHOH \\ | \\ CH_2OP \\ (PGA) \end{matrix} + 2H^+
$$

$$
\begin{matrix} CH_2OP \\ | \\ C(OH)COO^- \\ | \\ CHOH \\ | \\ CHOH \\ | \\ CH_2OP \\ (CRBP) \end{matrix} \tag{4}
$$

The formation of the hypothetical transition C_6-intermediate, 2-carboxy-3-ketoribitol-1,5-bisphosphate (3-keto-CRBP), during the enzyme reaction appears to be supported by two experimental approaches (SIEGEL and LANE, 1972, 1973; SJÖDIN and VESTERMARK, 1973). From the experimental evidences showing the involvement of an SH-group in the carboxylation reaction (see Sect. B), RABIN and TROWN (1964) have proposed a mechanism involving the formation of a hemiacetal linkage between the enzyme-SH and C-2 of RuBP. However, the experimental result dealing with the retention of the oxygen atom at C-2 excludes this mechanism. Furthermore, since the oxygen atom of C-3 is retained during the carboxylation it is unlikely that the Schiff base is formed between C-3 and the lysyl residue of the enzyme molecule (LORIMER, 1978; SUE and KNOWLES, 1978).

It has remained a long-standing question as to whether CO_2 or HCO_3^- is the actual molecular species in the carboxylase reaction. The enigmatic problem was solved by COOPER et al. (1969). Because the following reaction, Eq. (5), requires more than 60 s to attain equilibrium at temperatures below 15 °C, they were able to determine $^{14}CO_2$ incorporation into PGA

$$
H_2O + CO_2 \underset{\text{slow}}{\rightleftharpoons} H_2CO_3 \underset{\text{fast}}{\rightleftharpoons} H^+ + HCO_3^- \tag{5}
$$

catalyzed by purified spinach RuBP carboxylase at 10 °C, using either $(^{14}CO_2 + H^{12}CO_3^-)$ or $(^{12}CO_2 + H^{14}CO_3^-)$ as the initial substrate. At the same time the effect of carbonic anhydrase on the equilibrium of the reaction was examined. Results obtained were exactly those predicted if CO_2 was the active molecular species utilized.

Remarkable research achievements in recent years have solved the problem of the $K_m(CO_2)$ of RuBP carboxylase. From the time of investigations by WEISS-BACH et al. (1956) and RACKER (1957) up to relatively recently (PAULSEN and LANE, 1966), K_m values of the enzyme with respect to "CO_2" $(CO_2 + HCO_3^-)$ were invariably large, 10–30 mM, regardless of its origins, and could not account for the in vivo role of the enzyme. The concentration of CO_2 in leaf tissues may vary considerably, and in fact has not been measured adequately. However, analytical studies on photosynthetic CO_2 fixation by intact leaf tissues of tobacco, sugarcane, and maize (GOLDSWORTHY, 1968) or freshly prepared chloroplast preparations from spinach (JENSEN and BASSHAM, 1966) gave exceedingly small apparent $K_m("CO_2")$ values, nearly equal to the ambient atmospheric CO_2 concentration 0.03% ($\cong 10$ µM). These discrepant observations have been an enigma in photosynthesis research, and several mechanistic proposals were put forward to explain the regulatory mechanism of the enzyme reaction (for review, GIBBS, 1971; WALKER, 1973, 1976). Recent investigations have shown that the purified RuBP carboxylase preparations studied previously, which showed high $K_m(CO_2)$ values contained a kinetically inactive form of the enzyme. BAHR and JENSEN (1974a) demonstrated that upon rapid breakage of spinach chloroplasts with hypotonic medium, a form of RuBP carboxylase having a low $K_m(CO_2; 11–18$ µM, pH 7.8) can be isolated; this enzyme form is unstable, rapidly converting to a high $K_m(CO_2)$ form (20–25 mM HCO_3^-) similar to the purified enzyme. Independently, BADGER et al. (1974) isolated a high affinity (active) form [$K_m(CO_2)$, 16–42 µM] of the enzyme from spinach chloroplasts as well as from several other C_3, C_4 and CAM plant leaf tissues, and BOWES et al. (1975) reported basically similar observations with the spinach leaf enzyme. Subsequent thorough kinetic investigations dealing with the mechanism of interconversion of inactive–active forms using the purified spinach leaf enzyme by the Australian group (ANDREWS et al., 1975; LORIMER et al., 1976), and using the purified enzyme from soybean (*Glycine max*) leaves by LAING and his associates (1975, 1976) have demonstrated that upon preincubation with CO_2 and Mg^{2+}, the enzyme molecule is activated, accompanied by the formation of a ternary complex [Enzyme-CO_2-Mg^{2+}; Eq. (6)].

$$\text{Enzyme} + CO_2 \rightleftharpoons (\text{Enzyme-}CO_2)$$

$$(\text{Enzyme-}CO_2) + Mg^{2+} \rightleftharpoons (\text{Enzyme-}CO_2\text{-}Mg^{2+}) \tag{6}$$

Based on the results of kinetic analysis of the activation process as a function of pH, which indicate that CO_2 reacts with a group having a distinctly alkaline pK_a, LORIMER et al. (1976) have proposed the following sequential mechanism involving carbamate formation [Eqs. (7–9)].

$$\underset{\text{(inactive enzyme)}}{E—NH_3^+} \xrightarrow[(\varDelta pH)]{} E—NH_2 + H^+ \tag{7}$$
$$\uparrow$$
$$h\nu$$

$$E—NH_2 + CO_2 \underset{\text{slow}}{\rightleftharpoons} E—NH_2—COO^- + H^+ \tag{8}$$

$$E—NH_2—COO^- + Mg^{2+} \underset{\text{fast}}{\rightleftharpoons} \underset{\text{(active enzyme)}}{E—NH_2COO^-—Mg^{2+}} \tag{9}$$

The essential NH_2-group involved in CO_2-Mg^{2+} activation may well be different from the recently reported lysyl NH_2-groups in the substrate binding domain of the enzyme molecule (see Sect. B). Although chemical modifidation experiments have many inherent difficulties in interpretation of the results obtained, future investigations along such lines may provide useful information regarding the regulatory mechanism of the RuBP carboxylase reaction.

From these research advances it is now evident that there are two separate phases in the RuBP carboxylase reaction, i.e., activation and catalysis. These must be carefully taken into account during kinetic analyses, and in studies of the regulatory mechanism of the enzyme reaction. One can see that fresh leaf extracts prepared in the presence of CO_2 and Mg^{2+} contain the active form of the enzyme [low $K_m(CO_2)$], but most importantly, both the enzyme activation achieved by preincubation and the precise control of assay conditions such as the order of substrate additions, are required for the accurate measurement of enzyme activity (see also RuBP oxygenase in Sect. D). Therefore, while it is true that some of the contradictory findings or uncertainties pertaining to the properties of the enzyme reported in past literature may be explained in terms of inadequate control of these factors, there are many secrets yet to be uncovered by future investigations. Recent experiments by BAHR and JENSEN (1978) have shown that RuBP carboxylase in intact chloroplasts assumes various degrees of activation as a function of Mg^{2+} concentration, pCO_2, pH, and levels of sugar P.

For many years another important question has been whether or not the low maximal velocity of RuBP carboxylase in chloroplasts or leaf extracts could account for the in vivo rate of photosynthetic CO_2 fixation (PETERKOFSKY and RACKER, 1961). In this context LILLEY and WALKER (1975) reported that the enzyme from spinach chloroplasts having an apparent $K_m(CO_2)$ of 46 μM in air exhibits an activity of 1000 μmol CO_2 fixed per mg chl per h under the specific assay conditions. This observation, taken together with the finding of a high affinity-CO_2 form of the enzyme discussed above, makes it clear that RuBP carboxylase functioning in chloroplasts in situ can adequately account for photosynthetic CO_2 fixation in leaves in natural environments.

The requirement for Mg^{2+} in the enzyme catalysis observed in the very early research has been constantly confirmed by later workers. Although a part of the phenomena is likely to be attributed to the enzyme activation process described above, the intrinsic role of Mg^{2+} in enzyme catalysis remains unclear (MIZIORKO and MILDVAN, 1974). It is intriguing to speculate how Mg^{2+} might combine with the enzyme molecule at the level of subunit interactions. Some data have accumulated suggesting that the small subunit (B) is probably involved in the Mg^{2+}-effect of the enzyme reaction. It was previously found that by increasing Mg^{2+} concentrations in the assay mixture from 0 up to 10 mM, not only was the carboxylase activity displayed by enzymes from various sources enhanced, but also the optimal pH of the reaction shifted from alkaline (pH 9.0) to near neutral (about pH 7.5; SUGIYAMA et al., 1968). This property can be observed using various types of buffers and even by assaying the enzyme after activation. However, this specific phenomenon is not evident in the enzyme reaction catalyzed by the large catalytic subunit (A_8), depleted of subunit B (NISHIMURA and AKAZAWA, 1973; TAKABE and AKAZAWA, 1973a). The addition of the antiserum, anti-(B) to the native enzyme gives rise to the same effect, optimal pH remaining alkaline (pH 8.5) regardless of the presence or absence of Mg^{2+}. The reconstituted enzyme molecule regains the Mg^{2+}-induced optimal pH-shift (NISHIMURA and AKAZAWA, 1974a). It is noteworthy that kinetic studies by LAING and CHRISTELLER (1976) using the soybean enzyme have indicated that Mg^{2+} is not involved in catalysis, suggesting that the metal is bound at a regulatory site [Eq. (6)].

Regulatory mechanisms of the RuBP carboxylase reaction which operate in vivo are the most intriguing aspect of photosynthesis research. The light-induced pH increase in the stroma of chloroplasts first discovered by Neumann and Jagen-dorf (1964) may be significant in this respect. It was recently shown that upon illumination an uptake of H^+ across the thylakoid membranes occurs, which is accompanied by an alkalization of the stroma, giving rise to a total pH difference of 2.5 units, and increasing the pH in the stroma by 1 unit, to pH 8.1, which is the optimal pH of CO_2 fixation (Heldt et al., 1973; Werdan et al., 1975). These investigators further measured the light-dependent change of Mg^{2+} in thyla-koid membranes and stroma upon illumination of intact chloroplasts at pH 8.0 and found that in the stroma there was an increase of 1–3 mM Mg^{2+} concentration (Portis and Heldt, 1976); the mechanism being further reinforced by experiments of Bahr and Jensen (1978) concerning the light activation of photosynthetic CO_2 assimilation by spinach chloroplasts.

Unification of all such findings in relation to the regulatory mechanism of photosynthetic CO_2 fixation at the level of the light activation of RuBP carboxylase has been attempted by many workers (cf. Jensen and Bahr, 1977; Kelly et al., 1976; Walker, 1973, 1976; Werdan et al., 1975). It is likely that RuBP carboxylase molecules in chloroplasts oscillate between inactive/active states during dark/light transients; the enzyme activity being low in the dark due to the acidic stromal pH, and high in the light, due to the energized state of chloroplasts which leads to alkalization (pH 8.0) and enhanced Mg^{2+} concentrations in the stroma (Fig. 3). It should be noted in this figure that the treatment of the enzyme preparation by anti-(B) is apparently indifferent to the enhancement of Mg^{2+}-induced enzyme activity, and the nature of the mutual interplay of the two subunits in enzyme regulation in vivo remains to be uncovered by future investigations.

It has been reported that a number of metabolites such as 6-P-gluconate, FBP, and NADPH, among others, elicit allosteric effects on the RuBP carboxylase reaction and the implications of these findings have been considered in relation to regulatory mechanisms of photosynthesis at the cellular level (Chu and Bassham, 1974, 1975; Tabita and McFadden, 1972; Buchanan and Schürmann, 1973; Ryan and Tolbert, 1975; Chollet and Anderson, 1976). It can be seen, however, that the magnitude of such stimulatory or inhibitory effects of metabolites is gener-ally not very great and also not necessarily consistent among different investigators. Some of these observed effects appear to be attributable to the order of addition of substrates and/or effectors to the assay mixture. Although it is evident that the levels of intermediates or cofactors of the Calvin-Benson cycle in chloroplasts may fluctuate during light/dark transients and may be important factors modulating the activity of RuBP carboxylase, the validity of proposed allosteric models merely based on the kinetic behavior of purified enzyme preparations should be treated with great caution.

The crystalline RuBP carboxylase from tobacco leaves exhibits reversible cold inactivation (Kawashima et al., 1971), and unlike other oligomeric enzyme proteins it was found that dissociation–reassociation of subunits is not involved during the step of cold-inactivation and heat-reactivation (Chollet and Anderson, 1976). A more recent study indicates that the overall process of reversible cold inactivation

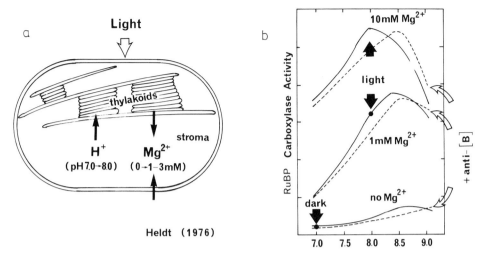

Fig. 3. Light-induced activation of RuBP carboxylase in chloroplasts. **a** modified after PORTIS and HELDT (1976); **b** modified after NISHIMURA and AKAZAWA (1974a)

is closely associated with a reversible change in the conformation of the enzyme protein (CHOLLET and ANDERSON, 1977).

D. RuBP Oxygenase

Since the classical study by BENSON and CALVIN (1950), it has been known that glycolate is a predominant product of the photosynthetic CO_2 fixation reaction in plants in an O_2-containing atmospheric environment. From the results of the O_2 effect on $^{14}CO_2$ fixation by *Chlorella* cells, BASSHAM and KIRK (1962) proposed that carbon atoms 1 and 2 of the RuBP molecule are oxidatively split to produce 2-P-glycolate. Indeed, this hypothetical view has led to the discovery of the RuBP oxygenase reaction catalyzed by RuBP carboxylase some 10 years later. Based on the competitive inhibitory effect of O_2 on the soybean RuBP carboxylase reaction, OGREN and BOWES (1971) hypothesized that the enzyme may regulate photorespiration. Subsequently, the enzymic production of 2-P-glycolate and PGA from RuBP, following Eq. (10), was demonstrated by the same workers (BOWES et al., 1971; BOWES and OGREN, 1972).

$$\text{RuBP} + O_2 \xrightarrow{\text{Mg}^{2+}} \text{2-P-glycolate} + \text{PGA} \tag{10}$$

The association of RuBP oxygenase activity with purified RuBP carboxylase preparations (fraction 1 protein) from spinach and soybean leaves was firmly established by ANDREWS et al. (1973). Their finding on the incorporation of one atom of ^{18}O into the carboxyl group of 2-P-glycolate, but not into PGA, was in fact consistent with the reaction mechanism requiring molecular O_2 (LORIMER et al., 1973).

In the beginning of studies on RuBP oxygenase, the specific activity of the oxygenase relative to carboxylase was extremely small and the optimal pH was approximately pH 9.0, at which pH the carboxylase activity was practically nil. Therefore, some reservations as to whether RuBP oxygenase plays a substantial role in the photorespiratory process prevailed for some time (cf. Tolbert, 1973). The demonstration for the presence of the oxygenase activity in the obligate anaerobe, *Chromatium* (Takabe and Akazawa, 1973b), can be interpreted on the basis of the oxygenolytic cleavage of the enzyme-bound enolate anion inherent to the RuBP carboxylase molecule as originally suggested by Lorimer and Andrews (1973). However, a sizable amount of experimental evidences has accumulated showing the functional role of the enzyme activity in the glycolate production of the bacterial cells under the photorespiratory environment (Takabe and Akazawa, 1977). Later kinetic as well as immunological studies showed that the large subunit moiety shares the catalytic site for both carboxylase and oxygenase reactions (Nishimura and Akazawa, 1974a; Takabe and Akazawa, 1975b). Since then a great many researches have emerged on this interesting enzyme reaction. Thus RuBP carboxylase is unique in its bifunctional nature, and also because, unlike most other oxygenase, it lacks characteristic oxygenase prosthetic groups, i.e., flavin, Cu, and Fe (Chollet et al., 1975); this activity of the enzyme can be categorized as an internal monooxygenase.

Recent studies on the physiological role of RuBP oxygenase on the photorespiratory processes have made significant progress. It was revealed that fresh extracts from spinach chloroplasts exhibit not only high specific oxygenase activities but also have a broad optimal pH range, 8.6–8.8, similar to that for the carboxylase reaction (Badger et al., 1974; Andrews et al., 1975; Bahr and Jensen, 1974b; Laing and Christeller, 1976). Further kinetic analyses on the competitive inhibition of RuBP carboxylase by O_2 and of RuBP oxygenase by CO_2, respectively, using the purified soybean enzyme, demonstrated the equality of the two kinetic constants, $K_m(CO_2)$ (carboxylase) = $K_i(CO_2)$ (oxygenase); $K_m(O_2)$ (oxygenase) = $K_i(O_2)$ (carboxylase) (Laing et al., 1974). On this basis a scheme was presented showing that the bifunctional RuBP carboxylase molecule regulates two counteracting reactions, photosynthesis (CO_2 fixation) on the one hand and the Warburg effect on the other (direct inhibition of photosynthesis by O_2 and photorespiratory production of glycolate and its catabolism; Chollet and Ogren, 1975; Laing et al., 1974, 1976). From the mutually competitive nature of CO_2 and O_2, it can be assumed that the carboxylase and oxygenase reaction which presumably take place at the same catalytic site are affected by the relative concentrations of CO_2/O_2, and will be the subject of control by some other factors. As presented in Figure 4, $K_i(O_2)$ values were determined to be nearly identical in the photosynthetic CO_2 fixation catalyzed by spinach leaf protoplasts (0.28 mM, pH 8.5) and the RuBP carboxylase reaction by the purified spinach enzyme preparations (0.25 mM, pH 8.0).

Although strong arguments have long persisted concerning multiple pathways for glycolate biosynthesis in many photosynthetic organisms (see Chap. II.25, this vol.), recent studies suggest that the photorespiratory carbon oxidation (PCO) cycle is potentially the important pathway (Krause et al., 1977; Lorimer et al., 1978b). According to this scheme, in the presence of O_2 RuBP is split oxygenolytically to produce 2-P-glycolate and PGA [Eq. (10)], the former being immediately

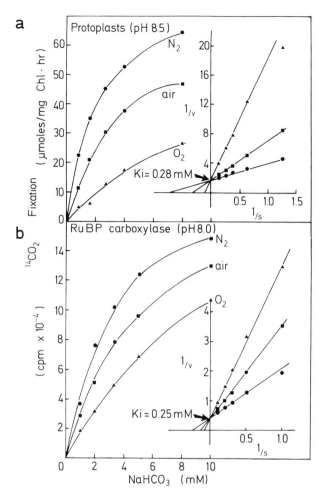

Fig. 4a and b. Inhibitory effect of O_2 on photosynthetic CO_2-fixation reactions. **a** spinach leaf protoplasts (NISHIMURA et al., 1975), **b** spinach leaf RuBP carboxylase. (NISHIMURA and AKAZAWA, unpublished data)

hydrolyzed in chloroplasts (RICHARDSON and TOLBERT, 1961). Glycolate is further metabolized in multi-organellar interactions (PCO cycle-glycolate pathway), so that two mol produce one CO_2 and one 3C-entity which eventually returns to the photosynthetic carbon reduction (PCR) cycle (Calvin-Benson cycle) as PGA. Two major experimental findings appear to support this formulation. Firstly, the glycolate formation from sugar-phosphate (R5P or Xu5P) fed to intact chloroplasts in light is specifically inhibited by uncoupling reagents (arsenate, FCCP and Dio-9), although one has to bear in mind that they will exert some unknown effects (KIRK and HEBER, 1976; KRAUSE et al., 1977, 1978). Secondly, the enrichment of ^{18}O in the carboxyl carbon of glycolate produced by *Chromatium, Chlorella* and spinach chloroplasts exposed to $^{18}O_2$ can be satisfactorily explained by the RuBP oxygenase reaction (LORIMER et al., 1977; 1978a). However, one cannot totally exclude other mechanism(s) such as that involving a transketolase-catalyzed reaction utilizing the superoxide anion in light-dependent glycolate formation (ASAMI and AKAZAWA, 1977; see Chap. III.2, this vol.).

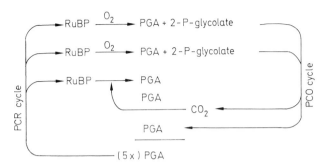

Fig. 5. The balance between the photosynthetic carbon reduction (PCR) cycle and photorespiratory carbon oxidation (PCO) cycle, involving RuBP carboxylase and RuBP oxygenase, respectively, at the CO_2 compensation point. (LORIMER et al., 1977b)

The sequence and stoichiometry of the in vivo incorporation of $^{18}O_2$ compensation point (LORIMER et al., 1978b) suggests that the PCR and PCO cycles are linked as shown in Figure 5. The ratio of carbon flow into the PCR cycle and the PCO cycle mediated by RuBP carboxylase/oxygenase reactions will be determined by the prevailing concentrations of CO_2 and O_2. Further tests under controlled O_2 and CO_2 tensions different from that at the CO_2 compensation point would add additional weight to this formulation (BERRY et al., 1977; 1978).

Currently much discussion has been given to the possible genetic or metabolic control of photorespiration, in particular relation to the specific inhibition of the RuBP oxygenase reaction leading to possible reduction of the glycolate formation. Naturally such a research endeavor has its own significance because of potential enhancement of plant productivity. However, extensive research by CHOLLET and ANDERSON (1976) has demonstrated that none of the chloroplast metabolites examined differentially regulate the carboxylase/oxygenase activities, results being quite predictable from the previously established notion concerning the same active site of enzyme molecule shares two opposing reactions. NELSON and SURZYCKI (1976) reported that RuBP carboxylase isolated from a Mendelian mutant strain of *Chlamydomonas reinhardtii* exhibits altered RuBP oxygenase activity and discussed the results in connection with glycolate metabolism in algae. Analogous experiments were undertaken by KUNG and MARSHO (1976) using mutant tobacco plants. However, careful inspection of data of all these authors invariably show that the enzyme assays conducted were not thorough enough. It is this writer's opinion that better understanding and critical analysis of activation and catalytic properties of these bifunctional enzyme reactions are prerequisites to further advances in our knowledge of the genetic and metabolic control of photorespiration.

E. Biosynthesis of RuBP Carboxylase

Since fraction 1 protein comprises the major portion of soluble protein in chloroplasts and occupies a key role in photosynthetic carbon metabolism, it is natural that many investigators have undertaken experiments to explore the mechanism of its biosynthesis. The advance of research in this field is very fast, and experimental results from different laboratories give many dramatic discoveries and new conceptual ideas concerning the synthesis of giant macromolecules in plant cells. In this review it is worthwhile to present a chronological account of the work for future studies.

Initially, the effects of several antibiotics on the pattern of fraction 1 protein synthesis during the greening of dark-grown plant leaves were tested and were related to the structural development of chloroplasts, but information obtained at this early stage of investigation was not particularly rewarding (MARGULIES, 1966; HUFFAKER et al., 1966; KELLER and HUFFAKER, 1967). Since the discovery of the presence of specific genetic information, as well as protein synthesizing machinery in chloroplasts and mitochondria, much recent interest has been focused on the autonomous nature of these cell organelles (BOARDMAN et al., 1971). Subsequent elucidation of the molecular mechanism of the biosynthesis of chloroplastic RuBP carboxylase is considered to be particularly relevant here because the enzyme molecule is composed of two structurally distinguishable components, differing in their functions (for review, see BOULTER et al., 1972; ELLIS et al., 1973; KUNG, 1976, 1977). KAWASHIMA (1970) reported the nonsynchronous incorporation of $^{14}CO_2$ into the two subunit components of RuBP carboxylase using photosynthesizing tobacco leaves and suggested that the synthesis of the enzyme may occur at two ribosomal sites or perhaps involve two DNA cistrons. This observation was extended by CRIDDLE et al. (1970) using intact barley plants, and it was suggested that subunit A is synthesized on the chloroplastic ribosomes (70s) and subunit B on the cytoplasmic ribosomes (80s), respectively. These conclusions were based on the differential inhibitory effects exerted by cycloheximide and chloramphenicol on the light-dependent synthesis of the enzyme. Independent genetic studies undertaken by CHAN and WILDMAN (1972) on the crossing of various strains of tobacco plants have revealed that the inheritance of subunit B of fraction 1 protein (RuBP carboxylase) is of the Mendelian type, whereas that of subunit A follows the maternal inheritance, suggesting that the synthesis of the two subunits is regulated on both transcriptional and translational levels.

Extensive research on the in vitro synthesis of RuBP carboxylase at the translational level has been carried out by ELLIS and his associates using intact chloroplasts or etioplasts isolated from developing pea plants (*Pisum sativum*), and has been corroborated by examining the inhibitory effects of various antibiotics (BLAIR and ELLIS, 1973; SIDDELL and ELLIS, 1975). By examining the tryptic finger print map of the labeled product, they demonstrated that one of the proteins synthesized on chloroplast ribosomes and requiring light energy is subunit A. Later similar observations were obtained by BOTTOMLEY et al. (1974) and MORGENTHALER et al. (1976), using spinach chloroplasts. Results of radioisotopic experiments using chloroplastic RNA (spinach) on the cell-free synthesis of subunit A in a heterologous system (*E. coli*) led HARTLEY et al. (1975) to conclude that the RNA fraction contains the mRNA for subunit A. More recent experiments by SAGHER et al. (1976) and HOWELL et al. (1977) revealed that the mRNA's for subunit A [lacking poly (A)], which were isolated from chloroplasts of *Euglena gracilis* and *Chlamydomonas reinhardtii*, were specifically translated on the chloroplast ribosomes. The latter workers reported the identification of the chloroplast gene coding for subunit A of RuBP carboxylase from *Chlamydomonas reinhardtii* by employing molecular hybridization with mRNA for this subunit isolated from the same organism (GELVIN et al., 1977). Independently, it was shown that a chloroplastic DNA fragment isolated from *Zea mays*, purified by molecular cloning in *E. coli*, can direct the synthesis of subunit A in the in vitro transcription–translation system (Coen et al., 1978). On the other hand, the in vitro studies on the synthesis of the small subunit (B) using French bean seedlings (Gray and Kekwick, 1974), wheat leaves (Gooding et al., 1973; Roy et al., 1976), barley leaves (Alscher et al., 1976), and pea leaves (Roy et al., 1977) have shown that this protein is specifically synthesized on the 80s cytoplasmic ribosomes.

The most intriguing question pertaining to the biosynthesis of the carboxylase molecule is the mechanism of the passage of the extraplastidic subunit B molecule synthesized across the chloroplast envelope and subsequent assembly with subunit A inside the organelle to make up the whole enzyme molecule. One relevant observation in this subject was the possible association of subunit B with a polypeptide component associated with the envelope membranes of spinach chloroplasts (Pineau and Douce, 1974). These results appear to be strengthened by the radioisotopic pulse-chase experiments of Cobb and Wellburn (1976), showing that the small subunit bound to the envelope of *Avena* chloroplasts is preferentially labeled in light. How will such experimental results be corroborated with the in vitro biosynthesis studies? A fascinating finding was reported recently by Dobberstein et al. (1977), who demonstrated the formation of a putative precursor (pS) of subunit B (MW 1.65×10^4, approximately 3500 larger than small subunit) in the in vitro system for RuBP carboxylase biosynthesis directed by the poly (A)-rich mRNA of *Chlamydomonas reinhardtii*. As it was found that endoprotease present in the postribosomal supernatant fraction cleaved pS to make subunit B, the above authors proposed that this small peptide plays a role in the transfer of subunit B into the chloroplast. Analogous experiments were conducted by Ellis and his associates (Ellis et al., 1978; Highfield and Ellis, 1978), who showed that the precursor (P-20) of subunit B was synthesized on the poly (A)-mRNA from a pea cytoplasmic system coupled with wheat germ ribosomes. Upon incubation with the intact pea chloroplasts, P-20 can be cleaved to make subunit B, and it has been hypothesized that the specific protease localized in the chloroplastic envelope plays an important role in the processing of the biosynthesis of RuBP carboxylase molecules (see also Cashmore et al., 1978; Chua and Schmidt, 1978).

In addition to the role of receptor protein(s) localized in chloroplast membranes, their hypothetical view is related to another important question concerning the regulatory mechanism which governs the biosynthesis of the oligomeric whole enzyme molecules occurring in different cell sites and its subsequent assembly to form active enzymes. From the experimental results dealing with the coordinated interaction of the two subunits during the enzyme synthesis described above, particularly on the uptake of the extraplastidic subunit B by intact chloroplasts, Ellis (1975, 1977) suggested that subunit B functions as a positive effector in the biosynthesis or translation of the mRNA for subunit A which takes place in the chloroplasts. He has thus proposed a hypothetical scheme for the biosynthesis of RuBP carboxylase as shown in Figure 6.

An extensive investigation, using synchronously growing cells of the green alga, *Chlamydomonas reinhardtii*, to examine the effects of various types of antibiotics on the synthesis and turnover of RuBP carboxylase, Iwanij et al. (1975) concluded that the biosynthesis of the two subunits is stringently controlled and that there is no pool of either large or small subunits in the algal cells. Although radioisotopic

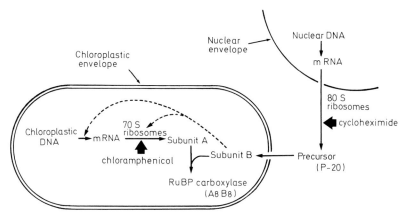

Fig. 6. Proposed biosynthetic mechanism of RuBP carboxylase molecules. Proposed sites of action of chloramphenicol and cycloheximide are indicated by *bold arrows*. (Modified after ELLIS et al., 1977)

experiments using whole tissues of pea seedlings (*Pisum sativum*) by CASHMORE (1976) as well as those using spinach leaf protoplasts (NISHIMURA and AKAZAWA, 1978) suggest the operation of a coordinated coupling mechanism in the biosynthesis of the whole enzyme molecule, the validity of the hypothetical scheme as shown in Figure 6 in higher plant cells remains to be established by future investigations. In this context it is interesting to note that there is a substantial experimental proof that the synthesis of the two subunits can be completely uncoupled in rye plants (*Secale cereale*) grown at an elevated temperature, 32°C (FEIERABEND, 1976). It was reported that the high temperature-induced deficiency of the chloroplastic 70s ribosomes causes the sole formation of subunit B on the cytoplasmic ribosomes (FEIERABEND and WILDNER, 1978).

Acknowledgments. The author is most grateful to his colleagues for many useful discussions and comments, and to Drs. R. Chollet, R.G. Jensen, C.B. Osmond, for critical reading of the manuscript and their kind help. The research carried out in the author's laboratory has been supported by the research grants from the Ministry of Education of Japan, the Toray Foundation (Tokyo) and the Nissan Science Foundation (Tokyo).

References

Akazawa, T.: Prog. Phytochem. *2*, 107–141 (1970)
Akazawa, T.: In: Proc. IV Int. Congr. Photosynthesis. Hall, D.O., Coombs, J., Goodwin, T.W. (eds.). pp. 447–456, London: Biochem. Soc. 1978
Alscher, R., Smith, M.A., Petersen, L.W., Huffaker, R.C., Criddle, R.S.: Arch. Biochem. Biophys. *174*, 216–225 (1976)
Andrews, T.J., Badger, M.R., Lorimer, G.H.: Arch. Biochem. Biophys. *171*, 93–103 (1975)
Andrews, T.J., Lorimer, G.H., Tolbert, N.E.: Biochemistry *12*, 11–18 (1973)
Asami, S., Akazawa, T.: Biochemistry *16*, 2202–2207 (1977)
Badger, M.R., Andrews, T.J., Osmond, C.B.: Proc. III Int. Congr. Photosynthesis. Avron, M. (ed.). pp. 1421–1429, Amsterdam: Elsevier Publ. Co. (1974)

Bahr, J.T., Jensen, R.G.: Plant Physiol. 53, 39–44 (1974a)
Bahr, J.T., Jensen, R.G.: Arch. Biochem. Biophys. 164, 408–413 (1974b)
Bahr, J.T., Jensen, R.G.: Arch. Biochem. Biophys. 185, 39–48 (1978)
Baker, T.S., Eisenberg, D., Eiserling, F.A., Weissman, L.: J. Mol. Biol. 91, 391–399 (1975)
Baker, T.S., Eisenberg, D., Eiserling, F.A.: Science 196, 293–295 (1977a)
Baker, T.S., Suh, S.W., Eisenberg, D.: Proc. Natl. Acad. Sci. USA 74, 1037–1041 (1977b)
Bassham, J.A., Kirk, M.: Biochem. Biophys. Res. Commun. 9, 376–380 (1962)
Benson, A.A., Calvin, M.: J. Exp. Bot. 1, 63–68 (1950)
Berry, J.A., Osmond, C.B., Lorimer, G.H.: Carnegie Inst. Yearb. 76, 307–313 (1977)
Berry, J.A., Osmond, C.B., Lorimer, G.H.: Plant Physiol. 62, 954–967 (1978)
Björkman, O., Gauhl, E.: Planta (Berl.) 88, 197–203 (1969)
Blair, C.E., Ellis, R.J.: Biochim. Biophys. Acta 319, 223–234 (1973)
Boardman, N.K., Linnane, A.W., Smillie, R.M.: Autonomy and biosynthesis of mitochondria and chloroplasts. Amsterdam: North-Holland 1971
Bottomley, W., Spencer, D., Whitfeld, P.R.: Arch. Biochem. Biophys. 164, 106–117 (1974)
Boulter, D., Ellis, R.J., Yarwood, A.: Biol. Rev. 47, 113–175 (1972)
Bowes, G., Ogren, W.L.: J. Biol. Chem. 247, 2171–2176 (1972)
Bowes, G., Ogren, W.L., Hageman, R.H.: Biochem. Biophys. Res. Commun. 45, 716–722 (1971)
Bowes, G., Ogren, W.L., Hageman, R.H.: Plant Physiol. 56, 630–633 (1975)
Bowien, B., Mayer, F., Codd, G.A., Schlegel, H.G.: Arch. Microbiol. 110, 157–166 (1976)
Buchanan, B.B., Schurmann, P.: Current Top. Cell. Regul. 7, 1–20 (1973)
Buchanan, B.B., Sirevag, R.: Arch. Microbiol. 109, 15–19 (1976)
Calvin, M.: J. Chem. Soc. 1895–1915 (1956)
Cashmore, A.R.: J. Biol. Chem. 251, 2848–2853 (1976)
Cashmore, A.R., Broadhurst, M.K., Gray, R.E.: Proc. Natl. Acad. Sci. 75, 655–659 (1978)
Chan, P.M., Wildman, S.G.: Biochim. Biophys. Acta 277, 677–680 (1972)
Chollet, R., Anderson, L.L.: Arch. Biochem. Biophys. 176, 344–351 (1976)
Chollet, R., Anderson, L.L.: Biochim. Biophys. Acta 482, 228–240 (1977)
Chollet, R., Anderson, L.L.: Biochim. Biophys. Acta 525, 455–467 (1978)
Chollet, R., Anderson, L.L., Hovsepian, L.C.: Biochem. Biophys. Res. Commun. 64, 97–107 (1975)
Chollet, R., Ogren, W.L.: Bot. Rev. 41, 137–179 (1975)
Christeller, J.T., Laing, W.A., Troughton, J.H.: Plant Physiol. 57, 580–582 (1976)
Chu, D.K., Bassham, J.A.: Plant Physiol. 54, 556–559 (1974)
Chu, D.K., Bassham, J.A.: Plant Physiol. 55, 720–726 (1975)
Chua, N.H., Schmidt, G.W.: Proc. Natl. Acad. Sci. USA 75, 6110–6114 (1978)
Cobb, A.H., Wellburn, A.R.: Planta (Berl.) 129, 127–131 (1976)
Codd, G.A., Stewart, W.D.P.: Planta (Berl.) 130, 323–326 (1976)
Codd, G.A., Stewart, W.D.P.: FEBS Lett. 1, 35–38 (1977a)
Codd, G.A., Stewart, W.D.P.: Arch. Microbiol. 113, 105–110 (1977b)
Coen, D.M., Bedbrook, J.R., Bogorad, L., Rich, A.: Proc. Natl. Acad. Sci. USA 74, 5487–5491 (1977)
Cooper, T.G., Filmer, D., Wishnick, M., Lane, M.D.: J. Biol. Chem. 244, 1081–1083 (1969)
Criddle, R.S., Dau, B., Kleinkopf, G.E., Huffaker, R.C.: Biochem. Biophys. Res. Commun. 41, 621–627 (1970)
Dobberstein, B., Blobel, G., Chua, N-H.: Proc. Natl. Acad. Sci. USA 74, 1082–1085 (1977)
Dorner, R.W., Kahn, A., Wildman, S.G.: Biochim. Biophys. Acta 29, 240–245 (1958)
Ellis, R.J.: Comment. Plant Sci. 4, 29–38 (1973)
Ellis, R.J.: Phytochemistry 14, 89–93 (1975)
Ellis, R.J.: Biochim. Biophys. Acta 463, 185–215 (1977)
Ellis, R.J., Blair, G.E., Hartley, M.R.: Biochem. Soc. Symp. 38, 137–162 (1973)
Ellis, R.J., Highfield, P.E., Silverthorne, J.: In: Proc. IV Int. Congr. Photosynthesis. Hall, D.O., Coombs, J., Goodwin, T.W. (eds.), pp. 497–506. London: Biochem. Soc. 1978
Feierabend, J.: In: Genetics and biogenesis of chloroplasts and mitochondria. Bucher, T., Neupert, W., Sebald, W., Werner, S. (eds.), pp. 99–102. Amsterdam: Elsevier/North-Holland Biomedical Press 1976
Feierabend, J., Wildner, G.: Arch. Biochem. Biophys. 186, 283–291 (1978)

Fuller, R.C., Gibbs, M.: Plant Physiol. *34*, 324–329 (1959)
Gelvin, S., Heizmann, P., Howell, S.H.: Proc. Natl. Acad. Sci. USA *74*, 3193–3197 (1977)
Gibbs, M.: In: Structure and function of chloroplasts. Gibbs, M. (ed.), pp. 169–214. Berlin, Heidelberg, New York: Springer 1971
Givan, A.L., Criddle, R.S.: Arch. Biochem. Biophys. *149*, 153–163 (1972)
Goldsworthy, A.: Nature (London) *217*, 62 (1968)
Goldthwaite, J.J., Bogorad, L.: Anal. Biochem. *41*, 57–66 (1971)
Gooding, L.R., Roy, H., Jagendorf, A.T.: Arch. Biochem. Biophys. *159*, 324–335 (1973)
Gray, J.C., Kekwick, R.G.O.: Eur. J. Biochem. *44*, 491–500 (1974)
Hartley, M.R., Wheeler, A., Ellis, R.J.: J. Mol. Biol. *91*, 67–77 (1975)
Hartman, F.G., Welch, M.H., Norton, I.L.: Proc. Natl. Acad. Sci. USA *70*, 3721–3724 (1973)
Haslett, B.G., Yarwood, A., Evans, I.M., Boulter, D.: Biochim. Biophys. Acta *420*, 122–132 (1976)
Hattersley, P.W., Watson, L., Osmond, C.B.: Aust. J. Plant Physiol. *4*, 523–539 (1977)
Heber, U., Pon, N.G., Heber, M.: Plant Physiol. *38*, 355–360 (1963)
Heldt, H.W., Werdan, K., Milovancev, M., Geller, G.: Biochim. Biophys. Acta *314*, 224–241 (1973)
Henriques, F., Park, R.B.: Arch. Biochem. Biophys. *176*, 472–478 (1976)
Highfield, P.E., Ellis, R.J.: Nature (London) *271*, 420–424 (1978)
Howell, S.H., Heizmann, P., Gelvin, S., Walker, L.L.: Plant Physiol. *59*, 464–470 (1977)
Huffaker, R.C., Obendorf, R.L., Keller, C.J., Kleinkopf, G.E.: Plant Physiol. *41*, 913–918 (1966)
Iwanij, V., Chua, N.-H., Siekevitz, P.: J. Cell Biol. *64*, 572–585 (1975)
Jensen, R.G., Bahr, J.T.: Annu. Rev. Plant Physiol. *28*, 379–400 (1977)
Jensen, R.G., Bassham, J.A.: Proc. Natl. Acad. Sci. USA *56*, 1095–1101 (1966)
Kannangara, C.G., Wyk, D. van, Menke, W.: Z. Naturforsch. *25 B*, 613–618 (1970)
Kawashima, N.: Biochem. Biophys. Res. Commun. *38*, 119–124 (1970)
Kawashima, N., Wildman, S.G.: Annu. Rev. Plant Physiol. *21*, 325–358 (1970)
Kawashima, N., Singh, S., Wildman, S.G.: Biochem. Biophys. Res. Commun. *42*, 664–668 (1971)
Keller, C.J., Huffaker, R.C.: Plant Physiol. *42*, 1277–1283 (1967)
Kelly, G.J., Latzko, E., Gibbs, M.: Annu. Rev. Plant Physiol. *27*, 181–205 (1976)
King, M.C., Wilson, A.C.: Science *188*, 107–116 (1975)
Kirk, M.R., Heber, U.: Planta (Berl.) *132*, 131–141 (1976)
Krause, G.H., Lorimer, G.H., Heber, U., Kirk, M.: In: Proc. IV Int. Congr. Photosynthesis. Hall, D.O., Coombs, J., Goodwin, T.W. (eds.), pp. 299–310. London: Biochem. Soc. 1978
Krause, G.H., Thorne, S.W., Lorimer, G.H.: Arch. Biochem. Biophys. *183*, 471–479 (1977)
Kung, S.D.: Science *191*, 429–434 (1976)
Kung, S.D.: Annu. Rev. Plant. Physiol. *28*, 401–437 (1977)
Kung, S.D., Marsho, T.V.: Nature (London) *259*, 325–326 (1976)
Kung, S.D., Sakano, K., Wildman, S.G.: Biochim. Biophys. Acta *365*, 138–147 (1974)
Laing, W.A., Christeller, J.T.: Biochem. J. *159*, 563–570 (1976)
Laing, W.A., Ogren, W.L., Hageman, R.H.: Plant Physiol. *54*, 678–685 (1974)
Laing, W.A., Ogren, W.L., Hageman, R.H.: Biochemistry *14*, 2269–2275 (1975)
Lerman, J.C.: In: Environmental and biological control of photosynthesis. Marcelle, R. (ed.), pp. 323–335. The Hague: Junk 1975
Lilley, R.M., Walker, D.A.: Plant Physiol. *55*, 1087–1092 (1975)
Lorimer, G.H.: Eur. J. Biochem. *89*, 43–50 (1978)
Lorimer, G.H., Andrews, T.J.: Nature (London) *243*, 359–360 (1973)
Lorimer, G.H., Andrews, T.J., Tolbert, N.E.: Biochemistry *12*, 18–23 (1973)
Lorimer, G.H., Badger, M.R., Andrews, T.J.: Biochemistry *15*, 529–536 (1976)
Lorimer, G.H., Krause, G.H., Berry, J.A.: FEBS Lett. *78*, 199–202 (1977)
Lorimer, G.H., Osmond, C.B., Akazawa, T., Asami, S.: Arch. Biochem. Biophys. *185*, 49–56 (1978a)
Lorimer, G.H., Woo, K.C., Berry, J.A., Osmond, C.B.: In: Proc. IV Int. Congr. Photosynthesis. Hall, D.O., Coombs, J., Goodwin, T.W. (eds.), pp. 311–322. London: Biochem. Soc. 1978b
Marchalonis, J.J., Weltman, J.K.: Comp. Biochem. Physiol. *38 B*, 609–625 (1971)
Margulies, M.M.: Plant Physiol. *41*, 992–1003 (1966)

Margulis, L.: Am. Sci. *59*, 230–235 (1971)

Matsumoto, K., Nishimura, M., Akazawa, T.: Plant Cell Physiol. *18*, 1281–1290 (1977)

McFadden, B.A.: Bacteriol. Rev. *37*, 289–319 (1973)

McFadden, B.A., Lord, J.M., Rowe, A., Dilks, S.: Eur. J. Biochem. *54*, 195–206 (1975)

McFadden, B.A., Tabita, F.R.: Biosystems *6*, 93–112 (1974)

Miziorko, H.M., Mildvan, A.S.: J. Biol. Chem. *249*, 2743–2750 (1974)

Morgenthaler, L.R.M., Morgenthaler, J.J., Price, C.A.: FEBS Lett. *62*, 96–100 (1976)

Nelson, P.E., Surzycki, S.J.: Eur. J. Biochem. *61*, 475–480 (1976)

Neumann, J., Jagendorf, A.T.: Arch. Biochem. Biophys. *107*, 109–119 (1964)

Nishimura, M., Akazawa, T.: Biochem. Biophys. Res. Commun. *54*, 842–848 (1973)

Nishimura, M., Akazawa, T.: Biochemistry *13*, 2277–2281 (1974a)

Nishimura, M., Akazawa, T.: J. Biochem. *76*, 169–176 (1974b)

Nishimura, M., Akazawa, T.: Biochem. Biophys. Res. Commun. *59*, 584–590 (1974c)

Nishimura, M., Akazawa, T.: Plant Physiol. *62*, 97–100 (1978)

Nishimura, M., Beevers, H.: Plant Physiol. *62*, 40–43 (1978)

Nishimura, M., Takabe, T., Sugiyama, T., Akazawa, T.: J. Biochem. *74*, 945–954 (1973)

Nishimura, M., Graham, D., Akazawa, T.: Plant Physiol. *56*, 718–722 (1975)

Nishimura, M., Graham, D., Akazawa, T.: Plant Physiol. *58*, 309–314 (1976)

Norton, I.L., Welch, M.H., Hartman, F.C.: J. Biol. Chem. *250*, 8062–8068 (1975)

Ogren, W.L., Bowes, G.: Nature New Biol. *230*, 159–160 (1971)

Osmond, C.B.: Annu. Rev. Plant Physiol. *29*, 375–414 (1978)

Osmond, C.B., Akazawa, T., Beevers, H.: Plant Physiol. *55*, 226–230 (1975)

Paech, C., Ryan, F.J., Tolbert, N.E.: Arch. Biochem. Biophys. *179*, 279–288 (1977)

Park, R.B., Pon, N.G.: J. Mol. Biol. *3*, 1–10 (1961)

Paulsen, J.M., Lane, M.D.: Biochemistry *5*, 2350–2357 (1966)

Peterkofsky, A., Racker, E.: Plant Physiol. *36*, 409–414 (1961)

Pineau, B., Douce, R.: FEBS Lett. *47*, 255–259 (1974)

Portis, A.R., Heldt, H.W.: Biochim. Biophys. Acta *449*, 434–446 (1976)

Poulsen, C., Strøbaek, S., Haslett, B.G.: In: Genetics and biogenesis of chloroplasts and mitochondria. Bucher, T., Neupert, W., Sebald, W., Werner, S. (eds.), Amsterdam: Elsevier/North-Holland, pp. 17–24, 1976

Purohit, K., McFadden, B.A.: J. Bacteriol. *129*, 415–421 (1977)

Purohit, K., McFadden, B.A., Cohen, A.L.: J. Bacteriol. *127*, 505–516 (1976a)

Purohit, K., McFadden, B.A., Shaykh, M.M.: J. Bacteriol. *127*, 516–522 (1976b)

Quayle, J.R., Fuller, R.C., Benson, A.A., Calvin, B.: J. Am. Chem. Soc. *76*, 3610–3611 (1954)

Rabin, B.R., Trown, P.W.: Nature *202*, 1290–1293 (1964)

Racher, E.: Arch. Biochem. Biophys. *69*, 300–310 (1957)

Richardson, K.E., Tolbert, N.E.: J. Biol. Chem. *236*, 1285–1290 (1961)

Roy, H., Patterson, R., Jagendorf, A.T.: Arch. Biochem. Biophys. *172*, 64–73 (1976)

Roy, H., Terrena, B., Cheong, L.C.: Plant Physiol. *60*, 532–537 (1977)

Roy, H., Valeri, A., Pope, D.H., Rueckert, L., Costa, K.A.: Biochemistry *17*, 665–668 (1978)

Rutner, A., Lane, M.D.: Biochem. Biophys. Res. Commun. *28*, 531–537 (1967)

Ryan, F.J., Tolbert, N.E.: J. Biol. Chem. *250*, 4229–4233 (1975)

Sagher, D., Grosfeld, H., Edelman, M.: Proc. Natl. Acad. Sci. USA *73*, 722–726 (1976)

Schloss, J.V., Hartman, F.C.: Biochem. Biophys. Res. Commun. *75*, 320–328 (1977a)

Schloss, J.V., Hartman, F.C.: Biochem. Biophys. Res. Commun. *77*, 230–236 (1977b)

Schloss, J.V., Stringer, C.D., Hartman, F.C.: J. Biol. Chem. *253*, 5705–5711 (1978)

Shively, J.M., Ball, F.L., Kline, B.W.: J. Bacteriol. *116*, 1405–1411 (1973)

Siddell, S.G., Ellis, R.J.: Biochem. J. *146*, 675–685 (1975)

Siegel, M.I., Lane, M.D.: Biochem. Biophys. Res. Commun. *48*, 508–516 (1972)

Siegel, M.I., Lane, M.D.: J. Biol. Chem. *248*, 5486–5498 (1973)

Siegel, M.I., Wishnick, M., Lane, M.D.: In: The enzymes. Vol. 6. Boyer, P.D. (ed.), pp. 169–192. New York: Academic Press 1972

Sjödin, B., Vestermark, A.: Biochim. Biophys. Acta *297*, 165–173 (1973)

Slack, C.R., Hatch, M.D.: Biochem. J. *103*, 660–665 (1967)

Slack, C.R., Hatch, M.D., Goodchild, D.J.: Biochem. J. *114*, 489–498 (1969)

Smillie, R.M.: Can. J. Bot. *41*, 123–146 (1963)

Staehelin, L.A.: J. Cell Biol. *71*, 136–158 (1976)

Stewart, R., Auchterlonie, C.C., Codd, G.A.: Planta *136*, 61–64 (1977)

Stringer, C.D., Hartman, F.C.: Biochem. Biophys. Res. Commun. *80*, 1043–1048 (1978)

Strøbaek, S., Gibbons, G.C.: Carlsberg Res. Commun. *41*, 57–72 (1976)

Strøbaek, S., Gibbons, G.C., Haslett, B., Boulter, D., Wildman, S.G.: Carlsberg Res. Commun. *41*, 335–343 (1976)

Strotmann, H., Hesse, H., Edelmann, K.: Biochim. Biophys. Acta *314*, 202–210 (1973)

Sue, J.M., Knowles, J.R.: Biochemistry *17*, 4041–4044 (1978)

Sugiyama, T., Akazawa, T.: Biochemistry *9*, 4499–4504 (1970)

Sugiyama, T., Ito, T., Akazawa, T.: Biochemistry *10*, 3406–3411 (1971)

Sugiyama, T., Nakayama, N., Akazawa, T.: Arch. Biochem. Biophys. *126*, 737–745 (1968)

Tabita, F.R., McFadden, B.A.: Biochem. Biophys. Res. Commun. *48*, 1153–1159 (1972)

Tabita, F.R., McFadden, B.A.: J. Biol. Chem. *249*, 3459–3464 (1974a)

Tabita, F.R., McFadden, B.A.: Arch. Microbiol. *99*, 231–240 (1974b)

Tabita, F.R., McFadden, B.A., Pfenning, N.: Biochim. Biophys. Acta *341*, 187–194 (1974a)

Tabita, F.R., Stevens, S.E., Quijano, R.: Biochem. Biophys. Res. Commun. *61*, 45–52 (1974b)

Takabe, T.: Agric. Biol. Chem. *41*, 2255–2260 (1977)

Takabe, T.: In: Taniguchi Int. Symp. Molecular evolution and polymorphism. Kimura, M. (ed.). Mishima: Natl. Int. Genetics, pp. 296–312, 1978

Takabe, T., Akazawa, T.: Arch. Biochem. Biophys. *157*, 303–308 (1973a)

Takabe, T., Akazawa, T.: Biochem. Biophys. Res. Commun. *53*, 1173–1179 (1973b)

Takabe, T., Akazawa, T.: Arch. Biochem. Biophys. *169*, 686–694 (1975a)

Takabe, T., Akazawa, T.: Biochemistry *14*, 46–50 (1975b)

Takabe, T., Akazawa, T.: Plant Cell Physiol. *16*, 1049–1060 (1975c)

Takabe, T., Akazawa, T.: Plant Cell Physiol. *18*, 753–765 (1977)

Takabe, T., Nishimura, M., Akazawa, T.: Biochem. Biophys. Res. Commun. *68*, 537–544 (1976)

Tolbert, N.E.: Current Top. Cell Regul. *7*, 21–50 (1973)

Troughton, J.H., Cord, K.A., Hendy, C.H.: Carnegie Inst. Yearb. *73*, 768–780 (1974)

Walker, D.A.: New Phytol. *72*, 209–235 (1973)

Walker, D.A.: Current Top. Cell Regul. *11*, 203–241 (1976)

Weissbach, A., Horecker, B.L., Hurwitz, J.: J. Biol. Chem. *218*, 795–810 (1956)

Weissbach, A., Smyrniotis, P.Z., Horecker, B.L.: J. Am. Chem. Soc. *76*, 3611 (1954)

Werdan, K., Heldt, H.W., Milovancev, M.: Biochim. Biophys. Acta *396*, 276–292 (1975)

Wildman, S.G., Bonner, J.: Arch. Biochem. *14*, 381–413 (1947)

Wishnick, M., Lane, M.D., Scrutton, M.C.: J. Biol. Chem. *245*, 4937–4939 (1970)

18. Carbonic Anhydrase

R.P. POINCELOT

A. Introduction

The hydration of CO_2 to HCO_3^- ($CO_2 + H_2O \rightleftharpoons H_2CO_3 \rightleftharpoons H^+ + HCO_3^-$) is relatively slow, especially in the forward direction. Carbonic anhydrase (EC 4.2.1.1.; carbonate hydro-lyase, carbonate dehydratase) is an efficient catalyst of this reaction. Since CO_2, HCO_3^-, and H^+ are involved in photosynthesis, and carbonic anhydrase is present in plants, it is logical to assume a photosynthetic role for carbonic anhydrase. However, the function of carbonic anhydrase still remains elusive.

The literature on plant carbonic anhydrase has been reviewed previously (WAY-GOOD, 1955; LINDSKOG et al., 1971; LAMB, 1977). The discussion here, unlike earlier reviews, will cover only carbonic anhydrase in plants with particular emphasis on areas of importance to photosynthesis.

B. Characterization

I. Histochemical and Other Detection

Hansson's medium has been used for the light microscopy detection of carbonic anhydrase in plant tissues (TRIOLO et al., 1974). The reaction mixture contains $CoSO_4$, H_2SO_4, and KH_2PO_4. As CO_2 is enzymically converted to HCO_3^-, a compound containing Co and P precipitates at the reaction site and is converted to CoS. The application of this medium for electron microscopic localization of carbonic anhydrase in animal systems (YOKOTA et al., 1975) suggests such a technique would be feasible with plants. Carbonic anhydrase can be detected on polyacrylamide gels after treatment with CO_2 and bromocresol blue (PATTERSON et al., 1971).

II. Occurrence

Carbonic anhydrase has been found in a large number of plants, which are too numerous to cite individually. It occurs in both monocotyledonous and dicotyledonous plants on land (EVERSON and SLACK, 1968; CHEN et al., 1970; ATKINS et al., 1972b; and LAMB, 1977). In water it is found in the various algal divisions and in marine angiosperms (GRAHAM and SMILLIE, 1976). Apparently there are no published reports on the occurrence of carbonic anhydrase in gymnosperms.

III. Location

1. In C_3 Species

There is general agreement that mesophyll chloroplasts isolated with aqueous or nonaqueous media from plants with less efficient photosynthesis (C_3 species) contain most of the carbonic anhydrase activity associated with the whole leaf (EVERSON and SLACK, 1968; EVERSON, 1970; POINCELOT, 1972a; CHANG, 1975a). Whether all the activity resides in the chloroplast, or a small part of it is associated with the cytoplasm is uncertain. These uncertainties result from the difficulty of isolating entirely intact chloroplasts or from completely separating the cytoplasm from chloroplasts. Isoenzymes of carbonic anhydrase have been isolated from several plants, and it has been suggested that they represent cytoplastic and chloroplastic forms (ATKINS et al., 1972a, 1972b; KACHRU and ANDERSON, 1974; KOSITSIN and KHALIDOVA, 1974; WALK, 1975). This writer found carbonic anhydrase was located in the stroma of spinach chloroplasts (POINCELOT, 1972a), and this was subsequently confirmed by others (WERDAN and HELDT, 1972; JACOBSON et al., 1975). Chloroplast envelope membranes of C_3 species, once thought to be the site of carbonic anhydrase, have no traces of activity (POINCELOT, 1973; POINCELOT and DAY, 1976).

Some authors have shown the presence of carbonic anhydrase in the roots of grapes (CHAMPAGNOL, 1976a) and legumes (ATKINS, 1974). The latter activity was localized in the root nodules as a soluble enzyme, with only traces being present in the bacteroids.

2. In C_4 Species

The location of carbonic anhydrase in plants with more efficient photosynthesis (C_4 plants) is unclear. Some reports indicate that the bulk of the activity is confined to the mesophyll cells (GRAHAM et al., 1971), and others describe a more even distribution between the mesophyll and bundle sheath cells (POINCELOT, 1972b; TRIOLO et al., 1974; RATHNAM and DAS, 1975). The location within these cells has been found to be cytoplastic and/or chloroplastic (EVERSON and SLACK, 1968; ATKINS et al., 1972a; POINCELOT, 1972b; RATHNAM and DAS, 1975). Carbonic anhydrase has been stated to be present and absent in the envelope membranes of mesophyll chloroplasts (RATHNAM and DAS, 1975; POINCELOT and DAY, 1976), and as being a stromal enzyme in bundle sheath chloroplasts (RATHNAM and DAS, 1975; POINCELOT, 1977). It is apparent that experiments are required in which cleanly separated mesophyll, bundle sheath cells and their respective chloroplasts are obtained, and that chloroplast fractions must be isolated with careful procedures that prevent inhibition from occurring. Such techniques have become available recently [see Chaps. II.2 (JENSEN) and II.7 (EDWARDS and HUBER), this vol.].

IV. Levels of Activity

Activities of the enzyme are similar in both the C_4 and C_3 crop and weed plants (GRAHAM et al., 1971; ATKINS et al., 1972a; POINCELOT, 1972b; RATHNAM, 1975). Earlier researchers, who showed lower levels of activity with C_4 plants (EVERSON

Table 1. Isolation and purification of carbonic anhydrase

Plant	Extent of Purifi- cation (-fold)	Overall Yield (%)	Reference
Cotton (*Gossypium hirsutum*)	33	–	CHANG, 1975b
Lettuce (*Lactuca sativa*)	900	1	WALK, 1975
Navy Bean (*Phaseolus vulgaris*)	3.5	90	ATKINS, 1974
Parsley (*Petroselinum crispum*)	133	80	TOBIN, 1970
Pea (*Pisum sativum*)	200	32	KISIEL and GRAF, 1972
Spinach (*Spinacia oleracea*)	330	7	POCKER and NG, 1973
Tomato (*Lycopersicon lycopersicum*)	6	40	KOSITSIN and KHALIDOVA, 1974
Tradescantia albiflora	24	30	ATKINS et al., 1972b

and SLACK, 1968; CHEN et al., 1970; CERFIGNI et al., 1971), were apparently in error in view of the above later work. Insufficient grinding or suboptimal extraction procedures probably accounts for the lesser enzymic activities. Many weeds, grasses, and the crop plants (beet, cotton, sunflower, spinach, maize, and pea) have activities of about 2000–6000 units/mg chl (CHEN et al., 1970; EVERSON, 1970; ATKINS et al., 1972a; POINCELOT, 1972a, 1972b; CHANG, 1975c; JACOBSON et al., 1975). Green, brown, red, or blue-green algae, phytoplankton, and marine angiosperms have comparable or lower values on a chlorophyll basis compared to the terrestrial plants (GRAHAM and REED, 1971; DOHLER, 1974; INGLE and COLMAN, 1975, 1976; GRAHAM and SMILLIE, 1976). Calculations (POINCELOT, 1972a) show that there is sufficient enzyme activity at $0°$ C and 4000 units to supply over 100 times more CO_2 or HCO_3^- to chloroplasts than is required by the rate of photosynthesis. The photosynthetic rate was assumed to be 200 μmol mg chl^{-1} h^{-1}. On this basis carbonic anhydrase would not be rate-limiting for leaf photosynthesis.

V. Isolation and Purification

Carbonic anhydrase has been isolated and purified to various degrees from a number of plants as shown in Table 1. The greatest purifications were with parsley, pea, spinach, and lettuce (Table 1). The criteria of purity for parsley carbonic anhydrase included its appearance as a single component in analytical polyacrylamide gel electrophoresis at pH 8.7 and 7.4, the finding of a single component in sedimentation velocity experiments (S $°_{20,\omega}=8.8$), no evidence of impurities with sedimentation equilibrium experiments, and showing a single precipitin line in immunodiffusion-antiserum experiments (TOBIN, 1970). While the supporting data for the criteria of purity with lettuce, pea and spinach carbonic anhydrase are not as complete, they would appear to be as pure as their parsley counterpart.

VI. Enzymic Parameters

1. Molecular Weight

The molecular weight of carbonic anhydrase has been described for spinach as 180,000 (POCKER and NG, 1973) and 148,000 (ROSSI et al., 1969), pea as 188,000

(ATKINS et al., 1972 b) and 194,000 (KISIEL and GRAF, 1972), parsley as 180,000 (TOBIN, 1970), lettuce as 195,000 (WALK, 1975), and navy bean as 205,000 (ATKINS, 1974). Since the electrophoretic migration of leaf carbonic anhydrases from species of 19 dicotyledonous families are all similar (ATKINS et al., 1972 a), it may be inferred that the molecular weight of carbonic anhydrase from dicots is around 180,000. Similarly the molecular weight of 42,000 for the monocot, *Tradescantia albiflora*, and supporting electrophoretic migration data suggest a molecular weight of 40,000 for carbonic anhydrase from monocots (ATKINS et al., 1972 a).

2. Stability

Carbonic anhydrase is highly thermostabile. Purified pea carbonic anhydrase retained 40% of its activity after incubation at 60 °C for 15 min at pH 8.1 (KISIEL and GRAF, 1972). Cotton carbonic anhydrase at pH 8.0 lost 50% of its activity after 5 min at 65 °C, but lost less than 50% after 20 min at 55 °C (CHANG, 1975 b).

Most studies with carbonic anhydrase have been conducted in the presence of a sulfhydryl reducing agent. With parsley carbonic anhydrase, activity was rapidly lost without the sulfhydryl reducing reagent (TOBIN, 1970). However, the reverse appears to be true for spinach and cotton carbonic anhydrase. With spinach the absence of the sulfhydryl reducing agent increased the storage life such that little loss of activity occurred at room temperature for 50 h or at 4° C for one year (POCKER and NG, 1973). With cotton the enzyme was quite stable at 4° C for 20 h without the protective sulfhydryl agent (CHANG, 1975 c). It appears the requirement of such protection will have to be established on a plant by plant basis.

Carbonic anhydrase from peas and *Tradescantia* exhibited maximal stability at pH 8.25 (ATKINS et al., 1972 b). The pH optimum for hydration of CO_2 and dehydration of HCO_3^- was 7.0 and 7.5, respectively, for pea carbonic anhydrase (KACHRU and ANDERSON, 1974). The half-life of activity at these pH values was about 6 days (ATKINS et al., 1972 b).

3. Metal Content

Zinc has been found to be associated with carbonic anhydrases from pea, lettuce, parsley, *Tradescantia*, and spinach (TOBIN, 1970; ATKINS et al., 1972 b; KISIEL and GRAF, 1972; POCKER and NG, 1973; WALK, 1975). There is also one claim that it is not associated with spinach carbonic anhydrase (ROSSI et al., 1969). The minimal protein/zinc ratio found in these purified enzymes is about 30,000 g protein per mol zinc, or 6 mol zinc per mol enzyme. Attempts to remove zinc without denaturing the protein were unsuccessful (TOBIN, 1970). Zinc deficiency in plants appears to correlate with carbonic anhydrase deficiency, which can be alleviated by applications of foliar sprays containing zinc (BAR-AKIVA et al., 1971; EDWARDS and MOHAMED, 1973; RANDALL and BOUMA, 1973).

4. Isoenzymes

Carbonic anhydrase isoenzymes have been found in a number of plants (ATKINS et al., 1972 a, 1972 b; KACHRU and ANDERSON, 1974; KOSITSIN and

KHALIDOVA, 1974; WALK, 1975). Also no isoenzymes have been observed for some of these same plants (ROSSI et al., 1969; TOBIN, 1970; KISIEL and GRAF, 1972; POCKER and NG, 1973). In the case of pea the two isoenzymes have isoelectric points of 5.75 and 6.30; these forms are thought to be cytoplasmic and chloroplastic, respectively (KACHRU and ANDERSON, 1974). More research will be necessary to determine whether carbonic anhydrase isoenzymes exist or are artifactual.

5. Physical Constants

Kinetic studies with carbonic anhydrase are most extensive for the enzyme from parsley and spinach (TOBIN, 1970; POCKER and NG, 1973). Some selected values at specific conditions for the catalytic and Michaelis constants (k_3 and K_m) of carbonic anhydrase from these and other plants are shown in Table 2.

The catalytic and Michaelis constants (CO_2 hydration) of parsley carbonic anhydrase, extrapolated to zero buffer concentration, show an upward trend with increasing pH from 6.5 to 7.5. The specific activity of the enzyme is independent of enzyme concentration between concentrations of 4–160 nM. The enzymic velocities with CO_2 concentrations from 3 to 16 mM at pH 6.5–7.5 are consistent with a normal Michaelis-Menten relationship. The data indicated that the basic form of a group with a pK about 7 participates in catalysis (TOBIN, 1970).

For spinach (POCKER and NG, 1973) the enzyme–substrate interaction seems to follow a Michaelis-Menten mechanism with a Hill constant of 1.02. This suggests that one CO_2 molecule binds to each enzyme molecule or to each of the active sites of the enzyme molecule, provided that the active sites are noninteracting with each other. Extrapolation to zero buffer concentration showed the K_m is constant for CO_2 hydration at 25° C over the pH range 6.0–9.0. The pH dependence of the catalytic constant suggests the involvement in catalysis of a basic group of pK 7.7.

6. Inhibitors

Acetazolamide, ethoxyzolamide, and azide inhibit carbonic anhydrase activity of pea, *Tradescantia*, spinach, and parsley (EVERSON, 1970; TOBIN, 1970; ATKINS et al., 1972b; KISIEL and GRAF, 1972; SWADER and JACOBSON, 1972; JACOBSON et al., 1975). At concentrations as low as 3.0–50.0 µM 50% inhibition has been reported (EVERSON, 1970; TOBIN, 1970; ATKINS et al., 1972b; JACOBSON et al., 1975). Other sulfonamides, particularly Diamox (5-acetamido-1,3,4-thiadiazole-2-sulfonamide), have been used as inhibitors of carbonic anhydrase in higher plants, *Chlorella*, and blue–green algae (EVERSON, 1970; GRAHAM and REED, 1971; GRAHAM et al., 1971; INGLE and COLMAN, 1975, 1976). Arsenite, nitrite, nitrate, iodide, chloride, Hg^{2+}, and Mg^{2+}, have been reported as inhibitors (ROSSI et al., 1969; EVERSON, 1970; ATKINS et al., 1972b; KISIEL and GRAF, 1972). With spinach, $H_2PO_4^-$ inhibits at pH 6.0–7.4 and HPO_4^{2-} stimulates at alkaline pH, suggesting that phosphate buffer should be avoided in studies involving carbonic anhydrase (POKER and NG, 1973). Polyvinylpyrrolidone-40, a phenol-trapping reagent, also inhibits carbonic anhydrase (POINCELOT, 1972b).

Many of these inhibitors also inhibit reactions required for photosynthesis (EVERSON, 1969, 1970; GRAHAM and REED, 1971; SWADER and JACOBSON, 1972;

Table 2. Physical constants of carbonic anhydrase

	Catalytic Constant of Hydration (k_3)	Michaelis Constant of Hydration (K_m)	Michaelis Constant of Dehydration (K_m)	Conditions		Reference
	s^{-1}	mM	mM	pH	°C	
Cotton	–	–	11	7.0	9	CHANG, 1975b
Parsley	$5.5 \pm 1.8^a \times 10^4$	11[a]	–	7.32	25	TOBIN, 1970
Pea	–	–	30	7.0	20	KISIEL and GRAF, 1972
	–	31 ± 8, 34 ± 6[b]	–	7.4	0	KACHRU and ANDERSON, 1974
Spinach	5.32×10^5	1.5[a]	–	9.0	25	POCKER and NG, 1973
	–	22	50	7.2	19	ROSSI et al., 1969

[a] Extrapolated to zero buffer concentration.
[b] Cytoplasmic and chloroplastic forms.

WERDAN and HELDT, 1972; JACOBSON et al., 1975), thus making it difficult to separate the effects of inhibition of carbonic anhydrase from the inhibition of photosynthesis. With azide, nitrite, magnesium, and ethoxyzolamide it was suggested that the decrease in CO_2 fixation by intact chloroplasts is a direct result of their inhibition of carbonic anhydrase activity (BAMBERGER and AVRON, 1975; JACOBSON et al., 1975).

The inhibitory action of Diamox on carbonic anhydrase is the basis for its recent use in affinity chromatography for partial purification of carbonic anhydrase (CHAMPAGNOL, 1976b).

7. Regulation

Light, CO_2, and zinc levels affect the biosynthesis of carbonic anhydrase. High CO_2 concentrations (three times ambient level) cause decreased activity of carbonic anhydrase in cotton plants for leaves 15 days old (25% loss) and 25 days old (75% loss), but young leaves (5–10 days) remained unaffected (CHANG, 1975c). However, compared to plants grown in air, maize and pea showed little change in carbonic anhydrase levels at 10% CO_2 (GRAHAM et al., 1971), but in another study showed a loss (about 30%) for pea and a gain (about 20%) for maize at 0.06% CO_2 (CERFIGNI et al., 1971). Similarly, 3% CO_2, 5% CO_2, and 1.5% CO_2 by volume in air decreased the activity of carbonic anhydrase in blue–green algae (DOHLER, 1974; INGLE and COLMAN, 1975, 1976), *Chlamydomonas* and *Chlorella* (GRAHAM and REED, 1971; GRAHAM et al., 1971), and *Scenedesmus* (FINDENEGG, 1974), respectively. Losses exceeded 90%. Carbonic anhydrase activity appears to be correlated with irradiance during growth of *Lolium*, as the activity increased about 25% on a dry weight basis and 100% on a leaf area basis when the irradiance went from 15 to 110 W/m² (REYSS and PRIOUL, 1975). Zinc deficiency results in a reduction of carbonic anhydrase in a number of plants (BAR-AKIVA et al., 1971; EDWARDS and MOHAMED, 1973; RANDALL and BOUMA, 1973).

C. Function

I. Chloroplast Envelope Membrane Permease

Various functions have been suggested for the role of carbonic anhydrase in plants, but none have been proven nor disproven completely. Carbonic anhydrase has been suggested to facilitate CO_2/HCO_3^- movement through the chloroplast envelope membrane (EVERSON and SLACK, 1968; CHEN et al., 1970; GRAHAM and REED, 1971; GRAHAM et al., 1971) based on model systems. However, the evidence for the absence of carbonic anhydrase in the chloroplast envelope membrane is such (see Sect. III) that this function appears unlikely.

II. Carbonic Anhydrase – RuBP Carboxylase Complex

The function of stromal carbonic anhydrase in C_3 plants is complicated by the uncertainty of which molecular species, CO_2 or HCO_3^-, crosses the membrane. The rates of transport of HCO_3^- (POINCELOT, 1974; POINCELOT and DAY, 1976) and CO_2 (WERDAN and HELDT, 1972) are both more than adequate to account for observed rates of photosynthesis.

Once past the envelope membrane, the proportion of molecular species is regulated by pH i.e., at pH 5.0, 98% would be present as CO_2, whereas at pH 8.0, 95% would appear as HCO_3^-. The pH of the stroma of illuminated spinach chloroplasts is close to pH 8.0 (WERDAN and HELDT, 1972), so if CO_2 is the entering species, the stromal carbonic anhydrase would rapidly convert it to HCO_3^-. If HCO_3^- enters, there is no further conversion at this point.

Since the active species for RuBP carboxylase is CO_2, it has been suggested that carbonic anhydrase is in close association with RuBP carboxylase and acts to increase the availability of CO_2 at the site of carboxylation (EVERSON and SLACK, 1968; GRAHAM and REED, 1971; POINCELOT, 1972a; WERDAN and HELDT, 1972). Similarly, in C_4 plants carbonic anhydrase may be associated with PEP carboxylase, and act to increase availability of HCO_3^- at the site of carboxylation (RATHNAM and DAS, 1975). These views are supported by the concomitant loss of RuBP carboxylase and carbonic anhydrase from spinach chloroplasts, similar distribution of carbonic anhydrase and PEP carboxylase in *Eleusine coracana*, and an overlap of isoelectric focusing patterns of pea carbonic anhydrase and RuBP carboxylase which suggest similar net charges (POINCELOT, 1972a; KACHRU and ANDERSON, 1974; RATHNAM and DAS, 1975). Mathematical calculations of the affinities of carbonic anhydrase and RuBP carboxylase for CO_2, based on K_m values, can either support or fault this view, since the literature values for K_m vary considerably (see Sect. 2.6.5, and Table 2).

III. Proton Source, Buffering Capacity and Ionic Flux Regulation

Other suggested functions of carbonic anhydrase have little evidence to support them at present. The reaction catalyzed by carbonic anhydrase is a source of protons and as such may be responsible for the proton gradient associated with the thylakoids and photophosphorylation (GRAHAM and REED, 1971; GRAHAM et al., 1971; TRIOLO et al., 1974). Alternately it may act in a buffering capacity to mediate localized pH jumps resulting from proton pumping and/or CO_2 incorporation in RuBP (GRAHAM and REED, 1971; JACOBSON et al., 1975). Finally, some evidence suggests it may be involved in regulation of ionic fluxes of HCO_3^- and Cl^- in the chloroplast (GRAHAM and REED, 1971; FINDENEGG, 1974a, 1974b).

References

Atkins, C.A.: Phytochemistry *13*, 93–98 (1974)
Atkins, C.A., Patterson, B.D., Graham, D.: Plant Physiol. *50*, 214–217 (1972a)
Atkins, C.A., Patterson, B.D., Graham, D.: Plant Physiol. *50*, 217–223 (1972b)
Bamberger, E.S., Avron, M.: Plant Physiol. *56*, 481–485 (1975)
Bar-Akiva, A., Gotfried, A., Lavon, R.: J. Hortic. Sci. *46*, 397–401 (1971)
Cerfigni, T., Teofani, F., Bassanelli, C.: Phytochemistry *10*, 2991–2994 (1971)
Champagnol, F.: C.R. Acad. Sci. Paris *289*, 1273–1275 (1976a)
Champagnol, F.: J. Chromatogr. *120* (2), 489–490 (1976)
Chang, C.W.: Phytochem. *14*, 119–121 (1975a)
Chang, C.W.: Plant Sci. Lett. *4*, 109–113 (1975b)
Chang, C.W.: Plant Physiol. *55*, 515–519 (1975c)
Chen, T.M., Brown, R.H., Black, C.C.: Weed Sci. *18*, 399–403 (1970)
Döhler, G.: Planta (Berl.) *117*, 97–99 (1974)
Edwards, G.E., Mohamed, A.K.: Crop Sci. *13*, 351–354 (1973)
Everson, R.G.: Nature (London) *222*, 876 (1969)
Everson, R.G.: Phytochemistry *9*, 25–32 (1970)
Everson, R.G., Slack, C.R.: Phytochemistry *7*, 581–584 (1968)
Findenegg, G.R.: Planta (Berl.) *116*, 123–131 (1974a)
Findenegg, G.R.: In: Membrane transport in plants. Zimmerman, U., Dainty, J. (eds.), pp. 192–196. Berlin, Heidelberg, New York: Springer 1974b
Graham, D., Reed, M.L.: Nature (London) *231*, 81–82 (1971)
Graham, D., Smillie, R.M.: Aust. J. Plant Physiol. *3*, 113–119 (1976)
Graham, D., Atkins, C.A., Reed, M.L., Patterson, B.D., Smillie, R.M.: In: Photosynthesis and photorespiration. Hatch, M.D., Osmond, C.B., Slatyer, R.O. (eds.), pp. 267–274. New York, London, Sydney, Toronto: John Wiley & Sons, Inc. 1971
Ingle, R.K., Colman, B.: Can J. Bot. *53*, 2385–2387 (1975)
Ingle, R.K., Colman, B.: Planta (Berl.) *128*, 217–223 (1976)
Jacobson, B.S., Fong, F., Heath, R.L.: Plant Physiol. *55*, 468–474 (1975)
Kachru, R.B., Anderson, L.E.: Planta (Berl.) *118*, 235–240 (1974)
Kisiel, W., Graf, G.: Phytochemistry *11*, 113–117 (1972)
Kositsin, A.V., Khalidova, G.B.: Fiziol. Rast. *21*, 1178–1181 (1974)
Lamb, J.E.: Life Sci. *20* (3), 393–406 (1977)
Lindskog, S., Henderson, L.E., Kannan, K.K., Liljas, A., Nyman, P.O., Strandberg, B.: In: The enzymes. Boyer, P.D. (ed.), Vol. V, pp. 587–665. New York: Academic Press 1971
Patterson, B.D., Atkins, C.A., Graham, D., Wills, R.B.H.: Anal. Biochem. *44*, 388–391 (1971)
Pocker, Y., Ng, J.S.Y.: Biochemistry *12*, 5127–5134 (1973)
Poincelot, R.P.: Biochem. Biophys. Acta. *258*, 637–642 (1972a)

Poincelot, R.P.: Plant Physiol. *50*, 336–340 (1972b)
Poincelot, R.P.: Arch. Biochem. Biophys. *159*, 134–142 (1973)
Poincelot, R.P.: Plant Physiol. *54*, 520–526 (1974)
Poincelot, R.P.: Plant Physiol. *60*, 767–770 (1977)
Poincelot, R.P., Day, P.R.: Plant Physiol. *57*, 334–338 (1976)
Randall, P.J., Bouma, D.: Plant Physiol. *52*, 229–232 (1973)
Rathnam, C.K.M., Das, V.S.R.: Z. Pflanzenphysiol. *75*, 360–364 (1975)
Reyss, A., Prioul, J.L.: Plant Sci. Lett. *5*, 189–195 (1975)
Rossi, C., Chersi, A., Cortivo, M.: In: CO_2: chemical, biochemical and physiological aspects.
    Forster, R.E., Edsall, J.T., Otis, A.B., Roughton, F.J.W. (eds.), pp. 131–138. Washington,
    D.C.: Office of Technology Utilization, National Aeronautics and Space Administration
    1969
Swader, J.A., Jacobson, B.S.: Phytochemistry *11*, 65–70 (1972)
Tobin, A.J.: J. Biol. Chem. *245*, 2656–2666 (1970)
Triolo, L., Bagnara, D., Anselmi, L., Bassanelli, C.: Physiol. Plant *31*, 86–89 (1974)
Walk, R.A., Metzner, H.: Z. Physiol. Chem. *356*, 1733–1741 (1975)
Waygood, E.R.: In: Methods in enzymology. Colowick, S.P., Kaplan, N.O. (eds.), Vol. II,
    pp. 836–845. New York: Academic Press 1955
Werdan, K., Heldt, H.W.: Biochem. Biophys. Acta *283*, 430–441 (1972)
Yokota, S., Waller, W.K.: Albrecht V. Graefes Arch. Klin. Exp. Ophthal. *197*, 145–152
    (1975)

19. Enzymes of the Reductive Pentose Phosphate Cycle

E. Latzko and G.J. Kelly

A. Introduction

Thirteen enzyme-catalyzed reactions constitute the Calvin cycle of CO_2 fixation (Fig. 1). Evidence for all these reactions was obtained in the 1950's, and in the last two decades each reaction has been studied in greater or lesser detail, emphasis being placed on the enzyme involved. Much has been learnt, but it might not be an overstatement to comment that the learning process has been, and still is, associated with a considerable amount of discussion over which of the observed molecular, kinetic, and regulatory properties of each enzyme are relevant to in vivo conditions in the chloroplast. The purpose of this chapter is to prepare, as far as is possible, a comprehensive picture of each Calvin cycle enzyme on the basis of investigations reported up to the beginning of 1978. The one omission is RuBP carboxylase, which is treated separately by T. Akazawa in Chapter II. 17 of this volume. In addition, light-mediated activation is only superficially considered since L.E. Anderson evaluates this aspect in Chapter II. 22.

B. Characteristics of Regulatory Enzymes

Several attributes are generally associated with regulatory enzymes, and should, as far as possible, considered together when evaluating the extent to which any one enzyme might govern the flux through a metabolic pathway such as the Calvin cycle. First, the dependence of enzyme activity on substrate concentration commonly departs from Michaelis-Menten kinetics, resulting in nonlinear double-reciprocal (Lineweaver-Burk) plots. Second, it is not unusual to find that the enzyme is composed of subunits, a phenomenon often associated with non-Michaelis-Menten kinetics. Third, the enzyme activity is often allosterically regulated, i.e., specifically affected by low concentrations of certain key metabolites which have no structural resemblance to the substrates for the enzyme. Fourth, an enzyme is more likely to have a regulatory role when its activity is of the same order as the rate of flux through the metabolic sequence in which it participates. Fifth, this relatively low activity can result in the catalyzed reaction being far from equilibrium in vivo, a situation which can be deduced from estimations of the levels of substrates and products in tissues; the detection of nonequilibrium reactions is, in turn, a valuable clue to the identification of regulatory enzymes (see Turner and Turner, 1975). Sixth, many reactions catalyzed by regulatory enzymes, including regulated steps of the Calvin cycle (Bassham and Krause, 1969), have

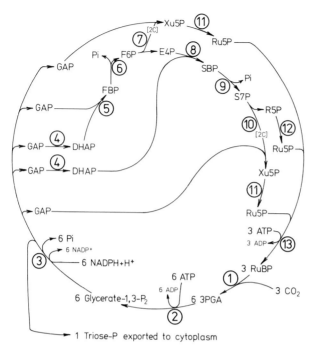

Fig. 1. Enzymes of the Calvin cycle of photosynthetic CO_2 fixation. Net synthesis of one triose-P from three molecules of CO_2 is shown. *Circled numbers* designate the participating enzymes: *1*, RuBP carboxylase; *2*, 3PGA kinase; *3*, GAP dehydrogenase; *4*, triose-P isomerase; *5*, FBP-aldolase; *6*, FBPase; *7*, transketolase; *8*, SBP-aldolase; *9*, SBPase; *10*, transketolase; *11*, pentose-P epimerase; *12*, pentose-P isomerase; *13*, Ru5P kinase. *Arrows* are in the direction of CO_2 fixation; in fact all reactions except *1*, *6*, *9*, and *13* are physiologically reversible. Additional abbreviations: *[2C]*, glycolaldehyde/transketolase complex; *F6P*, fructose-6-P

high free energy changes and are practically irreversible under physiological conditions. Finally, regulatory enzymes are usually strategically located near branch points in metabolic pathways and often catalyze the first committed step in a pathway, thereby determining the flux through that sequence.

C. Activities and Location of Calvin Cycle Enzymes

In early investigations, Peterkofsky and Racker (1961) and Latzko and Gibbs (1969) were able to confirm that, in both prokaryotic and eukaryotic organisms, the activities of most of the enzymes of the Calvin cycle were sufficient to accommodate observed rates of photosynthesis. However, the measured activities of RuBP carboxylase, FBPase, and SBPase were too low to support high rates of CO_2 fixation. Recent improvements in extraction and assay techniques have since enabled sufficient levels of RuPB carboxylase and FBPase to be demonstrated (Gar-

NIER and LATZKO, 1972; LILLEY and WALKER, 1975). SBPase remains the one enzyme for which a suitably high activity has not yet been consistently detected.

It is generally accepted that, in eukaryotes, the Calvin cycle operates only in the chloroplast stroma. This view is encouraged by the detection of all Calvin cycle enzymes in soluble extracts from isolated chloroplasts, and the restriction of RuBP carboxylase, NADP-GAP dehydrogenase, "alkaline" FBPase, and Ru5P kinase to these organelles (SMILLIE, 1963; LATZKO and GIBBS, 1968). Furthermore, with the exception of SBPase, reported enzyme activities were compatible with high rates of CO_2 fixation (LATZKO and GIBBS, 1968; LILLEY and WALKER, 1975).

C_4 plants constitute one major exception to these observations. Mesophyll cells from the leaves of C_4 plants contain chloroplasts reported to lack certain Calvin cycle enzymes (most notably RuBP carboxylase, "alkaline" FBPase, and Ru5P kinase) and the ability to achieve CO_2 fixation; however, other enzymes (including NADP-GAP dehydrogenase) are clearly present (see HATCH and OS-MOND, 1976). The presence of NADP-GAP dehydrogenase is in one sense unusual, since in all other instances this enzyme is associated with a functioning Calvin cycle.

D. Glycerate-3-P Kinase

Carboxylation of RuBP in the Calvin cycle produces two molecules of 3PGA which are phosphorylated to form glycerate-1,3-P_2 in a reaction (reaction 2, Fig. 1) catalyzed by 3PGA kinase. This enzyme differs from most kinases in catalyzing a reversible reaction. The pea chloroplast 3PGA kinase has been separated from its cytoplasmic counterpart by PACOLD and ANDERSON (1975) who found that the properties of the two enzymes were almost identical. With the chloroplast enzyme operating in the photosynthetic direction, strong inhibition by ADP (competitive with ATP; $K_i = 53$ μM) was observed, and related to earlier experiments (PACOLD and ANDERSON, 1973; LAVERGNE et al., 1974) which demonstrated the potential for energy charge regulation (i.e., activity controlled by the extent of phosphorylation of the adenylate pool). 3PGA kinase has been purified to homogeneity from the leaves of spinach (LAVERGNE and BISMUTH, 1973) and Beta vulgaris (CAVELL and SCOPES, 1976; see Table 1).

E. Glyceraldehyde-3-P Dehydrogenase

Reducing power from photosynthesis, in the form of NADPH, is utilized in the chloroplast to reduce glycerate-1,3-P_2 to GAP in reaction 3 (Fig. 1) which involves a GAP dehydrogenase able to use NADP as coenzyme. Apart from RuBP carboxylase, this enzyme has received more attention than any other of the Calvin cycle. Originally demonstrated in leaves by GIBBS (1952) and ARNON (1952) and shown present in several photosynthetic bacteria and algae (FULLER and GIBBS, 1959),

Table 1. Representative examples of purified preparations of enzymes of the reductive pentose phosphate cycle. Numbers before each enzyme correspond to the circled numbers shown in Figure 1

Enzyme	Plant	Molecular weight	Specific activity μmol $\min^{-1}$ mg protein^{-1}	References
1 RuBP carboxylase	see Chapter II.17			
2 3PGA kinase	*Beta vulgaris*	48,000	900	CAVELL and SCOPES, 1976
3 GAP dehydrogenase	*Pisum sativum*	140,000	130	McGOWAN and GIBBS, 1974
	Spinacia oleracea	145,000	39	PUPILLO and GIULIANI PICCARI, 1973, 1975
	Scenedesmus obliquus	140,000	25	O'BRIEN and POWLS, 1976
4 Triose-P isomerase	*Pisum sativum*	–	–	ANDERSON, 1971a
5 FBP-aldolase	*Spinacia oleracea*	142,000	44	BROOKS and CRIDDLE, 1966
	Anacystis nidulans	137,000	128	WILLARD and GIBBS, 1968b
6 FBPase	*Spinacia oleracea*	160,000	109	ZIMMERMANN et al., 1976
	Rhodopseudomonas palustris	130,000	8.8	SPRINGGATE and STACHOW, 1972a
7 Transketolase	*Spinacia oleracea*	–	152	HORECKER et al., 1956
8 SBP aldolase	*Spinacia oleracea*	142,000	25	BROOKS and CRIDDLE, 1966
9 SBPase	*Spinacia oleracea*	72,500	5.0	BUCHANAN et al., 1976
10 Transketolase	see 7 (above)			
11 Pentose-P epimerase	*Spinacia oleracea*	–	–	LATZKO and GIBBS, 1968
12 Pentose-P isomerase	*Spinacia oleracea*	53,200	2171	RUTNER, 1970
13 Ru5P kinase	*Spinacia oleracea*	46,000	320	LAVERGNE and BISMUTH, 1973
	Hydrogenomonas eutropha	237,000	89	ABDELAL and SCHLEGEL, 1974

NADP-GAP dehydrogenase activity increases substantially after plants are exposed to light (HAGEMAN and ARNON, 1955; BRAWERMAN and KONIGSBERG, 1960). This development of activity is independent of chlorophyll (FULLER and GIBBS, 1959), but rather appears to be mediated by the phytochrome system (MARCUS, 1960; MARGULIES, 1965).

In 1963 HEBER et al. (1963) and SMILLIE (1963) simultaneously reported that NADP-GAP dehydrogenase is restricted to chloroplasts, thus substantiating the initial proposal (GIBBS, 1952) that the enzyme participates only in photosynthesis. This view is further supported by the relatively greater affinity for NADPH and glycerate-1,3-P_2 as compared to NADP and GAP (SCHULMAN and GIBBS, 1968; GRISSOM and KAHN, 1975). However, chloroplasts also contain NAD-GAP dehydrogenase activity, and this fact, coupled with the presence of an NAD-GAP dehydrogenase in the cytoplasm, has complicated further investigations of the chloroplast enzyme. The cytoplasmic enzyme can be largely separated from that of the chloroplasts by ammonium sulfate fractionation (SCHULMAN and GIBBS, 1968; MELANDRI et al., 1970), but further purification of the chloroplast enzyme has not yet produced appreciable amounts of NADP-specific enzyme. The failure to separate NAD-linked activity from the NADP-linked activity of either spinach leaves or pea shoots led SCHULMAN and GIBBS (1968) to propose that these two activities in the chloroplast belong to the same enzyme molecule (or at least to almost identical molecules). Other investigators have arrived at essentially the same conclusion for the chloroplast enzyme from *Euglena gracilis* (VACCHI et al., 1973; THEISS-SEUBERLING, 1974), *Sinapis alba* (CERFF and QUAIL, 1974), as well as spinach (YONUSCHOT et al., 1970) and peas (MELANDRI et al., 1968; McGOWAN and GIBBS, 1974), and for the NADP-linked activity in the prokaryote *Anabaena variabilis* (HOOD and CARR, 1969). The function of the NAD-linked activity is not clear, although it may well participate in the glycolytic breakdown of chloroplast starch. It is noteworthy that the affinities of the spinach chloroplast enzyme for NAD and NADH are five to ten times lower than those for NADP and NADPH (PAWLIZKI and LATZKO, unpublished). Questions are raised by the conclusions of PUPILLO (1972) and GRISSOM and KAHN (1975) that the GAP dehydrogenases from the photosynthetic flagellates *Euglena* and *Ochromonas* are specific for NADP. In contrast to the oxygen-evolving photosynthetic organisms, anaerobic photosynthetic and chemosynthetic bacteria completely lack NADP-GAP dehydrogenase, but contain an NAD-linked enzyme presumably used in the Calvin cycle (SMILLIE and FULLER, 1960).

Like most Calvin cycle enzymes (BRADBEER, 1969), NADP-GAP dehydrogenase activity increases over a period of hours upon illumination of dark-grown plants; even a brief exposure to white or red light is sufficient to raise the enzyme level (see above). It has been questioned whether this enzyme synthesis involves a conversion of the cytoplasmic, NAD-specific GAP dehydrogenase into the chloroplast enzyme. Reciprocal changes of the NADP- and total NAD-linked activities in greening algal cells were consistent with this proposal for *Chlamydomonas*, but not for *Euglena* (BRAWERMAN and KONIGSBERG, 1960; HUDOCK and FULLER, 1965). More recently, CERFF and QUAIL (1974) followed changes in the two NAD-linked activities (cytoplasmic and chloroplast enzymes) and the NADP-linked activity (chloroplast enzyme) in the cotyledons of *Sinapis alba* grown under continuous far-red light and found that a rise in NADP-linked activity was quantitatively similar to a fall in that part of the NAD-linked activity attributable to the cytoplasmic enzyme. CERFF (1974) favors conversion of the cytoplasmic enzyme into the chloroplast enzyme as an explanation for these results. Quite the opposite conclusion was reached by McGOWAN and GIBBS (1974) after finding that, according to immunological criteria, pea and spinach chloroplast GAP dehydrogenases are structurally similar to each other but most dissimilar to pea root, pea seed, and pea leaf cytoplasmic GAP dehydrogenases. These authors consider that the two leaf GAP dehydrogenases are transcribed from different structural genes.

Interest in chloroplast GAP dehydrogenase was stimulated when ZIEGLER and ZIEGLER (1965) reported two- to eightfold increases in the NADP-linked (but not NAD-linked) activity of the enzyme over a period of 20 min in green leaves following transfer from darkness to illumination. This light-mediated activation, originally seen in ten species of the Angiospermae [but not yet observed in *Euglena* (THEISS-

SEUBERLING, 1974) or in some C_4 plants (STEIGER et al., 1971)], was absent in a chlorophyll-less *Pisum sativum* mutant (ZIEGLER and ZIEGLER, 1966). MÜLLER et al. (1969) subsequently reported that the enzyme in isolated chloroplasts was also activated following illumination, and that a similar activation was possible in vitro by incubating chloroplast extracts with NADPH; the curve of enzyme activity versus NADPH concentration was sigmoid, the most marked activation occurring between 0.5 and 1.5 mM NADPH. MÜLLER (1970) found ATP at 1 to 6 mM similarly activated the enzyme, and attributed the light-mediated activation in vivo to comparable effects of NADPH and ATP made available during photosynthesis. However, other mechanisms for achieving this activation are also under consideration (see Chap. II.22, this vol.). Nevertheless, effects of NADPH and ATP, including their ability to influence the extent to which the enzyme polymerizes (see below), might still be relevant to the light activation experiments.

The majority of reports indicate that chloroplast GAP dehydrogenase has a molecular weight of about 140,000 (PUPILLO and GIULIANI PICCARI, 1973; VACCHI et al., 1973; THEISS-SEUBERLING, 1974; McGOWAN and GIBBS, 1974; GRISSOM and KAHN, 1975; SCHWARZ et al., 1976; O'BRIEN and POWLS, 1976; see Table 1), but in these and other reports (YONUSCHOT et al., 1970; PAWLIZKI and LATZKO, 1974; WOLOSIUK and BUCHANAN, 1976), higher molecular weight values consistent with the association of two or four of the 140,000 molecular weight species were frequently mentioned. The enzyme is composed of subunits of molecular weight about 35,000 (McGOWAN and GIBBS, 1974; GRISSOM and KAHN, 1975; O'BRIEN and POWLS, 1976). The observation by PAWLIZKI and LATZKO (1974) that the spinach enzyme is composed of dissimilar subunits has received further support (CERFF, 1978; FERRI et al., 1978).

Of particular interest are the reports by PUPILLO and GIULIANI PICCARI (1973, 1975) describing dissociation of a GAP dehydrogenase polymer of molecular weight 600,000 into molecular weight 140,000 units by NADP, NADPH, or ATP. The dissociation by NADP was accompanied by a greatly increased affinity for this co-factor, although the catalytic activity measured with saturating levels of substrate was unchanged. As low as 30 µM NADP was sufficient to cause dissociation with low protein concentrations during gel filtration, but if 100 µM NAD was also included, the polymer remained intact. Activation of enzyme by several minutes preincubation with NADP, NADPH or ATP was observed (as also reported elsewhere; MÜLLER, 1970; WOLOSIUK and BUCHANAN, 1976) and correlated with a depolymerization, apparently from 600,000 to 140,000 molecular weight (PUPILLO and GIULIANI PICCARI, 1975). These authors noted that "...data are consistent with the assumption that the tetramers do not act themselves as catalysts for the NADP(H)-dependent activity, which arises only after depolymerization to protomers" (i.e., 140,000 molecular weight). Recently, dissociation of 600,000 molecular weight units to 140,000 molecular weight units of GAP dehydrogenase was reported to be caused by NADP with the *Euglena* enzyme (THEISS-SEUBERLING, 1974), and by cysteine plus a glycerate-1,3-P_2 generating system with the enzyme from *Scenedesmus* (and apparently also from *Chlorella* and *Chlamydomonas*, but not from higher plants; O'BRIEN et al., 1976); in this latter case dissociation was accompanied by a tenfold increase of the catalytic activity with NADPH.

It is worth noting that chloroplast GAP dehydrogenase binds NADP rather tightly (SCHULMAN and GIBBS, 1968; FERRI et al., 1978), and McGOWAN and GIBBS

(1974) have evidence that the strength of this binding dictates whether or not the nucleotide might be removed during purification, an event which probably contributes to polymerization. In addition to NADP, a binding fraction (in which a polypeptide similar to the enzyme subunit predominates) is suspected by CERFF (1978) to be involved in enzyme aggregation. One possible explanation for the tendency to polymerize is inherent in the observation by GRISSOM and KHAN (1975) that the exterior of *Euglena* chloroplast GAP dehydrogenase is probably hydrophobic.

F. Triose-P Isomerase and Aldolase

Triose-P isomerase is required to produce dihydroxyacetone-P from GAP (reaction 4, Fig. 1). ANDERSON (1971a) reported that the triose-P isomerase in chloroplasts is practically identical to the cytoplasmic enzyme, and that the chloroplast enzyme activity is competitively inhibited by P-enolpyruvate (K_i 1.3 mM), RuPB (K_i 0.56 mM), and most noticeably by glycolate-2-P (K_i 15 µM).

The Calvin cycle requires aldolase activity to condense dihydroxyacetone-P with GAP to form FBP (reaction 5, Fig. 1), and with erythrose-4-P to form SBP (reaction 8, Fig. 1). In 1966 BROOKS and CRIDDLE (1966) followed the activity of aldolase toward FBP and SBP during purification of FBP-aldolase from spinach chloroplasts and found that SBP-aldolase was purified in a parallel fashion. These authors concluded that the same enzyme catalyzes both FBP and SBP formation in the Calvin cycle. The two activities also coincided when chloroplast extracts were electrofocused by ANDERSON and PACOLD (1972). Values of V_{max} and K_m for the enzymic cleavage of FBP and SBP are of the same order of magnitude (BROOKS and CRIDDLE, 1966; ANDERSON and PACOLD, 1972) but kinetic constants for the reactions in the photosynthetic direction have not been reported.

Photosynthetic organisms contain three distinct types of aldolase. Certain green, red, and golden-brown algae, *Rhodospirillum rubrum*, and spinach contain aldolase (FEWSON et al., 1962) which does not require a divalent metal ion for activity. The enzyme in *Euglena*, unlike the spinach enzyme, is inhibited by EDTA and stimulated by KCl (LATZKO and GIBBS, 1969). Other organisms, such as blue-green algae (WILLARD and GIBBS, 1968a, b), and *Chromatium* (LATZKO and GIBBS, 1969), contain aldolase which requires cysteine and a divalent metal ion such as Fe^{2+}. The enzyme purified from *Anacystis nidulans* required a 10-min preincubation with cysteine and Fe^{2+} for maximal activity, and was found to cleave both FBP and SBP at comparable rates (WILLARD and GIBBS, 1968b). Interestingly, the green alga *Chlamydomonas mundana*, which contains the metal-independent aldolase of other green algae and higher plants when growing photosynthetically, synthesizes in addition the Fe^{2+}-dependent aldolase when adapted to media containing acetate (RUSSELL and GIBBS, 1967); in this instance the Fe^{2+}-dependent enzyme functions in glycolysis, rather than in photosynthetic CO_2 fixation as is the case for the *Anacystis* enzyme.

G. Fructosebisphosphatase and Sedoheptulosebisphosphatase

FBP and SBP formed by aldolase in the Calvin cycle are subsequently hydrolyzed at the C1 positions by specific phosphatases (FBPase and SBPase) to produce, respectively, fructose-6-P (reaction 6, Fig. 1) and sedoheptulose-7-P (reaction 9, Fig. 1). Most attention has been devoted to FBPase, although SBPase requires further investigation since it is the only Calvin cycle enzyme for which reported activities are too low to support observed rates of CO_2 fixation.

A Mg^{2+}-dependent FBPase with an alkaline optimum pH of 8.5, and inactive below pH 7.8, was described and purified from spinach by RACKER and SCHROEDER (1958), and shown by SMILLIE (1960) to occur in a wide variety of photosynthetic tissues where it is localized in the chloroplast. The first indication that the enzyme has regulatory properties came from PREISS et al. (1967) who found that curves of spinach leaf FBPase activity versus FBP concentration were sigmoid, and in addition that the optimum pH could be shifted from 8.5 to 7.5 by increasing the Mg^{2+} concentration from 5 to 40 mM. These properties have been confirmed for the spinach enzyme in other laboratories (MORRIS, 1968; GARNIER and LATZKO, 1972; EL-BADRY, 1974). The interactions between Mg^{2+} and FBP concentrations, pH, and activation by dithiothreitol (BUCHANAN et al., 1967), were further studied by GARNIER and LATZKO (1972) and BAIER and LATZKO (1975) and related to estimated conditions in the chloroplast stroma in light and darkness. Results were consistent with almost full activity of the enzyme during illumination but no activity in the dark.

Unlike most FBPases, the photosynthetic enzyme is unaffected by AMP (PREISS et al., 1967; SCALA et al., 1968; MORRIS, 1968). The enzyme from spinach chloroplasts (MORRIS, 1968), and that from *Rhodospirillum rubrum* (JOINT et al., 1972a) are inhibited by $MgATP^{2-}$. Chloroplast FBPase has a molecular weight of about 140,000 (SCALA et al., 1968; EL-BADRY, 1974; see Table 1) and consists of four apparently identical subunits (ZIMMERMANN et al., 1976). A characteristic of the enzyme is its spontaneous dissociation at pH 8.8 into two halves which retain catalytic activity (LÁZARO et al., 1974; ZIMMERMANN et al., 1976) and are reported to acquire SBPase activity (BUCHANAN et al., 1976).

In the studies of BAIER and LATZKO (1975), activation by dithiothreitol was more conspicuous at suboptimal pH. Shortly afterwards, ZIMMERMANN et al. (1976) showed that a similar activation was accompanied by a threefold increase in the number of freely available sulfhydryl groups on the enzyme molecule. This reduction might be analogous to the in vivo light-mediated activation which is detected only at suboptimal pH values (KELLY et al., 1976); results consistent with daylight activation of the enzyme (in tapioca leaves) were obtained as early as 1962 (VISWA-NATHAN and KRISHNAN, 1962). Mechanisms of light-dependent reductive activation are considered by ANDERSON (Chap. II.22).

Properties of alkaline FBPases from the photosynthetic microorganisms *Rhodospirillum rubrum* and *Rhodopseudomonas palustris* differ substantially from those of the spinach enzyme. The *R. rubrum* enzyme possesses hyperbolic (rather than sigmoid) curves for FBP, and a response to pH unaffected by Mg^{2+} concentration; interestingly, the enzyme co-purified with Ru5P kinase from this organism (JOINT et al., 1972a, b). FBPase from *R. palustris* (SPRINGGATE and STACHOW, 1972a,

b) is rather similar to FBPase from nonphotosynthetic tissues in that it can also utilize SBP, and is inhibited by high concentrations of substrate (FBP or SBP). However, like the photosynthetic enzyme, it has an optimum pH of 8.5 and easily dissociates into two active halves.

It remains to be clarified whether the principal SBPase of photosynthesis (ANDERSON, 1974) is a separate, specific SBPase or a dissociated FBPase: BUCHANAN et al. (1976) report that SBPase activity appears in chloroplast FBPase after dissociation into two halves at pH 8.5. A similar observation is that *R. palustris* FBPase, which also displays SBPase activity (SPRINGGATE and STACHOW, 1972a), might also be dissociated in reaction mixtures since, even at pH 7.4, the enzyme is split in the presence of FBP (SPRINGGATE and STACHOW, 1972b), and possibly also when SBP is present.

H. Transketolase, Pentose-P Epimerase, and Pentose-P Isomerase

Xylulose-5-P is synthesized in the Calvin cycle by transfer of a 2C (glycolaldehyde) unit from fructose-6-P (reaction 7, Fig. 1) and from sedoheptulose-7-P (reaction 10, Fig. 1) to GAP acceptor molecules. These reactions are catalyzed by transketolase and produce, in addition, the erythrose-4-P required by aldolase for SBP synthesis (see Sect. F), and ribose-5-P from which more CO_2 acceptor (RuBP) is generated (see below). Transketolase has been purified from spinach (HORECKER et al., 1956) and shown to be present in chloroplasts (LATZKO and GIBBS, 1968). Little further information on the chloroplast enzyme is available.

Pentose-Ps produced by transketolase are rearranged into Ru5P by pentose-P epimerase and pentose-P isomerase. The epimerase, which converts xylulose-5-P to Ru5P (reaction 11, Fig. 1) has been identified in chloroplasts (LATZKO and GIBBS, 1968), but not studied further. The formation of Ru5P from ribose-5-P (reaction 12, Fig. 1) is catalyzed by pentose-P isomerase which was orginally purified from alfalfa leaves (ALXELROD and JANG, 1954) and spinach (HURWITZ et al., 1956) and has since been purified from *Rhodospirillum rubrum* (ANDERSON and FULLER, 1969), spinach (RUTNER, 1970), peas (ANDERSON, 1971b), and tobacco (KAWASHIMA and TANABE, 1976). These reports indicate that the isomerase, like 3PGA kinase, is a relatively small enzyme (see Table 1). Competitive inhibition by AMP (K_i about 1.3 mM) was reported for the *R. rubrum* and pea isomerases. The tobacco enzyme rapidly lost activity at 40C, but this loss was prevented by the presence of 0.2% protein and 1 mM Mg^{2+}.

I. Ribulose-5-P Kinase

The Calvin cycle CO_2 acceptor RuBP is generated from Ru5P and ATP (reaction 13, Fig. 1) by a kinase identified and purified from spinach by RACKER (1955,

1957). Hurwitz et al. (1956) showed that this kinase was specific for ATP, and sensitive to sulfhydryl poisons; the latter property was reflected in the later demonstration of light-mediated activation involving a photosynthetically produced reductant (Latzko et al., 1970; Avron and Gibbs, 1974). There are conflicting reports as to whether or not Ru5P kinase from leaves is influenced by energy charge (Anderson, 1973; Lavergne et al., 1974).

Ru5P kinases from photosynthetic bacteria appear to have more regulatory properties than the higher plant enzyme. Following the report that *Thiobacillus thioparus* Ru5P kinase has a sigmoid ATP curve and is allosterically inhibited by AMP (MacElroy et al., 1968), stimulation by NADH was observed with the enzymes from *Hydrogenomonas facilis* (MacElroy et al., 1969) and *Rhodopseudomonas spheroides* (Rindt and Ohmann, 1969) and allosteric inhibition by P-enolpyruvate was found with the enzymes from *Chromatium* (Hart and Gibson, 1971) and *Pseudomonas facilis* (Ballard and MacElroy, 1971). These regulatory properties have since been reported for the *Rhodospirillum rubrum* (Joint et al., 1972b), *Thiobacillus neopolitanus* (MacElroy et al., 1972), and *Hydrogenomonas eutropha* (Abdelal and Schlegel, 1974) Ru5P kinases, and contrast with the lack of effect of AMP or P-enolpyruvate on the spinach leaf (MacElroy et al., 1972) and pea leaf (Anderson, 1973) enzymes. Ru5P kinase has been reported to be completely absent from *Chlorobium thiosulfatophilum* which is believed to achieve net CO_2 fixation by a mechanism other than the Calvin cycle (Buchanan and Sirevåg, 1976). Oddly enough, no algal Ru5P kinase has yet been studied in detail.

J. Concluding Remarks

Two trends become apparent from the present survey. First, the kinetic, regulatory, and molecular properties of each Calvin cycle enzyme vary according to the source of the enzyme, and this variation can be great when the evolutionary distance between two photosynthetic organisms is large. There is, therefore, considerable scope for studies in the comparative biochemistry of the Calvin cycle enzymes. Second, where potential regulatory properties have been observed for a particular enzyme, emphasis has been placed on the elaboration of a role for that enzyme in the control of photosynthetic CO_2 fixation. A cautionary note can be added here, since regulatory properties comprise but two of seven characteristics of a truly rate-determining enzyme (see Sect. B of this Chap.). Chloroplast GAP dehydrogenase might be viewed with this in mind: despite the prolific data available, the conclusion of Heber et al. (1963) that "…TPN-linked triosephosphate dehydrogenase is available in large excess as compared with the maximal efficiency of photosynthesis in the intact leaf" appears as relevant today as when written, and makes it unlikely that this enzyme has a regulatory role in vivo. Even before light-mediated activation, enzyme activity is sufficient to support CO_2 fixation at a rate in excess of 200 μmol per mg chlorophyll per h (Müller, 1970). On the other hand, the likelihood that FBPase can dictate rates of CO_2 fixation is real (see Latzko and Kelly, 1976) since this enzyme possesses most of the characteristics typical of rate-determining enzymes, and indeed recent studies by Portis et al. (1977) indicate a close connection between in vivo activities of FBPase and SBPase (but not GAP dehydrogenase) and rates of CO_2 fixation with intact spinach chloroplasts.

References

Abdelal, A.T.H., Schlegel, H.G.: Biochem. J. *139*, 481–489 (1974)
Anderson, L.E.: Biochim. Biophys. Acta *235*, 237–244 (1971a)
Anderson, L.E.: Biochim. Biophys. Acta *235*, 245–249 (1971b)
Anderson, L.E.: Biochim. Biophys. Acta *321*, 484–488 (1973)
Anderson, L.E.: Biochem. Biophys. Res. Commun. *59*, 907–913 (1974)
Anderson, L.E., Fuller, R.C.: In: Prog. Photosynthetic Res. Metzner, H. (ed.), Vol. III, pp. 1618–1623. Tübingen: Laupp, 1969
Anderson, L.E., Pacold, I.: Plant Physiol. *49*, 393–397 (1972)
Arnon, D.I.: Science *116*, 635–637 (1952)
Avron, M., Gibbs, M.: Plant Physiol. *53*, 136–139 (1974)
Axelrod, B., Jang, R.: J. Biol. Chem. *209*, 847–855 (1954)
Baier, D., Latzko, E.: Biochim. Biophys. Acta *396*, 141–147 (1975)
Ballard, R.W., MacElroy, R.D.: Biochem. Biophys. Res. Commun. *44*, 614–618 (1971)
Bassham, J.A., Krause, G.H.: Biochim. Biophys. Acta *189*, 207–221 (1969)
Bradbeer, J.W.: New Phytol. *68*, 233–245 (1969)
Brawerman, G., Konigsberg, N.: Biochim. Biophys. Acta *43*, 374–381 (1960)
Brooks, K., Criddle, R.S.: Arch. Biochem. Biophys. *117*, 650–659 (1966)
Buchanan, B.B., Kalberer, P.P., Arnon, D.I.: Biochem. Biophys. Res. Commun. *29*, 74–79 (1967)
Buchanan, B.B., Schürmann, P., Wolosiuk, R.A.: Biochem. Biophys. Res. Commun. *69*, 970–978 (1976)
Buchanan, B.B., Sirevåg, R.: Arch. Microbiol. *109*, 15–19 (1976)
Cavell, S., Scopes, R.K.: Eur. J. Biochem. *63*, 483–490 (1976)
Cerff, R.: Z. Pflanzenphysiol. *73*, 109–118 (1974)
Cerff, R.: Plant Physiol. *61*, 369–372 (1978)
Cerff, R., Quail, P.H.: Plant Physiol. *54*, 100–104 (1974)
El-Badry, A.M.: Biochim. Biophys. Acta *333*, 366–377 (1974)
Ferri, G., Comerio, G., Iadarola, P., Zapponi, M.C., Speranza, M.L.: Biochim. Biophys. Acta *522*, 19–31 (1978)
Fewson, C.A., Al-Hafidh, M., Gibbs, M.: Plant Physiol. *37*, 402–406 (1962)
Fuller, R.C., Gibbs, M.: Plant Physiol. *34*, 324–329 (1959)
Ganier, R.V., Latzko, E.: In: Proc. 2nd Int. Congr. Photosynthesis. Forti, G., Avron, M., Melandri, A. (eds.),Vol. III, pp. 1839–1845. The Hague: Dr. Junk 1972
Gibbs, M.: Nature (London) *170*, 164–165 (1952)
Grissom, F.E., Kahn, J.S.: Arch. Biochem. Biophys. *171*, 444–458 (1975)
Hageman, R.H., Arnon, D.I.: Arch. Biochem. Biophys. *57*, 421–436 (1955)
Hart, B.A., Gibson, J.: Arch. Biochem. Biophys. *144*, 308–321 (1971)
Hatch, M.D., Osmond, C.B.: In: Encyclopedia of plant physiology, Stocking, C.R., Heber, U. (eds.), Vol. III, pp. 144–184. Berlin, Heidelberg, New York: Springer 1976
Heber, U., Pon, N.G., Heber, M.: Plant Physiol. *38*, 355–360 (1963)
Hood, W., Carr, N.G.: Planta (Berl.) *86*, 250–258 (1969)
Horecker, B.L., Smyrniotis, P.Z., Hurwitz, J.: J. Biol. Chem. *223*, 1009–1019 (1956)
Hudock, G.A., Fuller, R.C.: Plant Physiol. *40*, 1205–1211 (1965)
Hurwitz, J., Weissbach, A., Horecker, B.L., Smyrniotis, P.Z.: J. Biol. Chem. *218*, 769–783 (1956)
Joint, I.R., Morris, I., Fuller, R.C.: J. Biol. Chem. *247*, 4833–4838 (1972a)
Joint, I.R., Morris, I., Fuller, R.C.: Biochim. Biophys. Acta *276*, 333–337 (1972b)
Kawashima, N., Tanabe, Y.: Plant Cell Physiol. *17*, 765–769 (1976)
Kelly, G.J., Zimmermann, G., Latzko, E.: Biochem. Biophys. Res. Commun. *70*, 193–199 (1976)
Latzko, E., Garnier, R.v., Gibbs, M.: Biochem. Biophys. Res. Commun. *39*, 1140–1144 (1970)
Latzko, E., Gibbs, M.: Z. Pflanzenphysiol. *59*, 184–194 (1968)

Latzko, E., Gibbs, M.: Plant Physiol. *44*, 295–300 (1969)
Latzko, E., Kelly, G.J.: Fortschr. Bot. *38*, 81–99 (1976)
Lavergne, D., Bismuth, E.: Plant Sci. Lett. *1*, 229–236 (1973)
Lavergne, D., Bismuth, E., Champigny, M.L.: Plant Sci. Lett. *3*, 391–397 (1974)
Lázaro, J.J., Chueca, A., Gorgé, J.L., Mayor, F.: Phytochemistry *13*, 2455–2461 (1974)
Lilley, R.McC., Walker, D.A.: Plant Physiol. *55*, 1087–1092 (1975)
MacElroy, R.D., Johnson, E.J., Johnson, M.K.: Biochem. Biophys. Res. Commun. *30*, 678–682 (1968)
MacElroy, R.D., Johnson, E.J., Johnson, M.K.: Arch. Biochem. Biophys. *131*, 272–275 (1969)
MacElroy, R.D., Mack, H.M., Johnson, E.J.: J. Bacteriol. *112*, 532–538 (1972)
Marcus, A.: Plant Physiol. *35*, 126–128 (1960)
Margulies, M.M.: Plant Physiol. *40*, 57–61 (1965)
McGowan, R.E., Gibbs, M.: Plant Physiol. *54*, 312–319 (1974)
Melandri, B.A., Baccarini, A., Pupillo, P.: Biochem. Biophys. Res. Commun. *33*, 160–164 (1968)
Melandri, B.A., Pupillo, P., Baccarini-Melandri, A.: Biochim. Biophys. Acta *220*, 178–189 (1970)
Morris, I.: Biochim. Biophys. Acta *162*, 462–464 (1968)
Müller, B.: Biochim. Biophys. Acta *205*, 102–109 (1970)
Müller, B., Ziegler, I., Ziegler, H.: Eur. J. Biochem. *9*, 101–106 (1969)
O'Brien, M.J., Powls, R.: Eur. J. Biochem. *63*, 155–161 (1976)
O'Brien, M.J., Easterby, J.S., Powls, R.: Biochim. Biophys. Acta *449*, 209–223 (1976)
Pacold, I., Anderson, L.E.: Biochem. Biophys. Res. Commun. *51*, 139–143 (1973)
Pacold, I., Anderson, L.E.: Plant Physiol. *55*, 168–171 (1975)
Pawlizki, K., Latzko, E.: FEBS Lett. *42*, 285–288 (1974)
Peterkofsky, A., Racker, E.: Plant Physiol. *36*, 409–414 (1961)
Portis, A.R., Chon, C.J., Mosbach, A., Heldt, H.W.: Biochim. Biophys. Acta *461*, 313–325 (1977)
Preiss, J., Biggs, M.L., Greenberg, E.: J. Biol. Chem. *242*, 2292–2294 (1967)
Pupillo, P.: Phytochemistry *11*, 153–161 (1972)
Pupillo, P., Giuliani Piccari, G.: Arch. Biochem. Biophys. *154*, 324–331 (1973)
Pupillo, P., Giuliani Piccari, G.: Eur. J. Biochem. *51*, 475–482 (1975)
Racker, E.: Nature (London) *175*, 249–251 (1955)
Racker, E.: Arch. Biochem. Biophys. *69*, 300–310 (1957)
Racker, E., Schroeder, E.A.R.: Arch. Biochem. Biophys. *74*, 326–344 (1958)
Rindt, K.-P., Ohmann, E.: Biochem. Biophys. Res. Commun. *36*, 357–364 (1969)
Russell, G.K., Gibbs, M.: Biochim. Biophys. Acta *132*, 145–154 (1967)
Rutner, A.C.: Biochemistry *9*, 178–184 (1970)
Scala, J., Patrick, C., Macbeth, G.: Arch. Biochem. Biophys. *127*, 576–584 (1968)
Schulman, M.D., Gibbs, M.: Plant Physiol. *43*, 1805–1812 (1968)
Schwarz, Z., Maretzki, D., Schönherr, J.: Biochem. Physiol. Pflanzen *170*, 37–50 (1976)
Smillie, R.: Nature (London) *187*, 1024–1025 (1960)
Smillie, R.M.: Can. J. Bot. *41*, 123–154 (1963)
Smillie, R.M., Fuller, R.C.: Biochem. Biophys. Res. Commun. *3*, 368–372 (1960)
Springgate, C.F., Stachow, C.S.: Arch. Biochem. Biophys. *152*, 1–12 (1972a)
Springgate, C.F., Stachow, C.S.: Biochem. Biophys. Res. Commun. *49*, 522–527 (1972b)
Steiger, E., Ziegler, I., Ziegler, H.: Planta (Berl.) *96*, 109–118 (1971)
Theiss-Seuberling, H.-B.: Ber. Dtsch. Bot. Ges. *87*, 465–471 (1974)
Turner, J.F., Turner, D.H.: Annu. Rev. Plant Physiol. *26*, 159–186 (1975)
Vacchi, C., Giuliani Piccari, G., Pupillo, P.: Z. Pflanzenphysiol. *69*, 351–358 (1973)
Viswanathan, P.N., Krishnan, P.S.: Nature (London) *193*, 166–167 (1962)
Willard, J.M., Gibbs, M.: Plant Physiol. *43*, 793–798 (1968a)
Willard, J.M., Gibbs, M.: Biochim. Biophys. Acta *151*, 438–448 (1968b)
Wolosiuk, R.A., Buchanan, B.B.: J. Biol. Chem. *251*, 6456–6461 (1976)
Yonuschot, G.R., Ortwerth, B.J., Koeppe, O.J.: J. Biol. Chem. *245*, 4193–4198 (1970)
Ziegler, H., Ziegler, I.: Planta (Berl.) *65*, 369–380 (1965)
Ziegler, H., Ziegler, I.: Planta (Berl.) *69*, 111–123 (1966)
Zimmermann, G., Kelly, G.J., Latzko, E.: Eur. J. Biochem. *70*, 361–367 (1976)

20. Enzymes of C_4 Metabolism

J. COOMBS

A. Introduction

All plant tissues contain enzymes which interconvert three-carbon (3-C) and four-carbon (4-C) organic acids by carboxylation and decarboxylation. This subject has been reviewed by WALKER (1962). At that time these reactions were regarded as important in dark (nonphotosynthetic) metabolism, restoring carbon lost from the tricarboxylic acid (TCA) cycle during synthesis of amino acids, fats and pigments. Walker quotes BASSHAM and KIRK (1960) as follows "the quantity of CO_2 introduced by the $C_3 + C_1$ reaction is about 3% of the total label incorporated in photosynthesis" and states that "It seems clear, then, that phosphoenolpyruvate (PEP) carboxylase plays a role in CO_2 fixation in the light and yet there appears at present to be no direct connection between this carboxylation and the photosynthetic cycle, except that the phosphoenolpyruvate concerned is derived from phosphoglycerate".

Since then it has been well established (HATCH and SLACK, 1970; COOMBS, 1971; BLACK, 1973) that C_4 carboxylation reactions are of particular importance in the photosynthetic assimilation of CO_2 in some higher plants, which are now known as C_4 plants for this reason (see Chap. II.6, this vol.).

In C_4 plants the photosynthetic carbon reduction (RPP) cycle (see Chap.II. 1, this vol.) remains the sole route of net reductive carbon assimilation. However, this carbon is not derived directly from the atmosphere, but trapped first into oxaloacetic acid (OAA), malate or aspartate, and subsequently released at the site of reductive reassimilation. This requires the cooperative interaction of the following sequence of reactions: (a) carboxylation (b) reduction or transamination (c) decarboxylation and (d) substrate regeneration.

The details of the individual reactions involved differ within various groups of C_4 plants, both in the importance of transamination reactions, and in the nature of the enzyme responsible for decarboxylation (GUTIERREZ et al., 1974; HATCH et al., 1975). It has been suggested that this C_4 pathway functions as a "carbon shuttle" transferring carbon from the mesophyll layers to the bundle sheath cells – although alternative views have been expressed (COOMBS, 1971, 1973, 1976).

Many of the studies on enzymes of C_4 metabolism have been concerned with the distribution of enzymes within the layers of photosynthetic tissue. This subject is considered in detail in Chapter II.7, this volume. This chapter deals with the isolation, purification, assay, kinetics, and regulation of the more important enzymes under headings corresponding to the four phases of the sequence.

B. Isolation of Enzymes from Tissues of C_4 Plants

C_4 plants have physical and biochemical properties which may cause difficulties during the isolation of enzymes. In many C_4 grasses the tough cuticle and high proportion of vascular tissue make total extraction difficult. In addition such plants may have high levels of o-diphenols (BALDRY et al., 1970 a, b) and an active o-diphenol oxidase (COOMBS et al., 1974 a) which cause inhibition of enzymes unless particular precautions are taken.

The use of sulfhydryl compounds such as dithiothreitol (DTT), mercaptoethanol and thioglycollate to overcome such problems is widespread. Inclusion of polymers such as polyvinylpyrrolidone (PVP) and polyethyleneglycol, or rapid clean-up procedures using gel chromatography on Sephadex G25 have also proved successful.

Conditions of illumination are also of importance. The levels of both NADP-malic enzyme and pyruvate P_i dikinase which can be detected in extracts of green leaves are increased by a period of light treatment (JOHNSON and HATCH, 1970; HATCH et al., 1969). These effects are over and above the increases seen during greening of etiolated material (GRAHAM et al., 1970).

C. Carboxylation – PEP Carboxylase

In C_4 plants the initial carboxylation is catalyzed by PEP carboxylase (orthophosphate: oxaloacetate carboxylase phosphorylating – E.C.4.1.1.31). The reaction, which requires Mg^{2+}, is essentially irreversible.

Although it has been suggested (WAYGOOD et al., 1969) that the enzyme from maize (Am: Corn) uses CO_2, other studies (COOPER et al., 1968; COOMBS et al., 1975) indicate that bicarbonate is the preferred substrate.

I. General Characteristics

This enzyme is probably ubiquitous in plant tissue, generally being located in the cytoplasm. It has been suggested (HATCH and SLACK, 1970) that it is associated with the mesophyll chloroplasts of C_4 plants. The alternative view that it is located in the cytoplasm (GIBBS et al., 1970; BALDRY et al., 1971; COOMBS, 1971) is now widely accepted. A detailed study of a variety of plant species (TING and OSMOND, 1973a,b) suggested that the enzyme exists in a number of distinct forms. These correspond to a C_4 photosynthetic form with high K_m for PEP and Mg^{2+} and a high V_{max}; a C_3 form with a low K_m for PEP and Mg^{2+} and a low V_{max}; a CAM form with a low K_m for PEP and a high V_{max}, and finally an enzyme associated with nongreen tissue having a low K_m for PEP and a low V_{max}. However, the situation does not appear that clear-cut. For instance MUKERJI and TING (1971) isolated three isoenzymes from leaves of the C_3 plant *Gossypium hirsutum* which differed in their requirement for cations and could be distinguished by gel electrophoresis, although their kinetic properties were similar.

During greening of etiolated leaves from C$_4$ plants the level of enzyme increases about eight fold (GRAHAM et al., 1970). At the same time the characteristics of the enzyme change from those associated with the dark form to those of the C$_4$ photosynthetic form (GOATLY and SMITH, 1974). This change does not appear to be associated with de novo synthesis of protein (GOATLY et al., 1975), but rather with a structural change – possibly association of subunits (VIDAL et al., 1976).

Two reports (PAN and WAYGOOD, 1971; LEBLOVA and MARES, 1975) suggest that a heat-stable form of the enzyme occurs in most plant tissues. However, the assay consists of the spectrophotometric determination of an unidentified breakdown product of the reaction rather than of OAA or other well-defined chemical product.

It has now been shown (WHELAN et al., 1973) that the low discrimination against the heavy (^{13}C) isotope of carbon associated with C$_4$ metabolism can be demonstrated at the enzyme level. In intact plants the carbon atoms of both glucose and malate were enriched by 2 to 3 parts per thousand with respect to the ^{12}CO$_2$/^{13}CO$_2$ fed. Synthesis of malate from PEP and HCO$_3$$^-$ using enzyme from *Sorghum* resulted in fractionation of a similar magnitude. In contrast the synthesis of phosphoglyceric acid (PGA) from RuBP and CO$_2$ in preparations from the same plant resulted in an enrichment of between 18 and 34 parts per thousand.

II. Physical Properties and Kinetics

The characteristics of this enzyme, isolated from a number of sources, has been reviewed in detail by UTTER and KOLENBRANDER (1972). More recent studies include the determination of the molecular weight (MW) at around 4×10^5 daltons (TING and OSMOND, 1973b; MIZIORKO et al., 1974 and UEDAN and SUGIYAMA, 1976). Dissociation of the enzyme and electrophoresis in dodecyl sulphate polyacrylamide gels yielded a single band corresponding to a subunit weight of about 1×10^5 daltons. Thus it appears in general that the native protein is composed of four similar polypeptide chains. This conclusion differs from that of SMITH (1968) who found an apparent MW of 2.7×10^5 daltons with no evidence of dissociation or aggregation. NOWAK et al. (quoted by UTTER and KOLENBRANDER, 1972) have suggested that the spinach enzyme has a MW of 7.5×10^5 daltons and is composed of 12 similar subunits.

All reports dealing with PEP carboxylase isolated from C$_4$ plants give the pH optimum as around 8.0 (with a range from 7.8 to 8.4 amongst different species). Reported K$_m$ values vary over a wide range; values from 0.006 to 2.6 mM PEP, 0.02 to 3.3 mM Mg^{2+} and 0.02 to 0.8 mM CO$_2$ or HCO$_3$$^-$ have been reported. On a leaf chlorophyll basis a typical value for V$_{max}$ is around 30 µmol per mg chl per min for C$_4$ plants and ten to twenty times lower for C$_3$ plants. On a protein basis values for V$_{max}$ are about 80 µmol per min per mg protein.

Numerical values recorded for kinetic constants depend to a large extent on the type of plants (C$_4$ or C$_3$), tissue and prehistory of illumination. In addition, purification procedure (or in some cases lack of it), assay used, inclusion or omission of sulfhydryl compounds during isolation or assay, and the treatment of kinetic data which may show sigmoid or nonlinear characteristics, may also affect the results.

III. Regulation, Activation and Inhibition

Although it has been established for some time that PEP carboxylase from bacteria has allosteric properties (CANOVAS and KORNBERG, 1966; CORWIN and FANNING, 1968), such regulatory function was not ascribed to the plant enzyme until the early 1970's. In 1972 UTTER and KOLENBRANDER put the plant enzymes into a separate class on the basis that they were not activated by allosteric effectors. In the same year COOMBS and BALDRY (1972) demonstrated the activation of the enzyme from *Pennisetum purpureum* by sugar phosphates. Similar observations were also made independently by other groups (WONG and DAVIES, 1973; TING and OSMOND, 1973).

Considerable variation is found in more recent reports from studies on activation of the enzyme from various plant tissues. For instance, the enzyme isolated from dark-grown leaves of sugar cane does not appear to show such properties (GOATLY and SMITH, 1974), whereas that from etiolated *Zea mays* (WONG and DAVIES, 1973) and potato (BONUGLI and DAVIES, 1977) responded to a variety of metabolites. Differences may once again reflect different isolation procedures and assay conditions. In several instances (COOMBS et al., 1972; WONG and DAVIES, 1973) sulfhydryl compounds have been shown to reduce allosteric effects, possibly by promoting the disassociation of enzyme into subunits (COOMBS et al., 1973).

In general the enzyme from C_4 plants is activated by sugar phosphates (COOMBS and BALDRY, 1972; COOMBS et al., 1973a; TING and OSMOND, 1973; GOATLY and SMITH, 1974) and inhibited by organic acids (TING, 1968; LOWE and SLACK, 1971; HUBER and EDWARDS, 1975; RAGHAVENDRA and DAS, 1976; BHAGWAT and SANE, 1976). Glycine is also an activator (NISHIKIDO and TAKANASNI, 1973; BHAGVAT and SANE, 1976; UEDAN and SUGIYAMA, 1976).

The most consistent effect noted is the activation by glucose 6-phosphate, usually associated with a change from sigmoid to hyperbolic Michaelis-Menten kinetics with respect to PEP as the variable substrate, also high concentration of glucose 6-phosphate may be inhibitory. Similar results have been obtained using enzyme purified from *P. purpureum* (COOMBS et al., 1973a) or *Z. mays* (UEDAN and SUGIYAMA, 1976). When PEP was the variable substrate cooperative rate/ concentration plots with a Hill number of two or greater were obtained. Cooperativity disappeared in the presence of the activator. These effects were greater at lower pH values; about half a pH unit below the optimum for the native enzyme. A possible reaction mechanism is shown in Figure 1.

Although other sugar phosphates can modify the activity of this enzyme, such observations are complicated by interaction between the apparent effectors and Mg^{2+} ions (COOMBS and BALDRY, 1975). Similar problems are encountered with adenylates, when considered in terms of energy charge (COOMBS et al., 1974b; the term energy charge is used in the context suggested by ATKINSON, 1968, and is defined by ATP+0.5 ADP / AMP+ADP+ATP). It was found that changes in assay conditions could variously produce apparent U-type, R-type or nil response when activity was plotted as a function of charge.

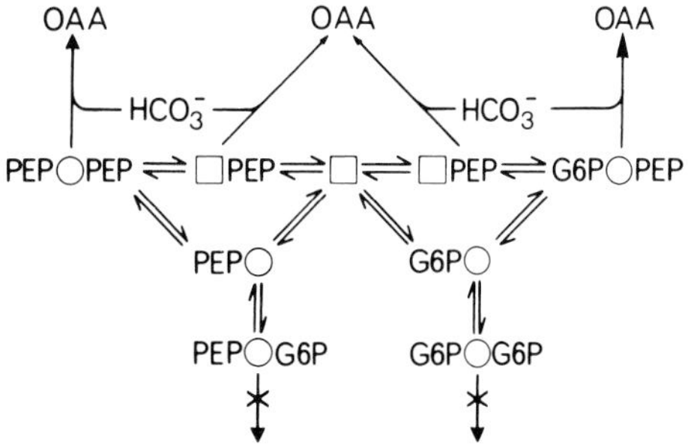

Fig. 1. Activation and inhibition of PEP carboxylase from a C$_4$ plant by glucose-6-phosphate (G6P). From COOMBS et al., 1973. In this figure the *square* represents a less active form of the enzyme which can be converted to the active form (represented by a *circle*) by binding of either G6P or PEP at the effector site to the left of the enzyme molecule. The maximum velocity of reaction will be catalyzed by the enzyme when the effector site is filled by G6P and the substrate site by PEP. Intermediate velocities will be observed when both sites are filled by PEP. Dead-end complexes may be formed when G6P fills the substrate site, thus inhibiting the enzyme

The possible fine regulation of C$_4$ photosynthesis by the interaction of feedback inhibition by 4-carbon organic acids from the C$_4$ pathway and activation by sugar phosphates produced in the RPP cycle has been discussed by COOMBS (1973, 1976) and by HUBER and EDWARDS, 1976. Using isolated PEP carboxylase an interaction could be demonstrated in vitro (COOMBS et al., 1976); see Figure 2. However, the problems remain in extrapolating from the in vitro system to a definitive regulatory system in vivo.

D. Formation of C$_4$ Acids by Reduction and Transamination

I. Reduction

It is generally assumed that OAA is reduced by the NADP malate dehydrogenase present in mesophyll chloroplasts of C$_4$ plants (BLACK, 1973), although all plant tissues contain high levels of the NAD-specific enzyme (ZSCHOCHE and TING, 1977).

1. NADP Malate Dehydrogenase (E.C.1.1.1.82)

This enzyme which was first observed in C$_4$ plants by JOHNSON and HATCH, 1970 catalyzes the reversible reaction: –

L-malate + NADP $\rightleftharpoons$ OAA + NADPH.

Isolation of the enzyme requires the presence of thiol compounds such as DTT in the extraction media. It has been partially purified by $(NH_4)_2SO_4$ precipitation

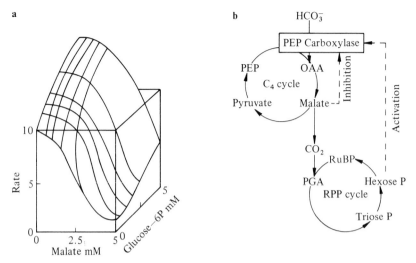

Fig. 2. a In vitro changes in activity of PEP carboxylase from a C_4 plant as a function of concentration of malate and glucose-6-phosphate in reaction mixtures. **b** Possible mechanism of control of C_4 photosynthesis in vivo by modulation of the activity of PEP carboxylase by malate (the product of the C_4 cycle) and glucose-6-phosphate (the product of the RPP cycle)

and gel chromatography on Sephadex G200. This procedure separates the NADP-specific enzyme from the NAD form.

The pH optimum varies with substrate (OAA) concentration from about 8.1 at 0.05 mM to 8.9 at 2 mM OAA. In the direction of oxidation of NADPH double reciprocal velocity/substrate plots are linear. Under these conditions the K_m for NADPH was 60 µM. With 0.2 mM NADPH the K_m for OAA was 25 mM. In the reverse direction kinetic plots were sigmoid. The concentrations of malate and NADP giving half maximal activity were about 2 mM and 0.25 mM respectively. When assayed in vitro the enzyme was not affected by the addition of sugar phosphates or other photosynthetic metabolites.

The enzyme can be isolated in several forms of differing MW (ZIEGLER, 1974b), ranging from 3.9×10^4 to 1.8×10^5 daltons. Addition of sulphite both inhibits the enzyme and favors the appearance of a low MW form. At the same time the reaction kinetics show a loss of negative cooperativity at low substrate or cofactor concentrations. These results suggest that SO_3^{2-} may split S-S bonds causing conformational changes.

2. NAD Malate Dehydrogenase (E.C.1.1.1.37)

This enzyme catalyzes a reaction similar to that of the NADP-specific form, although it has not been considered with reference to C_4 photosynthetic CO_2 fixation other than by COOMBS et al. (1973b). More recently a function has been suggested in the reduction of OAA to malate in the bundle sheath cells of those species having active NAD-malic enzyme (GUTIERREZ et al., 1974; HATCH et al., 1975).

This enzyme is specific for NAD and OAA with K$_m$ values similar to those of the NADP form. Both the C$_4$ enzyme from *P. purpureum* (COOMBS et al., 1973b) and the C$_3$ enzyme from spinach (ZIEGLER, 1974) show complex kinetic patterns. Again different MW forms (1.7×10^4 to 7×10^5 daltons) are found. A form of MW 1.27×10^5 daltons is stabilized in the presence of Mg^{2+} and NADH. Incubation with thiols causes splitting and partial reaggregation to multiple forms with MW between 3.5×10^4 and 1.8×10^5 daltons. It has been suggested that these conformational changes are responsible for the unusual kinetics observed with the spinach enzyme. With *P. purpureum* it would appear that malate reacts not only at the substrate binding site but also at a regulatory site.

3. Direct Reduction

Studies of a light-dependent activity of PEP carboxylase in isolated chloroplasts from sugar cane (BALDRY et al., 1969) led to the suggestion that OAA may also be reduced directly to malate by reduced ferredoxin, or an other primary photoreductant produced by photosystem I in mesophyll chloroplasts.

II. Transamination

1. Aspartate Aminotransferase (E.C.2.6.1.1.)

This enzyme catalyzes the reversible amination of OAA

L-aspartate + α-ketoglutarate ⇌ OAA + L-glutamate

Activity of this enzyme is higher in those C$_4$ species which possess high levels of PEP carboxykinase or NAD malic enzyme. Isoenzymes have been partially purified from leaves of *Atriplex spongiosa* (HATCH and MAU, 1973) using (NH$_4$)$_2$SO$_4$ fractionation, ion exchange chromatography and polyacrylamide gel electrophoresis. The activity of the enzyme increased from 2.4 μmol per g fr wt per min in dark-grown tissue to a tenfold higher rate in fully greened tissue. The two isoenzymes appeared to be located in different tissues of the leaf, one in the mesophyll cells and one in the bundle sheath cells. This activity was not associated with the chloroplasts. In the bundle sheath it appeared to be a mitochondrial enzyme, whereas in the mesophyll cells it was cytoplasmic.

The reaction velocities, activities, and apparent cellular location suggest that the enzyme in the mesophyll cells converts OAA to aspartate, whereas the isoenzyme in the bundle sheath favors the reverse reaction (HATCH, 1973a). The K$_m$ values which have been reported for the enzyme in the two forms are as follows: aspartate, 0.85 and 0.50 mM; α-ketoglutarate, 0.08 and 0.12 mM; OAA, 0.1 and 0.45 mM; glutamate, 1.6 and 0.75 mM. In the direction of deamination the enzyme has a broad pH optimum between 6.8 and 8.3. In the opposite direction the enzyme was optimal at pH 8.0, but nearly inactive at pH 7.0. Addition of 5 mM malate to reaction mixtures caused about 50% inhibition.

E. Decarboxylation

Atmospheric CO_2 incorporated into C_4 acids is subsequently released at the site of refixation in reactions catalyzed either by NADP malic enzyme (Andropogoneae, Maydeae, Paniceae and Aristidea), NAD malic enzyme (Chlorideae, Eragrosteae, Sporoboleae, Paniceae) or PEP carboxykinase (Chlorideae, Sporoboleae, Zoysieae).

I. NADP-Malic Enzyme (E.C.1.1.1.40)

This enzyme catalyzes the reaction

L-malate + NADP $\rightleftharpoons$ pyruvate + CO_2 + NADPH.

Two forms, of differing MW (2.3×10^5 and 4.6×10^5 daltons), can be obtained by Sephadex fractionation of enzyme from leaves of maize (ZIEGLER, 1974a), as described earlier by JOHNSON and HATCH (1970). The enzyme has a requirement for a divalent metal ion and shows no activity with NAD. In contrast with the malic enzyme of C_3 plants, the C_4 form has a low K_m for malate and a high pH optimum. However, the responses to change in pH vary with malate concentration. For instance JOHNSON and HATCH (1970) found that at 0.1 mM, 1 mM and 10 mM malate the optima were pH 7.4, 8.0, and 8.5, respectively. The apparent K_m for malate also varied with pH, being about 0.4 mM at pH 8.5 and 0.15 mM at pH 8.0. At pH 8.0 with 2 mM malate the K_m for NADP was 25 μM and the values for Mg^{2+} and Mn^{2+} were 55 and 4 μM respectively.

Results from other studies are generally in agreement. COOMBS et al. (1973) using enzyme partially purified, by $(NH_4)_2SO_4$ precipitation and DEAE cellulose chromatography, from leaves of *P. purpureum* found a pH optimum of 8.0. Under these conditions K_m values of 3.7 μM for NADP, 0.35 mM for malate and 8 μM for Mn^{2+} were recorded. Using enzyme from *Digitaria sanguinalis*, NISHIKIDO and WADA (1974) found a K_m for malate of 0.18 mM at pH 7.8.

Inhibition by sulphite has been studied in detail by ZIEGLER (1974). In malate decarboxylation, under conditions of saturating Mn^{2+}, SO_3^{2-} acts as a partially competitive inhibitor, in a manner similar to CO_2-saturated bicarbonate. In the direction of carboxylation SO_3^{2-} acts as a fully competitive inhibitor. This enzyme is also inhibited by phosphate, triose phosphates and AMP. In a study of 30 species DAVIES et al. (1974) found that malic enzyme from 27 out of 28 species had allosteric properties. No such activity could be demonstrated in members of the Gramineae.

II. NAD-Malic Enzyme (E.C.1.1.1.39)

High levels of this enzyme are found in the dicotyledon C_4 species such as *Amaranthus* and *Atriplex* in addition to the members of the Gramineae listed above. This enzyme is most active with NAD and Mn^{2+}; little activity is observed when Mn^{2+} is replaced by Mg^{2+} in assay mixtures (HATCH and KAGAWA, 1974).

Purification from leaves of *Atriplex spongiosa, Amaranthus edulis,* and *Panicum miliaceum* by $(NH_4)_2SO_4$ precipitation, gel filtration on Sephadex and chromatography on DEAE cellulose has been described by HATCH et al. (1974). The purified enzyme had an absolute requirement for Mn^{2+} which could not be replaced by Mg^{2+}. The enzyme was inhibited, in the direction of reductive decarboxylation, by bicarbonate. With the native enzyme the pH optimum was about 6.8. This increased towards 7.5 when activated by CoA or acetyl CoA. Plots of activity as a function of concentration of Mn^{2+}, SO_4^{2-}, CoA, or acetyl CoA were all sigmoidal. Maximal activity was observed with about 12 mM SO_4^{2-}, but only 20 to 25 µM CoA or acetyl CoA was required. With high concentrations of CoA plots of activity as a function of NAD content were hyperbolic. Double reciprocal rate/substrate plots of the same data indicated a rather high K_m for NAD of 0.3 mM. Half-maximal rates were obtained with concentrations of malate of about 1 to 2 mM.

III. PEP Carboxykinase (E.C.4.1.1.49)

PEP carboxykinase from C$_4$ plants catalyzes the following reversible reaction

$$OAA + ATP \rightleftharpoons PEP + CO_2 + ADP.$$

The enzyme also catalyzes a nucleotide-dependent exchange of CO_2 into OAA. A simple assay procedure, based on the change in absorbance at 280 nm, due to loss of OAA has been described by HATCH (1973). The activity may also be assayed in terms of the ATP-dependent decarboxylation, the ADP-dependent carboxylation, or the exchange reaction.

The enzyme has been partially purified from leaves of *Panicum maximum* and separated from PEP carboxylase activity using chromatography on DEAE-Sephadex A25 (RAY and BLACK, 1976). Using this preparation the following kinetic (K_m) values were obtained:
a) Decarboxylation reaction: OAA, 0.13 mM; ATP, 0.02 mM;
b) Carboxylation reaction; PEP, 0.38 mM; ADP, 0.05 mM; HCO_3^-, 11 mM;
c) Exchange reaction; OAA, 0.5 mM; ATP, 0.3 mM; HCO_3^-, 31 mM (equivalent to about 16 mM CO_2).
This enzyme has recently been studied in more detail by HATCH and MAU (1977).

F. Substrate Regeneration

Decarboxylation of malate by either the NAD- or the NADP-specific malic enzyme yields pyruvate. Decarboxylation of OAA by PEP carboxykinase will usually yield PEP but may also, in the presence of ADP, yield pyruvate. This may be reconverted directly to the substrate for CO_2 fixation (PEP) by the activity of the enzyme pyruvate P$_i$ dikinase, which is located in the mesophyll chloroplasts of C$_4$ plants. It has been suggested that in some C$_4$ species the pyruvate may first be converted to alanine in the bundle sheath, translocated as alanine, and reconverted to pyruvate in the mesophyll cells. Hence, in some C$_4$ plants alanine aminotransferase may also be of importance.

I. Pyruvate P_i Dikinase (E.C.2.7.9.1)

This enzyme was first reported from studies on C_4 plants (*Saccharum, Zea,* and *Sorghum*) by HATCH and SLACK (1968).

The enzyme requires Mg^{2+}. This cannot be replaced by either Mn^{2+} or Ca^{2+}. Without Mg^{2+}, mercaptoethanol or other thiol-compounds the enzyme is rapidly but reversibly inactivated. It is more unstable at $0°C$, or when frozen, than at $22°C$.

The reaction mechanism has been studied in detail using enzyme purified by $(NH_4)_2SO_4$ precipitation, hydroxyapatite and Sephadex G200 chromatography (ANDREWS and HATCH, 1969; HATCH and SLACK, 1969). An enzyme phosphate complex is formed first:

$$Enzyme + ATP + P_i \rightleftharpoons Enzyme\text{-}P + AMP + PP_i$$

The activated enzyme then reacts with pyruvate:

$$Enzyme\text{-}P + pyruvate \rightleftharpoons Enzyme + PEP.$$

Exchange reactions were used to show that the phosphate groups in the PP_i formed were derived from P_i and the terminal phosphate of ATP.

At pH 8.3 the initial velocity in the direction of PEP synthesis was about six times that in the reverse direction. Similar methods were used by SUGIYAMA (1973) to purify the enzyme from leaves of maize. Analytical centrifugation indicated a sedimentation coefficient of 8.86 s, suggesting a MW of 3.8×10^5 daltons. SDS polyacrylamide gel electrophoresis of this material gave a single band, suggesting that the enzyme was comprised of four equal subunits of 9.4×10^4 daltons.

Kinetic constants (K_m values) have been determined for both the enzyme from sugar cane (HATCH and SLACK, 1968) and maize (SUGIYAMA, 1973). In the direction of PEP formation the respective values determined for enzyme from the two plants were as follows: pyruvate, 0.11 and 0.25 mM; ATP, 0.09 and 0.015 mM; P_i, 0.5 and 1.5 mM. In the reverse direction of pyruvate formation values were: PEP, 0.11 and 0.14 mM; AMP, very low and 0.01 mM; PP_i, 0.04 mM for both plants.

HATCH and SLACK (1969) found that the extractable activity in leaves of *Amaranthus* or maize dropped rapidly when the plants were transferred to dark. Activation requires a thiol and P_i but was reversed by AMP, GMP, and partially by ADP and ATP. These and other observations suggest that the rate of regeneration of PEP may be regulated in vivo by the levels of ADP in the chloroplast.

II. Alanine Aminotransferase (E.C.2.6.1.2)

The reaction catalyzed is as follows

$$L\text{-alanine} + \alpha\text{-ketoglutarate} \rightleftharpoons pyruvate + L\text{-glutamate}.$$

Purification and assay are similar to those described above for the aspartate amino transferase (HATCH and MAU, 1973; HATCH, 1973a) except that the buffer pH was 7.5, aspartate was replaced by 10 mM alanine and lactate dehydrogenase was used as the coupling enzyme. Three isoenzymes, all apparently associated with the cytoplasm rather than chloroplast or mitochondria, could be demonstrated. One, in the mesophyll layer, favored the formation of pyruvate, whereas the other in the bundle sheath preferentially catalyzed the formation of alanine. The pH optimum was close to 8 for pyruvate formation but nearer to 7.5 for alanine synthesis. The third form was only detected when fractions had been incubated with 20 µg/ml of pyridoxal phosphate. Values of K$_m$ recorded for the three forms identified as ALA-DEAE 1, ALA-DEAE 2 and ALA-DEAE 3 were as follows: alanine, 0.25, 3.0 and 3.1 mM; α-ketoglutarate, 0.3, 0.09 and 0.03 mM; pyruvate, 0.45, 0.04 and 0.02 mM; glutamate, 0.55, 1.15 and 0.8 mM. The enzyme was inhibited by about 25% on addition of 5 mM malate to the reaction mixture.

G. Summary

The overall reactions comprising the C$_4$ cycle are now well established. However, considerably more attention has been paid to the distribution of the enzymes within leaf tissues than to studies of the purified enzymes. Some of the key reactions have only been studied in detail by one or two laboratories. The exception is PEP carboxylase. The demonstration that this enzyme has regulatory properties prompted a number of studies on kinetics in vitro. However, there is still no real evidence that such regulation is of importance in vivo, although it has been suggested (COOMBS et al., 1973c) that the change in ^{14}C incorporation into metabolites in short-term photosynthesis by leaves of *P. purpureum* could be due to inactivation of PEP carboxylase by high levels of sugar phosphate formed under these conditions.

References

Andrews, T.J., Hatch, M.D.: Biochem. J. *114*, 117–125 (1969)
Atkinson, D.E.: Biochemistry *7*, 4030–4034 (1968)
Baldry, C.W., Bucke, C., Coombs, J.: Biochem. Biophys. Res. Commun. *37*, 828–832 (1969)
Baldry, C.W., Bucke, C., Coombs, J.: Planta (Berl.) *97*, 310–319 (1971)
Baldry, C.W., Bucke, C., Coombs, J.: Planta (Berl.) *94*, 124–133 (1970a)
Baldry, C.W., Bucke, C., Coombs, J., Gross, D.: Planta (Berl.) *94*, 107–123 (1970b)
Bassham, J.A., Kirk, M.: Biochim. Biophys. Acta. *43*, 447–464 (1960)
Bhagwat, A.S., Sane, P.V.: Indian. J. Exp. Biol. *14*, 155–158 (1976)
Black, C.C.: Annu. Rev. Plant Physiol. *24*, 253–286 (1973)
Bonugli, K.J., Davies, D.D.: Planta (Berl.) *133*, 281–287 (1977)
Canovas, J.L., Kornberg, H.L.: Proc. R. Soc. Lond. *B 165*, 189–205 (1966)
Coombs, J.: Proc. R. Soc. Lond. *B 179*, 221–235 (1971)
Coombs, J.: Curr. Adv. Plant Sci. *1*, 1–10 (1973)
Coombs, J.: In: The Intact Chloroplast. J. Barber (ed.), pp. 279–313. The Netherlands: Elsevier 1976
Coombs, J., Baldry, C.W.: Nature New Biol. *238*, 268–270 (1972)
Coombs, J., Baldry, C.W.: Planta (Berl.) *124*, 153–158 (1975)

Coombs, J., Baldry, C.W., Bucke, C.: Biochem. J. *130*, 25 (1972)
Coombs, J., Baldry, C.W., Bucke, C.: Planta (Berl.) *110*, 95–107 (1973a)
Coombs, J., Baldry, C.W., Bucke, C.: Planta (Berl.) *110*, 109–120 (1973b)
Coombs, J., Baldry, C.W., Brown, J.E.: Planta (Berl.) *110*, 121–129 (1973c)
Coombs, J., Baldry, C.W., Bucke, C., Long, S.P.: Phytochem. *13*, 2703–2708 (1974a)
Coombs, J., Maw, S.L., Baldry, C.W.: Planta *117*, 279–292 (1974b)
Coombs, J., Maw, S.L., Baldry, C.W.: Plant. Sci. Lett. *4*, 97–102 (1975)
Coombs, J., Baldry, C.W., Bucke, C.: In: Perspectives in Experimental Biology. Vol. 2, Botany.
    Sunderland, N. (ed.), pp. 177–188. Oxford: Pergamon Press 1976
Cooper, T.G., Tchen, T.T., Wood, H.G., Benedict, C.R.: J. Biochem. *243*, 3857–3863 (1968)
Corwin, L.M., Fanning, G.R.: J. Biol. Chem. *243*, 3517–3525 (1968)
Davies, D.D., Nascimento, K.H., Patil, K.D.: Phytochem. *13*, 2417–2425 (1974)
Gibbs, M., Latzko, E., O'Neil, D., Hew, Choy-Sin: Biochem. Biophys. Res. Comm. *40*,
    1356–1361 (1970)
Goatly, M.B., Smith, H.: Planta *117*, 67–73 (1974)
Goatly, M.B., Coombs, J., Smith, H.: Planta *125*, 15–24 (1975)
Graham, D., Hatch, M.D., Slack, C.R., Smillie, R.M.: Phytochem. *9*, 521–532 (1970)
Gutierrez, M., Gracen, V.E., Edwards, G.E.: Planta *119*, 279–300 (1974)
Hatch, M.D.: Anal. Biochem. *52*, 280–283 (1973)
Hatch, M.D.: Arch. Biochem. Biophys. *156*, 207–214 (1973a)
Hatch, M.D., Kagawa, T.: Aust. J. Plant Physiol. *1*, 357–369 (1974)
Hatch, M.D., Kagawa, T., Craig, S.: Aust. J. Plant Physiol. *2*, 111–128 (1975)
Hatch, M.D., Mau, S.L.: Arch. Biochem. Biophys. *156*, 195–206 (1973)
Hatch, M.D., Mau, S.L.: Aust. J. Plant Physiol. *4*, 207–216 (1977)
Hatch, M.D., Mau, S.L., Kagawa, T.: Arch. Biochem. Biophys. *165*, 188–200 (1974)
Hatch, M.D., Slack, C.R.: Biochem. J. *106*, 141–146 (1968)
Hatch, M.D., Slack, C.R.: Biochem. J. *112*, 549–558 (1969)
Hatch, M.D., Slack, C.R.: Ann. Rev. Plant Physiol. *21*, 141–162 (1970)
Hatch, M.D., Slack, C.R., Bull, T.H.: Phytochemistry *8*, 697–706 (1969)
Huber, S.C., Edwards, G.E.: Can. J. Bot. *53*, 1925–1933 (1975)
Johnson, H.S., Hatch, M.D.: Biochem. J. *119*, 273–280 (1970)
Leblova, S., Mares, J.: Photosynthetica *9*, 177–184 (1975)
Lowe, J., Slack, C.R.: Biochim. Biophys. Acta *235*, 207–209 (1971)
Miziorko, H.M., Nowak, T., Mildvan, A.S.: Arch. Biochem. Biophys. *163*, 378–389 (1974)
Mukerji, S.K., Ting, I.P.: Arch. Biochem. Biophys. *143*, 297–317 (1971)
Nishikido, T., Takanashi, H.: Biochem. Biophys. Res. Commun. *53*, 126–133 (1973)
Nishikido, T., Wada, T.: Biochem. Biophys. Res. Commun. *61*, 243–249 (1974)
Pan, D., Waygood, E.R.: Can. J. Bot. *49*, 631–643 (1971)
Raghavendra, A.S., Das, V.S.R.: Z. Pflanzenphysiol. *78*, 434–437 (1976)
Ray, B.T., Black, C.C.: Plant Physiol. *58*, 603–607 (1976)
Smith, T.E.: Arch. Biochem. Biophys. *125*, 178–188 (1968)
Sugiyama, T.: Biochemistry *12*, 2862–2868 (1973)
Ting, I.P.: Plant Physiol. *43*, 1919–1924 (1968)
Ting, I.P., Osmond, C.B.: Plant. Sci. Lett. *1*, 123–128 (1973)
Ting, I.P., Osmond, C.B.: Plant Physiol *51*, 439–447 (1973a)
Ting, I.P., Osmond, C.B.: Plant Physiol. *51*, 448–453 (1973b)
Uedan, K., Sugiyama, T.: Plant Physiol. *57*, 906–910 (1976)
Utter, M.F., Kolenbrander, H.M.: Enzymes *6*, 117–168 (1972)
Vidal, J., Cavalie, G., Gadal, P.: Plant Sci. Lett. *7*, 265–270 (1976)
Walker, D.A.: Biol. Rev. *37*, 215–256 (1962)
Waygood, E.R., Mache, R., Tan, C.K.: Can. J. Bot. *47*, 1455–1458 (1969)
Whelan, T., Sackett, W.M., Benedict, C.R.: Plant. Physiol. *51*, 1051–1054 (1973)
Wong, K.F., Davies, D.D.: Biochem. J. *131*, 451–458 (1973)
Ziegler, I.: Phytochemistry *13*, 2403–2410 (1974)
Ziegler, I.: Biochim. Biophys. Acta *364*, 28–37 (1974a)
Ziegler, I.: Phytochemistry *13*, 2411–2416 (1974b)
Zschoche, W.C., Ting, I.P.: Plant Sci. Lett. *9*, 103–106 (1977)

21. Enzymes of Crassulacean Acid Metabolism

P. Dittrich

A. Introduction

Crassulacean acid metabolism (CAM), whose general features have been discussed by Kluge in Chapter II.8, requires the interaction of several groups of enzymes during alternate periods of the diurnal cycle of light and darkness. These enzymes, which must be subject to a sophisticated mode of regulation in order to bypass undesired competition, are the subject of this chapter.

B. Enzymes of Starch Metabolism

The conversion of starch to glucose can be catalyzed by α- and β-amylase in concert with α-glucosidase. The occurrence of these three enzymes in *Kalanchoë daigremontiana* was demonstrated by Schilling and Dittrich (1979). The degradation reaction of starch, however, which yields glucose-1-P, is carried out by phosphorylase. Despite the catabolic role of this enzyme, phosphorylase is generally assayed in the direction of starch synthesis, and consequently kinetic parameters must be considered with care. Sutton (1975b) investigated the starch-synthesizing properties of phosphorylase in an ammonium sulfate fraction from *K. daigremontiana* leaves, and found that P_i (1 mM) inhibited the synthesis of starch, reducing the V_{max} by 59%. Clearly, P_i can exert a mass action effect since it is a component of the reaction; removal of phosphate from the enzyme assay mixture by Al^{3+} (Schnabl, 1976) or Mn^{2+} (Schilling, 1976) results in a stimulation of starch formation. Other inhibitors reported by Sutton (1975b) were malate (8 mM) and PEP (2 mM) which inhibited the enzyme by 13% and 21% respectively. In contrast, addition of AMP or cAMP (1 mM each) brought about a stimulation of 24%. The total phosphorylase activity of 51 µmol per mg chl per h found in *K. daigremontiana* is well in accord with the maximum rate of starch degradation (51 µmol glucose per mg chl per h) observed by Vieweg and de Fekete (1977).

Starch metabolism in *K. daigremontiana* has been further tested with respect to enzymes synthesizing and interconverting starch and glucans (Schilling and Dittrich, 1979). The characteristic enzymes were similar to those in spinach: ADP-glucose-α-1-4 glucosyltransferase catalyzed the primed synthesis of starch, phosphorylase incorporated glucose-1-P into starch (as already reported by Sutton 1975b) and also degraded labeled starch to glucose-1-P, and α-1-4 glucosyl-glucan transferase (Linden et al., 1974) mediated an exchange between glucose, maltose, and higher dextrans, and finally a primer-free maltose phosphorylase (Schilling and KANDLER, 1975) converted labeled glucose in the presence of unlabeled glucose-1-P (and vice versa) into maltose and maltotriose. No detailed kinetic studies of these enzymes have yet been undertaken.

C. Glycolytic Enzymes

Operation of the glycolytic sequence in *K. daigremontiana* leaves was confirmed by the investigation of SUTTON (1975a) on the activities and kinetic properties of the enzymes involved. The data summarized in Table 1 provide ample evidence that the degradation of glucose follows the pathway present in other higher plants. 6-Phosphofructokinase, one enzyme which usually represents a cornerstone of glycolytic regulation, was investigated in detail (SUTTON, 1975b). This enzyme isolated from non-CAM plants usually exhibits cooperative kinetics (KELLY and TURNER, 1971). In *K. daigremontiana*, however, it appeared to follow regular Michaelis-Menten kinetics. ATP and PEP, which are known to inhibit PFK from other sources, also inhibited the *K. daigremontiana* enzyme. The substrate ATP (K_m 12 μM) inhibited the reaction at concentrations exceeding 0.24 mM. There was no interaction between ATP and fructose-6-P, since increasing concentrations of the latter could not alleviate the inhibitory action of the former. PEP was demonstrated to act in the mode of a noncompetitive inhibitor with a K_i of about 0.25 mM. More interesting, with respect to CAM, is the finding that PFK was also inhibited by relatively high levels of malate, the K_i being about 6 mM. It is known that PFK from other tissues is subject to allosteric inhibition by citrate (GARLAND et al., 1963). Since in CAM plants the accumulation of malate is sustained by the process of glycolysis, it appears reasonable that the level of malate should participate in the regulation of carbon flow through glycolysis. The same argument applies to the regulatory effect of PEP, whose accumulation would inhibit

Table 1. Activities of enzymes of starch metabolism and glycolysis in *Kalanchoë daigremontiana*

Enzyme	μmol g fr. wt^{-1} h^{-1}
Phosphorylase	51
Amylase	25
Hexokinase	9
Glucosephosphate isomerase	160
6-Phosphofructokinase	22
Fructosebisphosphate aldolase	472
Triosephosphate isomerase	9867
Glyceraldehydephosphate dehydrogenase	332
Phosphoglycerate kinase	1037
Phosphoglyceromutase-enolase	55

Data for amylase from VIEWEG and DE FEKETE (1977); other data from SUTTON (1975a).

its own formation and, consequently, diminish the substrate input for PEP carboxylase.

D. Gluconeogenic Enzymes

Carbon entry into the gluconeogenic sequence takes place either by phosphorylation of pyruvate, catalyzed by pyruvate orthophosphate dikinase or by phosphorylative decarboxylation of oxaloacetate, catalyzed by PEP carboxykinase. Since the latter plays a major role during daylight in supplying CO_2 for ribulosebisphosphate carboxylase, it is discussed below (Sect. F.II). According to Eq. (1), pyruvate orthophosphate dikinase converts pyruvate to PEP

$$Pyruvate + ATP + P_i \rightarrow PEP + AMP + PP_i \tag{1}$$

in an energetically unfavorable reaction. However, in concert with pyrophosphatase and adenylate kinase, accumulation of PEP becomes possible. The enzyme is considered to divert pyruvate from breakdown within the tricarboxylic acid cycle to synthesis of carbohydrates through gluconeogenesis. The presence of pyruvate orthophosphate dikinase has been reported in the CAM plants *Bryophyllum tubiflorum* and *Sedum praealtum* by KLUGE and OSMOND (1972) at levels of 21 μmol per mg chl per h. A more detailed investigation by SUGIYAMA and LAETSCH (1975) revealed enzyme activities of 96 μmol per mg chl per h, more in accord with the assigned role of this enzyme in *K. daigremontiana*. The partially purified enzyme not only depends on Mg^{2+} for catalytic activity, but also for protein stability. Removal of Mg^{2+} by gel filtration caused irreversible loss of 60% of the enzyme activity after incubation for 1 h at pH 7.5 and 22° C. Like several other plant enzymes associated with photosynthesis (KELLY et al., 1976), pyruvate orthophosphate dikinase from CAM plants exhibited a pronounced light activation (SUGIYAMA and LAETSCH, 1975). Transferring light-grown leaf tissue to darkness for 45 min resulted in a 90% loss of enzyme activity, the half-decay time being 15 min. Illumination of the leaves for about 30 min fully restored the initial enzyme activity. This phenomenon constitutes a basic mechanism for discrimination between glycolysis and gluconeogenesis.

E. Carboxylating Enzymes

I. The Formation of Malate

During the dark period of CAM, PEP carboxylase catalyzes the fixation of CO_2 according to the following Eq. (2)

$$PEP + HCO_3^- \rightarrow OAA + P_i. \tag{2}$$

The oxaloacetate formed is reduced by malate dehydrogenase to malate which is subsequently stored in the vacuole. Because of the central role of PEP carboxylase, a considerable number of investigations concerning this enzyme have accumulated. Two facts can be reliably filtered from the information obtained: PEP carboxylase in CAM plants is (a) inhibited by L-malate, and (b) activated by glucose 6-P. Both these properties have also been demonstrated for PEP carboxylases from other sources (Ting, 1968; Bonugli and Davies, 1977).

In 1967, Queiroz reported that L-malate inhibited PEP carboxylase in CAM plants. Since these plants contain malate in amounts dependent upon prevailing environmental factors, Kluge and Osmond (1972) investigated the effect of the malate present in crude extracts of CAM plants upon the enzyme. The authors showed that leaf tissue of *Bryophyllum tubiflorum* displayed only 30% of its potential PEP carboxylase activity when high levels of malate prevailed in the morning, while a 100% level of activity could be obtained with extracts from leaves with low content of malate in the evening. Removal of the endogenous malate by Sephadex G 25 filtration eliminated the differences between the morning and evening extracts. The effect of malate was attributed to a mixed type inhibition (K_i 3.6 mM). This inhibition of PEP carboxylase by malate can be overcome by addition of 2 mM glucose-6-P to the assay mixture as demonstrated by Ting and Osmond (1973a) with extracts of *K. daigremontiana*. This compound restores the parameters of the uninhibited control assay. In the absence of malate (filtered extracts), glucose-6-P reduces the K_m(PEP) 1.7-fold and increases V_{max} slightly. While malate appears to play the role of a feed-back inhibitor, glucose-6-P seems to effect a precursor activation. In concert with the inhibition of PFK by malate, ATP and PEP, glucose-6-P activation of PEP carboxylase provides an additional factor for regulation of the diurnal cycle of acidification and deacidification.

There exists a wide spectrum of extraction and assay procedures described in the literature for PEP carboxylase in CAM plants. The pH-optima range from 6.25 to 8.5; values of 6.25 and 7.8 have been reported for the same plant species by the same co-author (Kluge and Osmond, 1972; Ting and Osmond, 1973b), while the respective K_m values (PEP) were 0.22 and 0.7 mM. PEP carboxylase from *Bryophyllum blossfeldianum* has been extracted using a buffer with glutathione (Queiroz, 1969) or without any SH-reagent (Queiroz and Morel, 1974). Enzyme activities reported were obtained from plants either under a strictly controlled photoperiod (Wilkinson and Smith, 1976) or not (Ting and Osmond, 1973b) and crude leaf extracts were either filtered through Sephadex G 25 (Dittrich et al., 1973) or used without further treatment (Queiroz and Morel, 1974). The claim of Queiroz and Morel (1974) for periodically changing activities of PEP carboxylase in *Bryophyllum blossfeldianum* has not yet been confirmed elsewhere. In those experiments, plant extracts were prepared without SH-reagents and without gel filtration, although malate inhibition of PEP carboxylase was previously reported from the same laboratory. In contrast, Wilkinson and Smith (1976) obtained diurnal changes in PEP carboxylase activity opposite to those reported by Queiroz and Morel (1974). These authors used gel-filtered extracts without addition of SH-reagents. The properties and regulation of PEP carboxylase in CAM is obviously still unclear. A diurnal shift of the pH-optimum, such as reported for PEP carboxylase in potato (Bonugli and Davies, 1977), could be involved as well as in vivo activation or deactivation via SH-groups.

II. The Photosynthetic Fixation of CO_2

The carboxylase common to all green plants, RuBP carboxylase, was investigated in a comparative study in *K. daigremontiana* by BADGER et al., 1975. This enzyme catalyzes the following reaction (3)

$$RuBP + CO_2 + H_2O \rightarrow 2\,PGA. \tag{3}$$

The same enzyme also exhibits an oxygenase activity apparently according to Eq. (4):

$$RuBP + O_2 \rightarrow \text{P-glycolate} + PGA. \tag{4}$$

Much interest has been focused on RuBP carboxylase because of reported discrepancies in kinetic properties (see Chap. II.17, this vol.). Kinetic properties of the enzyme isolated from *K. daigremontiana* are quite similar to those of the enzyme from C_3 or C_4 plants. The K_m (CO_2) was 18 µM, and the K_i (O_2) for the carboxylating activity was 490 µM. The oxygenase activity required a concentration of 310 µM O_2 for half saturation. Upon comparing the V_{max} values for O_2 fixation and CO_2 fixation, the carboxylation was found to proceed about four times faster than the oxygenation. No unusual properties of RuBP carboxylase or oxygenase could be detected in this CAM plant.

Operation of the complete series of Calvin cycle enzymes was further demonstrated by LEVI and GIBBS (1975) in a study of CO_2 fixation by intact chloroplasts from *K. daigremontiana*.

F. Decarboxylating Enzymes

Two enzymes have been detected in CAM plants able to catalyze the decarboxylation of malate, and a further enzyme is known to decarboxylate oxaloacetate, each providing CO_2 for photosynthetic fixation and reduction by the Calvin cycle.

I. The Decarboxylation of Malate

The NADP-dependent malic enzyme described by ANDERSON et al. (1952), was studied in detail by BRANDON and BOEKEL-MOL (1973) and by GARNIER-DARDART and QUEIROZ (1974) in *K. tubiflora* and *K. daigremontiana*, respectively. The enzyme requires Mn^{2+} or Mg^{2+}; Mn^{2+} shows a slightly higher efficiency. The pH optimum lies between pH 6.8 and 7.2; the exact value appeared to depend upon the malate concentration present (WALKER, 1960). A malate concentration of 15 mM shifted the pH optimum from 7.2 to 7.9. This shift was accompanied by a rise in the K_m for malate, which was 0.59 mM at pH 7.2 and 2.1 mM at pH 7.9 (GARNIER-DARDART and QUEIROZ, 1974). The K_m for malate is also susceptible to temperature (BRANDON and BOEKEL-MOL, 1973): at temperatures below 17° C and above 39° C the affinity for malate was found to be lower than that between these temperatures. This peculiarity was considered to be involved in the regulation of malic enzyme during daylight, when temperatures up to 39° C prevail in the habitats of CAM plants. NADP-malic enzyme shows a remarkable temperature stability; the protein can be kept at 55° C for 30 min without loss of activity. Furthermore, the temperature optimum lies around 50° C. Other factors reported to affect malic enzyme were coenzyme A (CoA) and thiamine pyrophosphate, which at concentrations of 7 mM and 10 mM respectively completely inhibited the catalytic activity.

Certain previous studies have suggested that the NADP-malic enzyme was present in chloroplasts (GARNIER-DARDART, 1965; MUKERJI and TING, 1968 a), in mitochondria (BRANDON and BOEKEL-MOL, 1973) or in both organelles plus cytoplasm (MUKERJI and TING, 1968 b). However, recent experiments by DITTRICH and MEUSEL (1979) have clearly demonstrated that the NADP-malic enzyme was located entirely in the cytoplasm of cells of *K. daigremontiana*. This was established by analysis of fractions prepared from leaf homogenates. The NADP-malic enzyme was not found in either the chloroplast fraction, containing RuBP carboxylase, NADP-GAPDH and APS sulfotransferase, or the mitochondrial fraction, containing cytochrome c oxidase and fumarase. More than 90% of activity for the NADP-malic enzyme was found in the supernatant.

The NAD-dependent malic enzyme is located in the mitochondria (DITTRICH, 1976 a). This enzyme shows significant activities only after activation with CoA, acetyl-CoA or SO_4^{2-}, while thiamine pyrophosphate does not exhibit any apparent effect. The sigmoidal kinetics obtained indicate an allosteric nature of the protein. Activation by the above compounds resulted in a 14-fold stimulation of V_{max} with a concomitant fivefold reduction of the $S_{0.5}$ for malate. In contrast to NADP-malic enzyme, the NAD-dependent enzyme is specific for Mn^{2+} and does not operate with Mg^{2+}. Both types of malic enzyme occur at about similar activities (120 to 200 μmol per mg chl per h) in the Crassulaceae. The contrasting behavior toward CoA suggests that each enzyme fills a different function within CAM, but so far no explanation is at hand which would link the regulatory effects of CoA to specific metabolic or environmental situations.

II. The Decarboxylation of Oxaloacetate.

The third enzyme in CAM plants known to produce CO_2 for the Calvin cycle is PEP carboxykinase, which was reported to occur mainly in those plants which lack adequate activities of both malic enzymes (DITTRICH et al., 1973). PEP carboxykinase catalyzes reaction (5).

$$OAA + ATP \rightarrow CO_2 + PEP + ADP \tag{5}$$

The substrate for decarboxylation (OAA) is produced from malate by the action of malate dehydrogenase. It is not clear whether the NAD- or NADP-linked dehydrogenase is responsible, although KLUGE and OSMOND (1971) reported the presence of the apparently light-activated NADP enzyme in CAM plants.

DALEY et al. (1977) have recently purified PEP carboxykinase from pineapple leaves. The energy of activation at 15° C (night temperature) was about 80 kcal/

mol, while at day temperatures of 30° C the activation threshold was reduced to 13 kcal/mol. Taking into account the low affinities for PEP and bicarbonate (K_m 5 mM and 3.4 mM respectively) and high affinities for OAA and ATP (K_m 0.4 mM and 0.02 mM respectively) the authors concluded that PEP carboxykinase functions primarily as a decarboxylating enzyme under natural temperature regimes in CAM plants. The enzyme required both Mn^{2+} and Mg^{2+} for maximum activity. The PEP carboxykinase was also found to be active in a decarboxylation reaction in which ATP was replaced by ADP; in resemblance to avian liver PEP carboxykinase (NOCE and UTTER, 1975), the reaction product was thought to be pyruvate instead of PEP. This alternative route reflects the variability of CAM plants, which, according to environmental factors, are able to adjust their biochemical pathway (OSMOND et al., 1973). The utilization of ADP could be a measure triggered by a low energy charge to divert carbon from gluconeogenesis into the tricarboxylic acid cycle. The decarboxylation of OAA by PEP carboxykinase not only fulfils the demand for CO_2 by RuBP carboxylase, but also caters for the process of gluconeogenesis by providing PEP. Consequently pyruvate orthophosphate dikinase would be redundant in plants with PEP carboxykinase; indeed, the presence of both enzymes in the same CAM plant has not yet been reported. Bypassing pyruvate orthophosphate dikinase would save one mol of ATP per PEP formed, but the possible regulatory advantage of a light-activated dikinase would not be available in its absence.

G. Respiratory Enzymes

The uptake of oxygen during darkness and its mandatory presence for the formation and accumulation of malate points to operation of the tricarboxylic acid cycle in CAM plants. This contention is supported by experiments with labeled pyruvate (BRADBEER and RANSON, 1963), where no evidence for any irregularity of the tricarboxylic acid cycle in *K. crenata* was found. Activities of fumarase and cytochrome c oxidase were demonstrated in mitochondria of *K. daigremontiana* (DITTRICH, 1976a); fumarase was also shown to catalyze the equilibration of label in malate during CO_2 fixation in the dark (DITTRICH, 1976b). SUTTON (1975a) has measured pyruvate kinase in *K. daigremontiana*; he reported only low enzyme activities, but these were nevertheless sufficient to provide enough substrate for the tricarboxylic acid cycle, while not posing a problem of competition for PEP with regard to PEP carboxylase.

H. Conclusion

CAM is not an obligatory mechanism like the C_4 pathway of photosynthesis, though both mechanisms are based upon the same groups of enzymes. Environmental factors can either induce C_3 metabolism in CAM plants or require strict diurnal cycles of acidification and deacidification typical of CAM; all transitory forms between both alternatives are possible. This remarkable flexibility is in part a consequence of the participating enzymes which must be subject to specific modes of regulation, since they are not protected from mutual competition by intercellular compartmentation as in the C_4 pathway. Our present efforts in understanding interactions during CAM still concentrate on elucidating the basic pathways of carbon flow. Insights gained from the resolution of such enigmas as the mechanism of RuBP carboxylase will continue to contribute to studies of the complex nature of CAM and its associated enzymes.

References

Anderson, D.G., Stafford, H.A., Conn, E.E., Vennesland, B.: Plant Physiol. *27*, 576–684 (1952)
Badger, M.R., Andrews, T.J., Osmond, C.B.: In: Proc. 3rd Int. Congr. Photosynthesis. Avron, M. (ed.) pp. 1421–1429. Amsterdam: Elsevier Scientific Publ. Co. 1975
Bonugli, K.J., Davies, D.D.: Planta (Berl.) *133*, 281–287 (1977)
Bradbeer, J.W., Ranson, S.L.: Proc. R. Soc. B *157*, 258–278 (1963)
Brandon, P.C., Boekel-Mol, T.N.v.: Eur. J. Biochem. *35*, 62–69 (1973)
Daley, L.S., Ray, T.B., Vines, H.M., Black, C.C.: Plant Physiol. *59*, 618–622 (1977)
Dittrich, P.: Plant Physiol. *57*, 310–314 (1976a)
Dittrich, P.: Plant Physiol. *58*, 288–291 (1976b)
Dittrich, P., Meusel, M.: in preparation (1979)
Dittrich, P., Campbell, W.H., Black, C.C.: Plant Physiol. *52*, 357–361 (1973)
Garland, P.B., Randle, P.J., Newsholme, E.A.: Nature (London) *200*, 169–170 (1963)
Garnier-Dardart, J.: Physiol. Veg. *3*, 215–217 (1965)
Garnier-Dardart, J., Queiroz, O.: Phytochemistry *13*, 1695–1702 (1974)
Kelly, G.J., Turner, J.F.: Biochim. Biophys. Acta *242*, 559–565 (1971)
Kelly, G.J., Zimmermann, G., Latzko, E.: Biochim. Biophys. Res. Commun. *70*, 193–199 (1976)
Kluge, M., Osmond, C.B.: Naturwissenschaften *58*, 414–415 (1971)
Kluge, M., Osmond, C.B.: Z. Pflanzenphysiol. *66*, 97–105 (1972)
Levi, C., Gibbs, M.: Plant Physiol. *56*, 164–166 (1975)
Linden, C.J., Tanner, W., Kandler, O.: Plant Physiol. *54*, 752–757 (1974)
Mukerji, S.K., Ting, I.: Phytochemistry *7*, 903–911 (1968a)
Mukerji, S.K., Ting, I.: Biochim. Biophys. Acta *167*, 239–249 (1968b)
Noce, P.S., Utter, M.F.: J. Biol. Chem. *250*, 9099–9105 (1975)
Osmond, C.B., Allaway, W.G., Sutton, B.G., Troughton, J.H., Queiroz, O., Lüttge, U., Winter, K.: Nature (London) *246*, 41–42 (1973)
Queiroz, O.: C.R. Acad. Sci. (Paris) *265*, 1928–1931 (1967)
Queiroz, O.: Phytochemistry *8*, 1655–1663 (1969)
Queiroz, O., Morel, C.: Plant Physiol. *53*, 596–602 (1974)
Schilling, N.: personal communication (1976)
Schilling, N., Dittrich, P.: Planta (Berl.) (1979), in preparation
Schilling, N., Kandler, O.: Biochem. Soc. Transact. *3*, 985–987 (1975)
Schnabl, H.: Z. Pflanzenphysiol. *77*, 167–173 (1976)
Sugiyama, T., Laetsch, W.M.: Plant Physiol. *56*, 605–607 (1975)
Sutton, B.G.: Aust. J. Plant Physiol. *2*, 389–402 (1975a)
Sutton, B.G.: Aust. J. Plant Physiol. *2*, 403–411 (1975b)
Ting, I.P.: Plant Physiol. *43*, 1919–1924 (1968)
Ting, I.P., Osmond, C.B.: Plant Sci. Lett. *1*, 123–128 (1973a)
Ting, I.P., Osmond, C.B.: Plant Physiol. *51*, 448–453 (1973b)
Vieweg, G.H., de Fekete, M.A.R.: Z. Pflanzenphysiol. *81*, 74–79 (1977)
Walker, D.A.: Biochem. J. *74*, 216–223 (1960)
Wilkinson, M.J., Smith, H.: Plant Sci. Lett. *6*, 319–324 (1976)

22. Interaction Between Photochemistry and Activity of Enzymes

L.E. ANDERSON

A. Introduction

The photochemical apparatus affects the activity of chloroplastic and of cytoplasmic enzymes in several ways. Light-mediated activation or inactivation results in changes in maximal velocity of several enzymes. Within the chloroplast the light-modulated enzymes and other enzymes are also affected by pH and Mg^{2+} and/or by ATP and NADPH. Photochemistry then affects stromal chemistry at two levels: (1) by altering the catalytic properties of several enzymes; (2) through a superimposed regulation by effectors and substrates generated by the photochemical apparatus.

B. Light-Mediated Modulation

I. Occurrence

Light-mediated activation of a reductive pentose phosphate cycle enzyme, NADP-linked GAP dehydrogenase, was first observed by ZIEGLER and ZIEGLER (1965). Since that time four other C_3 pathway enzymes and two chloroplast enzymes involved in reactions related to transport have been reported to be light-activated, and two enzymes which participate in hexose breakdown have been shown to be light-inactivated (ANDERSON, 1974; see Fig. 1, Table 1). Light-mediated modulation of enzyme activity occurs in O_2-evolving plants, in species as diverse as blue-green algae and angiosperms, in plants which fix CO_2 by the C_3, C_4, or CAM pathway (Table 1). It has not been observed in the photosynthetic bacteria.

II. Metabolic Significance

The effect of light modulation on the catalytic properties of four of the modulatable enzymes has been examined. The pH optima of the more active and the less active forms of NADP-linked GAP dehydrogenase and of chloroplastic and cytoplasmic G6P dehydrogenases are similar or identical (ANDERSON and DUGGAN, 1976). Modulation affects maximal velocity of these enzymes and of FBPase (ANDERSON and DUGGAN, 1976; ANDERSON et al., 1976).

It follows that, when the levels of reductive pentose phosphate cycle intermediates are high, activation should only lead to about a two-fold increase in CO_2 fixation, since the increase in the rate of the cycle will be equal to the increase in the maximal velocity of the rate-limiting enzyme SBP phosphatase, (1.8-fold activated in peas). However, when cycle intermediates are low (at dawn, when

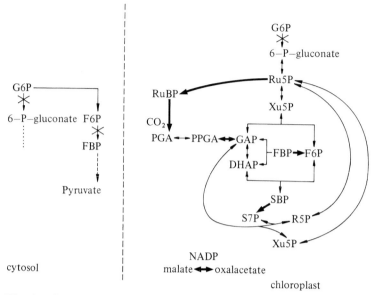

Fig. 1. Effect of light on carbon metabolism in C_3 plants. *Heavy arrows* indicate reactions which have been reported to be light-activated; *crossing out* indicates light inactivation

plant has been in dark), the increase in flow of carbon through the cycle will be related to the product of the activation factors. This follows from a consideration of the effect of substrate levels on enzyme activity: at low substrate levels enzyme activity is a function of the substrate level, and therefore, in a cycle, to the maximal velocity of each substrate-producing enzyme. The effect of activation will be damped off as intermediate levels rise, resulting in decrease in acceleration with increase in velocity. The inherent properties of enzyme-catalyzed reactions, which derive from saturation kinetics, in combination with light-mediated activation probably are responsible for the autocatalytic enhancement of CO_2 fixation in intact chloroplasts observed by WALKER (1967) and others.

Arsenite inhibits light-mediated enzyme activation (ANDERSON and AVRON, 1976) and photosynthetic CO_2 fixation (GIBBS and CALO, 1960). Arsenite inhibition of photosynthetic CO_2 fixation only occurs when the inhibitor is present prior to illumination; it can be partially overcome if FBP is added to chloroplast suspensions, in which case the rate of CO_2 uptake is 1.7-fold higher in the chloroplasts not treated with arsenite (GIBBS et al., 1967). Apparently in the presence of arsenite the lag period in CO_2 fixation is extended. When an intermediate is added CO_2 fixation is limited by the rate of the rate-limiting enzyme, which, not having been activated, is 1.7-fold less active than in the normal chloroplast.

III. Mechanism

1. The Light Receptor

The involvement of the photochemical apparatus in the activation of NADP-linked GAP dehydrogenase was first demonstrated by ZIEGLER et al. (1965) who monitored

Table 1. Light modulation in different types of plants. Values are for activation (x-fold). *Anacystis nidulans* data from DUGGAN and ANDERSON (1975); algal data from ZIEGLER and ZIEGLER (1967); *Pisum* malic dehydrogenase from ANDERSON and AVRON (1976), GAP dehydrogenase and G6P dehydrogenase from ANDERSON and DUGGAN (1976), Ru5P kinase and FBPase, SBPase from ANDERSON (1974) and ANDERSON, unpublished. P-fructokinase from KACHRU and ANDERSON (1975); *Spinacia* malic dehydrogenase from WOLOSIUK et al. (1977), GAP dehydrogenase and Ru5P kinase from STEIGER et al. (1971), FBPase and SBPase from CHAMPIGNY and BISMUTH (1976), G6P dehydrogenase from LENDZIAN and ZIEGLER (1970), *Zea* malic dehydrogenase from JOHNSON (1971), GAP dehydrogenase and Ru5P kinase from STEIGER et al., (1971), pyruvate Pi dikinase from HATCH and SLACK (1969), *Tidestromia* data from BJÖRKMAN and BADGER (1977); *Kalanchoë* data from GUPTA and ANDERSON (1978) except P-fructokinase which is unpublished data of GUPTA and ANDERSON

	Prokaryotes			Eukaryotes				
	Ana- cystis nidulans	Algae[a]		C₃ Plants		C₄ Plants		CAM
		Green	Brown	*Pisum*	*Spina- cia*	*Zea*	*Tides- tromia*	*Kalan- choë*
Light- activated enzymes	Light stimulation x-fold							
NADP-linked malic dehydrogenase				14[b]	∞	50	3.3	1.7
NADP-linked GAP dehydrogenase	Nil	~2	Nil	2.4[d]	5[d]	2	3.1	2
Ru5P kinase	2			7.7	3.2	1.6	4	4.4
FBPase	Nil			1.7	2.2[b]		Nil	Nil
SBPase	Nil			1.8	1.7[b]			1.7
Pyruvate Pi dikinase						~12[e]		
Dark activated enzymes	Dark stimulation x-fold							
G6P dehydrogenase	4.8			1.3[b] 1.4[c]	3			Nil
P-fructokinase				11				3

[a] Activation of glyceraldehyde-3-P dehydrogenase was observed in all (4) species of green algae tested, in some (4) species of red algae, and in none (2 species) of brown algae tested.
[b] Values from experiments with intact chloroplasts.
[c] Cytoplasmic enzyme.
[d] Glyceraldehyde-3-P dehydrogenase is activated in *Vicia faba, Nicotiana tabacum, Brassica napus, Valerianella olitoria, Beta vulgaris, Myrothamnus flabellifolia, Triticum aestivum, Avena sativa, Lemna gibba* (ZIEGLER and ZIEGLER, 1965) and *Saccharum* (STEIGER et al., 1971).
[e] Pyruvate Pi dikinase is activated in *Amaranthus palmeri* (HATCH and SLACK, 1969).

activation in *Vicia faba* leaves irradiated with light of selected wavelengths. Chlorophyll was implicated as the photoreceptor. These experiments were later extended to include the red alga *Ceramium rubrum*, in which the phycobilins also act as photoreceptors (ZIEGLER et al., 1968). A complete action spectrum showing clear correlation of activation of the NADP-linked GAP and malate dehydrogenases and O_2 evolution in duckweed (*Lemna gibba*) appeared in 1969 (SCHMIDT-CLAUSEN and ZIEGLER, 1969). In maize (*Zea mays*) leaves the action spectrum for activation of the mesophyll chloroplast enzyme pyruvate, phosphate dikinase, is essentially identical with the action spectrum of photosynthesis (YAMAMOTO et al., 1974).

The use of inhibitors of photosynthetic electron transport offered another means of implicating the photochemical apparatus in light-mediated modulation of chloroplast enzyme activity. DCMU at low levels inhibits photosynthetic electron transport and light-mediated activation of NADP-linked GAP dehydrogenase in *Lemna* (ZIEGLER and ZIEGLER, 1966) and of Ru5P kinase, NADP-linked malic dehydrogenase and NADP-linked GAP dehydrogenase in intact chloroplasts (AVRON and GIBBS, 1974; ANDERSON and AVRON, 1976; HATCH, 1977). Light-mediated inactivation of G6P dehydrogenase is DCMU-inhibited (ANDERSON and AVRON, 1976). CHAMPIGNY and BISMUTH (1976) likewise have reported inhibition of activation of GAP dehydrogenase and of FBPase in chloroplasts treated with CMU. Activation of pyruvate, phosphate dikinase, is markedly inhibited in maize leaf disks infiltrated with DCMU (YAMAMOTO et al., 1974).

Diquat, which shunts electron transport at the photosystem I electron acceptor level, interferes with light-mediated modulation of Ru5P kinase (AVRON and GIBBS, 1974), of the NADP-linked GAP and malate dehydrogenases and of G6P dehydrogenase (ANDERSON and AVRON, 1976). Photosystem I must then be involved in the activation of these four enzymes. Since the ferredoxin antagonist DSPD has no effect on activation of Ru5P kinase (AVRON and GIBBS, 1974) and actually stimulates light-mediated modulation of the activity of the three dehydrogenases (ANDERSON and AVRON, 1976), the electron transport chain component involved in the modulation of the activity of these enzymes must precede ferredoxin. The uncoupler FCCP has little or no effect on the modulation of the G6P and malic dehydrogenases (ANDERSON and AVRON, 1976) or on the activation of the kinase (AVRON and GIBBS, 1974) but does affect activation of GAP dehydrogenase (ANDERSON and AVRON, 1976). Modulation of the activity of the G6P and malic dehydrogenases and of Ru5P kinase, and possibly of NADP-linked GAP dehydrogenase, then involves component(s) of the electron transport chain located on the reducing side of photosystem I prior to ferredoxin.

Reduced ferredoxin is required for activation of FBPase (BUCHANAN et al., 1967, 1971) and SBPase (SCHÜRMANN and BUCHANAN, 1975) in broken chloroplast systems. The ferredoxin antagonist DSPD is a potent inhibitor of FBPase and SBPase activation in a related system (CHIN and ANDERSON, 1977; ANDERSON and AVRON, 1976).

Two sites in the electron transport chain are then implicated in the modulation of the activity of chloroplastic enzymes. One, on the reducing side of photosystem I, but before ferredoxin, is involved in the inactivation of G6P dehydrogenase and the activation of Ru5P kinase and NADP-linked malic dehydrogenase. The other, probably ferredoxin, is involved in the activation of the phosphatases. The primary electron acceptor of photosystem I is apparently an iron-sulfur protein

(THORNBER et al., 1977; GOLBECK et al., 1976) and may be the only component of the electron transport chain prior to ferredoxin on the reducing side of photosystem I. Both of the electron transport chain components involved in light modulation of enzyme activity may then be iron–sulfur proteins.

2. Mediators

Two systems have been proposed to account for light-mediated modulation of enzyme activity in chloroplasts. It has also been suggested that modulation is mediated by ATP and NADPH.

a) LEM System

Modulation of the activity of Ru5P kinase and of the NADP-linked malic, GAP, and G6P dehydrogenases in the pea leaf chloroplast system appears to be mediated by a membrane-bound vicinal-dithiol-containing factor, or light effect mediator (LEM; ANDERSON and AVRON, 1976). The dark or inactive form of the LEM is thought to be converted to the light, active form by reduction of a disulfide bond, the electrons being supplied by photosynthetic electron transport. The LEM is sensitive to sulfite in both light and dark, and therefore probably contains a disulfide group in the active as well as in the inactive form. There are at least two LEM's, LEM_I, which accepts electrons from the electron transport system at a site on the reducing side of photosystem I prior to ferredoxin, and LEM_{II}, which accepts electrons from ferredoxin or from a component beyond ferrodoxin (Fig. 2). There may be as many LEM's as there are light-modulated enzymes in the chloroplast (ANDERSON and AVRON, 1976). $LEM_{I(G6P-D)}$ has a sharp pH optimum around 7.3 and is NADP- and NAD-sensitive (ANDERSON and DUGGAN, 1976). $LEM_{II(FBPase)}$, likewise, has a sharp pH optimum around 7.3. In contrast to $LEM_{I(G6P-D)}$, $LEM_{II(FBPase)}$ requires Mg^{2+} for activity, 10 mM being optimal and higher concentrations being inhibitory (CHIN and ANDERSON, 1977).

DTNB treatment of stromal extracts, followed by removal of uncombined DTNB, results in inhibition of light-mediated modulation of pea leaf G6P and malic dehydrogenases in reconstituted chloroplast systems. DTT treatment, also, results in inhibition of modulation. The modulatable enzyme must then have both a free thiol group and a disulfide bond in order to undergo LEM-catalyzed modulation. These experiments, and the fact that the active form of the LEM apparently contains both dithiol and disulfide groups, suggest that light-mediated modulation involves thiol-disulfide exchange on the modulatable protein catalyzed by the reductively activated LEM (ANDERSON, et al., 1978; Fig. 3).

Experiments with the CAM plant *Kalanchoë* implicate a LEM_I system in the activation of NADP-linked GAP dehydrogenase and malic dehydrogenase in that plant. The pea leaf and *Kalanchoë* $LEM_{I(GAP-D)}$ systems are interchangeable (GUPTA and ANDERSON, unpublished).

b) Thioredoxin System

A soluble protein factor is required for ferredoxin-dependent activation of the phosphatases in broken spinach chloroplast preparations (BUCHANAN et al., 1971;

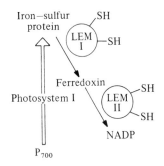

Fig. 2. Localization of LEM systems with respect to the photosynthetic electron transport system. The LEM's might either interact with the respective iron–sulfur proteins or with some component after the first iron–sulfur protein but prior to the second iron–sulfur protein (here labeled ferredoxin) in the case of LEM$_I$ and prior to NADP in the case of LEM$_{II}$

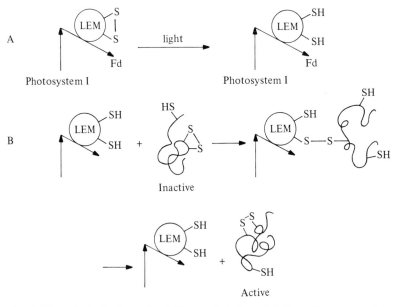

Fig. 3. Hypothetical scheme for LEM-mediated modulation of enzyme activity. **A** Reductive activation of LEM; **B** activation of stromal enzyme by thiol, disulfide exchange

SCHÜRMANN and BUCHANAN, 1975; BUCHANAN et al., 1976). This protein factor has now been resolved into two components: the enzyme ferredoxin-thioredoxin reductase, and thioredoxin (SCHÜRMANN et al., 1976; BUCHANAN and WOLOSIUK, 1976: HOLMGREN et al., 1977; WOLOSIUK and BUCHANAN, 1977; Fig. 4).

Thioredoxin is the reductant utilized almost ubiquitously by ribonucleotide reductase. In microbial and mammalian systems thioredoxin is reduced enzymatically by NADPH (REICHARD, 1968; HOLMGREN, 1977). Thioredoxin is also a potent general disulfide reductant: HOLMGREN and MORGAN (1976) have proposed that it be used instead of DTT in studies of the location and function of protein disulfide bridges. Reduced thioredoxin has now been shown to be capable of activating NADP-linked malic dehydrogenase (WOLOSIUK et al., 1977), NADP-linked GAP dehydrogenase and Ru5P kinase (SCHÜRMANN et al., 1976), as well as the phosphatases. Since DTT mimics light-treatment in increasing the activity

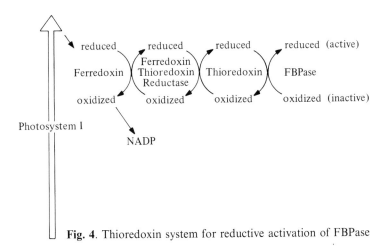

Fig. 4. Thioredoxin system for reductive activation of FBPase

of the light-activated enzymes and in decreasing the activity of the light-inactivated enzymes (ANDERSON, 1974), the similarity between modulation by light, by thioredoxin, and by DTT make the thioredoxin system a plausible alternative to the LEM system.

c) ATP, NADPH

Dissociation of a low-activity aggregate of spinach leaf GAP dehydrogenase into a lower molecular weight, higher specific activity form is observed when the low activity aggregate is incubated with ATP or NADPH. These observations, coupled with the observed correlation of light activation with the action spectrum of photosynthesis (see above), have led to the suggestion that light activation of this enzyme is mediated by low molecular weight substances, ATP and/or NADPH, generated by photosynthetic electron transport (MÜLLER et al., 1969; MÜLLER, 1972; PUPILLO and PICCARI, 1975). Metabolite activation appears to be hysteretic (WOLOSIUK and BUCHANAN, 1976). The experiments of ANDERSON and AVRON (1976) would seem to rule out NADPH as mediator of light activation, at least in pea leaf chloroplasts, since the ferredoxin antagonist DSPD, which blocks NADP reduction, stimulates activation. The uncoupler FCCP does inhibit activation of the dehydrogenase, which could point to the involvement of ATP in activation. But FCCP is also a sulfhydryl reagent and both the LEM and thioredoxin contain reactive dithiol groups. FCCP inhibition may simply indicate that the LEM for activation of GAP dehydrogenase is itself sensitive. A similar problem has been encountered in a bacterial sugar transport system (KABACK et al., 1974). Activation of GAP dehydrogenase does occur in broken chloroplast systems even after the crude stromal fraction has been filtered through G-25 Sephadex (ANDERSON, unpublished) and it seems most unlikely that ATP concentrations approaching those in intact chloroplasts are generated in such preparations. ATP, then, as well as NADPH, seems to be eliminated as a mediator of light activation of the pea leaf dehydrogenase. Unlike the spinach enzyme the pea leaf enzyme does not form aggregates (McGOWAN and GIBBS, 1974, ANDERSON and LIM, 1972).

Lendzian and Bassham (1975) and Wildner (1975) have suggested that NADPH is responsible for light-dependent inactivation of chloroplast G6P dehydrogenase. Since the LEM system is clearly involved in inactivation of this enzyme, and since the ferredoxin antagonist DSPD enhances light inactivation of this enzyme (Anderson and Avron, 1976), it seems highly unlikely that light inactivation of this chloroplastic dehydrogenase is the result of inhibition by NADPH. It seems reasonable, however, that a further, superimposed regulation of the activity of the enzyme is accomplished by NADPH (Anderson and Duggan, 1976).

IV. Special Cases

1. Cytoplasmic Enzymes

Light-mediated inactivation of cytoplasmic G6P dehydrogenase is inhibited in pea leaves which have been infiltrated in the dark with 10^{-4} M DCMU (Anderson and Nehrlich, 1977). The photochemical apparatus, then, appears to be involved in light modulation of the activity of this enzyme in the cytoplasm. Nothing further is known about this modulation, but clearly the membrane-bound LEM which inactivates the chloroplastic G6P dehydrogenase cannot interact directly with the cytoplasmic enzyme.

Phosphofructokinase, presumably the cytoplasmic form, is light-inactivated in pea (Kachru and Anderson, 1975) and in *Kalanchoë* (Gupta and Anderson, unpublished). Experiments to determine whether the photochemical apparatus is involved have not been done.

2. RuBP Carboxylase

Light activation of RuBP carboxylase in crude extracts of a number of autotrophic organisms ranging from photosynthetic bacteria to higher plants has been reported by Wildner and Criddle (1969). A light activating factor (LAF) has been prepared from tomato leaf tissue. It can be resolved into two components. One is probably a protein. The other, probably the light-absorbing component, is a low molecular weight compound which may be chlorogenic acid (Wildner et al., 1972).

Purified RuBP carboxylase is activated by Mg^{2+} and CO_2 (Pon et al., 1963; Badger and Lorimer, 1976; Lorimer et al., 1976; see Chap. II.17, this vol.). In effect this is light activation mediated by H^+ and Mg^{2+}, the levels of which change in the stroma when the chloroplast is illuminated (see Sect. D).

C. Dark Modulation

The dark reversal of light-mediated modulation has received little attention. Dark inactivation of Ru5P kinase in intact spinach chloroplasts is slow (Avron and Gibbs, 1974). Light-mediated modulation of the NADP-linked dehydrogenases and FBP phosphatase is readily reversible in broken chloroplast preparations and in intact chloroplasts (ANDERSON and AVRON, 1976; ANDERSON and DUGGAN, 1976; CHIN and ANDERSON, 1977). Inactivation of thioredoxin-activated NADP-linked malic dehydrogenase is mediated by a membrane-bound oxidant (WOLOSIUK et al., 1977). Inactivation of thioredoxin activated FBPase can be accomplished with soluble oxidants such as dehydroascorbate, tetrathionate, or oxidized glutathione (SCHÜRMANN and WOLOSIUK, 1978).

D. Thylakoid-Generated Effectors

The photochemical apparatus affects the activity of the enzymes of the reductive pentose phosphate cycle as a result of the uptake of protons into the thylakoid membranes in the light and the concomitant release of Mg^{2+} ions, and as a result of phosphorylation of ATP. In some cases these effects are superimposed on the effect of light-mediated modulation of enzyme activity.

I. pH

The pH of the chloroplast stroma shifts from 7.1 to 8.1 when chloroplasts are illuminated, as a result of the uptake of protons into the thylakoid membranes (HELDT et al., 1973). Three of the enzymes of the reductive pentose phosphate cycle, FBPase (BAIER and LATZKO, 1975), SBPase (SCHÜRMANN and BUCHANAN, 1975; ANDERSON, unpublished), and Ru5P kinase (HURWITZ et al., 1956; ANDERSON, 1973) are markedly pH-dependent (see Chap. II.16, this vol.). The chloroplast seems to be unique in that activity of enzymes is routinely controlled by pH changes within the organelle.

II. Mg^{2+}

Changes in Mg^{2+} levels in the stroma were first reported by DILLEY and VERNON (1965). The Mg^{2+} dependent enzymes of the reductive pentose phosphate cycle are RuBP carboxylase (WEISSBACH et al., 1956), PGA kinase (PACOLD and ANDERSON, 1975) FBPase (BAIER and LATZKO, 1975; PREISS et al., 1967) SBPase (ANDERSON, 1974), and Ru5P kinase (HURWITZ et al., 1956). These enzymes will be affected by light-induced changes in stromal Mg^{2+} (see Chap. II.16, this vol.).

III. Energy Charge

ATP levels increase in the chloroplast when it is illuminated, with the result that energy charge within the organelle is significantly higher in the light (PACOLD and ANDERSON, 1973, calculated from data of HEBER and SANTARIUS, 1970). PGA kinase is energy charge-sensitive and it seems possible that the activity of this enzyme is regulated by energy charge levels in vivo (PACOLD and ANDERSON, 1973;

Lavergne et al., 1974). Although Ru5P kinase from spinach reportedly is under energy charge control (Lavergne et al., 1974), Anderson (1973) was unable to demonstrate an energy charge effect on this enzyme isolated from pea leaves.

E. Conclusion

The photochemical apparatus apparently controls stromal carbon metabolism by altering the catalytic properties of certain enzymes and by generating effectors and substrates which affect the activity of these and other enzymes. Control is complex as a result of the interplay of these two regulatory mechanisms.

References

Anderson, L.E.: Biochim. Biophys. Acta *321*, 484–488 (1973)
Anderson, L.E.: Biochem. Biophys. Res. Commun. *59*, 907–913 (1974)
Anderson, L.E.: In: Proc. 3rd Int. Congr. Photosynthesis. Avron, M., (ed.), pp. 1393–1406. Amsterdam: Elsevier 1974
Anderson, L.E., Avron, M.: Plant Physiol. *57*, 209–213 (1976)
Anderson, L.E., Duggan, J.X.: Plant Physiol. *58*, 135–139 (1976)
Anderson, L.E., Lim, T.-C.: FEBS Lett. *27*, 189–191 (1972)
Anderson, L.E., Nehrlich, S.C.: FEBS Lett. *76*, 64–66 (1977)
Anderson, L.E., Duggan, J.X., Chin, N-M., Park, K.E.Y.: Fed. Proc. *35*, 1014 (1976)
Anderson, L.E., Nehrlich, S.C., Champigny, M-L.: Plant Physiol. *61*, 601–605 (1978)
Avron, M., Gibbs, M.: Plant Physiol. *53*, 136–139 (1974)
Badger, M.R., Lorimer, G.H.: Arch. Biochem. Biophys. *175*, 723–729 (1976)
Baier, D., Latzko, E.: Biochim. Biophys. Acta *396*, 141–148 (1975)
Björkman, O., Badger, M.: Annual report of the director. Dept. Plant Biol. Carnegie Inst. 1976–1977, 346–354 (1977)
Buchanan, B.B., Wolosiuk, R.A.: Nature (London) *264*, 669–670 (1976)
Buchanan, B.B., Kalberer, P.P., Arnon, D.I.: Biochem. Biophys. Res. Commun. *29*, 74–79 (1967)
Buchanan, B.B., Schürmann, P., Kalberer, P.P.: J. Biol. Chem. *246*, 5952–5959 (1971)
Buchanan, B.B., Schürmann, P., Wolosiuk, R.A.: Biochem. Biophys. Res. Commun. *69*, 970–978 (1976)
Champigny, M-L., Bismuth, E.: Physiol. Plant. *36*, 95–100 (1976)
Chin, H-M., Anderson, L.E.: Abstracts, 174th Nat. Am. Chem. Soc. Meet. Biol-155 (1977)
Dilley, R.A., Vernon, L.P..: Arch. Biochem. Biophys. *111*, 365–375 (1965)
Duggan, J.X., Anderson, L.E.: Planta (Berl.) *122*, 293–297 (1975)
Gibbs, M., Calo, N.: Biochim. Biophys. Acta *44*, 341–347 (1960)
Gibbs, M., Bamberger, E.S., Ellyard, P.W., Everson, R.G.: In: Biochemistry of chloroplasts. Goodwin, T.W. (ed.), Vol. II, pp. 3–38. New York: Academic Press 1967
Golbeck, J.H., Lien, S., San Pietro, A.: Biochem. Biophys. Res. Commun. *71*, 452–458 (1976)
Gupta, V.K., Anderson, L.E.: Plant Physiol. *61*, 469–471 (1978)
Hatch, M.D.: Plant Cell Physiol. Special Issue 311–314 (1977)
Hatch, M.D., Slack, C.R.: Biochem. J. *112*, 549–558 (1969)
Heber, U., Santarius, K.A.: Z. Naturforsch. *25b*, 718–728 (1970)
Heldt, H.W., Werdan, K., Milovancev, M., Geller, G.: Biochim. Biophys. Acta *314*, 224–241 (1973)

Holmgren, A.: J. Biol. Chem. *252*, 4600–4606 (1977)
Holmgren, A., Morgan, F.J.: Eur. J. Biochem. *70*, 377–383 (1976)
Holmgren, A., Buchanan, B.B., Wolosiuk, R.A.: FEBS Lett. *82*, 351–354 (1977)
Hurwitz, J., Weissbach, A., Horecker, B.L., Smyrniotis, P.Z.: J. Biol. Chem. *218*, 769–783 (1956)
Johnson, H.S.: Biochem. Biophys. Res. Commun. *43*, 703–709 (1971)
Kaback, R.R., Reeves, J.P., Short, S.A., Lombardi, F.J.: Arch. Biochem. Biophys. *160*, 215–222 (1974)
Kachru, R.B., Anderson, L.E.: Plant Physiol. *55*, 199–202 (1975)
Lavergne, D., Bismuth, E., Champigny, M-L.: Plant Sci. Lett. *3*, 391–397 (1974)
Lendzian, K., Bassham, J.A.: Biochim. Biophys. Acta *396*, 260–275 (1975)
Lendzian, K., Ziegler, H.: Planta (Berl.) *94*, 27–36 (1970)
Lorimer, G.H., Badger, M.R., Andrews, T.J.: Biochemistry *15*, 529–536 (1976)
McGowan, R.E., Gibbs, M.: Plant Physiol. *54*, 312–319 (1974)
Müller, B.: Z. Naturforsch. *27b*, 925–932 (1972)
Müller, B., Ziegler, I., Ziegler, H.: Eur. J. Biochem. *9*, 101–106 (1969)
Pacold, I., Anderson, L.E.: Biochem. Biophys. Res. Commun. *51*, 139–143 (1973)
Pacold, I., Anderson, L.E.: Plant Physiol. *55*, 168–171 (1975)
Pon, N.G., Rabin, B.R., Calvin, M.: Biochem. Z. *338*, 7–19 (1963)
Preiss, J., Biggs, M.L., Greenberg, E.: J. Biol. Chem. *242*, 2292–2294 (1967)
Pupillo, P., Piccari, G.: Eur. J. Biochem. *51*, 475–482 (1975)
Reichard, P.: Eur. J. Biochem. *3*, 259–266 (1968)
Schmidt-Clausen, H.J., Ziegler, I.: In: Progr. Photosynthesis Res. Metzner, H. (ed.), Vol. III, pp. 1646–1652. Tübingen: H. Laupp, Jr. 1969
Schürmann, P., Buchanan, B.B.: Biochim. Biophys. Acta *376*, 189–192 (1975)
Schürmann, P., Wolosiuk, R.A.: Biochim. Biophys. Acta *522*, 130–138 (1978)
Schürmann, P., Wolosiuk, R.A., Breazeale, V.D., Buchanan, B.B.: Nature (London) *263*, 257–258 (1976)
Steiger, E., Ziegler, I., Ziegler, H.: Planta (Berl.) *96*, 109–118 (1971)
Thornber, J.P., Alberte, R.S., Hunter, F.A., Shiozawa, J.A., Kan, K-S.: Brookhaven Symp. Biol. *28*, 132–148 (1977)
Walker, D.A.: In: Biochemistry of chloroplasts. Goodwin, T.W. (ed.), Vol. II, pp. 53–69. New York: Academic Press 1967
Weissbach, A., Horecker, B.L., Hurwitz, J.: J. Biol. Chem. *218*, 795–810 (1956)
Wildner, G.F.: Z. Naturforsch. *30c*, 756–760 (1975)
Wildner, G.F., Criddle, R.S.: Biochem. Biophys. Res. Commun. *37*, 952–960 (1969)
Wildner, G.F., Zilg, H., Criddle, R.S.: In: Proc. 2nd Int. Cong. Photosynthesis. Forti, G., Avron, M., Melandri, A., (ed.), pp. 1825–1830. The Hague: Dr. W. Junk, N.V. 1972
Wolosiuk, R.A., Buchanan, B.B.: J. Biol. Chem. *251*, 6456–6461 (1976)
Wolosiuk, R.A., Buchanan, B.B.: Nature (London) *266*, 565–567 (1977)
Wolosiuk, R.A., Buchanan, B.B., Crawford, N.A.: FEBS Lett. *81*, 253–258 (1977)
Yamamoto, E., Sugiyama, T., Miyachi, S.: Plant Cell Physiol. *15*, 987–992 (1974)
Ziegler, H., Ziegler, I.: Planta (Berl.) *65*, 369–380 (1965)
Ziegler, H., Ziegler, I.: Planta (Berl.) *69*, 111–123 (1966)
Ziegler, H., Ziegler, I.: Planta (Berl.) *72*, 162–169 (1967)
Ziegler, H., Ziegler, I., Schmidt-Clausen, H.J.: Planta (Berl.) *67*, 344–356 (1965)
Ziegler, H., Ziegler, I., Schmidt-Clausen, H.J.: Planta (Berl.) *81*, 169–180 (1968)

II E. Metabolism of Primary Products of Photosynthesis

23. Metabolism of Starch in Leaves

J. Preiss and C. Levi

A. Introduction

Starch is present in most green plants and in practically every type of tissue; leaves, fruits, pollen grains, roots, shoots and stems. Many early experiments have demonstrated the disappearance of starch in leaves either by growth in low light or in dark for 24–48 h (Sachs, 1887). If the leaf is brightly illuminated for an hour or more then the reappearance of starch granules in chloroplasts can easily be demonstrated by a number of means; electron microscopy (Badenhuizen, 1969) or I_2 staining (Meyer and Gibbons, 1951). Thus starch synthesis is an important end product of carbon fixation during photosynthesis. The starch formed in the light is degraded by respiration in the dark (Sachs, 1887). The biosynthetic and degradative processes of leaf starch are therefore more dynamic than the metabolism of starch in reserve tissues. In the development of the storage tissue the biosynthetic processes are predominant. The degradation of starch predominates during germination.

This chapter will review the present information known about leaf starch biosynthesis and degradation and the regulatory phenomena associated with these processes. Comparisons where possible will be made with similar systems in non-chlorophyllous plant tissue and algae. Recently reviews (Manners, 1974a; French, 1975) and a monograph (Banks and Greenwood, 1975) of the structure and composition of the starch granule and its components amylose and amylopectin have appeared. Thus, these subjects will not be reviewed here.

B. Starch Biosynthesis

I. Reactions Involved in Starch Biosynthesis

Prior to 1960 the only enzyme known to be responsible for syntheses of the α-1,4 glucosidic linkages in plants was phosphorylase [reaction (1); Hanes, 1940].

$$\text{glucose-1-P} + (\text{glucosyl})_n \rightleftharpoons P_i + (\text{glucosyl})_{n+1}. \tag{1}$$

However it is now believed that the starch synthase reaction [reaction (2)] is the main if not exclusive reaction involved in the synthesis of the starch α-1,4 glucosidic linkages (Leloir et al., 1961).

$$\text{ADP(UDP)glucose} + (\text{glucosyl})_n \rightarrow \text{ADP(UDP)} + (\text{glucosyl})_{n+1}. \tag{2}$$

This reaction was first described using UDPglucose as the sugar nucleotide donor. Subsequently it was shown by RECONDO and LELOIR that ADPglucose was a better substrate both in terms of V_{max} and affinity (RECONDO and LELOIR, 1961). In contrast leaf starch synthases (MURATA and AKAZAWA, 1964; GHOSH and PREISS, 1965; CARDINI and FRYDMAN, 1966) and bacterial glycogen synthases (GREENBERG and PREISS, 1964) are specific for ADPglucose. The starch synthase can be either bound to the starch granule or can be found in soluble form in the cytoplasm.

Both ADPglucose and UDPglucose are synthesized by classical pyrophosphorylase reactions [reaction (3); ESPADA, 1962] or by reversal of the sucrose synthase reaction [reaction (4); CARDINI et al., 1955; DE FEKETE and CARDINI, 1964].

$$ATP(UTP) + glucose-1-P \rightleftharpoons ADP(UDP)glucose + PP_i \qquad (3)$$

$$sucrose + ADP(UDP) \rightleftharpoons fructose + ADP(UDP)glucose. \qquad (4)$$

The synthesis of α-1,6 linkages found in amylopectin and phytoglycogen is catalyzed by an enzyme named branching enzyme or Q enzyme (BOURNE and PEAT, 1945; HOBSON et al., 1950). A distinction is usually made between Q enzyme, which is able to make a branched product from amylose similar to amylopectin, and branching enzyme, which is able to branch α-glucans to a greater degree to form glycogen-type molecules.

All the above enzymes appear to be ubiquitous both in plants and in bacteria (PREISS, 1969). Although an amylopectin type molecule has been reported to be synthesized directly from sucrose in a reaction [reaction (5)] catalyzed by amylosucrase, this enzyme has been found only in a restricted number of bacteria and never in plant extracts (OKADA and HEHRE, 1974).

$$sucrose \rightleftharpoons (glucosyl)_n + fructose \qquad (5)$$

II. The Predominant Pathway of Starch Synthesis

It is obvious from the foregoing discussion that there are a number of possible enzymatic routes from glucose-1-P to the biosynthesis of α-1,4 glucosidic linkages in plant systems: the phosphorylase route, via UDPglucose, or via ADPglucose. The question that may be asked is whether one pathway is predominant over another. At present there is no definitive answer to this question. However on the basis of available information it appears that the ADPglucose pathway is predominant. This view is based on the following data.

In leaves the starch synthase is solely specific for ADPglucose (MURATA and AKAZAWA, 1964; GHOSH and PREISS, 1965; CARDINI and FRYDMAN, 1966). UDPglucose is virtually inactive with either the starch bound or soluble starch synthases present in leaves. Similarly the soluble starch synthases in reserve tissues (CARDINI and FRYDMAN, 1966; FRYDMAN et al., 1966) also appear to be specific for ADPglucose. Only reserve tissue starch synthase bound to starch granules can utilize UDPglucose in addition to ADPglucose. However the rate of glucosyl transfer from UDPglucose is usually $^1/_3$ to $^1/_{10}$ of that observed from ADPglucose (LELOIR et al., 1961; CARDINI and FRYDMAN, 1966). Moreover the K_m for UDPglucose is about 15- to 30-fold higher than the K_m for ADPglucose for this starch-bound starch synthase (Cardini and Frydman, 1966). The K_m for ADPglucose for the bean starch-bound enzyme ranges from 2 to 4 mM and for UDPglucose the K_m is about 60 mM. These K_m values appear to be higher than what appears to be the physiological concentrations of these sugar nucleotides. In *Chlorella pyrenoidosa* the UDPglucose is estimated to be about 0.7 mM (Kanazawa et al., 1972) and most likely this is also the concentration found in plant tissues. From chromatographic patterns observed in rice grains, the ADP-glucose concentration appears to be equivalent to UDPglucose concentrations (Mutara et al., 1965). It is quite possible therefore that the bound starch synthase may be inoperative or much less active in vivo than the soluble ADPglucose specific starch synthase which has a K_m for the sugar nucleotide substrate of 0.1–0.3 mM (Frydman and Cardini, 1966; Lavintman and Cardini, 1968; Ozbun et al., 1971b; Ozbun et al., 1972). Furthermore in many tissues, for example maize or potato tubers, there is more activity of the soluble starch synthase than the bound starch synthase (Ozbun et al., 1973; Hawker and Downton, 1974; Downton and Hawker, 1975). All the foregoing strongly suggests that the ADPglucose pathway is by far preferred for starch synthesis in either leaf or reserve tissue.

The question remains whether phosphorylase is functional in the biosynthetic pathway. A number of observations would suggest that phosphorylase may not be operative in starch biosynthesis. It can be readily calculated that the P_i to α-glucose-1-P ratio is about 3–300 (Heber and Santarius, 1965; Bassham and Krause, 1969) in algae or leaves. Since the phosphorylase reaction in vitro has an equilibrium constant of 2.4 at pH 7.3 (Cohn, 1961), the formation of α-1,4 glucosyl linkages would not occur. Furthermore the concentration of glucose-1-P in plants is very low and usually cannot be detected. Its concentration can be estimated by assuming that it is in equilibrium with glucose-6-P via the P-glucomutase reaction ($K_{eq} = 17$; Colowick and Sutherland, 1942). Since the concentration of glucose-6-P in *C. pyrenoidosa* is about 0.7 mM (Bassham and Krause, 1969) the calculated concentration of glucose-1-P would be about 40 µM. The K_m values of glucose-1-P for phosphorylase in plants range from one to 50 mM (Alexander, 1973; Frydman and Slabnik, 1973; Burr and Nelson, 1975; Chen and Whistler, 1976) and are therefore far above the physiological glucose-1-P concentration. The in vitro activity of phosphorylase at saturating glucose-1-P may be 10- to 20-fold in excess of starch synthase activity measured at saturating ADPglucose concentrations in certain tissues (Ozbun et al., 1973). However if the physiological concentration of the substrate is taken into account the phosphorylase synthetic activity may be less than the starch synthase activity in vivo. There is always the possibility at the site of phosphorylase in the cell that glucose-1-P may be extraordinarily high compared to the rest of the cell. There is no evidence that this occurs.

There is strong suggestive evidence, however, that in maize endosperm 75% of the starch is synthesized via the ADPglucose pathway. Two maize endosperm mutants, shrunken-2 (sh-2) and brittle-2 (bt-2), have reduced levels of starch which are 25% of the normal (Cameron and Teas, 1954; Creech, 1965). Correlated with this are the levels of the ADPglucose pyrophosphorylase activity of normal starch maize endosperm (Tsai and Nelson, 1966; Dickinson and Preiss, 1969b). Thus at least 75% if not all the starch may be synthesized via the ADPglucose pyrophosphorylase reaction. Consistent with the above studies are the reports where during the development of reserve tissues, starch accumulation, the starch biosynthetic enzymes, ADPglucose pyrophosphorylase and starch synthase, and phosphorylase and UDPglucose pyrophosphorylase were measured. The increase in activity of both starch synthase and ADPglucose pyrophosphorylase could be correlated closely with the increase in rate of starch accumulation (MOORE and TURNER, 1969; TURNER, 1969; OZBUN et al., 1973; SOWOKINOS, 1976). Both phosphorylase and/or UDPglucose pyrophosphorylase activities could not be correlated as well with the starch accumulation rate.

Although not entirely conclusive, the above evidence strongly indicates that the ADPglucose pathway is the more important route to starch biosynthesis. However, some reports indicated that at the time where starch synthesis is initiated in the endosperm development there is no starch synthase activity but ample phosphorylase activity (TSAI et al., 1970; BAXTER and DUFFUS, 1971). Starch synthase activity appears at the latter stage of development. On the basis of these results the investigators have proposed that phosphorylase is involved in the initiation of starch synthesis and in the later stages starch synthase is the predominant activity for biosynthesis of starch. Close examination of these experiments reveals that relatively insensitive methods have been used to measure starch synthase activity. An example of this is the report suggesting that 12 days after pollination developing maize endosperm does not contain starch synthase activity (TSAI et al., 1970). A subsequent report however by other investigators showed substantial starch synthase activity at this stage of development if a highly sensitive radioactive assay was employed where incorporation of one nmol of glucose or less into starch from ADPglucose could be easily detected (OZBUN et al., 1973). Furthermore in many plant cell extracts potent amylase activities can interfere with the starch synthase assay. For example, in our laboratory we have assayed many crude leaf extracts for starch synthase and have found little or no activity. If, however, these crude extracts are fractionated with ammonium sulfate (to 40% saturation) and the obtained precipitate dissolved and dialyzed in buffer containing dithiothreitol, high amounts of starch synthase could be discerned. The ammonium sulfate fractionation has essentially separated most of the interfering amylase activity from the starch synthase activity. MANNERS (1974a) in his review has expressed similar sentiments about the difficulty of the starch synthase assay.

III. Regulation of Starch Biosynthesis

1. Leaf ADPglucose Pyrophosphorylases

The finding that starch granule accumulation in leaves or algae is dependent on light suggests some regulatory mechanism coordinating increased CO_2 fixation and photophosphorylation with starch biosynthesis. It therefore was of great significance to find that all the leaf and green algal ADPglucose pyrophosphorylases studied so far are subject to allosteric phenomena and are activated by 3-P-glycerate and inhibited by P_i (GHOSH and PREISS, 1966; PREISS et al., 1967; SANWAL and PREISS, 1967; SANWAL et al., 1968). Other glycolytic intermediates, such as P-enol pyruvate, fructose-1,6-P_2 and fructose-6-P, activate to lesser extents, and much

Table 1. Stimulation of spinach leaf ADPglucose pyrophosphorylase by metabolite activators at pH 8.5

Activator	Concentration mM	ADPglucose formed μmol mg protein^{-1} min^{-1}	Activation -fold
None		1.6	
3-P-glycerate	0.9	90.0	58
Fructose-6-P	1.6	36.9	24
Fructose-1,6-P$_2$	1.0	24.6	16
P-enol pyruvate	1.0	23.3	15
2,3-P$_2$-glycerate	1.0	22.7	14
Ribose-5-P	1.0	13.6	9
P-glycolate	1.0	5.1	3

The conditions of the experiment are described in GHOSH and PREISS (1966).

Table 2. Effect of 3-P-glycerate on the kinetic parameters of the substrates of spinach leaf ADPglucose pyrophosphorylase at pH 7.5

Substrate	no 3-P-Glycerate K_m mM	V_{max} μmol mg protein^{-1} min^{-1}	+1 mM 3-P-Glycerate K_m mM	V_{max} μmol mg protein^{-1} min^{-1}
ADPglucose	0.93	35.5	0.15	93.5
PP$_i$	0.50		0.04	
ATP	0.45	13.2	0.04	92.4
Glucose-1-P	0.07		0.04	
MgCl$_2$	1.4		1.6	

The data described above are obtained with homogenous enzyme (RIBEREAU-GAYON and PREISS, 1971).

higher concentrations are required to elicit the activation. Table 1 shows the activation of the spinach leaf enzyme by a number of metabolites. Of the lesser activators fructose-6-P is most active, giving about 40% of the activation noted for 3-P-glycerate. P-glycolate gives negligible activation. The enzyme studied in the greatest detail is that obtained from spinach leaf (GHOSH and PREISS, 1966; PREISS et al., 1967). 3-P-Glycerate increases the apparent affinity of all the substrates from two to 13-fold (Table 2). All substrate saturation curves are hyperbolic in the presence or absence of 3-P-glycerate. The MgCl$_2$ saturation curve is sigmoidal in the presence or absence of the activator, and its apparent affinity is not changed by 3-P-glycerate.

The stimulation of the spinach leaf enzyme by 3-P-glycerate is dependent on pH because of the different pH optima of the activated and unactivated reactions (GHOSH and PREISS, 1966; PREISS et al., 1967). Stimulation of ADPglucose synthesis can vary from 9- to 80-fold. The 3-P-glycerate activation curve is hyperbolic in shape at pH 7.0 and 7.5, but becomes progressively sigmoidal as the pH increases to 8.5 (GHOSH and PREISS, 1966). ADPglucose pyrophosphorylase has been isolated from ten other leaf sources, and the specificity of the activation is the same whether the enzyme was from a plant fixing CO$_2$ via the C$_3$ pathway

Table 3. 3-P-glycerate activation of ADPglucose synthesis catalyzed by various ADPglucose pyrophosphorylases

Enzyme source	$A_{0.5}$[a] μM	Activation -fold[b]
Turkish tobacco leaf	45	9.0
Red cherry tomato leaf	90	5.5
Barley leaf	7.0	13.3
Sorghum leaf	370	7.1
Sugar beet leaf	190	8.6
Rice leaf	180	14.6
Spinach leaf	20	9.3
Chlorella pyrenoidosa	400	18.0
Synechococcus 6301	112	25.0

The data are obtained from PREISS et al. (1976), SANWAL and PREISS (1967), SANWAL et al. (1968) and LEVI and PREISS (1976).
[a] Concentration required for 50% maximal stimulation of ADPglucose synthesis.
[b] Maximum stimulation seen at saturating 3-P-glycerate.

or the C_4 pathway (SANWAL et al., 1968). Table 3 shows some kinetic parameters for the activation of various leaf ADPglucose pyrophosphorylases by 3-P-glycerate. The concentration of 3-P-glycerate required for 50% of maximal activation varied from 7 μM for the barley enzyme to 370 μM for the sorghum leaf enzyme. Thus the leaf enzyme is quite sensitive to activation by 3-P-glycerate.

Inorganic phosphate is an effective inhibitor of ADPglucose synthesis for all the leaf and algal enzymes studied. ADPglucose synthesis catalyzed by the spinach leaf enzyme is inhibited 50% by 22 μM P_i in the absence of activator at pH 7.5 (GHOSH and PREISS, 1966). In the presence of 1 mM 3-P-glycerate, 50% inhibition requires 1.3 mM phosphate. Thus, the activator decreases sensitivity to P_i inhibition about 450-fold. As would be expected P_i at 0.5 mM increases the concentration of 3-P-glycerate needed for activation (GHOSH and PREISS, 1966). The 3-P-glycerate activation curve becomes sigmoidal in the presence of the P_i and conversely, 3-P-glycerate increases the sigmoidicity of the P_i inhibition curve. The interaction between 3-P-glycerate and P_i is clearly shown in Figure 1 where P_i decreases the apparent affinity of 3-P-glycerate for the tobacco leaf ADPglucose pyrophosphorylase and in Figure 2 where 3-P-glycerate decreases the apparent affinity of P_i in its inhibition of the tobacco leaf enzyme.

The sigmoidal saturation curves of activator and inhibitor cause the enzyme to be more responsive to smaller changes of effector molecules and therefore provide a more sensitive system for regulation by the fluxes of metabolite concentrations (GHOSH and PREISS, 1966; SANWAL et al., 1968). Phosphate is a noncompetitive or mixed inhibitor with respect to the substrates, ADPglucose, PP_i, ATP, and glucose-1-P.

The concentration of phosphate required for 50% inhibition of the other leaf ADPglucose pyrophosphorylases varies from 20 μM for the barley enzyme to 200 μM for the sorghum leaf enzyme (Table 4; SANWAL et al., 1968). In all cases

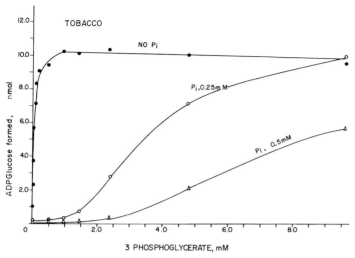

Fig. 1. Effect of 3-P-glycerate and P_i on the rate of ADP-glucose synthesis catalyzed by tobacco leaf ADPglucose pyrophosphorylase. The conditions ,are described in Sanwal et al. (1968)

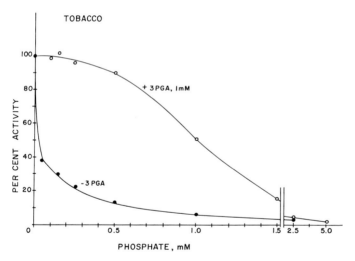

Fig. 2. Effect of 3-P-glycerate on P_i inhibition of tobacco leaf ADPglucose pyrophosphorylase. The conditions are described in Sanwal et al. (1968)

the concentration of phosphate required for 50% inhibition is increased by the presence of the activator 3-P-glycerate.

The spinach leaf ADPglucose pyrophosphorylase has been purified to homogeneity (Ribereau-Gayon and Preiss, 1971). The molecular weight as determined by disc gel electrophoresis is about 210,000. No information is currently known about its subunit molecular weight.

Table 4. P_i inhibition constants of ADPglucose pyrophosphorylases in presence and absence of 3-P-glycerate

Enzyme source	$I_{0.5}$[a] (−3-P-glycerate) µM	$I_{0.5}$ (+3-P-glycerate) µM	Conc. of 3-P-glycerate mM
Turkish tobacco	30	1010	1.0
Red cherry tomato	80	880	2.5
Barley	20	2300	1.4
Sorghum	190	410	2.2
Sugar beet	50	430	0.87
Rice	60	270	1.0
Spinach	60	1200	1.0
Chlorella pyrenoidosa	180	1000	2.0
Synechococcus 6301	72	1000	1.0

The data are obtained from SANWAL et al. (1968), PREISS et al. (1967), SANWAL and PREISS (1967) and LEVI and PREISS (1976). The assay conditions for measuring ADPglucose synthesis are indicated in the above references.
[a] Concentration of inhibitor required for 50% inhibition under conditions of the experiment.

2. *Chlorella pyrenoidosa* ADPglucose Pyrophosphorylase

The ADPglucose pyrophosphorylase from this green alga is very similar in properties to the leaf enzymes (SANWAL and PREISS, 1967). Some of the properties of the enzyme are summarized in Tables 3 and 4. The ATP and ADP glucose saturation curves for the *C. pyrenoidosa* enzyme are sigmoidal in the presence or absence of the activator 3-P-glycerate and this result differs from that for the spinach leaf enzyme.

The ADPglucose pyrophosphorylases of *Chlorella vulgaris, Scenedesmus obliquus,* and *Chlamydomonas reinhardii* are also activated by 3-P-glycerate and inhibited by orthophosphate (SANWAL and PREISS, 1967).

3. Physiological Significance of Regulation by 3-P-glycerate and P_i: Evidence for the Allosteric Phenomena in Vivo

Due to the great sensitivity of the leaf ADPglucose pyrophosphorylases to 3-P-glycerate, the primary CO_2 fixation product of photosynthesis, and P_i, it is suggested that they play a significant role in the regulation of starch biosynthesis. The level of P_i has been shown to decrease in leaves during photosynthesis because of photophosphorylation, and glycolytic intermediates are known to increase in the chloroplast in light (SANTARIUS and HEBER, 1965; HEBER, 1967). This would contribute to conditions necessary for optimal starch synthesis via the increased rate of formation of ADPglucose. In the light, the levels of ATP and reduced pyridine nucleotides are also increased, leading to the formation of sugar phosphates from 3-P-glycerate. In the dark, there is an increase in P_i concentration with concomitant decreases in the levels of glycolytic intermediates, ATP, and reduced pyridine nucleotides. This would lead to inhibition of ADPglucose synthesis and therefore starch synthesis.

The results obtained by a number of laboratories on the concentration of glycolytic intermediates, P_i, and ATP in the leaf chloroplast (the organelle where starch synthesis occurs) in the light and in the dark qualitatively support the hypothesis of regulation of starch synthesis by 3-P-glycerate and P_i levels. Santarius and Heber (1965) have shown that the concentration of P_i in the dark is 5–10 mM and decreases by 30%–50% in the light. In vitro kinetic studies with ADPglucose pyrophosphorylase show that at 5 mM 3-P-glycerate there is an increase of five-fold in the rate of ADPglucose synthesis when the P_i concentration is decreased from 10 to 7.5 mM (Sanwal et al., 1968). A 23-fold increase in rate is observed when P_i is decreased to 5 mM. Thus, under these conditions a decrease of phosphate concentration of only 30%–50% in the chloroplast may cause a rapid acceleration of ADPglucose synthesis and therefore of starch synthesis.

MacDonald and Strobel (1970) reported that wheat leaves infected with the fungus *Puccinia striiformis* accumulated more starch than noninfected leaves. They could correlate starch accumulation with the decrease observed in P_i levels in diseased leaves during the infection process. This decrease in P_i concentration is most probably due to synthesis of nucleic acids required for fungal growth. These data suggested that in diseased leaves the variations in the level of P_i and to a lesser extent variations in the level of activators of the wheat leaf ADPglucose pyrophosphorylase (3-P-glycerate, fructose-1,6-P_2, etc.) regulated the rate of starch synthesis via control of ADPglucose pyrophosphorylase activity.

Kanazawa et al. (1972) have shown in *C. pyrenoidosa* cells that starch and ADPglucose synthesis occur in the light. Starch synthesis abruptly ceases and the ADPglucose level drops to below detectable limits when the light is turned off. UDPglucose levels do not perceptibly change in the light to dark transition. ADPglucose is not detectable at any time later in the dark, despite the high steady state level of the substrates of the pyrophosphorylase reaction, ATP and hexose phosphate. Kanazawa et al. (1972) indicate that this observation provides strong support for the importance of the regulatory role of ADPglucose pyrophosphorylase in starch synthesis in vivo.

Finally recent interesting observations from a number of laboratories strongly suggest that P_i plays a role in regulating starch biosynthesis and degradation. Steup et al. (1976) have shown that during $^{14}CO_2$ fixation in spinach leaf chloroplasts in the light increasing levels of P_i (0.1–0.25 mM) increases the lag of starch synthesis slightly (<5 min) while having little effect on the rate of $^{14}CO_2$ fixation. Higher concentrations of P_i, 0.5 mM and 1.0 mM, not only induce a prolonged lag (about 15 min) in starch synthesis but also inhibit the maximum rate attained. Furthermore the amount of label in dihydroxy-acetone-P and sugar phosphates increases up to ten-fold under the P_i inhibiting conditions while the incorporation of label into 3-P-glycerate only differs about 25% between the 0 and 1.0 mM P_i conditions. Steup et al. (1976) conclude that inhibition of starch synthesis at high P_i concentrations can be explained by the allosteric inhibition of ADP-glucose pyrophosphorylase by P_i. The reason for only a short increase in lag at the low concentrations is also explained by the authors on the basis of the allosteric properties of the pyrophosphorylase. At low concentrations of P_i, the 3-P-glycerate concentration remains high while the P_i levels will be lowered due to photophosphorylation. Therefore only a small lag in starch synthesis is observed.

A similar study by HELDT et al. (1977) showed that P_i as low as 1 mM completely inhibited starch synthesis in intact chloroplasts. The P_i inhibition was overcome by adding 3-P-glycerate to the incubation mixture. The authors showed that a high ratio of 3-P-glycerate to P_i in the stroma could be correlated with a rate of CO_2 incorporated into starch; a low ratio correlated with a low rate. No correlation was seen with stromal hexose monophosphate levels. These authors compared their data on stromal concentration of P_i and 3-P-glycerate with the in vitro data of P_i and 3-P-glycerate effects on the spinach leaf ADPglucose pyrophosphorylase. The in vitro ADPglucose pyrophosphorylase activity at the different 3-P-glycerate/P_i ratios (GHOSH and PREISS, 1966; PREISS et al., 1967) correlated with the rates of starch synthesis observed in the chloroplast.

In another study SHEN-HWA et al. (1975) have shown that photosynthetic starch synthesis in leaf disks of spinach beet, sugar beet or spinach was increased more than ten-fold by the presence of exogenous 10 mM mannose. In the above tissues mannose is converted to mannose-6-P in the cytoplasm but cannot be incorporated into starch. SHEN-HWA et al. (1975) postulated two reasons for the mannose effect. They believed the cytoplasmic P_i levels are decreased because of the sequestration into mannose-6-P via phosphorylation. Thus triose phosphates and 3-P-glycerate could not be translocated from the chloroplast into the cytoplasm during photosynthesis as the translocation process requires P_i. Due to the increased levels of glycolytic intermediates in the chloroplast it was postulated that starch synthesis would be stimulated. SHEN-HWA et al. (1975) also suggest that the increased starch synthesis may also occur because the lower P_i concentration and higher 3-P-glycerate and other glycolytic intermediate concentrations in the chloroplast would increase the activity of the ADPglucose pyrophosphorylase. In a subsequent report (HEROLD et al., 1976) it was shown that leaf tissues from many C_3 species respond in the same way to exogenous mannose in the light by a marked stimulation of starch synthesis. HEROLD et al. (1976) also showed that leaf disks of plants grown in P_i-deficient media contained much greater amounts of starch than the leaf disks from control plants.

All these data strongly support the hypothesis that the allosteric phenomena observed in vitro for the leaf and algal ADPglucose pyrophosphorylases are operative in vivo and play important parts in the biosynthesis of starch. They also suggest the important role that ADPglucose pyrophosphorylase plays in the biosynthesis of starch. Since the level of 3-P-glycerate does not appreciably change in the dark to light transition, while the phosphate levels appear to increase in the dark and decrease in the light, it is suggested that the variation of P_i (the negative effector) is the more important control element.

4. ADPglucose Pyrophosphorylases of Nonchlorophyllous Plant Tissue

The ADPglucose pyrophosphorylases occurring in nonphotosynthetic plant tissues (maize endosperm and embryo: DICKINSON and PREISS, 1969a; PREISS et al., 1971; HANNAH and NELSON, 1975; wheat germ, etiolated peas, mung bean seedlings, potato tuber: PREISS et al., 1967; SOWOKINOS, 1976; carrot roots, and avocado mesocarp: PREISS et al., 1967) are similar to the leaf enzymes in that they are activated by 3-P-glycerate. The stimulation by 3-P-glycerate is abut 1.5–10-fold for these enzymes. Table 5 shows the stimulation of some of the above enzymes

Table 5. Activation of ADPglucose synthesis catalyzed by ADPglucose pyrophosphorylases from non-chlorophyllous plant tissues

Activator	Enzyme Source					
	Potato tuber	Maize endosperm	Carrot roots	Etiolated Alaska pea seedlings	Avocado mesocarp	Wheat germ
	nmol mg protein^{-1} min^{-1}					
None	0.34	143	0.67	0.13	0.55	0.25
3-P-glycerate	1.76	612	6.7	0.44	3.5	0.44
Fructose-6-P	0.57	420	2.2	0.09	0.55	–
Fructose-1,6-P$_2$	0.41	150	4.2	0.12	0.40	0.35
P-enol pyruvate	1.1	140	4.0	0.11	0.45	0.25
Ribose-5-P	–	291	–	–	–	–

The data for the maize endosperm enzyme are obtained from DICKINSON and PREISS (1969a); for the other enzymes the data are obtained from PREISS et al. (1967). The concentration of activators used is 10 mM for the maize endosperm enzyme and 1.5 mM for the other enzymes.

by various metabolites. The enzyme of this group studied in most detail is the one purified from maize endosperm (DICKINSON and PREISS, 1969a) and from sweet corn endosperm (AMIR and CHERRY, 1972). The stimulation by 3-P-glycerate is only about three- to four-fold and the concentration needed for activation is very high (2.2–10 mM) compared to that of the leaf enzymes. Fructose-6-P also stimulates the enzyme activity about three-fold at 10 mM. P$_i$ at 3 mM causes 50% inhibition in the absence of activator.

The lesser sensitivity of the maize enzyme to allosteric effects as compared to the leaf enzymes may reflect differences between leaf and endosperm cells with respect to intracellular levels of metabolites. It is also possible that in the nonphotosynthetic tissues, such as endosperm, there is no need for allosteric type regulation. The starch synthetic and degradative rates in leaves appear to be a more dynamic phemonena than what is observed in a reserve tissue such as endosperm or seed where the processes of synthesis and degradation are temporally separated during the development and germination of the tissue. Thus the endosperm enzyme may have evolved to a form which is comparatively insensitive to activation in inhibition. It appears in this regard that the most important phenomena in regulating starch biosynthesis in endosperm may be regulation of synthesis of the starch biosynthetic enzymes, ADPglucose pyrophosphorylase and ADPglucose: α-glucan-4-glucosyltransferase (MOORE and TURNER, 1969; TURNER, 1969; TSAI et al., 1970; BAXTER and DUFFUS, 1971; OZBUN et al., 1972; SOWOKINOS, 1976).

5. ADPglucose Pyrophosphorylases of Blue-green Algae (Bacteria)

The activation and inhibition of the partially purified ADPglucose pyrophosphorylase from the blue-green bacterium *Synechococcus 6301 (Anacystis nidulans)* resembles the regulation of ADPglucose pyrophosphorylases from higher plants and green algae but not those from photosynthetic bacteria i.e., 3-P-glycerate is a very effective activator and P_i is a very effective inhibitor (LEVI and PREISS, 1976). Some properties of the *Synechococcus* enzyme are shown in Tables 3 and 4. The blue-green bacteria grow very poorly if at all on organic compounds and thus must depend on CO_2 fixation of carbon assimilation. This is done via the ribulose-P_2 carboxylase reaction which has been demonstrated in blue-green bacteria (SMITH, 1973). Furthermore the obvious evolutionary relationship of the blue-green bacteria to the green algae and higher plants has been pointed out previously with respect to ability to evolve O_2 during photosynthesis and with respect to having similar photosynthetic pigments (SMITH, 1975).

IV. Properties of the ADPglucose: 1,4-α-D-glucan 4-α Glucosyltransferase

1. Starch Bound and Soluble Starch Synthase

As indicated previously the formation of the α-1,4-glucosidic linkages of starch was first observed by LELOIR's group (LELOIR et al., 1961). Synthesis occurred by the transfer of the glucosyl residue of UDPglucose to intact starch granules (reaction 2). The starch granules had been prepared from dwarf beans, young potatoes, and sweet corn. Subsequently, RECONDO and LELOIR (1961) reported that ADPglucose was a glucosyl donor far superior to UDPglucose. Other glucosyl nucleotides were found to be inactive. Since then many other α-4-glucosyl transferases adsorbed to the starch granules have been found (CARDINI and FRYDMAN, 1966). In all cases the rate of transfer of glucose from ADPglucose was three to ten times higher than from UDPglucose. Since the enzyme was intimately associated with the starch granules, it was not possible to show primer requirements. However, if oligosaccharides of the maltodextrin series were added to the reaction mixtures containing ADPglucose-^{14}C (or UDP-glucose-^{14}C) and the active starch granules, glucose transfer to these oligosaccharides could be observed.

Recent studies with the α-1,4-glucan synthases from a number of starch grains have shown that they can vary with respect to specificity of glucosyl donor. Both deoxy-TDPglucose and GDPglucose could act as glucosyl donors in certain α-1,4-glucan synthesizing systems (CARDINI and FRYDMAN, 1966; FRYDMAN and CARDINI, 1967). However, with all starch grain systems, ADPglucose was the superior glucosyl donor. Deoxy-ADPglucose also appears to be an effective glucosyl donor (GHOSH and PREISS, 1965; FRYDMAN and CARDINI, 1967). However, this sugar nucleotide does not appear to occur in nature.

Besides the α-1,4-glucan synthase associated with the starch granules, there is also present in the same tissues a soluble α-1,4-glucan synthase. In contrast to the particulate systems, the soluble synthase systems are active with only ADPglucose, and not with any other naturally occurring glucosyl nucleotides (GHOSH and PREISS, 1965; CARDINI and FRYDMAN, 1966; FRYDMAN and CARDINI, 1966). It has been suggested that perhaps the adsorption or entrapment of the α-1,4-glucan synthase in the starch granule may change its properties with respect to glucosyl donor specificity. CHANDORKAR and BADENHUIZEN (1966) and FRYDMAN and

Cardini (1967) showed that the properties of starch granule enzymes from a number of sources could be changed by mere mechanical disruption. First the activity of the starch synthase with ADPglucose is increased and second, in the case of the potato enzyme, the K_m of ADPglucose is decreased from 40 mM to 3.3 mM. Moreover the activity with UDPglucose is drastically diminished. Not all starch granule bound starch synthases can utilize UDPglucose. Murata and Akazawa (1964) showed that the bound starch synthase from soybean leaves is solely specific for ADPglucose. This has been shown to be true for other leaf granule bound starch synthases (Nomura et al., 1967). It has been shown that in certain varieties of rice or maize where the granules are devoid of the amylose portion of the starch granule, the α-1,4-glucan synthase is not present in the granule but present only in the soluble portion of the cell (Frydman, 1963, Murata et al., 1965). It appears therefore that the adsorption of the α-1,4-glucan synthase by the starch granule is due to the presence of amylose. In a number of cases it has been shown that amylose is capable of absorbing the synthase and thus transforming it into a particulate enzyme (Akazawa and Murata, 1965).

As indicated before (see Sect. III) it appears that the soluble ADPglucose specific starch synthase is in greater activity and has higher affinity for the sugar nucleotide than the granule bound enzyme. Therefore the soluble starch synthase found after disruption of the cell may be considered the active form of the enzyme and the synthase activity entrapped in the granule an inactive form. The soluble synthase inside the plant cell most probably is concentrated at the site of starch granule formation. During starch granule formation some synthase binds to the amylose portion of the starch granule becoming entrapped and is no longer active. The evidence suggesting that the starch bound α-1,4-glucan synthase is an inactive form is from the experiments cited above where grinding of the starch granules increases the starch synthase activity and also decreases the K_m for ADPglucose. Frydman and Cardini (1967) speculate that changes in conformation of the synthase occur when bound to the starch granule allowing it to be active with UDPglucose and causing decreased affinity with ADPglucose.

2. Requirements for Starch Synthase Activity

The requirements for starch synthase activity are sugar nucleotides and primers. The primer requirement may be satisfied by either amylose, amylopectin, glycogen, or starch granules (Cardini and Frydman, 1966). Various oligosaccharides of the maltodextrin series starting with maltose can also be utilized. The immediate product is the next higher oligosaccharide i.e., the product with maltose as the acceptor is maltotriose (Ghosh and Preiss, 1965). Longer incubations however will convert the maltotriose product to maltotetraose and maltopentaose.

A number of reports have indicated the requirement of univalent cations on the activity of starch synthases associated with the granule (Murata and Akazawa, 1968; Nitsos and Evans, 1969; Hawker et al., 1974a). The requirement is almost absolute when ADPglucose is the substrate for the sweet corn enzyme (Nitsos and Evans, 1969). K^+ is the most effective cation. When UDPglucose is the substrate, the K^+ requirement is not absolute, and stimulations of only twofold are seen in the sweet corn system. The cation requirement of the starch synthases of plant leaves were studied by HAWKER et al. (1974a). Both Na$^+$ and K$^+$ stimulate about 1.2–1.3-fold the soluble starch synthase activities from sugar beet, bean and saltbush leaves but the spinach leaf enzyme is not stimulated. The starch bound synthases with the exception of the spinach leaf enzyme are stimulated about 2–2.4-fold. Thus in general a univalent cation stimulates starch synthase activity but the effect varies from a minimal to an absolute effect depending on the source of the enzyme.

The starch synthases are not affected by glycolytic intermediates. ADP is a competitive inhibitor with ADPglucose (GHOSH and PREISS, 1965). The enzyme also appears to have a sulfhydryl group requirement for activity as it is quite sensitive to sulfhydryl group reagents (GHOSH and PREISS, 1965; CARDINI and FRYDMAN, 1966).

3. Multiple Forms of the Soluble Starch Synthase

Many reports since 1971 have reported the presence of multiple forms of the soluble starch synthases of leaves and reserve tissues (OZBUN et al., 1971a, b; 1972; HAWKER et al., 1972; HAWKER and DOWNTON, 1974; PISIGAN and DEL ROSARIO, 1976). In most cases these enzyme forms were separated by DEAE-cellulose chromatography. The number of forms found in the systems ranged from two to four with most of the activity residing in two peaks. The multiple enzyme forms studied in greatest detail are from spinach leaf and waxy maize endosperm (OZBUN et al., 1971b, 1972). The two major fractions in maize could be distinguished from each other with respect to their reaction rates for different primers (Table 6). The most distinct difference is that one fraction in each of the plant extracts (starch synthase I in maize) is able to catalyze the synthesis of an α-1,4-glucan with the slight amount of endogenous primer associated with the enzyme in the presence of 0.5 M citrate (OZBUN et al., 1971a, b; 1972; HAWKER et al., 1972; HAWKER and DOWNTON, 1974). The molecular weight of the starch synthase carrying out this "unprimed" synthesis is about 70,000 in sucrose density ultracentrifugation for all tissues studied (HAWKER et al., 1974b) while the molecular weight of the second starch synthase fraction is 92,000–95,000.

These data suggest the presence of at least two different classes of soluble starch synthases. Their precise function in starch synthesis is unknown. One may be involved in the biosynthesis of amylose and one in the biosynthesis of amylopectin. The 70,000 molecular weight starch synthase fraction which has high affinity for the endogenous α-1,4-glucan primer in the presence of citrate (the K_m for amylopectin in the spinach leaf system is reduced from 530 μg/ml to 1.9 μg/ml in the presence of citrate) is associated with branching enzyme activity (OZBUN et al., 1972, 1973; HAWKER and DOWNTON, 1974; HAWKER et al., 1974b). In spinach leaf extracts the branching enzyme activity is only resolved from the synthase via chromatography on the affinity resin ADP-hexanolamine-Sepharose 4B (HAWKER et al., 1974b). Under certain conditions the branching enzyme can stimulate the unprimed synthesis catalyzed by the starch synthase about 11–14-fold. The glucan product formed contains α-1,6 linkages as well as α-1,4 linkages and the molecular weight and absorption spectra of the I$_2$-glucan complex was typical of amylopectin.

Table 6. Properties of the multiple forms of waxy maize starch synthases

Substrate/primer	Concentration	Relative reaction rates	
		Synthase I	Synthase II
Amylopectin	5 mg ml^{-1}	100[a]	100
Amylose	5 mg ml^{-1}	96	73
Rabbit liver glycogen	5 mg ml^{-1}	380	78
Oyster glycogen	5 mg ml^{-1}	133	43
Maltose	1.0 M	160	122
Maltotriose	0.1 M	90	51

[a] The rate observed for amylopectin was arbitrarily set at 100. The data are from Ozbun et al. (1971 b).

V. α-1,4-Glucan: α-1,4-Glucan 6-Glycosyl Transferase (Branching or Q Enzyme)

Although publications on branching enzyme or Q enzyme date from the 1940's, very little is known about the mechanism of action and the nature of the glucan substrate. The enzyme that has been studied in most detail is that isolated from potato tubers (Drummond et al., 1972; Borovsky et al., 1975a). The enzyme has been purified to near homogeneity, has a molecular weight of 85,000 and is a monomer (Borovsky et al., 1975a). The enzyme can utilize as acceptors amyloses of chain lengths of 40 glucose units or greater (Nussenbaum and Hassid, 1952; Peat et al., 1953; Borovsky et al., 1975b, 1976). Experiments by Borovsky et al. (1976) which showed that the radioactive label could be transferred from one amylose chain to another of disparate length indicated that formation of the α-1,6 branch linkage occurred by interchain transfer. Intrachain transfer however was not eliminated. The pure enzyme contains no hydrolytic activity and this contrasts with an earlier report that Q enzyme consists of two components, a hydrolase with a molecular weight of 70,000 and a transferase with a molecular weight of 20,000 (Griffin and Wu, 1968).

Borovsky et al. (1976) postulate two possible models for the branching enzyme catalytic mechanism and invoke complexing of amylose chains in double helices. In model one, the branching enzyme reacts with the amylose molecule to form a covalent bond with a fragment from the donor chain. The branching enzyme oligosaccharide complex then interacts with acceptor amylose chain to form the branch linkage. In the second model, the two amylose chains form a double helix first which facilitates the branching enzyme action of transferring a portion of the oligosaccharide chain from one amylose molecule to another. Borovsky et al. (1976) believe that the minimum chain length requirement of 40 for the branching enzyme would be explained by the need for a double helix formation. They refer to the postulation by French (1972) that amylose chains which can adopt a helical configuration can also form double helices that are stabilized via hydrogen bonding. Some suggestive evidence of double helix formation of amylose chains is seen in the experiments where branching action can occur at a lower temperature (4° C) with amyloses of chain lengths lower than 40 (BOROVSKY et al., 1975 b). However there are alternative interpretations of this phenomenon; it is also possible that branching enzyme affinity for the lower chain lengths can increase at lower temperatures.

The most detailed report on leaf branching enzymes is by HAWKER et al. (1974 a). They have shown that on DEAE-cellulose chromatography spinach leaf contains multiple peaks of branching enzyme activity. One of these peaks is coincident with one of the starch synthase activity peaks. Both branching enzyme activities stimulate starch synthase activity when low or unsaturating concentrations of primer are used. The product formed resembles amylopectin with respect to chain length, I_2 spectra, molecular weight and resistance to α-amylase and α-amylase digestion.

The molecular weight of the spinach leaf enzymes is about 80,000 and therefore is similar to the potato tuber enzyme in this repsect. The only different property noted between the two enzyme forms in spinach leaf is their pH optima in citrate buffer. The in vitro stimulation of starch synthase activity by the branching enzyme has been observed in many other systems (DOI, 1967; FRYDMAN and CARDINI, 1967; HAWKER and DOWNTON, 1974) and could be the in vivo mechanism for amylopectin synthesis i.e., the continuous enlargement of oligosaccharide chains occurs via the starch synthase reaction. After the chain increases to a required length (40?), the branching enzyme catalyzes the transfer of a portion of the enlarged chain to form the new branch linkage. The stimulation of starch synthase activity by the branching enzyme is believed to be due to the creation of an increased number of nonreducing ends able to accept glycosyl residues from ADP-glucose (HAWKER et al., 1974 b).

At present the requirement for the presence of two different branching enzyme activities is not understood but multiple activities of branching enzyme have been observed in a number of systems (HAWKER et al., 1974 a; HAWKER and DOWNTON, 1974; BOYER and PREISS, 1978 a, b). In some maize endosperm mutant systems, where in addition to amylopectin a more highly branched α-glucan, phytoglycogen, is formed, two branching enzyme activities have been distinguished (LAVINTMAN, 1966; MANNERS et al., 1968; HODGES et al., 1969) One is able to catalyze the branching of amylose to form an amylopectin type product and is called the Q enzyme. The other enzyme could catalyze the branching of amylose or amylopectin to form a glycogen type product and is called the branching enzyme. However the occurrence of phytoglycogen in plants is limited (BLACK et al., 1966) and therefore cannot explain the presence of multiple forms of branching enzyme in the many tissues where it is not present.

The detection of multiple branching enzymes is interesting in the light of recent studies of the structure of amylopectins and glycogen. The unit chains from enzymatic debranching of amylopectin have been shown to differ from the unit chains of glycogen by having a greater length as well as a bimodal versus a unimodal distribution when fractionated by gel filtration (GUNJA-SMITH et al., 1970). The ratio of outer exterior chains to inner chains in amylopectins generated by enzymatic techniques was observed to be considerably higher than in glycogen; 2:1 compared to 1:1 (MARSHALL and WHELAN, 1974). These two observations considered together suggest an asymmetry in the structure of the amylopectin molecule as contrasted to glycogen. There has been no detection of multiple forms of branching enzyme either in mammals (Gibson et al., 1971), or in bacteria (Boyer and Preiss, 1977); systems where glycogen is found. It is therefore possible that the multiple forms of starch synthase and branching enzyme may participate in the formation of different sections of the asymmetric amylopectin structure. The verification of this speculation must await the further characterization of the starch synthases and branching enzymes with respect to their properties of elongation and chain transfer, respectively.

VI. Remaining Problems in Starch Synthesis

Despite considerable efforts in studying various aspects of starch biosynthesis, the complete formation of a starch granule in vitro has never been demonstrated. The formation of the particular submicroscopic structure of the native starch granule which is unique for each plant source remains unsolved. Likewise the various functions and interactions of the multiple forms of starch synthase and branching enzyme remain to be characterized. Neither are the relative importances of either the starch bound or soluble starch synthase known. In this connection it is not clear how each plant tissue produces in constant proportions the two polymers amylose and amylopectin. Theories such as compartmentation (Whelan, 1958), or that excess amylose is produced because branching enzyme activity is insufficient to branch all the α-1,4 polyglucose chains produced (Geddes and Greenwood, 1969), or that amylopectin is synthesized by a starch synthase-Q enzyme complex and amylose is synthesized by a starch synthase per se (Schiefer et al., 1973) have been presented. However, there is presently no evidence to lend support to any of these theories.

Finally another possible perplexing problem is the de novo synthesis of starch in the absence of primer; i.e., the initiation of starch synthesis. Some recent reports claim evidence for the formation of a glucoprotein via transfer of glucose from UDPglucose to an acceptor protein (Lavintman and Cardini, 1973; Lavintman et al., 1974). The presumptive glucoprotein which is thought to contain oligosaccharide chains in α-1,4 linkage (the nature of these glucosyl linkages has never been demonstrated) is believed to function as primer for synthesis of starch. The suggestive evidence is that the glucose transferred from UDPglucose is precipitable by 5% trichloroacetic acid and that the presumed glucoprotein is no longer precipitated after pronase action. However it is quite possible that amylose or a similar glucan can co-precipitate with protein in 5% trichloroacetic acid. This has been shown to occur in at least two purported de novo systems studied in detail (Fox et al., 1973; Hawker et al., 1974b). Furthermore the studies on the above de novo systems have not eliminated the possibility that endogenous maltodextrin primers are present. Indeed the presence of endogenous primer molecules has been found in some of the de novo systems (Schiefer, 1973; Hawker et al., 1974b). At present no direct transfer of a glucosyl group to a specific amino acid residue in protein has been demonstrated in the above de novo systems. This type of proof is required to establish definitively the formation of the presumed glucoprotein intermediate. Many other postulations have been made to explain the formation of new primer molecules such as amylases (Parodi, 1967) or D enzyme (Whelan, 1957) or even by unprimed synthesis by phosphorylase (Slabnik and Frydman, 1970; Frydman and SLABNIK, 1973). However the objections of phosphorylase involvement in starch synthesis have been cited (Sect. III). The mechanism for generating new primer molecules remains to be shown whether it be via glucoprotein formation or via other routes.

C. Starch Degradation

Starch degradation occurs in three phases: reduction of the granule to soluble maltodextrins, debranching and degradation of the larger maltodextrins to glucose and glucose-1-phosphate, and finally, further metabolism of glucose or glucose-1-phosphate and export of the products from the site of polysaccharide storage. Granule dissolution is thought to be the exclusive domain of α-amylase. However, maltodextrin degradation is accomplished by debranching enzymes and a series of enzymes which degrade linear glucans.

I. Reactions Involved in Starch Degradation

1. α-Amylase

The properties and distribution of α-amylase have been reviewed by FISCHER and STEIN (1961) and THOMA et al. (1971). α-Amylases are Ca^{2+} requiring metalloenzymes. Although removal of enzyme-bound Ca^{2+} from the α-amylases of bacterial, fungal and animal sources requires extensive dialysis, the Ca^{2+} of plant α-amylases is more loosely bound, the affinity varying with the plant source; the pea cotyledon enzyme is inhibited by 5 mM EDTA (SWAIN and DEKKER, 1966a). The pH optima of α-amylases are low, with cereal α-amylases having pH optima between 4.75 and 6 (FISCHER and STEIN, 1961; THOMA et al., 1971).

The hydrolysis of amylose by α-amylase is biphasic (FISCHER and STEIN, 1961). Amylose is initially subject to rapid fragmentation into maltodextrin chains (FISCHER and STEIN, 1961; GREENWOOD et al., 1965). The maltodextrin chains are more slowly hydrolyzed, the rate of hydrolysis of maltodextrins being inversely proportional to chain length. During the degradation of long chains by dilute α-amylase, glucose, maltose, maltotriose and maltodextrins of 6–8 glucosyl units accumulate, since the attack at the first five bonds from the nonreducing end of a dextrin is hindered (BIRD and HOPKINS, 1954; GREENWOOD et al., 1965). Although maltotriose is hydrolyzed by α-amylase, the rate of hydrolysis is insignificant compared to the rate of hydrolysis of maltotetrose and larger glucans (PARRISH et al., 1970).

α-Amylase readily hydrolyzes amylopectin but is hindered in the region of the α-1,6 branch points (THOMA et al., 1971). The products of amylopectin hydrolysis by malted rye α-amylase are glucose, maltose, maltotriose, and branched limit dextrins of greater than three glucosyl units (MANNERS and MARSHALL, 1971).

α-Amylase is strongly absorbed onto starch granules; absorption is inhibited by maltose and β-limit dextrins and maltose is an inhibitor of the hydrolysis

of starch granules of malt α-amylase (SCHWIMMER and BALLS, 1949). This property
has been suggested as the basis for a mechanism of regulation of α-amylase in
vitro. SIMPSON and NAYLOR (1962) have found that although α-amylase is present
in dormant *Avena fatua* endosperms, starch degradation does not begin until germination, at which time maltase appears, and further, that α-amylase does not degrade
raw endosperm starch until maltase is added. DE FEKETE and VIEWEG (1973) have
shown inhibition of the hydrolyses of amylopectin by maize leaf extract, but DUNN
(1974) found that 50 mg/ml maltose had no effect on the degradation of isolated
maize starch granules by barley α-amylase.

2. β-Amylase

β-Amylase is restricted to higher plants, where it has been found in seeds and
roots (THOMA et al., 1971), in pea seedlings (SWAIN and DEKKER, 1966 b) and
in *Vicia faba* leaves (CHAPMAN et al., 1972). The properties of β-amylase have
been well reviewed (FRENCH, 1961; THOMA et al., 1971).

β-Amylase hydrolyzes maltosyl residues from amylose, starting at the nonreducing end. As with α-amylase, large maltosaccharides are the preferred substrates.
The rate of maltotriose hydrolysis is insignificant compared to the rate of hydrolysis
of maltotetrose and larger dextrins (PARRISH et al., 1970; THOMA et al., 1971; CHAP-
MAN et al., 1972).

β-Amylase is a sulfhydryl containing enzyme, inhibited by sulfhydryl oxidizing
reagents (BERNFELD, 1955; FRENCH, 1961). Loss of catalytic activity is associated
with the oxidation of the sulfhydryl groups which are most reactive to iodoacetamide and N-ethylmaleimide (SPRADLIN and THOMA, 1970). Inhibition by DTNB
is completely reversible by cysteine and SPRADLIN et al. (1969) and SPRADLIN and
THOMA (1970) have proposed that reversible oxidation of the sulfhydryl groups
of β-amylase cause conformational changes that allow the enzyme to be stored
in an inactive state in ungerminated seeds.

The activation of β-amylase in cereal seeds has been extensively studied. β-Amylase occurs as an inactive zymogen, "latent β-amylase", in dormant wheat and
barley endosperm, where in wheat it is bound to the storage protein glutenin
(ROWSELL and GOAD, 1962 a, b). Latent β-amylase is converted to the active form
"free β-amylase" during germination (POLLOCK and POOL, 1958). Activation of
latent β-amylase can also be achieved in vitro by thiol reducing agents or by
a combination of thiol reducing agents and proteolysis (WHELAN, 1958; ROWSELL
and GOOD, 1962 a, b; THOMA et al., 1971). It has been suggested that, in vivo,
latent β-amylase is reduced by a disulfide reductase (THOMA et al., 1971; ap REES,
1974). A protein disulfide reductase from plants has been described (HATCH and
TURNER, 1960).

3. Phosphorylases

Potato and maize phosphorylases have been given major attention. Although a
number of isozymes have been reported for the phosphorylases of both sources
(TSAI and NELSON, 1968, 1969; GERBRANDY and DOORGEEST, 1972; FRYDMAN and
SLABNIK, 1973; GERBRANDY et al., 1975), at least one of the potato phosphorylase

isozymes is the product of proteolysis of the major potato isozyme (GERBRANDY et al., 1975).

a) Potato Phosphorylase

The properties of potato phosphorylase are the most well studied. It is a dimer of 180,000–207,000 mol wt (LEE, 1960a; KAMOGAWA et al., 1971). It contains 2 mol bound pyridoxal-P per mol enzyme (1 subunit) (LEE, 1960b; KAMOGAWA et al., 1971). Potato phosphorylase sulfhydryl groups are not as reactive as those of muscle phosphorylase (KAMOGAWA et al., 1971) and PCMB inhibition can be partially reversed by cysteine (LEE, 1960b).

Unlike mammalian phosphorylase, potato phosphorylase is not known to be regulated by phosphorylation and dephosphorylation. The plant enzyme does not contain serine phosphate and is unaffected by incubation with mammalian protein kinase, phosphatase or AMP (LEE, 1960a, b). In addition, neither ATP, dithiothreitol (GOLD et al., 1971), glucose-6-P, fructose-6-P, fructose-1,6-P_2, ribose-5-P, nor fructose-1-P (KAMOGAWA et al., 1968) activate the enzyme. DE FEKETE and CARDINI (1964) have reported that UDPglucose, but not ADPglucose, inhibits potato phosphorylase ($K_i = 2.9$ mM).

Potato phosphorylase readily degrades amylopectin with about 45% conversion to glucose-1-P (LIDDLE et al., 1961). Glycogen is degraded only very slowly, with 3–21% conversion to glucose-1-P (LIDDLE et al., 1961). The reported K_ms of glucose-1-P and P_i are 2–3.5 mM and 5.9–7.5 mM, respectively (LEE, 1960a; DE FEKETE and CARDINI, 1964; GOLD et al., 1971). The pH activity curve shows a sharp optimum at 6.5 (LEE, 1960a).

Maltotriose can serve as a primer for glucan synthesis by phosphorylase, but maltotetrose and larger glucans are much more effective (PARRISH et al., 1970). Apparent activity in the absence of primer by a nonprimer requiring potato phosphorylase could be abolished by glucoamylase preincubation (KAMOGAWA et al., 1968). SLABNIK and FRYDMAN (1970) have identified a nonprimer requiring isozyme, but did not report the results of glucoamylase treatment.

b) Maize Phosphorylase

TSAI and NELSON (1968, 1969) and TSAI et al. (1970) have described four phosphorylase isozymes from sweet corn which differ in the timing of their appearance in the seed and in their regulatory properties. Isozyme I is present in the dry seed, whereas II and III appear and increase rapidly after day 12 of development. Isozyme I also increases after pollination, but more slowly than II. Isozyme IV is only present in the germinating embryo (with scutellum). Although TSAI and NELSON (1968, 1969) observed inhibition by ATP and a series of purine and pyrimidine nucleotides and Mg^{2+} stimulation, later work from the same laboratory (BURR and NELSON, 1975) failed to find any effect of ATP on isozyme II, the major and most sensitive isozyme according to the first reports. The K_m of glucose-1-P of isozyme II is 1 mM, the K_m of P_i is 4.2 mM, and ADP-glucose, UDPglucose and 10 mM Mg^{2+} are inhibitory. The enzyme is a pyridoxal-P containing dimer, MW 223,000 (BURR and NELSON, 1975).

Lee and Braun (1973) also found that sweet corn phosphorylase is unaffected by ATP or Mg^{2+}; ADPglucose and 2,3-P_2-glycerate inhibit. The pH optimum is 5.8. The enzyme is a dimer, MW 315,000, containing bound pyridoxal-P. Glycogen is not a good substrate, while amylopectin is readily degraded: the K_m of amylopectin is 2.4×10^{-5} M and the K_m of glycogen is 500×10^{-5} M. The smallest primer required for synthesis is maltotriose and the smallest substrate for degradation is maltotetrose. The enzyme is not phosphorylated by rabbit phosphorylase kinase.

c) Blue-Green Algal (Bacterial) Phosphorylase

Fredrick (1968) has reported the presence of two phosphorylase isozymes from *Oscillatoria princeps*. The a_2 isozyme of *Oscillatoria* contains bound glucan which allows its glucan synthesizing activity to be independent of added primer (Fredrick, 1973). α-Amylase treatment converts the primer-independent isozyme into the primer-dependent isozyme and incubation of the primer-dependent isozyme with UDPglucose converts it to the primer-independent isozyme. The mechanism of the interconversion is unknown at present.

4. α-Glucosidase

α-Glucosidase hydrolyzes the α-1,4 linkages of dextrins, attacking from the nonreducing end and liberating glucose. The enzyme is widely distributed in plants. Manners (1974b) has reviewed the properties and occurrence of the multiple isozymic forms. The pH optima for α-glucosidases are between 3 and 5 (Hutson and Manners, 1965; Marshall and Taylor, 1971; Takahashi et al., 1971). The enzymes are most active in the hydrolysis of maltose, but they hydrolyze maltotriose, maltotetrose, nigerose and other dextrins with varying degrees of activity; amylopectin is hydrolyzed more readily than soluble starch, glycogen or amylose (Hutson and Manners, 1965; Manners and Rowe, 1969b; Takahashi et al., 1971; Chiba and Shimomura 1975).

5. D-Enzyme

D-enzyme catalyzes the reversible condensation of donor and acceptor α-1,4 glucan with the release of free glucose (reaction 6).

$$G_{donor} + G_{acceptor} \rightleftharpoons G_{(donor+acceptor-1)} + G. \tag{6}$$

Small donor groups are transferred most readily, and the rate of transfer is proportional to the size of the group transferred (Jones and Whelan, 1969). Maltotriose is the smallest donor substrate for D-enzyme. Jones and Whelan (1969) have shown that the nonreducing end linkage and the bond penultimate to the reducing end are "forbidden" linkages.

A probable function for D-enzyme in starch degradation, as suggested by Lee et al. (1970), is in the regeneration of dextrins which are substrates for further degradation by phosphorylase from the smaller linear limit dextrins which accumulate as a result of starch degradation. Lee and Braun (1973) have shown that maltotriose can appear to serve as a substrate for phosphorolysis if D-enzyme is present.

6. Glucosyltransferase

A glucosyltransferase which transfers the nonreducing glucose of maltose to an acceptor glucose was first identified in plants by EDELMAN and KEYS (1961). Recently, LINDEN et al. (1974) have described glucosyltransferase activity in spinach. Glucosyl, maltosyl and larger α-1,4 glucans are transferred to glucose from the nonreducing end of donor glucans. This enzyme differs from D-enzyme in that maltose functions as a donor.

Crude enzyme preparations from the green alga *Cladophora rupestris* (DUNCAN and MANNERS, 1958) and from green gram (NIGAM and GIRI, 1960) have been reported to catalyze the incorporation of maltose into larger dextrins, but both experimenters used high concentrations of maltose (32.4 g/250 ml and 10 g/25 ml) over extended incubation periods. Therefore, the transglucosylation activity they reported may have been due to α- or β-amylase, since HEHRE et al. (1969) have shown that both of these enzymes catalyze slow transglycosylation reactions when presented with high substrate concentrations.

7. Debranching Enzymes

The specificity of the enzymes which hydrolyze the α-1,6 linkages of plant glucans have been subject to considerable confusion. MANNERS' laboratory has described enzymes from malted barley (MANNERS et al., 1970; MANNERS and ROWE, 1971) and from sweet corn (MANNERS and ROWE, 1969a) which were only active on α-limit dextrins, amylopectin β-limit dextrins and pullulan, but not on amylopectin, glycogen or glycogen β-limit dextrins. The enzymes with this specificity were termed limit dextrinases, while the name R-enzyme was reserved for the enzymes which also hydrolyze the α-1,6 linkages of amylopectin. Through the work of DRUMMOND et al. (1970), LEE et al. (1971) and LEE and WHELAN (1971), it has become clear that when the enzymes from potato and sweet corn which hydrolyze amylopectin as well as pullulan and the α- and β-limits of amylopectin were diluted, their activity on amylopectin decreased to the limit of resolution of the assay method. Thus R-enzyme appeared to be converted to limit dextrinase. Consequently, LEE and WHELAN (1971) have classified all plant debranching enzymes as pullulanases. DUNN et al., (1973) and HARDIE et al. (1976) have confirmed the dilution effect, pointing out that increasing the sensitivity of the assay shows that the activity on amylopectin remains proportional to enzyme concentration.

In any case, the activity of amylopectin debranching by the debranching enzymes of rice, oats, sorghum and sweet corn is always much greater for limit dextrins than for amylopectin, although both types of substrate are debranched (LEE et al., 1971; DUNN et al., 1973; HARDIE et al., 1976).

Other types of debranching activity have been described in plants. MACWILLIAM and HARRIS (1959) identified an "R-enzyme" from malt which had limited activity on β-limit dextrins and amylopectin and did not degrade α-limit dextrins, and a "limit dextrinase" which did not degrade β-limit dextrins or amylopectin. MAN-

NERS and Rowe (1969a) identified an "R-enzyme" from sweet corn which had no activity on α-limit dextrins, but did degrade amylopectin, an isoamylase which degraded glycogen and amylopectin, and a pullulanase. Lee and Whelan (1971) have suggested that both of these "R-enzymes" may be isoamylases, rather than debranching enzymes of the pullulanase type.

II. Degradation of Intact Granules in Vitro

Of the three major α-1,4 glucan degrading enzymes (α-amylase, β-amylase and phosphorylase), only α-amylase can degrade intact maize or potato starch granules (Walker and Hope, 1963; Dunn, 1974; Manners, 1974a, b; Banks and Greenwood, 1975; see Whelan, 1958 for earlier references). Furthermore, degradation of native starch granules from reserve tissues is slow compared to degradation of gelatinized granules (Bailey and Macrae, 1973).

Leaf starch granules are rapidly degraded by α-amylase without prior gelatinization (Bailey and Macrae, 1973) so the possibility remains that leaf starch granules may also serve as substrates for phosphorylase or β-amylase. It should be kept in mind that isolated starch granules used in this type of experiment are stripped of adhering membranes and protein and may maintain a looser structure in vivo than when isolated. In addition, there may be a synergistic effect of debranching and amylolysis in vivo which has not been examined in experiments on isolated starch granules. Lee et al. (1971) have shown that sweet corn pullulanase stimulates degradation of amylopectin and glycogen by β-amylase.

III. Starch Degradation in Vivo: Germinating Seeds

The association of enzyme activity with the course of starch degradation has been most successfully studied in germinating seeds. The initial degradation and the maximum decrease in starch are associated with the activity of α-amylase. Subsequent degradation occurs in conjunction with increases in the activity or products of β-amylase, phosphorylase, maltase and debranching enzymes. This area has been recently reviewed (Briggs, 1973; Manners, 1974a, b; ap Rees, 1974).

IV. Starch Degradation in Vivo: Leaves

Starch metabolism in leaves and algae is intimately associated with photosynthesis. As Gibbs (1966) has pointed out, the enzymes of starch synthesis and degradation must be located in the chloroplasts, since that is the location of the starch granules. The coordination of starch synthesis and degradation must be finely regulated, since simultaneous operation of synthetic and degradative pathways would constitute a futile cycle. As important as it would be actually to establish the distribution of enzymes involved in leaf starch degradation and their regulatory properties, very little has been done to approach this subject.

It should be noted that in the case of germinating seeds, at least some of the enzymes of starch degradation are activated by enzyme synthesis, and they must be transported from their site of origin into the starch storing cells of the endosperm. Furthermore, during germination the seed undergoes extensive proteolysis and cell wall degradation as well as starch degradation (VARNER, 1965). Thus, the events which regulate starch degradation in seeds are not freely reversible and cannot be expected to pertain to the regulation of starch degradation in photosynthetic tissues.

1. Enzyme Localization

Intracellular localization studies have been reported for α-amylase, β-amylase and phosphorylase. β-Amylase is located outside the chloroplasts of *Vicia faba* (CHAPMAN et al., 1972) and pea shoots (LEVI and PREISS, 1978). α-Amylase has been found in isolated chloroplasts of maize bundle sheath and mesophyll cells (DAVIES, 1974). Phosphorylase has been found in isolated chloroplasts of tobacco (STOCKING, 1959), pea shoots (LEVI and PREISS, 1978), spinach (LATZKO and GIBBS, 1968), maize bundle sheath cells (HUBER et al., 1969; DAVIES, 1974) and maize mesophyll cells (DAVIES, 1974), sometimes in and sometimes out of the chloroplasts of tobacco (MADISON, 1956), and restricted to the cytoplasm of tobacco (BIRD et al., 1965). Thus the question of localization still needs to be clarified.

2. Pathway of Starch Degradation

Although little definitive work has been done on the localization or regulation of the enzymes of the first steps of starch degradation, the selective permeability of the chloroplast to metabolites has been thoroughly examined. Transport of large dextrins, sucrose, glucose, hexose monophosphates and hexose diphosphates by the spinach chloroplast inner envelope occurs slowly if at all, while ribose-5-P, 3-P-glycerate, dihydroxyacetone-P and glyceraldehyde-3-P are rapidly transported (HEBER et al., 1967; BASSHAM et al., 1968; NOBEL, 1969; HELDT and SAUER, 1971; WANG and NOBEL, 1971; HEBER, 1974; WALKER, 1974; POINCELOT, 1975; SCHAFER et al., 1977). In order to be used for cytoplasmic functions or intercellular transport, starch degradation must be coordinated with the synthesis of those phosphorylated sugars which can be transported out of the chloroplast.

The products of starch degradation comprise such a major portion of the metabolites of germinating seeds or other reserve tissue that it is relatively straightforward to determine the major products of starch degradation in these tissues. However, in photosynthetic tissues, where starch degradation accounts for only part of the sugar, sugar phosphate and organic acid transformations occurring in the dark, it is more difficult to assign the appearance of intermediates to starch breakdown.

Since 3-P-glycerate and triose-phosphate make up the major portion of the products of photosynthetic CO_2 fixation, it is possible to allow isolated chloroplasts to accumulate photosynthetically produced ^{14}C-starch and then wash out the bulk of the labeled compounds, retaining the ^{14}C-starch, ^{14}C-fructose diphosphate, and a small amount of labeled hexose monophosphate and maltose (LEVI and GIBBS, 1976; HELDT et al., 1977; PEAVEY et al., 1977). When chloroplasts prepared this way are incubated in the dark, starch degradation takes place. The major labeled products are 3-P-glycerate, maltose and minor amounts of glucose-6-P and ribose-5-P. There is no detectable 6-P-gluconate or CO_2 (Peavey et al., 1977) produced by the labeled starch (Levi and Gibbs, 1976). Under some conditions, glucose is produced (Heldt et al., 1977; Peavey et al., 1977). 3-P-Glycerate is readily transported by the chloroplast, so it is probably the terminal product of starch degradation in the organelle. Maltose and glucose may need to be further metabolized before they can either be transported to the cytoplasm or used in the chloroplast. Schafer et al. (1977) and Peavey et al. (1977) have shown glucose transport across the chloroplast membranes. Starch degradation by intact spinach chloroplasts is accelerated by P_i, 2.5–5 mM, which means that phosphorylase is probably involved in starch degradation under these conditions (Steup et al., 1976; Heldt et al., 1977; Peavey et al., 1977). Phosphorylase accounts for all glucose-6-P synthesis from amylopectin in pea chloroplast stroma (Levi and Preiss, 1978). Phosphorylase may initiate all the sugar phosphate synthesis from starch and α- or β-amylase may account for glucose or maltose production.

The production of 3-P-glycerate and absence of 6-P-gluconate are evidence for the operation of glycolysis rather than the oxidative pentose phosphate pathway in darkened chloroplasts. The finding of phosphofructokinase in chloroplasts (Kelly and Latzko, 1975) supports this conclusion.

3. Maltose Synthesis

Maltose synthesis may also be involved in sugar phosphate metabolism. Maltose appears as a product of CO_2 fixation in the light (Linden et al., 1975) and as a product of starch degradation in the dark in isolated chloroplasts (Levi and Gibbs, 1976). Recent reports from Kandler's laboratory (Linden et al., 1975; Schilling et al., 1976) have suggested that, rather than being a product of amylolytic degradation of starch, the maltose produced during photosynthesis in young leaves and isolated spinach chloroplasts is synthesized from glucose and glucose-1-P by a maltose phosphorylase, or from glucose-1-P through a maltose-phosphate intermediate during CO_2 fixation in spinach chloroplasts. Hydrolysis of maltose phosphate would prove to be a mechanism for maltose synthesis in chloroplasts.

4. Regulation of Starch Degradation

The mode of regulation of starch degradation and its initiation in chloroplasts is unknown. However, it is possible to compare the properties of the enzymes of starch degradation from other tissues with the changes which are known to regulate enzyme activity in leaves subjected to light/dark transitions.

Photosynthetically driven pH changes in photosynthetic systems may inhibit degradation in the light. α-Amylase (Fischer and Stein, 1961; Thoma et al., 1971), α-glucosidase (Hutson and Manners, 1965; Marshall and Taylor, 1971; Takahashi et al., 1971), D-enzyme (Manners and Rowe, 1969b) and limit dextrinase (Hardie et al., 1976) have pH optima at or below pH 6.0. The pH of a starch mobilizing tissue, germinating barley, varies between 4.5 and 5.5 (Briggs, 1968), but the stroma of chloroplasts maintains a higher pH range, particularly during photosynthesis. When the medium surrounding isolated chloroplasts is pH 7.6, the stromal pH is 6.9–7.3 in the dark and 8.0 in the light (Werdan et al., 1975).

It is possible that leaf starch degrading enzymes generally operate in a higher pH range than seed enzymes. The pH optimum of β-amylase from *Vicia faba* leaves is 6.0 (CHAPMAN et al., 1972), higher than that for β-amylase from sweet potato (pH 4–5; BERNFIELD, 1955).

Additionally, starch degradation in leaves may be regulated by phosphate concentration. The concentration of P_i in chloroplasts is 30–50% less in the light than in the dark (SANTARIUS and HEBER, 1965). The requirement of phosphorylase for P_i may allow it to be more active in the dark than in the light because of the differential in P_i concentrations. Indeed, STEUP et al. (1976) have shown a doubling in the rate of starch degradation in isolated spinach chloroplasts when the concentration of P_i in the medium was raised from 0 to 2.5 mM. However, since the in vivo concentration of P_i in chloroplasts is 4–25 mM (SANTARIUS and HEBER, 1965), and STEUP et al. (1976) found maximum rates of starch degradation at 2.5–5 mM P_i, this potential regulatory mechanism may not always operate in vivo.

Enzymes of the photosynthetic CO_2 fixation cycle (fructose diphosphatase, NADPH dependent glyceraldehyde-3-phosphate dehydrogenase, ribulose-5-phosphate kinase and sedeheptulose-1,7-diphosphatase) are light activated by a reductive mechanism which has been mimicked by dithiothreitol or reduced ferredoxin (BUCHANAN et al., 1971; PAWLIZKI and LATZKO, 1974; ANDERSON and AVRON, 1976; see Chap. II.22, this vol.). It is possible that dark oxidation of the light-activated enzymes accompanies dark reduction of β-amylase, or instead that the β-amylase of chloroplasts is inactivated by reduction in the light, as is glucose-6-phosphate dehydrogenase (ANDERSON and AVRON, 1976).

As has been already stated, potato tuber (DE FEKETE and CARDINI, 1964) and maize seed (LEE and BRAUN, 1973; BURR and NELSON, 1975) phosphorylases are inhibited by sugar nucleotides. However, the concentration of ADPglucose necessary for inhibition of maize seed phosphorylases ($K_i = 1.5$ mM) is much higher than the ADPglucose concentration (up to 0.08 mM) found in spinach chloroplasts (HELDT et al., 1977). Leaf phosphorylase could be more sensitive to ADPglucose than seed and tuber phosphorylase, or could be inhibited by another product of photosynthesis.

Reasoning that chloroplast phosphofructokinase activity is required for starch degradation in chloroplasts, KELLY and LATZKO (1975, 1976) have suggested that regulation of phosphofructokinase may control starch degradation. However, there is as yet no evidence that any of the enzymes of the primary events of starch degradation (i. e., α-amylase, β-amylase, phosphorylase, or debranching enzymes) are inhibited by glycolytic intermediates. Elevation of the ratios of glucose-1-P to P_i can shift the equilibrium of phosphorylase from degradation to synthesis or increase the level of the inhibitor, ADPglucose. However, as explained in Section II, starch synthesis by phosphorylase requires nonphysiological concentrations of glucose-1-P and P_i and small changes in glucose-1-P concentration would not cause significant changes in ADPglucose pyrophosphorylase activity in the dark.

5. Conclusion

Until there are systematic studies on the distribution and regulation of degradative enzymes in leaves, it will be impossible to say whether, as in seeds, α-amylase initiates starch degradation with β-amylase, phosphorylase and possibly debranching enzyme acting primarily on the maltodextrin products of α-amylolysis, or whether, instead, phosphorylase and β-amylase actually function in the primary degradation of leaf starch. If the latter is the physiological case, then it should be straightforward to find activators and inhibitors of these enzymes. If instead α-amylase is the primary degradative enzyme, there must be some mechanism in the chloroplast for sequestering or inhibiting α-amylase which has not yet been proposed.

References

Akazawa, T., Murata, T.: Biochem. Biophys. Res. Commun. *19*, 21–26 (1965)
Alexander, A.G.: Ann. N.Y. Acad. Sci. *210*, 64–80 (1973)
Amir, J., Cherry, J.H.: Plant Physiol. *49*, 893–897 (1972)
Anderson, L.E., Avron, M.: Plant Physiol. *57*, 209–213 (1976)
ap Rees, T.: In: MTP Int. Rev. Sci. Biochem. Ser. One Plant biochemistry. Northcote, D.H. (ed.), Vol. 11, pp. 89–127. London, Baltimore: Butterworths-University Park Press 1974
Badenhuizen, N.P.: The biogenesis of starch granules in higher plants. New York: Appleton-Century-Crofts 1969
Bailey, R.W., Macrae, J.C.: FEBS Lett. *31*, 203–204 (1973)
Banks, W., Greenwood, C.T.: Starch and its components. New York: J. Wiley and Sons 1975
Bassham, J.A., Krause, G.H.: Biochim. Biophys. Acta *189*, 207–221 (1969)
Bassham, J.A., Kirk, M., Jensen, R.G.: Biochim. Biophys. Acta *153*, 211–218 (1968)
Baxter, E.D., Duffus, C.M.: Phytochem. *10*, 2641–2644 (1971)
Bernfeld, P.: Methods Enzymol. *1*, 149–158 (1955)
Bird, I.F., Porter, H.K., Stocking, C.R.: Biochim. Biophys. Acta *100*, 366–375 (1965)
Bird, R., Hopkins, R.H.: Biochem. J. *56*, 86–99 (1954)
Black, R.C., Loerch, J.D., McArdle, F.J., Creech, R.G.: Genetics *53*, 661–668 (1966)
Borovsky, D., Smith, E.E., Whelan, W.J.: Eur. J. Biochem. *59*, 615–625 (1975a)
Borovsky, D., Smith, E.E., Whelan, W.J.: FEBS Lett. *54*, 201–205 (1975b)
Borovsky, D., Smith, E.E., Whelan, W.J.: Eur. J. Biochem. *62*, 307–312 (1976)
Bourne, E.J., Peat, S.: J. Chem. Soc. Lond. *1945*, 877–882
Boyer, C., Preiss, J.: Biochem. *16*, 3693–3699 (1977)
Boyer, C.D., Preiss, J.: Biochem. Biophys. Res. Commun., *80*, 169–175 (1978a)
Boyer, C., Preiss, J.: Carbohydrate Res., In press (1978b)
Briggs, D.E.: Phytochemistry *7*, 513–529 (1968)
Briggs, D.E.: In: Biosynthesis and its control in plants. Milborrow, B.C. (ed.), pp. 219–277. New York: Academic Press 1973
Brown, W.A., Brown, B.I., Brown, D.H.: Biochemistry *10*, 4253–4262 (1971)
Buchanan, B.B., Schurmann, P., Kalberer, P.P.: J. Biol. Chem. *246*, 5952–5959 (1971)
Burr, B., Nelson, O.E.: Eur. J. Biochem. *56*, 539–546 (1975)
Cameron, J.W., Teas, H.J.: Am. J. Bot. *41*, 50–55 (1954)
Cardini, C.E., Frydman, R.B.: Methods Enzymol. *8*, 387–394 (1966)
Cardini, C.E., Leloir, L.F., Chiriboga, J.: J. Biol. Chem. *214*, 148–155 (1955)
Chandorkar, K.R., Badenhuizen, N.P.: Staerke *18*, 91–95 (1966)
Chapman, G.W. Jr., Pallas, J.E. Jr., Mendicino, J.: Biochim. Biophys. Acta *276*, 491–507 (1972)
Chen, M., Whistler, R.L.: Int. J. Biochem. *7*, 433–437 (1976)
Chiba, S., Shimomura, T.: Agric. Biol. Chem. *39*, 1033–1040 (1975)

Cohn, M.: In: The enzymes. 2nd ed. Boyer, P.D., Lardy, H., Myrbäck, K. (eds.), Vol. V, pp. 179–206. New York, London: Academic Press 1961
Colowick, S.P., Sutherland, E.W.: J. Biol. Chem. *144*, 423–434 (1942)
Creech, R.G.: Genetics *52*, 1175–1186 (1965)
Davies, D.R.: In: Plant carbohydrate biochemistry. Pridham, J.B. (ed.), pp. 61–81. New York: Academic Press 1974
Dickinson, D.B., Preiss, J.: Arch. Biochem. Biophys. *130*, 119–128 (1969a)
Dickinson, D.B., Preiss, J.: Plant Physiol. *44*, 1058–1062 (1969b)
Doi, A.: Biochim. Biophys. Acta *184*, 477–485 (1967)
Downton, W.J.S., Hawker, J.S.: Phytochemistry *14*, 1259–1263 (1975)
Drummond, G.S., Smith, E.E., Whelan, W.J.: FEBS Lett. *9*, 136–140 (1970)
Drummond, G.S., Smith, E.E., Whelan, W.J.: Eur. J. Biochem. *26*, 168–176 (1972)
Duncan, W.A.M., Manners, D.J.: Biochem. J. *69*, 343–348 (1958)
Dunn, G.: Phytochem. *13*, 1341–1346 (1974)
Dunn, G., Hardie, D.G., Manners, D.J.: Biochem. J. *133*, 413–416 (1973)
Edelman, J., Keys, A.J.: Biochem. J. *72*, 12P (1961)
Espada, J.: J. Biol. Chem. *237*, 3577–3581 (1962)
Fekete, M.A.R. de, Cardini, C.E.: Arch. Biochem. Biophys. *104*, 173–184 (1964)
Fekete, M.A.R. de, Vieweg, G.H.: Ann. N.Y. Acad. Sci. *210*, 170–180 (1973)
Fischer, E.H., Stein, E.A.: In: The enzymes. 2nd ed. Boyer, P.D., Lardy, H., Myrbäch, K. (eds.), Vol. IV, pp. 313–343. New York: Academic Press 1961
Fox, J., Kennedy, L.D., Hawker, J.S., Ozbun, J.L., Greenberg, E., Lammel, C., Preiss, J.: Ann. N.Y. Acad. Sci. *210*, 90–103 (1973)
Fredrick, J.F.: Ann. N.Y. Acad. Sci. *151*, 413–423 (1968)
Fredrick, J.F.: Plant Sci. Lett. *1*, 457–462 (1973)
French, D.: In: The enzymes. 2nd ed. Boyer, P.D., Lardy, H., Myrbäch, K. (eds.), Vol. IV, pp. 345–368. New York: Academic Press 1961
French, D.: J. Japn. Soc. Starch Sci. *19*, 8–25 (1972)
French, D.: In: MTP Int. Rev. Sci. Biochem. Ser. One. Biochemistry of carbohydrates. Whelan, W.J. (ed.), Vol. V pp. 267–335. London, Baltimore: Butterworths-Univ. Park Press 1975
Frydman, R.B.: Arch. Biochem. Biophys. *102*, 242–248 (1963)
Frydman, R.B., Cardini, C.E.: Arch. Biochem. Biophys. *116*, 9–18 (1966)
Frydman, R.B., Cardini, C.E.: J. Biol. Chem. *242*, 312–317 (1967)
Frydman, R.B., DeSouza, B.C., Cardini, C.E.: Biochim. Biophys. Acta *113*, 620–623 (1966)
Frydman, R.B., Slabnik, E.: Ann. N.Y. Acad. Sci. *210*, 153–169 (1973)
Geddes, R., Greenwood, C.T.: Staerke *21*, 148–156 (1969)
Gerbrandy, S.J., Doorgeest, A.: Phytochem. *11*, 2403–2407 (1972)
Gerbrandy, S.J., Shankar, V., Shiraran, K.N., Stegemann, H.: Phytochem. *14*, 2331–2333 (1975)
Ghosh, H.P., Preiss, J.: Biochemistry *4*, 1354–1361 (1965)
Ghosh, H.P., Preiss, J.: J. Biol. Chem. *241*, 4491–4505 (1966)
Gibbs, M.: In: Plant physiology. Steward, F.C. (ed.), pp. 3–115. New York: Academic Press 1966
Gibson, W.B., Brown, B.I., Brown, D.H.: Biochemistry *10*, 4253–4262 (1971)
Gold, A.M., Johnson, R.M., Sanchez, G.R.: J. Biol. Chem. *246*, 3444–3450 (1971)
Greenberg, E., Preiss, J.: J. Biol. Chem. *239*, 4314–4315 (1964)
Greenwood, C.T., MacGregor, A.W., Milne, E.H.: Arch. Biochem. Biophys. *112*, 466–470 (1965)
Griffin, H.L., Wu, Y.V.: Biochemistry *7*, 3063–3172 (1968)
Gunja-Smith, Z., Marshall, J.J., Mercier, C., Smith, E.E., Whelan, W.J.: FEBS Lett. *12*, 101–104 (1970)
Hanes, C.S.: Proc. R. Soc. Lond. B *129*, 174–208 (1940)
Hannah, L.C., Nelson, O.E.: Plant Physiol. *55*, 297–302 (1975)
Hardie, D.G., Manners, D.J., Yellowlees, D.: Carbohydrate Res. *50*, 75–85 (1976)
Hatch, M.D., Turner, J.F.: Biochem. J. *76*, 556–562 (1960)
Hawker, J.S., Downton, W.J.S.: Phytochemistry *13*, 893–900 (1974)
Hawker, J.S., Ozbun, J.L., Preiss, J.: Phytochemistry *11*, 1287–1293 (1972)
Hawker, J.S., Marschner, H., Downton, W.J.S.: Austr. J. Physiol. *1*, 491–501 (1974a)

Hawker, J.S., Ozbun, J.L., Ozaki, H., Greenberg, E., Preiss, J.: Arch. Biochem. Biophys. *160*, 530–551 (1974b)
Heber, U.: In: Biochemistry of chloroplasts. Vol. II, pp. 71–78. London, New York: Academic Press 1967
Heber, U.: Annu. Rev. Plant Physiol. *25*, 393–421 (1974)
Heber, U., Santarius, K.A.: Biochim. Biophys. Acta *109*, 390–408 (1965)
Heber, U., Hallier, V.W., Hudson, M.A.: Z. Naturforsch. *226*, 1200–1215 (1967)
Hehre, E.J., Okada, G., Genghof, D.S.: Arch. Biochem. Biophys. *135*, 75–89 (1969)
Heldt, H.W., Sauer, F.: Biochim. Biophys. Acta *234*, 83–91 (1971)
Heldt, H.W., Chon, C.J., Maronde, D., Herold, A., Stankovic, Z.S., Walker, D.A., Kraminer, A., Kirk, M.R., Heber, U.: Plant Physiol. *59*, 1146–1155 (1977)
Herold, A., Lewis, D.H., Walker, D.A.: New Phytol. *76*, 397–407 (1976)
Hobson, P.N., Whelan, W.J., Peat, S.: J. Chem. Soc. Lond. *1950*, 3566–3573 (1950)
Hodges, H.F., Creech, R.G., Loerch, J.D.: Biochim. Biophys. Acta *185*, 70–79 (1969)
Huber, W., de Fekete, M.A.R., Ziegler, H.: Planta (Berl.) *87*, 360–364 (1969)
Hutson, D.H., Manners, D.J.: Biochem. J. *94*, 783–789 (1965)
Jones, G., Whelan, W.J.: Carbohydrate Res. *9*, 483–490 (1969)
Kamogawa, A., Fukui, T., Nikuni, Z.: J. Biochem. *63*, 361–369 (1968)
Kamogawa, A., Fukui, T., Nikuni, Z.: Agric. Biol. Chem. *35*, 248–254 (1971)
Kanazawa, T., Kanazawa, K., Kirk, M.R., Bassham, J.A.: Biochim. Biophys. Acta *256*, 656–669 (1972)
Kelly, G.J., Latzko, E.: Nature (London) *256*, 429–430 (1975)
Kelly, G.J., Latzko, E.: FEBS Lett. *68*, 55–58 (1976)
Latzko, E., Gibbs, M.: Z. Pflanzenphysiol. *59*, 184–194 (1968)
Lavintman, N.: Arch. Biochem. Biophys. *116*, 1–8 (1966)
Lavintman, N., Cardini, C.E.: Plant Cell Physiol. *9*, 587–592 (1968)
Lavintman, N., Cardini, C.E.: FEBS Lett. *29*, 43–46 (1973)
Lavintman, N., Tandecarz, J., Carceller, M., Mendiara, S., Cardini, C.E.: Eur. J. Biochem. *50*, 145–155 (1974)
Lee, E.Y.C., Braun, J.J.: Arch. Biochem. Biophys. *156*, 276–286 (1973)
Lee, E.Y.C., Marshall, J.J., Whelan, W.J.: Arch. Biochem. Biophys. *143*, 365–374 (1971)
Lee, E.Y.C., Whelan, W.J.: In: The enzymes. 3rd ed. Boyer, P.D. (ed.), Vol. V, pp. 191–234. New York: Academic Press 1971
Lee, E.Y.C., Smith, E.E., Whelan, W.J.: In: Miami winter symp. Vol. I, pp. 139–150. Amsterdam: North-Holland Publishing Co. 1970
Lee, Y.P.: Biochim. Biophys. Acta *43*, 18–24 (1960a)
Lee, Y.P.: Biochim. Biophys. Acta *43*, 25–30 (1960b)
Leloir, L.F., De Fekete, M.A.R., Cardini, C.E.: J. Biol. Chem. *236*, 636–641 (1961)
Levi, C., Gibbs, M.: Plant Physiol. *57*, 933–935 (1976)
Levi, C., Preiss, J.: Plant Physiol. *58*, 753–756 (1976)
Levi, C., Preiss, J.: Plant Physiol. *61*, 218–220 (1978)
Liddle, A.M., Manners, D.J., Wrights, A.: Biochem. J. *80*, 304–309 (1961)
Linden, J.C., Tanner, W., Kandler, O.: Plant Physiol. *54*, 752–757 (1974)
Linden, J.C., Schilling, N., Brackenhofer, H., Kandler, O.: Z. Pflanzenphysiol. *76*, 176–181 (1975)
MacDonald, P.W., Strobel, G.A.: Plant Physiol. *46*, 126–135 (1970)
MacWilliam, I.C., Harris, G.: Arch. Biochem. Biophys. *84*, 442–454 (1959)
Madison, H. Jr.: Plant Physiol. *31*, 387–392 (1956)
Manners, D.J.: Essays Biochem. *10*, 37–71 (1974a)
Manners, D.J.: In: Plant carbohydrate biochemistry. Pridham, J.B. (ed.), pp. 109–125. New York: Academic Press 1974b
Manners, D.J., Marshall, J.J.: Carbohydrate Res. *18*, 203–209 (1971)
Manners, D.J., Rowe, K.L.: Carbohydrate Res. *9*, 107–121 (1969a)
Manners, D.J., Rowe, K.L.: Carbohydrate Res. *9*, 441–450 (1969b)
Manners, D.J., Rowe, K.L.: J. Int. Brewing *77*, 358–365 (1971)
Manners, D.J., Rowe, J.J.M., Rowe, K.L.: Carbohydrate Res. *8*, 72–81 (1968)
Manners, D.J., Marshall, J.J., Yellowlees, D.: Biochem. J. *116*, 539–541 (1970)
Marshall, J.J., Taylor, P.M.: Biochem. Biophys. Res. Commun. *42*, 173–179 (1971)

Marshall, J.J., Whelan, W.J.: Arch. Biochem. Biophys. *161*, 234–238 (1974)
Meyer, K.H., Gibbons, G.C.: Adv. Enzymol. *12*, 341–377 (1951)
Moore, C.J., Turner, J.F.: Nature (London) *223*, 303–304 (1969)
Murata, T., Akazawa, T.: Biochem. Biophys. Res. Commun. *16*, 6–11 (1964)
Murata, T., Akazawa, T.: Arch. Biochem. Biophys. *126*, 873–879 (1968)
Murata, T., Suziyama, T., Akazawa, T.: Biochem. Biophys. Res. Commun. *18*, 371–376 (1965)
Nigam, V.N., Giri, K.V.: J. Biol. Chem. *235*, 947–950 (1960)
Nitsos, R.E., Evans, H.J.: Plant Physiol. *44*, 1260–1266 (1969)
Nobel, P.S.: Biochim. Biophys. Acta *172*, 134–143 (1969)
Nomura, T., Nakayama, N., Murata, T., Akazawa, T.: Plant Physiol. *42*, 327–332 (1967)
Nussenbaum, S., Hassid, W.Z.: J. Biol. Chem. *196*, 785–792 (1952)
Okada, G., Hehre, E.J.: J. Biol. Chem. *249*, 126–135 (1974)
Ozbun, J.L., Hawker, J.S., Preiss, J.: Biochem. Biophys. Res. Commun. *43*, 631–636 (1971a)
Ozbun, J.L., Hawker, J.S., Preiss, J.: Plant Physiol. *48*, 765–769 (1971b)
Ozbun, J.L., Hawker, J.S., Preiss, J.: Biochem. J. *126*, 953–963 (1972)
Ozbun, J.L., Hawker, J.S., Greenberg, E., Lammel, C., Preiss, J., Lee, E.Y.C.: Plant Physiol. *51*, 1–5 (1973)
Parodi, A.J.: Arch. Biochem. Biophys. *120*, 547–553 (1967)
Parrish, F.W., Smith, E.E., Whelan, W.J.: Arch. Biochem. Biophys. *137*, 185–189 (1970)
Pawlizki, K., Latzko, E.: FEBS Lett. *42*, 285–288 (1974)
Peat, S., Whelan, W.J., Bailey, J.M.: J. Chem. Soc. *1953*, 1422–1427 (1953)
Peavey, D.G., Steup, M., Gibbs, M.: Plant Physiol. *60*, 305–308 (1977)
Pisigan, R.A., Del Rosario, E.J.: Phytochemistry *15*, 71–73 (1976)
Poincelot, R.P.: Plant Physiol. *55*, 849–852 (1975)
Pollock, J.R.A., Pool, A.A.: J. Inst. Brewing *64*, 151–156 (1958)
Preiss, J.: In: Current topics of cellular regulation. Horecker, B.L., Stadtman, E.R. (eds.), Vol. I, pp. 125–160. New York: Academic Press 1969
Preiss, J., Ghosh, H.P., Wittkop, J.: In: Biochemistry of chloroplasts. Vol. II, pp. 131–153. London-New York: Academic Press 1967
Preiss, J., Lammel, C., Sabraw, A.: Plant Physiol. *47*, 104–108 (1971)
Recondo, E., Leloir, L.F.: Biochem. Biophys. Res. Commun. *6*, 85–88 (1961)
Ribereau-Gayon, G., Preiss, J.: Methods Enzymol. *23*, 618–624 (1971)
Rowsell, E.V., Goad, L.J.: Biochem. J. *84*, 73P (1962a)
Rowsell, E.V., Goad, L.J.: Biochem. J. *84*, 73P–74P (1962b)
Sachs, J.: In: Lectures of the physiology of plants (translated by H.M. Ward). pp. 304–325. Oxford: Clarendon Press 1887
Santarius, K.A., Heber, U.: Biochim. Biophys. Acta *102*, 39–54 (1965)
Sanwal, G.G., Preiss, J.: Arch. Biochem. Biophys. *119*, 454–469 (1967)
Sanwal, G.G., Greenberg, E., Hardie, J., Cameron, E., Preiss, J.: Plant Physiol. *43*, 417–427 (1968)
Schäfer, G., Heber, U., Heldt, H.W.: Plant Physiol. *60*, 286–289 (1977)
Schiefer, S.: Proc. Fed. Am. Soc. Exp. Biol. *32*, 602 (1973)
Schiefer, S., Lee, E.Y.C., Whelan, W.J.: FEBS Lett. *30*, 129–132 (1973)
Schilling, N., Scheibe, R., Beck, E., Kandler, O.: FEBS Lett. *61*, 192–193 (1976)
Schwimmer, S., Balls, A.K.: J. Biol. Chem. *180*, 883–894 (1949)
Shen-Hwa, C.-S., Lewis, D.H., Walker, D.A.: New Phytol. *74*, 383–392 (1975)
Simpson, G.M., Naylor, J.M.: Can. J. Bot. *40*, 1659–1673 (1962)
Slabnik, E., Frydman, R.B.: Biochem. Biophys. Res. Commun. *38*, 709–714 (1970)
Smith, A.J.: Biochem. Soc. Trans. *3*, 12–14 (1975)
Smith, A.J.: The biology of blue green algae. Carr, N.G., Whitton, B.A. (eds.), pp. 1–38. Berkeley, Los Angeles: Univ. of California Press 1973
Sowokinos, J.R.: Plant Physiol. *57*, 63–68 (1976)
Spradlin, J.E., Thoma, J.A.: J. Biol. Chem. *245*, 117–127 (1970)
Spradlin, J.E., Thoma, J.A., Filmer, D.: Arch. Biochem. Biophys. *134*, 262–264 (1969)
Steup, M., Peavey, D.G., Gibbs, M.: Biochem. Biophys. Res. Commun. *72*, 1554–1561 (1976)
Stocking, C.R.: Plant Physiol. *34*, 56–61 (1959)
Swain, R.R., Dekker, E.E.: Biochim. Biophys. Acta *122*, 75–86 (1966a)
Swain, R.R., Dekker, E.E.: Biochim. Biophys. Acta *122*, 87–105 (1966b)

Takahashi, N., Shimomura, R., Chiba, S.: Agric. Biol. Chem. *35*, 2015–2024 (1971)

Tanaka, Y., Akazawa, T.: Plant Cell Physiol. *12*, 493–495 (1971)

Thoma, J.A., Spradlin, J.E., Dygert, S.: In: The enzymes, 3rd ed. Boyer, P.D. (ed.), pp. 115–189. New York: Academic Press 1971

Tsai, C.Y., Nelson, O.E.: Science *151*, 341–343 (1966)

Tsai, C.Y., Nelson, O.E.: Plant Physiol. *43*, 103–112 (1968)

Tsai, C.Y., Nelson, O.E.: Plant Physiol. *44*, 159–167 (1969)

Tsai, C.Y., Salamini, F., Nelson, O.E.: Plant Physiol. *46*, 299–306 (1970)

Turner, J.F.: Austr. J. Biol. Sci. *22*, 1145–1151 (1969)

Varner, J.E.: In: Plant Biochemistry. Bonner, J., Varner, J.E. (eds.), pp. 763–792. New York: Academic Press 1965

Walker, D.A.: In: MTP Int. Rev. Sci. Biochem. Ser. One. Plant biochemistry. Northcote, D.H. (ed.), Vol. XI, London, Baltimore: Butterworths Univ. Park Press 1974

Walker, D.A., Kraminer, A., Kirk, M.R., Heber, U.: Plant Physiol. *59*, 1146–1155 (1977)

Walker, G.J., Hope, P..M.: Biochem. J. *86*, 452–462 (1963)

Wang, C.-T., Nobel, P.S.: Biochim. Biophys. Acta *241*, 200–212 (1971)

Werdan, K., Heldt, H.W., Milovancev, M.: Biochim. Biophys. Acta *396*, 275–292 (1975)

Whelan, W.J.: In: Encyclopedia of plant physiology. Ruhland, W. (ed.), Vol. VI, pp. 155–240. Berlin: Springer-Verlag 1958

Whelan, W.J.: Staerke *9*, 74–76, 98–101 (1975)

24. The Enzymology of Sucrose Synthesis in Leaves

C.P. WHITTINGHAM, A.J. KEYS and I.F. BIRD

A. Introduction

BROWN and MORRIS concluded in 1893 from experiments on photosynthesis by leaves of *Tropaeolum majus* that sucrose was the first carbohydrate formed and that free hexoses and starch were made from it. PARKIN (1912) also obtained evidence that monosaccharides were derived from sucrose and not precursors of it. However, leaves supplied exogenously with solutions of glucose or fructose synthesized sucrose (VIRTANEN and NORDLUND, 1934) by a mechanism that was not dependent on light, but which required oxygen (MCCREADY and HASSID, 1941; HARTT, 1943a). Plants deficient in phosphate accomplished the conversion less readily (SISAKYAN, 1936; KURSANOV and KRYUKOVA, 1939; HARTT, 1943b); also, sucrose formation was inhibited by phosphatases and inhibitors of phosphate metabolism added to the sugar solutions supplied (HARTT, 1943b, c). VITTORIO et al. (1954), PUTMAN and HASSID (1954), and PORTER and MAY (1955) supplied [^{14}C] hexoses and concluded that free hexoses were not the direct precursors of sucrose. CALVIN and BENSON (1949) studied the sequence in which metabolites were formed in *Chlorella* during photosynthesis from $NaH^{14}CO_3$ and after 90 s, sucrose was radioactive in both hexose moieties but the free hexoses were unlabeled. Hexose phosphates became radioactive before sucrose. BENSON and CALVIN (1950) found similar results during photosynthesis by barley leaves. Following the discovery of UDPG as a cofactor in the interconversion of glucose and galactose in a yeast (CAPUTTO et al., 1950), radioactivity was observed in UDPG before sucrose during photosynthesis (BUCHANAN et al., 1952). BUCHANAN (1953) found sucrose phosphate in sugar beet leaves and suggested that this compound arose from reaction of UDPG with fructose-1-phosphate. The sucrose phosphate was finally identified as sucrose-6'-phosphate (i.e., sucrose esterified with phosphate at the 6-position of the fructose moiety) by BUCHANAN et al. (1972). It is now generally accepted that sucrose phosphate is formed from UDPG and F6P and is the immediate precursor of sucrose.

The mechanism of sucrose synthesis is probably the same in all plants in spite of differences in photosynthetic metabolism. KORTSCHAK et al. (1965) and HATCH and SLACK (1966) showed that in leaves of sugarcane and other C_4 plants sucrose became radioactive after sugar phosphates but before free hexoses. Hence, the precursors of sucrose in C_4, as in C_3 plants, are phosphate esters and not free hexoses.

B. Physiological Relationships of Sucrose in Leaves

Sucrose and certain of its derivatives are the main forms in which carbon is transported in plants (Kursanov, 1961; Ziegler, 1975; Beck, 1975; Pontis, 1977). It serves as a storage carbohydrate, especially in leaves of certain monocotyledons (Parkin, 1899), increasing during the day and decreasing at night. Sucrose is imported into young, growing leaves to provide carbon for growth and respiration (Jones et al., 1959; Kursanov, 1961; Thrower, 1962; Joy, 1964). In partly grown leaves the mature distal region begins to export carbon while the basal region, which is still developing, imports carbon (Jones and Eagles, 1962; Fellows and Geiger, 1974). Mature leaves export most of the carbon they assimilate.

In young leaves diurnal changes occur both in growth rate (Bünning, 1956) and metabolism. Steer (1973) showed that while sucrose was a major product of photosynthesis early in the photoperiod, later, amino acids became major products. This was related to a periodicity of nitrate reductase activity. Steer (1974) investigated the controlling mechanism and suggested that NH_4^+ stimulates pyruvate kinase (E.C.2.7.1.40) activity allowing more carbon to enter the TCA-cycle. Stimulation of aldolase (E.C.4.1.2.13) activity and therefore of FBP production was observed when less NH_4^+ was available. If sucrose is synthesized in the cytoplasm (see Sect. D.I) then the cytoplasmic aldolase activity must be involved in this control. Alternatively, control may be through the cytoplasmic fructosebisphosphatase (E.C.3.1.3.11; Latzko et al., 1974). By condensation catalyzed by aldolase followed by hydrolysis catalyzed by fructose bisphosphatase, (E.C.3.1.3.11) triose phosphates from the chloroplast give rise to hexose monophosphates from which sucrose is made.

During photosynthesis, glycine and serine accumulate in the cytoplasm (Roberts et al., 1970), formed from gylcollate excreted from the chloroplasts. In C_3 plants, under near natural conditions, the amount of carbon metabolized by the glycollate pathway may equal the total assimilated (Mahon et al., 1974; Bird et al., 1975; Keys et al., 1975). One quarter is evolved as CO_2 when glycine is converted to serine; the remainder either recycles to glycollate by way of glyceric acid, PGA and triose phosphates or forms sucrose (Wang and Waygood, 1962; Waidyanatha et al., 1975a). It is not clear whether sucrose is synthesized from serine by a route located entirely in the cytoplasm, or whether the chloroplast is involved. Waidyanatha et al. (1975b) showed that the synthesis of sucrose from serine was dependent on concurrent photosynthetic CO_2 assimilation and on the presence of oxygen suggesting that sucrose synthesis from intermediates of the glycollate pathway may be supported by photorespiration in terms of a supply of energy and carbon. The glycollate pathway operates more slowly in C_4 than in C_3 plants.

Intermediates of the glycollate pathway and of C_4 metabolism provide carbon skeletons for synthesis of various compounds without operation of glycolysis or the TCA-cycle. Indeed, the amino acids that become labeled most rapidly in leaves during photosynthesis from $^{14}CO_2$ are related to glycollate or C_4 metabolism (Hellebust and Bidwell, 1963; Ongun and Stocking, 1965; Kennedy and Laetsch, 1973; Reynolds et al., 1974; Dickson and Larson, 1975; Blackwood and Miflin, 1976). Provision of specific amino acids is not, however, the main role of these ancilliary pathways; most of the carbon involved is metabolized to sucrose.

C. Enzymology

I. Sucrose Phosphate Synthetase E.C.2.4.1.14
and Sucrose Synthetase E.C.2.4.1.13

The reactions catalyzed by the two enzymes are:

1. UDPG + F6P $\rightleftharpoons$ UDP + sucrose phosphate

2. UDPG + D-fructose $\rightleftharpoons$ UDP + sucrose.

The equilibrium for the first reaction strongly favors sucrose phosphate synthesis with values for the apparent equilibrium constant estimated to be 3250 at pH 7.5 and 53 at pH 5.5 (MENDICINO, 1960). Estimated values for the equilibrium constant for reaction 2 range from 1.3 to 8.0 (CARDINI et al., 1955; AVIGAD, 1964; AVIGAD and MILNER, 1966). On thermodynamic grounds, reaction 1, especially when coupled to the hydrolysis of sucrose phosphate, is more favorable for sucrose synthesis than reaction 2. Sucrose synthetase may be involved in the utilization of sucrose. Consistent with this view, sucrose phosphate synthetase is associated especially with tissues that make much sucrose, while sucrose synthetase is found in tissues where sucrose is consumed (PRESSEY, 1969; DELMER and ALBERSHEIM, 1970; MAC-LACHLAN et al., 1970; ARAI and FUJISAKI, 1971; HAWKER, 1971).

Both sucrose phosphate synthetase and sucrose synthetase were first detected in wheat germ (LELOIR and CARDINI, 1953, 1955; CARDINI et al., 1955). Extraction of the enzymes and their purification from leaves has proved difficult (MENDICINO, 1960; ROREM et al., 1960; DUTTON et al., 1961; SKEWS, 1961; FRYDMAN and HASSID, 1963; BIRD et al., 1965; HAQ and HASSID, 1965). Amounts of the enzymes extracted were often insufficient (HAWKER, 1967b) to account for rates of sucrose synthesis in intact leaves. Rates necessary, if much of the carbon assimilated by photosynthesis is converted into sucrose, are in excess of 20 μmol $h^{-1}g$ fr. $wt.^{-1}$. Increased activities of sucrose phosphate synthatase were extracted from leaves of sugarcane, pea, tobacco and beet in a medium designed to suppress polyphenol oxidase (E.C.1.10.3.1; SLACK, 1966; HAWKER, 1967b). BIRD et al. (1974) found still higher activities by omitting F^- and EDTA, decreasing the ionic strength and including $MgCl_2$ in the reaction mixtures; desalted extracts of bean and spinach leaves in 0.01 M Tris buffer (pH 7.0) containing 0.1% bovine serum albumin had sucrose phosphate synthetase activities equivalent to 25 and 27.7 μmol $h^{-1}g$ fr. $wt.^{-1}$ of leaf, respectively and sucrose synthetase activities of 0.4 and 3.3 μmol $h^{-1}g$ fr. $wt.^{-1}$. DAVIES (1974) found activities of sucrose phosphate synthetase and sucrose synthetase in chloroplasts from sugarcane leaves (equivalent to approximately 15 and 30 μmol $h^{-1}g$ fr. $wt.^{-1}$ of leaf) that are much more than activities found in whole leaves by HAWKER (1967b). BUCKE and OLIVER (1975) studied leaves of *Pennisetum purpureum* and *Muhlenbergia montana*, and CHEN et al. (1974) leaves of *Cyperus rotundus* and found activities for sucrose phosphate synthetase in excess of 150 μmol $h^{-1}g$ fr. $wt.^{-1}$. These high activities are for C_4 plants and using media containing 0.33 M sucrose or sorbitol for extraction. The fast rates may reflect the potential for faster photosynthesis and translocation by C_4 plants (WARD-LAW, 1976).

1. Sucrose Phosphate Synthetase

Sucrose phosphate synthetase has been purified from wheat germ (LELOIR and CARDINI, 1955; MENDICINO, 1960; PREISS and GREENBERG, 1969) potato tubers (SLABNIK et al., 1968; MURATA, 1972b), sweet potato roots (MURATA, 1972b) and the scutella of rice seeds (NOMURA and AKAZAWA, 1974). pH optimum is in the range 6.4 to 7.5 and K_m values for UDPG and F6P range from 2.7 to 25.0 and 2.2 to 5.5 mM respectively. NOMURA and AKAZAWA found the molecular weight of the enzyme to be 4.5×10^5. The purified sucrose phosphate synthetase of LELOIR and CARDINI (1955), MENDICINO (1960), and SLABNIK et al. (1968) was assayed in the absence of Mg^{2+} but PREISS and GREENBERG (1969), MURATA (1972b) and NOMURA and AKAZAWA (1974) found activation by Mg^{2+}. Mg^{2+} also stimulated sucrose phosphate synthesis catalyzed by extracts of leaves and scutella of wheat seedlings (KEYS, 1959; BIRD et al., 1974). MURATA (1972b) reported that Mg^{2+} stimulated sucrose phosphate synthetase from potato tubers and sweet potato roots, but inhibited the activity of a partially purified barley leaf enzyme; in other respects the enzymes from the three sources were similar.

2. Sucrose Synthetase

Sucrose synthetase has been purified from wheat germ (CARDINI et al., 1955), soybean and broad bean seeds (NAKAMURA, 1959), artichoke tubers (AVIGAD, 1964), maize endosperm (FEKETE and CARDINI, 1964; TSAI, 1974), sugar beet roots (MILNER and AVIGAD, 1965; AVIGAD and MILNER, 1966), potato tubers (SLABNIK et al., 1968; PRESSEY, 1969; JAARMA and RYDSTROM, 1969; MURATA, 1972a), mung bean seedlings (GRIMES et al., 1970; DELMER, 1972a), tapioca tubers (SHUKLA and SAN-WAL, 1971), sweet potato roots (MURATA, 1971), and rice grains (MURATA, 1972a; NOMURA and AKAZAWA, 1973). The pH optimum for sucrose synthesis by the purified enzymes ranges from 7.0 to 8.8 and for the reverse reaction from 6.3 to 7.5. Estimates of Michaelis constants for the four substrates range for UDPG from 0.27 to 8.3 mM; fructose, 1 to 6.9 mM; UDP, 0.06 to 6.6 mM, and sucrose 17 to 290 mM. Since the Michaelis constants were not always measured under optimum conditions the true values probably lie near the lower value of the range given. Estimates of the molecular weight of sucrose synthetase range from 2.9×10^5 for the enzyme from potato tubers (PRESSEY, 1969) to 1×10^6 for the enzyme from mung bean seedlings (GRIMES et al., 1970).

Mg^{2+} and other divalent cations stimulate the reaction in the direction of sucrose synthesis but inhibit the reverse reaction (NAKAMURA, 1959; SLABNIK et al., 1968; DELMER, 1972b; TSAI, 1974; PRASOLOVA et al., 1976).

Unlike sucrose phosphate synthetase, which is highly specific for UDPG as the donor sugar nucleotide, sucrose synthetase is active with ADPG, less active with TDP-glucose and only slightly active with GDP-glucose (SLABNIK et al., 1968). In the reverse direction, MILNER and AVIGAD (1965) found TDP three times more effective than ADP as an acceptor for glucose from sucrose, but only half as effective as UDP.

II. Sucrose Phosphatase (E.C.3.1.3.24)

Sucrose phosphatase catalyzes the hydrolysis of sucrose phosphate by an essentially irreversible reaction forming sucrose and orthophosphate.

Sucrose phosphate + H_2O → sucrose + Pi.

The enzyme has not been purified from leaves, but activities up to 113 μmol $h^{-1}g$ fr. wt.$^{-1}$ have been measured in sugarcane leaves (HAWKER, 1966). Some activity, usually less than 10% of the total, remains associated with particles during extraction (HAWKER and HATCH, 1966; HAWKER, 1966).

Sucrose phosphatase purified from carrot root or sugarcane stem tissue has a pH optimum between 6.4 and 6.7 and a K_m for sucrose phosphate between 0.13 and 0.17 mM (HAWKER and HATCH, 1966). Activity requires Mg^{2+} (K_m 0.3 mM) or Mn^{2+} and is inhibited by EDTA, F^-, PP_i.

III. UDPglucose Pyrophosphorylase (E.C.2.7.7.9)

Using crude homogenates of leaves of sugar beet, BURMA and MORTIMER (1956) and GANGULI and HASSID (1957) demonstrated a UDPG pyrophosphorylase similar to that in yeast (MUNCH-PETERSEN et al., 1953) catalyzing the reversible reaction:

UTP + G1P ⇌ UDPG + PPi.

Adequate UDPG pyrophosphorylase activity has been detected in leaves to account for its role in sucrose synthesis; spinach leaves showed rates up to 250 μmol $h^{-1}g$ fr. wt.$^{-1}$ (BIRD et al., 1974). The electrophoretic mobility of the enzyme from oat leaves was similar to that from oat roots or coleoptiles (GORDON, 1973). UDPG pyrophosphorylase has been purified from mung bean seedlings (GINSBURG, 1958; TSUBOI et al., 1969), *Sorghum* seedlings (GANDER, 1966; GUSTAFSON and GANDER, 1972), wheat grain (TOVEY and ROBERTS, 1970), lily pollen (HOPPER and DICKENSON, 1972) and oat seedlings (GORDON, 1973). The pH optimum is between 7.5 and 8.5 and the K_m for UDPG, PPi, G1P and UTP range from 0.06 to 0.27, 0.05 to 0.25, 0.05 to 1.0 and 0.03 to 0.25 mM respectively. The molecular weight of the enzyme from *Sorghum* and oat seedlings was 50,000 and 53,000 respectively (GANDER, 1966; GORDON, 1973). Mg^{2+} was obligatory because UTP and PPi react only as Mg^{2+} complexes (GUSTAFSON and GANDER, 1972; GORDON, 1973). The equilibrium is dependent on pH and the concentration of Mg^{2+}; the values for the equilibrium constant reported by TURNER and TURNER (1958) range at 30° C from 0.12 at pH 7.9 to 0.29 at pH 7.0 with 2.5 mM Mg^{2+}, while GUSTAFSON and GANDER (1972) report a value at 37° C of 0.63 at pH 8.5 with 10 mM Mg^{2+}. The purified enzymes from *Sorghum* seedlings and lily pollen were specific for UDPG and UTP as nucleotide substrates. GINSBURG (1958) found that galactose-1-phosphate, xylose-1-phosphate, arabinose-1-phosphate, glucuronic acid-1-phosphate and galacturonic acid-1-phosphate were not alternative substrates to G1P for the mung bean enzyme.

IV. Sucrose Phosphorylase (E.C.2.4.1.7)

Doudoroff et al. (1943) discovered in the bacterium *Pseudomonas saccharophila* the enzyme sucrose phosphorylase. It catalyzed a reversible synthesis of sucrose from G1P and fructose with the elimination of orthophosphate.

$$G1P + \text{D-fructose} \rightleftharpoons \text{sucrose} + Pi.$$

Particular importance was attached to this enzyme as its mode of action resembled that of the starch-forming phosphorylase (E.C.2.4.1.1.) discovered in pea seeds and potato tubers by Hanes (1940a, b). Some reports purporting to show the presence of sucrose phosphorylase in higher plants, even in leaves, have been published (Pandya and Ramakrishnan, 1956; Alexander, 1964, 1965). However, Hatch et al. (1963) were unable to detect sucrose phosphorylase in sugarcane stem where rapid sucrose synthesis was taking place. Because the equilibrium of the reaction is unfavorable (Keq. $=0.05$) and free fructose, not a fructose phosphate (see Sect. A), is one of the substrates (Doudoroff, 1943) it is unlikely that sucrose phosphorylase is involved in sucrose synthesis in leaves.

V. Invertase (E.C.3.2.1.26)

Invertases hydrolyze sucrose to its constituent hexoses. Both acid and neutral invertases have been found in sugarcane stems, carrot roots, and pea roots (Hatch et al., 1963; Ricardo and ap Rees, 1970; Lyne and ap Rees, 1971). Acid invertase is often associated with cell walls in tissue homogenates but it is not always clear whether this reflects the situation in vivo or is an artifact (Little and Edelman, 1973). In immature sugarcane stem tissue there is a soluble acid invertase in the vacuoles of storage parenchyma cells and also associated with the cell walls. High acid invertase activities are associated with rapidly growing tissues (Hellebust and Forward, 1962). The neutral invertases are probably located in the cytoplasm, and high activities are associated with tissue containing or storing sucrose. Invertases are present in leaves (Krotkov and Bennet, 1952; Allen and Bacon, 1956; Pavlinova and Kursanov, 1956) and Manning and Maw (1975) showed that the activity of acid invertase in tomato leaves ranged from 223 to 369 μmol h^{-1}g fr. wt.$^{-1}$.

The pH optimum of acid invertase ranges from 3.8 for the bound to 5.5 for soluble forms. K_m values for sucrose range from 0.3 to 56 mM. The pH optimum for neutral invertase is close to 7.0 and K_m values for sucrose range from 13 to 25 mM.

VI. Enzyme Control Mechanisms

Some properties of UDPG pyrophosphorylase and sucrose phosphate synthetase may have controlling functions on sucrose synthesis from assimilated carbon. Rapid sucrose synthesis may be initiated as the amounts of sugar phosphates and UDPG increase. Preiss and Greenberg (1969) showed that there was a sigmoid relationship between concentrations of F6P or UDPG and the activity of sucrose phosphate synthetase. Free nucleotides, especially UDP (SLABNIK et al., 1968; MURATA, 1972b; SALERNO and PONTIS, 1976) inhibit the reaction. These properties together might form a basis for control of sucrose synthesis. FEKETE (1971) has reported that sucrose phosphate synthetase in *Vicia faba* cotyledons is associated with a natural activator which can be removed from the enzyme by freezing and thawing. The sigmoid response to UDPG and F6P is observed only with the enzyme separated from the natural activator, whereas the effect of UDP and other free nucleotides is found only when the natural activator is present. Citrate at low concentrations (20 mM) inhibits the activity of the enzyme plus natural activator, but at higher concentrations this inhibition is reversed and in the presence of 100 mM citrate the activity of the enzyme without the natural activator is as high as that of the enzyme plus activator in the absence of citrate. Citrate also overcomes the inhibitory effects of free nucleotides including UDP (FEKETE, 1971). In some leguminous seeds, where citrate concentrations are high, sucrose synthesis may be dependent only on the availability of UDPG and F6P. The nature of the natural activator and whether it occurs in leaves have not been established.

UDPG pyrophosphorylase from *Sorghum* seedlings was not activated by PGA, PEP, F6P, or FBP (GUSTAFSON and GANDER, 1972); in this respect UDPG pyrophosphorylase differs from ADPG pyrophosphorylase (GHOSH and PREISS, 1965). Inhibition by reaction products, i.e., UDPG in one direction and UTP in the other, is significant (TSUBOI et al., 1969; GUSTAFSON and GANDER, 1972; HOPPER and DICKENSON, 1972). The inhibition is competitive, or partly competitive and K_i values are 0.13, 0.05, and 16.0 mM for UDPG with the enzymes from lily pollen, *Sorghum* seedlings, and mung bean seedlings respectively. In the direction of UDPG pyrophosphorolysis the K_i values for UTP were 0.8 mM with mung bean and 0.1 mM with the *Sorghum* seedling enzyme.

Sucrose phosphatase was inhibited by sucrose, maltose, melezitose, and 6-kestose to an extent that depended on the source of the enzyme (HAWKER, 1967a). Glucose, fructose, 1-kestose, lactose, cellobiose, trehalose, gentiobiose, and raffinose did not inhibit the enzyme from immature sugarcane stem. Turanose inhibited sucrose phosphatase from carrot root, but not the enzyme from tobacco leaf and sugarcane stem or leaf; maltose was a stronger inhibitor of the enzyme from tobacco leaf than was sucrose. The inhibition by sugars may have an important function in regulation of sucrose metabolism.

The breakdown of sucrose by sucrose synthetase in the presence of UDP was inhibited by UTP (TSAI, 1974; DELMER, 1972b) and by glucose or sugar phosphates (FEKETE, 1969). PPi and NADP, on the other hand, stimulated the reaction with NADPH antagonizing the effect of NADP. While ADP or ADPG are substrates for sucrose synthetase, they are not as effective as UDP and UDPG, and reaction with the adenosine compounds is almost completely inhibited in the presence of the uridine compounds (FEKETE and CARDINI, 1964; GRIMES et al., 1970). Chloroplasts in tobacco leaves contain less uridine nucleotides than the cytoplasm (KEYS and WHITTINGHAM, 1969) so in these organelles conditions may be favorable for ADPG synthesis catalyzed by sucrose synthetase.

Like the better-known yeast enzyme, synthesis of the soluble acid invertase of surgarcane stem is subject to repression by glucose (GLASZIOU et al., 1967). Acid invertase activity in stem tissue is increased by gibberellic acid (GAYLER and GLASZIOU, 1972; DANIEL, 1975); natural inhibitors of acid invertase that are proteins of low molecular weight may have a central function in various roots and storage tissues (PRESSEY, 1968). The activities of invertases in plant tissues are related to whether the tissue is consuming, storing, or making sucrose.

D. Intracellular and Intercellular Site of Sucrose Synthesis in Leaves

I. Chloroplast or Cytoplasm?

Evidence as to the intracellular site of sucrose synthesis has been derived from experiments involving homogenization of leaves and subsequent fractionation. Two techniques have been used. One consists of homogenizing freeze-dried leaves in nonaqueous liquids e.g., CCl_4 or hexane followed by density gradient centrifugation (STOCKING, 1959; HEBER, 1960). The other consists of homogenizing fresh tissue in aqueous media of a suitable osmotic potential followed by a low-speed centrifugation to obtain a chloroplast-rich pellet (WALKER, 1971); the material remaining in suspension constitutes the chloroplast-depleted fraction. The nonaqueous procedure has the advantage that water-soluble components, both enzymes and metabolites, are not leached from organelles or re-distributed between fractions; disadvantages are that the organic solvents denature certain proteins and, by removing lipids, destroy the physical structure of the cell components.

STOCKING et al. (1963) found, during photosynthesis in tobacco leaves, that sucrose became radioactive first in the chloroplast and later in the rest of the cell. Similar experiments with bean and spinach leaves by HEBER and WILLENBRINK (1964) showed the opposite result. In these experiments, the nonaqueous procedure was used so that the chloroplast fraction was contaminated with cytoplasm (HEBER, 1960; STOCKING et al., 1968; BIRD et al., 1973). The cytoplasm adhering to the chloroplasts was probably that in contact with the chloroplasts in the intact leaf and is likely to contain substances recently exported from the chloroplast.

Using aqueous media, EVERSON et al. (1967) found [^{14}C] sucrose among the products of photosynthesis from $^{14}CO_2$ by chloroplast preparations from spinach leaves. However, many other chloroplast preparations obtained from leaves by similar methods made only insignificant amounts of sucrose (GRANT et al., 1972). LARSSON et al. (1971) have shown that spinach chloroplasts isolated in aqueous media can be separated by countercurrent distribution into classes of different degrees of purity. Only chloroplasts shown to be associated with cytoplasm made significant amounts of sucrose during photosynthesis (LARSSON and ALBERTSSON, 1974).

The sucrose phosphate synthetase activity detected in chloroplast fractions isolated by the nonaqueous method was low; equivalent to approximately 0.6 μmol $h^{-1}g$ fr. wt.$^{-1}$ of leaf (BIRD et al., 1965) which could be from contaminating cytoplasm. The calculated distribution of the enzyme showed almost equal activity in the chloroplasts and the chloroplast-depleted fraction but this may only reflect a higher concentration of an inactivator in the cytoplasm (WALKER, 1971).

Sucrose phosphate synthetase activity has been observed in intact chloroplasts isolated in aqueous media from sugarcane and mung bean leaves (HAQ and HASSID,

1965; Delmer and Albersheim, 1970; Davies, 1974) but Huber et al. (1969) and Bird et al. (1974) considered the activity present in chloroplasts from *Zea mays,* pea and spinach leaves as insignificant. In chloroplasts of sugarcane leaves the activity was 0.09 nmol $h^{-1}g$ fr. wt.$^{-1}$ of leaf (Haq and Hassid, 1965). Similarly, activities in chloroplasts from mung bean leaves were very low and the methods used were inadequate to detect activity in extracts of whole leaves (Delmer and Albersheim, 1970). Bird et al. (1974) assayed extracts of the intact tissue, isolated chloroplasts and chloroplast-depleted fractions obtained from leaves of pea and spinach. Assays of chlorophyll and RuBP carboxylase showed the distribution of sucrose phosphate synthetase was distinct from that of RuBP carboxylase and, of the total sucrose phosphate synthetase activity extracted, less than 4%, and usually less than 1% (0.25 µmol $h^{-1}g$ fr. wt.$^{-1}$ for spinach) was associated with the chloroplast.

The distribution in leaf fractions of sucrose phosphatase, UDPG pyrophosphorylase, and UDPG has also been studied. Hawker (1966) showed that most of the sucrose phosphatase in sugarcane leaves was soluble; less than 10% was associated with a particulate fraction containing mitochondria and there was no evidence of a significant amount of the enzyme in chloroplasts. UDPG pyrophosphorylase, like sucrose phosphate synthetase, was cytoplasmic and not present to a significant extent in intact chloroplasts of pea and spinach (Bird et al., 1974) or bean leaves (Koenigs and Heinz, 1974). Whilst some work with chloroplast fractions isolated in nonaqueous media suggested that the enzyme was in the chloroplasts of bean, rice (Nomura et al., 1967), and tobacco leaves (Bird et al., 1965), the activities detected were low, equivalent to less than 10 µmol $h^{-1}g$ fr. wt.$^{-1}$ of leaves and the basis on which Nomura et al. reached their conclusions is obscure. The results of analyses after nonaqueous fractionations (Keys, 1968; Keys and Whittingham, 1969) showed 80% of the total UDPG of tobacco was in the nonchloroplast fraction and only 20% in the chloroplasts with no allowance made for cytoplasmic contamination of the chloroplast fraction which may have been up to 20% (Bird et al., 1973). Thus, most evidence suggests that sucrose synthesis does not occur to a significant extent in the chloroplast. Even if sucrose were made in the chloroplast, it is doubtful whether it could be transported out. Sucrose solutions are used as an osmoticum during isolation of chloroplasts; suggesting that the membranes round the chloroplast are impermeable to sucrose. Edelman et al. (1971) showed that sucrose was taken up by sugarcane chloroplasts but Heldt and Sauer (1971) found that with spinach this was only into the space between the two envelope membranes. Poincelot (1975) also found that the inner chloroplast membrane of spinach was impermeable to sucrose.

II. Intercellular Localization of Sucrose Synthesis in C_4 Plants

C_4 plants have two distinct cell-types that contain chloroplasts, namely bundle sheath and mesophyll cells. Historical and recent studies have suggested a "division of labor" between the two cell types. Present evidence suggests that the enzyme sucrose phosphate synthetase occurs in both bundle sheath and mesophyll cells with possibly a higher activity in the latter (Edwards and Black, 1971; Downton and Hawker, 1973; Fekete and Vieweg, 1973; Chen et al., 1974; Bucke and

OLIVER, 1975). DOWNTON and HAWKER (1973) found sucrose phosphatase and UDPG pyrophosphorylase in both cell types; expressed as activity per mg of chlorophyll, more was in the bundle sheath but in terms of protein, more was associated with the mesophyll cells. BUCKE and OLIVER (1975) found UDPG pyrophosphorylase activity mainly in the mesophyll cells of *Pennisetum purpureum*; more activity was associated with bundle sheath cells of *Muhlenbergia montana*. Bundle sheath cell preparations usually include vascular tissue. Further studies are needed but it appears that sucrose may be made in both bundle sheath and mesophyll cells of C_4 plants.

III. Intercellular Distribution Between Cells Containing Chlorophyll and Vascular Tissue

UDPG was found in extracts of vascular tissue from sugar beet and cow parsnip, *Heracleum sphondylium* (PAVLINOVA and AFANASIEVA, 1962; PAVLINOVA, 1965) and low activities of sucrose synthetase and sucrose phosphate synthetase were detected in phloem exudates from *Robinia pseudoacacia* (KENNECKE et al., 1971). From studies of diurnal changes in sucrose concentration in mesophyll cells of cotton leaves (C_3), PHILLIS and MASON (1933) concluded that these cells were the site of sucrose synthesis but the tissue referred to consisted of the leaf lamina with the minor veins; only the midrib and lateral veins were removed. In mature leaves of sugar beet, ^{14}C from $^{14}CO_2$ assimilated by photosynthesis accumulated rapidly in the minor veins (phloem loading; FELLOWS and GEIGER, 1974) and in the light there were much higher concentrations of sugar in the minor veins than in the surrounding mesophyll cells (GEIGER et al., 1973; FELLOWS and GEIGER, 1974). GEIGER et al. (1974) showed that [^{14}C] sucrose infiltrated into the free space of sugar beet leaves quickly accumulated in the fine veins and that during photosynthesis from $^{14}CO_2$ some [^{14}C] sucrose could be washed from the free space. They concluded that sucrose was made in the mesophyll cells and transferred through the free space to the vascular tissue. CATALDO (1974) found that the minor veins separated from tobacco leaves absorbed sucrose more readily than glucose, although the rate of absorption was less than for minor veins in an intact leaf disk. He concluded that sucrose was made in the mesophyll cells but transported to the veins by a symplastic route. GEIGER (1975), from work with sucrose labeled with ^{14}C in one of the hexose moieties, concluded that during the loading of veins in mature leaves sucrose is accumulated without inversion. BROVCHENKO (1965, 1967, 1970) reported evidence contrary to this view.

Since vein loading involves accumulation of sucrose against a concentration gradient, an input of chemical energy is necessary. The mechanism is understood in sugarcane storage tissue. Sucrose is hydrolyzed in the free space by acid invertase, the hexoses released are phosphorylated and sucrose phosphate is synthesized in the cytoplasm of storage parenchyma cells. The sucrose phosphate is dephosphorylated by sucrose phosphatase, probably at the tonoplast membrane, so that sucrose is released into the vacuole (GLASZIOU and GAYLER, 1972). A neutral invertase in the cytoplasm may have a role in maintaining storate (GAYLER and GLASZIOU, 1972). No detailed studies of invertase in leaves permit one to judge whether a similar mechanism of storage operates in them. An ability of the mesophyll cells to make sucrose does not in itself preclude synthesis or resynthesis in the vascular tissue which would be necessary if vein loading proceeded by a mechanism similar to sucrose storage in the sugarcane stem.

References

Alexander, A.G.: J. Agric. Univ. Puerto Rico *48*, 265–283 (1964)
Alexander, A.G.: J. Agric. Univ. Puerto Rico *49*, 60–75 (1965)
Allen, P.J., Bacon, J.S.D.: Biochem. J. *63*, 200–206 (1956)
Arai, Y., Fujisaki, M.: Shokubutsugaku Zasshi *84*, 76–87 (1971)
Avigad, G.: J. Biol. Chem. *239*, 3613–3618 (1964)
Avigad, G., Milner, Y.: In: Methods in enzymology. Neufeld, E.F., Ginsburg, V. (eds.), Vol. VIII, pp. 341–345. New York, London: Academic Press 1966
Beck, E.: Prog. Bot. *37*, 121–132 (1975)
Benson, A.A., Calvin, M.: J. Exp. Bot. *1*, 63–68 (1950)
Bird, I.F., Porter, H.K., Stocking, C.R.: Biochim. Biophys. Acta *100*, 366–375 (1965)
Bird, I.F., Cornelius, M.J., Dyer, T.A., Keys, A.J.: J. Exp. Bot. *24*, 211–215 (1973)
Bird, I.F., Cornelius, M.J., Keys, A.J., Whittingham, C.P.: Phytochemistry *13*, 59–64 (1974)
Bird, I.F., Cornelius, M.J., Keys, A.J., Kumarasinghe, S., Whittingham, C.P.: In: Proc. 3rd Int. Congr. Photosynthesis Res. Avron, M. (ed.), pp. 1291–1301. Amsterdam, New York, London: Elsevier 1975
Blackwood, G.C., Miflin, B.J.: J. Exp. Bot. *27*, 735–747 (1976)
Brovchenko, M.I.: Fiziol. Rast. *12*, 270–279 (1965)
Brovchenko, M.I.: Fiziol. Rast. *14*, 415–424 (1967)
Brovchenko, M.I.: Fiziol. Rast. *17*, 31–39 (1970)
Brown, H.T., Morris, G.H.: J. Chem. Soc. *LXIII*, 604–677 (1893)
Buchanan, J.G.: Arch. Biochem. Biophys. *44*, 140–149 (1953)
Buchanan, J.G., Bassham, J.A., Benson, A.A., Bradley, D.F., Calvin, M., Daus, L.L., Goodman, M., Hayes, P.M., Lynch, V.H., Norris, L.T., Wilson, A.T.: In: Phosphorus metabolism. McElroy, W.D., Glass, B. (eds.), Vol. II, pp. 440–466. Baltimore: The Johns Hopkins Press 1952
Buchanan, J.G., Cummerson, D.A., Turner, D.M.: Carbohyd. Res. *21*, 283–292 (1972)
Bucke, C., Oliver, I.R.: Planta (Berl.) *122*, 45–52 (1975)
Bünning, E.: In: The growth of leaves. Milthorpe, F.L. (ed.), pp. 119–126. London: Butterworths Scientific Publications 1956
Burma, D.P., Mortimer, D.C.: Arch. Biochem. Biophys. *62*, 16–28 (1956)
Calvin, M., Benson, A.A.: Science *109*, 140–142 (1949)
Caputto, R., Leloir, L.F., Cardini, C.E., Paladini, A.C.: J. Biol. Chem. *184*, 333–350 (1950)
Cardini, C.E., Leloir, L.F., Chiriboga, J.: J. Biol. Chem. *214*, 149–155 (1955)
Cataldo, D.A.: Plant Physiol. *53*, 912–917 (1974)
Chen, T.N., Dittrich, P., Campbell, W.H., Black, C.C.: Arch. Biochem. Biophys. *163*, 246–262 (1974)
Daniel, M.: Planta (Berl.) *125*, 91–103 (1975)
Davies, D.R.: In: Plant carbohydrate biochemistry. Pridham, J.B. (ed.), pp. 60–81. London: Academic Press 1974
Delmer, D.P.: J. Biol. Chem. *247*, 3822–3828 (1972a)
Delmer, D.P.: Plant Physiol. *50*, 469–472 (1972b)
Delmer, D.P., Albersheim, P.: Plant Physiol. *45*, 782–786 (1970)
Dickson, R.E., Larson, P.R.: Plant Physiol. *56*, 185–193 (1975)
Doudoroff, M.: J. Biol. Chem. *151*, 351–361 (1943)
Doudoroff, M., Kaplan, N., Hassid, W.Z.: J. Biol. Chem. *148*, 67–75 (1943)
Downton, W.J.S., Hawker, J.S.: Phytochemistry *12*, 1551–1556 (1973)
Dutton, J.V., Carruthers, A., Oldfield, J.F.T.: Biochem. J. *81*, 266–272 (1961)

Edelman, J., Schoolar, A.I., Bonner, W.B.: J. Exp. Bot. *22*, 534–545 (1971)

Edwards, G.E., Black, C.C.: In: Photosynthesis and photorespiration. Hatch, M.D., Osmond, C.B., Slatyer, R.O. (eds.), pp. 153–168. New York, London, Sydney, Toronto: Wiley-Interscience 1971

Everson, R.G., Cockburn, W., Gibbs, M.: Plant Physiol. *42*, 840–844 (1967)

Fekete, M.A.R. de: Planta (Berl.) *87*, 311–323 (1969)

Fekete, M.A.R. de: Eur. J. Biochem. *19*, 73–80 (1971)

Fekete, M.A.R. de, Cardini, C.E.: Arch. Biochem. Biophys. *104*, 173–184 (1964)

Fekete, M.A.R. de, Vieweg, G.H.: Ber. Dtsch. Bot. Ges. *86*, 227–231 (1973)

Fellows, R.J., Geiger, D.R.: Plant Physiol. *54*, 877–885 (1974)

Frydman, R.B., Hassid, W.Z.: Nature (London) *199*, 382–383 (1963)

Gander, J.E.: Phytochemistry *5*, 405–410 (1966)

Ganguli, N.C., Hassid, W.Z.: Plant Physiol. Suppl. *32*, 34 (1957)

Gayler, K.R., Glasziou, K.T.: Physiol. Plant. *27*, 25–31 (1972)

Geiger, D.R.: In: Encyclopedia of plant physiology, New Ser. Pirson, A., Zimmermann, M.H. (eds.), Vol. I, pp. 395–431. Berlin, Heidelberg, New York: Springer 1975

Geiger, D.R., Giaquinta, R.T., Sovonick, S.A., Fellows, R.J.: Plant Physiol. *52*, 585–589 (1973)

Geiger, D.R., Sovonick, S.A., Shock, T.L., Fellows, R.J.: Plant Physiol. *54*, 892–898 (1974)

Ghosh, H.P., Preiss, J.: J. Biol. Chem. *240*, PC960–962 (1965)

Ginsburg, V.: J. Biol. Chem. *232*, 55–61 (1958)

Glasziou, K.T., Gayler, K.R.: Bot. Rev. *38*, 471–490 (1972)

Glasziou, K.T., Waldron, J.C., Most, B.H.: Phytochemistry *6*, 769–775 (1967)

Gordon, W.C.: Dissertation: Univ. of California, Riverside (1973)

Grant, B.R., Canvin, D.T., Fock, H.: In: Proc. 2nd. Int. Congr. Photosynthesis Res. Forti, G., Avron, M., Melandri, A. (eds.), pp. 1917–1925. The Hague: Dr. W. Junk, N.V. 1972

Grimes, W.J., Jones, B.L., Albersheim, P.: J. Biol. Chem. *245*, 188–197 (1970)

Gustafson, G.L., Gander, J.E.: J. Biol. Chem. *247*, 1387–1397 (1972)

Hanes, C.S.: Proc. R. Soc. Lond. *B 128*, 421–450 (1940a)

Hanes, C.S.: Proc. R. Soc. Lond. *B 129*, 174–208 (1940b)

Haq, S., Hassid, W.Z.: Plant Physiol. *40*, 591–594 (1965)

Hartt, C.E.: Hawaii. Plant. Rec. *47*, 113–132 (1943a)

Hartt, C.E.: Hawaii. Plant. Rec. *47*, 155–170 (1943b)

Hartt, C.E.: Hawaii. Plant. Rec. *47*, 223–255 (1943c)

Hatch, M.D., Slack, C.R.: Biochem. J. *101*, 103–111 (1966)

Hatch, M.D., Sacher, J.A., Glasziou, K.T.: Plant Physiol. *38*, 338–343 (1963)

Hawker, J.S.: Phytochemistry *5*, 1191–1199 (1966)

Hawker, J.S.: Biochem. J. *102*, 401–406 (1967a)

Hawker, J.S.: Biochem. J. *105*, 943–946 (1967b)

Hawker, J.S.: Phytochemistry *10*, 2313–2322 (1971)

Hawker, J.S., Hatch, M.D.: Biochem. J. *99*, 102–107 (1966)

Heber, U.: Z. Naturforsch. *15b*, 95–109 (1960)

Heber, U., Willenbrink, J.: Biochim. Biophys. Acta *82*, 313–324 (1964)

Heldt, H.W., Sauer, F.: Biochim. Biophys. Acta *234*, 83–91 (1971)

Hellebust, J.A., Bidwell, R.G.S.: Can. J. Bot. *41*, 985–994 (1963)

Hellebust, J.A., Forward, D.F.: Can. J. Bot. *40*, 113–126 (1962)

Hopper, J.E., Dickenson, D.B.: Arch. Biochem. Biophys. *148*, 523–535 (1972)

Huber, W., Fekete, M.A.R. de, Ziegler, H.: Planta (Berl.) *87*, 360–364 (1969)

Jaarma, M., Rydstrom, J.: Acta Chem. Scand. *23*, 3443–3450 (1969)

Jones, H., Eagles, J.E.: Ann. Bot. N.S. *26*, 505–510 (1962)

Jones, H., Martin, R.V., Porter, H.K.: Ann. Bot. N.S. *23*, 493–509 (1959)

Joy, K.W.: J. Exp. Bot. *15*, 485–494 (1964)

Kennecke, M., Ziegler, H., Fekete, M.A.R. de: Planta (Berl.) *98*, 330–356 (1971)

Kennedy, R.A., Laetsch, W.M.: Planta (Berl.) *115*, 113–124 (1973)

Keys, A.J.: Ph. D. Thesis, Univ. of London (1959)

Keys, A.J.: Biochem. J. *108*, 1–8 (1968)

Keys, A.J., Whittingham, C.P.: In: Prog. Photosynthesis Res. Metzner, H. (ed.), Vol. I, pp. 352–358. Tübingen 1969

Keys, A.J., Bird, I.F., Cornelius, M.J., Kumarasinghe, S., Whittingham, C.P.: In: Tracer techniques for plant breeding. pp. 13–18. Int. At. Energy Agency, Vienna 1975

Koenigs, B., Heinz, E.: Planta (Berl.) *118*, 159–169 (1974)

Kortschak, H.P., Hartt, C.E., Burr, G.O.: Plant Physiol. *40*, 209–213 (1965)

Krotkov, G., Bennet, W.C.G.: Can. J. Bot. *30*, 29–39 (1952)

Kursanov, A.L.: Endeavour *20*, 19–25 (1961)

Kursanov, A.L., Kryukova, N.N.: Biokhimiya *4*, 229–240 (1939)

Larsson, C., Albertsson, P.Å.: Biochim. Biophys. Acta *357*, 412–419 (1974)

Larsson, C., Collin, C., Albertsson, P.Å.: Biochim. Biophys. Acta *245*, 425–438 (1971)

Latzko, E., Zimmermann, G., Feller, U.: Hoppe Seyler's Z. Physiol. Chem. *355*, 321–326 (1974)

Leloir, L.F., Cardini, C.E.: J. Am. Chem. Soc. *75*, 6084 (1953)

Leloir, L.F., Cardini, C.E.: J. Biol. Chem. *214*, 157–165 (1955)

Little, G., Edelman, J.: Phytochemistry *12*, 67–71 (1973)

Lyne, R.L., ap Rees, T.: Phytochemistry *10*, 2593–2599 (1971)

McCready, R.M., Hassid, W.Z.: Plant Physiol. *16*, 599–610 (1941)

Maclachlan, G.A., Datko, A.H., Rollit, J., Stokes, E.: Phytochemistry *9*, 1023–1030 (1970)

Mahon, J.D., Fock, H., Canvin, D.T.: Planta (Berl.) *120*, 245–254 (1974)

Manning, K., Maw, G.A.: Phytochemistry *14*, 1965–1969 (1975)

Mendicino, J.: J. Biol. Chem. *235*, 3347–3352 (1960)

Milner, Y., Avigad, G.: Nature (London) *206*, 825 (1965)

Munch-Petersen, A., Kalckar, H.M., Cutolo, E., Smith, E.E.B.: Nature (London) *172*, 1036–1037 (1953)

Murata, T.: Agric. Biol. Chem. *35*, 1441–1448 (1971)

Murata, T.: Agric. Biol. Chem. *36*, 1815–1818 (1972a)

Murata, T.: Agric. Biol. Chem. *36*, 1877–1884 (1972b)

Nakamura, M.: Bull. Agric. Chem. Soc. Jpn. *23*, 398–405 (1959)

Nomura, T., Akazawa, T.: Arch. Biochem. Biophys. *156*, 644–652 (1973)

Nomura, T., Akazawa, T.: Plant Cell Physiol. *15*, 477–483 (1974)

Nomura, T., Nakayama, N., Murata, T., Akazawa, T.: Plant Physiol. *42*, 327–332 (1967)

Ongun, A., Stocking, C.R.: Plant Physiol. *40*, 819–824 (1965)

Pandya, K.P., Ramakrishnan, C.V.: Naturwissenschaften *43*, 85 (1956)

Parkin, J.: Phil. Trans. R. Soc. Lond. *191*, 35–79 (1899)

Parkin, J.: Biochem. J. *6*, 1–47 (1912)

Pavlinova, O.A.: Fiziol. Rast. *12*, 606–617 (1965)

Pavlinova, O.A., Afanasieva, T.P.: Fiziol. Rast. *9*, 133–141 (1962)

Pavlinova, O.A., Kursanov, A.L.: Fiziol. Rast. *3*, 539–546 (1956)

Phillis, E., Mason, T.G.: Ann. Bot. *CLXXXVII*, 585–634 (1933)

Poincelot, R.B.: Plant Physiol. *55*, 849–852 (1975)

Pontis, H.G.: In: Int. Rev. Biochem. Northcote, D.H. (ed.), Vol. XIII, pp. 80–117. Baltimore: Univ. Park Press 1977

Porter, H.K., May, L.H.: J. Exp. Bot. *6*, 43–63 (1955)

Prasolova, M.F., Mambetkulov, A., Pavlinova, O.A., Pechenov, V.A.: Fiziol. Rast. *23*, 292–299 (1976)

Preiss, J., Greenberg, E.: Biochem. Biophys. Res. Commun. *36*, 289–295 (1969)

Pressey, R.: Plant Physiol. *43*, 1430–1434 (1968)

Pressey, R.: Plant Physiol. *44*, 759–764 (1969)

Putman, E.W., Hassid, W.Z.: J. Biol. Chem. *207*, 885–902 (1954)

Reynolds, P.E., Raigosa, J., Trip, P.: Plant Physiol. (Suppl.) *53*, 62 (1974)

Ricardo, C.P.P., ap Rees, T.: Phytochemistry *9*, 239–247 (1970)

Roberts, G.R., Keys, A.J., Whittingham, C.P.: J. Exp. Bot. *21*, 683–692 (1970)

Rorem, E.S., Walker, H.G., McCready, R.M.: Plant Physiol. *35*, 269–272 (1960)

Salerno, G.L., Pontis, H.G.: FEBS Lett. *64*, 415–418 (1976)

Shukla, R.N., Sanwal, G.G.: Arch. Biochem. Biophys. *142*, 303–309 (1971)

Sisakyan, N.M.: Biokhimiya *1*, 301–320 (1936)

Skews, S.J.: M. Sc. Thesis, Univ. of London (1961)

Slabnik, E., Frydman, R.B., Cardini, C.E.: Plant Physiol. *43*, 1063–1068 (1968)

Slack, C.R.: Phytochemistry *5*, 397–403 (1966)

Steer, B.T.: Plant Physiol. *51*, 744–748 (1973)
Steer, B.T.: Plant Physiol. *54*, 758–761 (1974)
Stocking, C.R.: Plant Physiol. *34*, 56–61 (1959)
Stocking, C.R., Williams, G.R., Ongun, A.: Biochem. Biophys. Res. Commun. *10*, 416–421 (1963)
Stocking, C.R., Shumway, L.K., Weier, T.E., Greenwood, D.: J. Cell Biol. *36*, 270–275 (1968)
Thrower, S.L.: Aust. J. Biol. Sci. *15*, 629–649 (1962)
Tovey, K.C., Roberts, R.M.: Plant Physiol. *46*, 406–411 (1970)
Tsai, C.-Y.: Phytochemistry *13*, 885–891 (1974)
Tsuboi, K.K., Fukunga, K., Petricciana, J.C.: J. Biol. Chem. *244*, 1008–1015 (1969)
Turner, D.H., Turner, J.F.: Biochem. J. *69*, 448–452 (1958)
Virtanen, A.I., Nordlund, M.: Biochem. J. *28*, 1729–1732 (1934)
Vittorio, P.V., Krotkov, G., Reed, G.B.: Can. J. Bot. *32*, 369–377 (1954)
Waidyanatha, U.P. de S., Keys, A.J., Whittingham, C.P.: J. Exp. Bot. *26*, 15–26 (1975a)
Waidyanatha, U.P. de S., Keys, A.J., Whittingham, C.P.: J. Exp. Bot. *26*, 27–32 (1975b)
Walker, D.A.: In: Methods in enzymology. San Pietro, A. (ed.), Vol. XXIII, pp. 211–220. New York, London: Academic Press 1971
Wang, D., Waygood, E.R.: Plant Physiol. *37*, 826–832 (1962)
Wardlaw, I.F.: Aust. J. Plant Physiol. *3*, 377–387 (1976)
Ziegler, H.: In: Encyclopedia of plant physiology. New Ser. Pirson, A., Zimmermann, M.H. (eds.), Vol. 1, pp. 59–100. Berlin, Heidelberg, New York: Springer 1975

II F. Glycolic Acid and Photorespiration

25. Glycolate Synthesis

E. BECK

A. Introduction: Glycolate Formation, Photorespiration and the Warburg Effect

Photosynthetic glycolate formation is a complex and irreversible process which at least in intact systems cannot be uncoupled from photosynthetic carbon metabolism. Although some crucial breakthroughs were achieved recently, the biochemistry and regulation of this process has not yet been completely elucidated. Usually, glycolate synthesis is confined to the chloroplast, but its metabolism takes place outside this organelle. Consequently glycolate formation represents a substantial carbon sink with respect to the photosynthetic carbon cycle. Glycolate formation, being the beginning of the "wasteful" process of photorespiration, was often considered to cause the Warburg effect, i.e., the inhibition of photosynthesis by oxygen (WARBURG, 1920; COOMBS and WHITTINGHAM, 1966; for ref. see TURNER and BRITTAIN, 1962 and BJÖRKMAN, 1966). However, although both phenomena are certainly correlated, it is now clear that the Warburg effect comprises more metabolic features than the synthesis and metabolism of glycolate (ROBINSON and GIBBS, 1974; ROBINSON et al., 1977; WAH KOW et al., 1977).

In summary, glycolate formation is a feature of several aspects of plant carbon metabolism and since its discovery (BENSON and CALVIN, 1950) has permanently attracted the attention of plant physiologists. Available information concerning glycolate synthesis has been reviewed several times (TOLBERT, 1963; 1971; 1973a, b; TOLBERT and RYAN, 1976; GIBBS, 1969; ZELITCH, 1971; 1975; BECK, 1972; MERRETT and LORD, 1973).

B. Environmental Factors Affecting Glycolate Synthesis

Four environmental factors are known to influence more or less directly the extent of glycolate formation.

1. CO_2. At a constant O_2 partial pressure glycolate production is inversely proportional to the CO_2 concentration up to levels which saturate CO_2 fixation (TOLBERT, 1973b; BECK et al., 1975).

2. O_2. At saturating levels of CO_2, oxygen concentrations higher than that of air increase almost linearly the proportion of fixed carbon which is incorporated into glycolate, whereas at lower O_2 levels that proportion is unchanged. Thus a biphasic curve results (EICKENBUSCH and BECK, 1973). The position of its sharp bend depends directly on the CO_2 concentration (Beck et al., 1975). Therefore, at very low CO_2 levels the biphasic curve becomes a straight line and glycolate formation depends on the oxygen concentration alone.

3. Light Intensity. Glycolate synthesis follows the curve for light saturation of photosynthesis (Tolbert, 1973b; Beck et al., 1975; Gibbs, 1971), suggesting that light influences glycolate formation mainly by providing substrates. Sometimes an influence of the wavelength of the incident light on the extent of formation or excretion of glycolate has been described. These results, however, are rather contradictory (for a short review see Zelitch, 1975).

4. pH. With algae and isolated spinach chloroplasts, stimulation of glycolate formation by alkaline media has been observed (Tolbert, 1973b; Orth et al., 1966; Kirk and Heber, 1976). Since at least with isolated chloroplasts CO_2 fixation was identical at pH 7.8 and 8.5 the stimulation of glycolate formation with increasing pH of the medium is probably not due to trapping of CO_2 as bicarbonate/carbonate (Robinson et al., 1977).

C. Mechanisms of Glycolate Formation

A variety of mechanisms have been described by which glycolate could arise from intermediates of photosynthetic carbon metabolism. Several important ones are presented here. Their relevance in vivo is discussed below (Sect. D).

I. Reductive Glycolate Formation

1. The Glycolate–Glyoxylate Shuttle

Glyoxylate moves readily into chloroplasts (Kearney and Tolbert, 1962) where it is reduced by the NADPH-dependent glyoxylate reductase (Zelitch and Gotto, 1962; Tolbert et al., 1970) to glycolate. This in turn passes out of the organelle and is reoxidized to glyoxylate either by glycolate oxidase (Tolbert, 1973a) in the leaf peroxisome or by glycolate dehydrogenase in algae (Tolbert, 1976; Grodzinski and Colman, 1976; for earlier ref. see Tolbert, 1973a). A glycolate–glyoxylate shuttle between the chloroplast and its surroundings was suggested whereby reducing power from the chloroplast can be exported (Tolbert, 1971, 1973a, 1976; Thompson and Whittingham, 1968). It should be mentioned that this mechanism does not allow for a de novo synthesis of glycolate.

2. Reductive CO_2 Plus CO_2 Condensation?

From some in vivo experiments, glycolate synthesis by a reductive CO_2 plus CO_2 condensation has been indicated. However, conclusive evidence for such a mechanism is still lacking. The often cited experiment with *Chlorella* which, photosynthesizing under low CO_2 and high O_2 concentrations, excreted 92% of assimilated carbon (indirectly determined) as glycolate (WARBURG and KRIPPAHL, 1960) can also be interpreted in terms of glycolate formation from storage material (e.g., KIRK and HEBER, 1976). Furthermore, reduced rates of glycolate excretion, observed with *Chlorella* under manganese deficiency (TANNER et al., 1960; HESS and TOLBERT, 1967) are not sufficient to substantiate a manganese-dependent radical mechanism for glycolate synthesis by a C plus C condensation. On the other hand the finding that, in $^{14}CO_2$ assimilating tobacco leaf disks, the specific radioactivity of both glycolate carbons was significantly higher than that of the carboxyl group of PGA could indeed indicate glycolic acid synthesis by a C plus C condensation (ZELITCH, 1965). However, this finding could also be due to a mixture of labeled and unlabeled PGA pools upon tissue extraction (HESS and TOLBERT, 1966, see also ZELITCH, 1971), resulting in a decrease of the specific radioactivity of PGA. Such an explanation was also given for similar results obtained with *Chlorella* (FOCK et al., 1974). Evidence for glycolate being the first CO_2 fixation product has been presented from studies with photoheterotrophically grown *Rhodospirillum rubrum* (ANDERSON and FULLER, 1967). However, in order to suggest biosynthesis by a CO_2 condensation, the labeling pattern of the glycolate still has to be determined and shown to be 1:1.

II. Oxidative Glycolate Synthesis

1. Oxidation of Activated Glycolaldehyde by Hydrogen Peroxide and O_2^-

Precursor studies with ribose-1-^{14}C and glucose-1- and -2-^{14}C, administered to tobacco leaves or *Chlorella*, revealed that the hydroxymethyl group of glycolate originates from carbon 1 of a pentulose-, hexulose- or heptulose phosphate and the carboxyl group from the corresponding carbon atom 2 (GRIFFITH and BYERRUM, 1959; MARKER and WHITTINGHAM, 1966). From these types of sugar phosphate thiamine pyrophosphate-activated glycolaldehyde is produced by the transketolase reaction. For this reason, and since an oxidation of activated glycolaldehyde to glycolic acid has been demonstrated with ferricyanide (HOLZER and SCHRÖTER, 1962, BRADBEER and RACKER, 1961; GOLDBERG and RACKER, 1962; ASAMI and AKAZAWA, 1975a, 1977), the origin of glycolate from activated glycolaldehyde [2-(α,β-dihydroxyethyl, thiamine pyrophosphate)] was suggested (WILSON and CALVIN, 1955). In reconstituted chloroplast systems with ferredoxin and NADP as electron acceptors up to 10 µmoles glycolate per mg chl per h could be obtained from this carbon source (SHAIN and GIBBS, 1971; EICKENBUSCH et al., 1975). With photosynthesizing isolated *Chromatium* chromatophores glycolate formation from added F6P plus transketolase was observed at a rate of 4–5 nmoles per mg Bchl per h (ASAMI and AKAZAWA, 1977). With respect to the oxidant, in vitro studies have revealed that a normal (pH 7) redox potential of about 120–200 mV is necessary for the oxidation of activated glycolaldehyde (ASAMI and AKAZAWA, 1975b; SHAIN and GIBBS, 1971). Search for an appropriate oxidant now concentrates on two species of photosynthetically reduced oxygen: H_2O_2 and O_2^-.

Besides artificial electron acceptors, H_2O_2 ($E_0' = +270$ mV) was most effective in reconstituted chloroplast systems (SHAIN and GIBBS, 1971 and unpublished results from the author's laboratory). However, in vitro experiments with yeast transketolase or a crude extract of *Chromatium* cells, F6P and the xanthine oxidase system as a source for reduced oxygen, O_2^- was found to be more active than H_2O_2 (ASAMI and AKAZAWA, 1977). However, the concentrations of both oxidants were not comparable in these experiments, since O_2^- dismutates spontaneously by itself and, if present, by SOD (see Chap. III.2 by ELSTNER), yielding H_2O_2 and O_2 (MCCORD and FRIDOVICH, 1971). Equations (1) and (2) show that H_2O_2 should arise from the reaction of (activated) glycolaldehyde with O_2^- as is the case with the well-known oxidation of hydroxylamine by O_2^- (ELSTNER and HEUPEL, 1975).

$$CH_2OH\text{-}CHOH\text{-}R + H_2O + 2O_2^- + 2H^+ \rightarrow CH_2OH\text{-}COOH + RH + 2H_2O_2 \quad (1)$$

$$CH_2OH\text{-}CHOH\text{-}R + H_2O + H_2O_2 \rightarrow CH_2OH\text{-}COOH + RH + 2H_2O. \quad (2)$$

Thus, in photosynthesizing chloroplasts both oxidants could be simultaneously involved in glycolate formation from activated glycolaldehyde. The mechanism shown in Eq. (2) is a peroxidatic one. However, the recently described peroxidase of chloroplast membranes (GRODEN and BECK, 1977) could not use the activated glycolaldehyde as a substrate.

An unknown oxidant produced by photosystem II has also been suggested to act on this substrate (PLAUT and GIBBS, 1970).

2. Cleavage of Sugar Phosphates by H_2O_2

Some of the sugar phosphates of the reductive pentose phosphate cycle are readily attacked by oxygen. Exposure of RuBP to air resulted in the formation of P-glycolate PGA, glycolate, and other compounds (KAUSS and KANDLER, 1964). A similar cleavage is to be expected with FBP and SBP, all the more when oxygen is replaced by H_2O_2. Recent in vitro studies have confirmed the assailability of RuBP and F6P by H_2O_2 (GERSTER and TOURNIER, 1977).

It is noteworthy that some investigators have found FBP to be the best precursor for glycolate synthesis in isolated chloroplasts (BRADBEER and ANDERSON, 1967; VANDOR and TOLBERT, 1968)[1]. In addition, from other in vitro studies with a reconstituted chloroplast system, such a reaction is evident: more ^{14}C was incorporated from FBP-U-^{14}C than could be expected via C-1 dephosphorylation and subsequent oxidation of the activated glycolaldehyde produced in the transketolase reaction (EICKENBUSCH et al., 1975). The reaction should proceed by a four-electron transition mechanism and would normally lead to the formation of P-glycolate.

$$\begin{array}{c} R_1 \\ | \\ C{=}O \\ | \\ C\text{—}OH \\ | \\ R_2 \end{array} \rightleftharpoons \begin{array}{c} R_1 \\ | \\ C\text{—}OH \\ \| \\ C\text{—}OH \\ | \\ R_2 \end{array} \quad \xrightarrow[2H_2^*O]{2H_2^*O_2} \quad R_1\text{—}C^*OOH + R_2\text{—}C^*OOH \qquad (3)$$

---

[1]   In these cases, however, no allowance was made for different permeation velocities of the various precursors.

In this context it should be mentioned that RuBP, FBP, or SBP cannot be expected to be oxidized by solely electron-abstracting oxidants like Fe^{3+} (see Table 1 in ASAMI and AKAZAWA, 1975).

3. The RuBP Oxygenase Reaction (cf. Chap. II.17, this vol.)

The most important reaction for glycolate synthesis seems to be the oxidative cleavage of RuBP catalyzed by RuBP carboxylase/oxygenase. The ability of this enzyme to utilize O_2 instead of CO_2 and thereby to produce glycolate, the substrate of photorespiration, was predicted by A. HALL in 1971 (HALL, 1971) and simultaneously demonstrated by BOWES and coworkers (1971; 1972). TOLBERT'S group (ANDREWS et al., 1973; LORIMER et al., 1973) has identified the reaction products as P-glycolate and PGA (instead of 2 molecules of PGA as in the carboxylation reaction), evaluated the 1:1 stoichiometry of O_2 uptake and RuBP consumption (however, without demonstration of stoichiometric amounts of P-glycolate and PGA being produced) and established the identities of RuBP carboxylase and RuBP oxygenase. ^{18}O incorporation studies prompted them to propose a reaction mechanism closely analogous to that of the carboxylase reaction, with an intermediate peroxide and a cleavage of the carbon bond between C-2 and C-3.

$$
\begin{array}{ccccc}
CH_2OP & & CH_2OP & & CH_2OP{-}C^{*}OOH \\
| & & | & & \\
C{-}OH & & HO^{*}{-}{*}O{-}C{-}OH & OH^{-} & + \\
\| & \longrightarrow & | & \searrow & \\
C{-}OH & & C{=}O & \longrightarrow & {*}OH^{-} \quad (4) \\
| & {*}O_2 & | & & \\
CHOH & & CHOH & & + \\
| & & | & & \\
CH_2OP & & CH_2OP & & CH_2OP{-}CHOH{-}COOH
\end{array}
$$

From their data they ruled out any requirement of copper (as a prosthetic group) as well as any involvement of oxygen radicals in the reaction. However, from an energetic viewpoint, a quantitatively important reaction of the relatively inert triplet oxygen molecule with RuBP is difficult to understand. Thus, more recently, from oxidation and reduction studies of the enzyme's SH-groups, as well as from the finding that the oxygenase activity is inhibited by mild alkylation whilst the carboxylase activity is not, a participation of oxygen radicals in the reaction sequence was concluded (WILDNER, 1976). Activation of a sulfhydryl group by the OH· radical on the one hand, and formation of the peroxide by the superoxide radical on the other was proposed.

With respect to glycolate formation, three aspects from the large body of data on fraction I protein (ELLIS, 1973) should be emphasized:
a) A competitive inhibition of the carboxylase by oxygen and of the oxygenase by CO_2 (LAING et al., 1974) indicates that there is only one active center for both enzyme activities. Nevertheless, the observation that 5 mM glycidic acid (2,3-epoxypropionic acid), as well as iodoacetamide inhibit only the oxygenase reaction, suggests that perhaps an additional sulfhydryl group might be necessary for the reaction with oxygen (WILDNER and HENKEL, 1976).
b) Several inconsistent findings reported in the literature might have originated from the fact that the enzyme obviously changes some of its properties as well as its activity after isolation from chloroplasts (BADGER and LORIMER, 1976; LAING et al., 1975; LORIMER and BADGER, 1976). This might be also the case with the

pH optima of both activities which have now been definitively determined at 8.5–8.8 for the oxygenase reaction and at 8.2–8.3 for the carboxylase reaction, respectively (Badger and Lorimer, 1976; Jensen and Bahr, 1976; Badger et al., 1975).

c) Oxygenase activity of fraction 1 protein is also exhibited by the strictly anaerobic purple sulfur bacterium *Chromatium* (Takabe and Akazawa, 1973).

4. Phosphoglycolate Phosphatase

Both the nonenzymic oxidation of sugar bisphosphates and the enzymic oxidation with RuBP oxygenase result in the formation of P-glycolate, a very potent inhibitor of triosephosphate isomerase (Wolfenden, 1970). High activities of a specific P-glycolate phosphatase (Christeller and Tolbert, 1974) were found to be located predominantly in the chloroplasts of terrestrial plants (for ref. see Tolbert, 1973a) as well as in algal cells (Randall, 1976). The pH optima of the enzymes differ markedly from species to species and vary between about 6 [tobacco (Anderson and Tolbert, 1966) and *Chlorella* (Tolbert, 1973a)] and approximately 8 [pea (Kerr and Gear, 1974) and the marine alga *Halimeda* (Randall, 1976)]. Two isoenzymes have been found in *Phaseolus* leaves which differ markedly in the K_m values for the substrate and the requirement of divalent cations (Baldy et al., 1977).

D. Photosynthetic Glycolate Formation in Vivo; Which Reaction Predominates?

It has been shown that more than one glycolate-producing reaction exists, and it has therefore to be expected that all of them contribute to glycolate formation. In addition, it is reasonable that alterations in the external or internal environment will change the relative contributions of the individual reactions. Therefore the title of this section should perhaps read: to what extent may the described reactions contribute to total glycolate production? Obviously the number of possibilities will be proportional to the variety of conditions and plants. Most experiments have been performed with C_3 plants, therefore these will be discussed in more detail.

I. Glycolate Synthesis by C_3 Plants

1. Results of Labeling Experiments

Double labeling of glycolate from $^{14}CO_2$ assimilation by *Chlorella* in tritiated water revealed almost the same $^3H/^{14}C$ ratio in glycolate as in sugar phosphates. This observation suggested that glycolate formed by the glycolate–glyoxylate shuttle is quantitatively of limited importance compared to the net synthesis of glycolic acid (PLAMONDON and BASSHAM, 1966).

As mentioned above, application of position-labeled carbohydrates does not allow conclusive statements to be made with respect to the nature of the immediate precursor of glycolate. The same is true for short-time fixation experiments with $^{14}CO_2$ which yielded uniformly labeled glycolate when leaves were used (HESS and TOLBERT, 1966; SHOU et al., 1950; CALVIN and MASSINI, 1952) and predominantly C-2 labeled glycolic acid when the experiments were performed with algae (HESS and TOLBERT, 1967) or with intact chloroplasts (EICKENBUSCH, 1975) in which spreading of ^{14}C into photosynthetic intermediates is significantly delayed (PLAUT and GIBBS, 1970). In isolated chloroplasts, however, the dependence of the intramolecular ^{14}C distribution in glycolate on the oxygen concentration of the suspension indicated that different mechanisms for glycolate synthesis prevail at low and high oxygen concentrations (EICKENBUSCH, 1975).

More information can be obtained upon application of the isotope $^{18}O_2$. *Chromatium* fed $^{18}O_2$ excreted glycolic acid, the carboxyl group of which contained on a percent basis almost as much ^{18}O as was administered in the oxygen gas (LORIMER et al., 1976). This indicates an oxidative splitting of sugar phosphates yielding glycolate or phosphoglycolate rather than glycolate formation from activated glycolaldehyde. However, the results obtained with this anaerobic organism cannot be readily extended to aerobic organisms. This can be seen from the fact that neither the activities of RuBP oxygenase and P-glycolate phosphatase (ASAMI and AKAZAWA, 1975) nor the rate of glycolate formation from activated glycolaldehyde could account for the rate of glycolate production by *Chromatium* upon photosynthesis in O_2 (ASAMI and AKAZAWA, 1977).

With algae (*Chlorella, Euglena*) the rate of ^{18}O incorporation into the carboxyl group of glycolate was 65% to 80% of the ^{18}O applied; and in the presence of mM cyanide ^{18}O was still incorporated (GERSTER and TOURNIER, 1977), although CN^- is a potent inhibitor of the RuBP carboxylase-RuBP complex (SIEGEL et al., 1972). Both findings indicate that the RuBP oxygenase reaction is not the sole reaction responsible for ^{18}O incorporation into glycolate.

O_2 treatment of a whole leaf resulted in a 30% and 50% labeling of the carboxyl groups of serine and glycine, respectively (ANDREWS et al., 1971), both being products derived from glycolate. Although the different ^{18}O labeling of these amino acids indicates clearly the existence of more than one pool of serine, the relatively poor labeling of both carboxyl groups suggests participation of the RuBP oxygenase reaction in addition to a reaction which does not incorporate ^{18}O.

2. Kinetic Studies

When rates of P-glycolate hydrolysis were compared with the minimum rates of glycolate formation by *Chlorella* which had previously been exposed to CO_2 plus air, followed subsequently by 100% O_2 and 100% N_2, only 30%–50% of the formed glycolate could have been derived from P-glycolate and could therefore be mainly attributed to the oxidative cleavage of RuBP. The remainder of the glycolate might have originated from a sugar monophosphate (BASSHAM and KIRK, 1973).

3. Studies Employing Various O_2/CO_2 Concentrations

In photosynthesizing chloroplasts, even when kept exclusively under N_2, considerable oxygen concentrations originate from the photosynthetic process (Steiger et al., 1977). At oxygen pressures below air level, and with saturating CO_2, the extent of carbon incorporation into glycolate by intact chloroplasts was found to be independent of the oxygen concentration. However, with higher oxygen levels, that portion of the carbon that was incorporated into glycolate increased linearly with the increase in oxygen concentration (Eickenbusch and Beck, 1973) and could be attributed to the RuBP oxygenase reaction (Eickenbusch et al., 1975). Various sugar phosphates and, apparently, activated glycolaldehyde were shown to be the carbon sources for the "basic" reaction (i.e., at lower O_2 levels; Eickenbusch et al., 1975). Decrease of the CO_2 concentration shifted the first appearance of the RuBP oxygenase reaction to lower oxygen levels, as would be expected from competition between CO_2 and O_2 in this reaction (Eickenbusch, 1975). However, even at CO_2 concentrations so far above saturation that the oxygenase reaction should be totally eliminated, glycolate synthesis continued (Beck et al., 1975). The same can be seen from experiments with isolated chloroplasts under conditions which greatly favor the RuBP oxygenase reaction. Inhibition of RuBP regeneration resulted in a large decrease, but not complete abolition of further glycolate formation (Kirk and Heber, 1976, Fig. 11). Also, with isolated chloroplasts, inhibition of phosphatases with NaF resulted in a relative accumulation of P-glycolate, indicating glycolate synthesis predominantly from sugar bisphosphates (Larsson, 1975). However, such treatment also inhibits the formation of F6P which is a principal source of activated glycolaldehyde and therefore these results do not permit definite conclusions to be reached.

4. Quantitative Considerations

The question of whether a proposed reaction mechanism could account for rates of glycolate formation in vivo involves not only the velocity of the glycolate-producing reaction, but also that of the formation of carbon sources, oxidants, and competitors. At least with respect to the glycolate-producing reactions an enzymic reaction should predominate over all nonenzymic mechanisms. Hence, under normal environmental conditions, when only CO_2 is limiting photosynthesis, the RuBP oxygenase reaction should prevail over other glycolate-synthesizing reactions. Recently published calculations revealed that under these conditions about 40% of the fixed carbon would be incorporated into glycolate by means of the RuBP oxygenase reaction alone (Jensen and Bahr, 1976). Experiments with photosynthesizing tobacco leaf disks revealed somewhat higher values of glycolate formation from freshly fixed carbon (50%–60%, Zelitch, 1959). However, it should be emphasized that in vivo and especially under normal, that is, rate-limiting CO_2 concentrations, the carbon fluxes through glycolate may be considerably higher than those values and even exceed apparent CO_2, uptake (Mahon et al., 1974) indicating an involvement of storage material in glycolate synthesis. Such high rates of glycolate formation (up to 92 μmoles, i.e., 184 μatoms of C per mg chl per h) were also obtained with isolated chloroplasts upon illumination when supplied with FBP in a CO_2 free oxygen atmosphere and at pH 8.0 (Wah Kow et al., 1977). These chloroplast preparations exhibited a CO_2 fixation rate (at saturating CO_2 levels) of 34–40 µmoles per mg chl per h. Hence, quantitative considerations also indicate that, in addition to RuBP oxygenase, other mechanisms contribute to total glycolate synthesis. Since only part of the oxygen uptake observed during photosynthesis can be ascribed to the RuBP oxygenase reaction (GLIDEWELL and RAVEN, 1976; EGNEUS et al., 1975; HEBER et al., 1976), it seems likely that photosynthetically reduced oxygen also participates in glycolate formation under normal conditions. Taking into account the ubiquitous occurrence of SOD from strictly anaerobic organisms to the aerobic ones, O_2^- should be of minor importance compared with H_2O_2 in this respect.

II. Glycolate Formation by C_4 Plants

The question has often been raised whether or not C_4 plants, which do not exhibit photorespiration, produce glycolate upon photosynthesis. It is now quite clear that these plants form glycolate, but probably mostly in bundle sheath cells and at a lower rate than is found in C_3 plants (HATCH, 1971). This is plausible since decarboxylation of 4-C acids (e.g., malate) provides a higher CO_2 partial pressure in the bundle sheath cells (ZELITCH, 1973) thus inhibiting RuBP oxygenase (CHOLLET and OGREN, 1972). It is therefore reasonable that glycolate-producing mechanisms other than RuBP oxygenase could predominate in C_4 plants (HATCH, 1971; ZELITCH, 1973).

E. Conclusion: The Inhibition of Glycolate Formation by Some Common Metabolites – an Open Question

Glycidate has been reported to inhibit glycolate synthesis, block photorespiration, and significantly increase net CO_2 assimilation (ZELITCH, 1974a, b; WILDNER and HENKEL, 1976). Although stimulation of CO_2 fixation was confirmed with isolated chloroplasts, neither the inhibition of glycolate production (CHOLLET, 1976) nor a direct effect of glycidate upon the RuDP carboxylase/oxygenase reaction could be demonstrated (ZELITCH, 1976). In whole leaves carbon incorporation into glutamate and aspartate was significantly enhanced upon glycidate treatment (ZELITCH, 1974b). In this context the recent finding that supply of leaf disks with glutamate, aspartate, glyoxylate, and some other common metabolites resulted in the same effects on glycolate synthesis, inhibition of photorespiration, and increase of CO_2 fixation as that with glycidate is of particular interest (OLIVER and ZELITCH, 1977a, b). Since the chloroplast is not able to produce glutamate or glyoxylate from its own metabolites, the effect of glycidate on leaves could merely represent an increased supply of the chloroplast with these metabolites. This explanation, however, emphasizes the question for the mechanism of the inhibitory effect on glycolate formation exhibited by the metabolites mentioned above. It might be of some interest that glutamate and glyoxylate were found to inhibit P-glycolate phosphatase (BALDY et al., 1977).

References

Anderson, D.E., Tolbert, N.E.: Wood, W.A. (ed.). Methods Enzymol. 9, 645–650 (1966)
Anderson, L., Fuller, R.C.: Biochim. Biophys. Acta 131, 198–201 (1967)

Andrews, T.J., Lorimer, G.H., Tolbert, N.E.: Biochemistry *10*, 4777–4782 (1971)
Andrews, T.J., Lorimer, G.H., Tolbert, N.E., Biochemistry *12*, 11–18 (1973)
Asami, S., Akazawa, T.: Plant Cell Physiol. *16*, 631–642 (1975a)
Asami, S., Akazawa, T.: Plant Cell Physiol. *16*, 805–814 (1975b)
Asami, S., Akazawa, T.: Biochemistry *16*, 2202–2207 (1977)
Badger, M.R., Lorimer, G.H.: Arch. Biochem. Biophys. *175*, 723–729 (1976)
Badger, M.R., Andrews, T.J., Osmond, C.B.: In: Proc. 3rd Int. Congr. Photosynthesis. Rehovot, Israel. Avron, M. (ed.), Vol. II, pp. 1421–1429. Amsterdam: Elsevier Publ. Co. 1975
Baldy, P., Verin, C., Cavalié, G.: Abstr. 4th Int. Congr. Photosynthesis, Reading, 15–16 (1977)
Bassham, J.A., Kirk, M.: Plant Physiol. *52*, 407–411 (1973)
Beck, E.: Fortschr. Bot. *34*, 139–154 (1972)
Beck, E., Eickenbusch, J.D., Steiger, H.M.: Proc. 3rd Int. Congr. Photosynthesis Rehovot. Avron, M. (ed.), Vol. II, pp. 1329–1334. Amsterdam, Oxford, New York: Elsevier Publ. Comp. 1975
Benson, A.A.: Calvin, M.J.: J. Exp. Bot. *1*, 63–68 (1950)
Björkman, O.: Physiol. Plant. *19*, 618–633 (1966)
Bowes, G., Ogren, W.L.: J. Biol. Chem. *247*, 2171–2176 (1972)
Bowes, G., Ogren, W.L., Hagemann, R.H.: Biochem. Biophys. Res. Commun. *45*, 716–722 (1971)
Bradbeer, J.W., Anderson, C.M.A.: In: Biochemistry of chloroplasts II. Goodwin, T.W. (ed.), pp. 175–179. London, New York: Academic Press 1967
Bradbeer, J.W., Racker, E.: Fed. Proc. *20*, 88 (1961)
Calvin, M., Massini, P.: Experientia *8*, 445–457 (1952)
Chollet, R.: Plant Physiol. *57*, 237–240 (1976)
Chollet, R., Ogren, W.: Biochem. Biophys. Res. Commun. *46*, 2062–2066 (1972)
Christeller, J.T., Tolbert, N.E.: Plant Physiol. *53*, 28 S (1974)
Coombs, J., Whittingham, C.P.: Proc. R. Soc. Lond., *B 164*, 511–520 (1966)
Egneus, H., Heber, U., Mathiesen, U., Kirk, M.: Biochim. Biophys. Acta *408*, 252–268 (1975)
Eickenbusch, J.D.: Thesis, Univ. München (1975)
Eickenbusch, J.D., Beck, E.: FEBS Lett. *31*, 225–228 (1973)
Eickenbusch, J.D., Scheibe, R., Beck, E.: Z. Pflanzenphysiol. *75*, 375–380 (1975)
Ellis, R.J.: Curr. Adv. Plant Sci. *3*, 29–38 (1973)
Elstner, E.F., Heupel, A.: Planta (Berl.) *123*, 145–154 (1975)
Fock, H., Bate, G.C., Egle, K.: Planta (Berl.) *121*, 9–16 (1974)
Gerster, R., Tournier, P.: Abstr. 4th Int. Congr. Photosynthesis. Reading, 129–130 (1977)
Gibbs, M.: Ann. N.Y. Acad. Sci. *168*, 356–368 (1969)
Gibbs, M.: In: Photosynthesis and photorespiration. Hatch, M.D., Osmond, C.B., Slatyer, R.O. (eds.), pp. 433–441. New York, London, Sydney, Toronto: Wiley-Interscience 1971
Glidewell, S.M., Raven, J.A.: J. Exp. Bot. *27*, 200–204 (1976)
Goldberg, M.L., Racker, E.: J. Biol. Chem. *237*, PC 3841–PC 3842 (1962)
Griffith, T., Byerrum, R.U.: J. Biol. Chem. *234*, 762–764 (1959)
Groden, D., Beck, E.: Abstr. 4th Int. Congr. Photosynthesis, Reading 139 (1977)
Grodzinski, B., Colman, B.: Plant Physiol. *58*, 199–202 (1976)
Hall, A.E.: Plant Physiol. *47*, 10 S (1971)
Hatch, M.D.: Biochem. J. *125*, 425–432 (1971)
Heber, U., Andrews, T.J., Boardman, N.K.: Plant Physiol. *57*, 277–283 (1976)
Hess, J.L., Tolbert, N.E.: J. Biol. Chem. *241*, 5705–5711 (1966)
Hess, J.L., Tolbert, N.E.: Plant Physiol. *42*, 371–379 (1967)
Holzer, H., Schröter, W.: Biochim. Biophys. Acta *65*, 271–288 (1962)
Jensen, R.G., Bahr, J.T.: In: CO_2-metabolism and plant productivity. Burris, R.H., Black, C.C. (eds.), pp. 3–18. Baltimore, London, Tokyo: Univ. Park Press 1976
Kauss, H., Kandler, O.: Z. Naturforsch. *19b*, 439–441 (1964)
Kearney, P.C., Tolbert, N.E.: Arch. Biochem. Biophys. *98*, 164–171 (1962)
Kerr, M.W., Gear, C.F.: Biochem. Soc. Trans. *2*, 338–340 (1974)
Kirk, M.R., Heber, U.: Planta (Berl.), *132*, 131–141 (1976)
Laing, W.A., Ogren, W.L., Hageman, R.H.: Plant Physiol. *54*, 678–685 (1974)

Laing, W.A., Ogren, W.L., Hageman, R.H.: Biochemistry *14*, 2269–2275 (1975)
Larsson, C.: In: Proc. 3rd Int. Congr. Photosynthesis. Avron, M. (ed.), Vol. II, pp. 1321–1328. Amsterdam, Oxford, New York: Elsevier Publ. Comp. 1975
Lorimer, G.H., Andrews, T.J., Tolbert, N.E.: Biochemistry *12*, 18–23 (1973)
Lorimer, G.H., Badger, M.R., Andrews, T.J.: Biochemistry *15*, 529–536 (1976)
Lorimer, G.H., Osmond, C.B., Akazawa, T.: Plant Physiol. *57*, 6 S (1976)
Mahon, J.D., Fock, H., Canvin, D.T.: Planta (Berl.) *120*, 245–254 (1974)
Marker, A.H.F., Whittingham, C.P.: Proc. R. Soc. Lond. *B 165*, 473–485 (1966)
McCord, J.M., Fridovich, I.: J. Biol. Chem. *246*, 6886–6890 (1971)
Merrett, M.J., Lord, J.M.: New Phytol. *72*, 751–767 (1973)
Oliver, D.J., Zelitch, I.: Plant Physiol. *59*, 688–694 (1977a)
Oliver, D.J., Zelitch, I.: Science *196*, 1450–1451 (1977b)
Orth, G.M., Tolbert, N.E., Jiminez, E.: Plant Physiol. *41*, 143–147 (1966)
Plamondon, J.E., Bassham, J.A.: Plant Physiol. *41*, 1272–1275 (1966)
Plaut, Z., Gibbs, M.: Plant Physiol. *45*, 470–474 (1970)
Plaut, Z., Gibbs, M.: Plant Physiol. *46*, 488–490 (1970)
Randall, D.D.: Aust. J. Plant Physiol. *3*, 105–111 (1976)
Robinson, J.M., Gibbs, M.: Plant Physiol. *53*, 790–797 (1974)
Robinson, J.M., Gibbs, M., Cotler, D.N.: Plant Physiol. *59*, 530–534 (1977)
Shain, Y., Gibbs, M.: Plant Physiol. *48*, 325–330 (1971)
Shou, L., Benson, A.A., Bassham, J.A., Calvin, M.: Physiol. Plant (Berl.) *3*, 487–494 (1950)
Siegel, M.I., Wishnick, M., Lane, M.D.: In: The enzymes. Boyer, P.D. (ed.), Vol. VI, pp. 169–192. New York, London: Academic Press 1972
Steiger, H.M., Beck, E., Beck, R.: Plant Physiol. *60*, 903–906 (1977)
Takabe, T., Akazawa, T.: Biochem. Biophys. Res. Commun. *53*, 1173–1179 (1973)
Tanner, H.A., Brown, T.E., Eyster, C., Treharne, R.W.: Biochem. Biophys. Res. Commun. *3*, 205–210 (1960)
Thompson, C.M., Whittingham, C.P.: Biochim. Biophys. Acta *153*, 260–269 (1968)
Tolbert, N.E.: Natl. Acad. Sci. – Natl. Res. Council. Publ. *1145*, 648–662 (1963)
Tolbert, N.E.: Annu. Rev. Plant Physiol. *22*, 45–74 (1971)
Tolbert, N.E.: In: Symp. Soc. Biol. Vol. XXVII, pp. 474–504. Cambridge: Univ. Press 1973a
Tolbert, N.E.: In: Current topics in cellular regulation. Horecker, B.L., Stadtman, E.R. (eds.), pp. 21–50. New York: Academic Press 1973b
Tolbert, N.E.: In: Photorespiration in marine plants. Tolbert, N.E., Osmond, C.B. (eds.), pp. 129–132. Baltimore, CSIRO, Melbourne: Univ. Park Press 1976
Tolbert, N.E., Ryan, F.J.: In: CO$_2$-metabolism and plant productivity. Burris, R.H., Black, C.C. (eds.), pp. 141–159. Baltimore, London, Tokyo: Univ. Park Press 1976
Tolbert, N.E., Yamazaki, R.K., Oeser, A.: J. Biol. Chem. *245*, 5129–5136 (1970)
Turner, J.S., Brittain, E.G.: Biol. Rev. *37*, 130–170 (1962)
Vandor, S.L., Tolbert, N.E.: Plant Physiol. *43*, 12 S (1968)
Wah Kow, Y., Robinson, J.M., Gibbs, M.: Plant Physiol. *60*, 492–495 (1977)
Warburg, O.: Biochem. Z. *103*, 188–217 (1920)
Warburg, O., Krippahl, G.: Z. Naturforsch. *15b*, 197–199 (1960)
Wildner, G.F.: Ber. Dtsch. Bot. Ges. *89*, 349–360 (1976)
Wildner, G.F., Henkel, J.: Biochem. Biophys. Res. Commun. *69*, 268–275 (1976)
Wilson, A.T., Calvin, M.: J. Am. Chem. Soc. *77*, 5948–5957 (1955)
Wolfenden, R.: Biochemistry *9*, 3404–3405 (1970)
Zelitch, I.: J. Biol. Chem. *234*, 3077–3081 (1959)
Zelitch, I.: J. Biol. Chem. *240*, 1869–1876 (1965)
Zelitch, I.: Photosynthesis, photorespiration and plant productivity. pp. 173–212. New York, London: Academic Press 1971
Zelitch, I.: Plant Physiol. *51*, 299–305 (1973)
Zelitch, I.: Plant Physiol. *53*, 29 S (1974a)
Zelitch, I.: Arch. Biochem. Biophys. *163*, 367–377 (1974b)
Zelitch, I.: Annu. Rev. Biochem. *44*, 123–145 (1975)
Zelitch, I.: Plant Physiol. *57*, 54 S (1976)
Zelitch, I., Gotto, A.M.: Biochem. J. *84*, 541–546 (1962)

26. Glycolate Metabolism by Higher Plants and Algae

N.E. Tolbert

A. Introduction

Carbon flow into the photosynthetic carbon cycle or into the photorespiratory process of glycolate biosynthesis and metabolism is determined by the dual function of ribulose-1,5-bisphosphate carboxylase/oxygenase (reaction 1). The competition between a continuum of changing CO_2 and O_2 concentrations for these initial reactions is a major factor that produces different rates of photosynthesis and photorespiration. The fixation of CO_2 yields 3-P-glycerate, most of which is reduced to carbohydrates. Fixation of O_2 yields P-glycolate and P-glycerate. The P-glycolate is hydrolyzed by a specific chloroplastic phosphatase (reaction 2) and the glycolate excreted. The physiological conditions favoring glycolate biosynthesis and photorespiration are explainable on the basis of P-glycolate formation from ribulose-P_2. The competition between CO_2 and O_2 increases glycolate formation at lower CO_2 or higher O_2 concentrations than air. High temperature increases the O_2 to CO_2 ratio for the solubility of these gases in solution and thus favors the oxygenase reaction. High pH lowers the free CO_2 concentration. High light increases the amount of photosynthetic assimilatory power and increases the demand to dispose of it through photorespiration by lowering the available CO_2 through assimilation and an upward pH shift. At or above the compensation point sufficient P-glycerate is formed by both the carboxylase and the oxygenase reactions to regenerate ribulose-P_2. Below the compensation point storage carbohydrates would be catabolized to sustain carbon flow through the intermediates of the photosynthetic carbon reduction cycle for photorespiration.

Through an alternate pathway for serine and glycine synthesis, 3-P-glycerate can be hydrolyzed by a 3-P-glycerate phosphatase in chloroplasts (reaction 19). Thus the chloroplasts form two organic acids, glycolate and glycerate, which are further converted to glycine and serine in the peroxisomes. More details of these interlocking metabolic pathways associated with photorespiration and photosynthesis can be found in many reviews (GIBBS, 1969; TOLBERT, 1963, 1971, 1973, 1974; TOLBERT and RYAN, 1975; ZELITCH, 1971). One of the latest reviews, in

→

Fig. 1. The glycolate pathway. The numbers are for the name of the enzyme or reaction: *1*, ribulose-P_2 carboxylase/oxygenase; *2*, P-glycolate phosphatase; *3*, NADPH glyoxylate reductase; *4*, glyoxylate oxidation by any oxidant; *5*, glycolate oxidase; *6*, catalase; *7*, glutamate-glyoxylate aminotransferase; *8*, serine-glyoxylate aminotransferase; *9*, glycine oxidase; *10*, serine hydroxymethyl transferase; *11*, NADH-glutamate dehydrogenase; *12*, glutamate-hydroxypyruvate aminotransferase; *13*, glutamate-oxaloacetate aminotransferase or aspartate aminotransferase; *14*, NADPH-glutamate dehydrogenase; *15*, NADH-hydroxypyruvate reductase or glycerate dehydrogenase; *16*, NAD-malate dehydrogenase; *17*, NADP-malate dehydrogenase; *18*, glycerate kinase; *19*, P-glycerate phosphatase

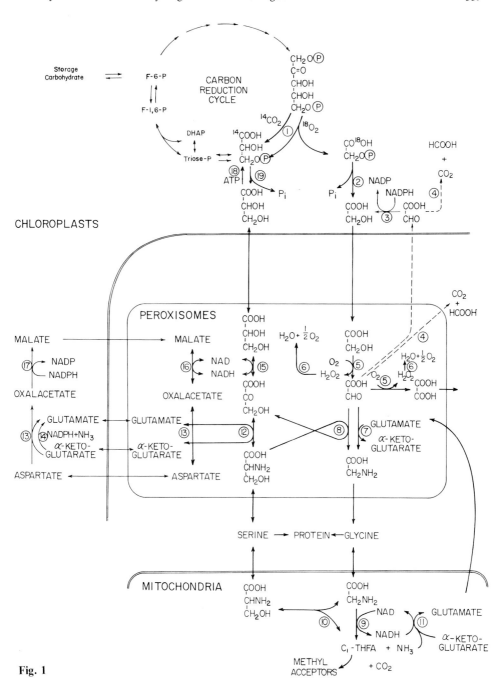

Fig. 1

Volume 3 of this series (SCHNARRENBERGER and FOCK, 1976) and Chap. II.27, this vol. should be examined for details about the pathway and measurements of photorespiration, whereas this contribution will update recent work on the enzymes involved. The reaction numbers cited referred to the glycolate pathway as outlined in Figure 1.

B. Glycolate Biosynthesis

I. Properties of Ribulose-P_2 Carboxylase/Oxygenase for Phosphoglycolate Biosynthesis

Some of the properties of ribulose-P_2 carboxylase/oxygenase are reviewed in chapters by Akazawa and Beck of this volume, and other reviews are available (MCFADDEN, 1973; JENSEN and BAHR, 1977). This protein catalyzes two reactions, the carboxylation which fixes CO_2 into the carboxyl group of 3-P-glycerate and the oxygenase which incorporates one oxygen atom at carbon 2 of ribulose-P_2 to form the carboxyl group of P-glycolate. The stoichiometry of the oxygenase reaction has been established with the purified enzyme and $^{18}O_2$ (LORIMER et al., 1973). For in vivo $^{18}O_2$ experiments the glycine and serine products were isolated and examined by mass spectrometry. The in vivo experiments with $^{18}O_2$ indicate that ribulose-P_2 oxygenase is the major, if not the sole pathway, for glycolate formation in plants and algae (ANDREWS et al., 1971; DIMON and GERSTER, 1976; LORIMER et al., 1977).

II. Phosphoglycolate Phosphatase and Phosphoglycerate Phosphatase

These two phosphatases of the chloroplast are present in sufficient amounts to hydrolyze the products from ribulose-P_2 carboxylase/oxygenase to the organic acids for further peroxisomal metabolism. P-glycolate phosphatase (reaction 2) with a K_m(P-glycolate) of about $6 \cdot 10^{-5}$M, is present in C_3 plants in larger amounts (ratio about 4:1) than P-glycerate phosphatase ($K_m = 8 \cdot 10^{-4}$ M). In C_4 plants there is more P-glycerate phosphatase (ratios of about 1:2; RANDALL et al., 1971; RANDALL and TOLBERT, 1971). This is due to less P-glycolate phosphatase in the bundle sheath cells of C_4-plants than in tissue from C_3-plants (consistent with less glycolate formation in C_4-plants) and more P-glycerate phosphatase in the cytoplasm of the C_4-mesophyll cells. P-glycerate phosphatase (reaction 3) is shown in Figure 1 as being in the chloroplast of C_3-plants based on unpublished work. When combined with glycerate kinase, also in the chloroplast (HATCH and SLACK, 1976), this system represents a potential system for ATP loss unless carefully regulated.

The properties of the two phosphatases for such similar substrates are remarkably different. That there is an enzyme for each indicates a need for differential regulation of their hydrolysis. P-glycolate phosphatase has a requirement for Mg^{2+}, whereas P-glycerate phosphatase has none and is not inactivated by EDTA. P-glycolate phosphatase requires a tricarboxylic acid (e.g., isocitrate) for stability (unpublished). Isolated P-glycolate phosphatase has a molecular weight of about 80,500 and is composed of four equal subunits (CHRISTELLER and TOLBERT, 1978). Ribose-5-P is a competitive inhibitor ($K_i \sim 5\,mM$) of P-glycolate phosphatase, while other phosphate esters of the photosynthetic carbon cycle are without effect. Since ribose-5-P is also an effective inhibitor of ribulose-P_2 carboxylase/oxygenase (RYAN and TOLBERT, 1975; CHOLLET and ANDERSON, 1976), which catalyze P-glycolate formation, these two consecutive reactions in glycolate biosynthesis appear to have a coordinated regulation. Further both enzymes are activated by increasing Mg^{2+} concentration. The activity pH curve for P-glycolate phosphatase from tobacco leaves exhibits two different profiles dependent upon the concentration of Mg^{2+} and indicates active components with pK's at pH 5.7 and 8.1 (CHRISTELLER and TOLBERT, 1978). P-glycolate phosphatase from pea leaves (KERR and GEAR, 1974) and from spinach leaves (CHRISTELLER and TOLBERT, 1978) have an extended pH range over 6 to 8.5. Studies with enzymes from tobacco leaves indicate that there are two distinct divalent cation binding sites, and it is the first that had previously been monitored for the phosphatase activity at pH 5.5 to 6.3 (RICHARDSON and TOLBERT, 1961; RANDALL et al., 1971). The second site, which is specific for Mg^{2+}, binds it in a negatively cooperative fashion about 100-fold greater at pH between 8 and 9. Consequently, enzyme activity in the chloroplast stroma in the light with the high concentrations of Mg^{2+} (10 mM) is extended into the physiological pH range of 8 to 9, although there is no change in total activity compared to that at pH 5 to 6. Thus the enzyme is able to function in the chloroplasts in the light, while in the dark when no P-glycolate is formed it is essentially inactive in vivo with a pH optimum of 5 to 6. This regulation represents a new mechanism for Mg^{2+} activation of chloroplast enzyme activity in the light.

Because of the almost absolute specificity of P-glycolate phosphatase, the mechanism of its reaction has been further investigated (CHRISTELLER and TOLBERT, 1978). Mass spectrometry studies with ($^{18}O_2$)-H_2O showed the cleavage to be between the O-P bond followed by an ordered release first of the glycolate and then the phosphate. The small active site excludes the catalysis of larger alkyl residues. Using small alternate substrates such as ethyl phosphate, the velocity of hydrolysis was found to be proportional to the pK_a of the corresponding alcohol product. The rate-limiting step may be the protonation of the bridge oxygen. The enzyme is inhibited by glycidol phosphate (2,3-epoxypropanol-phosphate) and by the thio analog of P-glycolate, $PO_3H_2\text{-S-}CH_2COOH$ (K_i of 0.4 mM).

The rate of synthesis and pool size of glycolate and glycerate during photosynthesis also can be compared. The observed pool size of glycolate is small, unless experiments are run at abnormally high O_2 concentration or the measurements include the glycolate excreted by algae. The rate of formation of uniformly labeled glycolate by $^{14}CO_2$ is extremely rapid, nearly as rapid as that for the sugar phosphates of the photosynthetic carbon cycle (SCHNARRENBERGER and FOCK, 1976, see also Chaps. II.27 and II.28, this vol.). It is the glycine and serine, products from the glycolate, which accumulate in the cell. On the other hand ^{14}C-glycerate, carboxy-labeled, is formed slowly, but in some plants there is a relatively large pool of free glycerate. Rapid glycolate synthesis can be explained by the great activity of ribulose-P_2 carboxylase/oxygenase, P-glycolate phosphatase and small pools by equally active peroxisomal glycolate oxidase. Slow accumulation of free glycerate reflects the fact that glycerate kinase activity (reaction 18) in the chloroplasts is low and that conversion of glycerate to hydroxypyruvate (reaction 15) in the peroxisomes is unfavorable. The equilibrium for the latter enzyme favors hydroxypyruvate reductase activity for glycerate formation, especially in the light with high levels of reducing capacity from photosynthesis.

C. Glycolate Pathway

I. Pathways in Peroxisomes

In previous reviews glycolate biosynthesis and metabolism have long been called "the glycolate pathway." The biosynthesis of P-glycolate and the metabolism of glycolate may also be trivalized to the "C_2 pathway", consistent with the terms for C_3 and C_4 pathways of photosynthetic carbon metabolism. The expression photorespiratory carbon oxidation cycle (Woo and Osmond, 1977) is descriptive but cumbersome.

The irreversible peroxisomal portion of the C_2 pathway which includes glycolate oxidase (reaction 5), catalase (reaction 6), glutamate-glyoxylate transaminase (reaction 7) and serine-glyoxylate (reaction 8), as shown in Figure 1, is well established. Oxidation of part of the glyoxylate by glycolate oxidase is a side reaction that may vary in magnitude among plants during different developmental stages or during nitrogen deficiency. The latter lowers the glutamate pool needed for the conversion of glyoxylate to glycine. The flow of carbon is ribulose-P_2 to glycolate and then from glycolate to glycine, as all of these reactions are physiologically irreversible. Glycolate oxidase is present in microbodies from all tissues that have been examined; in leaf peroxisomes this is a major pathway, but other sources of microbodies contain only low activity of glycolate oxidase. The relative activity of glycolate oxidase or α-hydroxy acid oxidase from leaves of C_3 plants when compared with glycolate equated to 100 range from 15 to 66 for L-lactate, 2 to 10 for α-hydroxybutyrate, but 100 for α-hydroxyisocaproate (McGroarty et al., 1974). The significance of activity with a C_8 α-hydroxyacid is unknown.

The reversible interconversion in the leaf peroxisome between glycerate and serine involves NAD:hydroxypyruvate reductase or glycerate dehydrogenase (reaction 15), and a transamination of hydroxypyruvate to serine (reaction 12). These two active enzyme systems in leaf peroxisomes may be detectable in low activity in other plant microbodies, but they are not found in mammalian microbodies. Plant, but not animal microbodies, contain a very active NAD-malate dehydrogenase (reaction 14) which has been theorized, similar to the mitochondrial system, to be a part of a C_4 acid shuttle for moving reducing capacity between peroxisomes and other subcellular compartments. Peroxisomal malate dehydrogenase is one of three isoenzymes, it represents about half of the total activity in the leaf, and it is a major component of the peroxisomes (Yamazaki and Tolbert, 1969; Rocha and Ting, 1971). The isoenzymes of malate dehydrogenase differ in cellular location, electrophoretic mobility, and the degree of product inhibition. Pools of malate and aspartate should be in the cytoplasm, peroxisomes, chloroplasts, and mitochondria. As shown in Figure 1 and reviewed in more detail in reference 52, the malate–aspartate shuttle can be linked with the chloroplasts in the light to move reducing capacity into the peroxisome as NADH for glycerate synthesis from serine. During the reduction of hydroxypyruvate to glycerate (reaction 15) malate would be oxidized to oxalacetate (reaction 16) which would be transaminated to aspartate (reaction 13). Leaf peroxisomes contain two isoenzymes of aspartate aminotransferase (REHFELD and TOLBERT, 1972). The conversion of aspartate back to malate-utilizing photosynthetic NADPH in the chloroplasts (reaction 13 and 17) is shown in Figure 1. Alternatively in the dark the synthesis of serine from glycerate could be coupled to NADH oxidation by a malate-aspartate shuttle through the mitochondria.

II. Mitochondrial Interconversion of Glycine and Serine

This process consists of several reactions involving tetrahydrofolate (C_1-THFA) bound to a mitochondrial enzyme system. Most biochemical investigations of the THFA complex system have been done with microorganisms and further work is needed with photosynthetic tissue (COSSINS and LOR, 1976), which is the primary source for this vitamin, folic acid, named after the Latin word, folium, for leaf. Oxidative glycine decarboxylation (reaction 9) is linked to NAD reduction and yields CO_2 from the carboxyl group of glycine, NH_3, and the C_1-THFA from C_2 of glycine. The CO_2 may be either refixed photosynthetically, or it may escape from the leaf as an indication of photorespiration. Because of the rapid interconversion of glycine and serine when either is given to a leaf, and because the label in the C_2 of glycine only slowly appears in CO_2 during photorespiration, the C_1-THFA must condense primarily with another glycine to form serine. The serine hydroxymethyltransferase (reaction 10) catalyzes the equilibrium exchange between serine plus THFA and glycine plus N^5, N^{10}-methylene THFA. The latter, if hydrolyzed, would be at the reduction level of a -CH_2OH group. It is possible that N^5, N^{10}-methylene THFA could be oxidized to N^5- or N^{10}-formyl THFA, and the formyl THFA could dissociate to formate. This pathway appears active for the incorporation of added ^{14}C-formate by leaves into the β-carbon of serine in light or dark (TOLBERT, 1955). Formate can be oxidized to CO_2 by a formate dehydrogenase in the mitchondria (HALLIWELL, 1974), and when ^{14}C-formate is given to leaves, much of it is oxidized to $^{14}CO_2$ and photosynthetically refixed (TOLBERT, 1955). The magnitude of CO_2 production through formate during photorespiration is judged small on the basis that label in the α carbon of glycine does not rapidly appear in CO_2 relative to the carboxyl group. Rather the rapid appearance of labeled serine after feeding labeled glycolate or glycine is consistent with the C_2 pathway as presented in Figure 1. The net products from two glycines are one serine, one CO_2, one NH_3, and the reduction of NAD.

Theoretically, as much NH_3 is produced during photorespiration as CO_2. However, because of NH_3 toxicity as an uncoupler of photosynthetic electron transport, its accumulation would injure the plant. Early plant scientists, such as PRIANISHNIKOV and CHIBNALL, emphasized that accumulation of NH_3 in starving leaves was a terminal phenomenon. Therefore during photorespiration, all the NH_3 must be immediately refixed by glutamate dehydrogenase (reaction 11) and glutamine synthetase (see Chap. III.5, this vol.). Refixation of NH_3 requires an equal amount of NADH, which is generated during glycine decarboxylation. Therefore it is reasonable that both the glycine decarboxylase and glutamate dehydrogenase are located together in the mitochondria, and this argument indicates why glycine decarboxylation should not occur in the peroxisomes. It is not known to what extent the NH_3 might also be used by the glutamate dehydrogenase of the chloroplasts.

During glycine decarboxylation neither NH_3 nor NADH should accumulate; there would be no excess NADH for ATP synthesis, nor is there any O_2 uptake from the oxidation of the NADH. During studies of glycine oxidation with isolated mitochondria, it is possible with the incomplete cellular system to observe NAD reduction (Woo and Osmond, 1976, 1977) and some ATP synthesis (Bird et al., 1971). Woo and Osmond have shown that the glycine decarboxylation system is located behind the inner membrane of the mitochondria from spinach leaves. They coupled the system to oxalacetate reduction by utilizing the mitochondrial malate dehydrogenase to oxidize the NADH, but no tests were run to couple glycine decarboxylase to glutamate dehydrogenase. With adequate regeneration of NAD^+, they observed rates of glycine decarboxylation that approached those required for CO_2 evolution during glycolate metabolism by the C_2 pathway. Whether the NADH generated during glycine decarboxylation contributes to a pool of NADH or is directly coupled to NH_3 fixation has not yet been evaluated.

NH_3 turnover during glycine formation and oxidation requires an equally large turnover of glutamate and α-ketoglutarate. The glutamate shuttles the nitrogen back to peroxisomes and presumably α-ketoglutarate shuttles from the peroxisomes to the mitochondria. This shuttle results in the oxidation of the C_2 compounds to CO_2, but there is no net energy gain to the cell nor O_2 uptake, and the α-ketoglutarate carbon skeleton does not change.

C_1-THFA generated from glycine or serine is at the reduction level of N^5, N^{10}-methylene THFA, which is the starting point for a wide variety of essential one-carbon transfers in cellular growth processes. Of greatest importance is its further reduction to N^5-methyl THFA, and then the transfer of this methyl group by way of methionine to all the other methyl acceptors. The reduction of the methylene THFA by NADH requires an additional mitochondrial generating system for the NADH other than the oxidation of glycine, since that NADH equivalent was used to refix the NH_3. On the basis that glycine and serine are necessary for protein synthesis and are the primary donors for all methyl groups, the biosynthesis of these two amino acids are essential peroxisomal reactions. The modification permitted is synthesis either from P-glycolate during photorespiration or from P-glycerate.

D. O_2 and CO_2 Exchange and Energy Balance

I. Sites of O_2 Uptake and CO_2 Release in the Glycolate Pathway

In the C_2 or glycolate pathway O_2 uptake occurs during ribulose-P_2 oxidation in the chloroplasts and during glycolate oxidation in the peroxisomes. Previous hypotheses also included O_2 uptake in the mitochondria associated with glycine decarboxylation. In view of the need to utilize the NADH generated during glycine decarboxylation in the mitochondria to fix NH_3, no O_2 uptake or ATP synthesis should occur during this process.

Other proposed mechanisms for CO_2 formation have been reviewed (BEEZLEY et al., 1976; SCHNARRENBERGER and FOCK, 1976; ZELITCH, 1971). According to the C_2 pathway on Figure 1, the primary source for CO_2 is from glycine decarboxylation. Recent research discussed in Section C.II indicates that this decarboxylation in the mitchondria is sufficiently active to account for the rate of CO_2 formation during photorespiration. In order to account for a larger efflux of CO_2 during photorespiration than predicted by the conversion of two glycines to one serine and CO_2, I have proposed that the serine may be converted to a glycine plus a C_1-THFA. The additional glycine would be further oxidized to CO_2 and another C_1-THFA. This can occur only to the extent that the excess C_1-THFA is used to form methyl groups or converted to formate and then to CO_2. Formate dehydrogenase is a mitochondrial enzyme and this C_1 unit has been reported to be oxidized to CO_2 by the mitchondria (HALLIWELL, 1974). Evidence for CO_2 arising from glyoxylate oxidation was well critiqued in the chapter of this series by SCHNARRENBERGER and FOCK (1976). With isolated subcellular organelles the oxidation of glyoxylate to CO_2 and formate (reaction 4) in the peroxisomes by H_2O_2 or in the chloroplasts by strong oxidants such as Mn^{3+}, is readily measured, but this does not seem to contribute significantly to CO_2 evolution during photorespiration. The main arguments against glyoxylate oxidation are the labeling kinetics for glycine and the absence of a formate pool or CO_2 formation from the C_2 of glycolate or glyoxylate.

II. O_2 Uptake During Photosynthesis

By definition photorespiration is considered to be O_2 uptake and CO_2 release in the light, over and above dark respiration. The biochemistry of the glycolate pathway has been able to explain the CO_2 release during photorespiration. Evaluation of the O_2 uptake has been most difficult without a radioactive label, but careful studies with manometry and $^{18}O_2$ have established a substantial O_2 uptake as superoxide and H_2O_2 production during auto-oxidation of reduced components of photosystem I (Mehler reaction). Superoxide dismutase of the chloroplasts rapidly converts any superoxide formed at this site to H_2O_2. This phenomenon is a direct competition between O_2 and CO_2 for the photosynthetically generated reducing power. Thus photorespiration, as O_2 uptake and as a mechanism to dissipate excess photosynthetic energy, is accomplished either by the glycolate pathway of carbon metabolism or by direct oxidation by O_2 of the electron acceptors of photosystem I. An integration of these two mechanisms of O_2 uptake can be proposed in reviews but is difficult to evaluate experimentally. For example, H_2O_2 in the chloroplast is toxic (KAISER, 1976), but its disposition in the chloroplasts without catalase (TOLBERT, 1971) is unclear. H_2O_2 may diffuse down a steep concentration gradient to the peroxisomal catalase. H_2O_2 may also oxidize such chloroplast components as ascorbate, reduced glutathione, or components of electron transport.

III. Energy Balance

The total carbon pathway for glycolate biosynthesis and metabolism consists of several integral parts. The reductive photosynthetic carbon cycle forms P-glycolate from ribulose-P_2, and reduces the other product of ribulose-P_2 oxidation, 3-P-glycerate, back to ribulose-P_2. After carbon flows through the C_2 glycolate pathway to glycine, two glycines are converted to serine and CO_2, and the serine reconverted to glycerate. This latter glycerate is reclaimed and reduced with photosynthetic energy. For each glycolate, one mol of O_2 is taken up in its formation and one half mol in its oxidation to glyoxylate. For two glycolates this is a total of 3 mol of oxygen uptake for 1 mol of CO_2 release, if only the carbon flows around the system as shown in Figure 1. However, this high O_2 uptake is balanced off by the O_2 evolved during the photosynthetic reduction of three mol of 3-P-glycerate regenerated in the formation of the one CO_2. For the photosynthetic carbon cycle the energy requirement for 6 CO_2 fixation to 12 3-P-glycerates and then reduction to ribulose-P_2 and fructose-6-P is 18 ATP and 12 NADPH. Reduction of the three 3-P-glycerates generated during one cycle of the glycolate pathway would require approximately $4^1/_2$ ATP and 3 NADPH. In addition 1 ATP is required to phosphorylate a glycerate and 1 NADPH equivalent to reduce hydroxy-pyruvate to glycerate. During the generation of this total of about 6 ATP and 4 NADPH, 2 mol of O_2 would have been formed by photosynthesis. This results in a net O_2 uptake for photorespiration of one mol per mol CO_2 loss.

Photorespiration does not function in the absence of any photosynthetic CO_2 fixation. The mol of CO_2 produced or its equivalent from air would be refixed photosynthetically with the evolution of 1 mol O_2. Net CO_2 and O_2 exchange would then be zero, which occurs at the compensation point. Thus the essence of photorespiration is a metabolic cyclic process for wasting excess photosynthetic capacity which is lost for net CO_2 fixation or other growth processes. Oxidation of the reduced components of photosystem I by O_2 accomplishes the same purpose during which O_2 uptake equals O_2 evolution, but without metabolic carbon turn-over. Both systems use photosynthetic reducing capacity and thus both are photorespiration by purpose, by this definition, and by O_2 uptake, but only the glycolate pathway involves CO_2 exchange.

E. Leaf Peroxisomal Membrane and Transport

Definitive data for transport between peroxisomes and other subcellular compartments are lacking (Tolbert, 1973). Since steps of the C_2 pathway, which are known to be sequential from ^{14}C-labeling experiments (Tolbert, 1963), are found in different parts of the cell (Tolbert, 1971) as shown in Figure 1, movement of large amounts of components in and out of the peroxisomes must occur. Since intact leaf peroxisomes exhibit latency for NAD^+-linked malate dehydrogenase (Yamazaki and Tolbert, 1969) and catalase activity, transport through the single bounding membrane may be rate-limiting. However, this single membrane should not represent as much of a barrier as the double membrane around the mitochondria and chloroplasts. Specific translocases in peroxisomal membranes are not known. In lipid composition and NADH-cytochrome c reductase activity the leaf peroxisomal membrane is similar to the endoplasmic reticulum (DONALDSON et al., 1972), from whence it comes. More definitive information is known about the glyoxysomal membrane composition (BEEVERS, 1975).

The following excretion or shuttle systems involving leaf peroxisomes have been proposed and discussed in some detail in previous reviews (TOLBERT, 1971; SCHNARRENBERGER and FOCK, 1976). (1) Glycolate excretion by the cholorplasts or by unicellular algae into their medium has so far not been coupled to any counter ion movement or to an enzymatic process. Glycolate excretion is highly specific and one wonders how this is accomplished. P-glycolate phosphatase does not seem to be involved, since it is in the chloroplast matrix (TOLBERT, 1971) and is not associated with isolated chloroplast membrane (DOUCE et al., 1973). However, this phosphatase in *Chlamydomonas* is lost by the intact cells when they are subjected to mild osmotic shock by suspension in distilled water. (2) The glycolate/glyoxylate shuttle between the chloroplasts and peroxisomes, proposed upon the basis of the chloroplast location of NADP-glyoxylate reductase (reaction 3), has not been substantiated. If it exists, it would be a potentially wasteful terminal oxidase and would have to be very regulated. (3) The several transaminases in the peroxisomes (REHFELD and TOLBERT, 1972) necessitate transport of both amino acids and possibly α-ketoacids across the peroxisomal membrane. In particular the irreversible glutamate-glyoxylate aminotransferase requires a large turnover of glutamate as the amino donor for glycine synthesis. (4) The malate-aspartate shuttle as shown in Figure 1 is proposed to function for moving reducing capacity in and out of the peroxisome as it does in the mitochondria.

F. Glycerate and Sucrose from Glycolate

Sucrose is synthesized in the cytoplasm from newly formed products of the photosynthetic carbon cycle. This pathway most probably involves the excretion of dihydroxyacetone phosphate (not shown in Fig. 1; HEBER, 1974) for transport of carbon and energy from the chloroplasts which is then converted to hexose phosphates in the cytoplasm. Several lines of research had suggested that glycerate from serine formed by the glycolate pathway might also be converted directly into cytoplasmic sucrose. However, this proposal has been deleted from Figure 1 for several reasons listed in the next paragraph. Instead the glycerate product of photorespiration probably returns to the chloroplast to be phosphorylated and reenters the 3-P-glycerate pool of the photosynthetic carbon cycle.

(1) ^{14}C labeling experiments showed that when glycolate, glycine or serine were given to leaves, ^{14}C rapidly appeared in glycerate and sucrose with a ^{14}C distribution pattern consistent with the glycolate pathway, as shown in Figure 1, followed by the combination of two glycerates to form one hexose for sucrose (TOLBERT, 1971). For example, α-labeled glycolate or glycine formed 2,3-labeled serine and glycerate and 1,2,5,6 labeled hexoses of the sucrose. Such data indicated that the glycerate from the glycolate pathway could be utilized for sucrose synthesis in the light, or that ^{14}C-glycerate could be exchanged into hexose phosphates by equilibrium reactions. These whole leaf experiments could not differentiate

among the sites where this might occur in the cell. (2) Glycerate kinase has been reported to be in the chloroplasts (Slack et al., 1969). The glycerate kinase has a low specific activity and is not sufficient for a large carbon flux through the glycolate pathway to sucrose. (3) That sucrose labeling only occurs in the light even from glycolate and glycerate, indicates that photosynthetic energy is involved and this could be explained by the formation and reduction of 3-P-glycerate from glycerate in the chloroplast. (4) Whereas glycolate, glycine and serine are uniformly labeled during short periods of $^{14}CO_2$ fixation, hexoses from sucrose are 3,4-labeled (Gibbs and Kandler, 1957) and could not arise quickly from uniformly labeled components of the C_2 pathway. (5) Free ^{14}C-glycerate accumulates very slowly during $^{14}CO_2$ fixation, relative to very rapidly labeling of sucrose. Likewise ^{18}O-labeling of glycerate from O_2 uptake into the carboxyl groups of P-glycolate, glycolate, glycine and serine is also slow, but it does occur (G. Lorimer, personal communication). The final glycerate pool size in such leaves as tobacco, wheat bean and spinach is large. Fast sucrose labeling is not kinetically possible from the glycerate pool. (6) Sucrose is formed rapidly in low O_2 atmospheres which slows down glycolate biosynthesis. These data support cytoplasmic sucrose synthesis from the triose-P pool of the photosynthetic carbon cycle. The data also indicate that carbon flows all the way through the glycolate pathway to glycerate, but the rate of glycerate synthesis relative to serine formation needs further evaluation.

G. The Glycolate Pathway in Algae

I. Introduction

Many aspects of glycolate biosynthesis and metabolism in algae are similar to those in higher plants (Tolbert, 1974). Algal ribulose-P_2 carboxylase/oxygenase has similar properties for P-glycolate synthesis (McFadden, 1973) and P-glycolate phosphatase is equally as ubiquitous and active (Randall et al., 1971). From prior photosynthetic research with $^{14}CO_2$, it is known that algae form uniformly labeled glycolate, glycine and serine in substantial amounts during photosynthesis. Glycolate biosynthesis by algae is increased by high O_2 and high pH and repressed by high CO_2, similar to higher plants. Therefore, it is concluded that algae have the glycolate pathway of photorespiration. This may not always be manifested by much CO_2 release for reasons to be discussed in G.III., but $^{18}O_2$ uptake during photosynthesis does occur (Radmer and Kok, 1976). However, in algal metabolism of glycolate there are some significant differences from higher plants which merit more research.

II. Glycolate Excretion

Whereas plant chloroplasts excrete glycolate into the cytoplasm from whence it is taken up by the peroxisomes and converted to glycine, some unicellular algae may excrete a significant amount (1% to 10%) of the newly fixed $^{14}CO_2$ as glycolate into their medium. The algal excretion represents a near total loss, as the glycolate in dilute solution is only slowly if at all reabsorbed (Tolbert, 1974), but rather it is apparently metabolized by microorganisms and other phytoplankton.

A few years ago this topic seemed to be of considerable ecological interest, but little new has been reported recently. The amount of glycolate in lakes and the ocean is small (0 to 10 ppm) and difficult to measure, and glycolate excretion may not be as ecologically important as that observed with algal cultures. The mechanism, specificity, and reason for glycolate excretion remain unknown. Although glycolate excretion is nearly unique, *Ankistrodesmus* and *Chlamydomonas* also excrete the stable lactone of isocitrate (CHANG and TOLBERT, 1970). Since the free acid was not excreted, we hypothesized that glycolate might also be excreted as the lactone, glycolide, which being unstable would be quickly hydrolyzed in the medium to reform glycolic acid. TOLBERT and ZILL (1956) upon discovering the dependency of glycolate excretion upon HCO_3^- postulated that the phenomenon might involve some ion exchange process for pumping HCO_3^- into the cells. This has recently been extended by further studies with *Chlamydomonas* grown in low CO_2 or air (BERRY et al., 1976). Such algae develop good rates of photosynthesis in the low CO_2 by some unknown mechanism, perhaps for concentrating CO_2 combined with glycolate excretion. The slow adaptation does not involve changes in ribulose-P_2 carboxylase/oxygenase (BERRY et al., 1976), but rather increases in carbonic anhydrase (GRAHAM and REED, 1971) and glycolate dehydrogenase (NELSON and TOLBERT, 1969). Whereas glycolate excretion is maximum when *Chlamydomonas,* adapted to growth on a CO_2 enriched atmosphere, are switched to air or low CO_2, glycolate excretion also occurs when these algae are grown on low CO_2, but excretion is readily suppressed by increasing amounts of CO_2 (BERRY et al., 1976). As the HCO_3^- concentration becomes limiting, there is an increase in the ratio of glycolate excreted to HCO_3^- assimilated.

III. Glycolate Dehydrogenase

A biochemical difference between unicellular algae and multicellular algae and angiosperms is in the enzyme which oxidizes glycolate to glyoxylate. In higher plants there is an active glycolate oxidase (~ 50 nmol per min per mg protein), and in algae there is a lower amount of a glycolate dehydrogenase. Glycolate oxidase is always located in the peroxisomes, has FMN as a cofactor, and takes up O_2 to form H_2O_2, which in turn is destroyed by the accompanying catalase, so the net O_2 uptake is one atom per glycolate (TOLBERT, 1971). Glycolate dehydrogenase has been found in most but not all unicellular freshwater green and blue green algae (FREDERICK et al., 1973), in some marine algae and two marine angiosperms (TOLBERT, 1976). Glycolate dehydrogenase is not linked, at least directly, to oxygen uptake (NELSON and TOLBERT, 1970), but like glycolate oxidase, it has been assayed by the artificial electron acceptor dichlorophenolindophenol. When glycolate dehydrogenase activity is present, the catalase content is about $^1/_{10}$ that found in tissue with glycolate oxidase (FREDERICK et al., 1973), presumably because the H_2O_2 generating system of glycolate oxidase is absent. *Euglena* without any catalase is one such example. Volcani's group (PAUL et al., 1975) reports that cytochrome b is reduced during glycolate oxidation by the diatom, *Cylindrotheca fusiformis.* By cytochemical and biochemical studies glycolate dehydrogenase has been located in the mitochondria of some algae including *Chlorella* (GRUBER et al., 1974), *Euglena* (COLLINS and MERRETT, 1975), *Chlamydomonas* (STABENAU, 1974; BEEZLEY et al., 1976), the diatom (PAUL et al., 1975) and in the photosynthetic lamella of blue green algae (GRODZINSKI and COLMAN, 1976). For comparison glycolate oxidase catalyzes an energy wasteful terminal oxidation in peroxisomes, whereas with glycolate dehydrogenase the energy may be conserved in an electron membrane transport system. Further investigations of glycolate dehydrogenase are needed, but being a membrane system, it becomes very unstable upon partial isolation and purification (NELSON and TOLBERT, 1970).

Glycolate oxidase oxidizes L-lactate, but not D-lactate, about $^1/_3$ to $^2/_3$ as rapidly as glycolate. Glycolate dehydrogenase oxidizes D-lactate but not L-lactate, equally as well or better than glycolate (GRUBER et al., 1974; NELSON and TOLBERT, 1970). Glycolate oxidase is not inhibited by 10^{-3} M cyanide, but in fact initial rates of O_2 uptake increase due to inhibition of contaminating catalase by the cyanide. Glycolate dehydrogenase is completely inhibited by 10^{-3} M cyanide. Therefore surveys of plant tissues for these two different glycolate oxidizing systems can be based on (a) O_2 uptake, (b) ratios of substrate specificities for glycolate, L-lactate and D-lactate with dichlorophenolindophenol as the electron acceptor, and (c) inhibition by cyanide. The evolutionary significance of the distribution of glycolate dehydrogenase among many unicellular green and blue green algae (FREDERICK et al., 1973; GRUBER et al., 1974) and glycolate oxidase in higher plants is not understood, but leads to some speculation. Both enzymes have never been found in the same tissue. One comparison has been that any unicellular alga that excretes glycolate has glycolate dehydrogenase, perhaps because its specific activity is low and under a condition (high O_2) for high P-glycolate production it can not oxidize all the glycolate. All higher land plants have glycolate oxidase, but two marine higher plants have been found to have glycolate dehydrogenase (TOLBERT, 1976). Multicellular algae will have glycolate oxidase, but all single cell algae do not necessarily have glycolate dehydrogenase. *Chlorella pyrenoidosa* (211/8p) contains glycolate dehydrogenase (NELSON and TOLBERT, 1970), but glycolate oxidase is found in a yellow *Chlorella vulgaris* mutant (KOWALLIK and SCHMID, 1971). In most algae, microbodies can be observed cytologically (TOLBERT, 1974), even in algae with glycolate dehydrogenase, but in most cases the enzymes in these microbodies have not been fully identified.

To consider further glycolate dehydrogenase-like activity in algae careful comparison with lactate dehydrogenases will be necessary to avoid confusion when evaluating evolutionary trends associated with photorespiration. As already mentioned, glycolate oxidase of peroxisomes readily oxidized L-lactate, but NAD$^+$ L-lactate dehydrogenase will not oxidize glycolate (TOLBERT, 1971). In the aerobic condition in leaves of higher plants L-lactate dehydrogenase is generally insignificant, so lactate cannot be formed and the glycolate oxidase substrate is primarily glycolate. However, in *Chlorella* there are two enzymes capable of oxidizing or forming D-lactate but not L-lactate (GRUBER et al., 1974). One is the particulate glycolate dehydrogenase coupled to a cytochrome. This dehydrogenase is similar to NAD$^+$-independent D-lactate dehydrogenases found in various other organisms, including the D-lactate dehydrogenase in bacterial plasma membranes (KABACK, 1974). The other is a widely-distributed cytoplasmic NAD$^+$-D-lactate dehydrogenase. The function of those two enzymes in bacteria has been detailed by KABACK (1974) for providing energy as D-lactate for transport across bacterial cytoplasmic membranes. The cytoplasmic NAD-D-lactate dehydrogenase reduces pyruvate to D-lactate and the D-lactate dehydrogenase (cytochrome-linked) in the membrane is coupled to the membrane transport. Similarities between the algal enzymes and bacterial enzymes indicate a similar research effort is needed on algal membrane transport involving glycolate metabolism. Both glycolate excretion described in 1956 (TOLBERT and ZILL, 1956) and D-lactate dehydrogenase first described in 1957 (WARBURG, 1957) for *Chlorella* have not been definitively associated with membrane transport. Algal membrane transport may be coupled to the oxidation of glycolate in the light and to the oxidation of D-lactate in the dark or under anaerobic conditions.

IV. Glycerate-Serine Pathway in Algae

From ^{14}C tracer research it is known that algae contain the same ability to intercon-vert glycine and serine and to biosynthesize serine from P-glycerate and glycerate as the higher plant. NADH-hydroxypyruvate reductase is active in algal homogen-ates, but the amino transferases, malic dehydrogenases, and the enzymes for glycine-serine interconversion have not been carefully examined in algae. During ^{14}CO$_2$ fixation in high light intensity by angiosperms uniformly-labeled glycine and serine are formed, as if insignificant carbon flows to serine from carboxyl-labeled P-glycerate. In algae on the other hand in short time periods the serine may be carboxyl-labeled and formed as fast as glycine (HESS and TOLBERT, 1967). This, together with the low level of glycolate dehydrogenase activity and glycolate excretion, suggests that glycerate metabolism contributes relatively more to serine in algae than in leaves. When glycine production by the glycolate pathway is limited, more serine must come from glycerate. Under these conditions CO$_2$ loss in the conversion of two glycines to a serine would be reduced. In leaves when the glycolate pathway is reduced in amount by lowered O$_2$ or light intensity, or higher CO$_2$, serine synthesis must also occur from glycerate. One way to evaluate the relative distribution of carbon flow from glycolate to glycine or glycerate to serine will be by heavy-isotope, mass spectrometric studies of the carbon distribu-tion in the products.

References

Andrews, T.J., Lorimer, G.H., Tolbert, N.E.: Biochemistry *10*, 4777–4785 (1971)
Andrews, T.J., Badger, M.R., Lorimer, G.H.: Arch. Biochem. Biophys. *171*, 93–103 (1975)
Beevers, H.: In: Recent advances in chemistry and biochemistry of plant lipids, pp. 287–299. New York: Academic Press 1975
Beezley, B.B., Gruber, P.J., Frederick, S.E.: Plant Physiol. *58*, 315–319 (1976)
Berry, J., Boynton, J., Kaplan, A., Badger, M.: Carnegie Inst. Yearb. *75*, 423–432 (1976)
Bird, I.F., Cornelius, M.J., Keys, A.J., Whittingham, C.P.: Biochem. J. *128*, 191–192 (1971)
Bowes, G., Ogren, W.L.: J. Biol. Chem. *247*, 2171–2176 (1972)
Chang, W-H., Tolbert, N.E.: Plant Physiol. *46*, 377–385 (1970)
Chollet, R., Anderson, L.L.: Arch. Biochem. Biophysics *176*, 344–351 (1976)
Christeller, J.T., Tolbert, N.E.: J. Biol. Chem. *253*, in press
Collins, N., Merrett, M.J.: Biochem. J. *148*, 321–328 (1975)
Cossins, E.A., Lor, K.L.: In: Chemistry and biology of pteridines, pp. 321–328. Berlin, New York: Walter de Gruyter 1976
Dimon, B., Gerster, R.: C. R. Sér. *D 283*, 507–510 (1976)
Donaldson, R.P., Tolbert, N.E., Schnarrenberger, C.: Arch. Biochem. *152*, 199–215 (1972)

Douce, R., Holtz, R.B., Benson, A.A.: J. Biol. Chem. *248*, 7215–7222 (1973)
Frederick, S.E., Gruber, P., Tolbert, N.E.: Plant Physiol. *52*, 318–323 (1973)
Gibbs, M.: Ann. N.Y. Acad. Sci *168*, 356–368 (1969)
Gibbs, M., Kandler, O.: Proc. Natl. Acad. Sci. USA *43*, 446–451 (1957)
Graham, D., Reed, M.L.: Nature (London) *231*, 81–83 (1971)
Grodzinski, B., Colman, B.: Plant Physiol. *58*, 199–202 (1976)
Gruber, P.J., Frederick, S.E., Tolbert, N.E.: Plant Physiol. *53*, 167–170 (1974)
Halliwell, B.: Biochem. J. *138*, 77–85 (1974)
Hatch, M.D., Osmond, C.B.: In: Encyclopedia of plant physiology. New Ser. Heber, U., Stocking, C.R. (eds.), pp. 144–184 1976
Heber, U.: Annu. Rev. Plant Physiol. *25*, 393–421 (1974)
Hess, J.L., Tolbert, N.E.: Plant Physiol. *42*, 371–379 (1967)
Jensen, R.G., Bahr, J.T.: Annu. Rev. Plant Physiol. *28*, 379–400 (1977)
Kaback, H.R.: Science *186*, 882–892 (1974)
Kaiser, W.: Biochem. Biophysica Acta *440*, 476–482 (1976)
Kerr, M.W., Gear, C.F.: Biochemical Soc. Trans. *2*, 338–340 (1974)
Kowallik, W., Schmid, G.H.: Planta (Berl.) *96*, 224–237 (1971)
Ku, S-B., Edwards, G.E.: Plant Physiol. *59*, 986–990 (1977)
Lorimer, G.H., Andrews, T.J., Tolbert, N.E.: Biochemistry *12*, 18–23 (1973)
Lorimer, G.H., Badger, M.R., Andrews, T.J.: Biochemistry *15*, 529–536 (1976)
Lorimer, G.H., Berry, J.A., Krause, G.H., Osmond, C.B.: Plant Physiol. (Suppl.) *59*, 43 (1977)
McFadden, B.A.: Bacteriol. Rev. *37*, 289–319 (1973)
McGroarty, E., Hsieh, B., Wied, D., Tolbert, N.E.: Arch. Biochem. Biophys. *161*, 194–210 (1974)
Nelson, E.B., Tolbert, N.E.: Biochem. Biophys. Acta *184*, 263–270 (1969)
Nelson, E.B., Tolbert, N.E.: Arch Biochem. Biophysics *141*, 102–110 (1970)
Paul, J.S., Sullivan, C.W., Volcani, B.E.: Arch. Biochem. Biophys. *169*, 152–159 (1975)
Radmer, R.J., Kok, B.: Plant Physiol. *58*, 336–340 (1976)
Randall, D.D., Tolbert, N.E.: J. Biol. Chem. *246*, 5510–5517 (1971)
Randall, D.D., Tolbert, N.E., Gremel, D.: Plant Physiol. *48*, 480–487 (1971)
Rehfeld, D.W., Tolbert, N.E.: J. Biol. Chem. *247*, 4803–4811 (1972)
Richardson, K.E., Tolbert, N.E.: J. Biol. Chem. *236*, 1285–1290 (1961)
Rocha, V., Ting, I.P.: Arch. Biochem. Biophys. *147*, 114–122 (1971)
Ryan, F.J., Tolbert, N.E.: J. Biol. Chem. *250*, 4234–4238 (1975)
Schnarrenberger, C., Fock, H.: In: Encyclopedia of plant physiology. New Ser. Vol. III, pp. 185–234 1976
Slack, C.R., Hatch, M.D., Goodchild, D.J.: Biochem. J. *114*, 489–498 (1969)
Stabenau, H.: Planta (Berl.) *118*, 35–42 (1974)
Tolbert, N.E.: J. Biol. Chem. *215*, 27–34 (1955)
Tolbert, N.E.: In: Photosynthetic mechanisms in green plants. Publication 1145. Natl. Acad. Sci. USA, pp. 648–662. Natl. Res. Council 1963
Tolbert, N.E.: Annu. Rev. Plant Physiol. *22*, 45–74 (1971)
Tolbert, N.E.: In: Current topics in cellular regulation. Horecker, B.L., Stadtman, E.R. (eds.), pp. 21–50. New York: Academic Press 1973
Tolbert, N.E.: In: Algal physiology and biochemistry. Stewart, W.D.P. (ed.), pp. 474–504. Oxford, England.: Blackwell Scientific Publication Ltd. 1974
Tolbert, N.E.: Soc. Exp. Biol. Symp. *27*, 215–239 (1973)
Tolbert, N.E.: Aust. J. Plant Physiol. *3*, 129–132 (1976)
Tolbert, N.E., Ryan, F.J.: In: CO_2 metabolism and plant productivity. Burris, R.H., Black, C. (eds.), pp. 141–159. Univ. Park Press 1975
Tolbert, N.E., Zill, L.P.: J. Biol. Chem. *222*, 895–906 (1956)
Warburg, O., Gewitz, H.S., Völker, W.: Naturforschung *126*, 722–726 (1957)
Woo, K.C., Osmond, C.B.: Aust. J. Plant Physiol. *3*, 771–785 (1976)
Woo, K.C., Osmond, C.B.: Plant Cell Physiol. Spec. Issue, 315–323 (1977)
Yamazaki, R.K., Tolbert, N.E.: Biochem. Biophys. Acta *178*, 11–20 (1969)
Zelitch, I.: Photosynthesis, photorespiration, and plant productivity. p. 347. New York: Academic Press

27. Photorespiration: Studies with Whole Tissues

I. Zelitch

A. Discovery of Photorespiration

The discovery of the process of photorespiration, the rapid evolution of CO_2 in light in photosynthetic tissues, was long delayed, and the history of the failure of scientists to recognize the importance of this process provides an example that may be instructive. This has been discussed in greater detail (Zelitch, 1971) and will only be summarized here. Rabinowitch (1945) in his comprehensive treatise on photosynthesis considered the possibility that an acceleration of dark respiration might occur in photosynthetic tissues in the light. However, he and others failed to consider that photorespiration might be derived from biochemical reactions completely different from those of dark respiration and might possess distinctly different properties. Thus experiments were frequently carried out at high concentrations of CO_2, low levels of O_2, weak light, and low temperatures, conditions which inhibit photorespiration and interfere with its detection.

The existence and importance of photorespiration was first clearly described by Decker and Tió (1959) based on the fully reproducible post-illumination CO_2 burst they observed in many C_3 species. The burst decelerated for several minutes following darkening after a period of photosynthesis. The CO_2 outburst was first described in tobacco leaves by Decker (1955). He attributed the phenomenon to a brief overshoot resulting because the substrate for photorespiration was synthesized only in the light at low concentration and because synthesis of the substrate was promptly cut off when photosynthesis ceased in darkness. Krotkov (1963) later confirmed Decker's observations and extended them by showing high O_2 levels were required to obtain the CO_2 outburst (Tregunna et al., 1966; Forrester et al., 1966a). The remarkable similarities between the properties of photorespiration and the synthesis and metabolism of glycolate first prompted the suggestion that glycolate was the primary substrate of photorespiration (Zelitch, 1964).

The use of an electrical analog of gaseous diffusion into and within a leaf has been helpful in explaining the relation between photosynthesis and photorespiration (Zelitch, 1971; Ludlow and Jarvis, 1971; Canvin and Fock, 1972; Schnarrenberger and Fock, 1976). Such models assist in understanding the difficulties associated with obtaining precise measurements of rates of photorespiration. Photorespiration occurs in photosynthetic systems where the main flux of CO_2 is from the atmosphere surrounding the leaf into the chloroplast, hence all assays of photorespiration will underestimate the rate because a portion of the released CO_2 will be refixed and will not escape to the atmosphere where it can be measured. Some of the more common methods used to demonstrate the process of photorespiration and to measure it will be described, and the advantages and limitations of each method will be indicated.

B. Assays of Photorespiration in Leaves

I. Post-Illumination CO_2 Outburst

DECKER (1955, 1959) used this assay to show that photorespiration increased with increasing irradiance during the photosynthetic period immediately preceding the outburst, it increased with higher leaf temperatures between 14.5 and 33.5° C, and it was unchanged when the leaf was maintained either at the CO_2 compensation point (45 µl/l CO_2) or in normal air. At high CO_2 levels (1200 µl/l) the CO_2 outburst was completely eliminated in bean and sunflower leaves (EGLE and FOCK, 1967).

Maize (corn) leaves do not show a typical post-illumination outburst, an indication that photorespiration is slow in C_4 species (FORRESTER et al., 1966b), and no outburst was observed in maize even in an atmosphere of 100% O_2.

Although the assay is relatively easy and rapid to perform, there are several disadvantages associated with it. The rate of CO_2 release decreases with time, and determination of the maximal rate is subjective (LUDLOW and JARVIS, 1971). In addition, if the stomatal diffusive resistance increases during the assay, photorespiration will be underestimated even more. It is also possible that a part of the substrate remaining from the photosynthetic period will be metabolized to products that do not yield as much CO_2 in subsequent darkness. This assay is probably less sensitive to smaller changes in photorespiration than other assays, since two tobacco varieties that differed in photorespiration when measured by the rate of CO_2 efflux in CO_2-free air were no different when assayed by the post-illumination outburst (HEICHEL, 1973).

II. Inhibition of Net CO_2 Assimilation by Oxygen

Increases in net photosynthesis of 33% to 50% are obtained in C_3 species when the O_2 content in the atmosphere is decreased from 21% to 1%–3% (Table 1; Table 8.5 in ZELITCH, 1971). The synthesis of glycolate and its oxidation to produce photorespiratory CO_2 are both highly dependent on O_2 concentration; hence it is often assumed that the decrease in net photosynthesis with increasing O_2 is a function of the rate of photorespiration. In sharp contrast to the effect in C_3 species, net photosynthesis in maize is unchanged up to O_2 concentrations of 50% (FORRESTER et al., 1966b). It has been suggested that the O_2 inhibition of net photosynthesis overestimates photorespiration because some of the inhibition is caused by an inhibition of the carboxylation reaction (D'AOUST and CANVIN, 1973). It seems more likely that this assay underestimates the magnitude of photorespiration because glycolate metabolism occurs even at low atmospheric levels of O_2 (EICKENBUSCH and BECK, 1973; ROBINSON and GIBBS, 1974). Oxygen is always produced by chloroplasts during photosynthesis and may therefore be enriched at the site of glycolate synthesis so that photorespiration cannot be inhibited completely. This assay gave similar values for photorespiration in wheat leaves compared with measurement of the rate of CO_2 release in CO_2-free air (Table 1; KEYS et al., 1977), and presumably both methods underestimate photorespiration.

Table 1. Minimal rates of photorespiration in leaves of various species[a]

Species	Method of assay	Temperature (°C)	Photorespiration, % of net photosynthesis in normal air	Reference
Soybean[b]	CO_2 release, CO_2-free air	26	46	SAMISH et al., 1972
Soybean	Post-illumination CO_2 burst	25	75	BULLEY and TREGUNNA, 1971
Soybean	CO_2 release, CO_2-free air	30	42	HOFSTRA and HESKETH, 1969
Sunflower	Short-time uptake, $^{14}CO_2$ minus $^{12}CO_2$, 300 ppm CO_2	25	31	BRAVDO and CANVIN, 1974
Sunflower	Same, 600 ppm	25	45	BRAVDO and CANVIN, 1974
Sunflower	Short-time uptake, $^{14}CO_2$ minus $^{12}CO_2$, 21% O_2	25	33	LAWLOR and FOCK, 1975
Sunflower	Same, 1.5% O_2	25	14	LAWLOR and FOCK, 1975
Sunflower	Short-time uptake, $^{14}CO_2$ minus $^{12}CO_2$	25	27	LUDWIG and CANVIN, 1971
Sunflower	Short-time uptake, $^{14}CO_2$, minus $^{12}CO_2$, 421 ppm CO_2	25	27	FOCK and PRZYBYLLA, 1976
Sunflower	Same, 190 ppm CO_2	25	57	FOCK and PRZYBYLLA, 1976
Sugar beet	CO_2 release, CO_2-free air	25	34	TERRY and ULRICH, 1973a
Sugar beet	CO_2 release, CO_2-free air	25	47	TERRY and ULRICH, 1973b
Sugar beet	CO_2 release, CO_2-free air	25	43	TERRY and ULRICH, 1974
Tobacco	CO_2 release, CO_2-free air	25	55	KISAKI, 1973
Tobacco	Short-time uptake, $^{14}CO_2$ minus $^{12}CO_2$	25	47	FOCK and PRZYBYLLA, 1976
Tobacco	Extrapolation of net photosynthesis to "zero" CO_2	25	25	DECKER, 1957
Tobacco	Post-illumination CO_2 burst	25.5	45	DECKER, 1959
Tobacco	Same	33.5	66	DECKER, 1959
Wheat	Net CO_2 uptake 21% O_2 vs 2% O_2	28	53	KEYS et al., 1977
Wheat	CO_2 release, CO_2-free air	28	69	KEYS et al., 1977
Wheat	Short-time uptake, $^{14}CO_2$ minus $^{12}CO_2$	28	17	KEYS et al., 1977
Alfalfa	Calculated from CO_2 compensation point	25	36	HODGKINSON, 1974
Potato	Net CO_2 uptake, 21% O_2 vs 2.5% O_2	25	50	KU et al., 1977
Tall Fescue	Net CO_2 uptake, 21% O_2 vs 0% O_2	27	47, 36[c]	NELSON et al., 1975
Maize	Output of $^{13}CO_2$	28–34	5.7	VOLK and JACKSON, 1972
Maize	CO_2 release, air passed through leaf	30	0	TROUGHTON, 1971
Maize	CO_2 release, CO_2-free air	35	0	MOSS, 1966
Maize	CO_2 release, CO_2-free air	35	0	HOFSTRA and HESKETH, 1969

[a] These values are minimal and underestimates because photorespiration is assayed under conditions that favor rapid net photosynthesis and the main flux of CO_2 is into the leaf while the measurements are made of necessity outside the leaf. Dark respiration may contribute somewhat to the photorespiration measured. Descriptions of these assays and their limitations are discussed by ZELITCH (1971).

[b] Recalculated by the authors considering internal diffusive resistances; the results are the mean values of 20 varieties.

[c] Values obtained with two different varieties of tall fescue.

III. CO_2 and $^{14}CO_2$ Efflux in CO_2-Free Air

When a leaf is placed in light in a rapid stream of CO_2-free air, the rate of CO_2 released depends upon the rate of photorespiration, the diffusive resistance to CO_2 fixation by the chloroplasts, and the stomatal diffusive resistance (BRAVDO, 1968). This method of assay therefore reveals only a portion of the photorespiration even when stomata are wide open. As indicated previously, an atmosphere of CO_2-free air probably does not change photorespiration compared with rates in normal air, since the post-illumination CO_2 burst was not detectably altered. The rate of glycolate synthesis by tobacco leaf disks was also similar in "zero" CO_2 and normal air (ZELITCH and WALKER, 1964). Thus the use of CO_2-free air does not seem to introduce complications and merely facilitates the efflux of a higher proportion of the photorespiratory CO_2. This assay has been applied to a number of C_3 species, and the photorespiration rate was often found to be about 50% of net photosynthesis (Table 1). Photorespiration was about twice as great in 100% O_2 as in 21% O_2 in sunflower leaves with this method of measurement (HEW and KROTKOV, 1968).

This assay is a convenient one, but the rate of CO_2 efflux is determined in an open system with rapid flow rates and at concentrations of CO_2 close to zero. These conditions limit the sensitivity of the measurement with infrared CO_2 gas analyzers and hence the accuracy is limited. To increase the sensitivity a modification was introduced whereby leaf disks were first allowed to assimilate $^{14}CO_2$ in a closed system, the ^{14}C-labeled products were permitted to recycle at the CO_2 compensation concentration for a period of time, and then the $^{14}CO_2$ released in a rapid stream of CO_2-free air was collected and measured (GOLDSWORTHY, 1966; ZELITCH, 1968). The results were usually expressed as the ratio of the $^{14}CO_2$ released in the light to that in darkness during 30-min periods when the rates were constant. Photorespiration was three to five times faster than dark respiration in leaves of C_3 species. The specific radioactivity of $^{14}CO_2$ obtained in the light was not greatly different from that in darkness (GOLDSWORTHY, 1966).

The ^{14}C-assay was used to show that having open stomata was essential to detect maximal rates of photorespiration, that a large increase in photorespiration occurred at higher temperature, that photorespiration was strongly inhibited in 3% O_2 compared with 21% O_2, and that photorespiration was blocked by inhibitors of photosynthetic electron transport such as CMU, while dark respiration was unaffected (ZELITCH, 1968). Maize showed barely detectable photorespiration with this assay, and it was possible to observe differences in photorespiration among plants of several tobacco varieties (ZELITCH and DAY, 1968, 1973).

The ^{14}C-assay is usually carried out under conditions where the leaf disks completely fix the $^{14}CO_2$ supplied within 15 min in a closed system, and a further 30 min is allowed to elapse before CO_2-free air is passed over the disks at zero time. If the collection of the released $^{14}CO_2$ is started after the first 15 min, the ratio of the $^{14}CO_2$ released in the light/dark is about twice as great as the ratio observed after a wait of 30 or the usual 45 min. The ratio is constant, however, after 30 min incubation, and the release of $^{14}CO_2$ is linear with time. This suggests that after about 15 min of recycling within the leaf disks the specific radioactivity of the $^{14}CO_2$ released into the CO_2-free air is constant. In sunflower leaves the rate of $^{14}CO_2$ evolution also became constant, as did the specific radioactivity after about 10 min of supplying $^{14}CO_2$ in the light (D'AOUST and CANVIN, 1972).

The ^{14}C-assay does not measure photorespiration in absolute units because the specific radioactivity of the $^{14}CO_2$ is usually not determined, but it measures photorespiration relative to dark respiration. The assay has been criticized because "total CO_2 production" cannot be determined (CANVIN and FOCK, 1972). For reasons discussed in Sect. A, "total" CO_2 production in photorespiration cannot be measured by any method. Hence in spite of its limitations the ^{14}C-assay has been found useful in studies in this laboratory on the genetic variation of photorespiration and the biochemical control of this process (Sect. E II–EIV).

IV. Short-Time Uptake of $^{14}CO_2$ and $^{12}CO_2$

Under steady-state conditions of photosynthesis, if a leaf is suddenly supplied with $^{14}CO_2$, the initial rate of uptake of $^{14}CO_2$ represents the gross photosynthesis and it will be greater than the previous $^{12}CO_2$ uptake which measures the net photosynthesis. The difference between the $^{14}CO_2$ and $^{12}CO_2$ uptake should be equal to the CO_2 evolved by photorespiration. (Some workers still use the awkward expression "true" photosynthesis for gross photosynthesis and "apparent" photosynthesis for net photosynthesis in spite of the fact that the measurements are neither "true" nor "apparent".) This assay would seem to be the most reliable one on theoretical grounds, but recently fixed $^{14}CO_2$ is respired and recycled so rapidly within the leaf cells that $^{14}CO_2$ released can be detected *outside* the leaf within 15 to 45 s (LUDWIG and KROTKOV, 1967). Therefore considerable internal CO_2 release and refixation must occur in periods less than 15 s within the leaf. This method then also underestimates photorespiration because this recycling will cause gross photosynthesis to be underestimated even in the short times used to make the necessary measurements.

The method as initially described required 30 s for the simultaneous measurement of $^{14}CO_2$ and $^{12}CO_2$ uptake (LUDWIG and CANVIN, 1971a). Later the apparatus was modified and the time was decreased to 20 s (CANVIN and FOCK, 1972), and finally to 15 s (BRAVDO and CANVIN, 1974). The specific radioactivity of the $^{14}CO_2$ available to the leaf is calculated from the average of the specific radioactivity of the gas stream entering the leaf chamber and the specific radioactivity of the exiting gas stream (CANVIN and FOCK, 1972). Large errors may be introduced by this calculation especially with high CO_2 concentrations and low specific radioactivity. This may explain why it appears by this method of assay that photorespiration *increases* at higher CO_2 concentrations (BRAVDO and CANVIN, 1974; FOCK and PRZYBYLLA, 1976). This result is inconsistent with a vast array of data showing that photorespiration decreases at high CO_2 levels. Also by this method photorespiration was unaffected between 15° and 35°C in sunflower and tobacco leaves (FOCK and PRZYBYLLA, 1976), and this is contrary to all other data in the literature. Some examples of estimates of photorespiration obtained by this assay are shown in Table 1. This assay has been used with exposure times of several minutes (FRASER and BIDWELL, 1974), and photorespiration would be even more greatly underestimated under such conditions (LUDWIG and CANVIN, 1971b).

V. The Magnitude of Photorespiration in Leaves

Various assays show that at about 25 °C photorespiration in C_3 species occurs at rates at least 50% of net CO_2 assimilation (Table 1). At higher temperatures photorespiration will account for a still higher proportion of net CO_2 uptake (ZELITCH, 1971). These values of photorespiration obtained from the literature contradict the hypothesis that photorespiration results from a stoichiometric relation between glycolate metabolism and the photosynthetic carbon reduction cycle such that photorespiration occurs in a fixed ratio of 15% to 25% of net CO_2 assimilation (LORIMER and ANDREWS, 1973; LAING et al., 1974). Laboratory experiments have also shown that net photosynthesis in tobacco leaf disks can be increased at least 50% by slowing photorespiration. This can be done by blocking glycolate oxidation with an α-hydroxysulfonate (ZELITCH, 1966), by inhibiting glycolate synthesis with glycidate (ZELITCH, 1974), or by increasing the pool size of some common metabolites (OLIVER and ZELITCH, 1977a, b). These results further confirm that at least 50% of the carbon fixed in photosynthesis may be lost by photorespiration in C_3 species. In C_4 species, however, photorespiration is slow, and respiration is probably no faster than the expected rates of dark respiration (Table 1). In maize leaves the rate of glycolate synthesis is only about 10% of that in tobacco tissue (ZELITCH, 1973, 1974). Net photosynthesis was not increased in maize by inhibition of photorespiration. The slow rate of photorespiration in maize probably results from the decreased synthesis of the photorespiratory substrate, glycolate, because of the high CO_2 concentration in the bundle sheath cells (HATCH, 1971) and possibly by regulation of synthesis brought about by aspartate in these cells (OLIVER and ZELITCH, 1977a) or perhaps by phosphorylated intermediates (KOW et al., 1977).

The existence of C_4 species suggests that photorespiration is a wasteful process that might be regulated with beneficial effects on net photosynthesis. The finding of examples of C_3 species and varieties that have intermediate rates of photorespiration also supports that view. ZELITCH and DAY (1968, 1973) have described tobacco varieties with slower than normal photorespiration and more rapid rates of net photosynthesis than is normal for this species. The grass *Panicum milioides* is a C_3 species in which net CO_2 uptake is less sensitive to O_2 and the rate of CO_2 evolution in CO_2-free air is slower than usual for C_3 species (BROWN and BROWN, 1975; GOLDSTEIN et al., 1976; KECK and OGREN, 1976). Leaves of carpetweed, *Mollugo verticillata*, had morphological characteristics and photorespiration rates intermediate between C_3 and C_4 species (KENNEDY and LAETSCH, 1974).

C. Photorespiration in Algae and Submerged Aquatic Plants

The assay of photorespiration in aqueous systems is subject to even greater underestimation than for leaves of land plants, because the diffusion of CO_2 in water is several orders of magnitude slower than in air, and thus the recycling of photorespired CO_2 will be even greater. In 1920 WARBURG observed an increasing inhibition of photosynthesis in *Chlorella* with increasing O_2, so that presumably these cells have a rapid photorespiration. A faster $^{14}CO_2$ release in the light than in darkness was observed with the ^{14}C-assay with cells of *Chlorella* and *Chlamydomonas* grown in normal air (ZELITCH and DAY, 1968). Lower values for photorespiration were obtained for these algae by CHENG and COLMAN (1974) who observed still lower rates for blue-green algae. A pronounced inhibition of photosynthesis in an atmosphere of air was found compared with N_2, and a post-illumination burst of CO_2 was observed in seven species of marine algae and one freshwater species, indicating they all photorespired (BURRIS, 1977). The synthesis and metabolism of glycolate by algae has been thoroughly reviewed (MERRETT and LORD, 1973). There is recent evidence that some algae possess an additional pathway of glycolate metabolism previously found only in bacteria whereby glyoxylate is converted to CO_2 and tartronic semialdehyde which is reduced to yield glycerate (BADOUR and WAYGOOD, 1971; PAUL and VOLCANI, 1976). Such a pathway has not yet been demonstrated in higher plants.

Recent papers show that when algae are placed in thin layers on nylon cloth or filter paper they no longer possess characteristics associated with a rapid photorespiration. Several algal species did not exhibit a conventional photorespiration when assayed by the difference between $^{14}CO_2$ and $^{12}CO_2$ uptake (BIDWELL, 1977). There was little difference between rates of $^{14}CO_2$ and $^{12}CO_2$ uptake in either 21% or 1% O_2, and net CO_2 uptake was the same in 2%, 21%, or 50% O_2 in four species of algae (LLOYD et al., 1977). The reasons for these unexpected results obtained with algae examined in thin layers is not known.

The ^{14}C-assay of photorespiration has been applied to submerged aquatic angiosperms, and in spite of extensive refixation of CO_2 in these environments the existence of photorespiration was established by the strong dependence of $^{14}CO_2$ release on the dissolved O_2 concentration (HOUGH and WETZEL, 1972; HOUGH, 1974).

D. Photorespiration in Callus, Isolated Plant Cells, and Protoplasts

The ability of undifferentiated callus cells to produce chlorophyll in the light occurs when the callus is grown on sucrose, and it was shown by LAETSCH and STETLER (1965) that such callus carries out photosynthesis. Photoautotrophic growth was demonstrated for tobacco callus and suspension culture (BERGMANN, 1967; CHANDLER et al., 1972; NEUMANN and RAAFAT, 1973). Continuous autotrophic growth in elevated CO_2 concentrations has also been achieved for callus from several varieties of tobacco (BERLYN and ZELITCH, 1975) and for a suspension culture of *Chenopodium rubrum* (HÜESEMANN and BARZ, 1977).

Photorespiration was measured in *Portulaca oleracea* (C_4) and *Streptanthus tortuosus* callus (C_3) grown on sucrose (KENNEDY, 1976). Only the C_3 callus released more $^{14}CO_2$ in the light than in the dark, and 100% O_2 greatly stimulated photorespiration only in the C_3 callus. Although callus from C_4 species lacks the specialized anatomy characteristic of leaves of these species, tissue cultures of *Froelichia gracilis* (LAETSCH and KORTSCHAK, 1972) and *Portulaca oleracea* (KENNEDY et al., 1977) also produced predominantly C_4 acids on short-term exposure

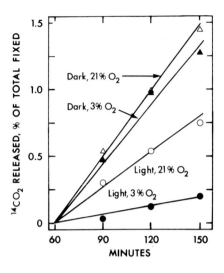

Fig. 1. Direct demonstration of photorespiration in autotrophically grown tobacco callus. Data from BERLYN et al. (1978). About 1.0 g of callus in each flask was permitted to fix $^{14}CO_2$ photosynthetically under aseptic conditions in a closed system for about 22 h. At zero time, a rapid stream of CO_2-free air or 3% O_2 in N_2 was passed through the flasks in light or in darkness and the $^{14}CO_2$ released was collected in traps containing ethanolamine solution for determination of ^{14}C

to $^{14}CO_2$. In carrot callus an inhibition of CO_2 fixation with increasing O_2 levels, and a demonstration that glycine and serine were heavily labeled from $^{14}CO_2$ in O_2 and almost unlabeled in N_2 (HANSON and EDELMAN, 1972) provide an indication that this callus had rapid rates of photorespiration. In this laboratory (BERLYN et al., 1978) photorespiration has been demonstrated as an oxygen-dependent release of $^{14}CO_2$ in a stream of CO_2-free air in autotrophically grown tobacco callus previously allowed to fix $^{14}CO_2$ for 18 to 22 h (Fig. 1). The rate of release was 3.8 times faster in 21% O_2 than in 3% O_2. Dark respiration was hardly affected by the change in O_2 level in this assay. The apparently faster dark respiration undoubtedly results from refixation within the bulky tissue of $^{14}CO_2$ released in the light. We have also found that these cells synthesize glycolate rapidly in the light relative to net photosynthesis. The rate of glycolate synthesis is about three times higher for autotrophically than heterotrophically grown cells. The autotrophic cells metabolize [$1\text{-}^{14}C$] glycolate in the light to yield $^{14}CO_2$, ^{14}C-glycine, and ^{14}C-serine. Thus there is little doubt that photosynthetically active callus of C_3 species has a rapid system of photorespiration.

The isolation of mesophyll cells of soybean leaf has permitted studies on photosynthesis and photorespiration to be carried out in a somewhat simpler system than intact leaf tissue, although these cells have rates of net photosynthesis in air about one-third of intact leaves on a chlorophyll basis. The oxygen sensitivity of CO_2 assimilation and its reversal by high concentrations of CO_2 indicate that rapid rates of photorespiration occur in such systems (SERVAITES and OGREN, 1977a). Further studies using biochemical inhibitors of the glycolate pathway on isolated mesophyll cells showed that when the further metabolism of glycine was blocked, about 50% as much ^{14}C accumulated in glycine as in glycolate when the oxidation of glycolate was blocked (SERVAITES and OGREN, 1977b). This suggests that about 50% of the glycolate was metabolized to $^{14}CO_2$ before it was converted to glycine, presumably by the decarboxylation of glyoxylate.

Because they lack cell walls, protoplasts may provide an even simpler system for the study of photorespiration than isolated cells. Photosynthesis in protoplasts isolated from spinach leaves was stable for up to 9 h (NISHIMURA and AKAZAWA, 1975a) and for 20 h in protoplasts from various grasses (RATHNAM and EDWARDS, 1976; see Chap. II.2, by JENSEN). Photosynthesis by protoplasts was sensitive to O_2 in C_3 species (NISHIMURA and AKAZAWA, 1975b; HUBER and EDWARDS, 1975), indicating that they also have an active photorespiratory system.

E. The Control of Photorespiration

I. The Energetics and Possible Origins of Photorespiration

Since at least 50% of the net CO_2 assimilated during photosynthesis in C_3 plants is lost by photorespiration, this results in decreased CO_2 uptake and yield. Photorespiratory CO_2 almost certainly arises from the glycolate pathway of carbohydrate synthesis. As usually depicted this pathway converts four molecules of glycolate to yield one of glucose and two of CO_2, the CO_2 all released during the conversion of glycine to serine (ZELITCH, 1971). It was shown that formation of photorespiratory CO_2 is much greater than 25% of the glycolate synthesized (Table 1), hence CO_2 must also be released elsewhere in the pathway. Most of the additional CO_2 is probably produced from the oxidation of glyoxylate (ZELITCH, 1972; ELSTNER and HEUPEL, 1973; HALLIWELL and BUTT, 1974; GRODZINSKI and BUTT, 1976; HALLIWELL, 1976).

BIRD et al. (1972) and MOORE et al. (1977) have shown that ATP may be generated during the conversion of glycine to serine and CO_2 by mitochondrial preparations, and implied that the formation of ATP might provide some benefit to the plant. However, additional ATP's would be required to convert serine back to carbohydrate, and still more ATP's would be necessary to refix the CO_2 lost by photorespiration. GOLDSWORTHY (1975) pointed out that this small gain in ATP during photorespiration was analogous to a merchant buying back at one-half the price merchandise previously stolen from him by a thief.

Several "futile" biochemical cycles have been described in which cyclic transformations occur and the balance consists of the hydrolysis of ATP to ADP and Pi with an apparently deliberate waste of ATP (HERS, 1976). Perhaps photorespiration might be considered an example of a "futile" cycle. However, in those "futile" cycles that are understood, the transformations produce metabolites with powerful regulatory properties that effectively control metabolic flow. No such regulatory function is known for any metabolite of the glycolate pathway, and the essential products of the pathway such as glycine and serine can be synthesized by alternative routes without invoking photorespiration. These amino acids may be produced in C_4 species by an alternative pathway, and they may be synthesized at least in some C_3 species and under somes conditions (low irradiance or high CO_2, for example) from glycerate rather than glycolate (PLATT et al., 1977).

It has often been suggested that a function of photorespiration is to protect the chloroplast against photooxidative destruction in an environment of high irradiance and low CO_2 concentration (SCHNARRENBERGER and FOCK, 1976). This implies that carbon compounds are oxidized to CO_2 during photorespiration in order to protect the photochemical apparatus. The hypothesis fails to consider that mesophyll chloroplasts of C_4 species are surely protected against photoxidative damage without having a photorespiratory mechanism. It is well established that carotenoids fulfill such a protective function, and still other mechanisms are available to plant cells to minimize photoxidative damage.

A common product of reactions involving oxygen uptake is the superoxide anion, O_2^- (FRIDOVICH, 1977). This highly toxic radical is converted to H_2O_2 and O_2 by the enzyme superoxide dismutase which is present in photosynthetic organisms (ASADA et al., 1977). The superoxide anion may be the oxidant in the light-dependent formation of glycolate from the two-carbon fragment generated in the transketolase reaction (ASAMI and AKAZAWA, 1977). Hydrogen peroxide and superoxide anion may also serve as the oxidants in the production of photorespiratory CO_2 from glyoxylate (ZELITCH, 1972; ELSTNER and HEUPEL, 1973). CLAYTON (1960) observed that the catalase content of photosynthetic bacteria increased 20-fold when they were transferred from anaerobic conditions to (toxic) air. Thus an overproduction of enzyme activities that destroy the superoxide radical and hydrogen peroxide may help to decrease the wasteful process of photorespiration.

Laboratory experiments described below (Sect. E.III and E.IV) show that slowing various steps in the glycolate pathway results in a decreased rate of glycolate synthesis and increased CO_2 assimilation, and that these effects can be sustained for hours. These results suggest that alternative mechanisms are available to plant cells in whole tissues that enable toxic oxidants to be degraded without the mandatory wasteful respiration of products of photosynthesis. The effectiveness of conserving susceptible carbon compounds may therefore determine the ratio of the rate of photorespiration to the rate of net photosynthesis.

II. Biochemical Inhibition of Glycolate Oxidation

In order to be metabolized, glycolate must first be oxidized to glyoxylate by the flavoprotein glycolate oxidase. Long-term blocking of the glycolate oxidase reaction would not therefore seem a practical solution for the regulation of photorespiration, since glycolate would simply accumulate and ultimately reach toxic concentrations. Biochemical inhibitors of glycolate oxidase are useful in showing the importance of the glycolate pathway in photorespiration.

α-Hydroxysulfonates, aldehyde-bisulfite addition compounds, are effective inhibitors of glycolate oxidase (ZELITCH, 1957). When α-hydroxy-2-pyridinemethanesulfonic acid was supplied to tobacco leaf disks in the light, the enzymatic oxidation was blocked and glycolate accumulated in the tissue at initial rates sufficiently rapid to account for photorespiration in tobacco and sunflower (ZELITCH, 1973). This sulfonate inhibited photorespiration, but not dark respiration, in the ^{14}C-assay (ZELITCH, 1968). Under suitable conditions of temperature and short times of exposure, large increases in photosynthetic CO_2 uptake were obtained (ZELITCH, 1966). The α-hydroxysulfonates usually became toxic to leaf tissues and inhibited photosynthesis after about 15 min of exposure, and C_4 species were even more sensitive to toxic effects of the inhibitor (LÜTTGE et al., 1972). Results of an investigation of a number of compounds for their possible effectiveness as inhibitors of glycolate oxidase has been described (CORBETT and WRIGHT, 1971); none tested were as effective as the α-hydroxysulfonates.

JEWESS et al. (1975) described a new class of irreversible inhibitor of glycolate oxidase. These are acetylenic substrate analogs of glycolate. One of these, 2-hydroxy-3-butynoate, was supplied to pea leaf disks in the light and the accumulation of glycolate was measured. The action of this inhibitor was much slower than that observed with sulfonates; maximal rates of glycolate accumulation obtained were 1.8 µmol compared with rates of 70 µmol per g fr. wt. per h often observed with a sulfonate in C_3 tissues. It is therefore not surprising that they report no detectable effect on the respiration of leaf disks under the same conditions that glycolate accumulation was measured.

III. Biochemical Inhibition of Glycolate Synthesis

Inhibitors of glycolate synthesis were sought that would specifically block photorespiration in C_3 species and increase CO_2 assimilation in leaf disks (ZELITCH, 1974). The concentration of glycolate in leaf tissue is low and its turnover is very rapid. An assay was developed in which tobacco leaf disks were floated on solutions of the inhibitor to be tested for 60 min in the light. The inhibitor solution was then removed and replaced with a solution of a sulfonate for 3 min, and glycolate accumulation was measured to determine the rate of synthesis.

With such an assay 20 mM glycidate (2,3-expoxypropionate) was found to inhibit glycolate synthesis and photorespiration about 50% and to increase photosynthetic $^{14}CO_2$ uptake about 50% (ZELITCH, 1974). Glycidate also inhibited glycolate synthesis in maize leaf disks, but net photosynthesis was not affected, probably because the rate of glycolate formation and photorespiration are already so low in maize that altering glycolate metabolism had little effect on net CO_2 uptake.

The products of $^{14}CO_2$ assimilation in the presence and absence of glycidate were examined, and as expected if glycolate synthesis were inhibited, the pool sizes of glycine and serine were decreased (Table 2). It was also found that the pool sizes of aspartate and glutamate were about two-fold greater in leaf disks treated with glycidate. Although the significance of the latter observation was not fully appreciated at the time, as discussed later (Sect. E.IV), these changes in pool size may largely account for the regulatory effect of glycidate.

[1-^{14}C]Glycidate was synthesized and its binding to proteins in leaf disks was studied under conditions where glycolate synthesis in the tissue was inhibited at least 50% (ZELITCH, 1978). Glycidate did not combine with or inhibit ribulose bisphosphate carboxylase or affect the inhibition of this enzyme by oxygen. This might have been expected if glycidate directly inhibited the ribulose bisphosphate oxygenase reaction, a possible source of glycolate synthesis in vivo (BOWES et al., 1971). Glycidate also had no effect on the activities of glycolate oxidase, P-glycolate phosphatase, and NADH-glyoxylate reductase, and slightly inhibited NADPH-glyoxylate reductase (ZELITCH, 1978). A strong inhibition of glutamate:glyoxylate aminotransferase activity by glycidate was found in particulate preparations of tobacco leaf and callus (LAWYER and ZELITCH, 1978). When $^{14}CO_2$ was supplied to leaf disks treated with glycidate under conditions where net photosynthesis

Table 2. Effect of glycidate on distribution of ^{14}C in tobacco leaf disks supplied $^{14}CO_2$ in the light. Data from ZELITCH, 1974. After 60 min on water at 2000 ft-c and 28 °C, leaf disks were either floated on water or 20 mM glycidate for 60 min before measurement of $^{14}CO_2$ uptake was made for 4.0 min

Fraction or compound	Leaf disks on water	Leaf disks on glycidate
	% of total ^{14}C fixed	
Neutral compounds	6.0	11.5
Aspartate	3.1	5.4
Glutamate	0.18	0.68
Phosphoglycerate	17.5	17.6
Phosphoglycolate	2.3	2.9
Fructose bisphosphate, phosphoenolpyruvate, ribulose bisphosphate fraction	2.0	3.0
Glycolate	0.28	0.34
Glycerate	3.1	5.2
Glycine	15.3	7.6
Serine	22.8	6.8
Photosynthetic $^{14}CO_2$ fixation, μmol g fr. wt.$^{-1}$ h^{-1}	57.1	81.3 (+42%)

increased, the pool size of glyoxylate increased, as well as that of aspartate and glutamate (Table 2).

IV. The Metabolic Regulation of Photorespiration

The metabolic regulation of photorespiration by feedback or other type of inhibition of glycolate synthesis (or any other step of the glycolate pathway) had not been demonstrated until recently. To assist in laying a biochemical foundation for genetically altering photorespiration (ZELITCH, 1975; BERLYN and ZELITCH, 1975), our laboratory investigated whether floating leaf disks on solutions of common metabolites would inhibit glycolate biosynthesis. We used the leaf disk assay with which the inhibitory effect of glycidate was found earlier. These studies revealed that about 30 mM solutions of L-glutamate, L-aspartate, phosphoenolpyruvate, or glyoxylate effectively inhibited glycolate synthesis in the tobacco leaf disk assay (OLIVER and ZELITCH, 1977a, 1977b).

Floating leaf disks on 30 mM glutamate increased the total tissue concentration of glutamate from approximately 2 mM to 5 mM. Glycolate synthesis was inhibited about 35%, photorespiration was inhibited about 60%, and there was no substantial effect on the dark respiration. Photosynthetic $^{14}CO_2$ fixation by the leaf disks was increased about 25% by glutamate treatment. This regulation is probably brought about by a metabolite of glutamate rather than by glutamate itself (OLIVER and ZELITCH, 1977a).

Fig. 2. Effect of glyoxylate concentration on net CO_2 fixation in tobacco leaf disks. From OLIVER and ZELITCH (1977b). Disks in large Warburg flasks were kept for 1 h in the light on water or the solution shown at 30°C, and a continuous stream of $^{14}CO_2$ was passed through the flasks for 5 min before the disks were killed and the amount of ^{14}C fixed was determined. The photosynthetic rates for disks on water varied in different experiments, shown by the various symbols, from 38.0 to 60.3 μmol of $^{14}CO_2$ fixed per g fresh weight per hour

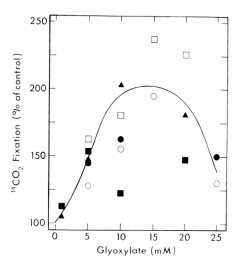

Floating leaf disks on 20 mM glyoxylate solution inhibited glycolate synthesis about 50% and inhibited photorespiration similarly. Net photosynthesis was increased two-fold under optimal conditions (Fig. 2). Floating the disks for 1 h on 20 mM glyoxylate increased the total glyoxylate concentration in the disks about 40%, from 0.6 mM to 0.9 mM. Thus modest alterations of the pool sizes of several common metabolites can increase photosynthesis by inhibiting glycolate synthesis and photorespiration probably by a feedback type mechanism. The effect of glycidate on inhibiting glycolate synthesis (Sect. E.III) is thus probably an indirect one brought about by increases in the pool sizes of glutamate, aspartate, and glyoxylate. These results suggest that chemical or genetic regulation of some commonly occurring metabolites can produce plants with higher rates of net photosynthesis. It is conceivable that species or varieties of C_3 species with intermediate rates of photorespiration (Sect. B.V) are metabolically regulated in a similar manner, and this type of regulatory mechanism may also help to explain the low rates of glycolate synthesis and photorespiration in C_4 species (OLIVER and ZELITCH, 1977a).

Acknowledgment. Helpful discussions with Mary B. Berlyn and David J. Oliver are gratefully acknowledged.

References

Asada, K., Kanematsu, S., Uchida, K.: Arch. Biochem. Biophys. *179*, 243–256 (1977)
Asami, S., Akazawa, T.: Biochemistry *16*, 2202–2207 (1977)
Badour, S.S., Waygood, E.R.: Biochim. Biophys. Acta *242*, 493–499 (1971)
Bergmann, L.: Planta (Berl.) *74*, 243–249 (1967)
Berlyn, M.B., Zelitch, I.: Plant Physiol. *56*, 752–756 (1975)
Berlyn, M.B., Zelitch, I., Beaudette, P.D.: Plant Physiol. *61*, 606–610 (1978)
Bidwell, R.G.S.: Can. J. Bot. *55*, 809–818 (1977)
Bird, I.F., Cornelius, M.J., Keys, A.J., Whittingham, C.P.: Biochem. J. *128*, 191–192 (1972)

Bowes, G., Ogren, W.L., Hageman, R.H.: Biochem. Biophys. Res. Commun. *45*, 716–722 (1971)

Bravdo, B.-A.: Plant Physiol. *43*, 479–483 (1968)

Bravdo, B., Canvin, D.T.: In: Proc. 3rd. Int. Congr. Photosynthesis. pp. 1277–1283. Amsterdam: Elsevier 1974

Brown, R.H., Brown, W.V.: Crop Sci. *15*, 681–685 (1975)

Bulley, N.R., Tregunna, E.B.: Can. J. Bot. *49*, 1277–1284 (1971)

Burris, J.E.: Mar. Biol. *39*, 371–379 (1977)

Canvin, D.T., Fock, H.: In: Methods in enzymology. Part B. San Pietro, A. (ed.), Vol. XXIV, pp. 246–260. New York: Academic Press 1972

Chandler, M.T., deMarsac, N.T., deKouchkovsky, Y.: Can. J. Bot. *50*, 2265–2270 (1972)

Cheng, K.H., Colman, B.: Planta (Berl.) *115*, 207–212 (1974)

Clayton, R.K.: Biochim. Biophys. Acta *37*, 503–512 (1960)

Corbett, J.R., Wright, B.J.: Phytochem. *10*, 2015–2024 (1971)

D'Aoust, A.L., Canvin, D.T.: Photosynthetica *6*, 150–157 (1972)

D'Aoust, A.L., Canvin, D.T.: Can. J. Bot. *51*, 457–464 (1973)

Decker, J.P.: Plant Physiol. *30*, 82–84 (1955)

Decker, J.P.: J. Sol. Energy Sci. Eng. *1*, 30–33 (1957)

Decker, J.P.: Plant Physiol. *34*, 100–102 (1959)

Decker, J.P., Tió, M.A.: J. Agric. Univ. Puerto Rico *43*, 50–55 (1959)

Egle, K., Fock, H.: In: The biochemistry of chloroplasts. Goodwin, T.W. (ed.), Vol. II, pp. 79–87. New York: Academic Press 1967

Eickenbusch, J.D., Beck, E.: FEBS Lett. *31*, 225–228 (1973)

Elstner, E.F., Heupel, A.: Biochim. Biophys. Acta *325*, 182–188 (1973)

Fock, H., Przybylla, K.-R.: Ber. Dtsch. Bot. Ges. *89*, 643–650 (1976)

Forrester, M.L., Krotkov, G., Nelson, C.D.: Plant Physiol. *41*, 422–427 (1966a)

Forrester, M.L., Krotkov, G., Nelson, C.D.: Plant Physiol. *41*, 428–431 (1966b)

Fraser, D.E., Bidwell, R.G.S.: Can. J. Bot. *52*, 2561–2570 (1974)

Fridovich, I.: Bioscience *27*, 426–466 (1977)

Goldstein, L.D., Ray, T.B., Kestler, D.P., Mayne, B.C., Brown, R.H., Black, C.C.: Plant Sci. Lett. *6*, 85–90 (1976)

Goldsworthy, A.: Phytochemistry *5*, 1013–1019 (1966)

Goldsworthy, A.: In: Physiological aspects of dryland farming. Gupta, V.S. (eds.), pp. 328–349. New Delhi: Oxford and IBH 1975

Grodzinski, B., Butt, V.S.: Planta (Berl.) *128*, 225–231 (1976)

Halliwell, B.: FEBS Lett. *64*, 266–270 (1976)

Halliwell, B., Butt, V.S.: Biochem. J. *138*, 217–224 (1974)

Hanson, A.D., Edelman, J.: Planta (Berl.) *102*, 11–25 (1972)

Hatch, M.D.: Biochem. J. *125*, 425–432 (1971)

Heichel, G.H.: Plant Physiol. *51S*, 42 (1973)

Hers, H.G.: Biochem. Soc. Trans. *4*, 985–988 (1976)

Hew, C.-S., Krotkov, G.: Plant Physiol. *43*, 464–466 (1968)

Hodgkinson, K.C.: Aust. J. Plant Physiol. *1*, 561–578 (1974)

Hofstra, G., Hesketh, J.D.: Planta (Berl.) *85*, 228–237 (1969)

Hough, R.A.: Limnol. Oceanogr. *19*, 912–927 (1974)

Hough, R.A., Wetzel, R.G.: Plant Physiol. *49*, 987–990 (1972)

Huber, S., Edwards, G.: Biochem. Biophys. Res. Commun. *67*, 28–34 (1975)

Hüesemann, W., Barz, W.: Physiol. Plant *40*, 77–81 (1977)

Jewess, P.J., Kerr, M.W., Whitaker, D.P.: FEBS Lett. *53*, 292–296 (1975)

Keck, R.W., Ogren, W.L.: Plant Physiol. *58*, 552–555 (1976)

Kennedy, R.A.: Plant Physiol. *58*, 573–575 (1976)

Kennedy, R.A., Laetsch, W.M.: Science *184*, 1087–1089 (1974)

Kennedy, R.A., Barnes, J.E., Laetsch, W.M.: Plant Physiol. *59*, 600–603 (1977)

Keys, A.J., Sampaio, E.V.S.B., Cornelius, M.J., Bird, I.F. J. Exp. Bot. *28*, 525–533 (1977)

Kisaki, T.: Plant Cell Physiol. *14*, 505–514 (1973)

Kow, Y.W., Robinson, J.M., Gibbs, M.: Plant Physiol. *60*, 492–495 (1977)

Krotkov, G.: In: Photosynthetic mechanisms in green plants. Washington, D.C.: Natl. Acad. pp. 452–454. Sci.-Natl. Res. Council 1963

Ku, S.-B., Edwards, G.E., Tanner, C.B.: Plant Physiol. *59*, 868–872 (1977)
Laetsch, W.M., Kortschak, H.P.: Plant Physiol. *49*, 1021–1023 (1972)
Laetsch, W.M., Stetler, D.A.: Am. J. Bot. *52*, 798–804 (1965)
Laing, W.A., Ogren, W.L., Hageman, R.H.: Plant Physiol. *54*, 678–685 (1974)
Lawlor, D.W., Fock, H.: Planta (Berl.) *126*, 247–258 (1975)
Lawyer, A.L., Zelitch, I.: Plant Physiol. *61*, 242–247 (1978)
Lloyd, N.D.H., Canvin, D.T., Culver, D.A.: Plant Physiol. *59*, 936–940 (1977)
Lorimer, G.H., Andrews, T.J.: Nature (London) *243*, 359–360 (1973)
Ludlow, M.M., Jarvis, P.G.: In: Plant photosynthetic production. Manual of methods. Šesták, Z., Čatský, J., Jarvis, P.G. (eds.), pp. 294–315. The Hague: W. Junk 1971
Ludwig, L.J., Canvin, D.T.: Can. J. Bot. *49*, 1299–1313 (1971a)
Ludwig, L.J., Canvin, D.T.: Plant Physiol. *48*, 712–719 (1971b)
Ludwig, L.J., Krotkov, G.: Plant Physiol. *42S*, 47 (1967)
Lüttge, U., Osmond, C.B., Ball, E., Brinckmann, E., Krinze, G.: Plant Cell Physiol. *13*, 505–514 (1972)
Merrett, M.J., Lord, J.M.: New Phytol. *72*, 751–767 (1973)
Moore, A.L., Jackson, C., Halliwell, B., Dench, J.E., Hall, D.O.: Biochem. Biophys. Res. Commun. *78*, 483–491 (1977)
Moss, D.N.: Crop Sci. *6*, 351–354 (1966)
Nelson, C.J., Asay, K.H., Patton, C.D.: Crop Sci. *15*, 629–633 (1975)
Neumann, K., Raafat, A.: Plant Physiol. *51*, 685–690 (1973)
Nishimura, M., Akazawa, T.: Plant Physiol. *55*, 712–716 (1975a)
Nishimura, M., Akazawa, T.: Plant Physiol. *55*, 718–722 (1975b)
Oliver, D.J., Zelitch, I.: Plant Physiol. *59*, 688–694 (1977a)
Oliver, D.J., Zelitch, I.: Science *196*, 1450–1451 (1977b)
Paul, J.S., Volcani, B.E.: Arch. Microbiol. *110*, 247–252 (1976)
Platt, S.G., Plaut, Z., Bassham, J.A.: Plant Physiol. *60*, 230–234 (1977)
Rabinowitch, E.I.: Photosynthesis and related processes. Vol. 1. New York: Wiley Interscience 1945
Rathnam, C.K.M., Edwards, G.E.: Plant Cell Physiol. *17*, 177–186 (1976)
Robinson, J.M., Gibbs, M.: Plant Physiol. *53*, 790–797 (1974)
Samish, Y.B., Pallas, J.C., Jr., Dornhoff, G.M., Shibles, R.M.: Plant Physiol. *50*, 28–30 (1972)
Schnarrenberger, C., Fock, H.: In: Encyclopedia of plant physiology. New Ser. Stocking, C.R., Heber, U. (eds.), Vol. III, pp. 185–234. Berlin, Heidelberg, New York: Springer 1976
Servaites, J.C., Ogren, W.L.: Plant Physiol. *60*, 693–696 (1977a)
Servaites, J.C., Ogren, W.L.: Plant Physiol. *60*, 461–466 (1977b)
Terry, N., Ulrich, A.: Plant Physiol. *51*, 43–47 (1973a)
Terry, N., Ulrich, A.: Plant Physiol. *51*, 783–786 (1973b)
Terry, N., Ulrich, A.: Plant Physiol. *54*, 379–381 (1974)
Tregunna, E.B., Krotkov, G., Nelson, C.D.: Can. J. Bot. *42*, 989–997 (1966)
Troughton, J.H.: Planta (Berl.) *100*, 87–92 (1971)
Volk, R.J., Jackson, W.A.: Plant Physiol. *49*, 218–223 (1972)
Zelitch, I.: J. Biol. Chem. *224*, 251–260 (1957)
Zelitch, I.: Annu. Rev. Plant Physiol. *15*, 121–142 (1964)
Zelitch, I.: Plant Physiol. *41*, 1623–1631 (1966)
Zelitch, I.: Plant Physiol. *43*, 1829–1837 (1968)
Zelitch, I.: Photosynthesis, photorespiration, and plant productivity. p. 347. New York: Academic Press 1971
Zelitch, I.: Arch. Biochem. Biophys. *150*, 698–707 (1972)
Zelitch, I.: Plant Physiol. *51*, 299–305 (1973)
Zelitch, I.: Arch. Biochem. Biophys. *163*, 367–377 (1974)
Zelitch, I.: Science *188*, 626–633 (1975)
Zelitch, I.: Plant Physiol. *61*, 236–241 (1978)
Zelitch, I., Day, P.R.: Plant Physiol. *43*, 1838–1844 (1968)
Zelitch, I., Day, P.R.: Plant Physiol. *52*, 33–37 (1973)
Zelitch, I., Walker, D.A.: Plant Physiol. *39*, 856–862 (1964)

28. Photorespiration: Comparison Between C₃ and C₄ Plants

D.T. CANVIN

A. Introduction

The "nightmare" of "photorespiration" that RABINOWITCH (1945) mentioned has now become reality, but it is now even more appropriate to say that "the relation between photosynthesis and respiration ... has become even less clear and the data even more controversial" (RABINOWITCH, 1956).

In the last *Encyclopedia of Plant Physiology* chapters by EGLE (1960) and ROSENSTOCK and RIED (1960) were devoted to the problems of measuring photorespiration and to discussing whether the process existed. Since then, great strides have been made in the methods of measurement (LUDLOW and JARVIS, 1971; CANVIN and FOCK, 1972), many characteristics of the process have been elucidated, and several reviews have appeared (FOCK, 1970; GOLDSWORTHY, 1970; WOLF, 1970; JACKSON and VOLK, 1970; MOSS and MUSGRAVE, 1971; TOLBERT, 1971, 1973, 1974; ZELITCH, 1971, 1973b, 1975; BLACK, 1973; BLACK et al., 1973; COOMBS, 1973; CHOLLET and OGREN, 1975; SCHNARRENBERGER and FOCK, 1976). There is little doubt that, in C₃ plants, normal dark respiration is replaced or supplemented by a light-dependent CO₂ evolution from green leaves. This CO₂ evolution or photorespiration is closely linked to photosynthesis and results from the metabolism of compounds in the glycolate pathway (TOLBERT, 1963, 1971; TOLBERT et al., 1969). In C₄ plants the situation is not as clear; manifestations of photorespiration external to the leaf are absent and there is considerable difference of opinion regarding the occurrence of hidden photorespiration in the leaf.

At the present time all the data on photorespiration cannot be reconciled. Vagueness and generalization do not contribute positively either to understanding or to constructive experimentation. Thus, before a comparison of photorespiration in C₃ and C₄ plants can be undertaken it is necessary to establish, as precisely as one can, the meaning and characteristics of photorespiration in C₃ plants. The first part of this article will thus be devoted to photorespiration in C₃ plants, but the majority of the article will deal with an evaluation of the results obtained with C₄ plants.

I have intentionally adopted a restricted treatment of photorespiration because many aspects of it will be dealt with more fully in this series and because I believe a restricted treatment is necessary to lessen some of the confusion and generalizations that are now extant.

B. Terminology and Perception

Photorespiration has been used "to describe all respiratory activity in the light, regardless of the pathways by which CO₂ is released or O₂ consumed", (JACKSON and VOLK, 1970) or as "light-dependent O_2 uptake and/or CO_2 output", (BEEVERS, 1971). Under normal atmospheric conditions photosynthesis occurs at the same time and involves the opposite fluxes of the same gases. Pictures of varying completeness of the complex situation of gas exchange in the leaf are provided by a number of models (LAKE, 1967; BRAVDO, 1968; WAGGONER, 1969; ZELITCH, 1971). Although these models include fairly complete descriptions of all presumed events and sites, their analytical usefulness would seem limited (BRAVDO, 1971) because of the necessity to include many unmeasurable terms. For the purposes of this article a picture based on SCHNARRENBERGER and FOCK, (1976) with the accompanying terminology will be sufficient (Fig. 1).

Apparent photosynthesis (APS) is the net exchange of CO_2 between the leaf and the atmosphere. True photosynthesis (TPS) is the total CO_2 uptake at the fixation site of photosynthesis and photorespiration (PR) is the difference between these two processes.

$$TPS - APS = PR \qquad (1)$$

As several authors (LUDWIG and CANVIN, 1971a, JACKSON and VOLK, 1970; ZELITCH, 1971; SCHNARRENBERGER and FOCK, 1976) have pointed out, TPS cannot be accurately measured because it is impossible to measure that portion of PR that is reassimilated (RA) in photosynthesis. Thus, only a portion of true photosynthesis (TPS') can be measured. For the same reason it is not possible to measure PR but only PR', or that portion of PR that escapes from the leaf. As SCHNARRENBERGER and FOCK (1976) have shown, because the term RA cancels out, the measured values of TPS' and PR' can be inserted in Eq. (1) and it is these values that will be referred to in this article.

$$TPS' - APS = PR' \qquad (2)$$

The degree of reassimilation (RA) of PR cannot be measured but indirect arguments suggest that it might be equivalent to as much as 30% to 40% of PR' (D'AOUST and CANVIN, 1974).

Photorespiration (PR') in this context is entirely consistent with the earlier definitions and it could consist of CO_2 from conventional dark respiration (R_D) or CO_2 from an additional process of light respiration (R_L) that is now attributed to the metabolism of compounds in the glycolate pathway (TOLBERT, 1963, 1973; WANG and WAYGOOD, 1962).

Dark respiration rates are largely insensitive to changes in the oxygen concentration above 2% O_2 (FORRESTER et al., 1966a; D'AOUST and CANVIN, 1974) and the $^{14}CO_2$ that is evolved from the leaf in the dark has a specific activity which is much lower than that of the $^{14}CO_2$ supplied in an immediately preceding photosynthetic period (Fig. 2; D'AOUST and CANVIN, 1972). Measured photorespiration from C₃ plant leaves, however, is sensitive to the oxygen concentration and is virtually eliminated in an atmosphere of 2% O_2 at moderate temperatures (D'AOUST and CANVIN, 1973, 1974; CORNIC, 1974). It also becomes rapidly labelled with ^{14}C when $^{14}CO_2$ is supplied to the leaf and, in fact, reaches a specific activity equal to that of the $^{14}CO_2$ supplied (LUDWIG and CANVIN, 1971b; D'AOUST and CANVIN, 1972; CANVIN et al., 1976). Such characteristics suggest that photorespira-

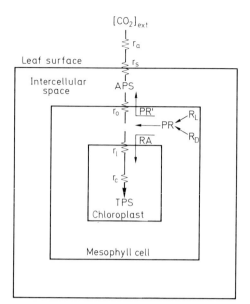

Fig. 1. Diagrammatic presentation of leaf gas exchange. *APS*, apparent photosynthesis; *TPS*, true photosynthesis; *PR*, photorespiration; R_D, dark respiration; R_L, light respiration; *PR'*, portion of PR that is released to intercellular spaces; *RA*, portion of PR that is reassimilated; r_a and r_s, boundary layer and stomatal resistances; the mesophyll or residual resistance r_m is the sum of r_o, r_i and r_c. The latter three resistances cannot be measured individually. (Adapted from SCHNARRENBERGER and FOCK, 1976)

tion (PR') is almost completely derived from light respiration (R_L) and that dark respiration (R_D) contributes, at most, only a small portion of the CO_2 (CANVIN et al., 1976). These same characteristics lead one to suggest that photorespiration (PR') should be used in a more limited context to refer *to the light-dependent CO_2 evolution from green leaves that is sensitive to the oxygen concentration and that originates largely from the metabolism of compounds through the glycolate pathway.* Only by including these properties of photorespiration in a definition is it possible to differentiate photorespiratory CO_2 evolution from other sources of CO_2 evolution, including dark respiratory processes, that may go on in the light under certain conditions.

Oxygen uptake that occurs in the light (JACKSON and VOLK, 1970) will be discussed in this article but its relationship to CO_2 evolution or photorespiration, as defined above, is not completely clear and it is by no means established that such oxygen uptake is a better means of measuring photorespiration than CO_2 evolution.

C. Measurement of Photorespiration

An accurate measurement of photorespiration cannot be obtained because of reassimilation. Within that shortcoming, however, several methods of measuring photorespiration have been described (ZELITCH, 1971; LUDLOW and JARVIS, 1971; CANVIN and FOCK, 1972). The advantages and disadvantages of these methods are discussed elsewhere (see Chap. II.27, this vol.).

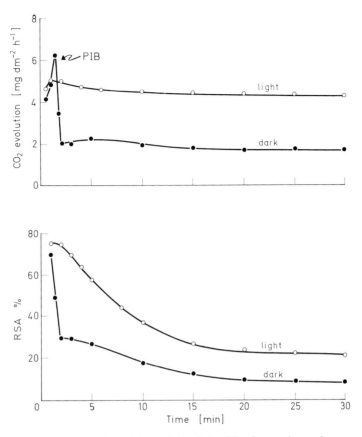

Fig. 2. CO_2 evolution and the relative specific activity (RSA) of the CO_2 from tobacco leaves flushed with CO_2-free air in the light or dark. Tobacco (Connecticut Havana 38) leaves were allowed to photosynthesize for 30 min in $^{14}CO_2$ with a total CO_2 concentration of 159 µl l^{-1}, 21% O_2, a temperature of 25° C and quantum flux of 270 µE m^{-2} s^{-1}. APS was 11 mg CO_2 dm^{-2} h^{-1}. After 30 min the leaves were flushed with CO_2-free air in the light or dark and CO_2 and $^{14}CO_2$ evolution measured. Relative specific activity (*RSA*) is the specific activity of the $^{14}CO_2$ evolved expressed as a percentage of the specific activity of the $^{14}CO_2$ supplied during the photosynthetic period. (Data from D'AOUST, 1970)

D. Characteristics of Photorespiration in C_3 Plants

At the moment, except for approximately 500 species of C_4 plants (DOWNTON, 1975; KRENZER et al., 1975) and 250 species of CAM plants (BLACK and WILLIAMS, 1976) that have been identified, the remaining 300,000 species of plants (BOLD, 1970) are viewed as C_3 plants. Of course, additional C_4 and CAM plants will be discovered, but a few additions will not greatly change the relationship between the numbers of species.

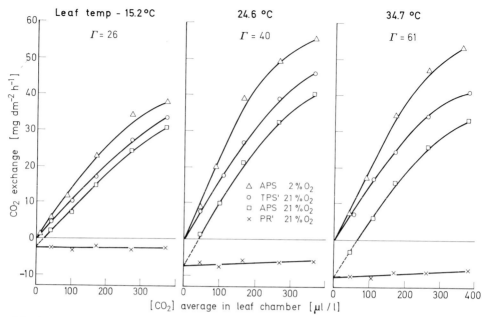

Fig. 3. The effect of CO_2 concentration, O_2 concentration, and temperature on TPS′, APS, PR′ and the compensation point in sunflower leaves. Quantum flux was $800 \, \mu E \, m^{-2} \, s^{-1}$ and temperature as indicated. Γ is the compensation point

I. Rates of Photorespiration

Under normal photosynthetic conditions the rates of photorespiration (PR′) of most higher C_3 plants (Fig. 3) range from 3 to 10 mg $CO_2 \, dm^{-2} \, h^{-1}$ (Hew et al., 1969a; Ludwig and Canvin, 1971b; Zelitch, 1971; D'Aoust and Canvin, 1974; Lloyd and Canvin, 1977). In all cases, these rates exceed dark respiration by factors of 1.2 to 4.0 times and are 15% to 20% of the rate of APS. Earlier results (see reviews of Jackson and Volk, 1970; Zelitch, 1971) that show photorespiration from C_3 plant leaves to be less or equal to dark respiration can be ascribed to poorly designed gas exchange systems or to poor experimental technique. Other results which show photorespiration to be much greater in relation to dark respiration are based on a technique that does not measure photorespiration but rather compares $^{14}CO_2$ evolution from a leaf in the light or dark after a period of photosynthesis in $^{14}CO_2$ (Zelitch, 1968). Since the specific activity of the $^{14}CO_2$ evolved in the light is greater than the specific activity of the $^{14}CO_2$ evolved in the dark (Fig. 2; Goldsworthy, 1966; Ludwig and Canvin, 1971b; D'Aoust and Canvin, 1972, 1974) the ratio of the $^{14}CO_2$ evolved in the two conditions is greater than the ratio of the CO_2 evolved and the magnitude of photorespiration (PR′) in relation to dark respiration is overestimated.

There are, however, two groups of C_3 plants that are exceptions to the above generalizations about rates of photorespiration. The first exception is that of submerged aquatic vascular plants which exhibit rates of photorespiration that are lower than dark respiration (Lloyd et al., 1977b). This would seem to be due to a high resistance to gas diffusion with resulting higher reassimilation. The second group of plants are air-grown algae which show little CO_2 evolution in the light (LLOYD et al., 1977a), a result in contradiction to much earlier work which implied the presence of photorespiration (MERRETT and LORD, 1973; TOL-BERT, 1974). The explanation for the lack of photorespiration in algae is unclear, but it has been suggested that ribulose bisphosphate oxygenase activity may be suppressed due to the formation of high CO_2 concentrations in the cells (BERRY et al., 1976).

II. The Post-Illumination Burst

The accelerated CO_2 evolution in the dark period immediately following the light has been called the post-illumination burst (PIB). The PIB has been viewed as a remnant of photorespiration that continues for a short period in the dark (DECKER, 1955, 1959) and evidence in favour of that view was provided when it was shown that the specific activity of the $^{14}CO_2$ evolved in the PIB was identical to that of the $^{14}CO_2$ evolved in photorespiration in the light (Fig. 3; LUDWIG and CANVIN, 1971b; D'AOUST and CANVIN, 1972; D'AOUST and CANVIN, 1974). The PIB is rapid, lasting less than 2 min (BULLEY and TREGUNNA, 1971; D'AOUST and CANVIN, 1974) and is often followed by a second burst of CO_2 (Fig. 2) that is associated with dark respiration (HEICHEL, 1971; CORNIC, 1973; D'AOUST and CANVIN, 1974) and has nothing to do with photorespiration.

Overall, the properties of the PIB are very similar to those of steady-state photorespiration (CORNIC et al., 1969), but it does not follow that the rate of photorespiration can be calculated from the PIB. Presumably the PIB would be comprised of carbon from the emptying of the pools from the precursor of glycolate to the substrate for CO_2 evolution. The total size of the PIB would, on that basis, provide an estimate of the pool sizes and not an estimate of steady rates of photorespiration. Rates of photorespiration computed from the PIB within a measuring system, however, seem to provide a comparative estimate of the rate of photorespiration. Computation and comparison of rates between systems is questionable, however, since the maximum rate measured from any burst will depend upon the resolution of the system (volume of plant chamber, etc.) and the resolution of the measuring cell (i.e., volume) in the infra-red gas analyzer.

III. Compensation Point

When a C_3 plant is placed in a closed container in the light, the CO_2 will be depleted to a stable CO_2 concentration where net photosynthesis is zero – this concentration of CO_2 is termed the CO_2 compensation point. At a temperature of 25° C, normal oxygen and high light intensity the compensation point of C_3 plants is about 40 µl CO_2 l^{-1} or higher (ZELITCH, 1971; KRENZER et al., 1975). The compensation point is not a static situation but rather a dynamic situation where photosynthesis is equal to photorespiration (LUDWIG, 1968; BULLEY and TREGUNNA, 1970). The compensation point increases linearly with the oxygen concentration (FORRESTER et al., 1966a; TREGUNNA et al., 1966; TREGUNNA and

Downton, 1967; Poskuta, 1968b; Jolliffe and Tregunna, 1973) and with temperature (Egle and Schenk, 1953; Heath and Orchard, 1957; Whiteman and Koller, 1967; Hew et al, 1969b; Björkman et al., 1970; Voskresenskaya et al., 1970 and Nelson et al., 1975), but is independent of light intensity above a relatively low intensity (Heath et al., 1967; Whiteman and Koller, 1967; Bulley et al., 1969). Reports of variations in the compensation point due to age, light intensity, or nitrogen nutrition (Fair et al., 1972, 1973, 1974a,b) and season (Smith et al., 1976) have appeared, but usually the compensation point is remarkably constant within (Moss et al., 1969; Cannell et al., 1969; Moss, 1971; Lloyd and Canvin, 1977) and between (Krenzer et al., 1975) most species.

Some variation in the compensation point has been reported for the progeny from a cross of *Lolium* sp. (Wilson, 1972) and three other plants, *Molluga verticillata* (Sayre and Kennedy, 1977), *Moricandia arvensis* (L.) DC. and *Panicum milioides* Nees ex Trin (Krenzer et al., 1975) have been reported to have compensation points between 15 and 25 μl 1^{-1}. *Panicum milioides* is a C_3 plant (Brown and Brown, 1975; Kestler et al., 1975; Kanai and Kashiwagi, 1975; Goldstein et al., 1976; Ku et al., 1976) which seems to have a reduced rate of photorespiration (Brown and Brown, 1975; Keck and Ogren, 1976; Quebedeaux and Chollet, 1977). It has been proposed that the lower compensation point in this C_3 species is due to a combination of lower photorespiratory activity and higher PEP carboxylase activity (Kestler, et al., 1975; Black et al., 1976) but this is not firmly established as it has not been shown that the malate and aspartate formed in this *Panicum* sp. turn over as they do in C_4 plants (Black et al., 1976).

IV. Effect of CO_2 Concentration

CO_2 concentrations up to 400 μl 1^{-1} seem to have little effect on the measured rates of photorespiration (Fig. 3; Ludwig and Canvin, 1971b). Above these CO_2 concentrations a larger rate of photorespiration has been reported as a result of improved measurement (Bravdo and Canvin, 1974) but this requires further examination. Early work (Fock and Egle, 1966; Egle and Fock, 1967) reported a lack of PIB at high CO_2 concentrations and the competition that exists between CO_2 and O_2 at the RuBP carboxylase/oxygenase would result in lower photorespiration at high CO_2 concentrations (Laing et al., 1974; Chollet and Ogren, 1975).

V. Effect of O_2

A large number of reports shows that photosynthesis of leaves (McAllister and Meyers, 1940; Forrester et al., 1966a; D'Aoust and Canvin, 1973; see also review articles cited earlier), protoplasts (Nishimura et al., 1975), protoplast extracts (Huber and Edwards, 1975), and chloroplasts (Ellyard and Gibbs, 1969; Gibbs, 1969) are inhibited with increasing oxygen concentrations. With intact leaves this is an effect on the mesophyll or residual resistance as there is no effect of oxygen on stomatal aperture (Gauhl and Björkman, 1969; Bull, 1969; Cornic, 1974). The algae appear to be an exception among C_3 plants as, in spite of a vast earlier literature showing inhibition of photosynthesis by oxygen (Turner

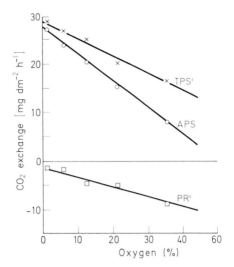

Fig. 4. Effect of oxygen concentration on the CO_2 gas exchange of sunflower leaves at 150 ppm CO_2, 270 μE m^{-2} s^{-1} and 25.5° C. (D'AOUST, 1970)

and BRITTAIN, 1962), later results show no effect of oxygen on CO_2 fixation by air-grown algal cells (FOCK et al., 1971; BERRY et al., 1976; LLOYD et al., 1977a).

In the work with leaves it was concluded that increasing oxygen concentrations had two effects, namely, an inhibition of TPS and a stimulation of photorespiration (FORRESTER et al., 1966a; TREGUNNA et al., 1966; CURTIS et al., 1969; LUDWIG and CANVIN, 1971b) and this was subsequently shown to be the case, (Fig. 4; D'AOUST and CANVIN, 1973).

VI. Effect of Temperature

Photorespiration increases with temperature (Fig. 3), with a maximum somewhat over 30° C (JOLLIFFE and TREGUNNA, 1968; HEW et al., 1969b; HOFSTRA and HES-KETH, 1969; CORNIC, 1974). Indirect conclusions on the changes in the compensation point (see earlier section) are consistent with these results.

VII. Interaction of Oxygen, Carbon Dioxide, and Temperature

Oxygen and carbon dioxide interact competitively with RuBP carboxylase/oxygen-ase to yield either products of CO_2 fixation or substrate for the glycolate pathway (BOWES et al., 1971; BOWES and OGREN, 1972). Since the interaction of oxygen and carbon dioxide on leaf photosynthesis was competitive, OGREN and BOWES (1971) suggested that the interaction of oxygen and carbon dioxide on the enzyme governed both the rate of photorespiration and photosynthesis in leaves. This was developed more fully in a later paper (LAING et al., 1974) where the effect of oxygen and carbon dioxide on the isolated enzyme was correlated with the effect of these same gases on leaf photosynthesis. The increase of photorespiration with increasing temperature was assigned to an increased affinity of the enzyme for O_2 and a decreased affinity for CO_2. Both carboxylase and oxygenase reactions had a similar energy of activation. BADGER and ANDREWS (1974) initially obtained different results but later confirmed LAING's findings (BADGER and COLLATZ, 1977).

More recently KU and EDWARDS (1977a, b) proposed that the effect of temperature on the inhibition of photosynthesis is partly due to the differential effect of temperature on the O_2/CO_2 solubility ratio in the leaf. Certainly there will be a general correlation (KU and EDWARDS, 1977a) between percentage inhibition and solubility ratio, but it is not established that the change in the solubility ratio can entirely account for the changes in percentage inhibition of photosynthesis. EHLERINGER and BJÖRKMAN (1977), for instance, have shown that changes in the O_2/CO_2 solubility ratio do not account for the decreases in quantum yield of C_3 plants with temperature (Fig. 12).

The rate of photorespiration does not show a correlation with solubility ratio (Fig. 3) and as the solubility ratio changes the proportion of inhibition that is due to photorespiration and that is due to the competitive effect of O_2 changes rather dramatically (Fig. 5; LUDWIG and CANVIN, 1971b; KU and EDWARDS, 1977b). Since both effects are supposedly due to the single effect of O_2 on the RuBP carboxylase/oxygenase, it is not apparent why photorespiration should not increase in concert with the competitive effect. It is also not apparent why the relative proportion of photorespiration and competition should be changed by temperature at the same O_2/CO_2 solubility ratio (KU and EDWARDS, 1977b). Thus, while the solubility ratio of O_2/CO_2 may in part explain the effect of temperature on photorespiration it does not seem likely that it is the entire explanation.

VIII. Effect of Light Intensity

At low light intensities the rate of CO_2 evolution in the light is less than that in darkness (HOLMGREN and JARVIS, 1967; HEW et al., 1969b) but as the light intensity increases photorespiration increases and exceeds dark respiration (HOLMGREN and JARVIS, 1967; HEW et al., 1969b; CARLSON et al., 1971; CORNIC and MOUSSEAU, 1969). Further increases to very high light intensities do not appear to increase photorespiration even though photosynthesis may still increase.

IX. The Glycolate Pathway

As previously mentioned, the term, "photorespiration" refers to the evolution of CO_2 from the metabolism of compounds in the glycolate pathway (see Chap. II. 26, this vol).

In order for CO_2 to be continuously evolved in photorespiration there must be a flux of carbon through the glycolate cycle equivalent to four times the rate of CO_2 evolution. Since the carbon for the pathway originates in the photosynthetic carbon reduction cycle, it means that if $^{14}CO_2$ is supplied in photosynthesis, then the intermediates of the glycolate pathway and the CO_2 evolved will rapidly become labelled and will eventually reach the specific activity of the $^{14}CO_2$ that was supplied. Similarly, if after $^{14}CO_2$ supply, $^{12}CO_2$ is supplied, the label should be rapidly flushed from the intermediates of the glycolate pathway and the evolution of $^{14}CO_2$ should rapidly cease (GOLDSWORTHY, 1966; LUDWIG and CANVIN, 1971b).

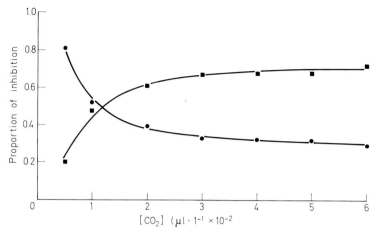

Fig. 5. The proportion of the difference between APS (2% O_2) and APS (21% O_2) that can be assigned to photorespiration (●) and to competition (■). (Data replotted from Fig. 4, sunflower 24.9° C, 790 µE m^{-2} s^{-1}. [CO_2] is the average in the leaf chamber

If no CO_2 is supplied after $^{14}CO_2$ supply, then the intermediates of the glycolate pathway must continue to be labelled from the previously labelled intermediates of the photosynthetic carbon reduction cycle and the evolution of $^{14}CO_2$ would continue. In other words, evolution of $^{14}CO_2$ from a previously labelled leaf will be greater in the absence than in the presence of CO_2.

If data on the labelling of glycolate cycle intermediates are to be used as evidence for photorespiration then it is necessary to show that, when $^{14}CO_2$ is supplied, the intermediates become labelled in the correct order. Further, the intermediates should eventually reach the same specific activity as the $^{14}CO_2$ supplied if there are not multiple pools of the compounds in the tissue. Evidence must be presented that there is carbon flow through the cycle or, at least, in my estimation, carbon flow from glycine to serine which is the CO_2-evolving step. The accumulation of ^{14}C only in glycolate or glycine cannot be used as evidence for photorespiration as no CO_2 would be evolved. The accumulation of ^{14}C in serine which can arise from 3-phosphoglyceric acid by the reverse reactions (TOLBERT, 1971) can also not be used as evidence for photorespiration. Thus, the simple detection of ^{14}C in some glycolate cycle intermediates after $^{14}CO_2$ is supplied to the leaf does not demonstrate that photorespiration is present. At the very least a flux of carbon through glycine and serine should be demonstrated.

To demonstrate a flux conclusively a period when $^{14}CO_2$ is supplied must be followed by a chase period with $^{12}CO_2$, and enough times sampled to provide a picture of the changes in glycolate cycle intermediates with time. Without the chase one cannot demonstrate that carbon is flowing through the compounds involved.

Concerted changes in the specific activity of glycine and serine (Fig. 6) will only occur if there are single pools of the compounds, but concerted changes in total radioactivity must occur even if there are multiple pools of the compounds. Only this type of concerted change in the intermediates provides evidence for the operation of the cycle and the possible evolution of CO_2. A further discussion

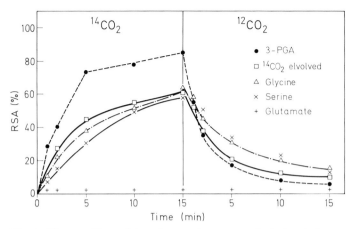

Fig. 6. The specific activity of several compounds isolated from sunflower leaves exposed to $^{14}CO_2$ and 400 ppm CO_2. Oxygen concentration, 21%; light intensity, 3500 ft-c. RSA is relative specific activity and is the specific activity of the compounds as a percentage of the specific activity of the supplied $^{14}CO_2$. (From Canvin, 1976)

of this approach can be found in earlier publications (Mahon et al., 1974b, c; Fock, et al., 1974; Canvin et al., 1976; Schnarrenberger and Fock, 1976).

Oxygen interacts at two and possibly three sites in the cycle. Increasing oxygen concentration should thus stimulate photorespiration, and decreasing oxygen concentration, or increasing CO_2 concentration should diminish it.

E. Photorespiration in C_4 Plants

C_4 plants are characterized by an initial fixation of CO_2 into C_4 acids in the mesophyll cells, a transfer to and subsequent decarboxylation of this acid in the bundle sheath cells and a final fixation of the CO_2 by the photosynthetic carbon reduction cycle in these cells (Hatch, 1976; Hatch and Osmond, 1976). Usually well-delineated chlorophyllous bundle sheaths and mesophyll cells are apparent in the leaves (see Chap. II. 6, this vol.).

I. Photorespiration as CO_2 Evolution

The first indications that the CO_2 gas exchange of C_4 plants was different from C_3 plants is found in the results of Meidner (1962) and Moss (1962) who showed that the compensation point of maize was near zero. These results suggested that CO_2 evolution was absent from these leaves in the light and since then many investigators have confirmed this conclusion. Thus, except for a few reports that will be discussed below, CO_2 evolution in the light or photorespiration cannot be demonstrated in C_4 plants by determinations on the external gas phase. No difference between $^{14}CO_2$ uptake and $^{12}CO_2$ uptake (LUDWIG, 1968; HEW et al., 1969a; YEMM and BIDWELL, 1969; FOCK et al., 1970; D'AOUST and CANVIN, 1973) can be shown when the carbon isotope technique (LUDWIG and CANVIN, 1971a) is used over an oxygen concentration from 2% to 60% (D'AOUST and CANVIN, 1973). No, or only slight, evolution from leaves of C_4 plants in the light can be detected when the leaves are flushed with CO_2-free air (MOSS, 1966; EL SHARKAWY et al., 1967; HEW et al., 1969a; HOFSTRA and HESKETH, 1969; OSMOND et al., 1969; MEIDNER, 1970) or even when CO_2-free air is forced directly through the leaf (TROUGHTON, 1971).

Some $^{14}CO_2$ evolution from leaves of C_4 plants was detected after the plants were allowed to photosynthesize in a $^{14}CO_2$ atmosphere (IRVINE, 1970; LAING and FORDE, 1971; RATHNAM and DAS, 1974; KENNEDY, 1976a, b). In all cases these results were interpreted as evidence for photorespiration in C_4 plants, but a critical examination of the data does not support this position. In C_3 plants the specific activity of the $^{14}CO_2$ evolved from labelled leaves in the dark is less than that in the light (Fig. 2; GOLDSWORTHY, 1966; LUDWIG and CANVIN, 1971b; CANVIN and FOCK, 1972; D'AOUST and CANVIN, 1974) and if this is also true for C_4 plants the observed CO_2 evolution in the light is only a small portion of the CO_2 evolution in the dark. In addition, deficiencies in experimental procedure or differences in the characteristics of the $^{14}CO_2$ evolved can be identified that make it unlikely that the observed $^{14}CO_2$ was of a photorespiratory origin. IRVINE (1970) measured the transfer of ^{14}C from labelled to unlabelled plants and observed a rate of transfer of ^{14}C in the light which was 0.7% to 15% of the rate of ^{14}C transfer in the dark. Since the system included much non-photosynthetic tissue and shaded photosynthetic tissue, the $^{14}CO_2$ evolution in the light can be entirely attributed to dark respiration.

Negligible $^{14}CO_2$ evolution in the light was observed with maize leaves in CO_2-free air (ZELITCH, 1968; YEMM and BIDWELL, 1969; LAING and FORDE, 1971) or CO_2-free oxygen (ZELITCH, 1968) but considerable evolution from the leaves of *Amaranthus lividus* L. cn. *lividus* was observed (LAING and FORDE, 1971). There is, however, an important deficiency in the experiments and several significant differences between the properties of the $^{14}CO_2$ evolution reported and those that are well established for C_3 plants. The experimental deficiency is the lack of results showing that the $^{14}CO_2$ evolution is eliminated in 2% O_2. The important differences are that 100% O_2 did not stimulate $^{14}CO_2$ evolution over that observed in air (although it did stimulate it over CO_2-free air) and that the presence of CO_2 in the air stream did not eliminate the $^{14}CO_2$ evolution. In C_3 plants CO_2 in the flushing stream quickly suppresses $^{14}CO_2$ evolution (GOLDSWORTHY, 1966; LUDWIG and CANVIN, 1971b; LAING and FORDE, 1971) and increased oxygen always stimulates evolution (D'AOUST and CANVIN, 1974). While the origin of the evolved $^{14}CO_2$ cannot be identified, the important differences in properties between it and $^{14}CO_2$ evolution in C_3 plants shows that it is very unlikely to be arising from recent photosynthetic products via a photorespiratory mechanism. RATHNAM and DAS (1974) measured $^{14}CO_2$ release from *Eleusine coracana* L. in the light at rates $^1/_{10}$ to $^1/_{20}$ of that of dark respiration, but provided no evidence to establish that the CO_2 was of photorespiratory origin. KENNEDY (1976b), using ZELITCH's technique (1968), showed that $^{14}CO_2$ evolution from senescent *Portulaca oleraceae* L. leaves was greater in the light than $^{14}CO_2$ evolution in the dark.

Pure oxygen in place of CO_2-free air, however, only stimulated the evolution 10% and pure nitrogen only inhibited it 35%. No other characteristics of the $^{14}CO_2$ evolution were reported. While it is not possible to explain the effect of light on $^{14}CO_2$ evolution in these experiments, the slight effect of oxygen on $^{14}CO_2$ evolution makes it unlikely that the $^{14}CO_2$ arose from a photorespiratory pathway.

Several authors have used inhibitors of photosynthesis to try to demonstrate photorespiration in C_4 plants. DCMU has most commonly been employed (EL SHARKAWY et al., 1967; DOWNTON and TREGUNNA, 1968b; HEICHEL, 1972; RATHNAM and DAS, 1974), and when photosynthesis has been completely inhibited, CO_2 evolution equivalent in rates to dark respiration has been observed. DOWNTON and TREGUNNA (1968b), however, have shown that the CO_2 evolution observed with DCMU treatment is completely insensitive to changes in the oxygen concentration (POSKUTA, 1968a, reports opposite results) and on that basis it is probably dark respiration which is also insensitive to changes in oxygen concentration above 2% O_2. TREGUNNA (1966) induced high compensation points, which were decreased to normal values by 2% O_2, in maize by supplying FMN. The FMN, however, inhibited photosynthesis about 60% and it would appear that the low oxygen also prevented this inhibition and thus the high compensation points in air can be attributed to a lower photosynthetic capacity. The explanation that the lack of FMN was inhibiting glycolic acid oxidase in maize was later shown to be erroneous (DOWNTON and TREGUNNA, 1968b). The most recent inhibitor study is that of RAY and BLACK (1976), who used 3-mercaptopicolinic acid, an inhibitor of phosphoenolpyruvate carboxykinase in *Panicum maximum* L. leaves. They found that treatment of the *Panicum* leaves for 4 h with the inhibitor changed the CO_2 compensation point from near zero to between 18 and 45 µl l^{-1} and interpreted this result as showing that the photorespiration of the bundle sheath cells was being made apparent.

Subsequently, however, the authors tested the effect of oxygen on net photosynthesis and found no effect of oxygen concentration. The increased compensation points are thus not due to an interaction of C_3 photosynthesis and photorespiration, (RAY and BLACK, 1977) but apparently due to interference with the CO_2-fixing reactions.

The CO_2 compensation point of C_4 plants has been shown to be zero or less than 10 µl l^{-1} by a large number of authors (MOSS, 1962; MEIDNER, 1962; FORRESTER et al., 1966b; TREGUNNA and DOWNTON, 1967; MANSFIELD, 1968; DOWNTON and TREGUNNA, 1968a; MOSS et al., 1969; OSMOND et al., 1969; POSKUTA, 1969; KRENZER and MOSS, 1969; BJÖRKMAN et al., 1970; CHEN et al., 1970; LAING and FORDE, 1971; MOSS et al., 1971; KRENZER et al., 1975; LONG et al., 1975; SHOMER-ILAN et al., 1975). Moreover, the compensation point was not affected by temperature (BJÖRKMAN et al., 1970; LONG et al., 1975) or by oxygen concentration (FORRESTER et al., 1966b; MEIDNER, 1967; TREGUNNA and DOWNTON, 1967; DOWNTON and TREGUNNA, 1968b; POSKUTA, 1969). Atypical compensation points of 9–25 µl CO_2 l^{-1} for field-grown maize were reported by HEICHEL and MUSGRAVE (1969), but the effect of oxygen on these compensation points was not tested and MOSS (1971) and MOSS et al. (1971) were not able to confirm these high compensation points, nor did they find any variation in the compensation points of a wide selection of maize genotypes. POSKUTA (1969) has reported that young maize plants had a compensation point of 10–15 µl CO_2 l^{-1}, but oxygen concentration had no effect on the compensation point and this elevated compensation point is probably due to a combination of dark respiration and a poorly developed photosynthetic apparatus in the young leaves.

LESTER and GOLDSWORTHY (1973) have reported high compensation points in *Amaranthus* sp. These compensation points were sensitive to oxygen concentration but they remain unexplained as high compensation point plants or leaves seemed to occur at random and could not be repeatably produced. Large increases in the compensation points of several C_4 plants have been reported as a result of supplying a high level of ammonia to the plants (GROSSMAN and CRESSWELL, 1974 a, b). These compensation points were further increased by oxygen concentration but they remain unexplained as high compensation point plants or leaves seemed to occur at random and could not be repeatably produced. Large increases in the compensation points of several C_4 plants have been reported as a result of supplying a high level of ammonia to the plants (GROSSMAN and CRESSWELL, 1974a, b). These compensation points were further increased by oxygen concentrations above ambient and the authors proposed that ammonia increases photorespiration to such an extent that the CO_2 could no longer be refixed by the mesophyll cells. The authors, however, did not show that these compensation points could be returned to a low value by 2% oxygen or that they are affected by temperature as are the compensation points of C_3 plants. Thus, while the observations are interesting, there is not enough evidence on the properties of these compensation points really to evaluate their significance.

In early work with C_3 plants the post-illumination burst was extensively used to demonstrate photorespiration (DECKER, 1955) and several authors have looked for such bursts in C_4 plants. In maize, a post-illumination burst is not present (TREGUNNA et al., 1964; FORRESTER et al., 1966b; MOSS, 1966; MEIDNER, 1967; BJÖRKMAN, 1968; D'AOUST, 1970; DOWNTON, 1970; HEICHEL, 1972) and, in fact, no CO_2-evolution can be detected from the leaves in the first minute of darkness (TREGUNNA et al., 1964; FORRESTER et al., 1966b; MEIDNER, 1967; D'AOUST, 1970; HEICHEL, 1972). In other C_4 plants, however, post-illumination bursts have been observed (BJÖRKMAN, 1968; DOWNTON, 1970; BROWN and GRACEN, 1972; RATHNAM and DAS, 1974) and DOWNTON (1970) has described three types of CO_2 transients that can be observed in C_4 plants upon transfer from light to darkness (Fig. 7). Some lists of plant species belonging to each type can be found in DOWNTON (1970) or BROWN and GRACEN (1972). Where this property has been investigated the bursts of Types II and III are oxygen-insensitive (BJÖRKMAN, 1968; BJÖRKMAN et al., 1970; DOWNTON, 1970). Type I and Type III transients are very different from the post-illumination burst observed in C_3 plants (Fig. 2) and provide no support for the presence of photorespiration in C_4 plants. The Type II transient superficially resembles a C_3 post-illumination burst (Fig. 2) but it is not sensitive to the oxygen concentration and also is saturated with $^{14}CO_2$ after a much shorter fixation time (WYNN et al., 1973) than is the burst from C_3 plants (D'AOUST and CANVIN, 1972). These important differences in the properties of the post-illumination burst indicate that the burst observed in C_4 plants is not of photorespiratory origin but is more likely arising from the decarboxylation of recently formed C_4 acids (BERRY et al., 1970; BROWN and GRACEN, 1972; WYNN et al., 1973; RATHNAM and DAS, 1974). The very delayed increases of dark respiration in maize

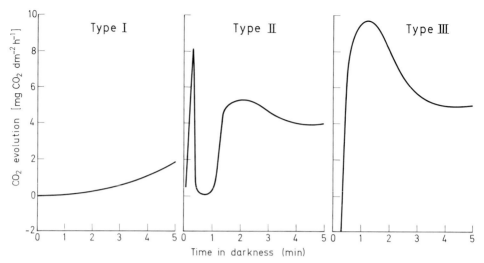

Fig. 7. The patterns of CO_2 gas exchange observed in C_4 plants on transfer from light to darkness. Type I is typical for *Zea mays* L., Type II for *Panicum bergii* Arech. and Type III for *Paspalum notatum* Flugge. Type I redrawn from D'Aoust (1970), and Types II and III from Brown and Gracen (1972)

observed by Heichel (1970, 1972) can certainly not have any connection with photorespiration.

Apparent photosynthesis of C_4 plant leaves is insensitive to changes in the oxygen concentration from 2% to 60% O_2 (Björkman, 1966; Forrester et al., 1966b; Hesketh, 1967; Downton and Tregunna, 1968b; Downes and Hesketh, 1968; Bull, 1969; Hofstra and Hesketh, 1969; Osmond et al., 1969; Björkman et al., 1970; Fock et al., 1970; Ludlow, 1970; Troughton, 1971; D'Aoust and Canvin, 1973; Voskresenskaya et al., 1974). Pure oxygen inhibits apparent photosynthesis in C_4 plants (Forrester et al., 1966b; Poskuta, 1969; Grishina et al., 1975), and while it has been contended that this inhibition is a manifestation of photorespiration, CO_2 evolution has not been shown under those circumstances nor has it been shown that increased CO_2 concentration will overcome the inhibition. To test the CO_2/O_2 interaction in C_4 plants may be difficult because, at least in corn, stomatal closure is rapidly obtained with high CO_2 concentrations (Raschke, 1972; Akita and Moss, 1973).

The conclusion, then, can be safely reached that with whole leaves of C_4 plants there is no evidence for any CO_2 evolution or photorespiration with properties similar to those that are well established in C_3 plants.

II. Photorespiration as Oxygen Uptake

According to current formulations of the glycolate pathway it should be possible to measure "photorespiration" as the oxygen uptake that must occur for glycolate formation and/or for glycolate and possibly glycine oxidation. Although the technique is not without its problems (Decker, 1958; Jackson and Volk, 1970) the oxygen uptake is determined using the oxygen isotope ^{18}O and the mass spectrometer. As the mass spectrometer will also simultaneously measure changes in carbon isotopes the use of this technique might be expected to provide the best insight into the gas exchange that occurs in an illuminated leaf. Unfortunately, the experiments that have been done with C_3 and C_4 plants do not provide much illumination. One would expect that the observed oxygen uptake under various conditions would be consistent with the effect of these conditions on CO_2 evolution if photorespiration in the sense defined earlier for C_3 plants was involved. However, the most remarkable feature of the reported results is the degree to which they differ from any measurements that have been done on CO_2 exchange with other systems.

Equal uptake and evolution of oxygen has been shown to occur at the compensation point of maize (JACKSON and VOLK, 1969; GERSTER et al., 1974). This oxygen uptake was about 20 $\mu mol\ dm^{-2}\ h^{-1}$ at 0.1% O_2 and increased to over 100 μmol $dm^{-2}\ h^{-1}$ at 8% O_2 (JACKSON and VOLK, 1969; VOLK and JACKSON, 1972), a rate which was three times that of dark respiration, (JACKSON and VOLK, 1969). If these were the only data presented, and if it were assumed that any evolved CO_2 was continuously refixed, this light-stimulated oxygen uptake at the compensation point would be entirely consistent with showing the presence of conventional photorespiration in maize. Further results, however, were presented and these, because of their unusual nature in comparison to what we know of C_3 photorespiration, show that the O_2 uptake in maize is not due only to photorespiration. The first unusual feature of the experiments is that a steady constant rate of CO_2 uptake, O_2 evolution and O_2 uptake is obtained from 1.9% CO_2 to zero CO_2 (VOLK and JACKSON, 1972). This is unusual because the stomata of corn are very sensitive to CO_2 concentration (PALLAS, 1965; AKITA and MOSS, 1972; RASCHKE, 1972) and are completely closed at 0.2% CO_2 (PALLAS, 1965). The closed stomata should considerably hamper any gas exchange and one might expect that, as the stomata open upon CO_2 depletion, the rate of photosynthesis would increase, which was not observed. The second variant feature is that the O_2 uptake is only decreased 30% in 1.9% CO_2 compared to zero CO_2.

In C_3 plants such as sunflower no effect of from 2% to 21% O_2 on photosynthesis was observed at 0.25% CO_2 (Fig. 8) or in bean at 0.15% CO_2 (VIIL and PARNICK, 1974). If the interpretation of a lack of effect of oxygen on photosynthesis indicates no photorespiration in a C_3 plant at those CO_2 concentrations, how can photorespiration proceed in maize at 1.9% CO_2. Also surprising about the results – if the process is photorespiration – is the fact that there is no change in the rate of O_2 uptake from 1.9% CO_2 to zero CO_2 – in other words, there is no competitive interaction between CO_2 and O_2.

VOLK and JACKSON (1972) also determined CO_2 exchange using $^{13}CO_2$ and $^{12}CO_2$. They showed in maize that from 1.9% to zero CO_2 the $^{13}CO_2$ was assimilated faster than $^{12}CO_2$ with a continual decline in specific isotope content. This again is at variance with $^{14}CO_2$ isotope experiments where no change in specific activity of $^{14}CO_2$ was observed from 0.025% to zero CO_2 (HEW et al., 1969a) or even an increase in specific activity was observed (YEMM and BIDWELL, 1969). Even in a C_3 plant where the specific activity rapidly declines (HEW et al., 1969a) there was no change in specific activity above 0.05% CO_2 (VOZNESENSKII, 1965). Using 95% atom percent $^{13}CO_2$ the authors further showed that $^{12}CO_2$ was evolved at a rate of 15 $\mu mol\ dm^{-2}\ h^{-1}$ for 15 min but by 30 min net $^{12}CO_2$ uptake was

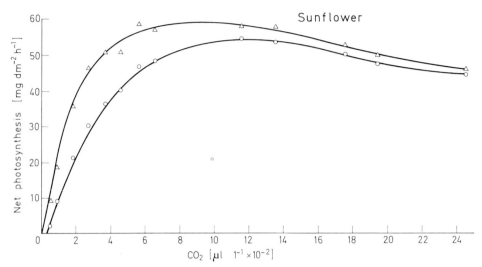

Fig. 8. Photosynthesis of sunflower leaves as a function of CO_2 concentration in 21% and 2% O_2. Temperature was 25 °C and quantum flux 800 µE m^{-2} s^{-1}. (△) 2% O_2; (○) 21% O_2

occurring. Even when $^{12}CO_2$ was being taken up, however, the atom % ^{13}C continued to decline. The continuous decline in atom % ^{13}C shows that any CO_2 evolution which occurred in the system must have arisen from unlabelled endogenous sources. This is entirely inconsistent with origin in recent photosynthesis. In C_3 plants the evolved CO_2 rapidly becomes equal in specific activity to the $^{14}CO_2$ supplied (LUDWIG and CANVIN, 1971b; D'AOUST and CANVIN, 1972) and then there is no change in the specific activity of $^{14}CO_2$ in the gas stream. If a similar photorespiration occurred in the maize leaves of VOLK and JACKSON (1972), the photorespiratory CO_2 should have rapidly become saturated with $^{13}CO_2$ and then no further decline in atom % ^{13}C should have occurred.

The striking differences between the properties of the CO_2 evolution or O_2 uptake determined in maize and that observed in C_3 plants leads to a firm conclusion that the CO_2 evolution and O_2 uptake observed in maize does not show photorespiration similar to that in C_3 plants. In fact, the large rates of O_2 uptake at 2% or 3% O_2 in C_3 plants (OZBUN et al., 1964; MULCHI et al., 1971) do not correlate with the lack of CO_2 evolution or no difference between $^{14}CO_2$ and $^{12}CO_2$ uptake under such conditions. While the oxygen exchange can be partly assigned to CO_2 recycling, it is clear that the rates are considerably in excess of any measured CO_2 exchange (BULLEY and TREGUNNA, 1970).

JACKSON and VOLK (1969) and VOLK and JACKSON (1972) mentioned the possibility that the O_2 uptake they observed could have been due to a Mehler reaction (MEHLER, 1951; WHITEHOUSE et al., 1971; EGNEUS et al., 1975) with oxygen acting as a terminal electron acceptor (FORTI and JAGENDORF, 1961; KRALL and BASS, 1962). That oxygen can act in such a role has been conclusively demonstrated with mesophyll cell preparations from C_4 plants (HUBER and EDWARDS, 1975; EDWARDS et al., 1976), where pyruvate-dependent CO_2 fixation was greatly stimulated via pseudocyclic ATP formation. This stimulation was dependent on the concentration of oxygen and was not saturated at even 100% O_2. High rates of oxygen exchange that are unrelated to photorespiration have also been observed in algae (RADMER and KOK, 1976).

Whether this mechanism of O_2 uptake occurs in the intact leaf is unknown, but if one adheres to the usual extrapolation of in vitro results to in vivo conditions, there is no reason to believe that it does not (ALLEN, 1975; EGNEUS et al., 1975; FORTI and GEROLA, 1977). It cannot be emphasized too much that this O_2 uptake has nothing to do with metabolism of glycolate in the glycolate pathway or with CO_2 evolution (the primary base on which photorespiration must be defined).

The explanation, then, that is most consistent with the reported results (JACKSON and VOLK, 1969; VOLK and JACKSON, 1972; GERSTER et al., 1974) is that most of the O_2 uptake in maize is due to a Mehler reaction and the low CO_2 evolution that is observed is due to residual dark respiration. In that regard, the O_2 uptake that was observed might better be termed, "oxygen photoreduction" (JACKSON and VOLK, 1969) to avoid further confusion of the term photorespiration and such O_2 uptake would not be a better measure of photorespiration than CO_2 evolution.

III. Photorespiration in C_4 Plants – Indirect Evidence

As presented in the previous section there is no entirely consistent or definitive direct evidence to establish the presence of photorespiration in C_4 plants. In early work this led to the conclusion that photorespiration was absent or that any evolved CO_2 was re-utilized before it reached the external atmosphere (FORRESTER et al., 1966b). With the elucidation of C_4 photosynthesis (HATCH and SLACK, 1966) and the recognition of the role that mesophyll and bundle sheath cells played in this photosynthesis (HATCH, 1976), and with the discovery that glycolate was a product of RuBP carboxylase/oxygenase (BOWES et al., 1971; CHOLLET and OGREN, 1975) and with further discoveries about the glycolate pathway (TOL-BERT, 1971), DOWNTON et al. (1969) and BLACK (1973) contended that photorespiration occurred in the bundle sheath cells but the respired CO_2 was refixed and recycled via the mesophyll cells and thus did not appear in the external atmosphere. A contemporary alternate suggestion (BJÖRKMAN, 1971; HATCH, 1971a) was that photorespiration was indeed suppressed because of an elevated CO_2 concentration in the bundle sheath cells brought about by the transport of C_4 acids from the mesophyll cells and their decarboxylation in the bundle sheath cells. Both of these proposals require an evaluation of the internal metabolism and capabilities of the leaf cells of C_4 plants, and reports pertinent to that question will be discussed in this section.

1. The Mesophyll Cells

The possibility of photorespiration in mesophyll cells does not seem to exist because of the apparent inability of these cells to synthesize glycolate by any of the proposed methods. Glycolate synthesis from CO_2 (ZELITCH, 1965), from a two-carbon fragment of the transketolase reaction (SHAIN and GIBBS, 1971) or from RuBP via RuBP oxygenase (BOWES et al., 1971; KIRK and HEBER, 1976) require the presence of RuBP carboxylase/oxygenase and the photosynthetic carbon reduction cycle. Studies based on several techniques have elegantly shown that RuBP carboxylase and the photosynthetic carbon reduction cycle are confined to the bundle sheath cells (HATCH, 1976; HATCH and OSMOND, 1976). Since the mesophyll cells apparently lack the ability to synthesize glycolate, the significance of peroxisomes (FREDERICK and NEWCOMB, 1971; HILLIARD et al., 1971) and glycolate cycle enzymes (REHFELD et al., 1970; OSMOND and HARRIS, 1971; LIU and BLACK, 1972; EDWARDS and GUTIERREZ, 1972; HUANG and BEEVERS, 1972; KU and EDWARDS, 1975) in these cells is not clear. Certainly, the presence of peroxisomes and a few of the enzymes of the glycolate pathway cannot be used as evidence for photorespiration unless glycolate is formed by an, as yet, undiscovered means, or it is provided by the bundle sheath cells.

2. The Bundle Sheath Cells

It is well established that RuBP carboxylase and the photosynthetic carbon reduction cycle are contained in the bundle sheath cells of C_4 plants (HATCH and OSMOND, 1976; see Chap. II.6, this vol.). As with the RuBP carboxylase from C_3 plants, the RuBP carboxylase from C_4 plants shows oxygenase activity (BAHR and JENSEN, 1974) and the cells might then be expected to be unable to prevent the production of glycolate in the presence of oxygen (LORIMER and ANDREWS, 1973). If so, then the glycolate pathway and photorespiration could be present to prevent the accumulation of glycolate and to recover the carbon that would be diverted to it.

Peroxisomes, a cell organelle that contains some of the key enzymes of the glycolate pathway (TOLBERT, 1971) are present in C_4 plants at 10% to 50% of the frequency found in C_3 plants (FREDERICK and NEWCOMB, 1971; HILLIARD et al., 1971). They are from 1.5 to 12 times more numerous in the bundle sheath cells compared to the mesophyll cells. In crabgrass and bermuda grass, peroxisome frequency in bundle sheath cells was equivalent to that found in C_3 mesophyll cells, but the frequency of peroxisomes in the C_4 mesophyll cells was only one-third to one-sixth of that in the bundle sheath cells (LIU and BLACK, 1972). The enzymes of the glycolate pathway are, in general, concentrated in the bundle sheath (TOLBERT et al., 1969; REHFELD et al., 1970; LIU and BLACK, 1972; OSMOND and HARRIS, 1971; EDWARDS and GUTIERREZ, 1972; HUANG and BEEVERS, 1972; KU and EDWARDS, 1975; RATHNAM and DAS, 1974; OSMOND, 1972; EDWARDS and BLACK, 1971; CHOLLET and OGREN, 1973; WOO and OSMOND, 1977), but the distribution for each enzyme is not similar, and several of them appear to be equally distributed between the bundle sheath and the mesophyll cells. Total activity of the enzymes in C_4 leaves was much lower than the activity observed in C_3 plants, with sorghum being particularly deficient (OSMOND, 1972; HUANG and BEEVERS, 1972).

Maize leaves were able to decarboxylate ^{14}C-labelled glycolate in the light (DOWNTON and TREGUNNA, 1968 b; HEICHEL, 1972; ZELITCH, 1966) or ^{14}C-labelled glycolate or glycine in the dark (KISAKI and TOLBERT, 1970; KISAKI et al., 1972) but why glycine decarboxylation in darkness should be stimulated by increased oxygen (KISAKI et al., 1972) is not clear. A C_4 *Atriplex* actively metabolized ^{14}C-labelled glyoxylate or glycine to CO_2 and other products in the light or dark (OSMOND and HARRIS, 1971) but the light decarboxylation rate of glycine-1-^{14}C

could not be made equal to the dark even when refixation via PEP carboxylase was inhibited with glyoxal bisulfite. Bundle sheath cells from crabgrass (LIU and BLACK, 1972) and from maize (CHOLLET, 1974) also metabolized glycolate to CO_2 and other products. The glycolate oxidation in maize (HEICHEL, 1972) and in maize bundle sheaths (CHOLLET, 1974) was stimulated by oxygen and inhibited by αHPMS, an inhibitor of glycolate oxidase.

Maize leaves, when treated with αHPMS, produced glycolate, but only at about 10% of the rate of glycolate production that was observed in tobacco (ZELITCH, 1966, 1973a) and it was suggested that the origin of glycolate in maize was different from that in tobacco. Glycolate formation from radioactive sugar phosphates in the presence of glyoxal bisulfite was equal in *Atriplex spongiosa* (C_4) and *Atriplex hastata* (C_3) and was at least ten fold the dark rate of formation (OSMOND and HARRIS, 1971). Bundle sheath strands from maize produced glycolate from ^{14}C sugar phosphates (CHOLLET, 1974) or from $^{14}CO_2$ (CHOLLET and OGREN, 1972b) and this production was suppressed by high concentrations of CO_2 or low concentrations of O_2. Phosphoglycolate formation from malate ^{14}C was reported for bundle sheath strands of *Digitaria sanguinalis* (L) Scop (DITTRICH et al., 1973).

CHOLLET and OGREN (1972a) and CHOLLET (1974) studied the effect of O_2 concentration on photosynthesis of bundle sheath strands of maize. They showed that increased O_2 concentration inhibited photosynthesis, that the O_2 effects were rapidly reversible and that increased CO_2 concentrations overcame the oxygen inhibition. Since these effects were similar to those observed in leaves of C_3 plants they concluded that the effect of oxygen was on the RuBP carboxylase/oxygenase and it would appear that maize bundle sheath strands are capable of a C_3 photorespiration, but it does not follow that photorespiration is taking place in the maize leaves under normal conditions. In fact, CHOLLET and OGREN (1972a) suggest that oxygenase activity of intact leaves is suppressed due to high CO_2 concentrations in the bundle sheath.

In C_3 leaves supplied with $^{14}CO_2$, glycine and serine, two intermediates of the glycolate pathway are rapidly and heavily labelled (MAHON et al., 1974b,c). Not only do they label in concert, but when $^{12}CO_2$ is supplied the label disappears from them, also in concert (Fig. 9). Such behaviour is completely consistent with a rapid flow of carbon through the glycolate pathway (MAHON et al., 1974b, c; CANVIN et al., 1976), and it has been calculated that the flux of carbon through the pathway is sufficient for the measured rates of photorespiration (SCHNARREN-BERGER and FOCK, 1976). In C_4 leaves supplied with $^{14}CO_2$ there are many reports of ^{14}C-labelled glycine and serine (OSMOND et al., 1969; OSMOND and AVADHANI, 1970; OSMOND, 1972; SHOMER-ILAN et al., 1975; STAMIESZKIN et al., 1972; KENNEDY and LAETSCH, 1973; GRISHINA et al., 1974, 1975; CHAPMAN and LEECH, 1976) as minor components, but none of these studies shows data in support of a flux of carbon through these two compounds. Oxygen does have an effect on the amount of label in glycine and serine with 1% O_2 resulting in a decrease in the amount of radioactivity (OSMOND and BJÖRKMAN, 1972; MAHON et al., 1974a) and 100% O_2 resulting in an increase in the amount of radioactivity in these compounds (GRISHINA et al., 1974, 1975; KUBOWICZ et al., 1977).

Only one study has been conducted with ^{18}O incorporation in maize. In 30% O_2 and CO_2-free conditions the half-time for labelling of glycine was 1 to 2 min, and for serine 3 to 4 min (DIMON et al., 1977). This would be consistent with

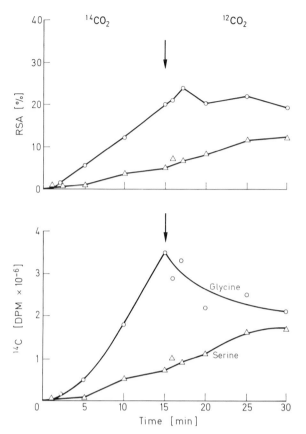

Fig. 9. The total radioactivity (*lower*) and relative specific activity (*upper*) of glycine and serine from corn leaves (cv. Golden Bantam) exposed to $^{14}CO_2$ (1.18×10^5 dpm μg^{-1} of C) 400 ppm of CO_2, 21% O_2, 25 °C, and 109 W m^{-2} for 15 min, followed by exposure to the same conditions and gas phase but without $^{14}CO_2$ for a further 15 min. (CANVIN et al., 1976)

rapid synthesis of these compounds from glycolate produced by the RuBP oxygenase reaction (ANDREWS et al., 1971; LORIMER et al., 1973, 1977) and would indicate rapid photorespiration in maize. Confirmation of these data would be helpful as the incorporation of ^{18}O and rapid flux through the glycolate pathway is inconsistent with flux experiments using $^{14}CO_2$.

The general picture that emerges from studies such as those detailed above is that the bundle sheath cells of C_4 plants, with the possible exception of sorghum (OSMOND and HARRIS, 1971; OSMOND, 1972; HUANG and BEEVERS, 1972), are capable of photorespiration in the C_3 pattern but this photorespiration is not displayed external to the leaf. Is this because the CO_2 is refixed by the mesophyll cells and subsequently recycled, or is the capacity for photorespiration inhibited and not expressed (OSMOND, 1971 a)?

IV. Evidence Against Photorespiration in C_4 Plants

The evidence from two types of experiment strongly suggests that photorespiration is inhibited and not expressed and that there is little recycling of CO_2 via the mesophyll cells. The first type of experiment has only been performed with maize (MAHON et al., 1974a; BLACKWOOD and MIFLIN, 1976) and shows a lack of flow of carbon through glycine and serine (Fig. 9). When $^{14}CO_2$ was supplied to a maize leaf, glycine and serine became labelled but when $^{12}CO_2$ was subsequently supplied the radioactivity in serine was not flushed or decreased. Such patterns on total radioactivity or specific activity (MAHON et al., 1974a; BLACKWOOD and MIFLIN, 1976) are not consistent with any significant flow through the glycolate pathway.

The second type of experiment has been performed with maize, *Amaranthus*, sugarcane (BULL, 1969), several C_4 *Atriplex* species and *Tidestromia oblongifolia* (BJÖRKMAN et al., 1970; OSMOND and BJÖRKMAN, 1972; EHLERINGER and BJÖRKMAN, 1977) and concerns the determination of the energy requirement for CO_2 evolution or refixation. In C_3 plants there is a demonstrable energy cost for photorespiration, and treatments that inhibit photorespiration such as 2% O_2 or high CO_2 concentrations (EHLERINGER and BJÖRKMAN, 1977; BJÖRKMAN, 1976) increase the quantum yield (Figs. 10 and 11). If photorespiration similar to that in C_3 plants proceeded in C_4 plants, there would also be an energy cost for the process. If all the CO_2 were refixed there would be a further energy cost. If photorespiration in C_4 plants were inhibited by low oxygen or high CO_2, as it is in C_3 plants, then an increase in quantum yield would be observed. No change, however, in quantum yield has been observed in any C_4 plant as a result of changing the oxygen concentration (Fig. 10; BULL, 1969; BJÖRKMAN et al., 1970; OSMOND and BJÖRKMAN, 1972; BJÖRKMAN, 1976; EHLERINGER and BJÖRKMAN, 1977), changing the CO_2 concentration or changing the temperature (Fig. 12; EHLERINGER and BJÖRKMAN, 1977). These extensive measurements on quantum yield demonstrate conclusively that photorespiration and concurrent refixation of the CO_2 does not proceed in C_4 plants.

If that is so, then we are left with the last possibility that the capacity for photorespiration is in some way suppressed in C_4 plants. At the present time all RuBP carboxylases that have been isolated have been shown by in vitro measurements to have oxygenase activity (JENSEN and BAHR, 1977) and current views are that, unless it is prevented in some way, this oxygenase activity will be expressed in vivo if RuBP is present. The suppression of oxygenase activity and the subsequent lack of glycolate formation would, of course, not allow the capacity of photorespiration to be expressed in C_4 plants. That oxygenase activity could be suppressed by high CO_2 concentrations in the bundle sheath cell was suggested by BJÖRKMAN (1971) and HATCH (1971a). HATCH (1971b) measured the size of the CO_2 pool in maize and *Amaranthus* and showed that it was ten times greater in the light than in darkness, and that it was larger than the concentration which would be in equilibrium with the ambient gas phase. Subsequently HATCH and OSMOND (1976) have developed earlier work of OSMOND (1971b) and HATCH (1971b, 1976) to show that the flux and decarboxylation rate of C_4 acids into the bundle sheath cells is sufficient to maintain a CO_2 concentration in the cells two to five times higher than the CO_2 concentration that would be in equilibrium with air. Even though the bundle sheath cells are permeable to CO_2 (GOLDSWORTHY and DAY, 1970), the calculated back flux of CO_2 to the mesophyll cells would be negligible (HATCH and OSMOND, 1976). While there is a need for further verification of the assumptions and a need to apply the approach of HATCH and OSMOND (1976) to other C_4 plants, it would seem possible that elevated CO_2 levels are maintained in the bundle sheath cells.

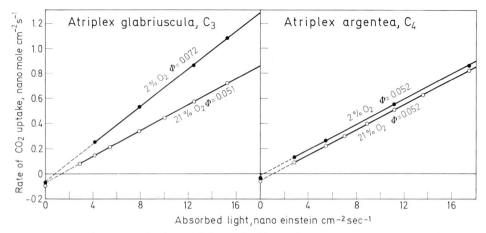

Fig. 10. Rate of CO_2 uptake in *A. glabriuscula* (C_3) and *A. argentea* (C_4) versus absorbed quantum flux in 21% and 2% O_2. Leaf temperature was 30 °C and CO_2 pressure was 325 µbar. (Ehleringer and Björkman, 1977)

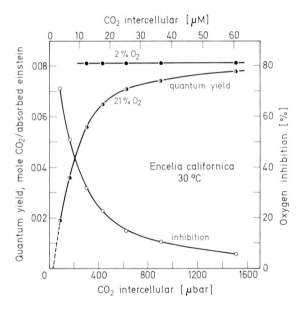

Fig. 11. Quantum yield for CO_2 uptake in *E. californica* (C_3) determined as a function of intercellular CO_2 pressure in 21% and 2% O_2 (●) and O_2 inhibition of quantum yield in 21% O_2 as a function of intercellular CO_2 pressure (○). Leaf temperature was 30 °C. (Ehleringer and Björkman, 1977)

F. Concluding Remarks

While some data on the external gas exchange of C_4 plants are suggestive of photorespiration in C_4 plants there is no definitive consistent evidence from the gas exchange data that demonstrates the existence of photorespiration in these plants. Because glycolate cannot be synthesized in the mesophyll cells it is most unlikely

Fig. 12. Quantum yield for CO_2 uptake in *E. californica* (C_3) and *A. rosea* (C_4) as a function of leaf temperature. *Curve* A of *E. californica* represents the measured quantum yields and *curve B* represents the quantum yields adjusted for changes in liquid phase solubilities of CO_2 and O_2. The CO_2 pressure was held constant at 325 µbar and O_2 concentration was 21%. (EHLERINGER and BJÖRKMAN, 1977)

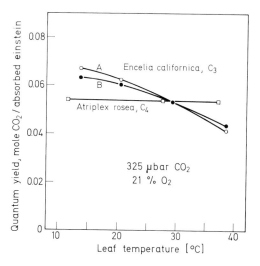

that photorespiration will occur in these cells. The presence of peroxisomes and glycolate pathway enzymes in the bundle sheath cells and experiments with isolated bundle sheath strands demonstrate that these cells appear to have the capacity for photorespiration. A flux of carbon through the glycolate pathway in C_4 plants under normal conditions has, however, not been demonstrated. In addition, treatments which suppress or enhance photorespiration in C_3 plants, when applied to C_4 plants, have no effect on the quantum yield. It thus seems that the capacity of photorespiration in the bundle sheath cells is inhibited or suppressed. At the moment, it seems likely that this suppression of photorespiration in C_4 plants is a result of the maintenance of elevated CO_2 levels in the bundle sheath cells.

References

Akita, S., Moss, D.N.: Crop Sci. *12*, 789–793 (1972)
Akita, S., Moss, D.N.: Crop Sci. *13*, 234–237 (1973)
Allen, J.F.: Nature (London) *256*, 599–600 (1975)
Andrews, T.J., Lorimer, G.H., Tolbert, N.E.: Biochemistry *10*, 4777–4782 (1971)
Badger, M.R., Andrews, T.J.: Biochem. Biophys. Res. Commun. *60*, 204–210 (1974)
Badger, M.R., Collatz, G.J.: Carnegie Inst. Yearb. *76*, 355–361 (1977)
Bahr, J.T., Jensen, R.G.: Biochem. Biophys. Res. Commun. *57*, 1180–1185 (1974)
Beevers, H.: In: Photosynthesis and photorespiration. Hatch, Osmond, Slatyer (eds.), pp. 541–543. New York: John Wiley and Sons 1971
Berry, J.A., Downton, W.J.S., Tregunna, E.B.: Can. J. Bot. *48*, 777–786 (1970)
Berry, J.A., Boynton, J., Kaplan, A., Badger, M.: Carnegie Inst. Yearb. *75*, 423–432 (1976)
Björkman, O.: Physiol. Plant. *19*, 618–633 (1966)
Björkman, O.: Carnegie Inst. Yearb. *66*, 220–228 (1968)
Björkman, O.: In: Photosynthesis and photorespiration. Hatch, Osmond, Slatyer (eds.), pp. 18–32. New York: John Wiley and Sons 1971
Björkman, O.: In: CO_2 metabolism and plant productivity. Burris, Black (eds.), pp. 287–309. Baltimore, Md.: Univ. Park Press 1976

Björkman, O., Gauhl, E., Nobs, M.A.: Carnegie Inst. Yearb. *68*, 620–623 (1970)
Black, C.C., Annu. Rev. Plant Physiol. *24*, 253–286 (1973)
Black, C.C., Williams, S.: In: CO_2 metabolism and plant productivity. Burris, Black (eds.), pp. 407–424. Baltimore, Md.: Univ. Park Press 1976
Black, C.C., Campbell, W.H., Chen, T.M., Dittrich, P.: Q. Rev. Biol. *48*, 299–313 (1973)
Black, C.C., Goldstein, L.C., Ray, T.B., Kestler, D.P., Mayne, B.C.: In: CO_2 metabolism and plant productivity. Burris, Black (eds.), pp. 113–139. Baltimore, Md.: Univ. Park Press 1976
Blackwood, G.C., Miflin, B.J.: J. Exp. Bot. *27*, 735–737 (1976)
Bold, H.C.: The plant kingdom. Englewood Cliffs, N.J.: Prentice-Hall Inc. 1970
Bowes, G., Ogren, W.L.: J. Biol. Chem. *247*, 2171–2176 (1972)
Bowes, G., Ogren, W.L., Hageman, R.H.: Biochem. Biophys. Res. Commun. *45*, 716–722 (1971)
Bravdo, B.A.: Plant Physiol. *43*, 479–483 (1968)
Bravdo, B.A.: Plant Physiol. *48*, 607–613 (1971)
Bravdo, B.A., Canvin, D.T.: Proc. 3rd Int. Congr. Photosynthesis. Avron (ed.), pp. 1277–1284. Amsterdam: Elsevier 1974
Brown, R.H., Brown, W.V.: Crop Sci. *15*, 681–685 (1975)
Brown, R.H., Gracen, V.E.: Crop Sci. *12*, 30–33 (1972)
Bull, T.A.: Crop Sci. *9*, 726–729 (1969)
Bulley, N.R., Tregunna, E.B.: Can. J. Bot. *48*, 1271–1277 (1970)
Bulley, N.R., Tregunna, E.B.: Can. J. Bot. *49*, 1277–1284 (1971)
Bulley, N.R., Nelson, C.O., Tregunna, E.B.: Plant Physiol. *44*, 678–685 (1969)
Cannell, R.Q., Brun, W.A., Moss, D.N.: Crop Sci. *9*, 840–842 (1969)
Canvin, D.T.: In: Genetic improvement of seed proteins. pp. 172–191. Washington, D.C.: Nat. Acad. Sci. 1976
Canvin, D.T., Fock, H.: In: Methods in enzymology. Part B. San Pietro (ed.), Vol. XXIV, pp. 246–260. New York: Academic Press 1972
Canvin, D.T., Lloyd, N.D.H., Fock, H., Przybylla, K.: In: CO_2 metabolism and plant productivity. Burris, Black (eds.), pp. 161–176. Baltimore, Md.: Univ. Park Press 1976
Carlson, G.E., Pearce, R.B., Lee, D.R., Hart, R.H.: Crop Sci. *11*, 35–38 (1971)
Chapman, D.J., Leech, R.M.: FEBS Lett. *68*, 160–163 (1976)
Chen, T.M., Brown, R.H., Black, C.C.: Weed Sci. *18*, 399–403 (1970)
Chollet, R.: Arch. Biochem. Biophys. *163*, 521–529 (1974)
Chollet, R., Ogren, W.L.: Biochem. Biophys. Res. Commun. *46*, 2062–2066 (1972a)
Chollet, R., Ogren, W.L.: Biochem. Biophys. Res. Commun. *48*, 684–688 (1972b)
Chollet, R., Ogren, W.L.: Plant Physiol. *51*, 787–792 (1973)
Chollet, R., Ogren, W.L.: Bot. Rev. *41*, 137–179 (1975)
Coombs, J.: Curr. Adv. Plant Sci. *2*, 1–10 (1973)
Cornic, G.: Physiol. Veg. *11*, 663–679 (1973)
Cornic, G.: Physiol. Veg. *12*, 83–94 (1974)
Cornic, G., Mousseau, M.: C. R. Acad. Sci. Ser. D. (Paris) *269*, 1774–1776 (1969)
Cornic, G., Chopin, G., Mousseau, M.: C. R. Acad. Sci. Ser. D. (Paris) *269*, 1194–1197 (1969)
Curtis, P.E., Ogren, W.L., Hageman, R.H.: Crop Sci. *9*, 323–327 (1969)
D'Aoust, A.L., Canvin, D.T.: Photosynthetica *6*, 150–157 (1972)
D'Aoust, A.L., Canvin, D.T.: Can. J. Bot. *51*, 457–464 (1973)
D'Aoust, A.L., Canvin, D.T.: Physiol. Veg. *12*, 545–560 (1974)
Decker, J.P.: Plant Physiol. *30*, 82–84 (1955)
Decker, J.P.: Plant Sci. Bull. *4*, 3–4 (1958)
Decker, J.P.: Plant Physiol. *34*, 100–102 (1959)
Dimon, B., Gerster, R., Tournier, P.: C. R. Acad. Sci. Ser. D. (Paris) *284*, 297–299 (1977)
Dittrich, P., Salin, M.L., Black, C.C.: Biochem. Biophys. Res. Commun. *55*, 104–110 (1973)
Downes, R.W., Hesketh, J.D.: Planta (Berl.) *78*, 79–84 (1968)
Downton, W.J.S.: Can. J. Bot. *48*, 1795–1801 (1970)
Downton, W.J.S.: Photosynthetica *9*, 96–105 (1975)
Downton, W.J.S., Tregunna, E.B.: Can. J. Bot. *46*, 207–215 (1968a)
Downton, W.J.S., Tregunna, E.B.: Plant Physiol. *43*, 923–929 (1968b)

Downton, W.J.S., Berry, J., Tregunna, E.B.: Science *163*, 78–79 (1969)
Edwards, G.E., Black, C.C.: In: Photosynthesis and photorespiration. Hatch, Osmond, Slayter (eds.). pp. 153–168. New York: John Wiley and Sons 1971
Edwards, G.E., Gutierrez, M.: Plant Physiol. *50*, 728–732 (1972)
Edwards, G.E., Huber, S.C., Ku, S.B., Rathnam, C.K.M., Gutierrez, M., Mayne, B.C.: In: CO_2 metabolism and plant productivity. Burris, Black (eds.), pp. 83–112. Baltimore, Md.: Univ. Park Press 1976
Egle, K.: In: Encyclopedia of plant physiology. Ruhland (ed.), Vol. V/I, pp. 182–210. Berlin, Heidelberg, New York: Springer 1960
Egle, K., Fock, H.: In: Biochemistry of chloroplasts. Goodwin (ed.), Vol. II, pp. 79–87. New York: Academic Press 1967
Egle, K., Schenk, W.: Planta (Berl.) *43*, 83–97 (1953)
Egneus, H., Heber, U., Matthiesen, V., Kirk, M.: Biochim. Biophys. Acta *408*, 252–269 (1975)
Ehleringer J., Bjorkman, O.: Plant Physiol. *59*, 86–90 (1977)
Ellyard, P.W., Gibbs, M.: Plant Physiol. *44*, 1115–1122 (1969)
El-Sharkawy, M.A., Loomis, R.S., Williams, W.A.: Physiol. Plant. *20*, 171–186 (1967)
Fair, P., Tew, J., Cresswell, C.: J. S. Afr. Bot. *38*, 81–95 (1972)
Fair, P., Tew, J., Cresswell, C.: Ann. Bot. *37*, 831–844 (1973)
Fair, P., Tew, J., Cresswell, C.: Ann. Bot. *38*, 39–44 (1974a)
Fair, P., Tew, J., Cresswell, C.: Ann. Bot. *38*, 45–52 (1974b)
Fock, H.: Biol. Zentralbl. *89*, 545–572 (1970)
Fock, H., Egle, K.: Beitr. Biol. Pflanz. *42*, 213–239 (1966)
Fock, H., Becker, J.D., Egle, K.: Can. J. Bot. *48*, 1185–1191 (1970)
Fock, H., Canvin, D.T., Grant, B.R.: Photosynthetica *5*, 389–394 (1971)
Fock, H., Mahon, J.D., Canvin, D.T., Grant, B.R.: In: Mechanisms of regulation of plant growth. Bieleski, Ferguson, Cresswell (eds.), Bull. 12, pp. 235–242. Wellington: Royal Soc. N. Z. 1974
Forrester, M.L., Krotkov, G., Nelson, C.D.: Plant Physiol. *41*, 422–427 (1966a)
Forrester, M.L., Krotkov, G., Nelson, C.D.: Plant Physiol. *41*, 428–431 (1966b)
Forti, G., Gerola, P.: Plant Physiol. *59*, 859–862 (1977)
Forti, G., Jagendorf, A.T.: Biochim. Biophys. Acta *54*, 322–330 (1961)
Frederick, S.E., Newcomb, E.H.: Planta (Berl.) *96*, 152–175 (1971)
Gauhl, E., Bjorkman, O.: Planta (Berl.) *88*, 187–191 (1969)
Gerster, R., Dimon, B., Peybernes, A.: Proc. 3rd Int. Congr. Photosynthesis. pp. 1589–1600. 1974
Gibbs, M.: In: How crops grow. A century later. Conn. Agric. Exp. State Bull. *708*, 63–79 (1969)
Goldstein, L.D., Ray, T.B., Kestler, D.P., Mayne, B.C., Brown, R.H., Black, C.C.: Plant Sci. Lett. *6*, 85–91 (1976)
Goldsworthy, A.: Phytochemistry *5*, 1013–1019 (1966)
Goldsworthy, A.: Bot. Rev. *36*, 321–340 (1970)
Goldsworthy, A., Day, P.R.: Nature (London) *228*, 687–688 (1970)
Grishina, G.S., Maleszewski, S., Frankevicz, A., Voskresenskaya, N.P., Poskuta, J.: Z. Pflanzenphysiol. *73*, 189–198 (1974)
Grishina, G.S., Maleszewski, S., Frankevicz, A., Poskuta, J., Voskresenskaya, N.P.: Sov. Plant Physiol. (Engl. transl.) *22*, 18–22 (1975)
Grossman, D., Cresswell, C.F.: Proc. Grassland Soc. S. Afr. *9*, 89–94 (1974a)
Grossman, D., Cresswell, C.F.: Proc. Grassland Soc. S. Afr. *9*, 95–104 (1974b)
Hatch, M.D.: In: Photosynthesis and photorespiration. Hatch, Osmond, Slatyer (eds.), pp. 139–152. New York: John Wiley and Sons Inc. 1971a
Hatch, M.D.: Biochem J. *125*, 425–433 (1971b)
Hatch, M.D.: In: Plant biochemistry. Bonner, Varner (eds.), pp. 797–844. New York: Academic Press 1976
Hatch, M.D., Osmond, C.B.: In: Encyclopedia of plant physiology. Stocking, Heber (eds.), Vol. III, pp. 144–184. Berlin, Heidelberg, New York: Springer 1976
Hatch, M.D., Slack, C.R.: Biochem. J. *101*, 130–111 (1966)
Heath, O.J.S., Orchard, B.: Nature (London) *180*, 180–181 (1957)

Heath, O.J.S., Meidner, H., Spanner, D.C.: J. Exp. Bot. *18*, 746–751 (1967)
Heichel, G.H.: Plant Physiol. *46*, 359–363 (1970)
Heichel, G.H.: Plant Physiol. *48*, 178–182 (1971)
Heichel, G.H.: Plant Physiol. *49*, 490–496 (1972)
Heichel, G.H., Musgrave, R.B.: Plant Physiol. *44*, 1724–1729 (1969)
Hesketh, J.: Planta (Berl.) *76*, 371–374 (1967)
Hew, C.S., Krotkov, G., Canvin, D.T.: Plant Physiol. *44*, 662–671 (1969a)
Hew, C.S., Krotkov, G., Canvin, D.T.: Plant Physiol. *44*, 671–678 (1969b)
Hilliard, J.H., Gracen, V.E., West, S.H.: Planta (Berl.) *97*, 93–106 (1971)
Hofstra, G., Hesketh, J.D.: Planta (Berl.) *85*, 228–238 (1969)
Holmgren, P., Jarvis, P.: Physiol. Plant. *20*, 1045–1051 (1967)
Huang, A.H.C., Beevers, H.: Plant Physiol. *50*, 242–248 (1972)
Huber, S., Edwards, G.: Biochem. Biophys. Res. Commun. *67*, 28–35 (1975)
Irvine, J.E.: Physiol. Plant. *23*, 607–613 (1970)
Jackson, W.A., Volk, R.J.: Nature (London) *222*, 269–271 (1969)
Jackson, W.A., Volk, R.J.: Annu. Rev. Plant Physiol. *21*, 385–432 (1970)
Jensen, R.G., Bahr, J.T.: Annu. Rev. Plant Physiol. *28*, 379–400 (1977)
Jolliffe, P.A., Tregunna, E.B.: Plant Physiol. *43*, 902–906 (1968)
Jolliffe, P.A., Tregunna, E.B.: Can. J. Bot. *51*, 841–853 (1973)
Kanai, R., Kashiwagi, M.: Plant Cell Physiol. *16*, 669–679 (1975)
Keck, R.W., Ogren, W.L.: Plant Physiol. *58*, 552–555 (1976)
Kennedy, R.A.: Plant Physiol. *58*, 573–575 (1976a)
Kennedy, R.A.: Planta (Berl.) *128*, 149–155 (1976b)
Kennedy, R.A., Laetsch, W.M.: Planta (Berl.) *115*, 113–124 (1973)
Kestler, D.P., Mayne, B.C., Ray, T.B., Goldstein, L.D., Brown, R.H., Black, C.C.: Biochem.
    Biophys. Res. Commun. *66*, 1439–1447 (1975)
Kirk, M.R., Heber, U.: Planta (Berl.) *132*, 131–141 (1976)
Kisaki, T., Tolbert, N.E.: Plant Cell Physiol. *11*, 247–258 (1970)
Kisaki, T., Yano, N., Hirabayaski, S.: Plant and Cell Physiol. *13*, 581–584 (1972)
Krall, A.R., Bass, E.R.: Nature (London) *196*, 791–792 (1962)
Krenzer, E.G., Moss, D.N.: Crop Sci. *9*, 619–621 (1969)
Krenzer, E.G., Moss, D.N., Crookston, R.K.: Plant Physiol. *56*, 194–206 (1975)
Ku, S.B., Edwards, G.E.: Z. Pflanz. *77*, 16–33 (1975)
Ku, S.B., Edwards, G.E.: Plant Physiol. *59*, 986–990 (1977a)
Ku, S.B., Edwards, G.E.: Plant Physiol. *59*, 991–999 (1977b)
Ku, S.B., Edwards, G.E., Kanai, R.: Plant Cell Physiol. *17*, 615–620 (1976)
Kubowicz, D., Maleszewski, S., Poskuta, J.: Z. Pflanzenphysiol. *81*, 141–146 (1977)
Laing, W.A., Forde, B.J.: Planta (Berl.) *98*, 221–231 (1971)
Laing, W.A., Ogren, W., Hageman, R.H.: Plant Physiol. *54*, 678–686 (1974)
Lake, J.: Aust. J. Biol. Sci. *20*, 487–493 (1967)
Lester, J.W., Goldsworthy, A.J.: J. Exp. Bot. *24*, 1031–1034 (1973)
Liu, A.Y., Black, C.C.: Arch. Biochem. Biophys. *149*, 269–281 (1972)
Lloyd, N.D.H., Canvin, D.T.: Can. J. Bot. *55*, 3006–3012 (1977)
Lloyd, N.D.H., Canvin, D.T., Culver, D.A.: Plant Physiol. *59*, 936–940 (1977a)
Lloyd, N.D.H., Canvin, D.T., Bristow, J.M.: Can. J. Bot. *55*, 3001–3005 (1977b)
Long, S.P., Incoll, L.D., Woolhouse, H.D.: Nature (London) *257*, 622–625 (1975)
Lorimer, G.H., Andrews, T.J.: Nature (London) *243*, 359 (1973)
Lorimer, G.H., Andrews, T.J., Tolbert, N.E.: Biochemistry *12*, 18–23 (1973)
Lorimer, G.H., Krause, G.H., Berry, J.A.: FEBS Lett. *78*, 199–202 (1977)
Ludlow, M.: Planta (Berl.) *91*, 285–291 (1970)
Ludlow, M.M., Jarvis, P.G.: In: Plant photosynthetic production. Sestak, Catsky, Jarvis (eds.),
    pp. 294–315. The Hague: Dr. W. Junk, N.V. 1971
Ludwig, L.J.: Ph.D. Thesis, Queen's Univ. Kingston, Ontario (1968)
Ludwig, L.J., Canvin, D.T.: Can. J. Bot. *49*, 1299–1319 (1971a)
Ludwig, L.J., Canvin, D.T.: Plant Physiol. *48*, 712–719 (1971b)
Mahon, J.D., Fock, H., Höhler, T., Canvin, D.T.: Planta (Berl.) *120*, 113–123 (1974a)
Mahon, J.D., Fock, H., Canvin, D.T.: Planta (Berl.) *120*, 125–134 (1974b)
Mahon, J.D., Fock, H., Canvin, D.T.: Planta (Berl.) *120*, 245–254 (1974c)

Mansfield, T.A.: Physiol. Planta. *21*, 1159–1162 (1968)

McAllister, E.D., Meyers, J.: Smithson. Misc. Collect. *99*, 6–26 (1940)

Mehler, A.H.: Arch. Biochem. Biophys. *33*, 65–77 (1951)

Meidner, H.: J. Exp. Bot. *13*, 284–293 (1962)

Meidner, H.: J. Exp. Bot. *18*, 177–185 (1967)

Meidner, H.: J. Exp. Bot. *21*, 1067–1075 (1970)

Merrett, M.J., Lord, J.M.: New Phytol. *72*, 751–768 (1973)

Moss, D.N.: Nature (London) *193*, 587 (1962)

Moss, D.N.: Crop Sci. *6*, 351–354 (1966)

Moss, D.N.: In: Photosynthesis and photorespiration. Hatch, Osmond, Slatyer (eds.), pp. 120–123. New York: John Wiley and Sons Inc., 1971

Moss, D.N., Musgrave, R.B.: Adv. Agron. *23*, 317–336 (1971)

Moss, D.N., Krenzer, E.G., Brun, W.A.: Science *164*, 187–188 (1969)

Moss, D.N., Willmer, C.M., Crookston, R.K.: Plant Physiol. *47*, 847–848 (1971)

Mulchi, C.L., Volk, R.J., Jackson, W.A.: In: Photosynthesis and photorespiration. Hatch, Osmond, Slatyer (eds.), pp. 35–50. New York: John Wiley and Sons Inc., 1971

Nelson, C.J., Asay, K.N., Patton, L.D.: Crop Sci. *15*, 629–633 (1975)

Nishimura, M., Graham, D., Akazawa, T.: Plant Physiol. *56*, 718–722 (1975)

Ogren, W.L., Bowes, G.: Nature New Biol. *230*, 159–160 (1971)

Osmond, C.B.: In: Photosynthesis and photorespiration. Hatch, Osmond, Slatyer (eds.), pp. 472–482. New York: John Wiley and Sons Inc. 1971 a

Osmond, C.B.: Aust. J. Biol. Sci. *24*, 159–163 (1971 b)

Osmond, C.B.: In: Proc. 2nd Int. Congr. Photosynthesis. Forti, Avron, Melandri (eds.), pp. 2233–2239. The Hague: Dr. W. Junk N.V. 1972

Osmond, C.B., Avadhani, P.N.: Plant Physiol. *45*, 228–231 (1970)

Osmond, C.B., Björkman, O.: Carnegie Inst. Yearb. *71*, 141–148 (1972)

Osmond, C.B., Harris, B.: Biochim. Biophys. Acta *234*, 270–287 (1971)

Osmond, C.B., Troughton, J.H., Goodchild, D.J.: Z. Pflanzenphysiol. *61*, 218–237 (1969)

Ozbun, J.L., Volk, R.J., Jackson, W.A.: Plant Physiol. *39*, 523–527 (1964)

Pallas, J.E.: Science *147*, 171–173 (1965)

Poskuta, J.: Sept. Exp. *24*, 344 (1968 a)

Poskuta, J.: Physiol. Plant. *21*, 1129–1135 (1968 b)

Poskuta, J.: Physiol. Plant. *22*, 76–85 (1969)

Quebedeaux, B., Chollet, R.: Plant Physiol. *59*, 42–44 (1977)

Rabinowitch, E.I.: Photosynthesis and related processes. Vol. I. Vol. II, Parts 1 and 2. New York: Interscience Publishers Inc., 1945, 1951, 1956

Radmer, R.J., Kok, B.: Plant Physiol. *58*, 336–340 (1976)

Raschke, K.: Plant Physiol. *49*, 229–234 (1972)

Rathnam, C.K.M., Das, R.: Indian, J. Biochem. Biophys. *11*, 303–306 (1974)

Ray, T.B., Black, C.C.: J. Biol. Chem. *251*, 5824–5826 (1976)

Ray, T.B., Black, C.C.: Plant Physiol. *60*, 193–196 (1977)

Rehfeld, D.W., Randall, D.D., Tolbert, N.E.: Can. J. Bot. *48*, 1219–1226 (1970)

Rosenstock, G., Ried, A.: In: Encyclopedia of plant physiol. Ruhland (ed.), Vol. XII/2, pp. 259–233. Berlin, Heidelberg, New York: Springer 1960

Sayre, R.T., Kennedy, R.A.: Planta (Berl.) *134*, 257–262 (1977)

Schnarrenberger, C., Fock, H.: In: Encyclopedia of plant physiology. New Ser. Stocking, Heber (eds.), Vol. III, pp. 185–234. Berlin, Heidelberg, New York: Springer 1976

Shain, Y., Gibbs, M.: Plant Physiol. *48*, 325–330 (1971)

Shomer-Ilan, A., Beer, S., Waisel, Y.: Plant Physiol. *56*, 676–680 (1975)

Smith, E.W., Tolbert, N.E., Ku, H.S.: Plant Physiol. *58*, 143–146 (1976)

Stamieszkin, E., Maleszewski, S., Poskuta, J.: Z. Pflanzenphysiol. *67*, 180–182 (1972)

Tolbert, N.E.: In: Photosynthetic mechanisms in green plants. Publication 1145. Natl. Acad. Sci. USA pp. 648–662. Natl. Res. Council 1963

Tolbert, N.E.: Annu. Rev. Plant Physiol. *22*, 45–74 (1971)

Tolbert, N.E.: In: Current topics in cellular regulation. Horecker, Stadtman (eds.), pp. 21–50. New York: Academic Press 1973

Tolbert, N.E.: In: Algal physiology and biochemistry. Stewart (ed.), pp. 474–504. Oxford: Blackwell Scientific Publ. 1974

Tolbert, N.E., Oeser, A., Yamazaki, R.K., Hageman, R.H., Kisaki, T.: Plant Physiol. *44*, 135–147 (1969)
Tregunna, E.B.: Science *151*, 1239–1241 (1966)
Tregunna, E.B., Downton, J.: Can. J. Bot. *45*, 2385–2387 (1967)
Tregunna, E.B., Krotkov, G., Nelson, C.D.: Can. J. Bot. *42*, 989–997 (1964)
Tregunna, E.B., Krotkov, G., Nelson, C.O.: Physiol. Plant. *19*, 723–733 (1966)
Troughton, J.H.: Planta (Berl.) *100*, 87–92 (1971)
Turner, J.S., Brittain, E.G.: Biol. Rev. *37*, 130–170 (1962)
Volk, R.J., Jackson, W.A.: Plant Physiol. *49*, 218–224 (1972)
Viil, J., Parnik, T.: Photosynthetica *8*, 208–215 (1974)
Voskresenskaya, N.P., Grishina, G.S., Chmora, S.N., Poyarkova, N.M.: Sov. Plant Physiol. (Engl. transl.) *17*, 195–202 (1970)
Voskresenskaya, N.P., Polyakov, M.A., Karpushkin, L.T.: Sov. Plant Physiol. (Engl. translation) *21*, 367–372 (1974)
Voznesenskii, V.L.: Sov. Plant Physiol. (Engl. translation) *12*, 652–659 (1965)
Waggoner, P.: Crop Sci. *9*, 315–321 (1969)
Wang, D., Waygood, E.R.: Plant Physiol. *37*, 826–832 (1962)
Whitehouse, D.G., Ludwig, L.J., Walker, D.A.: J. Exp. Bot. *22*, 772–792 (1971)
Whiteman, P.C., Koller, D.: New Phytol. *66*, 463–473 (1967)
Wilson, D.: J. Exp. Bot. *23*, 517–524 (1972)
Wolf, F.T.: Adv. Frontiers Plant Sci. *26*, 161–231 (1970)
Woo, K.C., Osmond, C.B.: In: Photosynthetic organelles. Miyachi, Katoh, Fujita, Shibata (eds.), pp. 315–323. Tokyo: Japan Soc. Pl. Physiol. Ctr. Acad. Publ. 1977
Wynn, T., Brown, H., Campbell, W.H., Black, C.C.: Plant Physiol. *52*, 288–291 (1973)
Yemm, E.W., Bidwell, R.G.S.: Plant Physiol. *44*, 1328–1335 (1969)
Zelitch, I.: J. Biol. Chem. *240*, 1869–1876 (1965)
Zelitch, I.: Plant Physiol. *41*, 1623–1631 (1966)
Zelitch, I.: Plant Physiol. *43*, 1829–1837 (1968)
Zelitch, I.: Photosynthesis, photorespiration and plant productivity. New York: Academic Press 1971
Zelitch, I.: Plant Physiol. *51*, 299–305 (1973a)
Zelitch, I.: Curr. Adv. Plant Sci. *3*, 44–54 (1973b)
Zelitch, I.: Annu. Rev. Biochem. *44*, 123–145 (1975)

III. Ferredoxin-Linked Reactions

1. Transhydrogenase

P. Böger

A. Introduction and Definitions

This chapter deals with nicotinamide nucleotide transhydrogenases, E.C.1.6.1.1 (pyridine nucleotide oxidoreductase) which catalyze hydrogen transfer between two pyridine nucleotides:

$$NADPH + NAD^+ \rightleftharpoons NADP^+ + NADH \tag{1}$$

According to the source of the enzyme the reaction may or may not be reversible, and since definite specificity is lacking, the hydrogen acceptor may be generally substituted for by various nucleotide analogs such as thionicotinamide $NAD(P)^+$ [$=TN-NAD(P)^+$] or 3-acetylpyridine-adenine dinucleotide (phosphate). TN-NAD(P)H has an absorption peak at 400 nm ($\varepsilon = 11.3 \, \text{mM}^{-1} \cdot \text{cm}^{-1}$), which often allows convenient determination of the reduced mother compounds in addition to the analog (COHEN and KAPLAN, 1970b; BÖGER, 1972a). A transhydrogenase reaction as defined here does not imply a third substrate mediating hydrogen transfer as, for example, malate in malate/lactate "transhydrogenase" (ALLEN and PATIL, 1972) or glutamate in glutamic dehydrogenases which are unspecific for pyridine nucleotides.

Nicotinamide nucleotide transhydrogenases can be divided into two groups (ERNSTER, 1964; RYDSTRÖM et al., 1976):
a) Soluble flavoproteins, apparently all containing FAD. The hydrogen transfer is stereospecific from and to the B-side of the pyridine ring of the nucleotide, without involving proton exchange with water (LOUIE and KAPLAN, 1970; HOEK et al., 1974). Specificity for the nucleotides is often not very great, although transfer from NADPH to NAD^+ is predominant. The equilibrium constant is close to unity (BÖGER, 1972a) provided the transhydrogenation is reversible. The reaction is under allosteric control.
b) Membrane-bound hydrophobic proteins specific for the 4A-hydrogen of NADH and the 4B-hydrogen of NADPH. Again, no proton exchange with water is observed (LEE et al., 1965; GRIFFITHS and ROBERTSON, 1966; FISHER and GUILLORY, 1971b). Their complete purification has been achieved only in two cases recently and it appears that they are not flavoproteins. Reduction of $NADP^+$ by NADH (or NADPH) is an energy-linked process in mitochondria, while the reaction NADPH $\rightarrow NAD^+$ generally is unaffected by the energized state of the membrane and attains an equilibrium of about unity, as expected from the midpoint-redox potentials of the two nucleotides involved (nonenergy-linked reaction, see RYDSTRÖM et al., 1976). The reverse reaction (NADH$\rightarrow NADP^+$ or TN-$NADP^+$) is slow in nonenergized membranes but can be linked to electron transport and energy-

conserving mechanisms (FISHER and GUILLORY, 1971 b), thereby permitting an "energization" of the reaction which accelerates it several-fold and markedly increases the (apparent) equilibrium constant. Both energy- and nonenergy-linked transhydrogenase reactions are catalyzed by the same enzyme (KAWASAKI et al., 1964; LEE and ERNSTER, 1964; FISHER and KAPLAN, 1973).

This article will not review all aspects of transhydrogenases and many molecular details are omitted (see e.g., RYDSTRÖM et al., 1976; RYDSTRÖM, 1977), but a brief scope of general facts, emphasizing photosynthetic organisms and ferredoxin-NAD(P)$^+$ reductase-containing bacteria is included. References should primarily serve as guide-lines to additional literature.

B. Soluble Flavoproteins with Transhydrogenase Activity

I. Bacterial Enzymes

Elaborate work has been done on the bacterial transhydrogenase by KAPLAN and coworkers with *Pseudomonas* sp. (for refs. see RYDSTRÖM et al., 1976; WIDMER and KAPLAN, 1976), and by CHUNG (MIDDLEDITCH and CHUNG, 1971) and Veeger (VAN DEN BROEK et al., 1971a) with *Azotobacter vinelandii*. The enzymes from different bacteria show a remarkable divergence in their properties. The transhydrogenase from both species mentioned and that from *Chromatium vinosum* (KEISTER and HEMMES, 1966) catalyze the reversible oxidation of NADPH by NAD$^+$ with a pH optimum around 7, although the degree of reversibility depends on the concentration of the nucleotides and may be influenced by certain activators. The *Pseudomonas* reaction appears to be irreversible (NADPH→NAD$^+$), since NADP$^+$ is an efficient inhibitor of the enzyme and therefore prevents the reaction from attaining equilibrium. 2'-AMP acts as an activator for the forward reaction, alleviates (product) inhibition by NADP$^+$ and allows the reverse reaction (NADH →NADP$^+$) to proceed. Furthermore, this activator is necessary when transhydrogenation occurs between diphosphopyridine nucleotides and its analogs (COHEN and KAPLAN, 1970b; WIDMER and KAPLAN, 1976). 2'-AMP has no influence on the (forward) reaction NADPH→NAD$^+$ with the *Azotobacter* enzyme (CHUNG, 1970); it was inhibitory with the *Chromatium* transhydrogenase, as was NADP$^+$, both competing with NADPH or NADH at the same donor site (KEISTER and HEMMES, 1966). 2'-AMP may be replaced by Ca^{2+} or Mn^{2+} (RYDSTRÖM et al., 1973).

Activation phenomena are superimposed by competition and inhibition effects of hydrogen donor and acceptor (COHEN and KAPLAN, 1970b); this has prevented elucidation of all details of the reaction mechanisms. The K$_m$ and K$_i$ data are between 10 to 125 μM (*Azotobacter* enzyme; VAN DEN BROEK and VEEGER, 1971b; see also COHEN and KAPLAN, 1970b; RYDSTRÖM et al., 1976) which can be considered as physiological concentrations, but reports on the metabolic role of these bacterial transhydrogenases are speculative.

Furthermore, transhydrogenases of this type have "diaphorase" activities, which allows for a transfer of hydrogen (or electrons) from NAD(P)H to unphysiological acceptors such as dichlorophenol indophenol or potassium ferricyanide (COHEN and KAPLAN, 1970a; VAN DEN BROEK et al., 1971a) although this does not hold for the *Chromatium* enzyme (KEISTER and HEMMES, 1966). Diaphorase activity is also often observed with FMN-containing flavoproteins (PALMER and MASSEY, 1968) which have no transhydrogenase activity.

II. Ferredoxin-NAD(P)$^+$ Reductases

Noteworthy are those transhydrogenases which also exhibit ferredoxin-NAD(P)$^+$ reductase activity (E.C.1.6.7.1). In the last decade, this enzyme was isolated and characterized from higher plants (first described by KEISTER et al., 1960a, b; SHIN et al., 1963), and from algae (YAMANAKA and KAMEN, 1966; BÖGER, 1971a) and was later demonstrated in ferredoxin-containing anaerobic bacteria. The general characteristics of the (higher) plant reductase have been described by FORTI, 1977. Consequently, only essential characteristics and the interaction of the reductase with ferredoxin, will be given here. In (aerobic) photosynthetic eukaryotes and blue-green algae this protein functions as the terminal enzyme of the photosynthetic redox chain. It was partially characterized in crude extracts from *Chromatium* (BUCHANAN and BACHOFEN, 1968) and *Clostridia* (THAUER et al., 1971; PETITDE-MANGE et al., 1973); in the latter case it occurs together with an NAD(H)-specific reductase (JUNGERMANN et al., 1971). Purified preparations were obtained from *Bacillus polymyxa* (YOCH, 1973), *Rhodopseudomonas palustris* (YAMANAKA and KA-MEN, 1967) and *Chlorobium thiosulfatophilum* (KUSAI and YAMANAKA, 1973).

As far as assayed, all these transhydrogenases catalyze an *irreversible* hydrogen transfer from NADPH to NAD$^+$ in vitro (KEISTER et al., 1960a; YAMANAKA and KAMEN, 1967; BÖGER, 1971a, b; THAUER et al., 1971; BÖGER, 1972b; KUSAI and YAMANAKA, 1973; YOCH, 1973) and a reversible transfer between NADPH and a corresponding analog of the mother compound (BÖGER, 1972a). The enzyme from the photosynthetic bacteria mentioned above uses both NADP(H) and NAD(H) in various redox reactions, e.g., reduction of cytochrome(s) c or photosynthetic reduction of pyridine nucleotides (KUSAI and YAMANAKA, 1973).

More detailed kinetic studies with these types of transhydrogenases have been performed with the plant and algae enzyme only (KEISTER et al., 1960a; BÖGER, 1971b; BÖGER, 1972b). NADP$^+$ and 2'-AMP compete with both the donor and acceptor nucleotides at a common binding site and cannot reverse the transhydrogenase reaction. It could be shown that the irreversibility of the reaction NADPH →NAD$^+$ is apparently due to the high K_m for NADH being in the same range (10^{-4} M) as the K_i, whereas the K_m for NADPH (approx. 10^{-7} M) is well below the corresponding K_i for NADPH (about 10^{-4} M). The substrates for the diaphorase reaction (e.g., dichlorophenolindophenol, methylviologen) occupy a reaction site different from that of the pyridine nucleotides; they do not compete with the diaphorase substrates. As was shown in detail for the reductase from the eukaryotic alga *Bumilleriopsis*, this site is seemingly identical with that for ferredoxin, which is the natural "diaphorase" substrate. Apparently, the same holds true for the spinach enzyme. One site is specific for ferredoxin and a second one for the pyridine nucleotides, as can be concluded from kinetic data (BÖGER, 1971a, b; BÖGER, 1972b), from fluorescence behavior in the presence of NADP$^+$

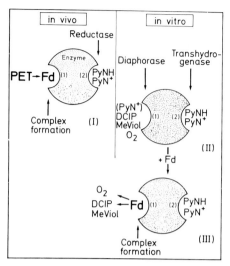

Fig. 1. Tentative scheme of binding sites on plastidic ferredoxin-NADP$^+$ reductase according to Böger (1971b, 1972b). **I** Photosynthetic NADP$^+$ reduction (*PET*, photosynthetic electron transport system). The enzyme forms a complex with ferredoxin (*Fd*) at site *1* and may bind reduced (*PyNH*) or oxidized (*PyN$^+$*) pyridine nucleotides at reaction site *2*. PyNH competes with PyN$^+$ (=reductase activity). **II** A transhydrogenase reaction (NADPH→NAD$^+$ or NADPH→TN-NADP$^+$ takes place at site *2* via a ping-pong mechanism. Rates depend on the plant species. For freshly prepared *Bumilleriopsis* transhydrogenase initial rates are 0.19 mmol NADH formed/μmol enzyme per min and 4.2 mmol TN-NADPH formed/μmol enzyme per min, both at pH 8.0 (reaction conditions in Böger, 1971a). Diaphorase substrates react at site *1*; these are dichlorophenolindophenol (DCIP) or methylviologen (MeViol, 1,1-dimethyl-4,4'-dipyridylium dichloride), oxygen, etc. Diaphorase activity of the *Bumilleriopsis* enzyme is about 2 mmol DCIP reduced/μmol enzyme per min at pH 8.0 (see Böger, 1971a, for assay). Apparently, pyridine nucleotides may also be bound at site *1* under certain conditions (see Nelson and Neumann, 1969; Böger, 1972a). **III** With low salt concentrations present (about 0.1 M) ferredoxin forms a complex with the reductase which enhances both the transhydrogenase and diaphorase activities. The affinity of ferredoxin to the enzyme is high, the diaphorase substrates may then react with bound ferredoxin. Under conditions of very low ionic strength (0.01) the bound ferredoxin is partially competitive with NADPH (see Nakamura and Kimura, 1971a, and text)

and ferredoxin (Böger et al., 1973), or further from studies with modified reductase. Butanedione inactivates the enzyme by reacting with arginine residues, one of which is responsible for the binding of nucleotides (Bookjans and Böger, 1978). The treated enzyme, however, still forms a complex with ferredoxin but not with NADP$^+$ (Bookjans and Böger, 1979).

The reaction site for pyridine nucleotides in the transhydrogenase reaction is identical with that of the (photosynthetic) ferredoxin-NADP$^+$ reductase activity. This was concluded by comparative investigation of the kinetic parameters of photosynthetic NADP$^+$ reduction versus the NADP→TN-NADP$^+$ transhydrogenase reaction. Figure 1 schematically demonstrates the conclusions.

The enzyme is easily reduced by NAD(P)H (SHIN and ARNON, 1965); in the reduced state it binds the oxidized nucleotide more strongly than the reduced nucleotide, and vice versa (BÖGER, 1971b). Consequently, as long as the enzyme is kept reduced by ferredoxin under physiological photosynthetic conditions, transhydrogenase activity is not likely to occur, but the enzyme acts as a reductase. Furthermore, the high K_m for NAD^+ (about 10^{-3} M; see also YOCH, 1973) renders it most unlikely that the reducing power is dissipated by a transhydrogenase reaction as proposed by other authors (GREEN and ISRAELSTAMM, 1970; CHOPOWICK and ISRAELSTAMM, 1973; KRAWETZ and ISRAELSTAMM, 1978), since physiological concentrations of the nucleotides in the chloroplast are in the 50–200 µM range (LENDZIAN and BASSHAM, 1976) or even smaller in the cell (KUSAI and YAMANAKA, 1973). The occurrence of NADH in addition to NADPH in the chloroplasts is most probably related to secondary reactions, e.g., the oxidation of malate (KRAUSE and HEBER, 1976).

Of particular interest is the three to four fold stimulation by oxidized ferredoxin of the V_{max} of the transhydrogenase (and diaphorase) activity, as observed with the clostridial enzyme (THAUER et al., 1971) and with the enzyme from higher plants (FREDERICKS and GEHL, 1971) and algae (*Bumilleriopsis*; BÖGER, 1971a; BÖGER, 1972b). This nonheme iron protein acts as a modulator upon the flavoprotein and furthermore it functions as its electron donor in the ferredoxin-NADP$^+$ reductase reaction. This dual role may explain the failure to replace ferredoxin by different dipyridylium salts (having an E_0' of < -400 mV) in the photosynthetic redox transport system, even under strictly anaerobic conditions and with photosynthetic oxygen evolution excluded (BÖGER, unpubl.). This agrees, for example, with the observation that the reduction of the enzyme from *Navicula* could be achieved much better (by sodium dithionite, $Na_2S_2O_4$) via ferredoxin than via benzylviologen (YAMANAKA and KAMEN, 1966). Ferredoxin appears to act rather specifically (BÖGER, 1971b), and stimulations by phytoflavin (from *Chlorella*) or thiol reagents could not be found in this laboratory. Under low ionic strength polylysine exerts a stimulation of activity (SCHNEEMAN and KROGMANN, 1975).

The stimulation by ferredoxin is observed under conditions of complex formation between both proteins (e.g., SHIN and SAN PIETRO, 1968; BÖGER, 1971a; higher concentrations of ferredoxin may inhibit the transhydrogenase reaction by competing with the nucleotides (NELSON and NEUMANN, 1969) or by mediating NADPH oxidation with molecular oxygen; the irreversibility of the transhydrogenase reaction is not altered by ferredoxin (BÖGER, 1971b).

As was first shown by KIMURA and coworkers and confirmed in this laboratory, under low ionic strength ($=0.01$) ferredoxin inhibits (diaphorase) activity, whereas stimulation is attained with about 0.1 M NaCl (NAKAMURA and KIMURA, 1971a, b). This inhibition is partially competitive with respect to NADPH (with a K_i of $2 \cdot 10^{-7}$ M; NAKAMURA and KIMURA, 1971a). Under these conditions ferredoxin seemingly interferes with the access or binding of NADP(H) to its reaction site. Complex formation between ferredoxin-NADP$^+$ reductase and ferredoxin is necessary for photosynthetic NADP$^+$ reduction (for refs. see FORTI, 1977 and DAVIS and SAN PIETRO, 1977a). The latter authors presented direct proof by using trinitrophenylated spinach ferredoxin which was unable to complex with the reductase but could still mediate, for example, photosynthetic cytochrome c reduction. Complex formation in vitro should be clarified further since data from various laboratories are not in accordance: by following complex formation spectrophotometrically, a decrease of the binding constant was found with increasing ionic strength by some workers (e.g., SHIN and SAN PIETRO, 1968; FOUST et al., 1969; NELSON and NEUMANN, 1969; BÖGER, 1971a), whereas others (NAKAMURA and KIMURA, 1971a), by kinetic studies, observed an increase of the binding constant up to 0.15 ionic strength which remained stable with higher salt concentrations.

NADPH oxidase activity (NADPH→O_2, yielding H_2O_2) is low (0.10 μmol of oxygen reduced/nmol reductase per min at pH 8.0 in an NADPH-regenerating system with freshly prepared *Bumilleriopsis* reductase (BÖGER, 1971b; SPILLER et al., 1976; cf. also NAKAMURA and KIMURA, 1971a) but is markedly enhanced by ferredoxin or methylviologen. In the assay system just mentioned, ferredoxin stimulates about 15-fold at a saturating concentration of 6-8 μM. In the presence of ferredoxin, polyamines (NAKAMURA et al., 1972), histones, and mammalian cytochrome c further enhance reactivity. With cytochrome c plus ferredoxin, water is the reaction product (NAKAMURA and KIMURA, 1972), which may be due to reoxidation of the cytochrome by peroxide. Mammalian cytochrome c and ferredoxin form a 2:1 complex which apparently acts as the diaphorase substrate (DAVIS and SAN PIETRO, 1977a, b). As with diaphorase reactions, the oxidase activity of the pure reductase is inhibited by ferredoxin at low ionic strength.

Little work has been done on the stereospecificity of this transhydrogenase type; for the spinach enzyme BB specificity was reported (AMMERAAL et al., 1965), which could be confirmed in this laboratory (BOOKJANS, unpubl.). No energized hydrogen transfer has been observed with the plant enzyme while still bound to isolated thylakoids (BÖGER, 1972a). An apoprotein from spinach which yields an active transhydrogenase upon reconstitution has been prepared recently (BOOKJANS et al., 1978).

The plant ferredoxin-NADP$^+$ reductase is bound to the thylakoid membrane although, as was proven with spinach chloroplasts by immunological methods (BÖHME, 1977, 1978), located peripherally on the outer thylakoid membrane. Surprisingly, the extent of solubilization depends on the species. Homogenization of the blue-green alga *Nostoc muscorum* with glass beads solubilizes about 90% of the activity, but only 5% when the eukaryotic alga *Bumilleriopsis filiformis* is used and about 10%–20% when spinach is treated that way. While about 50% can be solubilized by ageing of isolated spinach chloroplast material, this is not possible with *Bumilleriopsis* where it requires, for example, 1% Triton X-100 *and* strong sonication for a 90% solubilization. This amount of solubilization could be achieved with spinach chloroplasts by simple freeze-thawing (our findings and BÖHME, 1978). As investigated with spinach, the ferredoxin-NADP$^+$ reductase appears to be the only plastidic FAD-containing flavoprotein present.

Some data are available on the stoichiometric occurrence of the reductase in relation to plastidic ferredoxin. A 1:2 molar ratio (HASLETT and CAMMACK, 1976) was reported for higher plants, whereas a ratio of 3:5 was recently measured in this laboratory in spinach chloroplasts (BÖHME, 1978) and one of 1:2–4 for *Bumilleriopsis*. In the case of the alga the ratio was found to vary with the chlorophyll content of the cell (BÖGER, unpubl. results).

Soluble flavoproteins with properties as described in this section have not yet been isolated from animal or insect tissues.

C. Membrane-Bound (Particulate) Transhydrogenases

I. Mitochondrial and Bacterial Transhydrogenases; General Aspects

The occurrence of membrane-bound transhydrogenases is ubiquitous throughout bacteria, animals, and plants (see pp. 64–66 of RYDSTRÖM et al., 1976 for refs. on distribution). Most detailed work has been done with preparations from animal mitochondria (submitochondrial particles from the inner membranes) and photosynthetic bacteria. Purification up to only 25-fold has been achieved (KAPLAN, 1967), but is complicated by the high sensitivity of this enzyme to organic solvents or detergents (KRAMAR and SALVENMOSER, 1966; SALVENMOSER and KRAMAR, 1971). Consequently, mechanistic details of the enzyme reaction are still provisional; inhibitor studies on the enzyme have been presented (RYDSTRÖM et al., 1971a; RYDSTRÖM, 1972; SWEETMAN et al., 1974).

In the energy-independent transhydrogenase system NADH→NADP$^+$ the forward reaction rate is several-fold smaller than that of the backward reaction (RYDSTRÖM et al., 1970; TEIXEIRA DA CRUZ et al., 1971); therefore, this transhydrogenation is generally measured via the reverse reaction (SINGH and BRAGG, 1974). However, energization of the membrane by respiration (in the presence of succinate) or ATP hydrolysis accelerates the rate of NADP$^+$ reduction four to six fold, with an increase of the apparent equilibrium constant by up to 500 (LEE and ERNSTER, 1966; KEISTER and YIKE, 1967; FISHER and GUILLORY, 1971b). As found by Ernster's group (RYDSTRÖM et al., 1971b) the K$_m$ values of all nucleotides involved did not change appreciably except for the K$_m$ of NADP$^+$ which decreased from 40 μM under nonenergy-linked conditions to 6.5 μM under energy-linked conditions. Thus, energized membranes may provide for a special affinity of the oxidized NADP$^+$ to the transhydrogenase. A functional model has been published by HATEFI and GALANTE (1977). Energy-transfer inhibitors (oligomycin) only inhibit the ATP-driven reactions (e.g., with submitochondrial particles) whereas electron transport inhibitors (like antimycin) prevent the respiration-driven transhydrogenation (LEE and ERNSTER, 1964; LEE et al., 1964). The energy expenditure has been demonstrated by several laboratories (e.g., TAGER et al., 1969) and found to be 1 ATP per NADP$^+$ reduced; some reversibility could be shown (VAN DE STADT et al., 1971; OSTROUMOV et al., 1973).

The mode of energy input to the transhydrogenase system is unknown. This problem is intimately connected to the general mechanism of energy conservation in biological membranes. The nucleotides are suggested to react with an energy-rich intermediate X~Y (ERNSTER, 1964; ERNSTER et al., 1973; WIDMER and KAPLAN, 1976):

$$\text{NADH} + \text{NADP}^+ + \text{X} \sim \text{Y} \rightleftharpoons \text{NAD}^+ + \text{NADPH}$$
$$\underset{\text{ATP}}{\uparrow} \tag{2}$$

X~Y indicates an unknown compound, or an energized state coupled to a proton gradient across the membrane. Concurrent with transhydrogenation an inward flux of positively charged species, possibly protons, was observed (Dontsov et al., 1972). The energized state may represent energy-driven conformational changes of the transhydrogenase (system) itself (Rydström et al., 1970; Rydström et al., 1971b; Blazyk and Fisher, 1975). The mitochondrial transhydrogenase appears to be located in complex I (NADH-ubiquinone reductase, E.C.1.6.5.3; Djavadi-Oha-niance and Hatefi, 1975; Ragan and Widger, 1975; Hatefi and Galante, 1977), although transhydrogenase is separate from the NADH dehydrogenase (Zahl et al., 1978). The ATP-driven reaction requires the coupling factor (ATPase; Erns-ter et al., 1973), as was elegantly demonstrated with *E. coli* (Bragg and Hou, 1972; Cox et al., 1973; Singh and Bragg, 1974; Sabet, 1977). In this bacterium the ATP-linked transhydrogenase was found to be exclusively associated with the cytoplasmic membrane fractions (Sabet, 1977).

It has been claimed for the mitochondrial system that besides the NADH/NADP$^+$ transhydrogenase there is an NADH/NAD$^+$ enzyme present (Kaufman and Kaplan, 1961; Kramar and Salvenmoser, 1966; Salvenmoser and Kramar, 1971; Djavadi-Ohaniance and Hatefi, 1975; Köhler and Saz, 1976). However, the existence of two separate enzymes still has to be substantiated. A possible connection of the two transhydrogenations is discussed by Fisher and Guillory (1971b).

The transhydrogenase is, at least, an excellent tool to study formation and properties of the energized state. For its possible physiological function see Rydström et al. (1976).

II. Transhydrogenase of Photosynthetic Bacteria

Basically, the same transhydrogenase system as in submitochondrial particles oper-ates in chromatophore preparations of photosynthetic bacteria. A slow nonenergy-linked reaction can be energized. ATP (1 molecule hydrolyzed for 1 molecule of NADPH formed) may be replaced by pyrophosphate and light (Keister and Yike, 1967). Electron transport and energy transfer inhibitors exhibit the same effect as with submitochondrial particles (Sect. C.I). The enzyme could be demon-strated in *Rhodospirillum rubrum* and other photosynthetic bacteria (Keister and Yike, 1967). Photoreduction of NADP$^+$ with isolated thylakoids from *Rhodopseu-domonas spheroides* may depend on an energized transhydrogenase (Orlando et al., 1966) as does the thiosulfate-linked NAD$^+$ reduction in *Rhodopseudomonas palustris* (Knobloch, 1975). The latter energized reaction was reported to proceed with isolated particles not formed into vesicles. Isolated chromatophores from *R. rubrum*, depleted of the transhydrogenase activity, were fully able to perform the light-dependent NAD$^+$ reduction (Thomas et al., 1970), a finding reminiscent of the unimpaired respiration-driven NAD$^+$ reduction by succinate while the en-ergized transhydrogenase was blocked by an antibody (Kawasaki et al., 1964).

The physiological role of the energized transhydrogenases, and also the question whether they share with ferredoxin-NAD(P)$^+$ reductases in the direct reduction of pyridine nucleotides through the electron transport system (Yamanaka and Kamen, 1967; Buchanan and Bachofen, 1968; Kusai and Yamanaka, 1973) are far from elucidated. In contrast to plants and algae, a generalization is hampered by obvious differences between bacterial subgroups.

Of particular interest is the isolation of a "transhydrogenase factor" from R. *rubrum* (FISHER and GUILLORY, 1969). This factor (molecular weight of 70,000 daltons) has no flavin and no transhydrogenase activity (KONINGS and GUILLORY, 1973) and can be separated from the (insoluble) chromatophores by mild washings which concurrently remove bound pyridine nucleotides (FISHER and GUILLORY 1971a). NADP(H), but not NAD(H), protects the isolated protein against heat or trypsin attack (FISHER et al., 1975). Added back to the depleted chromatophore membrane the factor is readily bound in the presence of NADP(H) with complete reconstituion of both the nonenergy-linked and energy-linked reaction rates (FISHER and GUILLORY, 1971b). The ATP- and pyrophosphate-driven reaction required the presence of either the coupling factor (ATPase) or pyrophosphatase, respectively, as could be demonstrated by elegant reconstitution experiments (GUILLORY and FISHER, 1969). The bound transhydrogenase was solubilized by extraction of chromatophores by lysolecithin. The solubilized membrane component reconstituted transhydrogenase activity upon additon of the transhydrogenase factor. The solubilized enzyme has to await molecular characterization (JACOBS et al., 1977). Three binding sites for pyridine nucleotides were reported for the complete transhydrogenase complex (McFADDEN and FISHER, 1978). Also, chromatophores from *Rh. spheroides* could be depleted of light-driven transhydrogenase activity, which was restored by a (partly) purified proteinaceous factor of 4000–7000 daltons molecular weight (BERGER and ORLANDO, 1973).

Note Added in Proof. Recently, the membrane-bound transhydrogenase from bovine heart mitochondria has been purified to homogeneity (HÖJEBERG and RYDSTRÖM, 1977; ANDERSON and FISHER, 1978). The molecular weight is between 97 and 120 kilodaltons and represents a single polypeptide containing neither flavin nor cytochromes. It has been proposed that the transhydrogenase operates as a direct (reversible) proton pump in submitochondrial particles: $NADH + NADP^+ + nH_{in}^+ \rightleftharpoons NAD^+ + NADPH + nH_{out}^+$ (see C.I). The authors were able to incorporate transhydrogenase into artificial liposomes where – by quenching of 9-aminoacridine – a proton uptake could be shown during hydrogenation of NAD^+ which was sensitive to the uncoupler F-CCP. Presence of this uncoupler enhanced both the forward and reverse transhydrogenation, while it did not affect the reaction with "nonvesicular" transhydrogenase.

References

Allen, S.H.G., Patil, J.R.: J. Biol. Chem. *247*, 909–916 (1972)
Ammeraal, R.N., Krakow, G., Vennesland, B.: J. Biol. Chem. *240*, 1824–1828 (1965)
Anderson, W.M., Fisher, R.R.: Arch. Biochem. Biophys. *187*, 180–190 (1978)
Berger, T.J., Orlando, J.A.: Arch. Biochem. Biophys. *159*, 25–31 (1973)
Blazyk, J.F., Fisher, R.R.: FEBS Lett. *50*, 227–232 (1975)
Böger, P.: Planta (Berl.) *99*, 319–338 (1971a)
Böger, P.: Z. Naturforsch. *26b*, 807–815 (1971b)
Böger, P.: Z. Naturforsch. *27b*, 826–833 (1972a)
Böger, P.: In: Proc. 2nd Int. Congr. Photosynthesis. Forti, G. et al. (eds.), Vol. I, pp. 449–458. The Hague: Junk N.V. Publ. 1972b
Böger, P., Lien, S.S., San Pietro, A.: Z. Naturforsch. *28c*, 505–510 (1973)
Böhme, H.: In: Membrane bioenergetics. Packer, L. et al. (eds.), pp. 329–337. Amsterdam: Elsevier-North Holland 1977
Böhme, H.: Eur. J. Biochem. *83*, 137–141 (1978)

Bookjans, G., Böger, P.: Arch. Biochem. Biophys. *190*, 459–465 (1978)
Bookjans, G., Böger, P.: Arch. Biochem. Biophys. (1979), in press
Bookjans, G., San Pietro, A., Böger, P.: Biochem. Biophys. Res. Commun. *80*, 759–765 (1978)
Bragg, P.D., Hou, C.: FEBS Lett. *28*, 309–312 (1972)
Buchanan, B.B., Bachofen, R.: Biochim. Biophys. Acta *162*, 607–610 (1968)
Chopowick, R., Israelstamm, G.F.: Planta (Berl.) *101*, 171–173 (1973)
Chung, A.E.: J. Bacteriol. *102*, 438–447 (1970)
Cohen, P.T., Kaplan, N.O.: J. Biol. Chem. *245*, 2825–2836 (1970a)
Cohen, P.T., Kaplan, N.O.: J. Biol. Chem. *245*, 4666–4682 (1970b)
Cox, G.B., Gibson, R., McCann, L.M., Butlin, J.D., Crane, F.L.: Biochem. J. *132*, 689–695 (1973)
Davis, D.J., San Pietro, A.: Biochem. Biophys. Res. Commun. *74*, 33–40 (1977a)
Davis, D.J., San Pietro, A.: Arch. Biochem. Biophys. *182*, 266–272 (1977b)
Djavadi-Ohaniance, L., Hatefi, Y.: J. Biol. Chem. *250*, 9397–9403 (1975)
Dontsov, A.E., Grinius, L.L., Jasaitis, A.A., Severina, I.I., Skulachev, C.P.: J. Bioenerg. *3*, 277–303 (1972)
Earle, S.R., Anderson W.M., Fisher, R.R.: FEBS Lett. *91*, 21–24 (1978)
Ernster, L.: Annu. Rev. Biochem. *33*, 729–788 (1964)
Ernster, L., Juntti, K., Asami, K.: J. Bioenerg. *4*, 149–159 (1973)
Fisher, R.R., Guillory, R.J.: J. Biol. Chem. *244*, 1078–1079 (1969)
Fisher, R.R., Guillory, R.J.: J. Biol. Chem. *246*, 4679–4686 (1971a)
Fisher, R.R., Guillory, R.J.: J. Biol. Chem. *246*, 4687–4693 (1971b)
Fisher, R.R., Kaplan, N.O.: Biochemistry *12*, 1182–1188 (1973)
Fisher, R.R., Rampey, S.A., Sadighi, A., Fisher, K.: J. Biol. Chem. *250*, 819–825 (1975)
Forti, G.: Encyclopedia of plant physiology. Photosynthesis I, Trebst, A., Avron, M. (eds.), Vol. 5, pp. 222–226. Berlin, Heidelberg, New York: Springer 1977
Foust, G.P., Mayhew, S.G., Massey, V.: J. Biol. Chem. *244*, 964–970 (1969)
Fredericks, W.W., Gehl, J.M.: J. Biol. Chem. *246*, 1201–1205 (1971)
Green, W.G.E., Israelstamm, G.F.: Physiol. Plant. *23*, 217–231 (1970)
Griffiths, D.E., Robertson, A.M.: Biochim. Biophys. Acta *118*, 453–464 (1966)
Guillory, R.J., Fisher, R.R.: FEBS Lett. *3*, 27–30 (1969)
Haslett, B.G., Cammack, R.: New Phytol. *76*, 219–226 (1976)
Hatefi, Y., Galante, Y.M.: Proc. Natl. Acad. Sci. USA *74*, 846–850 (1977)
Höjeberg, B., Rydström, J.: Biochem. Biophys. Res. Commun. *78*, 1183–1190 (1977)
Hoek, J.B., Rydström, J., Höjeberg, B.: Biochim. Biophys. Acta *333*, 237–245 (1974)
Jacobs, E., Heriot, K., Fisher, R.R.: Arch. Microbiol. *115*, 151–156 (1977)
Jungermann, K., Rupprecht, E., Ohrloff, C., Thauer, R., Decker, K.: J. Biol. Chem. *246*, 960–963 (1971)
Kaplan, N.O.: Methods Enzymol. *10*, 317–322 (1967)
Kaufman, B., Kaplan, N.O.: J. Biol. Chem. *236*, 2133–2139 (1961)
Kawasaki, T., Satoh, K., Kaplan, N.O.: Biochem. Biophys. Res. Commun. *17*, 648–654 (1964)
Keister, D.L., Hemmes, R.B.: J. Biol. Chem. *241*, 2820–2825 (1966)
Keister, D.L., San Pietro, A., Stolzenbach, F.E.: J. Biol. Chem. *235*, 2989–2996 (1960a)
Keister, D.L., San Pietro, A., Stolzenbach, F.E.: Arch. Biochem. Biophys. *98*, 235–244 (1960b)
Keister, D.L., Yike, N.J.: Biochemistry *6*, 3847–3857 (1967)
Knobloch, K.: Z. Naturforsch. *30c*, 771–776 (1975)
Köhler, P., Saz, H.J.: J. Biol. Chem. *251*, 2217–2225 (1976)
Konings, A.W.T., Guillory, R.J.: J. Biol. Chem. *248*, 1045–1050 (1973)
Kramar, R., Salvenmoser, F.: Hoppe Seyler's Z. Physiol. Chem. *346*, 310–313 (1966)
Krause, G.H., Heber, U.: In: The intact chloroplast. Barber, J. (ed.), pp. 171–214. Amsterdam, New York, Oxford: Elsevier 1976
Krawetz, S.A., Israelstamm, G.F.: Plant Sci. Lett. *12*, 323–326 (1978)
Kusai, A., Yamanaka, T.: Biochim. Biophys. Acta *292*, 621–633 (1973)
Lee, C.P., Ernster, L.: Biochim. Biophys. Acta *81*, 187–190 (1964)
Lee, C.P., Ernster, L.: In: Regulation of metabolic processes in mitochondria. Tager, J.M. et al. (eds.), pp. 218–234. Amsterdam: Elsevier 1966
Lee, C.P., Azzone, G.F., Ernster, L.: Nature (London) *201*, 152–155 (1964)

Lee, C.P., Simard-Duquesne, N., Ernster, L.: Biochim. Biophys. Acta *105*, 397–409 (1965)
Lendzian, K., Bassham, J.A.: Biochim. Biophys. Acta *430*, 478–489 (1976)
Louie, D.D., Kaplan, N.O.: J. Biol. Chem. *245*, 5691–5698 (1970)
McFadden, B.J., Fisher, R.R.: Arch. Biochem. Biophys. *190*, 820–828 (1978)
Middleditch, L.E., Chung, A.E.: Arch. Biochem. Biophys. *146*, 449–453 (1971)
Nakamura, S., Kimura, T.: J. Biol. Chem. *246*, 6235–6241 (1971a)
Nakamura, S., Kimura, T.: FEBS Lett. *15*, 352–354 (1971b)
Nakamura, S., Kimura, T.: J. Biol. Chem. *247*, 6462–6468 (1972)
Nakamura, S., Kazim, A.L., Wang, H.P., Chu, J.W., Kimura, T.: FEBS Lett. *28*, 209–212 (1972)
Nelson, N., Neumann, J.: J. Biol. Chem. *244*, 1926–1931 (1969); *244*, 1932–1936
Orlando, J.A., Sabo, D., Curnyn, C.: Plant. Physiol. *41*, 937–945 (1966)
Ostroumov, S.A., Samuilov, V.P., Skulachev, V.P.: FEBS Lett. *31*, 27–29 (1973)
Palmer, G., Massey, M.: In: Biological oxidations. Singer, Th.P. (ed.), pp. 263–300. New York, London, Sydney: Interscience Publ. 1968
Petitdemange, H., Bengone, J.M., Bergere, J.L., Gay, R.: Biochimie *55*, 1307–1310 (1973)
Ragan, C.I., Widger, W.R.: Biochem. Biophys. Res. Commun. *62*, 744–749 (1975)
Rydström, J.: Eur. J. Biochem. *31*, 496–504 (1972)
Rydström, J.: Biochim. Biophys. Acta *463*, 155–184 (1977)
Rydström, J., Teixeira da Cruz, A., Ernster, L.: Eur. J. Biochem. *17*, 56–62 (1970)
Rydström, J., Panov, A.V., Paradies, G., Ernster, L.: Biochem. Biophys. Res. Commun. *45*, 1389–1397 (1971a)
Rydström, J., Teixeira da Cruz, A., Ernster, L.: Eur. J. Biochem. *23*, 212–219 (1971b)
Rydström, J., Hoek, J., Höjeberg, B.: Biochem. Biophys. Res. Commun. *52*, 421–429 (1973)
Rydström, J., Hoek, J.B., Ernster, L.: In: The enzymes. 3rd ed. Boyer, P.D. (ed.), Vol. XIII, pp. 51–88. New York, London: Academic Press 1976
Sabet, S.F.: J. Bacteriol. *129*, 1397–1406 (1977)
Salvenmoser, F., Kramar, R.: Enzymologia *40*, 322–327 (1971)
Schneeman, R., Krogmann, D.W.: J. Biol. Chem. *250*, 4965–4971 (1975)
Shin, M., Arnon, D.I.: J. Biol. Chem. *240*, 1405–1411 (1965)
Shin, M., San Pietro, A.: Biochem. Biophys. Res. Commun. *33*, 38–42 (1968)
Shin, M., Tagawa, K., Arnon, D.I.: Biochem. Z. *338*, 84–96 (1963)
Singh, A.P., Bragg, P.D.: J. Gen. Microbiol. *82*, 237–246 (1974)
Spiller, H., Bookjans, G., Böger, P.: Z. Naturforsch. *31c*, 565–568 (1976)
Sweetman, A.J., Green, A.P., Hooper, M.: Biochem. Biophys. Res. Commun. *58*, 337–343 (1974)
Tager, J.M., Groot, G.S.P., Roos, D., Papa, S., Quagliariello, E.: In: The energy level and metabolic control in mitochondria. Papa, S. et al. (eds.), pp. 453–466. Bari: Adriatica Editrice 1969
Teixeira da Cruz, A., Rydström, J., Ernster, L.: Eur. J. Biochem. *23*, 203–211 (1971)
Thauer, R.K., Rupprecht, E., Ohrloff, C., Jungermann, K., Decker, K.: J. Biol. Chem. *246*, 954–959 (1971)
Thomas, J.O., Fisher, R.R., Guillory, R.J.: Biochim. Biophys. Acta *223*, 204–206 (1970)
Van den Broek, H.W.J., Veeger, C.: Eur. J. Biochem. *24*, 72–82 (1971b)
Van den Broek, H.W.J., Santema, J.S., Wassink, J.H., Veeger, C.: Eur. J. Biochem. *24*, 31–45 (1971a); see also *24*, pp. 46–82
Van de Stadt, R.J., Niewenhuis, F.J.R.M., van Dam, K.: Biochim. Biophys. Acta *234*, 173–176 (1971)
Widmer, F., Kaplan, N.O.: Biochemistry *15*, 4693–4698 (1976)
Yamanaka, T., Kamen, M.D.: Biochim. Biophys. Acta *112*, 436–447 (1966)
Yamanaka, T., Kamen, M.D.: Biochim. Biophys. Acta *131*, 317–339 (1967)
Yoch, D.C.: J. Bacteriol. *116*, 384–391 (1973)
Zahl, K.J., Rose, C., Hanson, R.L.: Arch. Biochem. Biophys. *190*, 598–602 (1978)

2. Oxygen Activation and Superoxide Dismutase in Chloroplasts

E.F. ELSTNER

A. Introduction

Recently several groups of workers have focused on oxygen toxicity. Since the discovery of the catalytic function of erythrocuprein as a superoxide dismutase (McCORD and FRIDOVICH, 1969), a whole set of compounds and reactions involved in "oxygen activation" and "deactivation" of toxic oxygen species has been described. Several books and reviews on this topic have appeared (HAYAISHI, 1974; HALLIWELL, 1974; BORS et al., 1974; PRYOR, 1976; ELSTNER and KONZE 1976).

B. Principles of Oxygen Activation

Molecular oxygen is relatively stable and shows only very slow reaction kinetics with most organic materials at physiological temperatures. Oxygen can be activated, however, by forming complexes with metal ions (e.g., the perferryl, or the percupryl ion) or by the addition of one electron (usually coming from organic radicals), yielding the superoxide free radical ion (E_0 $O_2/O_2^{\cdot-} = -0.33$ V; ILAN et al., 1976). Further electrons are added to the superoxide free radical ion in successive exergonic reactions (ULLRICH and STAUDINGER, 1968; HAMILTON, 1974). Thus, the addition of one electron to oxygen by metal ions or by organic radicals represents the important endergonic step in oxygen activation, yielding three types of actively oxidizing or hydroxylating agents, namely (a) metal–oxygen complexes (b) organic peroxides or (c) oxygen-free radicals.

C. Superoxide Anion and Superoxide Dismutase

I. Dismutation of the Superoxide Anion ($O_2^{\cdot-}$); Superoxide Dismutase in Plants

Monovalent oxygen reduction yields $O_2^{\cdot-}$, which spontaneously dismutates into H_2O_2 and O_2 [Eq. (1)]

$$O_2^{\cdot-} + O_2^{\cdot-} + 2H^+ \rightarrow H_2O_2 + O_2 \tag{1}$$

Fig. 1. Scheme of successive oxygen reduction. (After HA-MILTON, 1974)

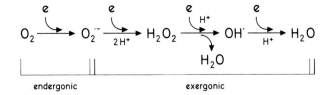

H_2O_2 and $O_2{}^{\cdot-}$ (under the catalysis of iron-ions) can react in turn according to the Haber-Weiss reaction [HABER and WEISS 1934; Eq. (2)]

$$H_2O_2 + O_2{}^{\cdot-} \rightarrow OH^- + OH^\cdot + O_2, \tag{2}$$

forming the very reactive hydroxyl radical (OH$^\cdot$). These reactions are summarized in Figure 1.

The rate constant for the spontaneous dismutation of $O_2{}^{\cdot-}$ [Eq. (1)] is about $1.2 \times 10^5\ M^{-1}\ s^{-1}$. In the presence of SOD, an enzyme which seems to be obligatory for all aerobic or aerotolerant organisms (FRIDOVICH, 1976), the rate constant of the dismutation is about $2 \times 10^9\ M^{-1}\ s^{-1}$.

Three different types of SOD are known to date. In addition to the plasmatic enzyme from eukaryotes, containing 2 g atoms copper (active center) and 2 g atoms zinc (stabilizing function) per mol apoprotein, iron-containing enzymes from bacteria and blue-green algae and mangano-enzymes from bacteria and mitochondria have been isolated. The metal content and molecular weight may be slightly different for the individual enzymes from different organisms (PUGET and MICHELSON, 1974; FRIDOVICH, 1976).

Of outstanding importance is the observation that one product of the spontaneous dismutation of $O_2{}^{\cdot-}$ is the most aggressive singlet oxygen ($^1O_2$; KHAN, 1970), while the product of the SOD-catalyzed dismutation seems to be triplet oxygen (GODA et al., 1974; MICHELSON, 1974). Isolated SOD has been widely used for the inhibition and thus identification of $O_2{}^{\cdot-}$ – utilizing reactions: knowing $O_2{}^{\cdot-}$ – dependent reactions, one can search for $O_2{}^{\cdot-}$ – producing reactions in chemical and biological systems with the aid of SOD.

II. Superoxide Dismutase in Chloroplasts

Chloroplasts from higher plants (spinach, *Spinacia oleracea;* sugar beet, *Beta vulgaris*) contain SOD activities in the stroma as well as bound to the chloroplast lamellae (ASADA et al., 1973; ELSTNER and HEUPEL, 1975, 1976a). Both the soluble and the bound SOD are of the same cyanide-sensitive type (ASADA et al., 1974), containing copper and zinc (ASADA et al., 1973). Approximately one third (e.g., about 30 enzyme units per mg chl, cf. ELSTNER and HEUPEL, 1975) of the chloroplastic SOD-activity appears to be bound to the lamellae. Chloroplasts contain soluble and bound manganese, but there seems to be no cyanide-insensitive SOD-activity in the chloroplasts of higher plants, although a manganese-dependent "SOD-like" activity has been observed in subchloroplast particles after Triton treatment (LUMSDEN and HALL, 1975). These observations are in contrast to algae and bacteria, where cyanide-insensitive mangano- and iron-SOD seem to predominate (ASADA et al., 1977; LUMSDEN and HALL, 1975).

III. Monovalent Oxygen Reduction in Chloroplasts

Photosynthetic oxygen reduction (also described as "pseudocyclic electron transport") and light-dependent peroxide production have been observed in intact algae (PATTERSON and MYERS, 1973; GLIDEWELL and RAVEN, 1976; RADMER and KOK, 1976), isolated intact chloroplasts from higher plants (EGNEUS et al., 1975) and isolated chloroplast lamellae (ARNON et al., 1967; BOTHE 1969; TELFER et al., 1970; ELSTNER and HEUPEL, 1973). Photosynthetic oxygen reduction is thought to contribute to over-all photophosphorylation, thus representing a vital necessity for higher plants if a P/2e ratio of 1.33 (TREBST, 1974) or less is implicated for photosynthetic NADP reduction (HEBER, 1973; ELSTNER und KRAMER, 1973; EGNEUS et al., 1975; ZIEGLER and LIBERA, 1975).

Oxygen reduction by photosystem I has been shown to proceed by one-electron steps, yielding the superoxide radical anion (NELSON et al., 1972; ASADA and KISO, 1973; ELSTNER and KRAMER, 1973; EPEL and NEUMANN, 1973). Ferredoxin and eventually other compounds located on the reducing side of photosystem I are involved (ELSTNER and HEUPEL, 1974a, b) as electron donors for molecular oxygen.

In isolated chloroplast lamellae, oxygen reduction is switched on as soon as approximately 80% of the NADP pool is reduced (ELSTNER and HEUPEL, 1973). On the other hand, oxygen radical formation is also possible in isolated chloroplast lamellae in the dark at the expense of NADPH, via NADP-ferredoxin reductase and ferredoxin (ELSTNER and HEUPEL, 1974a; ELSTNER and KONZE, 1974). The above reactions are summarized in Figure 2:

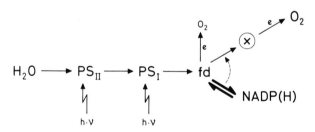

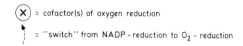

$\textcircled{x}$ = cofactor(s) of oxygen reduction

= "switch" from NADP - reduction to O_2 - reduction

Fig. 2. Photosynthetic oxygen reduction

D. Determination of the Products of Oxygen Reduction

According to Eqs. (1) and (2) and Figure 1, monovalent oxygen reduction yields at least three different reduced oxygen species, namely $O_2^{\cdot-}$, H_2O_2 and $OH^{\cdot}$.

These three species have been determined separately by the following methods:

1. Photosynthetic $O_2^{\cdot-}$-production can be followed by several methods, using either its oxidizing or reducing power. The oxidation of hydroxylamine, yielding nitrite, has been shown to be convenient for certain photosynthetic experiments (ELSTNER et al., 1975). This method is also useful for the location and quantitation of SOD in different cell compartments (ELSTNER and HEUPEL, 1975, 1976b).

2. H_2O_2-production can be determined spectrophotometrically (cf. LOSCHEN et al., 1971; BOVERIS et al., 1976) or by following the decarboxylation of a 1-^{14}C-α-keto acid (ZELITCH, 1972; ELSTNER and HEUPEL, 1973, 1976b).

3. The formation of the hydroxyl radical can be followed by measuring the production of ethylene from 3-mercaptomethyl propanal (methional; BEAUCHAMP and FRIDOVICH, 1970; ELSTNER and KONZE, 1975; see *noted added in proof*).

E. Possible Functions of Reduced Oxygen Species in Chloroplasts

Reduced oxygen species ("activated oxygen") might play a role in the following reactions, although their direct involvement has not yet been demonstrated in most cases.

I. Desaturation of Fatty Acids

The formation of oleyl-CoA from stearyl-CoA in chloroplasts seems to be dependent on an electron donor (NADPH or photosynthetic electron transport), ferredoxin, oxygen, and a stearyl-acyl carrier protein-desaturase (BLOCH, 1969; JACOBSON et al., 1974).

The above system (without the desaturase) has been shown to "activate" oxygen, yielding $O_2^{\cdot-}$, H_2O_2 and $OH^{\cdot}$ (ELSTNER and KONZE, 1974; KONZE and ELSTNER, 1976).

II. Hydroxylation of Aromatic Compounds

A similar mechanism of oxygen activation as described for fatty acid desaturation seems to play a role for the hydroxylation of cinnamic acid, yielding o-coumaric acid (GESTETNER and CONN, 1974) and p-coumaric acid (CZICHI and KINDL, 1975), or of p-coumaric acid, yielding caffeic acid (BARTLETT et al., 1972; HALLIWELL, 1975) by isolated chloroplasts.

III. Photorespiration

H_2O_2 as well as oxygen radicals have been discussed as "initiators" of photorespiration. While H_2O_2 has been proposed directly to split a transketolase complex, yielding glycolate (Plaut and Gibbs, 1970), oxygen radicals are discussed as "activating" ribulose-bisphosphate carboxylase which acting as an oxygenase, splits ribulose-bisphosphate, yielding phosphoglycerate and phosphoglycolate (Wildner, 1976).

IV. Ethylene Formation

Isolated chloroplasts can synthesize methionine (Shah and Cossins, 1970) and evolve ethylene from methionine in the presence of pyridoxalphosphate and ferredoxin (Konze and Elstner, 1976). Ethylene as a plant hormone is produced by various plant organs under a variety of conditions (Abeles, 1973). The production of ethylene by green leaves is inhibited by light, but ethylene formation is stimulated by light in "wounded" sugar beet leaves. A model reaction has been described for ethylene production in wounded green leaves, including the functions of photosynthetic electron transport and a phenoloxidase-induced quinone–semiquinone cycle as an oxygen-activating system (Elstner et al., 1976).

Apart from this indirect evidence, there is presently no experimental proof that chloroplasts (or generally plastids) represent a major or sole source of ethylene in plants.

F. Conclusions

There is good evidence that oxygen reduction by the photosynthetic electron transport system (or parts of the photosynthetic electron transport system, e.g., NADPH →NADP-ferredoxin reductase→ferredoxin) plays an important role in the overall synthesizing capacity of plastids. The plastids contain superoxide dismutase in order to avoid the production of a surplus of $O_2^{\cdot-}$ and of toxic singlet oxygen.

Note Added in Proof. The formation of *free* OH-radicals in chloroplasts is very questionable (Elstner et al., Eur. J. Biochem. *89*, 61–66, 1978; Elstner and Zeller, Plant Sci. Lett. *13*, 15–20, 1978).

References

Abeles, F.B.: Ethylene in plant biology. London: Academic Press 1973
Arnon, D.I., Tsujimoto, H.Y., McSwain, B.D.: Nature (London) *214*, 562–566 (1967)
Asada, K., Kiso, K.: Agr. Biol. Chem. *37*, 453–454 (1973)
Asada, K., Urano, M., Takahashi, M.: Eur. J. Biochem. *36*, 257–266 (1973)
Asada, K., Takahashi, M., Nagate, M.: Agric. Biol. Chem. *38*, 471–473 (1974)
Asada, K., Kanematsu, S., Uchida, K.: Arch. Biochem. Biophys. *179*, 243–256 (1977)
Bartlett, D.J., Poulton, J.E., Butt, V.S.: FEBS Lett. *23*, 265–267 (1972)
Beauchamp, C., Fridovich, I.: J. Biol. Chem. *345*, 4641–4646 (1970)
Bloch, K.: Accounts Chem. Res. *2*, 193–202 (1969)

Bors, W., Saran, M., Lengelder, E., Spöttl, R., Michel, C.: Curr. top. radiat. res. Q. *9*, 247–309 (1974)
Bothe, H.: Z. Naturforsch. *24b*, 1574–1582 (1969)
Boveris, A., Cadenas, E., Stoppani, A.O.M.: Biochem. J. *156*, 435–444 (1976)
Czichi, U., Kindl, H.: Hoppe-Seyler's Z. Physiol. Chem. *356*, 475–485 (1975)
Egneus, H., Heber, U., Matthiesen, U., Kirk, M.: Biochim. Biophys. Acta *408*, 252–268 (1975)
Elstner, E.F., Heupel, A.: Biochim. Biophys. Acta *325*, 182–188 (1973)
Elstner, E.F., Heupel, A.: Z. Naturforsch. *29c*, 564–571 (1974a)
Elstner, E.F., Heupel, A.: Z. Naturforsch. *29a*, 559–563 (1974b)
Elstner, E.F., Heupel, A.: Planta (Berl.) *123*, 145–154 (1975)
Elstner, E.F., Heupel, A.: Planta, (Berl.) *123*, 145–154 (1975)
Elstner, E.F., Heupel, A.: Anal. Biochem. *70*, 616–620 (1976a)
Elstner, E.F., Heupel, A.: Planta (Berl.) *130*, 175–180 (1976)
Elstner, E.F., Konze, J.R.: FEBS Lett. *45*, 18–21 (1974)
Elstner, E.F., Konze, J.R.: Z. Naturforsch. *30c*, 58–63 (1975)
Elstner, E.F., Konze, J.R.: Ber. Dtsch. Bot. Ges. *90*, (1976)
Elstner, E.F., Kramer, R.: Biochim. Biophys. Acta *314*, 340–353 (1973)
Elstner, E.F., Stoffer, C., Heupel, A.: Z. Naturforsch. *30c*, 53–56 (1975)
Elstner, E.F., Konze, J.R., Selman, B.R., Stoffer, C.: Plant Physiol. *58*, 163–168 (1976)
Epel, B.L., Neumann, J.: Biochim. Biophys. Acta, *325*, 520–529 (1973)
Fridovich, I.: In: Radicals in biology, Pryor, W.A. (ed.), Vol. I, p. 239. New York: Academic Press 1976
Gestetner, B., Conn, E.E.: Arch. Biochem. Biophys. *163*, 617–624 (1974)
Glidewell, S.M., Raven, J.A.: J. Exp. Bot. *27*, 200–204 (1976)
Goda, K., Kimura, T., Thayer, A.L., Kees, K., Schaap, A.P.: Biochem. Biophys. Res. Commun. *58*, 660–666 (1974)
Haber, F., Weiss, J.: Proc. R. Soc. Ser. A *147*, 322 (1934)
Halliwell, B.: New Phytol. *73*, 1075–1086 (1974)
Halliwell, B.: Eur. J. Biochem. *55*, 355–360 (1975)
Hamilton, G.A.: In: Molecular mechanisms of oxygen activation. Hayaishi, O., ed., pp. 405–451. New York, London: Academic Press 1974
Hayaishi, O.: Molecular mechanisms of oxygen activation. New York, London: Academic Press 1974
Heber, U.: Ber. Dtsch. Bot. Ges. *86*, 187–195 (1973)
Ilan, Y.A., Czapski, G., Meisel, D.: Biochim. Biophys. Acta *430*, 209–224 (1976)
Jacobson, B.S., Jaworski, J.G., Stumpf, P.K.: Plant Physiol. *54*, 484–486 (1974)
Khan, A.U.: Science *168*, 476–477 (1970)
Konze, J.R., Elstner, E.F.: FEBS Lett. *66*, 8–11 (1976)
Loschen, G., Flohé, L., Chance, B.: FEBS Lett. *18*, 261–264 (1971)
Lumsden, J., Hall, D.O.: Biochem. Biophys. Res. Commun. *64*, 595–602 (1975a)
Lumsden, J., Hall, D.O.: Nature (London) *257*, 670–672 (1975b)
McCord, J.M., Fridovich, J.: J. Biol. Chem. *244*, 6049–6055 (1969)
Michelson, A.M.: FEBS Lett. *44*, 97–100 (1974)
Nelson, N., Nelson, H., Racker, E.: Photochem. Photobiol. *16*, 481–489 (1972)
Patterson, C.O.P., Myers, J.: Plant Physiol. *51*, 104–109 (1973)
Plaut, Z., Gibbs, M.: Plant Physiol. *45*, 470–474 (1970)
Pryor, W.A.: Free radicals in biology. Vol. I, II, New York, London: Academic Press 1976
Puget, K., Michelson, A.M.: Biochem. Biophys. Res. Commun. *58*, 830–838 (1974)
Radmer, R.J., Kok, B.: Plant Physiol. *58*, 336–340 (1976)
Shah, S.P.J., Cossins, E.A.: FEBS Lett. *7*, 267–269 (1970)
Telfer, A., Cammack, R., Evans, M.C.W.: FEBS Lett. *10*, 21–24 (1970)
Trebst, A.: Annu. Rev. Plant Physiol. *25*, 423–458 (1974)
Ullrich, V., Staudinger, H.J.: In: Biochemie des Sauerstoffs. Hess, B., Staudinger, J.J. (Hrsg.), p. 229. Berlin, Heidelberg, New York: Springer
Wildner, G.F.: Ber. Dtsch. Bot. Ges. *89*, 349–360 (1976)
Zelitch, I.: Arch. Biochem. Biophys. *150*, 698–707 (1972)
Ziegler, I., Libera, W.: Z. Naturforsch. *30c*, 534, 637 (1975)

3. Ferredoxin-Linked Carbon Dioxide Fixation in Photosynthetic Bacteria

B.B. BUCHANAN

A. Introduction

Over three decades ago, LIPMANN (1946) suggested that the formation of pyruvate via a reductive carboxylation of what is now known as acetyl coenzyme A would be an ideal reaction for the fixation of carbon dioxide in photosynthesis. However, despite one report that sodium dithionite could drive pyruvate synthesis from acetyl phosphate and carbon dioxide in a cell-free preparation from the fermentative bacterium *Clostridium butyricum* (MORTLOCK and WOLFE, 1959), there was for years general agreement that pyruvate synthesis by reversal of α-decarboxylation was at best of questionable importance in carbon dioxide fixation (WOOD and STJERNHOLM, 1963). It was not until ferredoxin had been isolated and certain of its properties determined (MORTENSON et al., 1962; TAGAWA and ARNON, 1962) that pyruvate synthesis by this reaction gained recognition as a physiological mechanism for the fixation of carbon dioxide.

In 1964, BACHOFEN et al. found that reduced ferredoxin[1] can bring about in cell-free extracts of the fermentative bacterium *Clostridium pasteurianum* a reversal of the "phosphoroclastic" splitting of pyruvate – the reaction that MORTENSON et al. (1962) had shown earlier to depend on ferredoxin. Shortly thereafter, ferredoxin-dependent pyruvate synthesis was demonstrated in the photosynthetic bacterium *Chromatium vinosum* (BUCHANAN et al., 1964).

Enzymes catalyzing ferredoxin-dependent carbon dioxide fixation have since been found in numerous other bacteria, both photosynthetic and nonphotosynthetic; but, except for blue-green algae (LEACH and CARR, 1971; BOTHE et al., 1974), it has not been found in photosynthetic cells that evolve oxygen. The ferredoxin-linked carboxylases demonstrated in these various organisms include not only the one yielding pyruvate, but also enzymes subsequently found that may or may not utilize acyl coenzyme A derivatives other than acetyl coenzyme A.

This chapter summarizes evidence for these reductive carboxylation reactions and describes recent findings pertaining to the cyclic pathway of carbon dioxide assimilation driven by reduced ferredoxin (the reductive carboxylic acid cycle; EVANS et al., 1966; BUCHANAN et al., 1967). Certain of these topics have been touched on in earlier reviews on this subject (BUCHANAN and ARNON, 1970; BUCHANAN, 1972, 1973).

---

[1] For a detailed description of the properties of ferredoxin, the reader is referred to Chapter II.9 of Volume 5, Photosynthesis I.

B. Ferredoxin-Linked Carboxylation Reactions

The nomenclature followed below will be that used in the original description of ferredoxin-linked carbon dioxide fixation (BUCHANAN et al., 1964). Accordingly, the carboxylase enzymes are collectively referred to as "synthases" and are identified for individual reactions by the name of the α-keto acid product formed – for example, pyruvate or α-ketoglutarate synthase. In reactions for which other names have been applied, those names will also be given. In the equations below, reversibility of the reactions is not indicated.

I. Synthesis of Pyruvate

$$\text{Acetyl-CoA} + CO_2 + Fd_{red} \xrightarrow[\text{synthase}]{\text{pyruvate}} \text{Pyruvate} + \text{CoA-SH} + Fd_{ox}. \tag{1}$$

Pyruvate synthase (pyruvate:ferredoxin oxidoreductase) is widely distributed in anaerobic organisms. It has been found in each of the three main types of photosynthetic bacteria: purple sulfur, *Chromatium vinosum* (BUCHANAN et al., 1964); green sulfur, *Chlorobium thiosulfatophilum* (EVANS and BUCHANAN, 1965) and a mixed culture of *Prostheocochloris aestuarii* (EVANS, 1968; SHIOI et al., 1976); and purple nonsulfur, *Rhodospirillum rubrum* (BUCHANAN et al., 1967). As noted above, pyruvate synthase has also been described for two species of blue-green algae and for various types of fermentative bacteria (BUCHANAN, 1972, 1973). In many of these organisms, pyruvate synthase appears to be important in the assimilation of exogenous acetate and carbon dioxide to form pyruvate-based biosynthetic products such as alanine and aspartate. As discussed below, pyruvate synthase is also a key enzyme in reactions leading from the reductive carboxylic acid cycle of carbon dioxide assimilation. There is evidence that carbon dioxide rather than bicarbonate is the active species fixed by pyruvate synthase (THAUER et al., 1975a).

Pyruvate synthase has been purified from two photosynthetic bacteria (*C. thiosulfatophilum* and *Chromatium vinosum*) and, following specific treatment to release cofactor from the enzyme, was shown to require thiamine pyrophosphate (BUCHANAN et al., 1965). A similar enzyme has been highly purified from *Clostridium acidi-urici* and shown to contain an iron–sulfur chromophore in addition to thiamine pyrophosphate (UYEDA and RABINOWITZ, 1971a, 1971b). This chromophore appears to couple directly to ferredoxin in both the synthesis and breakdown of pyruvate. It is noteworthy that a corresponding enzyme recently purified from mixed cultures of rumen microorganisms showed only synthase activity (BUSH and SAUER, 1976; SAUER et al., 1976). A second enzyme, also partially purified, was required for pyruvate breakdown. These findings raise the possibility that at least some organisms may use one enzyme for pyruvate synthesis and another enzyme for pyruvate breakdown. Organisms that utilize pyruvate synthase (and other ferredoxin-linked carboxylases) contain only trace amounts of lipoic acid – the cofactor associated with the oxidation of pyruvate in aerobic cells (BOTHE, 1975).

II. Synthesis of α-Ketoglutarate

$$\text{Succinyl-CoA} + CO_2 + Fd_{red} \xrightarrow[\text{synthase}]{\text{α-ketoglutarate}} \text{α-Ketoglutarate} + \text{CoA-SH} + Fd_{ox}. \quad (2)$$

α-Ketoglutarate synthase (α-ketoglutarate:ferredoxin oxidoreductase) was discovered in *C. thiosulfatophilum* (BUCHANAN and EVANS, 1965) and later was found in *R. rubrum* (BUCHANAN et al., 1967) and in a mixed culture of *Prostheocochloris aesturii* (EVANS, 1968). So far there is no evidence for this enzyme in photosynthetic purple sulfur bacteria, such as *Chromatium vinosum*.

Apart from its role in the reductive carboxylic acid cycle described below, α-ketoglutarate synthase appears to function in the assimilation of exogenous succinate and carbon dioxide (SHIGESADA et al., 1966). The principal products formed from succinate and carbon dioxide by this reaction are amino acids – especially glutamate, which is derived directly from α-ketoglutarate by transamination.

α-Ketoglutarate synthase has been purified from *C. thiosulfatophilum* (GEHRING and ARNON, 1972). The enzyme at its highest state of purity was free of pyruvate synthase activity and catalyzed the breakdown as well as the synthesis of α-ketoglutarate. Like pyruvate synthase, α-ketoglutarate synthase shows a requirement for thiamine pyrophosphate (BUCHANAN and ARNON, 1969).

III. Synthesis of α-Ketobutyrate

$$\text{Propionyl-CoA} + CO_2 + Fd_{red} \xrightarrow[\text{synthase}]{\text{α-ketobutyrate}} \text{α-Ketobutyrate} + \text{CoA-SH} + Fd_{ox}. \quad (3)$$

α-Ketobutyrate synthase (α-ketobutyrate:ferredoxin oxidoreductase) occurs in both photosynthetic and fermentative bacteria (BUCHANAN, 1969; BUSH and SAUER, 1976). In these organisms α-ketobutyrate synthase appears to function in a novel pathway for the biosynthesis of isoleucine and α-aminobutyrate (BUCHANAN, 1969). The newly found pathway is independent of threonine and threonine deaminase – the two components characteristic of the established path of isoleucine synthesis in aerobic cells (BUCHANAN, 1972). It is not known whether the threonine-threonine deaminase pathway exists in addition to the newly found α-ketobutyrate synthase pathway in those anaerobes known to contain α-ketobutyrate synthase.

α-Ketobutyrate synthase has not been extensively purified. Nevertheless, although certain purified preparations of pyruvate synthase show also α-ketobutyrate synthase, there are several lines of evidence to indicate that at least in some organisms the two activities reside on separate proteins (BUCHANAN, 1969; BUSH and SAUER, 1976).

IV. Synthesis of Phenylpyruvate

$$\text{Phenylacetyl-CoA} + CO_2 + Fd_{red} \xrightarrow[\text{synthase}]{\text{phenylpyruvate}} \text{Phenylpyruvate} + \text{CoA-SH} + Fd_{ox}. \quad (4)$$

The phenylpyruvate synthase reaction was first proposed by ALLISON and ROBINSON (1967) on the basis of experiments with whole cells of *Chromatium vinosum* and *R. rubrum*. These authors described ^{14}C-labeling data which were consistent with a synthesis of phenylalanine via phenylpyruvate from a condensation of phenylacetate and carbon dioxide.

Direct evidence for this reaction came several years later with the demonstration of ferredoxin-dependent phenylpyruvate synthesis from phenylacetyl CoA and carbon dioxide in cell-free extracts from purple and green photosynthetic sulfur bacteria (GEHRING and ARNON, 1971). The enzyme appears to function in the synthesis of aromatic amino acids via a pathway (ALLISON, 1965; ALLISON and ROBINSON, 1967) that is independent of the shikimate pathway established for aerobic cells (BUCHANAN, 1972). It is not known whether such cells also utilize the shikimate pathway for the synthesis of aromatic amino acids.

Although phenylpyruvate synthase has not been purified, it appears that this activity is due to a specific enzyme (GEHRING and ARNON, 1971).

V. Synthesis of α-Ketoisovalerate

$$\text{Isobutyryl-CoA} + CO_2 + Fd_{red} \xrightarrow[\text{synthase}]{\text{α-ketoisovalerate}} \text{α-Ketoisovalerate} + \text{CoA-SH} + Fd_{ox}.$$

$$(5)$$

The ferredoxin-dependent reductive carboxylation of isobutyryl coenzyme A to α-ketoisovalerate was described by ALLISON and PEEL (1971) for cell-free extracts from two different fermentative bacteria. The α-ketoisovalerate formed in this reaction, which is dependent on thiamine pyrophosphate, is converted to valine by transamination. As with the other ferredoxin-linked carboxylation reactions that lead to amino acids, the α-ketoisovalerate synthase mechanism for valine biosynthesis does not involve steps of the α,β-dihydroxyisovalerate mechanism previously established. It is not known whether fermentative bacteria that utilize the α-ketoisovalerate synthase pathway of valine biosynthesis also utilize the earlier established pathway (BUCHANAN, 1972). The presence of α-ketoisovalerate synthase in photosynthetic cells has not been reported.

VI. Synthesis of Formate

$$CO_2 + Fd_{red} \xrightarrow[\text{reductase}]{\text{carbon dioxide}} \text{Formate} + Fd_{ox}$$

Carbon dioxide reductase (reduced ferredoxin:carbon dioxide oxidoreductase) was discovered by JUNGERMANN et al. (1970) in cell-free extracts of *Clostridium pasteurianum*, and has so far not been reported to occur in photosynthetic bacteria. Like pyruvate synthase, the active species fixed by carbon dioxide reductase is carbon dioxide rather than bicarbonate (THAUER et al., 1975a). Growth and inhibitor studies suggest that, despite its reversibility, carbon dioxide reductase functions in the synthesis of formate rather than in its degradation (THAUER et al., 1974) and that molybdenum is an essential component of the enzyme (THAUER et al., 1973). Carbon dioxide reductase is the only known case in which reduced ferredoxin specifically promotes the fixation of carbon dioxide via a reaction that does not involve an acyl coenzyme A derivative.

C. The Reductive Carboxylic Acid Cycle

The reductive carboxylic acid cycle was proposed in 1966 as a new cyclic pathway for the assimilation of carbon dioxide by the photosynthetic bacteria *C. thiosulfatophilum* (EVANS et al., 1966) and *R. rubrum* (BUCHANAN et al., 1967). On the basis of the influence of different colors of light on photosynthetic products, it has recently been suggested that the reductive carboxylic acid cycle functions also in higher plants (PUNNETT, 1976a, 1976b). The confirmation of this proposal awaits a demonstration in leaves of the enzymes associated with the cycle.

The reductive carboxylic acid cycle is in effect a reversal of the oxidative citric acid cycle of Krebs and in one turn yields one molecule of acetyl coenzyme A from two molecules of carbon dioxide. Reduced ferredoxin is needed to form (by reversal of α-decarboxylation) α-ketoglutarate, a key intermediate of the cycle, and pyruvate, a product made from acetyl coenzyme A and carbon dioxide. The above-discussed enzymes, α-ketoglutarate synthase and pyruvate synthase, catalyze these two carboxylation steps.

As depicted in Figure 1, the reductive carboxylic acid cycle was initially visualized to function in the synthesis of amino acids, principal products of bacterial photosynthesis (BUCHANAN and ARNON, 1970). The products formed by *C. thiosulfatophilum* under physiological concentrations of carbon dioxide are in accord with this view (BUCHANAN et al., 1972; SIREVÅG, 1974).

The possibility that green bacteria could utilize reactions leading from the reductive carboxylic acid cycle for the synthesis of carbohydrates from acetate and carbon dioxide arose from the assimilation experiments of SIREVÅG and ORMEROD with whole cells (SIREVÅG and ORMEROD, 1970a, 1970b). The capability of *C. thiosulfatophilum* to convert acetate produced by the cycle to carbohydrate, as indicated in Figure 2, was further substantiated by more recent labeling and enzyme studies (BUCHANAN et al., 1972; SIREVÅG, 1975). In a related study, *C. thiosulfatophilum* was shown to contain, in addition to pyruvate synthase and the pertinent enzymes of glycolysis, the enzyme puruvate, P_i dikinase (BUCHANAN, 1974), which functions in carbon dioxide assimilation by "C_4-plants" (HATCH and SLACK, 1968). Pyruvate, P_i dikinase was also found in cell-free preparations from *Chromatium vinosum* and *R. rubrum* (BUCHANAN, 1974).

Apart from ATP, reduced ferredoxin and reduced NAD(P), which would be supplied by light (EVANS and BUCHANAN, 1965; BUCHANAN and EVANS, 1969; KUSAI and YAMONAKA, 1973), the operation of the reductive carboxylic acid cycle requires the presence of the enzymes associated with the cycle (acetyl coenzyme A synthetase; pyruvate synthase; pyruvate P_i dikinase; phosphoenolpyruvate carboxylase; malate dehydrogenase; fumarate hydratase; succinate dehydrogenase; succinyl coenzyme A synthetase, α-ketoglutarate synthase; isocitrate dehydrogenase; aconitate hydratase; and citrate lyase). This laboratory presented evidence for the occurrence of these enzymes in cell-free extracts of *C. thiosulfatophilum*

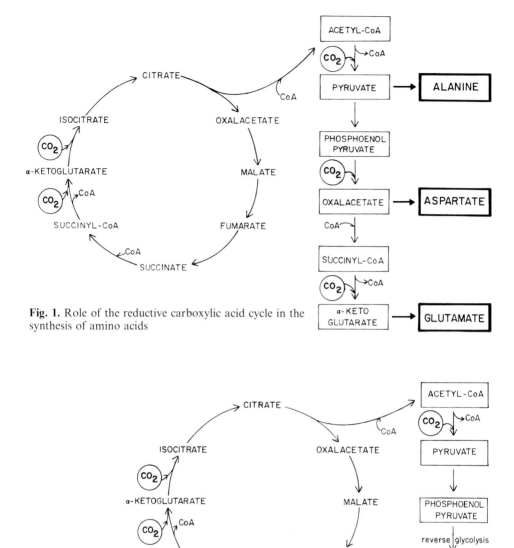

Fig. 1. Role of the reductive carboxylic acid cycle in the synthesis of amino acids

Fig. 2. Role of the reductive carboxylic acid cycle in the synthesis of carbohydrates. *PGA, 3-phosphoglycerate*

and *R. rubrum* (EVANS et al., 1966; BUCHANAN et al., 1967; BUCHANAN, 1974). The presence of citrate lyase has been questioned (BEUSCHER and GOTTSCHALK, 1972). As a consequence, we are currently reexamining evidence for the existence of this or a related enzyme that cleaves citrate to oxalacetate and a C_2-unit in these organisms.

In light of the widely held view that the reductive pentose phosphate cycle is present in all types of photosynthetic cells, a comment on the existence of this pathway in the photosynthetic green bacteria seems appropriate. In 1962, SMILLIE et al. reported that all enzymes of the cycle (including RuBPcarboxylase and ribulose-5-P kinase) were present in one strain of *C. thiosulfatophilum*. Some years later, CALLELY et al. (1968) concluded that a mixed culture of *Prostheocochloris aestuarii* was devoid of RuBPcarboxylase; the absence of the carboxylase in that organism was considered to be the basis for its requirement of a C_2-compound such as ethanol or acetate. It has since been established that the C_2-compound was required not by *Prostheocochloris aestuarii* but by its fermentative symbiont *Desulfuromonas acetoxidans* (SHIOI et al., 1976; PFENNIG and BIEBL, 1976).

In more recent studies, neither BUCHANAN (1972) nor SIREVÅG (1974) could find significant RuBPcarboxylase activity in cell-free extracts of the two strains of *C. thiosulfatophilum* examined. By contrast, TABITA et al. (1974) reported that low carboxylase activity was readily demonstrable in cell-free preparations of one of these strains following transatlantic air shipment of the parent cells. TABITA et al. described several characteristics of the purified enzyme including its stability, molecular weight, and quaternary structure.

In a recent collaborative study, BUCHANAN and SIREVÅG (1976) again were unable to detect significant amounts of RuBPcarboxylase in either of two strains of *C. thiosulfatophilum* even when following in detail the procedure of TABITA et al. (1974). These results prompted an attempt to demonstrate the carboxylase in whole cells by a mass spectrometric method that is based on the ability of growing cells to discriminate between the ^{12}C and ^{13}C isotopes of carbon dioxide (TROUGHTON et al., 1974). The results of that investigation (SIREVÅG et al., 1977), which were recently confirmed and extended to other photosynthetic green bacteria (QUANDT et al., 1977), indicate that RuBPcarboxylase and associated reactions of the reductive pentose phosphate cycle play at most a minor role in *C. thiosulfatophilum*. The reported absence (SIREVÅG, 1974; BUCHANAN and SIREVÅG, 1976) in this organism of ribulose-5-P kinase, an enzyme which, like RuBPcarboxylase, is peculiar to the cycle, is in accord with this conclusion.

Thus, in short, while there is not unanimity on this issue, recent findings tend to support the view (BUCHANAN et al., 1972) that the green bacteria may represent a group of photosynthetic organisms that has survived without the carbon reduction mechanism that is otherwise considered to be present universally in autotrophic cells.

D. Concluding Remarks

The discovery of a role for ferredoxin in the reductive synthesis of pyruvate from acetyl coenzyme A and carbon dioxide has led to the finding of a number of other carboxylation reactions dependent on reduced ferredoxin. Except for carbon dioxide reductase, these reactions involve a reductive carboxylation of an acyl coenzyme A derivative to an α-keto acid. The available evidence indicates that each different reaction is catalyzed by a specific enzyme. It has been reported that the flavoproteins, flavodoxin (ALLISON and PEEL, 1971) and phytoflavin (BOTHE, 1969), can substitute for ferredoxin in certain of the reactions.

The ferredoxin-linked carboxylation enzymes function in the joint assimilation of carbon dioxide and organic acids during growth of both photosynthetic and fermentative bacteria. The α-keto acids formed are key metabolic intermediates of particular importance in the synthesis of amino acids. The routes of amino acid biosynthesis involve in each case a newly found pathway that differs at least in part from the one previously established for other organisms.

The contribution of the ferredoxin-linked carboxylation enzymes seems not to be restricted to the assimilation of exogenous organic substrates in the presence of carbon dioxide. Pyruvate and α-ketoglutarate synthase form the basis for the operation of the reductive carboxylic acid cycle – a pathway of carbon dioxide assimilation proposed for photosynthetic bacteria. There is evidence that the carboxylation reactions associated with this cycle may be particularly important in the photosynthetic green bacteria (QUANDT et al., 1978).

References

Allison, M.J.: Biochem. Biophys. Res. Commun. *18*, 30–35 (1965)
Allison, M.J., Peel, J.L.: Biochem. J. *121*, 431–437 (1971)
Allison, J.J., Robinson, I.M.: J. Bacteriol. *93*, 1269–1275 (1967)
Bachofen, R., Buchanan, B.B., Arnon, D.I.: Proc. Natl. Acad. Sci. USA *51*, 690–694 (1964)
Beuscher, N., Gottschalk, G.: Z. Naturforsch. *27b*, 967–973 (1972)
Bothe, H.: In: Prog. Photosynthesis Res. Metzner, H. (ed.), Vol. III, pp. 1483–1491. Tübingen: Laupp 1969
Bothe, H., Falkenberg, B., Nolteernsting, U.: Arch. Microbiol. *96*, 291–304 (1974)
Bothe, H., Nolteernsting, U.: Arch. Microbiol. *102*, 53–57 (1975)
Buchanan, B.B.: J. Biol. Chem. *244*, 4218–4223 (1969)
Buchanan, B.B.: In: The enzymes, 3rd ed. Boyer, P.D. (ed.), Vol. VI, pp. 193–216. New York: Academic Press 1972
Buchanan, B.B.: In: Iron-sulfur proteins. Lovenberg, W. (ed.), Vol. I, pp. 129–150. New York: Academic Press 1973
Buchanan, B.B.: J. Bacteriol. *119*, 1066–1068 (1974)
Buchanan, B.B., Arnon, D.I.: In: Methods in enzymology. (Citric Acid Cycle). Lowenstein, J.M. (ed.), Vol. XIII, pp. 170–181. New York: Academic Press 1969
Buchanan, B.B., Arnon, D.I.: Adv. Enzymol. *33*, 119–176 (1970)
Buchanan, B.B., Evans, M.C.W.: Proc. Natl. Acad. Sci. USA *54*, 1212–1218 (1965)
Buchanan, B.B., Evans, M.C.W.: Biochim. Biophys. Acta *180*, 123–129 (1969)
Buchanan, B.B., Sirevåg, R.: Arch. Microbiol. *109*, 15–19 (1976)
Buchanan, B.B., Bachofen, R., Arnon, D.I.: Proc. Natl. Acad. Sci. USA *52*, 839–847 (1964)
Buchanan, B.B., Evans, M.C.W., Arnon, D.I.: In: Non-heme iron proteins. San Pietro, A. (ed.), pp. 175–188. Yellow Springs, Ohio: Antioch Press 1965
Buchanan, B.B., Evans, M.C.W., Arnon, D.I.: Arch. Mikrobiol. *59*, 32–40 (1967)
Buchanan, B.B., Schürmann, P., Shanmugam, K.T.: Biochim. Biophys. Acta *283*, 136–145 (1972)
Bush, R.S., Sauer, F.D.: Biochem. J. *157*, 325–331 (1976)
Callely, A.G., Rigopoulos, N., Fuller, R.C.: Biochem. J. *106*, 615–622 (1968)
Evans, M.C.W.: Biochem. Biophys. Res. Commun. *33*, 146–150 (1968)
Evans, M.C.W., Buchanan, B.B.: Proc. Natl. Acad. Sci. USA *53*, 1420–1425 (1965)
Evans, M.C.W., Buchanan, B.B., Arnon, D.I.: Proc. Natl. Acad. Sci. USA *55*, 928–934 (1966)
Gehring, U., Arnon, D.I.: J. Biol. Chem. *246*, 4518–4522 (1971)

Gehring, U., Arnon, D.I.: J. Biol. Chem. *247*, 6963–6969 (1972)

Hatch, M.D., Slack, C.R.: Biochem. J. *106*, 141–146 (1968)

Jurgermann, K., Kirchniawy, H., Thauer, R.K.: Biochem. Biophys. Res. Commun. *41*, 682–689 (1970)

Kusai, A., Yamanaka, T.: Biochim. Biophys. Acta *292*, 621–633 (1973)

Leach, C.K., Carr, N.G.: Biochim. Biophys. Acta *245*, 165–174 (1971)

Lipmann, F.: Adv. Enzymol. *6*, 231–267 (1946)

Mortenson, L.E., Valentine, R.C., Carnahan, J.E.: Biochem. Biophys. Res. Commun. *7*, 448–452 (1962)

Mortlock, R.P., Wolfe, R.S.: J. Biol. Chem. *234*, 1657–1658 (1959)

Pfennig, N., Biebl, H.: Arch. Microbiol. *110*, 3–12 (1976)

Punnett, T.: Fed. Proc. *35*, 1597 (1976a)

Punnett, T., Kelly, J.H.: Plant Physiol. Abstracts 59 (1976b)

Quandt, L., Gottschalk, G., Ziegler, H., Stichler, W.: FEMS Microbiol. Lett. *1*, 125–128 (1977)

Quandt, L., Pfennig, N., Gottschalk, G.: FEMS Microbiol. Lett. *3*, 227–230 (1978)

Sauer, F.D., Bush, R.S., Stevenson, L.L.: Biochim. Biophys. Acta *445*, 518–520 (1976)

Shigesada, K., Hidaka, K., Katsuki, H., Tanaka, S.: Biochim. Biophys. Acta *112*, 182–185 (1966)

Shioi, Y., Takamiya, K., Nishimura, M.: J. Biochem. (Tokyo) *79*, 361–371 (1976)

Sirevåg, R.: Arch. Microbiol. *98*, 3–18 (1974)

Sirevåg, R.: Arch. Microbiol. *104*, 105–111 (1975)

Sirevåg, R., Ormerod, J.G.: Science *169*, 186–188 (1970a)

Sirevåg, R., Ormerod, J.G.: Biochem. J. *120*, 399–408 (1970b)

Sirevåg, R., Buchanan, B.B., Berry, J.A., Troughton, J.H.: Arch. Microbiol., *112*, 35–38 (1977)

Smillie, R.M., Rigopoulos, N., Kelly, H.: Biochim. Biophys. Acta *56*, 612–614 (1962)

Tabita, R.F., McFadden, B., Pfennig, N.: Biochim. Biophys. Acta *341*, 187–194 (1974)

Tagawa, K., Arnon, D.I.: Nature (London) *195*, 537–543 (1962)

Thauer, R.K., Fuchs, G., Schnitker, U., Jungermann, K.: FEBS Lett. *38*, 45–48 (1973)

Thauer, R.K., Fuchs, G., Jungermann, K.: J. Bacteriol. *118*, 758–760 (1974)

Thauer, R.K., Käufer, B., Fuchs, G.: Eur. J. Biochem. *55*, 111–117 (1975a)

Thauer, R.K., Käufer, B., Scherer, P.: Arch. Microbiol. *104*, 237–240 (1975b)

Troughton, J.H., Card, K.A., Hendy, C.H.: Carnegie Inst. Yearb. *73*, 768–780 (1974)

Uyeda, K., Rabinowitz, J.C.: J. Biol. Chem. *246*, 3111–3119 (1971a)

Uyeda, K., Rabinowitz, J.C.: J. Biol. Chem. *246*, 3120–3125 (1971b)

Wood, H.G., Stjernholm, R.: The bacteria. Gunsalus, I.C., Stanier, R.Y. (eds.), pp. 41–117. New York: Academic Press 1963

4. Reduction of Nitrate and Nitrite

B. Vennesland and M.G. Guerrero

A. Introduction

For most plants in their natural environment, nitrate is the immediate source of nitrogen. This nitrate must be converted to ammonia before it is combined with carbon compounds to form the various nitrogenous components of the cell. Since the nitrogen of nitrate has a formal charge of $+5$, whereas the nitrogen of ammonia has a formal charge of -3, an 8-electron reduction is required to convert nitrate to ammonia. This occurs in two steps. First, nitrate is reduced to nitrite in a 2-electron reduction catalyzed by the enzyme nitrate reductase, a flavoprotein containing heme and molybdenum. After many early efforts to identify intermediates in the further reduction of nitrite to ammonia, a consensus has been reached that this 6-electron reaction is catalyzed by a single enzyme, the iron-protein nitrite reductase, without liberation of free intermediates (Hewitt, 1975; Losada, 1976a; Zumft, 1976).

The ability to use nitrate as a nitrogen source is shared with higher plants and algae by many fungi, yeasts, and bacteria. The scope of the present article is limited primarily to nitrate reduction in algae and higher plants, but in view of the advantages of a comparative approach, occasional reference will be made to studies with other microorganisms, especially to fungal nitrate reduction, which has been studied intensively.

In bacterial metabolism, two types of nitrate reduction are distinguished: (1) assimilatory, as in plants, and (2) respiratory (in the latter type of nitrate reduction, nitrate serves as an electron acceptor in place of molecular oxygen). Assimilatory nitrate reductase is generally easily soluble. An important exception is the assimilatory nitrate reductase of the blue-green algae (see Sect. B.II), which is membrane-bound.

Because of space limitations, comprehensive coverage even of the recent literature is impossible.[1] Many reviews of nitrate reduction are available (e.g., Kessler, 1959, 1964, 1971, 1976; Beevers and Hageman, 1969, 1972; Hewitt, 1970, 1975; Morris, 1974; Losada, 1976; Zumft, 1976). Most studies of bacterial nitrate reductase have been concerned with the respiratory enzyme, not covered here, but see Payne (1973), Brown et al. (1974), Stouthamer (1976).

---

[1] A more extensive chapter will be found in Volume 12 (*Inorganic Plant Nutrition*) of this Encyclopedia.

B. Reduction of Nitrate to Nitrite

I. Assimilatory Nitrate Reductase of Eukaryotes

1. Classification

The enzyme common to algae and higher plants and molds catalyzes the reduction of nitrate by reduced pyridine nucleotides [Eq. (1)]

$$NAD(P)H + NO_3^- + H^+ \rightarrow NAD(P)^+ + NO_2^- + H_2O \tag{1}$$

Three subclasses of NAD(P)H: nitrate oxidoreductase have been distinguished: E.C.1.6.6.1 is specific for NADH; E.C.1.6.6.2 operates with both NADH and NADPH; and E.C.1.6.6.3 is specific for NADPH. This classification is not always easy to apply, because many pyridine nucleotide nitrate reductases react to some extent with both pyridine nucleotides, and the boundaries between the groups are fuzzy. However, the commonest enzyme in green algae and higher plants has a strong preference for NADH, whereas the enzyme in fungi and yeast prefers NADPH, though not very exclusively. This latter enzyme has sometimes been termed the fungal type, but it occurs to some extent in higher plants also (Shen et al., 1976; Jolly et al., 1976; Campbell, 1976).

Although a particulate nitrate reductase preparation from the yeast *Candida utilis* was reported to have B stereospecificity for the reduced position of the pyridine nucleotide (Davies et al., 1972), all of the purified, soluble, assimilatory nitrate reductases examined to date were found to have A stereospecificity (Guerrero and Vennesland, 1975; Guerrero et al., 1977).

2. Enzyme Purification and Characterization

Because of its instability, assimilatory nitrate reductase has resisted numerous purification attempts, though many partial purifications have been reported (Hewitt, 1975). The enzyme from *Chlorella vulgaris* can be obtained in homogeneous form in a few simple steps mainly by affinity chromatography on a blue-dextran Sepharose column (Solomonson, 1975). A high level of purity has also been obtained for the enzyme from the yeast *Rhodotorula glutinis* (Guerrero and Gutierrez, 1977). A claim has been made that the nitrate reductase of *Ankistrodesmus braunii* has also been purified to homogeneity (Ahmed and Spiller, 1976), but it is difficult to accept this claim because of the low specific activity of the preparation.

The physical properties of some of the more highly purified assimilatory nitrate reductases from eukaryotes are summarized in Table 1. The enzyme from *Chlorella* contains at least three subunits of about the same size and about two mol each of heme, FAD, and Mo per mol of enzyme (Solomonson et al., 1975). The enzyme from *Rhodotorula* contains two subunits of about equal size and one mol cytochrome per mol of enzyme (Guerrero and Gutierrez, 1977). It was in *Neurospora* that the pyridine nucleotide-requiring assimilatory nitrate reductase was first clearly characterized as a molybdo-flavoprotein containing cytochrome b-557 (Nason and Evans, 1953; Nicholas and Nason, 1954; Nicholas et al., 1954; Kinsky and

Table 1. Physical properties of assimilatory nitrate reductases from eukaryotes

	$S_{20,w}$	Stokes radius (Å)	Molecular weight	Subunits
Chlorella vulgaris[a]	9.7	89	356,000	3 (MW 100,000)
Spinacea oleracea[b]	8.1	60	197,000	nr
Aspergillus nidulans[c, d]	7.8[c]	64[c]	197,000[c]	nr[c]
	7.6[d]	63[d]	190,000[d]	nr[d]
Neurospora crassa[e]	8.0	70	228,000	nr
Rhodotorula glutinis[f]	7.9	70.5	230,000	2 (MW 118,000)

nr = not reported
[a] Solomonson et al. (1975).
[b] cited from Notton et al. (1977).
[c] Downey (1971).
[d] McDonald and Coddington (1974).
[e] Garrett and Nason (1969).
[f] Guerrero and Gutierrez (1977).

McElroy, 1958; Garret and Nason, 1967, 1969). A cytochrome b-557 has been detected in all nitrate reductases which have been adequately purified from chlorophyll-containing tissues. This includes the nitrate reductase of spinach (Guerrero et al., 1977; Notton et al., 1977). In contrast to the *Chlorella vulgaris* enzyme, which contains rather firmly bound flavin, many nitrate reductases lose their flavin component rather readily, so that added flavin is necessary in the assay procedure. Thus, the flavin requirement of the enzyme was early recognized. Molybdenum is regarded as an important component of all nitrate reductases (see also Sects. B.I.4 and B.I.5).

3. Reactions Catalyzed and Reaction Mechanism

Like many flavoproteins, nitrate reductase catalyzes a so-called diaphorase or NAD(P)H dehydrogenase reaction in addition to the reduction of nitrate by reduced pyridine nucleotide. In the diaphorase reaction, the reduced pyridine nucleotide is oxidized by cytochrome c, ferricyanide, or other oxidants. Just as the reduced pyridine nucleotide-activating moiety of nitrate reductase can be measured independently of nitrate reduction in the diaphorase reaction, so also the nitrate-reducing activity can be measured independently of pyridine nucleotide by substituting reduced flavins or viologens for the natural reductant. The nitrate-activating Mo-containing moiety of the enzyme can be completely inhibited without inhibition of diaphorase, and the diaphorase moiety can be inhibited without affecting the activity of the terminal nitrate-activating moiety (Sorger, 1966; Garret and Nason, 1969; Losada et al., 1969; Notton and Hewitt, 1971c; Schrader et al., 1968; Losada, 1976).

The electron transport chain of nitrate reductase has been pictured as

$$\text{NAD(P)H} \rightarrow (\text{FAD} \rightarrow \text{cyt b-557} \rightarrow \text{Mo}) \rightarrow \text{NO}_3^- \tag{2}$$

In the case of the *Neurospora* enzyme, evidence has been presented that an SH group is interposed between the NAD(P)H and flavin (Amy et al., 1977). Though it has been shown that the cytochrome of nitrate reductase is reduced by added NAD(P)H and reoxidized on addition of nitrate (Garrett and Nason, 1969; Vennesland and Jetschmann, 1971; Guerrero et al., 1977; Guerrero and Gutierrez, 1977; Notton et al., 1977), rapid kinetic measurements of cytochrome reduction and oxidation have to our knowledge not yet been made, and are required to elucidate the function of the cytochrome.

Stiefel (1973) has proposed that Mo(IV) is oxidized to Mo(VI) when nitrate is reduced to nitrite, but did not exclude other redox changes of Mo (Stiefel et al., 1977). Orme-Johnson et al. (1977) cited EPR experiments with *N. crassa* nitrate reductase suggesting that the sequence Mo(VI)→Mo(V)→Mo(IV)→Mo(III)→Mo (II) occurs when the enzyme is reduced by NADPH. Complexes of Mo(III) can reduce nitrate to nitrite nonenzymatically and their use as models for nitrate reductase has been discussed (Ketchum et al., 1977). The main redox changes of Mo during enzyme action remain undefined, however.

Nitrate can be reduced to nitrite by ultraviolet radiation and it has been suggested that this could contribute significantly to nitrate utilization by plants growing under tropical conditions (Naik et al., 1976).

4. Effect of Tungsten Incorporation

Tungsten salts antagonize or compete with molybdenum. In barley it was shown that a nitrate-induced diaphorase activity appeared when plants were treated with tungsten, and nitrate utilization was impaired (Wray and Filner, 1970). Similar effects were noted in *Chlorella* (Vega et al., 1971). This is due to the fact that tungsten is built into nitrate reductase in place of Mo, to give a molecule incapable of activating nitrate, but with full diaphorase activity. The process has been studied in tobacco cells (Heimer and Filner, 1971; Heimer et al., 1969), spinach leaves (Notton and Hewitt, 1971a, b, 1972; Notton et al., 1972), cauliflower (Notton et al., 1974), *Chlorella fusca* (Paneque et al., 1972), and *Neurospora* (Subramanian and Sorger, 1972; Sorger et al., 1974; Lee et al., 1974a).

Nitrate reductase is also inhibited in vivo or in vitro by vanadium salts (Buczek, 1973; Hewitt, 1975), but it is not clear whether or not the vanadium is acting similarly to tungsten.

5. Genetic Studies

The genetics of fungal nitrate reduction has been extensively investigated, both in *Neurospora crassa* and in *Aspergillus nidulans* (see Coddington, 1976). Among the studies of *Neurospora* which have particular biochemical interest are those which show that extracts of the nit-1 mutant, inactive in nitrate reductase, can be reconstituted in vitro to give a fully active normal enzyme, by addition of a component from any of a wide variety of molybdo-enzymes. Sorger (1966) first showed that the nit-1 mutant contained a nitrate-inducible diaphorase activity, but no nitrate reductase. The conversion of this diaphorase to active nitrate reductase requires an organic molybdenum cofactor which appears to be common to many molybdenum-containing enzymes (see Zumft, 1976). The nit-1 locus is thought to specify this cofactor which presumably combines with two diaphorase subunits to give active enzyme (NASON et al., 1970; KETCHUM et al., 1970; NASON et al., 1971; LEE et al., 1974b).

6. Inhibitors

a) Chlorate, an Alternate Substrate

Chlorate, a weed killer, is toxic to cells growing on nitrate, but not to cells growing on ammonia. ÅBERG (1947) first suggested that the toxicity of chlorate is due to its reduction to chlorite, a much more toxic substance, by the enzyme nitrate reductase. The earlier literature has been reviewed by LILJESTRÖM and ÅBERG (1966).

Chlorate can in fact replace nitrate as a substrate for isolated nitrate reductase, each of the two compounds acting as a competitive inhibitor of the reduction of the other (SOLOMONSON and VENNESLAND, 1972a; VEGA et al., 1972).

During the reduction of chlorate by nitrate reductase, there is a progressive and irreversible degradation of the enzyme by the chlorite produced (SOLOMONSON and VENNESLAND, 1972a). These facts can explain many, though not all, of the biological effects of chlorate (TROMBALLA and BRODA, 1971; COVE, 1976).

b) HCN

Nitrate utilization in vivo is very sensitive to cyanide, and the enzyme nitrate reductase is strongly inhibited by HCN (SOLOMONSON and VENNESLAND, 1972b; GARRETT and GREENBAUM, 1973; SOLOMONSON, 1974; BAREA et al., 1976a; VEGA et al., 1972).

The inhibition of the nitrate reductase of *Chlorella vulgaris* by HCN has been studied in considerable detail, because it has been shown that an inactive nitrate reductase formed in vivo is in fact an HCN complex of the reduced enzyme (LORIMER et al., 1974).

In the presence of its reducing substrate, NADH, nitrate reductase of *Chlorella vulgaris* combines stoichiometrically with HCN to form a product which is inactive for nitrate reduction but which retains full diaphorase activity. This HCN complex is relatively stable. It is formed in a second-order reaction, the rate being proportional to the concentration of reduced enzyme and to the concentration of HCN. Removal of free HCN and NADH by gel filtration does not lead to activation, but the enzyme can be rapidly activated by oxidation with ferricyanide, and more slowly by tervalent manganese pyrophosphate (FUNKHOUSER and ACKERMANN, 1976), or other oxidants such as nitrate, O_2, and redox dyes. The HCN-inactivated enzyme can also be reactivated by added thiosulfate and rhodanese (thiosulphate: cyanide sulfur transferase E.C.2.8.1.1) without the addition of an oxidizing agent (TOMATI et al., 1976).

On reactivation in the presence of nitrate and O_2, HCN is released from inactive enzyme in proportion to the amount of enzyme reactivated (LORIMER et al., 1974). Nitrate inhibits the reaction of reduced enzyme with HCN, presumably by keeping the enzyme in the oxidized form (SOLOMONSON et al., 1973; BAREA et al., 1976a). HCN also apparently behaves as a competitive inhibitor of nitrate, when initial velocities are measured (VEGA et al., 1972; GARRETT and GREENBAUM, 1973), but dissociation constants determined in this way do not give a true measure of the high affinity of the reduced enzyme for HCN. The dissociation constant K_D for the reaction.

$$(E_R \cdot HCN) \rightleftharpoons (E_R) + (HCN) \tag{3}$$

where E_R represents reduced enzyme, was estimated to be $3.6 \cdot 10^{-10}$ M for the enzyme from *Chlorella vulgaris* (LORIMER et al., 1974).

Since the substitution of tungsten for Mo results in loss of nitrate reduction activity, but no loss of diaphorase, and the HCN-inhibited enzyme likewise has no nitrate-reducing capacity, but full diaphorase activity, it is thought that the HCN reacts at the Mo site, but the nature of the chemical binding is not known.

c) Other Inhibitors

Azide, cyanate, and thiocyanate are strong inhibitors, all competitive with nitrate. They do not inhibit the diaphorase reaction. The latter, however, is inhibited by SH reagents such as pHMB which in turn have no effect on the reduction of nitrate by reduced benzyl viologens (GARRETT and NASON, 1967; MORRIS and SYRETT, 1963; SCHRADER et al., 1968; SOLOMONSON and VENNESLAND, 1972b; RELIMPIO et al., 1971; SOLOMONSON, 1974; VEGA et al., 1972).

A reversible competitive inhibition of nitrate reduction by nitrite has been described for the enzyme from *Chlorella vulgaris* (SOLOMONSON and VENNESLAND, 1972b), *Aspergillus* (McDONALD and CODDINGTON, 1974) and *Rhodotorula glutinis* (GUERRERO and GUTIERREZ, 1977). In the case of *Chlorella*, the affinity of the enzyme for nitrite is lower than that for nitrate, but in the case of the fungal and the yeast enzymes, the affinities for nitrite and nitrate are about equal.

II. Assimilatory Nitrate Reduction in Prokaryotes

The prokaryotic blue-green algae *Anabaena* and *Anacystis* have been shown to contain a particle-bound nitrate reductase, which can be solubilized to give an enzyme which operates with artificial reductants but not with reduced pyridine nucleotides (HATTORI and MYERS, 1966, 1967; HATTORI, 1970; HATTORI and UE-SUGI, 1968a; MANZANO et al., 1976). Reconstituted systems from *Anabaena cylin-drica* (HATTORI and UESUGI, 1968b) and chlorophyll-containing particles from *Ana-cystis nidulans* and *Nostoc muscorum* which contained tightly bound nitrate reduc-tase (MANZANO et al., 1976; ORTEGA et al., 1976, 1977) could carry out a ferredoxin-dependent photoreduction of nitrate. The enzyme from *Anacystis nidulans* has been solubilized and purified to homogeneity by use of affinity chromotography on a feredoxin-Sepharose gel as the main step. This ferredoxin-dependent nitrate reductase has only one polypeptide chain with a molecular weight of 75,000 daltons (MANZANO and CANDAU, unpubl.). In the blue-green algae, nitrate reduction is intimately linked to photosynthesis (CANDAU et al., 1976; ORTEGA et al., 1976, 1977).

The presence of a NADH-dependent nitrate reductase in extracts of the photo-synthetic bacterium *Rhodospirillum rubrum* has been reported by KATOH (1963), but it is not clear whether this enzyme has an assimilatory or a respiratory function. In contrast, a ferredoxin-dependent nitrate reductase has been reported in *Ectothio-rhodospira shaposhnikovii* (MALOFEEVA et al., 1975).

C. Reduction of Nitrite to Ammonia

I. Nitrite Reductase of Photosynthetic Cells

1. Classification

The enzyme responsible for the reduction of nitrite to ammonia in photosynthesizing cells is ferredoxin:nitrite oxidoreductase (E.C.1.7.7.1), and the reaction catalyzed is shown in Eq. (4)

$$NO_2^- + 6e^- + 8H^+ \rightarrow NH_4^+ + 2H_2O \tag{4}$$

The six electrons are furnished by reduced ferredoxin, and there are no intermediates released in the reaction.

2. Enzyme Purification and Characterization

Nitrite reductase is usually readily solubilized. An exception is the prolonged disruption procedure required to solubilize the enzyme from the blue-green alga *Anacystis nidulans* (GUERRERO et al., 1974; MANZANO et al., 1976).

Purification to homogeneity has been achieved for the nitrite reductases from *Chlorella fusca* (ZUMFT, 1972) and from leaves of spinach, *Spinacea oleracea* (CÁRDENAS et al., 1972a; SHIMIZU and TAMURA, 1974), squash and marrow, *Cucurbita pepo* (CÁRDENAS et al., 1972b; HUCKLESBY et al., 1976), and from the red alga *Porphyra yezoensis* (HO et al., 1976a). The use of a ferredoxin-Sepharose column as an effective purification step for spinach nitrite reductase has been recently reported (IDA et al., 1976). Only partial purification has been reported for the enzyme from blue-green algae (HATTORI and UESUGI, 1968a). A molecular weight of 63,000 daltons was measured for *Chlorella*, marrow, spinach, and squash. IDA and MORITA (1973) have reported a molecular weight of 72,000 daltons for the spinach nitrite reductase, and the presence of two subunits, but SHIMIZU and TAMURA (1974) found no subunits in the spinach enzyme, as was also reported for *Chlorella fusca* (ZUMFT, 1972).

The absorption spectra of all purified nitrite reductases show similar peaks in the regions 380–390 and 570–580 nm. MURPHY et al., (1974) have shown that this absorption spectrum is primarily due to siroheme, an iron–porphyrin prosthetic group previously shown to be a component of the enzyme sulfite reductase (MURPHY et al., 1973; MURPHY and SIEGEL, 1973). Siroheme contains a reduced porphyrin with eight carboxylate side chains. APARICIO et al. (1975) presented evidence for the presence, in addition, of an iron–sulfur center in the spinach enzyme. A recent, extensive study of the spinach enzyme showed that it was composed of one polypeptide chain of 61,000 daltons, containing siroheme plus two additional iron atoms and two labile sulfides (Vega and Kamin, 1977).

It has also been demonstrated that a cytochrome c-553 after heat inactivation or chemical oxidation has nitrite reductase activity (Ho et al., 1976b).

3. Reactions Catalyzed and Reaction Mechanism

Flavodoxin, a physiological substitute for ferredoxin under conditions of limited iron supply, can substitute for ferredoxin as an electron source for nitrite reduction in the presence of nitrite reductase (Bothe, 1969; Zumft, 1972). Artificial one-electron reductants such as chemically reduced methyl or benzyl viologens are also effective electron donors, and have been used routinely for the assay of enzyme activity (Losada and Paneque, 1971).

In addition to nitrite, hydroxylamine can also be reduced by nitrite reductase, but this hydroxylamine reductase activity is very low in the purified enzyme (Zumft, 1972; Hucklesby and Hewitt, 1970).

Sulfite cannot be utilized as a substrate by nitrite reductase, but there are striking similarities between nitrite reductase and sulfite reductase (E.C.1.8.1.2). The latter enzyme can use both nitrite and hydroxylamine as alternate substrates to sulfite. Both enzymes catalyze six-electron reductions, and both have similar prosthetic groups. The $(Fe_2-S_2^*)$ (iron–sulfur center) plus a siroheme of nitrite reductase is an analog of the electron transport chain of the iron-containing subunit of sulfite reductase, which can transport one electron at a time through a $(Fe_4-S_4^*)$ cluster plus one siroheme (Siegel and Davis, 1974).

The binding of nitrite to the spinach enzyme markedly alters both the absorption spectrum and the high-spin heme EPR signal (Aparicio et al., 1975; Vega et al., 1976; Vega and Kamin, 1977), suggesting that the siroheme is the substrate binding site. One mol of nitrite is bound per mol of enzyme. It is though that the path of electron transfer is from iron–sulfur to siroheme to nitrite substrate, as shown:

$$Fd_{red} \rightarrow [Fe_2-S_2^*] \rightarrow (Siroheme) \rightarrow NO_2^- \tag{5}$$

The possibility that the iron–sulfur center might also donate electrons to substrate has, however, not been excluded (Vega and Kamin, 1977).

4. Inhibitors

Cyanide is a potent inhibitor of nitrite reductase from all the various sources studied. The inhibition is competitive with nitrite (Cárdenas et al., 1972b). Carbon monoxide is also an effective inhibitor which reacts only with reduced siroheme. Nitrite protects against CO inhibition (Vega and Kamin, 1977). When either cyanide or carbon monoxide are added to the enzyme under reducing conditions, EPR signals corresponding to the reduced iron–sulfur center are observed (Aparicio et al., 1975; Vega et al., 1976), which suggests that complete reduction of the iron–sulfur center only occurs when inhibitors bind to the siroheme and prevent the oxidation of the reduced iron–sulfur center.

Sulfhydryl reagents such as pHMB and sodium mersylate also inhibit nitrite reductase (Hucklesby et al., 1976; Zumft, 1972; Ho and Tamura, 1973). The pHMB effect seems to be due to a destruction of the siroheme prosthetic group (VEGA and KAMIN, 1977).

II. Nitrite Reductase of Nonphotosynthetic Cells

The nitrite reductase of *Neurospora crassa* has been characterized as an NAD(P)H: nitrite oxidoreductase (E.C.1.6.6.4) of 290,000 daltons, i.e., much larger than the nitrite reductase of leaves, but it has the same siroheme prosthetic group (LAFFERTY and GARRETT, 1974; VEGA et al., 1975a, b). These pyridine-nucleotide dependent nitrite reductases are flavoproteins.

D. Control of Nitrate Reduction

BEEVERS and HAGEMAN (1969) concluded that the rate-limiting enzyme in the reduction of nitrate is nitrate reductase. Nitrite and ammonia are normally present in plant cells in only very small amounts, except under exceptional circumstances, but intracellular nitrate concentrations can reach very high levels, presumably in the vacuole. Thus, studies of the mechanism of regulation of nitrate reduction have mainly been focused on nitrate reductase.

Such studies have been hampered by the questionable significance of enzyme assay procedures applied to crude extracts. There have been reports of unidentified inhibitors of nitrate reductase (e.g., STULEN et al., 1971; DALLING et al., 1972a; ASLAM, 1977), or of an initial lag in the enzyme reaction (DUSKY and GALITZ, 1977; TISCHNER, 1976). An "in vivo" assay procedure (JAWORSKI, 1971; STREETER and BOSLER, 1972; KLEPPER et al., 1971; BRUNETTI and HAGEMAN, 1976; JONES et al., 1977) which involves measurements of nitrite excreted by intact tissue after various prior treatments, has not entirely solved this problem. It has received considerable attention, since there is evidence that the ultimate protein content of the seeds of various crop plants can be predicted from assay of the nitrate reductase activity of seedling tissues (BRUNETTI and HAGEMAN, 1976; JOHNSON et al., 1976, 1977; BUTZ, 1977). Freezing and thawing has been used to measure nitrate and nitrite reductase in otherwise intact *Chlorella* cells (SYRETT, 1973; SYRETT and THOMAS, 1973).

I. Synthesis and Degradation of Enzymes

1. Variations in Enzyme Level

In higher plants, nitrate reductase activity has been shown to vary with light intensity, CO_2 levels, temperature, water availability, and nitrate supply (BEEVERS and HAGEMAN, 1969). In general the enzyme is formed in cells or plants growing on nitrate and is almost absent in cells growing on ammonium. Nitrite reductase, usually but not always, develops and disappears together with nitrate reductase (see e.g., GARRETT, 1972; KELKER and FILNER, 1971; DILWORTH and KENDE, 1974a, b; MORRIS, 1974).

Circadian rhythms of nitrate reductase activity have been described in higher plants, with the activity reaching highest levels in the light period, and declining in the dark (Upcroft and Done, 1972, 1974; Cohen and Cumming, 1974; Steer, 1974, 1976). There is evidence that plant hormones may be involved in such changes in enzyme level (Lips and Roth-Bejerano, 1969; Roth-Bejerano and Lips, 1970; Knypl, 1973; Steer, 1976). Phytochrome participation has been reported in the effect of light on induction of nitrate reductase in peas, maize, and mustard (Jones and Sheard, 1972, 1975, 1977; Johnson, 1976).

In synchronized *Chlorella* cultures growing in approximately circadian successive light and dark periods, a marked increase of both nitrate and nitrite reductase occurred at the beginning of the light period with activity passing through a maximum in the light and then declining, to reach a low level at the end of the dark period (Hodler et al., 1972; Tischner, 1976; see also Kanazawa et al., 1970). The increase required new protein synthesis, being sensitive to actidione. Both enzymes were so affected, and the conclusion was drawn that both enzymes were synthesized on 80 S ribosomes in the cytoplasm (Tischner, 1976). In bean leaves, cycloheximide inhibited the synthesis of both enzymes, whereas chloramphenicol was without effect (Sluiter-Scholten, 1973). The stimulatory effect of light (and CO_2) on enzyme synthesis could be duplicated by added carbohydrate in both bean leaves and *Chlorella*, as well as in pea roots (Sahulka et al., 1975) and corn roots (Aslam and Oaks, 1975). The metabolism of nitrogenous substances likewise plays a role (Filner, 1966; Smith and Thompson, 1971; Liu and Hellebust, 1974; Oaks, 1974; Oaks et al., 1977).

Little is known about the mechanism of control of nitrate or nitrite reductase synthesis at the level of transcription or translation. Cells starved of nitrogen may develop high enzyme levels (Morris and Syrett, 1963; Kessler and Oesterheld, 1970; Oesterheld, 1971; Syrett and Hipkin, 1973; Spiller et al., 1976b), as may cells grown on urea or certain amino acids (Syrett and Hipkin, 1973; Rigano et al., 1974) or nitrite (Lips et al., 1973; Kaplan et al., 1974).

A bewildering variety of apparently unrelated chemicals can elicit nitrate reductase synthesis in certain plant species (Dilworth and Kende, 1974a, b; Knypl, 1974; Knypl and Ferguson, 1975; Shen, 1972; Shen et al., 1976). Subtoxic concentrations of the herbicide simazin, and related compounds, cause an increased level of nitrate reductase in seedlings of certain species (rye, peas, rice, and maize), and such compounds have been used to increase the protein content of food plants (Ries et al., 1967; Tweedy and Ries, 1967; Sawhney and Naik, 1972).

2. Proteolytic Degradation of Nitrate Reductase

Travis et al., (1969) reported that cycloheximide prevented the normal decline of nitrate reductase which occurred in barley leaves in darkness. This suggested that another protein was necessary for enzyme degradation. An enzyme which specifically inactivates nitrate reductase but not a range of other enzymes, has been isolated from the roots of maize seedlings (Wallace, 1973, 1974). This enzyme acts on nitrate reductases from *Neurospora* and higher plants, but not on bacterial nitrate reductases. It appears to be a proteolytic enzyme, and particularly attacks the diaphorase moiety of the nitrate reductase (Wallace, 1975). Inactivation of a possibly similar nature has been noted in rice roots (Kadam et al., 1974) and in suspension cultures of rice cells (YAMAYA and OHIRA, 1976). In *Neurospora*, the presence of nitrate in the culture medium has been shown to slow down the rate of the in vivo degradation of nitrate reductase (SORGER et al., 1974).

II. Utilization of Nitrate

The control of nitrate utilization cannot be considered comprehensively if the topics of nitrate uptake and transport are ignored, but these topics cannot be covered here.

There is a close relationship between the development of nitrate transport activity and nitrate reductase activity. This has led to the suggestion that both processes are functions of the same molecule (see e.g., BUTZ and JACKSON, 1977). It should be pointed out, however, that nitrate uptake has been shown to occur in cells which contain no active nitrate reductase (HEIMER and FILNER, 1971; SCHLOEMER and GARRETT, 1974; RAO and RAINS, 1976), so the transport system is not dependent on the presence of the active enzyme, a fact which clearly refutes the speculative relationship suggested by BUTZ and JACKSON (1977).

Nitrate assimilation has been studied intensively in the green algae *Chlorella fusca* and *Chlamydomonas reinhardii* (MORRIS, 1974; SYRETT and LEFTLEY, 1976). Such cells show particularly clearly the strong stimulation of nitrate utilization by light and CO_2. Added carbohydrates can substitute for CO_2, indicating that the effect of CO_2 is due to its conversion to carbohydrates. The addition of ammonia to algae assimilating nitrate results in a prompt cessation of nitrate utilization. Cells starved of carbohydrates lose their normal capacity to control nitrate assimilation and can reduce nitrate to ammonia in the light in the absence of CO_2, ammonia being then excreted into the medium. From these facts it was first concluded that a product of nitrate assimilation inhibits nitrate utilization by affecting either the nitrate uptake system or the nitrate reductase itself (SYRETT and MORRIS, 1963; MORRIS and SYRETT, 1963; THACKER and SYRETT, 1972b; MORRIS, 1974). A consensus has later been reached that the nitrate uptake system seems to be the locus of the rapid ammonia action on algae, and that the uptake system is the first control point of nitrate utilization in both green algae (PISTORIUS et al., 1976; SYRETT and LEFTLEY, 1976; PISTORIUS and FUNKHOUSER, unpubl.) and higher plants (CHANTAROTWONG et al., 1976).

III. Reversible Inactivation of Nitrate Reductase

LOSADA et al. (1970) first observed that addition of ammonia to *Chlorella fusca* cells grown on nitrate resulted in a reversible inactivation of nitrate reductase, but not of its diaphorase activity or of nitrite reductase. This inactivation of nitrate reductase has been extensively investigated in the Seville laboratory both with *Chlorella fusca* and with *Chlamydomonas reinhardii* which behaves rather similarly. It seemed that this phenomenon might account for the prompt inhibition of algal nitrate utilization caused by added ammonia. In vitro, the inactivation requires reduced pyridine nucleotide and is potentiated by ADP. Experiments and interpretations have been reviewed (LOSADA 1974, 1976a, 1976b), and it has been proposed that ammonia exerts its effect on the activation level of nitrate reductase by uncoupling photosynthetic phosphorylation. This leads to increased levels of NADH and of the uncharged forms of the adenine nucleotides (Chaparro et al., 1976). The reversibly inactivated enzyme is regarded as an "overreduced" form.

Reversible inactivation of nitrate reductase has also been observed in *Chlorella vulgaris* both in vivo and in vitro. As in the case of *Chlorella fusca*, addition of ammonia to cells growing on nitrate resulted in accumulation of inactive enzyme, which was identified as the HCN complex of the reduced enzyme (Solomonson et al., 1973; Solomonson, 1974; Gewitz et al., 1974; Lorimer et al., 1974). [Later work showed, however, that it is the inhibition of the nitrate uptake system and not the inhibition of nitrate reductase which causes the prompt cessation of nitrate utilization after ammonia addition (Pistorius et al., 1976; Pistorius and Funk-houser, unpubl.).]

With *Chlorella vulgaris*, light, O_2, nitrate, and CO_2 all affect nitrate reductase inactivation interdependently. Maximum inactivation was observed in the absence of nitrate in O_2 at high light intensities. CO_2 prevented inactivation in the presence of nitrate in the light, but had little effect in the absence of nitrate. There was little inactivation in Mn^{2+}-deficient cells (Pistorius et al., 1976). Although the level of enzyme activity is not rate-limiting for nitrate reduction when the process is abruptly stopped by ammonia addition to *Chlorella vulgaris*, there may well be other physiological situations where the level of active enzyme is indeed rate-limiting.

It is not possible at present to integrate the results and interpretations of the Berlin group and the Seville group. Different algal species have been used, and many of the facts reported are different. Among the more striking differences are the effects of anaerobiosis, which favors inactivation in *Chlorella fusca* and *Chlamydomonas reinhardii*, whereas high O_2 tensions favor inactivation in *Chlorella vulgaris*. Furthermore, ADP has no potentiating effect on the in vitro inactivation of nitrate reductase of *Chlorella vulgaris*, such as has been found for *Chlorella fusca*.

Chlorella vulgaris can grow on ammonia just as well as on nitrate. It is difficult to see how ammonia could be regarded as an uncoupling agent for these cells, since the term uncoupling implies impairing ATP synthesis. It is possible, however, that ammonia might lead to a drop in ATP and increase in ADP levels, because of the rapid conversion of glutamic acid to glutamine in the glutamine synthetase reaction. One also can probably regard HCN as an uncoupler. It is not known yet whether the reversibly inactivated nitrate reductases of *Chlorella fusca* or of *Chlamydomonas reinhardii* contain HCN. Nevertheless, the strong similarity of the enzyme inactivated by NADH and HCN in vitro, to the in vivo inactivated enzyme of *Chlamydomonas reinhardii* has been stressed (Barea et al., 1976b).

Possible precursors of HCN in *Chlorella vulgaris* include the amino acid histidine and also glyoxylic acid oxime (Gewitz et al., 1976a, 1976b; Solomonson and Spehar, 1977). Both histidine and glyoxylic acid oxime give HCN when incubated with *Chlorella* extracts under suitable conditions. The formation of HCN from histidine has been shown to be due to an amino acid oxidase acting together with peroxidase, or cell particulates, or a redox cation such as Mn^{2+} (Pistorius et al., 1977; Pistorius and Voss, 1977). Solomonson and Spehar (1977) have proposed that the combined effects of O_2, CO_2, and light on the reversible inactivation of nitrate reductase in *Chlorella vulgaris* can be explained by the competition of O_2 and CO_2 for ribulose-bisphosphate in the presence of ribulose-bisphosphate carboxylase-oxygenase. They assume that the glyoxylate – formed from glycolate, the product of the oxidation of ribulose-bisphosphate – combines with hydroxylamine – formed when nitrite is reduced in the presence of ammonia (LOUSSAERT and HAGEMAN, 1976) – to produce the glyoxylate oxime. In actual fact, two different paths of HCN formation are not incompatible.

The greatest experimental weakness of the glyoxylic oxime route to HCN is the uncertainty about the formation of sufficient hydroxylamine. The greatest uncertainty in the histidine route to HCN is that D-histidine appears to be a far better HCN precursor than L-histidine, and nothing is known about the formation of D-histidine in *Chlorella*.

It is known that HCN can be converted to the carboxamide group of asparagine by at least one *Chlorella* species as well as by other cells (BLUMENTHAL et al., 1963; FOWDEN and BELL, 1965).

After the nitrate reductase of *Chlorella vulgaris* had been inactivated in vivo by illumination in absence of nitrate, the addition of nitrate resulted in the restoration of the initial level of active enzyme. This restoration of activity was partially inhibited by cycloheximide (PISTORIUS et al., 1976), but SOLOMONSON and SPEHAR (1977) found that the inactivated enzyme could be completely reactivated in vivo in the presence of sufficient tungstate to prevent all new synthesis of active enzyme. Tungstate does not activate the inactive enzyme in vitro.

A report has appeared showing that reversibly inactivated nitrate reductase from *Chlorella fusca* and from spinach leaves can be very rapidly activated by blue light (APARICIO et al., 1976). It was suggested that such an effect might account for the fact that blue light enhances the formation of protein and nucleic acids in photosynthetic organisms (VOSKRESENSKAYA, 1972; ANDERSAG and PIRSON, 1976).

The ferredoxin-dependent photoreduction of nitrate to nitrite by particles prepared from the blue-green alga *Nostoc muscorum* can also be inhibited by ammonia and other uncouplers of photophosphorylation (ORTEGA et al., 1977).

IV. Localization of Enzymes and Effect of Light and Carbohydrate on Nitrate Utilization

In the blue-green algae, no clear anatomical separation of nitrate and nitrite reductases has been demonstrated, both of the enzymes being membrane-bound and capable of interacting with ferredoxin. Thus, LOSADA and his associates have shown that unsupplemented illuminated lamellar preparations from blue-green algae can photoreduce nitrate to ammonia with concurrent O_2 evolution (CANDAU et al., 1976). In a particle preparation from *Nostoc* (ORTEGA et al., 1976) photoreduction of nitrate to nitrite (but not to ammonia) has been reported.

In leaves, nitrite reductase has definitely been localized in the chloroplast, and intact chloroplasts can photoreduce added nitrite without enzyme additions, but the localization of nitrate reductase is controversial (DALLING et al., 1972a; LIPS and AVISSAR, 1972; MIFLIN, 1974; RATHNAM and DAS, 1974). It seems not to be within the chloroplast-like nitrite reductase, but it is not clear whether it may not at least in part be particle-bound. Though Dalling et al. (1972a) concluded that most of the nitrate reductase of tobacco and spinach leaves was present in soluble form in the cytoplasm, evidence has also been presented that nitrate reductase is either attached to the chloroplast membrane (Rathnam and Das, 1974) or loosely attached to microsomes or microbody-like organelles, which may attach themselves to chloroplasts during illumination (Lips and Avissar, 1972; Lips, 1975; Kagan-Zur and Lips, 1975). This is of interest in connection with the model of HCN production proposed by Solomonson and Spehar (1977) since the enzymes of photorespiration are also present in the microbodies.

In the C_4 plants, there is a further anatomical separation. The leaf mesophyll cells, which are the site of malate and aspartate synthesis, contain the major portion of the nitrate and nitrite reductases (Rathnam and Das, 1974; Rathnam and Edwards, 1976; Harel et al., 1977). The bundle sheath cells contain relatively little of these enzymes.

In wheat roots, nitrate reductase was present in soluble form in the cytoplasm but nitrite reductase was particle-bound, possibly in proplastids (Dalling et al., 1972b; see also Blevins et al., 1976).

It has long been known that photosynthesis stimulates nitrate utilization in algae and higher plants (Warburg and Negelein, 1920; Kessler, 1959; Syrett, 1962; Hewitt, 1970; Beevers and Hageman, 1972; Ullrich, 1974; Ullrich-Eberius, 1973; Thomas et al., 1977). The stimulatory role of carbohydrate in enzyme synthesis has already been mentioned, but there is a more immediate, short-term effect of light which is more closely related to the generation of reducing power for the enzyme reactions. It makes physiological sense, that carbohydrate can provide the reductant for nitrate (in the form of NADH) and can also provide the carbon building blocks for making amino acids and other nitrogenous products. In the case of nitrate reductase, with its likely cytoplasmic localization, the effect of light is probably indirect, and is due mainly to the reducing power transmitted to the cytoplasm by way, e.g., of triose phosphates or of malate. The view that nitrite reduction by green algae is more closely linked to the light reactions of photosynthesis than nitrate reduction was advanced particularly by Kessler (1955, 1959) and can easily be accounted for by the fact that nitrite reductase is located in the chloroplast and uses ferredoxin as a reductant. The closer relation of nitrite than nitrate to photosynthetic light reactions is also demonstrated by the fact that nitrite causes greater quenching of chlorophyll fluorescence in green algae than does nitrate (Kessler and Zumft, 1973; Kulandaivelu et al., 1976). No direct demonstration has been made of the localization of nitrite reductase in algal chloroplasts, but it can be inferred from these experiments.

Ferredoxin transmits photon-generated reducing power to NADP by way of ferredoxin:NADP oxido-reductase (E.C.1.6.99.4), and probably to glutamate synthase as well as to nitrite reductase. Thus, there are potentially at least three key systems, one involving carbohydrate synthesis from CO_2, the other two involving reduction of nitrite and incorporation of the ammonia formed into glutamate (by way of glutamine and α-ketoglutarate), all using the same reducing agent, ferredoxin (in the blue-green algae, reduction of nitrate to nitrite is also competing for ferredoxin). It has been suggested that ferredoxin:NADP reductase has an important regulatory function in maintaining a balance between nitrogen and carbon metabolism by directing reducing power towards oxidized nitrogen when excess NADPH and carbohydrate are available (SPILLER et al., 1976a). Competition between nitrate reduction and carbohydrate synthesis has been demonstrated (Thomas et al., 1977). A study of nitrite disappearance and $^{14}CO_2$ uptake by isolated, intact spinach chloroplasts has led to the suggestion that specific intermediates of the carbon reduction cycle may have a regulatory role in nitrite reduction by illuminated chloroplasts (PLAUT et al., 1977).

V. Conclusions

In spite of intensive research efforts and substantial recent progress in our understanding of the molecular mechanism of nitrate reduction, no compelling general answer can yet be provided about the specific chemical mechanisms whereby the reduction of nitrate is controlled and integrated with the carbohydrate metabolism of the green cell. The earlier conclusion of BEEVERS and HAGEMAN (1969) that nitrate reductase is the rate-limiting enzyme should perhaps be modified to include the provision that nitrite reductase is just as likely to be rate-limiting in the dark in green cells (STULEN and LANTING, 1976). The location of the enzymes ensures that, in the light, nitrite has preference (over nitrate) for photochemically generated reducing power. However, in the dark, nitrite reduction is presumably limited by NADPH levels generated from carbohydrate oxidation (GUERRERO et al., 1971) with the NADPH reducing ferredoxin by way of ferredoxin: NADP oxidoreductase. Either the NADH for nitrate reduction or the NADPH for nitrite reduction could be in short supply.

The availability of reductants is only one of several possible controls on nitrate assimilation, however. The nitrate uptake system must be regarded as the first possible point of control. In vacuolated cells, the transfer of nitrate to the vacuole must also be considered. Then the reversible inactivation and reactivation of the nitrate reductase can play a role, as well as the degradation and resynthesis of the enzyme. Accumulated nitrite can exercise a feedback inhibition, either at the level of the nitrate uptake system (FERGUSON and BOLLARD, 1969; HEIMER and FILNER, 1971; SCHLOEMER and GARRETT, 1974) or as a competitive inhibitor of nitrate reductase. Other as yet unidentified enzyme inhibitors may be operating. The heart of the control center for nitrite reduction may lie in the chloroplast, where nitrite and CO_2 reduction both ultimately depend on photosynthetically generated reducing power.

Acknowledgments. We are indebted to L. Solomonson, M. Losada, and J.M. Vega for manuscripts prior to publication. We also acknowledge the helpful criticism and discussion which we have enjoyed with E. Pistorius, F. Wissing, C. Manzano and P. Candau.

References

Åberg, B.: Kungl. Lantbruckshögskolans Ann. *15*, 37–107 (1947)
Ahmed, J., Spiller, H.: Plant Cell Physiol. *17*, 1–10 (1976)
Amy, N.K., Garrett, R.H., Anderson, B.M.: Biochim. Biophys. Acta *480*, 83–95 (1977)
Andersag, R., Pirson, A.: Biochem. Physiol. Pflanz. *169*, 71–85 (1976)

Aparicio, P.J., Knaff, D.B., Malkin, R.: Arch. Biochem. Biophys. *169*, 102–107 (1975)
Aparicio, P.J., Roldán, J.M., Calero, F.: Biochem. Biophys. Res. Commun. *70*, 1071–1077 (1976)
Aslam, M.: Plant Sci. Lett. *9*, 89–92 (1977)
Aslam, M., Oaks, A.: Plant Physiol. *56*, 634–639 (1975)
Barea, J.L., Maldonado, J.M., Cárdenas, J.: Physiol. Plant. *36*, 325–332 (1976a)
Barea, J.L., Sosa, F., Cárdenas, J.: Z. Pflanzenphysiol. *79*, 237–245 (1976b)
Beevers, L., Hageman, R.H.: Annu. Rev. Plant Physiol. *20*, 495–522 (1969)
Beevers, L., Hageman, R.H.: Photophysiol. *7*, 85–113 (1972)
Blevins, D.G., Lowe, R.H., Staples, L.: Plant Physiol. *57*, 458–459 (1976)
Blumenthal, S.G., Butler, G.W., Conn, E.E.: Nature (London) *197*, 718–719 (1963)
Bothe, H.: Prog. Photosyn. Res. *3*, 1483–1491 (1969)
Brown, C.M., MacDonald-Brown, D.S., Meers, J.L.: Adv. Microb. Physiol. *11*, 1–52 (1974)
Brunetti, N., Hageman, R.H.: Plant Physiol. *58*, 583–587 (1976)
Buczek, J.: Acta Soc. Bot. Poliae *42*, 223–232 (1973)
Butz, R.G.: Nature (London) *266*, 383 (1977)
Butz, R.G., Jackson, W.A.: Phytochemistry *16*, 409–417 (1977)
Campbell, W.H.: Plant Sci. Lett. *7*, 239–247 (1976)
Candau, P., Manzano, C., Losada, M.: Nature (London) *262*, 715–717 (1976)
Cárdenas, J., Barea, J.L., Rivas, J., Moreno, C.G.: FEBS Lett. *23*, 131–135 (1972a)
Cárdenas, J., Rivas, J., Barea, J.L.: Rev. Acad. Ciencias (Spain) *66*, 565–577 (1972b)
Chantarotwong, W., Huffaker, R.C., Miller, B.L., Ganstedt, R.C.: Plant Physiol. *57*, 519–522 (1976)
Chaparro, A., Maldonado, J.M., Diez, J., Relimpio, A.M., Losada, M.: Plant Sci. Lett. *6*, 335–342 (1976)
Coddington, A.: Mol. Gen. Genet. *145*, 195–206 (1976)
Cohen, A.S., Cumming B.G.: Can. J. Bot. *52*, 2351–2360 (1974)
Cove, D.J.: Mol. Gen. Genet. *146*, 147–159 (1974)
Dalling, M.J., Tolbert, N.E., Hageman, R.H.: Biochim. Biophys. Acta *28*, 505–512 (1972a)
Dalling, M.J., Tolbert, N.E., Hageman, R.H.: Biochim. Biophys. Acta *28*, 513–519 (1972b)
Davies, D.D., Texeira, A., Kenworthy, P.: Biochem. J. *127*, 335–343 (1972)
Dilworth, M.F., Kende, H.: Plant Physiol. *54*, 821–825 (1974a)
Dilworth, M.F., Kende, H.: Plant Physiol. *54*, 826–828 (1974b)
Downey, R.J.: J. Bacteriol. *105*, 759–768 (1971)
Dusky, J.A., Galitz, D.S.: Physiol. Plant. *39*, 215–220 (1977)
Ferguson, A.R., Bollard, E.G.: Planta (Berl.) *88*, 344–352 (1969)
Filner, P.: Biochim. Biophys. Acta *118*, 299–310 (1966)
Fowden, L., Bell, E.A.: Nature (London) *206*, 110–112 (1965)
Funkhouser, E.A., Ackermann, R.: Eur. J. Biochem. *66*, 225–228 (1976)
Garrett, R.H.: Biochim. Biophys. Acta *264*, 481–489 (1972)
Garrett, R.H., Greenbaum, P.: Biochim. Biophys. Acta *302*, 24–32 (1973)
Garrett, R.H., Nason, A.: Proc. Natl. Acad. Sci. USA *58*, 1603–1610 (1967)
Garrett, R.H., Nason, A.: J. Biol. Chem. *244*, 2870–2882 (1969)
Gewitz, H.-S., Lorimer, G.H., Solomonson, L.P., Vennesland, B.: Nature (London) *249*, 79–81 (1974)
Gewitz, H.-S., Pistorius, E.K., Voss, H., Vennesland, B.: Planta (Berl.) *131*, 145–148 (1976a)
Gewitz, H.-S., Pistorius, E.K., Voss, H., Vennesland, B.: Planta (Berl.) *131*, 149–153 (1976b)
Guerrero, M.G., Gutierrez, M.: Biochim. Biophys. Acta *482*, 272–285 (1977)
Guerrero, M.G., Jetschmann, K., Völker, W.: Biochim. Biophys. Acta *482*, 19–26 (1977)
Guerrero, M.G., Manzano, C., Losada, M.: Plant Sci. Lett. *3*, 273–278 (1974)
Guerrero, M.G., Vennesland, B.: FEBS Lett. *51*, 284–286 (1975)
Guerrero, M.G., Rivas, J., Paneque, A., Losada, M.: Biochem. Biophys. Res. Commun. *45*, 82–89 (1971)
Harel, E., Lea, P.J., Miflin, B.J.: Planta (Berl.) *134*, 195–200 (1977)
Hattori, A.: Plant Cell Physiol. *11*, 975–978 (1970)
Hattori, A., Myers, J.: Plant Physiol. *41*, 1031–1036 (1966)
Hattori, A., Myers, J.: Plant Cell Physiol. *8*, 327–337 (1967)
Hattori, A., Uesugi, I.: Plant Cell Physiol. *9*, 689–699 (1968a)

Hattori, A., Uesugi, I.: In: Comparative biochemistry and biophysics of photosynthesis. Shibata, K., Takamiya, A., Jagendorf, A.T., Fuller, R.C. (eds.), pp. 201–205. Tokyo: Univ. of Tokyo Press 1968b
Heimer, Y.M., Folner, P.: Biochim. Biophys. Acta 230, 362–372 (1971)
Heimer, Y.M., Wray, J.L., Filner, P.: Plant Physiol. 44, 1197–1199 (1969)
Hewitt, E.J., In: Nitrogen nutrition of the plant. Kirkby, E.A. (ed.), pp. 78–103. Univ. of Leeds Agric. Chem. Symp. 1970
Hewitt, E.J.: Annu, Rev. Plant Physiol. 26, 73–100 (1975)
Ho, C.H., Ikawa, T., Nisizawa, K.: Plant Cell Physiol. 17, 417–430 (1976a)
Ho, C.H., Ikawa, T., Nisizawa, K.: Plant Cell Physiol. 17, 431–438 (1976b)
Ho, C.H., Tamura, G.: Agric. Biol. Chem. 37, 37–44 (1973)
Hodler, M., Morgenthaler, J.J., Eichenberger, E., Grob, E.C.: FEBS Lett. 28, 19–21 (1972)
Hucklesby, D.P., Hewitt, E.J.: Biochem. J. 119, 615–627 (1970)
Hucklesby, D.P., James, D.M., Banwell, M.J., Hewitt, E.J.: Phytochemistry 15, 599–603 (1976)
Ida, S., Kobayakawa, K., Morita, Y.: FEBS Lett. 65, 305–308 (1976)
Ida, S., Morita, Y.: Plant Cell Physiol. 14, 661–671 (1973)
Jaworski, E.G.: Biochem. Biophys. Res. Commun. 43, 1274–1279 (1971)
Johnson, C.B.: Planta (Berl.) 128, 127–131 (1976)
Johnson, C.B., Whittington, W.J., Blackwood, G.C.: Nature (London) 262, 133–134 (1976)
Johnson, C.B., Whittington, W.J., Blackwood, G.C., Wallace, W.: Nature (London) 266, 383 (1977)
Jolly, S.O., Campbell, W.H., Tolbert, N.E.: Arch. Biochem. Biophys. 174, 431–439 (1976)
Jones, R.W., Abbott, A.J., Hewitt, E.S., James, D.M., Best, G.R.: Planta (Berl.) 133, 27–34 (1977)
Jones, R.W., Sheard, R.W.: Nature (London) 238, 221–222 (1972)
Jones, R.W., Sheard, R.W.: Plant Physiol. 55, 954–959 (1975)
Jones, R.W., Sheard, R.W.: Plant Sci. Lett. 8, 305–311 (1977)
Kadam, S.S., Gandhi, A.P., Sawhney, S.K., Naik, M.S.: Biochim. Biophys. Acta 350, 162–170 (1974)
Kagan-Zur, V., Lips, S.H.: Eur. J. Biochem. 59, 17–23 (1975)
Kanazawa, T., Kanazawa, K., Kirk, M.R., Bassham, J.A.: Plant Cell Physiol. 11, 445–452 (1970)
Kaplan, D., Roth-Bejerano, N., Lips, H.: Eur. J. Biochem. 49, 393–398 (1974)
Katoh, T.: Plant Cell Physiol. 4, 13–28 (1963)
Kelker, H.C., Filner, P.: Biochim. Biophys. Acta 252, 69–82 (1971)
Kessler, E.: Nature (London) 176, 1069–1070 (1955)
Kessler, E.: Symp. Soc. Exp. Biol. 13, 87–105 (1959)
Kessler, E.: Annu. Rev. Plant. Physiol. 15, 57–72 (1964)
Kessler, E.: Prog. Bot. 33, 95–103 (1971)
Kessler, E.: Prog. Bot. 38, 108–117 (1976)
Kessler, E., Oesterheld, H.: Nature (London) 228, 287–288 (1970)
Kessler, E., Zumft, W.G.: Planta (Berl.) 111, 41–46 (1973)
Ketchum, P.A., Combier, H.Y., Frazier III, W.A., Madansky, C.H., Nason, A.: Proc. Natl. Acad. Sci. USA 66, 1016–1023 (1970)
Ketchum, P.A., Johnson, D., Taylor, R.C., Young, D.C., Atkinson, A.W., In: Bioinorganic chemistry II. Advances in chemistry series 162. Raymond, K.N. (ed.), pp. 408–420. Washington DC: American Chemical Society 1977
Kinsky, S.C., McElroy, W.D.: Arch. Biochem. Biophys. 73, 466–483 (1958)
Klepper, L.A., Flesher, D., Hageman, R.H.: Plant Physiol. 48, 580–590 (1971)
Knypl, J.S.: Z. Pflanzenphysiol. 70, 1–11 (1973)
Knypl, J.S.: Z. Pflanzenphysiol. 71, 37–48 (1974)
Knypl, J.S., Ferguson, A.R.: Z. Pflanzenphysiol. 74, 434–439 (1975)
Kulandaivelu, G., Spiller, H., Böger, P.: Plant Sci. Lett. 7, 225–231 (1976)
Lafferty, M.A., Garrett, R.H.: J. Biol. Chem. 249, 7555–7567 (1974)
Lee, K.Y., Erickson, R.H., Pan, S.S., Jones, G., May, F., Nason, A.: J. Biol. Chem. 249, 3953–3959 (1974a)
Lee, K.Y., Pan, S.S., Erickson, R., Nason, A.: J. Biol. Chem. 249, 3941–3952 (1974b)

Liljeström, S., Åberg, B.: Kungl. Lantbrukshögskolans Ann. *32*, 93–107 (1966)
Lips, S.H.: Plant Physiol. *55*, 598–601 (1975)
Lips, S.H., Avissar, Y.: Eur. J. Biochem. *29*, 20–24 (1972)
Lips, S.H., Kaplan, D., Roth-Bejerano, N.: Eur. J. Biochem. *37*, 589–592 (1973)
Lips, S.H., Roth-Bejerano, N.: Science *166*, 109–110 (1969)
Liu, M.S., Hellebust, J.A.: Can. J. Bot. *20*, 1119–1125 (1974)
Lorimer, G.H., Gewitz, H.-S., Völker, W., Solomonson, L.P., Vennesland, B.: J. Biol. Chem. *249*, 6074–6079 (1974)
Losada, M., In: 3rd. Int. Symp. Metab. Interconvers. Enzymes, Seattle. pp. 257–270. Berlin, Heidelberg, New York: Springer 1974
Losada, M.: J. Mol. Catal. *1*, 245–264 (1976a)
Losada, M., In: Reflections on biochemistry. pp. 73–84. Oxford, New York, Frankfurt, Paris: Pergamon Press 1976b
Losada, M., Paneque, A.: Methods Enzymol. *23*, 487–491 (1971)
Losada, M., Aparicio, P.J., Paneque, A.: Prog. Photosyn. Res. *3*, 1504–1509 (1969)
Losada, M., Paneque, A., Aparicio, P.J., Vega, J.M., Cárdenas, J., Herrera, J.: Biochem. Biophys. Res. Commun. *38*, 1009–1010 (1970)
Loussaert, D., Hageman, R.H.: Plant Physiol. (Suppl.) *57*, 38 (1976)
Malofeeva, I.V., Kondratieva, E.N., Rubin, A.B.: FEBS Lett. *53*, 188–189 (1975)
Manzano, C., Candau, P., Gómez-Moreno, C., Relimpio, A.M., Losada, M.: Mol. Cell. Biochem. *10*, 161–169 (1976)
Mc Donald, D.W., Coddington, A.: Eur. J. Biochem. *46*, 169–178 (1974)
Miflin, B.J.: Plant Physiol. *54*, 550–555 (1974)
Morris, I., In: Algal physiology and biochemistry. Botanical monographs. Stewart, W.D.P. (ed.), Vol. X, pp. 583–609. Oxford, London, Edinburgh, Melbourne: Blackwell 1974
Morris, I., Syrett, P.J.: Biochim. Biophys. Acta *77*, 649–650 (1963)
Murphy, M.J., Siegel, L.M.: J. Biol. Chem. *248*, 6911–6919 (1973)
Murphy, M.J., Siegel, L.M., Kamin, H., Rosenthal, D.: J. Biol. Chem. *248*, 2801–2814 (1973)
Murphy, M.J., Siegel, L.M., Tove, S.R., Kamin, H.: Proc. Natl. Acad. Sci. USA *71*, 612–616 (1974)
Naik, M.S., Sardhambal, K.V., Prakash, S.: Nature (London) *262*, 396–397 (1976)
Nason, A., Evans, H.J.: J. Biol. Chem. *202*, 655–673 (1953)
Nason, A., Antoine, A.D., Ketchum, P.A., Frazier III, W.A., Lee, D.K.: Proc. Natl. Acad. Sci. USA *65*, 137–144 (1970)
Nason, A., Lee, K.Y., Pan, S.S., Ketchum, P.A., Lamberti, A., DeVries, T.: Proc. Natl. Acad. Sci. USA *68*, 3242–3246 (1971)
Nicholas, D.J.D., Nason, A.: J. Biol. Chem. *207*, 353–360 (1954)
Nicholas, D.J.D., Nason, A., McElroy, W.D.: J. Biol. Chem. *207*, 341–351 (1954)
Notton, B.A., Hewitt, E.J.: Biochem. Biophys. Res. Commun. *44*, 702–710 (1971a)
Notton, B.A., Hewitt, E.J.: FEBS Lett. *18*, 19–22 (1971b)
Notton, B.A., Hewitt, E.J.: Plant Cell Physiol. *12*, 465–477 (1971c)
Notton, B.A., Hewitt, E.J.: Biochim. Biophys. Acta *275*, 355–357 (1972)
Notton, B.A., Hewitt, E.J., Fielding, A.H.: Phytochemistry *11*, 2447–2449 (1972)
Notton, B.A., Graf, L., Hewitt, E.J., Povey, R.C.: Biochim. Biophys. Acta *364*, 45–58 (1974)
Notton, B.A., Fido, R.J., Hewitt, E.J.: Plant Sci. Lett. 8, 165–170 (1977)
Oaks, A.: Biochim. Biophys. Acta *372*, 122–126 (1974)
Oaks, A., Aslam, M., Boesel, I.: Plant Physiol. *59*, 391–394 (1977)
Oesterheld, H.: Arch. Microbiol. *79*, 25–43 (1971)
Orme-Johnson, W.H., Jacob, G.S., Henzl, M.T., Averill, B.A.: In: Bioinorganic chemistry II. Advances in chemistry series 162. Raymond, K.N. (ed.), pp. 389–401. Washington DC: American Chemical Society 1977
Ortega, T., Castillo, F., Cárdenas, J.: Biochem. Biophys. Res. Commun. *71*, 885–891 (1976)
Ortega, T., Castillo, F., Cárdenas, J., Losada, M.: Biochem. Biophys. Res. Commun. *75*, 823–831 (1977)
Paneque, A., Vega, J.M., Cárdenas, J., Herrera, J., Aparicio, P.J., Losada, M.: Plant Cell Physiol. *13*, 175–178 (1972)
Payne, W.J.: Bacteriol. Rev. *37*, 409–452 (1973)
Pistorius, E.K., Voss, H.: Biochim. Biophys. Acta *481*, 395–406 (1977)

Pistorius, E.K., Gewitz, H.-S., Voss, H., Vennesland, B.: Planta (Berl.) *128*, 73–80 (1976)
Pistorius, E.K., Gewitz, H.-S., Voss, H., Vennesland, B.: Biochim. Biophys. Acta *481*, 384–394 (1977)
Plaut, Z., Lendzian, K., Bassham, J.A.: Plant Physiol. *59*, 184–188 (1977)
Rao, K.P., Rains, D.W.: Plant Physiol. *57*, 59–62 (1976)
Rathnam, C.K.M., Das, V.S.R.: Can. J. Bot. *52*, 2599–2605 (1974)
Rathnam, C.K.M., Edwards, G.E.: Plant Physiol. *57*, 881–885 (1976)
Relimpio, A.M., Aparicio, P.J., Paneque, A., Losada, M.: FEBS Lett. *17*, 226–230 (1971)
Ries, S.K., Chmiel, H., Dilley, D.R., Filner, P.: Proc. Natl. Acad. Sci. USA *58*, 526–532 (1967)
Rigano, C., Aliotta, G., Violante, U.: Plant Sci. Lett. *2*, 277–281 (1974)
Roth-Bejerano, N., Lips, S.H.: New Phytol. *69*, 165–169 (1970)
Sahulka, J., Gandinova, A., Hadacova, V.: Z. Pflanzenphysiol. *75*, 392–404 (1975)
Sawhney, S.K., Naik, M.S.: Biochem. J. *130*, 475–485 (1972)
Schloemer, R.H., Garrett, R.H.: J. Bacteriol. *118*, 259–269 (1974)
Schrader, L.E., Ritenour, C.L., Eilrich, G.L., Hageman, R.H.: Plant Physiol. *43*, 930–940 (1968)
Shen, T.C.: Plant Physiol. *49*, 546–549 (1972)
Shen, T.C., Funkhouser, E.A., Guerrero, M.G.: Plant Physiol. *58*, 292–294 (1976)
Shimizu, J., Tamura, G.: J. Biochem. *75*, 999–1005 (1974)
Siegel, L.M., Davis, P.S.: J. Biol. Chem. *249*, 1587–1598 (1974)
Sluiter-Scholten, C.M.Th.: Planta (Berl.) *113*, 229–240 (1973)
Smith, F.W., Thompson, J.F.: Plant Physiol. *48*, 224–227 (1971)
Solomonson, L.P.: Biochim. Biophys. Acta *334*, 297–308 (1974)
Solomonson, L.P.: Plant Physiol. *56*, 853–855 (1975)
Solomonson, L.P., Spehar, A.M.: Nature (London) *265*, 373–375 (1977)
Solomonson, L.P., Vennesland, B.: Plant Physiol. *50*, 421–424 (1972a)
Solomonson, L.P., Vennesland, B.: Biochim. Biophys. Acta *267*, 544–557 (1972b)
Solomonson, L.P., Jetschmann, K., Vennesland, B.: Biochim. Biophys. Acta *309*, 32–43 (1973)
Solomonson, L.P., Lorimer, G.H., Hall, R.L., Borchers, R., Bailey, J.L.: J. Biol. Chem. *250*, 4120–4127 (1975)
Sorger, G.J.: Biochim. Biophys. Acta *118*, 484–494 (1966)
Sorger, G.J., Debanne, M.T., Davies, J.: Biochem. J. *140*, 395–403 (1974)
Spiller, H., Bookjans, G., Böger, P.: Z. Naturforsch. *31c*, 565–568 (1976a)
Spiller, H., Dietsch, E., Kessler, E.: Planta (Berl.) *129*, 175–181 (1976b)
Steer, B.T.: Plant Physiol. *54*, 762–765 (1974)
Steer, B.T.: Plant Physiol. *57*, 928–932 (1976)
Stiefel, E.I.: Proc. Natl. Acad. Sci. USA *70*, 988–992 (1973)
Stiefel, E.I., Newton, W.E., Watt, G.D., Hadfield, K.L., Bulen, W.A.: In: Bioinorganic chemistry II. Advances in chemistry series 162. Raymond, K.N. (ed.), pp. 353–388. Washington DC: American Chemical Society 1977
Stouthamer, A.H.: Adv. Microb. Physiol. *14*, 315–375 (1976)
Streeter, J.G., Bosler, M.E.: Plant Physiol. *49*, 448–450 (1972)
Stulen, I., Koch-Bosma, T., Koster, A.: Acta Bot. Neerl. *20*, 389–396 (1971)
Stulen, I., Lanting, L.: Physiol. Plant. *37*, 139–142 (1976)
Subramanian, K.N., Sorger, G.J.: Biochim. Biophys. Acta *256*, 533–543 (1972)
Syrett, P.J., In: Physiology and biochemistry of algae. Lewin, R.A. (ed.), pp. 171–188. New York, London: Academic Press 1962
Syrett, P.J.: New Phytol. *72*, 37–46 (1973)
Syrett, P.J., Hipkin, C.R.: Planta (Berl.) *111*, 57–64 (1973)
Syrett, P.J., Leftley, J.W., In: Perspect. Exp. Biol. Vol. 2, pp. 221–234. Oxford, New York: Pergamon Press 1976
Syrett, P.J., Morris, I.: Biochim. Biophys. Acta *67*, 566–575 (1963)
Syrett, P.J., Thomas, E.M.: New Phytol. *71*, 1307–1310 (1973)
Thacker, A., Syrett, P.J.: New Phytol. *71*, 423–433 (1972a)
Thacker, A., Syrett, P.J.: New Phytol. *71*, 435–441 (1972b)
Thomas, R.J., Hipkin, C.R., Syrett, P.J.: Planta (Berl.) *133*, 9–13 (1977)
Tischner, R.: Planta (Berl.) *132*, 285–290 (1976)

Tomati, U., Giovannozzi-Sermanni, G., Depret, S., Cannella, C.: Phytochem. *15*, 597–598 (1976)
Travis, R.L., Jordan, W.R., Huffaker, R.C.: Plant Physiol. *44*, 1150–1156 (1969)
Tromballa, H.W., Broda, E.: Arch. Microbiol. *78*, 214–223 (1971)
Tweedy, J.A., Ries, S.K.: Plant Physiol. *42*, 280–282 (1967)
Ullrich, W.R.: Planta (Berl.) *116*, 143–152 (1974)
Ullrich-Eberius, C.J.: Planta (Berl.) *115*, 25–36 (1973)
Upcroft, J.A., Done, J.: FEBS Lett. *21*, 142–144 (1972)
Upcroft, J.A., Done, J.: J. Exp. Bot. *25*, 503–508 (1974)
Vega, J.M., Kamin, H.: J. Biol. Chem. *252*, 896–909 (1977)
Vega, J.M., Herrera, J., Aparicio, J.P., Paneque, A., Losada, M.: Plant Physiol. *48*, 294–299 (1971)
Vega, J.M., Herrera, J., Relimpio, A., Aparicio, P.J.: Physiol. Veg. *10*, 637–652 (1972)
Vega, J.M., Garrett, R.H., Siegel, L.M.: J. Biol. Chem. *250*, 7980–7989 (1975a)
Vega, J.M., Greenbaum, P., Garrett, R.H.: Biochim. Biophys. Acta, *377*, 251–257 (1975b)
Vega, J.M., Kamin, H., Orme-Johnson, W.H.: 67th Annu. Meet. Soc. Biol. Chem. Abstr. 1215. San Francisco (1976)
Vennesland, B., Jetschmann, C.: Biochim. Biophys. Acta *227*, 554–564 (1971)
Voskresenskaya, N.P.: Annu. Rev. Plant Physiol. *23*, 219–234 (1972)
Wallace, W.: Plant Physiol. *52*, 197–201 (1973)
Wallace, W.: Biochim. Biophys. Acta *341*, 265–276 (1974)
Wallace, W.: Biochim. Biophys. Acta *377*, 239–250 (1975)
Warburg, O., Negelein, E.: Biochem. Z. *110*, 66–115 (1920)
Wray, J.L., Filner, P.: Biochem. J. *119*, 715–725 (1970)
Yamaya, T., Ohira, K.: Plant Cell Physiol. *17*, 633–641 (1976)
Zumft, W.G.: Biochim. Biophys. Acta *276*, 363–375 (1972)
Zumft, W.G.: Naturwissenschaften *63*, 457–464 (1976)

5. Photosynthetic Ammonia Assimilation

P.J. LEA and B.J. MIFLIN

A. Introduction

Ammonia assimilation is defined as the incorporation of ammonia into organic compounds, and for the purpose of this chapter will be considered to include the formation of amides and amino acids from keto acids and NH_3.

The pathway of nitrogen assimilation in plants has been reviewed previously by the authors (MIFLIN and LEA, 1976a, b; 1977). Two possible routes of ammonia assimilation have been postulated, the first involving one enzyme, and the second involving two enzymes:

Route 1: The reductive amination of a keto acid to yield an amino acid directly. The enzyme widely distributed in plant tissues is glutamate dehydrogenase (E.C.1.4.1.3) which catalyzes reaction (1):

$$\alpha\text{-Ketoglutarate} + NH_3 + NAD(P)H + H^+ \rightleftharpoons \text{glutamate} + NAD(P)^+ + H_2O \quad (1)$$

Similar enzymes that catalyze the formation of alanine from pyruvate and aspartate from oxaloacetate have also been reported, although exact characterization of the products has not been carried out.

Route 2: The initial incorporation of ammonia into the amide position of glutamine by the enzyme glutamine synthetase (E.C.6.3.1.2) which catalyzes reaction (2):

$$\text{Glutamate} + NH_3 + ATP \rightarrow \text{glutamine} + ADP + Pi \quad (2)$$

The amide-amino group can then be transferred to the α position of α-ketoglutarate by the enzyme glutamine: α-ketoglutarate amino transferase NAD(P)H oxidizing (E.C.2.6.1.53) or reduced ferredoxin oxidizing (E.C.1.4.7.1), reaction (3):

$$\text{Glutamine} + \alpha\text{-ketoglutarate} + \text{reduced ferredoxin or } NAD(P)H + H^+$$
$$\rightarrow 2 \text{ glutamate} + \text{oxidized ferredoxin or } NAD(P)^+ \quad (3)$$

The enzyme has been given the trivial name glutamate synthase or the acronym GOGAT. The term GOGAT will be used in this review to prevent any confusion between glutamine synthetase and glutamate synthase.

Whichever route is utilized, a minimum of two reducing equivalents are required for the net formation of one glutamate unit, plus an extra ATP molecule if route (2) is used. In addition to this, a carbon skeleton is required to accept the ammonia assimilated. If the assimilated nitrogen is maintained as glutamate, then five carbon atoms are required, but if the storage compound asparagine is predominant, then only two carbon atoms are required for each molecule of ammonia incorporated (LEA and FOWDEN, 1975a). Inorganic nitrogen is normally supplied to the leaf as nitrate in the xylem stream. In the chapter by VENNESLAND and GUERRERO (Chap. III.4, this vol.), the reduction of nitrate and nitrite to ammonia has been discussed, however, neither nitrite nor ammonia accumulate in plants to any great extent, and experiments which measure the incorporation of nitrate into amino acids must involve ammonia assimilation. Indeed, it is arguable that such experiments may be more valid than those in which ammonia is fed to leaves, since there is no necessary reason why ammonia supplied externally is assimilated in the same place as that generated internally from nitrate reduction.

In the following sections evidence will be presented to support the hypothesis that ammonia, generated internally by nitrate, nitrite, or dinitrogen reduction in photosynthetic tissue, is assimilated by an energy-dependent reaction in which the energy is derived directly from the photo-chemical process. This will be based on studies on:

1) Intact unicellular and multicellular organisms
2) The distribution of enzymes known to be involved in ammonia assimilation
3) Intact chloroplasts.

B. Photosynthetic Ammonia Assimilation in Intact Organisms

Their experiments with $^{15}NH_4^+$ tracer studies on *Chlorella* suggested to BASSHAM and KIRK (1964) that the reductive amination of α-ketoglutarate was the primary route of incorporation of ammonia during photosynthesis. However, they "were unable to preclude the possibility that there was a small actively turning-over pool of glutamine which saturated very quickly and accounted for substantial amounts of $^{15}NH_4^+$ incorporation". In contrast, previous experiments by BAKER and THOMPSON (1961) had shown that $^{15}NH_3$ was rapidly incorporated into the amide group of glutamine in *Chlorella* and only later into the amino groups of glutamate and glutamine. However, neither series of experiments was sufficiently detailed to determine exactly which of the two assimilatory pathways were operating.

The pathway of nitrogen assimilation in the N_2-fixing blue-green alga *Anabaena cylindrica* has been elegantly described utilizing ^{13}N-labeled nitrogen gas obtained from a cyclotron by WOLK et al. (1976). Glutamine was rapidly labeled; after 15 s fixation nearly 90% of the label in organic compounds was in glutamine and of that the vast majority was present in the amide group. The kinetics of labeling and pulse chase experiments showed that ^{13}N entered glutamate secondarily via glutamine. Methionine sulphoximine, a potent inhibitior of glutamine synthetase (TATE and MEISTER, 1973) caused $^{13}NH_3$ to build up and prevented the incorporation of ^{13}N into amino acids. Azaserine, an inhibitor of plant GOGAT (WALLSGROVE et al., 1977), and all other glutamine amide transfer reactions allowed the formation of ^{13}N-glutamine but no other amino acids. Similar results were obtained by STEWART and ROWELL (1975), who showed that methionine sulphoximine caused the excretion of ammonia into the external medium of nitrogen-fixing

Anabaena. Detailed discussions of the action of amino acid analogs (LEA and NORRIS, 1976) with particular emphasis on ammonia assimilation (MIFLIN and LEA, 1976a) have previously been published.

In barley leaves the conversion of $^{15}NO_3^-$, $^{15}NO_2^-$, and $^{15}NH_4^+$ to amino acids is greatly stimulated by light. The light response is independent of carbon fixation as it takes place in CO_2-free air (CANVIN and ATKINS, 1974). In *Chlamydomonas* light stimulates the assimilation of ammonia which in normal cells does not occur in the dark, but acetate is able to replace light to a certain extent. Acetate also stimulates ammonia assimilation in the light, thus demonstrating that it requires both light and an available carbon source (THACKER and SYRETT, 1972). ^{15}N feeding studies of LEWIS and PATE (1973) to illuminated pea leaves are consistent with the assimilation of NO_3^- via the amide group of glutamine and glutamate (MIFLIN and LEA, 1976a). Further studies by LEWIS and BERRY (1975) with photosynthesizing leaves of *Datura* demonstrated a considerable routing of $^{15}NO_3^-$ into glutamine. Very recent studies with $^{15}NO_3^-$ feeding to leaves by LEWIS (1978) suggested that in certain species glutamate was the initial compound formed. However further studies with methionine sulphoximine showed that if glutamine formation was inhibited no glutamate synthesis occurred. In the aquatic plant *Lemna*, the internal pools of metabolites were measured in the presence of amino acid analogs (STEWART and RHODES, 1976). Methionine sulphoximine caused ammonia to accumulate from externally supplied NO_3^- and increase the pool size of glutamate and α-ketoglutarate. Azaserine concurrently increased the levels of glutamine and α-ketoglutarate whilst depressing the level of glutamate. These observations are very similar to those previously made with *Chlorella* (VAN DER MEULEN and BASSHAM, 1959) and *Anabaena* (STEWART and ROWELL, 1975; WOLK et al., 1976) and present strong evidence for the functioning of the glutamine synthetase/GOGAT pathway in plants.

C. Localization of Enzymes Involved in Ammonia Assimilation

If ammonia assimilation is to be directly light-driven, then the enzymes involved in the metabolic sequence must be present in the chloroplast at levels capable of catalyzing the observed in vivo rates of ammonia assimilation. The distribution of the following enzymes in leaves has been studied:

I. Glutamate Dehydrogenase

In crude extracts of plants the NADH-dependent activity of glutamate dehydrogenase frequently exceeds the NADPH activity by a factor of 10. The majority of activity is located in the mitochondria, which readily rupture on extraction and release their contents into the soluble fraction (LEA and THURMAN, 1972). A number of workers have failed to detect any glutamate dehydrogenase activity in the chloroplast RITENOUR et al., 1967; EHMKE and HARTMANN, 1976); even when sophisticated gradient techniques for the isolation of intact chloroplasts were employed, only very low levels of glutamate dehydrogenase were detected (MIFLIN, 1974b).

The first demonstration of a plastid glutamate dehydrogenase was by Leech and Kirk (1968) from *Vicia faba,* who described an NADPH-dependent enzyme which was tightly bound to the chloroplast lamellae. Santarius and Stocking (1969) also demonstrated glutamate dehydrogenase activity in chloroplasts from the same plant, but this was shown to be NADH-dependent. An NADPH-dependent glutamate dehydrogenase isolated from lettuce chloroplasts (Lea and Thurman, 1972) was found to have properties distinct from that located in the mitochondria. A similar enzyme was isolated from spinach chloroplasts (Magalhaes et al., 1974), and the authors calculated that sufficient enzyme activity was present to account for the in vivo rates of 2-amino nitrogen production. However, Magalhaes et al. assayed glutamate dehydrogenase at 100 mM NH_4^+, although such concentrations are unlikely to be reached in vivo since broken chloroplasts are uncoupled at 2 mM NH_4^+ (Good, 1960). Almost all reported K_m's for NH_4^+ for glutamate dehydrogenase from higher plants are high with values up to 100 mM and the only recorded value for glutamate dehydrogenase isolated from higher plant chloroplasts is 5.8 mM (Lea and Thurman, 1972). If these K_m's reflect the in vivo characteristics of glutamate dehydrogenase, then it is unlikely that chloroplastic glutamate dehydrogenase will be able to operate anywhere approaching its maximum efficiency and this, together with the generally low levels of activity, make it improbable that it is functioning in photosynthetic NH_4^+ assimilation. Recently an NADP-dependent glutamate dehydrogenase has been isolated from the siphonous marine alga *Caulerpa simpliciuscula* which has an apparent K_m for NH_4^+ of 0.4 to 0.7 mM (Gayler and Morgan, 1976).

In the blue-green algae levels of glutamate dehydrogenase are low (Haystead et al., 1973; Batt and Brown, 1974); considerable activities of alanine dehydrogenase have been reported, although on purification the enzyme was found to have a high K_m for ammonia which was dependent upon pH (Rowell and Stewart, 1976). Aspartate dehydrogenase has also been reported in blue-green algae (Batt and Brown, 1974) and in chloroplasts (Santarius and Stocking, 1969). Care should be taken in interpreting data on aspartate dehydrogenase as plant extracts contain large amounts of malate dehydrogenase which catalyzes a reaction between oxaloacetate and $NAD(P)H$ which can readily confuse spectrophotometric assays (Miflin and Lea, 1975). Anderson and Done (1977b) also demonstrated rapid evolution of O_2 in the presence of oxaloacetate.

II. Glutamine Synthetase

Unfortunately no reliable assay is available for glutamine synthetase in crude plant extracts (for a detailed discussion of the assay methods of glutamine synthetase see Miflin and Lea, 1977). The most widely used method involves the formation of γ-glutamyl hydroxamate, either by the "transferase" assay in the presence of glutamine and ADP or the "synthetase" assay in the presence of glutamate and ATP. As the "transferase" assay gives higher values, it is more frequently used, although there is no evidence that it has any physiological significance in plants or that it is specific to glutamine synthetase.

The localization of glutamine synthetase in chloroplasts was independently demonstrated by Haystead (1973) and O'Neal and Joy (1973a). There is no evidence for the presence of the enzyme in any other organelle, but it could be present in the cytoplasm since the amount in the "soluble fraction" of spinach was greater than could be explained by plastid breakage (MIFLIN, 1974b). However, RATHNAM and EDWARDS (1976) have isolated chloroplasts from protoplasts of maize mesophyll cells and showed that at least 74% of the glutamine synthetase activity was of plastid origin. By a careful comparison of the localization of known chloroplast enzymes isolated from pea protoplasts on sucrose density gradients (WALLSGROVE et al., 1979) have shown that only 50% of the leaf glutamine synthetase is located in the chloroplast, the remainder is in the cytoplasm. Besides its location in maize mesophyll tissue, glutamine synthetase is also present in bundle sheath tissue, probably in the chloroplasts (HAREL et al., 1977).

Highly purified glutamine synthetase has been prepared from pea leaves (O'NEAL and JOY, 1973b, 1974, 1975) and its properties described. The enzyme has also been purified from a blue-green alga (DHARMAWARDENE et al., 1973), *Chlorella* (KRETOVICH et al., 1974) and *Lemna minor* (RHODES, 1976). Plant glutamine synthetase has one important property in that it has a very high affinity for ammonia, with K_m values of 19 µM for pea leaf, 39 µM for pea seeds and 10 µM for *Lemna* reported. Unfortunately the affinity of a purely chloroplastic enzyme for ammonia has not been determined, but values of this order would mean that the internal concentration need not rise much above 100 µM, thus preventing any toxic effects of ammonia.

MIFLIN (1977) has discussed in detail mechanisms by which the activity of glutamine synthetase may be controlled in the chloroplast by the action of light. Increases in magnesium concentration (HIND et al., 1974), energy charge (KEYS, 1968) and pH (WERDAN et al., 1975) have all been demonstrated in chloroplasts upon illumination. The results of O'NEAL and JOY (1973b, 1974, 1975) suggest that in vitro glutamine synthetase activity would be greatly increased by these changes. Further evidence for light as a controlling factor in glutamine synthetase activity comes from studies which show that if *Lemna* plants, grown on ammonia, are placed in the dark, there is a rapid reduction in the glutamine synthetase level. Subsequently, the glutamate concentration increases and that of glutamine decreases. Following the transfer back to light, the level of glutamine synthetase increases rapidly and accompanying this is a re-adjustment in the levels of glutamine and glutamate. These changes in glutamine synthetase apparently involve a form of reversible inactivation. The enzyme can be re-activated in vitro by the addition of ATP, glutamate, Mg^{2+}, and NADPH (STEWART and RHODES, 1977). Presumably this mechanism prevents the cell depleting stocks of ATP and glutamate (required for the synthesis of glutamine) which cannot be replenished in the absence of light. Thus, the effect of light, which also acts as a source of reductant for NO_2^- reduction, would be to stimulate the incorporation of ammonia into nontoxic glutamine by activating glutamine synthetase as well as by the formation of ATP required for the enzyme reaction.

III. Glutamate Synthase (GOGAT)

Ferredoxin-dependent GOGAT was first isolated from pea chloroplasts by LEA and MIFLIN (1974), and recent studies using chloroplasts isolated from pea protoplasts (with 80% recovery of intact plastids) and separated on density gradients, have shown that the enzyme is solely located in the chloroplast (WALLSGROVE et al., 1979). Subsequently the enzyme has been purified from *Vicia faba* leaves (WALLSGROVE and MIFLIN, 1977). The enzyme is specific for glutamine (K_m 330 µM) and has no action on asparagine. Classical inhibitors of glutamine-amide transferases, such as albizziine and azaserine, are strong inhibitors of the enzyme. Only α-ketoglutarate (K_m 150 µM) is able to act as an acceptor; glyoxylate, pyruvate and oxaloacetate are inactive (WALLSGROVE et al., 1977).

The enzyme is apparently widely distributed, having been found in blue-green algae (LEA and MIFLIN, 1975a), green algae (LEA and MIFLIN, 1975b), *Lemna minor* (RHODES et al., 1976) many higher plants (NICKLISH et al., 1976; RATHNAM and EDWARDS, 1976; WALLSGROVE et al., 1977). In maize leaves higher levels of GOGAT were detected in the bundle sheath cells than in the mesophyll (HAREL et al., 1977). Nitrate and nitrite reduction, however, take place in the mesophyll cells where there is a greater capacity for the production of reducing power (MAYNE et al., 1971). RATHNAM and EDWARDS (1976) in a similar study were unable to detect any GOGAT activity in bundle sheath cells; this may be due to the presence of inhibitors in maize leaves. GOGAT in the bundle sheath chloroplasts would be an advantage when a large proportion of the nitrogen arrives in the leaves as glutamine after ammonia assimilation has taken place in the roots. GOGAT and glutamine synthetase in the mesophyll chloroplasts could assimilate the ammonia produced by leaf nitrate reduction.

In maize leaves GOGAT and glutamine synthetase are present in dark-grown and nitrogen-starved plants, and no increase in activity is detected on the addition of nitrate and transfer to the light (HAREL et al., 1977), however, nitrate and nitrite reductase activity increases rapidly during this period. Thus the plant always has the ability to assimilate ammonia in the light, whether supplied by nitrate reduction, protein breakdown, or external application.

GOGAT activity, which is able to use either NAD(P)H or ferredoxin as a source of reductant, has been detected in root plastids (MIFLIN and LEA, 1975), root nodule tissue and developing legume seeds (LEA and FOWDEN, 1975b; BEEVERS and STOREY, 1976); whether they are two separate enzymes has not yet been investigated.

D. Photosynthetic Ammonia Assimilation in Isolated Intact Chloroplasts

MITCHELL and STOCKING (1975) have provided evidence that intact chloroplasts can assimilate ammonia into glutamine, by showing that they can convert ^{14}C-glutamate to glutamine in a reaction that is dependent on ammonia (Fig. 1) and light. Concentrations greater than 1 mM are inhibitory to the reaction, probably due to the uncoupling effect of ammonia depleting ATP supply. The fact that the reaction responds to 10^{-4} M external ammonia suggests that the internal pool of ammonia within the chloroplast is less than this which is far below the reported K_m for NH_4^+ of glutamate dehydrogenase. Chloroplasts have been shown to convert

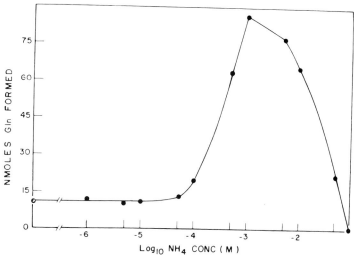

Fig. 1. Effect of NH_4^+ concentration on glutamine synthesis by isolated whole pea chloroplasts in the light. The incubation mixture contained 6.25 µmol of monosodium glutamate, 0.34 µCi of ^{14}C-glutamate, and chloroplasts equivalent to 25.2 µg of chlorophyll. (From MITCHELL and STOCKING, 1975)

^{14}C-α-ketoglutarate to glutamate in the light (GIVAN et al., 1970; TSUKAMOTO, 1970), but evidence was not provided for the action of glutamate dehydrogenase, as the process was not stimulated by ammonia. It is possible that this reaction was due to the action of transaminases which are present in the chloroplast (KIRK and LEECH, 1972).

Evidence for the presence of GOGAT has been demonstrated in a number of preparations of isolated chloroplasts (LEA and MIFLIN, 1974; WALLSGROVE et al., 1977; Table 1). There is a net production of two molecules of glutamate from one molecule of α-ketoglutarate and one molecule of glutamine by a light-dependent process. The reaction is not inhibited by aminooxyacetate, suggesting that transamination is not taking place. Further confirmation of the ability of chloroplasts to assimilate ammonia via the glutamine synthetase/GOGAT pathway in the light has come from experiments carried out in the authors' laboratories by ANDERSON and DONE in which the reductant-dependent steps have been coupled to oxygen evolution under conditions in which CO_2 fixation has been blocked by the addition of glyceraldehyde. Oxygen evolution rates of 8.5–13.9 µmol per mg chl per h, dependent on the presence of both glutamine and α-ketoglutarate have been recorded. The reaction only takes place with intact chloroplasts in the light and is inhibited by 0.5 µM DCMU (ANDERSON and DONE, 1977a). The concentrations of glutamine and α-ketoglutarate required for half maximal rates were similar to the K_m values of GOGAT reported by WALLSGROVE et al. (1977). The oxygen evolution was inhibited by azaserine, an inhibitor of GOGAT, but not by methionine sulphoximine, an inhibitor of glutamine synthetase. Illuminated pea chloroplasts also catalyze an ammonia plus α-ketoglutarate-dependent oxygen evolution at rates of 6.2–10.6 µmol per mg chl per h (ANDERSON and DONE, 1977b). The presence of PPi, ADP, and $MgCl_2$ is essential for the reaction to take place.

Table 1. Light-dependent glutamate synthesis by chloroplasts isolated from various plants

Plant	Rate (μmol mg chl^{-1} h^{-1})		
	Dark	Light	
	complete	complete	minus α-ketoglutarate
Barley	0.270	0.768	0.18
Maize	0.480	1.542	0.342
Bean	0.408	1.530	0.318
Pea	0.420	4.560	0.312

Isolated chloroplasts were incubated with α-ketoglutarate in a small tube, and the amount of glutamate formed over a 15-min period determined (WALLSGROVE et al., 1977).

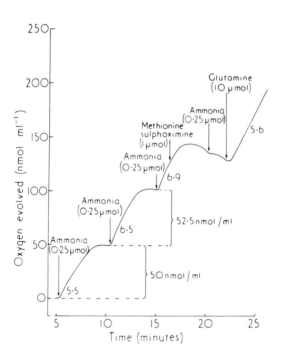

Fig. 2. Oxygen evolution by illuminated isolated pea chloroplasts. Chloroplasts were pre-incubated in the light for 10 min in the presence of 2.5 mM α-ketoglutarate, 4.5 mM ADP, 5 mM PPi, 10 mM MgCl$_2$ and 10 mM DL-glyceraldehyde. Substrates and inhibitors were added as shown. (From ANDERSON and DONE, 1977b)

A lag period of 10 min was required before oxygen evolution took place. Ammonia at a concentration of 0.2 mM supported half maximal rates, but concentrations above 1 mM were inhibitory. The reaction was inhibited by DCMU, methionine sulphoximine, and azaserine (Fig. 2). These results are consistent with the suggestion that chloroplasts, which generally have a high endogenous glutamate concentration, are able to synthesize glutamine from ammonia, which then reacts with α-ketogluta-

Table 2. Production of glutamate and glutamine by isolated chloroplasts

Added nitrogen source	Light-dependent rate (nmol h^{-1})	
	Glutamine	Glutamate
None	0	+53
5 mM KNO$_3$	0	+61
1 mM NaNO$_2$	0	+674
1 mM (NH$_4$)$_2$SO$_4$	+29	+992
1 mM (NH$_4$)$_2$SO$_4$ +2.5 mM methionine sulphoximine	0	+136
1 mM (NH$_4$)$_2$SO$_4$ +1 mM azaserine	+330	−231

Assays contained, in a volume of 0.5 ml, 2.25 μmol ADP, 1.25 μmol α-ketoglutarate, 2.5 μmol Na pyrophosphate, 5 μmol MgCl$_2$ and 200 μl chloroplasts containing 0.123 mg chlorophyll. All reagents were made up in 50 mM HEPES pH 7.6 incubation buffer as described by ANDERSON and DONE (1977a).

rate in the GOGAT reaction to oxidize ferredoxin and lead to oxygen evolution. The sensitivity of the oxygen evolution reaction to methionine sulphoximine excludes the possibility that glutamate dehydrogenase is involved. Indirect evidence is, therefore, provided that ammonia is converted to 2-amino acids by a reduction process dependent on light. This is further confirmed by the fact that two groups of workers have shown that illuminated chloroplasts are capable of converting nitrite to 2-amino nitrogen (MIFLIN, 1974a; MAGALHAES et al., 1974). Further evidence for ammonia assimilation in chloroplasts has been produced by WALLSGROVE et al. (1978). Illuminated chloroplasts are able to convert nitrite and ammonia to glutamate (Table 2). The addition of methionine sulphoximine inhibits the reaction while azaserine causes an accumulation of glutamine, as has been shown in other systems.

E. Photorespiratory Ammonia Evolution and Reassimilation

During the process of photorespiration glycine is converted to serine in the mitochondria in an ATP-liberating reaction (BIRD et al., 1972). Carbon dioxide is evolved but the fact that one molecule of ammonia is also formed is usually ignored.

$$2 \begin{array}{c} \text{CH}_2\text{—NH}_2 \\ | \\ \text{COOH} \end{array} \xrightarrow{1/2\,\text{O}_2} \text{CH}_2\text{OH—CH—NH}_2 + \text{CO}_2 + \text{NH}_3 . \qquad (4)$$
$$\underset{\text{COOH}}{|}$$

If the rate of photorespiration is 20% that of carbon fixation, then the ammonia produced would have to be assimilated at rates considerably higher than required from nitrate reduction.

Studies by Woo et al. (1977) feeding ^{15}N-glycine to isolated spinach leaf cells suggested that free ^{15}N-ammonia is produced, which is rapidly reassimilated into other amino acids.

Studies at Rothamsted utilizing ^{15}N-^{14}C-glycine have shown that mitochondria evolve stoichiometric amounts of CO_2 and NH_3. The ammonia is not reassimilated by the mitochondria. The addition of purified glutamine synthetase promoted the conversion of the liberated ammonia to glutamine. Isolated protoplasts also convert glycine to serine, the reassimilation of ammonia is inhibited by the presence of methionine sulphoximine (Keys et al., 1978). These results suggest that glutamate dehydrogenase is not operating in the mitochondria, but that the ammonia liberated during photorespiration is transported out of the mitochondria and reassimilated by glutamine synthetase in the cytoplasm.

F. Conclusions

Results from studies on the fate of labeled nitrogen compounds, the effect of specific inhibitors and the characteristics and distribution of the relevant enzymes all suggest that in photosynthetic organisms or tissues, ammonia is assimilated first into the amide position of glutamine and then transferred to the 2-amino position of glutamate as shown in Figure 3. The evidence presented in this paper also suggests that the requirement for ATP and reducing energy for this pathway is met directly by ATP and ferredoxin produced by the light reactions of photosynthesis. The rates of light-dependent ammonia assimilation that have been measured via the conversion of glutamate to glutamine (Mitchell and Stocking, 1975), or by coupling the overall process of O_2 evolution, are more than sufficient to cope with the rates of nitrite reduction measured in chloroplasts (Miflin, 1974a; Magalhaes et al., 1974) or in intact leaves (Miflin, 1972). This work emphasizes that it is insufficient to consider the dark reactions of photosynthesis as solely consisting of CO_2 fixation; the reactions of nitrogen metabolism should also be included. The relative potential of isolated chloroplasts to carry out CO_2 fixation and ammonia assimilation are more than sufficient to account for the in vivo rates of these processes and the balance of C and N in the cell (Anderson and Done, 1977b). However, it must be emphasized that there is little evidence to suggest that chloroplasts can independently synthesize the carbon skeletons to accept the reduced nitrogen, and it is likely that fixed carbon leaves the chloroplast as DHAP and PGA (Walker, 1974) and is converted in the mitochondria and cytoplasm to certain α-ketoacids, prior to re-entry. Elias and Givan (1977) have demonstrated the presence of an NADP-dependent isocitrate dehydrogenase in chloroplasts, thus α-ketoglutarate could be synthesized from isocitrate, although the authors could find no evidence for other enzymes required for citrate synthesis, as found in the mitochondria, being present in the chloroplast.

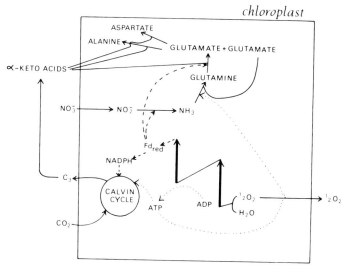

Fig. 3. Proposed pathway of nitrogen metabolism in chloroplasts

References

Anderson, J.W., Done, J.: Plant Physiol. *60*, 354–359 (1977a)

Anderson, J.W., Done, J.: Plant Physiol. *60*, 504–508 (1977b)

Baker, J.E., Thompson, J.F.: Plant Physiol. *36*, 208–212 (1961)

Bassham, J.A., Kirk, M.: Biochim. Biophys. Acta *90*, 553–562 (1964)

Batt, T., Brown, D.H.: Planta (Berl.) *116*, 27–37 (1974)

Beevers, L., Storey, R.: Plant Physiol. *57*, 862–866 (1976)

Bird, I.F., Cornelius, M.J., Keys, A.J., Whittingham, C.P.: Biochem. J. *128*, 191–192 (1972)

Canvin, D.T., Atkins, C.A.: Planta (Berl.) *116*, 207–224 (1974)

Davies, D.D., Teixeira, A.N.: Phytochemistry *14*, 647–656 (1975)

Dharmawardene, M.W.N., Haystead, A., Stewart, W.D.P.: Arch. Mikrobiol. *90*, 281–295 (1973)

Ehmke, A., Hartmann, T.: Phytochemistry *15*, 1611–1617 (1976)

Elias, R.T., Givan, C.V.: Plant Physiol. *59*, 738–740 (1977)

Gayler, K.R., Morgan, W.R.: Plant Physiol. *58*, 283–287 (1976)

Givan, C.V., Givan, A., Leech, R.N.: Plant Physiol. *45*, 624–630 (1970)

Good, N.E.: Biochim. Biophys. Acta *40*, 502–517 (1960)

Harel, E., Lea, P.J., Miflin, B.J.: Planta (Berl.) *134*, 195–200 (1977)

Haystead, A.: Planta (Berl.) *III*, 271–274 (1973)

Haystead, A., Dharmawardene, M.W.N., Stewart, W.D.P.: Plant Sci. Lett. *1*, 439–445 (1973)

Hind, G., Nakatini, H.Y., Izawa, S.: Proc. Natl. Acad. Sci. USA *71*, 1484–1488 (1974)

Keys, A.J.: Biochem. J. *108*, 1–8 (1968)

Keys, A.J., Bird, I.F., Cornelius, M., Wallsgrove, R.M., Lea, P.J., Miflin, B.J.: Nature *275*, 741–743 (1978)

Kirk, P.R., Leech, R.M.: Plant Physiol. *50*, 228–234 (1972)

Kretovich, W.L., Evstigneeva, Z.G., Bakh, A.N.: In: The organ of life and evolutionary biochemistry. Rose, A., Fox, S.W., Deborin, G.A., Pavlavskoya, T.E. (eds.), pp. 245–262. Plenum Press, 1974

Lea, P.J., Fowden, L.: Biochem. Physiol. Pflanz. *168*, 3–14 (1975a)

Lea, P.J., Fowden, L.: Proc. R. Soc. B. *192*, 13–26 (1975b)
Lea, P.J., Miflin, B.J.: Nature (London) *251*, 614–616 (1974)
Lea, P.J., Miflin, B.J.: Biochem. Soc. Trans. *3*, 381–383 (1975a)
Lea, P.J., Miflin, B.J.: Biochem. Biophys. Res. Commun. *64*, 856–861 (1975b)
Lea, P.J., Norris, R.D.: Phytochemistry *15*, 585–595 (1976)
Lea, P.J., Thurman, D.A.: J. Exp. Bot. *23*, 440–449 (1972)
Leech, R.M., Kirk, P.R.: Biochem. Biophys. Res. Commun. *64*, 685–690 (1968)
Lewis, O.A.M.: In: Nitrogen assimilation in plants. 6th Long Ashton Symposium. London: Academic Press in press 1979
Lewis, O.A.M., Berry, M.J.: Planta (Berl.) *125*, 77–80 (1975)
Lewis, O.A.M., Pate, J.S.: J. Exp. Bot. *24*, 596–606 (1973)
Magalhaes, A.C., Neyra, C.A., Hageman, R.H.: Plant Physiol. *53*, 411–415 (1974)
Mayne, B.C., Edwards, G.E., Black, C.C.: In: Photosynthesis and photorespiration. Hatch, M.D., Osmond, C.B., Slayter, R.O. (eds.), pp. 361–371. New York: Wiley Interscience 1971
Miflin, B.J.: Planta (Berl.) *105*, 225–233 (1972)
Miflin, B.J.: Planta (Berl.) *116*, 187–196 (1974a)
Miflin, B.J.: Plant Physiol. *34*, 550–555 (1974b)
Miflin, B.J.: In: Regulation of enzyme synthesis and activity in higher plants. Smith, H. (ed.), pp. 23–40. London: Academic Press 1977
Miflin, B.J., Lea, P.J.: Biochem. J. *149*, 403–409 (1975)
Miflin, B.J., Lea, P.J.: Phytochemistry *15*, 873–885 (1976a)
Miflin, B.J., Lea, P.J.: Trends Biochem. Sci. *1*, 103–106 (1976b)
Miflin, B.J., Lea, P.J.: Annu. Rev. Plant Physiol. *28*, 299–329 (1977)
Mitchell, C.A., Stocking, C.R.: Plant Physiol. *55*, 59–63 (1975)
Nicklisch, A., Geske, W., Kohl, J.G.: Biochem. Physiol. Pflanz. *170*, 85–90 (1976)
O'Neal, D., Joy, K.W.: Nature New Biol. *246*, 61–62 (1973a)
O'Neal, D., Joy, K.W.: Arch. Biochem. Biophys. *159*, 113–122 (1973b)
O'Neal, D., Joy, K.W.: Plant Physiol. *54*, 773–779 (1974)
O'Neal, D., Joy, K.W.: Plant Physiol. *55*, 968–974 (1975)
Rathnam, C.K.M., Edwards, G.E.: Plant Physiol. *57*, 881–883 (1976)
Rhodes, D.: Ph.D. Thesis, Univ. of Manchester U.K. (1976)
Rhodes, D., Rendon, G.A., Stewart, G.R.: Planta (Berl.) *129*, 203–210 (1976)
Ritenour, G.L., Joy, K.W., Bunning, J., Hageman, R.H.: Plant Physiol. *42*, 233–237 (1967)
Rowell, P., Stewart, W.D.P.: Arch. Microbiol. *107*, 115–121 (1976)
Santarius, K.A., Stocking, C.R.: Z. Naturforsch. *24b*, 1170–1179 (1969)
Stewart, G.R., Rhodes, D.: FEBS Lett. *64*, 296–299 (1976)
Stewart, G.R., Rhodes, D.: In: Regulation of enzyme synthesis and activity. Smith, H. (ed.), Vol. XIV, pp. 1–22. London: Academic Press 1977
Stewart, W.D.P., Rowell, P.: Biochem. Biophys. Res. Commun. *65*, 846–856 (1975)
Tate, S.S., Meister, A.: In: The enzymes of glutamine metabolism. Prusiner, S., Stadtman, E.R. (eds.), pp. 77–127. New York: Academic Press 1973
Thacker, A., Syrett, P.J.: New Phytol. *71*, 423–433 (1972)
Tsukamoto, A.: Plant Cell Physiol. *11*, 221–230 (1970)
van der Meulen, P.V.F., Bassham, J.A.: J. Am. Chem. Soc. *81*, 2233–2239 (1959)
Walker, D.A.: In: Med. Tech. Publ. Int. Rev. Sci. Biochem. Ser. I. Northcote, D.H. (ed.), Vol. II, pp. 1–49. London: Butterworths 1974
Wallsgrove, R.H., Lea, P.J., Miflin, B.J.: In: Nitrogen assimilation in plants. 6th Long Ashton Symposium. London: Academic Press in press 1979
Wallsgrove, R.M., Miflin, B.J.: Biochem. Soc. Trans. *5*, 269–271 (1977)
Wallsgrove, R.M., Harel, E., Lea, P.J., Miflin, B.J.: J. Exp. Bot. *28*, 588–596 (1977)
Wallsgrove, R.M., Lea, P.J., Miflin, B.J.: Plant Physiol. in press 1979
Werdan, K., Heldt, H.W., Milovancev, M.: Biochim. Biophys. Acta *396*, 276–292 (1975)
Wolk, C.P., Thomas, J., Schaffer, P.W., Austin, S.M., Galonsky, A.: J. Biol. Chem. *251*, 5027–5034 (1976)
Woo, K.C., Berry, J.A., Osmond, C.B., Lorimer, G.H.: In: Abstr. 4th Int. Congr. Photosynthesis, p. 415. 1977, Reading England

6. N$_2$ Fixation and Photosynthesis in Microorganisms

W.D.P. STEWART

A. Introduction

The inter-relationships between N$_2$ fixation and photosynthesis, particularly in agriculturally important legumes, has for long attracted attention (see WILSON, 1940; QUISPEL, 1974; BURNS and HARDY, 1975; NUTMAN, 1976; HARDY, 1977; PATE, 1977). In legumes and in other symbiotic systems, except those involving blue-green algae (see FOGG et al., 1973; MILLBANK, 1974; STEWART et al., 1978a), the inter-dependency of N$_2$ fixation and photosynthesis is an indirect one in that the function of the photosynthetic host is primarily to provide fixed carbon for the production of ATP and reductant by the heterotrophic N$_2$-fixing partner, and to provide a sink for the nitrogen which is fixed. Within recent years and particularly since the world energy crisis, interest has been aroused in the possibility of achieving, by genetic manipulation, plants which can photosynthesise and fix N$_2$ without the involvement of a symbiont (see e.g., POSTGATE, 1974; HOLLAENDER et al., 1977). To date only two groups of organisms are known which can carry out both N$_2$ fixation and photosynthesis simultaneously. These are the photosynthetic bacteria and blue-green algae (cyanobacteria). The ways in which they do so are considered in this chapter.

B. Distribution of Nitrogenase Among Photosynthetic Prokaryotes

Nitrogenase is widely distributed among photosynthetic prokaryotes, with over 250 strains having an active nitrogenase (see SIEFERT, 1976; RIPPKA and WATERBURY, 1977; STEWART et al., 1978b). The most comprehensive study on the range of photosynthetic bacteria is that of SIEFERT (1976) who tested 62 strains belonging to the Rhodospirillaceae: *Rhodospirillum* (10), *Rhodopseudomonas* (13), *Rhodomicrobium* (2), *Rhodocyclus* (1), Chromatiaceae: *Chromatium* (10), *Thiocystis* (3), *Thiocapsa* (4), *Amoebobacter* (2), and Chlorobiaceae: *Chlorobium* (12), *Prosthecochloris* (2) and *Pelodictyon* (3). Anaerobic nitrogenase activity occurred in 92% of the Rhodospirillaceae, in 86% of the Chromatiaceae and in 77% of the Chlorobiaceae.

All the major groups of blue-green algae have N$_2$-fixing members. The first demonstration of N$_2$ fixation in axenic culture by a unicellular cyanophyte (WYATT and SILVEY, 1969) has been expanded upon (NEILSON et al., 1971), and RIPPKA and WATERBURY (1977) summarised findings which showed acetylene reduction by the five *Gloeothece* strains which they tested. These strains also reduced acetylene aerobically and included two strains which were previously classified as species of *Gloeocapsa*. Of 50 other unicellular strains tested only three (marine strains of *Synechococcus*) reduced acetylene, but under anaerobic conditions only. Genera with no N_2-fixing representatives were *Gloeobacter, Gloeocapsa, Synechocystis*, and *Chamaesiphon*.

Certain pleurocapsalean strains of *Dermocarpa, Xenococcus, Myxosarcina, Chroococcidiopsis*, and of the *Pleurocapsa* group, reduce acetylene, but only anaerobically. The one strain of *Dermocarpella* tested does not reduce acetylene (RIPPKA and WATERBURY, 1977).

The ability of non-heterocystous filamentous forms, to reduce acetylene and fix N_2 under anaerobic conditions only (STEWART and LEX, 1970), has been confirmed (STEWART, 1971; KENYON et al., 1972; STEWART et al., 1977), and in an extensive survey RIPPKA and WATERBURY (1977) tested 44 strains of oscillatoriacean algae and found an active nitrogenase, under anaerobic conditions only, in 25 strains of the genera *Oscillatoria, Pseudanabaena* and the LPP (*Lyngbya-Plectonema-Phormidium*) group. The two *Spirulina* strains tested did not reduce acetylene.

The heterocystous algae, first shown to fix N_2 by DREWES (1928) have been confirmed as the most ecologically important group of N_2-fixing algae fixing N_2 under aerobic conditions (see FOGG et al., 1973; STEWART, 1975). Virtually every strain tested has an active nitrogenase. RIPPKA and STANIER (1978) have studied the non-heterocystous *Anabaena variabilis* strain 7118. This non-heterocystous form does not fix N_2 aerobically, although it does so anaerobically, but revertants of this alga which develop heterocysts fix N_2 aerobically.

There is no evidence of an active nitrogenase in any eukaryotic organism, whether photosynthetic or not. Possible reasons for this are discussed by HELINSKI (1977). In terms of photosynthetic prokaryotes, the cyanelle of *Cyanophora paradoxa*, which may be intermediate between a blue-green alga and a eukaryotic plant chloroplast (see FOGG et al., 1973; HERDMAN and STANIER, 1977), does not reduce acetylene aerobically or anaerobically (STEWART and BOTTOMLEY, 1976). In terms of genome size, that of even the smallest blue-green alga is at least 10 times larger than that of the largest eukaryotic plant chloroplast (see HERDMAN and STANIER, 1977) and it may be that the *nif* genes, or genes responsible for their expression, have been lost during the evolution of chloroplasts.

C. Oxygen Sensitivity and Protection of Algal Nitrogenase

An important factor which distinguishes the different groups of N_2-fixing organisms considered above is their differing capacities to prevent O_2 inactivation of nitrogenase. This inhibitory effect of O_2 is primarily via enzyme inactivation rather than via an inhibition of nitrogenase synthesis by O_2. Thus BONE (1971) demonstrated that when N_2-fixing filaments of *Anabaena*, which had been placed under high O_2 to inhibit their nitrogenase, were replaced in air, nitrogenase activity recovered in the absence but not in the presence of inhibitors of protein synthesis. The data suggest that at any one time the level of nitrogenase in the alga may be regulated, in part at least, by the extent to which the rate of nitrogenase synthesis exceeds the rate of O$_2$ inactivation of nitrogenase. STEWART and LEX (1970) and RIPPKA and STANIER (1978) have shown similarly in non-heterocystous forms exposed to O$_2$ that an immediate inhibition of nitrogenase occurs, followed by a lengthy period of recovery of nitrogenase activity when O$_2$ is removed, indicative of an immediate inactivation of existing nitrogenase followed by nitrogenase resynthesis. Secondary indirect effects of O$_2$ also occur. Thus high O$_2$ stimulates photorespiration in blue-green algae (LEX et al., 1972) through ribulose 1,5-bisphosphate (RuBP) carboxylase acting as an oxygenase (CODD and STEWART, 1977a). In this way reductant, which may otherwise have been available for N$_2$ reduction, is used up. Even in anaerobic N$_2$-fixing photosynthetic bacteria RuBP carboxylase may also act as an oxygenase (see TAKABE and AKAZAWA, 1973; MCFADDEN, 1974; RYAN et al., 1974).

In photosynthetic bacteria there is no protective mechanism against O$_2$ inactivation, the bacteria fixing N$_2$ anaerobically, although some may grow aerobically on combined nitrogen. The mechanism of O$_2$ protection in unicellular algae which fix N$_2$ aerobically may result from a temporal separation of O$_2$-sensitive nitrogenase activity and O$_2$-evolving photosynthesis (GALLON et al., 1975). Among the pleurocapsalean algae and oscillatoriacean algae likewise, where there is little protection against O$_2$ in excess of 2%; evidence of a temporal separation of photosynthetic O$_2$ evolution and nitrogenase activity has again been obtained (WEARE and BENEMANN, 1974). It is likely that in these non-heterocystous algae the products of nitrogen metabolism regulate photosynthesis and vice-versa. Thus an ATP:ADP ratio below 10:1 may inhibit nitrogenase activity (ORME-JOHNSON, 1978) and unpublished data from our laboratory show that CO$_2$ fixation in cyanophytes may be affected by products of nitrogen metabolism.

The major factor which allows heterocystous algae to fix N$_2$ in air is that the nitrogenase within heterocysts (Fig. 1) is protected from O$_2$ inactivation (FAY et al., 1968; STEWART et al., 1969; WOLK and WOJCIUCH, 1971; WEARE and BENEMANN, 1973; FLEMING and HASELKORN, 1973, 1974; PETERSON and BURRIS, 1976; TEL-OR and STEWART, 1976; WOLK et al., 1976). Under anaerobic conditions the vegetative cells of some heterocystous algae may also fix N$_2$ (see STEWART, 1971). Evidence for this comes from several sources. First, various non-heterocystous algae have an active nitrogenase under anaerobic conditions (see above). Second, VAN GORKOM and DONZE (1971) used phycocyanin, which acts as a rapidly mobilised nitrogen reserve, as an indicator of the nitrogen status of the filaments of *Anabaena cylindrica*. They measured phycocyanin fluorescence microscopically and found that when nitrogen-starved heterocystous filaments which showed little fluorescence were exposed to air, phycocyanin re-appeared first in the cells adjacent to the heterocysts and there was a decreasing phycocyanin gradient away from the heterocysts. On the other hand, when the nitrogen-starved filaments were exposed to N$_2$ under anaerobic conditions phycocyanin fluorescence developed equally in all the vegetative cells. Such evidence suggests that only the heterocysts fix N$_2$ aerobically, but that the vegetative cells can also do so anaerobically. Third, the finding that, although the non-heterocystous *Anabaena variabilis* 7118 fixes N$_2$ only under anaerobic conditions, revertants which develop heterocysts can fix N$_2$ aerobically (RIPPKA and STANIER, 1978), strongly suggests anaerobic vegetative cell nitrogenase activity, but with activity occurring only in the hetero-

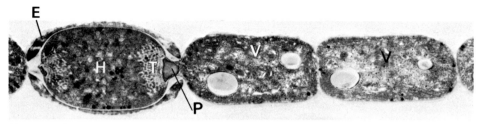

Fig. 1. An electron micrograph of *Anabaena cylindrica* showing a heterocyst (*H*) and vegetative cells (*V*). The heterocyst has highly contorted, densely packed thylakoids (*T*), a thick surrounding envelope (*E*) and a dense, osmiophilic polar plug (*P*) (×6800)

cysts in air. It would be useful, now that consistently high rates of nitrogenase activity in heterocyst extracts have been obtained, to compare the relative distribution of activity between extracts of vegetative cells and heterocysts from aerobic and anaerobically grown cultures. It cannot be emphasised too strongly in this connexion that it is the relative distribution of activities between vegetative cell and heterocyst extracts under any one set of assay conditions which is important. Comparisons of the activities of cell-free extracts of heterocysts or vegetative cells (which take time to prepare, and which may be incubated under H_2 in the presence of $Na_2S_2O_4$ and with an ATP-generating system) with the activities of intact filaments are of limited value in attempting to assess the distribution of the enzyme between the two cell types.

D. Requirements for an Active Nitrogenase

The nitrogenase complex of photosynthetic prokaryotes, like that of other N_2-fixing agents, is a polymeric complex composed of a large component (Component I or the FeMo protein) and a smaller component (Component II or the Fe protein). Of particular interest has been the discovery of factors which activate these components. Thus from *Rhodospirillum rubrum* Ludden and Burris (1976) have isolated a factor which is O_2- and trypsin-sensitive, and which in the presence of ATP and a divalent metal ion, can activate inactive component II, and a molybdenum-containing co-factor has been detected in *Azotobacter* which is associated with, and essential for, the activity of the FeMo protein (see Bishop et al., 1977). Although nitrogenase has been purified from various microorganisms (see Winter and Burris, 1976; Orme-Johnson and Davis, 1978) only partial purifications of the enzymes from photosynthetic prokaryotes have been achieved e.g., from *Rhodospirillum rubrum* (Burns and Bulen, 1966; Munson and Burris, 1969), *Chromatium vinosum* (formerly *Chromatium* strain D; Winter and Arnon, 1970; Evans et al., 1973) and *Anabaena cylindrica* (Smith et al., 1971b; Tsai and Mortenson, 1978). From the available data there is nothing to suggest that the enzymes of photosynthetic prokaryotes vary markedly from heterotrophic N_2-fixing agents where the mechanism proceeds via (1) the reduction of the Fe-protein by the electron donor (2) the assembly of a ternary complex containing both proteins and MgATP (3) a transfer of electrons from the Fe protein to the FeMo protein within the complex (4) the transfer of electrons from the FeMo protein to the reducible substrate such as N$_2$ or C$_2$H$_2$ (see THORNELEY et al., 1978). Photosynthetic N$_2$-fixing prokaryotes differ from heterotrophic N$_2$-fixing prokaryotes mainly in how the ATP and reductant are provided.

E. The Provision of Reductant and ATP in Photosynthetic Prokaryotes

Blue-green algae exhibit a higher plant type of photosynthesis with water acting as electron donor, with electron transfer occurring via two photosystems linked in series and with carbon being fixed by a typical Calvin cycle pathway. Certain blue-green algae also grow in the presence of H$_2$S as HINZE (1903), NAKAMURA (1937, 1938), SETLIK (1957) KNOBLOCH (1966) and STEWART and PEARSON (1970) reported in early work, HINZE (1903) and NAKAMURA (1937) in fact reporting on the presence of S droplets within the algal cells. More recently it has been shown that 11 out of 21 blue-green algae tested (GARLICK et al., 1977) including *Oscillatoria limnetica* (COHEN et al., 1975a, b) could use Na$_2$S in the presence of 3'-(3,4 dichlorophenol)-1',1'-dimethylurea (DCMU) as electron donor and the ability of *Anacystis nidulans* to use Na$_2$S$_2$O$_3$ has also been demonstrated (UTKILEN, 1976). The inhibition of thiosulphate oxidation by 2,5-dibromo-3-methyl-6-isopropyl-*p*-benzoquinone (DBMIB) but not by DCMU (UTKILEN, 1976) suggests that thiosulphate feeds into the electron transport chain after photosystem II but on the photosystem II side of plastoquinone. BELKIN and PADAN (1978) have shown both with *Oscillatoria limnetica* and *Aphanothece halophytica* that H$_2$ can also act as electron donor for the photofixation of CO$_2$ in a photosystem I-dependent reaction although, compared with Na$_2$S, H$_2$ is a poor source of reductant. The photosynthetic bacteria do not photolyse water, but use reduced S compounds or organic compounds as sources of reductant in photosynthesis.

I. Electron Donation

Early studies showed that in blue-green algae electrons required for N$_2$ reduction were not generated directly on the photolysis of water, e.g., DCMU had no immediate effect on N$_2$ reduction if endogenous carbon reserves or exogenous fixed carbon were available in the light (COX, 1966; COX and FAY, 1969; BOTHE, 1972; KENYON et al., 1972; LEX and STEWART, 1973; BOTTOMLEY and STEWART, 1977; RIPPKA and STANIER, 1978), the action spectrum of nitrogenase activity corresponded to that of photosystem I, not photosystem II (FAY, 1970), the absence of photosystem II light had no effect on nitrogenase activity (BOTHE and LOOS, 1972) and there was no Emerson enhancement of N$_2$ fixation (LYNE and STEWART, 1973). The reductant for nitrogenase in blue-green algae comes from fixed carbon compounds.

The electron donor to nitrogenase in blue-green algae has been variously suggested as NADPH, ferredoxin, or under iron-deficient conditions flavodoxin, and all may be important depending on the conditions (see STEWART, 1974). However ferredoxin is usually regarded as the natural electron donor, and evidence supporting this view is: (1) in a chloroplast-coupled system a ferredoxin- or flavodoxin-dependent electron transfer from ascorbate/dichlorophenol indophenol to *Anabaena* nitrogenase has been demonstrated (BOTHE, 1970; SMITH et al., 1971a); (2) plant ferredoxin and *Anabaena* ferredoxin function in electron transfer from glucose-6-phosphate to cytochrome c or nitrogenase in extracts of *Anabaena* (BOTHE, 1970; APTE et al., 1978); (3) H_2 can donate electrons to nitrogenase via clostridial ferredoxin in *Anabaena* extracts (HAYSTEAD and STEWART, 1972). The ferredoxins of blue-green algae have been reviewed by ANDREW et al. (1975) and HALL et al. (1975). HUTSON and RODGERS (1975) and SHIN et al. (1977) have reported the isolation and characterization of two plant-type ferredoxins from *Nostoc* MAC and *Nostoc verrucosum* respectively. In photosynthetic bacteria ferredoxin is again considered as electron donor to nitrogenase. Ferredoxins have been characterized from *Chromatium, Chlorobium* and *Rhodospirillum rubrum* (see YOCH and VALENTINE, 1972; MORTENSON and NAKOS, 1973; LOVENBERG, 1974), the last species having two extractable ferredoxins in phototrophically grown cultures, but only one in heterotrophically grown cultures (SHANMUGAM et al., 1972). YOCH and ARNON (1974) suggest that there is a non-cyclic photosynthetic electron flow to ferredoxin and nitrogenase from reduced S compounds or organic compounds, and although it is possible that this may be the main route of electron donation to nitrogenase, SCHICK (1971) has shown continued nitrogenase activity in the dark by *Rhodospirillum rubrum*. BENNETT et al. (1964) suggest that a dark reaction involving the phosphoroclastic cleavage of pyruvate provides electrons to nitrogenase.

II. The Production of ATP

ATP can be generated in blue-green algae by: non-cyclic photophosphorylation, cyclic photophosphorylation, oxidative phosphorylation, fermentation, and by a Knallgas reaction (see BOTTOMLEY and STEWART, 1976, 1977; BOTHE et al., 1977a, b). Both non-cyclic and cyclic photophosphorylation alone can support maximum nitrogenase activity. The ability of cyclic photophosphorylation alone to do so is shown by the fact that inhibition of photosystem II has no inhibitory effect on nitrogenase activity when fixed carbon is available (for other evidence see Sect. E.I above).

Oxidative phosphorylation alone cannot support optimum nitrogenase activity. First, it is of limited significance in the light. Thus, BOTTOMLEY and STEWART (1976) showed that when *Anabaena cylindrica* is transferred from light to dark aerobic conditions, the ATP pool drops to 50% of the light level and then recovers in the absence but not in the presence, of m-chlorocarbonylcyanide phenylhydrazone (CCCP). If oxidative phosphorylation was making a substantial contribution to the ATP pool in the light, such a drop in the ATP pool would not have been expected. Also, the oxidative pentose phosphate pathway is the main route of NADPH generation in blue-green algae, apart from the photosynthetic electron transport chain, and this pathway is unimportant in vegetative cells in the light since enzymes such as glucose-6-phosphate dehydrogenase are photoinhibited

(PELROY et al., 1972; LEX and CARR, 1974; DUGGAN and ANDERSON, 1975; GROSS-MAN and McGOWAN, 1975; APTE et al., 1978). Thus, oxidative phosphorylation is unlikely to be important in ATP generation in the light in non-heterocystous algae. In heterocysts, on the other hand, the oxidative pentose phosphate pathway functions both in the light and dark (see below). Second, in the dark the rate of oxidative phosphorylation, in vivo, is only about 10%–20% of the rate of photophosphorylation. The maximum reported prolonged rate of nitrogenase activity in the dark by *Anabaenopsis circularis,* which can grow heterotrophically, is 70% of the light rate (see BOTTOMLEY and STEWART, 1976, 1977).

Substrate level phosphorylation has been demonstrated in *Anabaena cylindrica* by BOTTOMLEY and STEWART (1976) but the data suggest, as others have inferred (see e.g., PELROY et al., 1972; RIPPKA, 1972; PELROY and BASSHAM, 1973; STANIER, 1973; GROSSMAN and McGOWAN, 1975; SANCHEZ et al., 1975), that it is of limited significance in cyanophytes.

The presence of a Knallgas reaction in blue-green algae has been demonstrated by BOTHE et al. (1977 b) who observed, in *Anabaena cylindrica,* that there is present in N$_2$-fixing cultures a H$_2$-dependent O$_2$ uptake which proceeds in the light and dark and which transfers electrons along the respiratory electron transport chain, as evidenced by its inhibition by KCN, generating ATP in the process. The extent to which ATP generation by this process alone can support nitrogenase activity in vivo is uncertain.

F. The Nitrogen-Fixing System of Heterocysts of *Anabaena cylindrica*

The ways in which the heterocysts of *Anabaena cylindrica* are adapted to perform a N$_2$-fixing function have been extensively studied using isolated heterocysts or their extracts. The data obtained corroborate the whole filament studies discussed above, and may be summarised briefly. First, heterocysts do not photolyse water, in agreement with the whole cell finding that DCMU does not immediately inhibit N$_2$-fixation. Thus, there is no photosynthetically evolved O$_2$ in heterocysts to inhibit their nitrogenase. This results from the facts that heterocysts are deficient in the accessory light-harvesting pigments of photosystem II (phycocyanin, allophycocyanin, and chlorophyll a 670, see THOMAS, 1970) and have a Mn content which, when compared with other photosynthesising cells (see CHENIAE, 1970), is too low for the photolysis of water to occur (TEL-OR and STEWART, 1975). It will be of interest to know whether heterocysts contain superoxide dismutase (Fe- and Mn-containing types in blue-green algae) which scavenges O$_2^{\cdot-}$ radicals on the photolysis of water in the vegetative cells of blue-green algae (see ASADA et al., 1976).

Despite the lack of a water-splitting capacity, the components of the photosynthetic electron transport chain of heterocysts appear to be intact from cytochrome b-559 to ferredoxin (TEL-OR and STEWART, 1977). There is evidence, obtained using difference spectroscopy, of cytochrome b-559, and of plastoquinone, based on the fact that in heterocyst preparations the addition of plastoquinone antagonist

DBMIB to heterocyst extracts inhibits light-dependent electron transport from diphenyl carbazide to methylviologen. DBMIB also inhibits light-dependent acetylene reduction by whole filaments of *Anabaena cylindrica*. Cytochrome c-554 has been detected by difference spectroscopy and there is evidence of plastocyanin, based on the fact that KCN inhibits the transfer of electrons from ascorbate/dichlorophenol indophenol to methylviologen by lamellar fragments of isolated heterocysts. An EPR signal of oxidised heterocyst samples also occurs near $g = 2.00$, which is typical of Cu^{2+} and which may be attributable in part at least to the presence of plastocyanin. A $g = 2.05$ signal which does not completely disappear on the reduction of heterocyst samples with ascorbate/dichlorophenol indophenol suggests the presence of a second form of copper (possibly non-protein) in heterocysts (Cammack et al., 1976). The presence of cytochrome b-563 has been detected by difference spectroscopy (see Tel-Or and Stewart, 1977). The presence of P-700 in heterocysts has been detected both by difference spectroscopy (Donze et al., 1972), and by EPR spectroscopy (Tel-Or and Stewart, 1977). Hiyama et al. (1977) have recently presented evidence for the presence of two types of P-700 in membrane fragments from whole filaments of *Nostoc muscorum*. Whether they occur in *Anabaena cylindrica*, or its heterocysts, is unknown.

Electron donation to soluble ferredoxin, which is present in heterocysts, occurs via membrane-bound iron–sulphur proteins and EPR signals indicative of the photo-oxidation of P-700 and the reduction of the bound iron–sulphur centres have been obtained using heterocyst preparations. Evidence for the presence of such membrane-bound components has been obtained in reduced heterocyst preparations examined at 16 K. Then, a spectrum with signals at $g = 2.05$, 1.94, 1.92, and 1.89 is observed (Cammack et al., 1976). This spectrum is typical of preparations containing the membrane-bound iron–sulphur proteins A and B of the primary acceptor complex of photosystem I of chloroplasts (Evans et al., 1972; Ke, 1973; Ke et al., 1974; Bearden and Malkin, 1975; Evans, Sihra and Cammack, 1976). Evidence has been obtained using extracts of chloroplasts (McIntosh et al., 1975; Evans, Sihra and Cammack, 1976; McIntosh and Bolton, 1976) and of whole filaments of *Chlorogloea fritschii* (Evans, Cammack and Evans, 1976) of a third component which is not a ferredoxin but which may be the primary electron acceptor from P-700 and which passes electrons on to the iron–sulphur proteins. This electron-accepting species has not yet been conclusively demonstrated in heterocysts (see Cammack et al., 1976).

Ferredoxin-NADP$^+$ oxidoreductase is also active in heterocysts (Tel-Or and Stewart, 1976, 1977) but despite this, and the fact that heterocysts photophosphorylate, heterocysts do not fix CO_2 photosynthetically. This is due in part at least to the lack of detectable RuBP carboxylase in heterocysts. The lack of RuBP carboxylase has been shown enzymatically (Winkenbach and Wolk, 1973; Stewart and Codd, 1975; Codd and Stewart, 1977b), immunologically (Codd and Stewart, 1977b) and indirectly by electron microscopy in that the polyhedral bodies, which are packets containing RuBP carboxylase in vegetative cells (Codd and Stewart, 1976), are absent from heterocysts (Stewart and Codd, 1975). The fact that the enzyme was detected immunologically in vegetative cells, but not in heterocysts, suggests that the enzyme protein is not synthesised in heterocysts, or is present in a conformational state which is different from the catalytically active enzyme of the vegetative cells.

The lack of RuBP carboxylase in heterocysts raises two questions: (1) how do they obtain the necessary fixed carbon for metabolism and (2) what is the function of ferredoxin-NADP$^+$ oxidoreductase in heterocysts? It is now clear that the fixed carbon is provided from the photosynthesising vegetative cells, possibly as maltose, since JUTTNER and CARR (1976) observed that after exposing whole filaments of *Anabaena cylindrica* to $^{14}CO_2$ for 10 s and then separating vegetative cell and heterocyst fractions rapidly at low temperatures, 62% of the ^{14}C-labelling in heterocysts was present in a disaccharide, probably maltose. Other compounds such as glutamate, which are required for glutamine synthesis, may also enter the heterocysts from vegetative cells, but are of lesser importance as sources of fixed carbon.

In heterocysts high activities of the enzymes of the oxidative pentose phosphate pathway occur but glyceraldehyde-3-phosphate dehydrogenase activity is extremely low (WINKENBACH and WOLK, 1973; LEX and CARR, 1974). Hexokinase is enriched in heterocysts (WINKENBACH and WOLK, 1973; LEX and CARR, 1974) and the necessary glucose-6-phosphate required for the oxidative pentose phosphate pathway can be easily provided. This, together with 6-phosphogluconate dehydrogenase, can generate NADPH.

The suggestion that glucose-6-phosphate serves as electron donor for N$_2$ reduction was made first by BOTHE (1970) and later by SMITH et al. (1971a) on the basis of whole filament, or filament extract, studies. APTE et al. (1978) have studied this pathway of electron donation in detail in heterocysts of *Anabaena cylindrica*. They observed (1) in heterocyst extracts, unlike in whole filament or vegetative cell extracts, there is no photoinhibition of G6P dehydrogenase and thus in heterocysts the oxidative pentose phosphate pathway can function equally well in the light and dark (2) ferredoxin-NADP$^+$ oxidoreductase, measured as cytochrome c reduction or as reduction of DCPIP (dichlorophenolindophenol), occurs in heterocyst extracts in the light and dark, although in whole filament or vegetative cell extracts it functions in electron transfer from NADPH to ferredoxin only in the dark (3) in heterocysts G6P, acting via G6P dehydrogenase and ferredoxin-NADP$^+$ oxidoreductase, can donate electrons to cytochrome c in a pathway inhibited by antibody to ferredoxin (4) electron transfer from G6P to ferredoxin can be controlled by the NADPH/NADP$^+$ ratio or reduction charge value with G6P dehydrogenase showing decreased activities above a NADPH/NADP$^+$ ratio of 0.3 and ferredoxin-NADP$^+$ oxidoreductase being most active above 0.3 and decreasing in activity at lower values. Thus when the entire system is poised at a reduction charge value of around 0.3, electron flow to ferredoxin from glucose-6-phosphate can occur with minimum wastage of electrons. Other components shown to inhibit *Anabaena cylindrica* glucose-6-phosphate dehydrogenase are ATP and dithiothreitol, but RuBP, ADP, phosphoenolpyruvate, glutamine and glutamate do not affect activity. Ferredoxin-NADP$^+$ oxidoreductase is inhibited, as expected, by NADP$^+$, but not by G6P, RuBP, ATP, ADP, glutamine, and glutamate (APTE et al., 1978).

Other possible electron donors to ferredoxin are known. Thus, pyruvate via pyruvate-ferredoxin oxidoreductase (BOTHE and FALKENBERG, 1973; BOTHE et al., 1974) donates electrons to nitrogenase in whole filament extracts of *Anabaena cylindrica* (CODD et al., 1974) but this flavoprotein has yet to be detected in heterocysts. In addition, MURAI and KATOH (1975) have shown that lamellar fragments from whole filaments of *Anabaena variabilis* can, in the presence of CMU [3′(p-chlorophenyl)-1′, 1′-dimethylurea], use glycollate, malate, succinate and isocitrate in decreasing order of effectiveness as electron donors for the photoreduction of methylviologen. LOCKAU et al. (1978) have considered various reductant sources in isolated heterocysts. H_2 can serve as electron donor to nitrogenase (WOLK and WOJCIUCH, 1971; HAYSTEAD and STEWART, 1972). This is due to the fact that heterocysts possess an uptake hydrogenase which is sensitive to high O_2, but which can, in aerobic cultures, function in heterocysts where O_2 levels are low (see BOTHE et al., 1977b; TEL-OR et al., 1977a). Via a Knallgas reaction (unlike the oxyhydrogen reaction it is insensitive to 2,4-dinitrophenol) and the respiratory electron transport chain electrons can thus be transferred from H_2 to nitrogenase with the concurrent generation of ATP (BOTHE et al., 1977a, b). Under anaerobic conditions in the light, H_2 can also donate electrons to nitrogenase independent of the photosynthetic electron transport chain, although the exact mechanism whereby this proceeds is uncertain (BOTHE et al., 1977b). This H_2 uptake mechanism plays an important role because the nitrogenase of blue-green algae, like that of other N_2-fixing systems (see EVANS et al., 1977) acts as an ATP-dependent hydrogenase (see BENEMANN and WEARE, 1974a, b; JONES and BISHOP, 1976; PETERS et al., 1976; BOTHE et al., 1977a, b; KOSIAK, 1977; TEL-OR et al., 1977a, b) so that even under normal conditions of N_2-fixation, a proportion of the available electrons, and ATP, for N_2 reduction is wasted as H_2 is produced. In root nodules wastage of ATP and reductant in this way may represent up to a third of the total available to nitrogenase (EVANS et al., 1977). The presence of an uptake hydrogenase thus helps to recycle H_2, conserve reductant and generate ATP. ATP-dependent H_2-production can be stimulated artificially by incubating the algae in the absence of a reducible substrate, or by adding CO which blocks electron transfer from nitrogenase to the reducible substrate. The inducible uptake hydrogenase is sensitive to high O_2 levels, C_2H_2, and also to CO. The nitrogenase also acts as an ATP-dependent hydrogenase in N_2-fixing photosynthetic bacteria (see GEST and KAMEN, 1949; YOCH and ARNON, 1974; GRUZINSKY and GOGOTOV, 1977).

ATP production in heterocysts may be by cyclic photophosphorylation, oxidative phosphorylation, or by a Knallgas reaction. SCOTT and FAY (1972) provided preliminary direct evidence of photophosphorylation by isolated heterocysts of *Anabaena variabilis*, the observed rate of photophosphorylation being 2.1 times that of oxidative phosphorylation. TEL-OR and STEWART (1977) observed rates of photophosphorylation by heterocysts of 64 µmol ATP per mg chl per h compared with values of 55–159 µmol ATP per mg chl per h for ruptured filaments. The rates of photophosphorylation by heterocysts were 2.6–7.1 times higher than the rates of oxidative phosphorylation. It may be that a major role of the photosynthetic electron transport chain in heterocysts is to generate ATP by cyclic photophosphorylation with reductant being transferred to ferredoxin from fixed carbon compounds via G6P and ferredoxin-NADP$^+$ oxidoreductase. In this connexion it is of interest that ARNON and CHAIN (1975) using spinach chloroplasts noted that cyclic photophosphorylation could be regulated by NADPH in the presence of ferredoxin, presumably with ferredoxin-NADP$^+$ oxidoreductase also being implicated (see also BÖHME, 1977). ATP generation linked to H_2 uptake is discussed by PACKER et al. (1977) and PETERSON and BURRIS (1978).

In heterocysts, with the availability of ATP and reductant, nitrogenase activity occurs at high rates, PETERSON and BURRIS (1976) having reported acetylene reduction rates of 3260 nmoles C$_2$H$_4$ mg N^{-1} 20 min^{-1} using heterocysts isolated with colloidal silica and incubated under H$_2$ in the presence of Na$_2$S$_2$O$_4$ and an ATP-generating system.

The pathway of assimilation of newly fixed NH$_3$ in blue-green algae has been the subject of detailed studies, and it is clear that in heterocystous algae the glutamine synthetase-glutamate synthase pathway is a major route of NH$_3$ assimilation (DHARMAWARDENE et al., 1973; LEA and MIFLIN, 1975; STEWART et al., 1975; THOMAS et al., 1975; LAWRIE et al., 1976; WOLK et al., 1976). Evidence suggests that whereas glutamine synthetase is active in the heterocysts, as well as in vegetative cells, there is little or no glutamate synthase activity in heterocysts (WOLK et al., 1976). Glutamate synthase in blue-green algae is ferredoxin-dependent (LEA and MIFLIN, 1975) and this compartmentation means that glutamate synthase does not compete with nitrogenase in heterocysts for reduced ferredoxin, and that the glutamine produced in heterocysts must be transported to the vegetative cells where it is utilised by glutamate synthase. Early ^{15}N studies suggested that glutamine was exported from heterocysts in this way (STEWART et al., 1975) and elegant kinetic studies using ^{13}N as tracer (WOLK et al., 1976) confirm that glutamine is in fact the major transported compound. Glutamate, from vegetative cells, may be transferred back to heterocysts to pick up this newly fixed NH$_3$ via glutamine synthetase (WOLK et al., 1976).

G. Nitrogenase and Its Possible Regulation by Glutamine Synthetase

Glutamine synthetase is of particular interest in blue-green algae, not only because of its role in NH$_3$ assimilation, but because it has been implicated in metabolic regulation in prokaryotes (see PRUSINER and STADTMAN, 1973), including N$_2$-fixing species (STREICHER et al., 1972; TUBB, 1974; SHANMUGAM et al., 1976). The most extensive early studies which implicate it in the regulation of N$_2$ fixation have been carried out with *Klebsiella* (STREICHER et al., 1974; SHANMUGAM et al., 1976). Such studies suggested that glutamine synthetase is regulated indirectly by NH$_4$$^+$ which affects the degree of adenylylation of the enzyme. In blue-green algae there is no evidence of activation/deactivation of glutamine synthetase via a deadenylylation/adenylylation mechanism, but enzymic activity may be subject to feedback inhibition by potential products of glutamine metabolism, by AMP, CTP, and by a Mn^{2+}:ATP ratio greater than 1 (DHARMAWARDENE et al., 1973; STEWART et al., 1975; ROWELL et al., 1977) and light-dark regulation may also occur (ROWELL et al., 1978). The generation of ATP through photophosphorylation is likely to be of importance for this ATP-dependent enzyme. Although glutamine synthetase is involved in the regulation of nitrogen metabolism and heterocyst differentiation in blue-green algae (STEWART and ROWELL, 1975; WOLK et al., 1976) its effect may not necessarily be a direct one, since changes in nitrogenase activity may precede changes in levels of catalytically active glutamine synthetase (ROWELL et al., 1977).

References

Andrew, P.W., Delaney, M.E., Rogers, L.J., Smith, A.J.: Phytochemistry *14*, 931–935 (1975)
Apte, S.K., Rowell, P., Stewart, W.D.P.: Proc. R. Soc. Lond. B. *200*, 1–25 (1978)
Arnon, D.I., Chain, R.K.: Proc. Natl. Acad, Sci. USA *72*, 4961–4965 (1975)
Asada, K., Kanematsu, S., Takahashi, M., Kono, Y.: In: Iron and copper proteins. Yasunobo, K.T., Mower, H.F., Hayaishi, O. (eds), pp. 551–564. New York, London: Plenum Press 1976
Bearden, A.J., Malkin, R.: Q. Rev. Biophys. *7*, 131–177 (1975)
Belkin, S., Padan, E.: Arch. Microbiol. *116*, 109–111 (1978)
Benemann, J.R., Weare, N.M.: Science *184*, 174–195 (1974a)
Benemann, J.R., Weare, N.M.: Arch. Microbiol., *101*, 401–408 (1974b)
Bennett, R., Rigopoulos, N., Fuller, R.C.: Proc. Natl. Acad. Sci. USA *52*, 762–768 (1964)
Bishop, P.E., Gordon, J.K., Shah, V.K., Brill, W.J.: In: Genetic engineering for nitrogen fixation. Hollaender, A. et al. (eds.), pp. 67–76. New York, London: Plenum Press 1977
Böhme, H.: Eur. J. Biochem. *72*, 283–289 (1977)
Bone, D.H.: Arch. Mikrobiol. *80*, 242–251 (1971)
Bothe, H.: Ber. Dtsch. Bot. Ges. *83*, 421–432 (1970)
Bothe, H.: Proc. 2nd. Int. Congr. Phot. Res. Stresa, Italy. *3*, 2169–2178 (1972)
Bothe, H., Falkenberg, B.: Plant Sci. Lett. *1*, 151–156 (1973)
Bothe, H., Loos, E.: Arch. Mikrobiol. *86*, 241–254 (1972)
Bothe, H., Falkenberg, B., Nolteernsting, U.: Arch. Mikrobiol. *96*, 291–304 (1974)
Bothe, H., Tennigkeit, J., Eisbrenner, G., Yates, M.G.: Planta (Berl.) *133*, 237–242 (1977a)
Bothe, H., Tennigkeit, J., Eisbrenner, G.: Arch. Microbiol. *114*, 43–49 (1977b)
Bottomley, P.J., Stewart, W.D.P.: Arch. Microbiol. *108*, 249–258 (1976)
Bottomley, P.J., Stewart, W.D.P.: New Phytol. *79*, 625–638 (1977)
Burns, R.C., Bulen, W.A.: Arch. Biochem. Biophys. *113*, 461–463 (1966)
Burns, R.C., Hardy, R.W.F.: Nitrogen fixation in bacteria and higher plants. p. 189. Berlin, Heidelberg, New York: Springer 1975
Cammack, R., Tel-Or, E., Stewart, W.D.P.: FEBS Lett. *70*, 241–244 (1976)
Cheniae, G.M.: Annu. Rev. Plant Physiol. *21*, 467–498 (1970)
Codd, G.A., Stewart, W.D.P.: Planta (Berl.) *130*, 323–326 (1976)
Codd, G.A., Stewart, W.D.P.: Arch. Microbiol. *113*, 105–110 (1977a)
Codd, G.A., Stewart, W.D.P.: FEMS Lett *2*, 247–249 (1977b)
Codd, G.A., Rowell, P., Stewart, W.D.P.: Biochem. Biophys. Res. Commun. *61*, 424–431 (1974)
Cohen, Y., Jorgensen, B.B., Padan, E., Shilo, M.: Nature (London) *257*, 489–492 (1975a)
Cohen, Y., Padan, E., Shilo, M.: J. Bacteriol. *123*, 855–861 (1975b)
Cox, R.M.: Arch. Mikrobiol. *53*, 263–276 (1966)
Cox, R.M.: Fay, P.: Proc. R. Soc. Lond. B. *172*, 357–366 (1969)
Dharmawardene, M.W.N., Haystead, A., Stewart, W.D.P.: Arch. Mikrobiol. *90*, 281–295 (1973)
Donze, M., Haveman, J., Schiereck, P.: Biochim. Biophys. Acta *256*, 157–161 (1972)
Drewes, K.: Zentralb. Bakteriol. Parasitenkd. (Abt II) *76*, 88–101 (1928)
Duggan, J.X., Anderson, L.E.: Planta (Berl.) *122*, 293–297 (1975)
Evans, E.H., Cammack, R., Evans, M.C.W.: Biochem. Biophys. Res. Commun. *68*, 1212–1218 (1976)
Evans, H.J., Ruiz-Argüeso, T., Jennings, N., Hanus, J.: In: Genetic engineering for nitrogen fixation. Hollaender, A. et al. (eds.), pp. 333–354. New York, London: Plenum Press 1977
Evans, M.C.W., Telfer, A., Lord, A.V.: Biochim. Biophys. Acta *267*, 530–537 (1972)

Evans, M.C.W., Telfer, A., Smith, R.V.: Biochim. Biophys. Acta *310*, 344–352 (1973)
Evans, M.C.W., Sihra, C.K., Cammack, R.: Biochem. J. *158*, 71–77 (1976)
Fay, P.: Biochim. Biophys. Acta *216*, 353–356 (1970)
Fay, P., Stewart, W.D.P., Walsby, A.E., Fogg, G.E.: Nature (London) *220*, 810–812 (1968)
Fleming, H., Haselkorn, R.: Proc. Natl. Acad. Sci. USA *70*, 2727–2731 (1973)
Fleming, H., Haselkorn, R.: Cell *3*, 159–170 (1974)
Fogg, G.E., Stewart, W.D.P., Fay, P., Walsby, A.E.: The blue-green algae. p. 459. London, New York: Academic Press 1973
Gallon, J.R., Kurz, W.G.W., LaRue, T.A.: In: Nitrogen fixation by free-living micro-organisms. Stewart, W.D.P. (ed.), I.B.P. Vol. VI, pp. 159–174. Cambridge: Cambridge Univ. Press 1975
Garlick, S., Oren, A., Padan, E.: J. Bacteriol. *129*, 623–629 (1977)
Gest, H., Kamen, M.D.: Science *109*, 558–559 (1949)
Grossman, A., McGowan, R.E.: Plant Physiol. *55*, 658–662 (1975)
Gruzinsky, I.V., Gogotov, I.N.: In: Abstr. 2nd. Int. Symp. Microb. Growth C$_1$-Compounds, pp. 143–144. Puschino: U.S.S.R. Acad. Sci. 1977
Hall, D.O., Cammack, R., Rao, K.K., Evans, M.C.W., Mullinger, R.: Biochem. Soc. Trans. *3*, 361–368 (1975)
Hardy, R.W.F.: In: Genetic engineering for nitrogen fixation. Hollaender, A. et al. (eds.), pp. 369–397. New York, London: Plenum Press 1977
Haystead, A., Stewart, W.D.P.: Arch. Mikrobiol. *82*, 325–336 (1972)
Helinski, D.R.: In: Genetic engineering for nitrogen fixation. Hollaender, A. et al. (eds.), pp. 19–50. New York, London: Plenum Press 1977
Herdman, M., Stanier, R.Y.: FEMS Lett. *1*, 7–12 (1977)
Hinze, G.: Ber. Dtsch. Bot. Ges. *21*, 394–398 (1903)
Hiyama, T., McSwain, B.D., Arnon, D.I.: Biochim. Biophys. Acta *460*, 76–84 (1977)
Hollaender, A., Burns, R.H., Day, P.R., Hardy, R.W.F., Helinski, D.R., Lamborg, M.R., Owens, L., Valentine, R.C.: Genetic engineering for nitrogen fixation. p. 538. New York, London: Plenum Press 1977
Hutson, K.G., Rogers, L.: Biochem. Soc. Trans. *3*, 277–279 (1975)
Jones, L.W., Bishop, N.I.: Plant Physiol. *257*, 659–665 (1976)
Juttner, F., Carr, N.G.: In: Proc. 2nd. Int. Symp. Photos. Prokaryotes, pp. 121–123. Dundee 1976
Ke, B.: Biochim. Biophys. Acta *301*, 1–33 (1973)
Ke, B., Hansen, R.E., Beinert, H.: Proc. Natl. Acad. Sci. USA *70*, 2941–2945 (1974)
Kenyon, C.N., Rippka, R., Stanier, R.Y.: Arch. Mikrobiol. *83*, 216–236 (1972)
Knobloch, K.: Planta (Berl.) *70*, 73–86 (1966)
Kosiak, A.V.: Abstr. 2nd. Int. Symp. Microb. Growth C$_1$-Compounds, pp. 135–136. Puschino: U.S.S.R. Acad. Sci. 1977
Lawrie, A.C., Codd, G.A., Stewart, W.D.P.: Arch. Microbiol. *107*, 15–24 (1976)
Lea, P.J., Miflin, B.J.: Biochem. Soc. Trans. *3*, 381–384 (1975)
Lex, M., Carr, N.G.: Arch. Microbiol. *101*, 161–167 (1974)
Lex, M., Stewart, W.D.P.: Biochim. Biophys. Acta *292*, 436–443 (1973)
Lex, M., Silvester, W.B., Stewart, W.D.P.: Proc. R. Soc. Lond. B. *180*, 87–102 (1972)
Lockau, W., Peterson, R.B., Wolk, C.P., Burris, R.H.: Biochim. Biophys. Acta *502*, 298–308 (1978)
Lovenberg, W.: In: Microbial iron metabolism. Neilands, J.B. (ed.), pp. 161–186. New York, London: Academic Press 1974
Ludden, P.W., Burris, R.H.: Science *194*, 424–426 (1976)
Lyne, R., Stewart, W.D.P.: Planta (Berl.) *109*, 27–38 (1973)
McFadden, B.A.: Biochem. Biophys. Res. Commun. *60*, 312–317 (1974)
McIntosh, A.R., Bolton, J.R.: Biochim. Biophys. Acta *430*, 555–559 (1976)
McIntosh, A.R., Chu, M., Bolton, J.R.: Biochim. Biophys. Acta *376*, 308–314 (1975)
Millbank, J.W.: In: The biology of nitrogen fixation. Quispel, A. (ed.), pp. 239–264. Amsterdam: North Holland Publishing Company 1974
Mortenson, L.E., Nakos, G.: In: Iron-sulfur proteins. Lovenberg, W. (ed.), Vol. I, pp. 37–64. New York, London: Academic Press 1973
Munson, T.O., Burris, R.H.: J. Bacteriol. *97*, 1093–1098 (1969)

Murai, T., Katoh, T.: Plant Cell Physiol. *16*, 789–797 (1975)

Nakamura, H.: Bot. Mag. Tokyo *51*, 529–533 (1937)

Nakamura, H.: Acta Phytochim. Jpn. *10*, 271–281 (1938)

Neilson, A., Rippka, R., Kunisawa, R.: Arch. Mikrobiol. *76*, 139–150 (1971)

Nutman, P.S.: Symbiotic nitrogen fixation in plants. I.B.P. Vol. VII, p. 548. Cambridge: Cambridge Univ. Press 1976

Orme-Johnson, W.H.: In: Genetic engineering for nitrogen fixation. Hollaender, A. et al. (eds.), pp. 317–332. New York, London: Plenum Press 1977

Orme-Johnson, W.H., Davis, L.C.: In: Iron-sulphur proteins. Lovenberg, W. (ed.), Vol. 3. Ch. 2. New York: Academic Press 1978 (in press)

Packer, L., Tel-Or, E., Luijk, L.M.: Fed. Proc. *36*, 81 (1977)

Pate, J.S.: In: A treatise on dinitrogen fixation. Section III Biology. Hardy, R.W.F., Silver, W.S. (eds.), pp. 473–518. New York: John Wiley and Sons Inc. 1977

Pelroy, R.A., Bassham, J.A.: J. Bacteriol. *115*, 937–942 (1973)

Pelroy, R.A., Rippka, R., Stanier, R.Y.: Arch. Mikrobiol. *87*, 303–322 (1972)

Peters, G.A., Evans, W.R., Toia, Jr., R.E.: Plant Physiol. *58*, 119–126 (1976)

Peterson, R.B., Burris, R.H.: Arch. Microbiol. *108*, 35–40 (1976)

Peterson, R.B., Burris, R.H.: Arch. Microbiol. *116*, 125–132 (1978)

Postgate, J.R.: J. Appl. Bacteriol. *37*, 185–202 (1974)

Prusiner, S., Stadtman, E.R. (eds.): The enzymes of glutamine metabolism. p. 615. New York: Academic Press 1973

Quispel, A. (ed.): The biology of nitrogen fixation. p. 769. Amsterdam: North Holland Publishing Company 1974

Rippka, R.: Arch. Mikrobiol. *87*, 93–98 (1972)

Rippka, R., Stanier, R.Y.: J. Gen Microbiol. *105*, 83–94 (1978)

Rippka, R., Waterbury, J.B.: FEMS Lett. *2*, 83–86 (1977)

Rowell, P., Enticott, S., Stewart, W.D.P.: New Phytol. *79*, 41–54 (1977)

Rowell, P., Sampaio, M.J.A.M., Stewart, W.D.P.: Proc. Soc. gen. Microbiol. *5*, 103–104 (1978)

Ryan, F.J., Jolly, S.O., Tolbert, N.E.: Biochem. Biophys. Res. Commun. *59*, 1233–1241 (1974)

Sanchez, J.J., Palleroni, N.J., Doudoroff, M.: Arch. Microbiol. *104*, 57–65 (1975)

Schick, H.J.: Arch. Mikrobiol. *75*, 89–101 (1971)

Scott, W.E., Fay, P.: Br. Phycol. J. *7*, 283–284 (1972)

Setlik, I.: Biochim. Biophys. Acta *24*, 436–437 (1957)

Shanmugam, K.T., Buchanan, B.B., Arnon, D.I.: Biochim. Biophys. Acta *256*, 477–486 (1972)

Shanmugam, K.T., Streicher, S.L., Morandi, C., Ausubel, F., Goldberg, R.B., Valentine, R.C.: In: Proc. 1st Int. Symp. N_2 fixation. pp. 313–326. Washington: Pullman 1976

Shin, M., Sukenobu, M., Reiko, O., Kitazume, Y.: Biochim. Biophys. Acta *460*, 85–93 (1977)

Siefert, E.: Dissertation zur Erlangung des Doktorgrades der Mathematisch-Naturwissenschaftlichen Fakultät der Georg-August-Universität zu Göttingen, Göttingen p. 149. 1976

Smith, R.V., Evans, M.C.W.: J. Bacteriol. *105*, 913–917 (1971)

Smith, R.V., Noy, R.J., Evans, M.C.W.: Biochim. Biophys. Acta. *253*, 104–109 (1971a)

Smith, R.V., Telfer, A., Evans, M.C.W.: J. Bacteriol. *107*, 574–575 (1971b)

Stanier, R.Y.: In: The biology of blue-green algae. Carr, N.G., Whitton, B.A. (eds.), pp. 501–578. Oxford: Blackwell Scientific Publications 1973

Stewart, W.D.P.: In: Biological nitrogen fixation in natural and agricultural habitats. Lie, T.A., Mulder, E.G. (eds.), Plant Soil Spec. Vol., pp. 377–391, 1971

Stewart, W.D.P.: In: The biology of nitrogen fixation. Quispel, A. (ed.), pp. 696–718. Amsterdam: North Holland Publishing Company 1974

Stewart, W.D.P. (ed.) Nitrogen fixation by free-living micro-organisms. I.B.P. Vol. VI, p. 471. Cambridge: Cambridge Univ. Press 1975

Stewart, W.D.P., Bottomley, P.J.: Proc. 1st Int. Symp. N_2 fixation. pp. 257–273. Washington: Pullman 1976

Stewart, W.D.P., Codd, G.A.: Br. Phycol. J. *10*, 273–278 (1975)

Stewart, W.D.P., Lex, M.: Arch. Mikrobiol. *73*, 250–260 (1970)

Stewart, W.D.P., Pearson, H.W.: Proc. R. Soc. Lond. B. *175*, 293–311 (1970)

Stewart, W.D.P., Rowell, P.: Biochem. Biophys. Res. Commun. *65*, 846–856 (1975)

Stewart, W.D.P., Haystead, A., Pearson, H.W.: Nature (London) *224*, 226–228 (1969)

Stewart, W.D.P., Haystead, A., Dharmawardene, M.W.N.: In: Nitrogen fixation by free-living micro-organisms, I.B.P., Vol. VI, Stewart, W.D.P. (ed.), pp. 129–158. Cambridge: Cambridge Univ. Press 1975

Stewart, W.D.P., Rowell, P., Apte, S.K.: Proc. 2nd Int. Symp. N$_2$ fixation, pp. 287–307. London-New York: Academic Press, 1977

Stewart, W.D.P., Rowell, P., Lockhart, C.M.: In: Nitrogen assimilation of plants. 6th Long Ashton Symp. Bristol 1977. (in press 1978a)

Stewart, W.D.P., Rowell, P., Codd, G.A., Apte, S.K.: Proc. 4th Int. Congr. Photosynthesis. Hall, D.O., Coombs, J., Goodwin, T.W. (eds.), pp. 133–146. London: The Biochemical Society 1978b

Streicher, S.L., Gurney, E.G., Valentine, R.C.: Nature (London) 239, 495–499 (1972)

Streicher, S.L., Shanmugam, K.T., Ausubel, F., Morandi, C., Goldberg, R.B.: J. Bacteriol. 120, 815–821 (1974)

Takabe, T., Akazawa, T.: Biochem. Biophys. Res. Commun. 53, 1178–1180 (1973)

Tel-Or, E., Stewart, W.D.P.: Nature (London) 258, 715–716 (1975)

Tel-Or, E., Stewart, W.D.P.: Biochim. Biophys. Acta 423, 189–195 (1976)

Tel-Or, E., Stewart, W.D.P.: Proc. R. Soc. Lond. B. 198, 61–86 (1977)

Tel-Or, E., Luijk, L., Packer, L.: In: Abstr. 2nd Int. Symp. Microb. Growth C$_1$-Compounds. pp. 147–150. Puschino: U.S.S.R. Acad. Sci. 1977a

Tel-Or, E., Luijk, L., Packer, L.: FEBS Lett. 78, 49–52 (1977b)

Thomas, J., Nature (London) 228, 181–183 (1970)

Thomas, J., Wolk, C.P., Shaffer, P.W., Austin, S.M., Galonsky, A.: Biochem. Biophys. Res. Commun. 67, 501–507 (1975)

Thorneley, R.N.F., Eady, R.R., Smith, B.E., Lowe, D.J., Yates, M.G., O'Donnell, M.J., Postgate, J.R.: In: Nitrogen assimilation of plants. 6th Long Ashton Symp. Bristol 1977 in press 1978

Tsai, L.B., Mortenson, L.E.: Biochem. Biophys. Res. Commun. 81, 280–287 (1978)

Tubb, R.S.: Nature (London) 251, 481–485 (1974)

Utkilen, H.C.: J. Gen. Microbiol. 95, 177–180 (1976)

Van Gorkom, H.J., Donze, M.: Nature (London) 234, 231–232 (1971)

Weare, N.M., Benemann, J.R.: Arch. Mikrobiol. 90, 323–332 (1973)

Weare, N.M., Benemann, J.R.: J. Bacteriol. 119, 258–265 (1974)

Wilson, P.W.: The biochemistry of symbiotic nitrogen fixation. Madison: Univ. of Wisconsin Press 1940

Winkenbach, F., Wolk, C.P.: Plant Physiol. 52, 480–483 (1973)

Winter, H.C., Arnon, D.I.: Biochim. Biophys. Acta 197, 170–179 (1970)

Winter, H.C., Burris, R.H.: Annu. Rev. Biochem. 45, 409–426 (1976)

Wolk, C.P., Thomas, J., Shaffer, P.W., Austin, S.M., Galonsky, A.: J. Biol. Chem. 251, 5027–5034 (1976)

Wolk, C.P., Wojciuch, E.: Planta (Berl.) 97, 126–134 (1971)

Wyatt, J.T., Silvey, J.K.G.: Science 165, 908–909 (1969)

Yoch, D.C., Arnon, D.I.: In: The biology of nitrogen fixation. Quispel, A. (ed.), pp. 687–695. Amsterdam: North Holland Publishing Company 1974

Yoch, D.C., Valentine, R.C.: Annu. Rev. Microbiol. 26, 139–162 (1972)

7. Symbiotic N_2 Fixation and Its Relationship to Photosynthetic Carbon Fixation in Higher Plants

B. QUEBEDEAUX

A. Introduction

Photosynthetic carbon reduction is a major process regulating symbiotic N_2 fixation. Photosynthetic carbon fixation provides the basic energy requirements for nitrogenase, the central N_2-fixing enzyme which couples ATP hydrolysis (to ADP and Pi) with electron transfer from a reductant, ferredoxin and/or flavodoxin, to molecular nitrogen which is reduced to ammonia. The ammonia, in turn, is converted to amino acids in a series of enzymatic reactions utilizing photosynthate from carbon fixation as the source of carbon skeletons. Sucrose, the major form in which carbon is translocated from the leaves to the root nodules in legumes, is further metabolized to hexose units and short-chain respiratory substrates and transferred to the bacteroid, the site of nitrogenase synthesis and activity. Photosynthetic CO_2 fixation rates regulate the supply of carbon assimilates to the bacteroids of plant root nodules.

In legumes, photosynthate supply is established as a major limitation for enhancing N_2 fixation (HARDY and HAVELKA, 1976; QUEBEDEAUX et al., 1975) and it is clear that in order to achieve a major increase in N_2 fixation in higher plants attention must be focused on photosynthetic efficiency and improvement of photosynthate availability. Most N_2-fixing symbioses in higher plants are with plants which are C_3 and photosynthetically inefficient. Major improvement in N_2 fixation could be achieved by extending the process to photosynthetically efficient (C_4) plants. Therefore, enhancing nitrogen fixation becomes a problem in carbon assimilation and partitioning of newly fixed carbon through the process of translocation. The N_2 fixation process requires a large energy supply in the form of ATP and reducing power as reduced ferredoxin and/or flavodoxin. Photosynthate provides this energy and supports root and nodule growth. Thus, as emphasized here, the photosynthetic carbon cycle is a significant part of the general mechanism of symbiotic N_2 fixation in higher plants. Several important features of the biochemistry and physiology of the photosynthetic carbon cycle are well covered in recent reviews (CHOLLET and OGREN, 1975; ZELITCH, 1975; BURRIS and BLACK, 1976; KELLY et al., 1976) and here I will attempt only to integrate knowledge of both fundamental processes. Results and problems of N_2 fixation will be resumed in Volume 12 (*Inorganic Plant Nutrition*) of this Encyclopedia.

B. Relationship of N_2 Fixation to Carbon Assimilation

I. Nitrogenase

Nitrogenase, the central enzyme that catalyzes the conversion of molecular nitrogen to ammonia is found only in N_2-fixing organisms. Nitrogenase preparations from all N_2-fixing organisms are surprisingly similar. The partially purified nitrogenase from *Rhizobium* bacteroids of soybean root nodules (WHITING and DIL-WORTH, 1974) is in no way different from the nitrogenase of free-living organisms or of other nitrogen-fixing organisms from which it has been isolated. Nitrogenase is an enzyme complex composed of two protein components containing molybdenum (Mo), iron (Fe), and inorganic sulfide. The large protein component, called the Mo-Fe protein, contains four subunits composed of two molybdenums and 24–32 iron and sulfide atoms with a molecular weight of 220,000. The small protein component, called the Fe protein, contains two identical subunits composed of four Fe and sulfur atoms with a molecular weight of 60,000. The nitrogenase protein is very labile; this has caused difficulty in obtaining highly purified preparations because of its O_2- and cold-sensitivity. All nitrogenases, both in vivo and in vitro, are extremely O_2-sensitive (BERGERSON, 1971) and their activity is quickly destroyed if exposed to ambient or higher concentrations of oxygen. Legume nodules are structurally organized to enable facilitated oxygen diffusion to the nitrogen-fixing bacteroids. According to this concept the nodules contain leghemoglobin, a red heme pigment that binds gaseous oxygen with high affinity, facilitating high O_2 diffusion to mitochondria and bacteroids for ATP production via oxidative phosphorylation (BERGERSON, 1971; TJEPKEMA and YOCUM, 1973; WITTENBURG et al., 1975). This physiological protective mechanism provides a high respiratory rate for high ATP levels and for the removal of O_2 continuously from the N_2-fixing enzyme site. This system is extremely complex and soybean nodules have developed a capability to adapt to a wide variety of external pO_2 by an as yet unidentified mechanism (CRISWELL et al., 1976, 1977).

Nitrogenase is a versatile enzyme and can reduce a variety of substrates in addition to N_2. It reduces $2H^+$ to H_2, and recent studies by SCHUBERT and EVANS (1976) suggest that the reduction of $2H^+$ to H_2 occurs in situ in soybean nodules and may be a competitive inhibitor of N_2 reduction reducing the energy efficiency of the nitrogenase reaction. In addition nitrogenase may reduce a wide variety of artificial substrates of triple-bonded nitrogen-carbon and oxygen-containing compounds (HARDY et al., 1971) with a number of these substrates serving as the basis of the nitrogenase assay. The substrate, acetylene, has become established as the most widely used and accepted assay for nitrogenase activity. The reaction is the reduction of acetylene to ethylene and the determination of ethylene through gas chromatography coupled with flame ionization. This nitrogenase assay is used extensively and techniques utilizing excised nodules or noduled roots (HARDY et al., 1968) as well as intact plants (QUEBEDEAUX et al., 1975) have been described.

Nitrogenase, photosynthetic, and photorespiratory activities are all extremely oxygen-sensitive. Low O_2 concentration and high CO_2 favor carboxylation and photosynthesis, while high O_2 and low CO_2 levels favor oxygenation, glycolate synthesis, and photorespiration. Photosynthesis is limited by the high external atmospheric O_2 concentration (21%) and the very low CO_2 atmospheric concentration (0.03%) for photosynthetic carbon fixation that is inadequate to saturate RuBP, while the nitrogen concentration is high (80%) and able to completely saturate nitrogenase which has K_m for N_2 of 0.056 atmosphere in soybean nodule bacteroids (KOCK et al., 1967). The oxygen interactions and the light-dependent release of CO_2 with glycolate biosynthesis during carbon fixation is controversial, and several mechanisms and pathways have been proposed. One is the synthesis of glycolate from "active glycolaldehyde" with the generation of an oxidant, presumably H_2O_2 from the oxidation of reduced ferredoxin (SHAIN and GIBBS, 1971; see Chap. II.25 by BECK). Another is based on the oxygenase reaction with P-glycolate produced in an oxygen-dependent reaction catalyzed by RuBP carboxylase. Oxygen competes with CO_2 for RuBP at the carboxylation site (BOWES et al., 1971). In addition, other investigators have suggested multiple pathways for glycolate synthesis during photorespiration (EICHENBUSCH et al., 1975; ZELITCH, 1975). Still another mechanism involves O_2 uptake not coupled to carbon chemistry in which O_2 and CO_2 compete for photosynthetically generated reducing power, with O_2 being the main electron acceptor during the induction process and under conditions in which CO_2 reduction cannot keep pace with O_2 evolution (RADMER and KOK, 1976). In spite of the different proposed mechanism described for photosynthesis and photorespiration, there is general agreement that an attractive approach for increasing nitrogenase activity and symbiotic N_2 fixation is decreasing photorespiration and increasing photosynthesis.

II. ATP and Reductant

Nitrogenase activity requires a large energy supply in the form of ATP and reducing power as ferredoxin and/or flavodoxin. ATP concentration and the ratio of ATP to ADP regulates enzyme activity. Soybean nodule ATP concentration is about 0.5 mM with a K_m of nitrogenase for ATP of 0.1–0.3 mM (CHING et al., 1975). Lowering of the nodule ATP concentration by a 24-h dark treatment gave an immediate decline in N_2 fixation and demonstrates the high ATP requirement and its significant energetic role in maintaining stable N_2 fixation rates. The calculated energy for reducing N_2 to 2 NH_3 is reported to be approximately 15 molecules of ATP and 6 electrons, equivalent to a total of 24 ATP's for each molecule of N_2 fixed (BURNS and HARDY, 1975). These calculations assume 9 ATP molecules are used for three pairs of electrons for reductant and 15 ATP molecules for the nitrogenase reaction with 4 ATP molecules for each electron pair and an 80% efficiency in coupling electrons to nitrogen.

Reduced ferredoxins and flavodoxins are the key physiological electron donors, as well as the compatible electron transfer agents which interact with but are not part of the nitrogenase enzyme complex (Fig. 1). These electron-transferring proteins used by nitrogenase are supplied directly by photoreduction or by dark reactions associated with photosynthesis. Ferredoxin was first discovered in the nitrogen-fixing organism *Clostridium pasteurianum* (MORTENSON et al., 1962) and has now been isolated and purified from a large number of photosynthetic organisms, bacteria, algae, and higher plants. In contrast to nitrogenase the isolation of purified ferredoxin and flavodoxin indicates that many properties including molecular weight, their amino acid, iron and sulfide content, redox properties and biological activity varies with the nitrogen-fixing organism. Rhizobium ferredoxin, isolated from bacteroids of soybean root nodules, is difficult to purify because of its extreme oxygen sensitivity. The ferredoxins and flavodoxins have many functions in bacterial and plant metabolism in addition to their specific role in the oxidation-reduction processes of both photosynthesis and N_2 fixation.

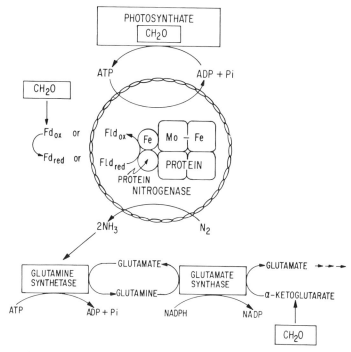

Fig. 1. Schematic of ammonia assimilation during symbiotic N_2 fixation and relationship to photosynthate, nitrogenase, ATP, ferredoxin and flavodoxin

III. Ammonia Assimilation

Ammonia is the key intermediate and the first stable product of N_2 fixation. Further enzymatic conversion of ammonia to amino acids utilizes photosynthate as a source of carbon skeletons. Results of $^{15}N_2$ tracer studies on root nodules (KENNEDY, 1966; ZELITCH et al., 1951) suggest that glutamine is one of the initial products of ammonia assimilation; further synthesis to other amino acids occurs through a series of reactions regulated by plant enzymes rather than the bacteroids (ROBERTSON et al., 1975; SCOTT et al., 1976). Glutamine synthetase and glutamate synthase are the key enzyme suggested for ammonia incorporation (DILWORTH, 1974; NAGATANI et al., 1971; THOMAS et al., 1975). Glutamine synthetase forms glutamine from ammonia, ATP and glutamate and subsequently glutamate synthase catalyzes the formation of two molecules of glutamate from glutamine, α-ketoglutarate and NADPH (Fig. 1). It is essential that the newly fixed NH_3-assimilates be metabolized and transported to utilization sites in plant vegetative and reproductive structures. The ammonia-assimilating enzymes are not part of the nitrogenase reaction, and failure to remove ammonia may inhibit or repress nitrogenase synthesis (TUBB and POSTGATE, 1973; O'GARA and SHANMUGAM, 1976). The high affinity of glutamine synthetase for its substrate ammonia and its relative low K_m of 0.2 mM (DENTON et al., 1970; MILLER and STADTMAN, 1972) is favorable for the maintenance of low cellular ammonia concentration, and allows for rapid transport from the nitrogenase complex. The main nitrogen compound exported from the xylem is usually species-specific with asparagine, glutamine, aspartic acid, and glutamic acid as major components in legumes (Pate, 1976). The exact assimilation pathway for each reaction of each component is unclear.

Recent ^{15}N studies with isolated spinach leaf cells suggests that, during photorespiration, C_3 plants generate NH_3 from glycine and serine in leaf mitochondria at rates equivalent to photorespiratory CO_2 production (Woo et al., 1977). It appears that this flux of free ammonia could result in losses of NH_3 and/or uncoupling of electron transport. The NH_3 generated during photosynthesis under conditions which permit photorespiration is fixed by the glutamate synthase system of the chloroplasts, and its direct influence on the nitrogenase system has not yet been established.

The addition of fertilizer nitrogen either as root or foliar applications (Hardy, 1973) reduces ammonia incorporation during N_2 fixation with greater depression effects from fertilizer nitrogen occurring as ammonia than with other nitrogen sources. High ammonia levels can inhibit nitrogenase activity completely by altering its synthesis, or by competition between nitrogenase and ammonia-assimilating enzymes for ATP and reductant.

IV. Photosynthate as the Limiting Factor

Recent pCO_2 and pO_2 studies with soybeans have identified photosynthate availability to the nodule as a major limitation of symbiotic N_2 fixation (Hardy and Havelka, 1976; Quebedeaux et al., 1975). An elevated CO_2/O_2 ratio decreases N_2 fixation. For example, N_2 fixation is increased 125% by a canopy exposure to 5% O_2, whereas, 30% O_2 decreases N_2 fixation by 50% (Quebedeaux et al., 1975). CO_2 enrichment of field-grown soybean increases N_2 fixation fivefold from 75 to 475 kg/ha (Hardy and Havelka, 1976). This multifold increase in N_2 fixation is due to an increase in specific N_2-fixing activity of the nodules, an increase in nodule mass, and a delay in the loss of the exponential phase due to a delay in plant senescence. The direct relationship of pO_2 and CO_2-enrichment to N_2 fixation is also attributed to the effect of the CO_2/O_2 ratio on CO_2 fixation in C_3 plants and to a decrease in photorespiration. Plant growth-regulating chemicals that decrease photorespiration or plant selections with decreased photorespiration should lead to increased N_2 fixation in C_3 legumes. According to this concept, decreasing photorespiration or increasing photosynthesis appears to be an attractive approach for increasing photosynthate availability for symbiotic N_2 fixation.

Recently, naturally occurring rhizosphere associations with free-living N_2-fixing bacteria have been identified in C_4 species, e.g., *Spirillum lipoferum-Digitaria decumbens* (Dobereiner and Day, 1975), *Spirillum-Zea mays* (Dobereiner et al., 1972), *Azotobacter-Paspalum* (Dobereiner, 1966). Some of these associations are extracellular while others appear intracellular. The amount of nitrogen fixed by these associations is suggested to be high, based on measurements following pre-incubation in low O_2, however, in situ measurements over a longer period of time or growth cycle are not yet available. Enhancing the activity of these rhizosphere associations may require improved photosynthate distribution at the site of N_2

fixation and selecting nitrogen-fixing symbionts that efficiently utilize the energy provided by photosynthesis. These associations suggest that extending the symbiotic process to photosynthetically efficient C_4 plants is an attractive approach and could result in a major enhancement of N_2 fixation.

A variety of physiological factors that influence photosynthate production or availability to the nodule produce similar effects on N_2 fixation. Factors such as increased light intensity (HARDY et al., 1968; LAWN and BRUN, 1974; SLOGER et al., 1975), increased source size by grafting additional foliage (STREETER, 1974), low planting density (HARDY and HAVELKA, 1976), decreased demand of competitive sinks by pod removal (LAWN and BRUN, 1974) all increase N_2 fixation. Factors such as decreased light intensity from shading and darkness (HARDY et al., 1968; LAWN and BRUN, 1974; SLOGER et al., 1975; VIRTANEN et al., 1955), decreased source size by defoliation (HARDY et al., 1968); high plant density and lodging (HARDY and HAVELKA, 1976), cessation of translocation to nodule by girdling and increased demand of competitive reproductive sinks (LAWN and BRUN, 1974; WHEELER, 1971) and water stress (SPRENT, 1972), all decrease the amount of photosynthate to the nodule with a decrease in N_2 fixation.

V. Translocation and Partitioning of Nitrogen and Carbon Assimilates

Translocation involving a source–sink relationship provides a high carbon and nitrogen transfer rate in symbiotic nitrogen-fixing systems. N_2 fixation, and translocation of photosynthate from leaves to nodules is generally higher during the day than at night (MINCHIN and PATE, 1973, 1974). Root nodules contain and store pools of starch and sugars to supplement supplies of photosynthate during hours of darkness or periods of unfavorable conditions. In soybeans, the lower leaves contribute the major source of photosynthate for the nodules. Efficient partitioning of both newly fixed carbon, as well as nitrogen assimilates, is necessary to maintain high rates of N_2 fixation. Little is known about factors that control sink activity of nodules and translocation involving import and export of carbon and nitrogen assimilates in nodules. Recent studies by QUEBEDEAUX and HARDY (1973, 1974, 1975, 1976) and QUEBEDEAUX et al. (1975) with intact C_3 and C_4 plants suggest a difference between vegetative and reproductive sinks in response to pO_2 with sink intensity of reproductive sinks regulated by pO_2. Alterations in nodule sink activity have not been explored and deserve further consideration.

Sucrose is the major form in which carbon is translocated from leaves to root nodules. Evidence for this observation is supported by results of $^{14}CO_2$ feeding studies to the leaf (BACH et al., 1958) and phloem exudate analysis (PATE, 1975). A source–sink transport process as shown diagramatically in Figure 2 is involved with the transfer and unloading of sucrose to nodules from transfer cells (PATE and GUNNING, 1972) of the plant vascular system to the bacteroids. The bacteroids do not appear directly to utilize sucrose, which must first be hydrolyzed to hexose units by plant invertases and further metabolized to short chain substrates in an aerobic respiration process. The transport process and the biochemical reactions of sucrose unloading during N_2 fixation in higher plants has not yet been established and deserves further consideration.

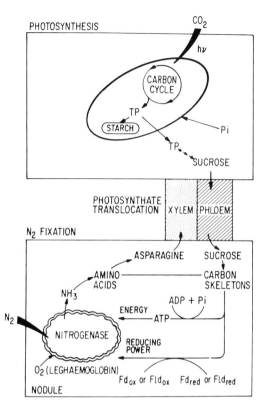

Fig. 2. Schematic of photosynthesis and symbiotic N_2 fixation in nodules and relationship to photosynthate, nitrogenase, ATP and reductant, ammonia assimilation, nitrogen and carbon transport

The export of nitrogen compounds from root nodules during N_2 fixation is an equally important process involving translocation. Although ammonia is considered the first stable product of N_2 fixation, ^{15}N tracer studies with legumes (OGHOGHORIE and PATE, 1972) indicate that ammonia is not transported but rather enzymatically converted to amino acids with export to the shoot xylem as amide nitrogen. The nitrogen compounds exported in the xylem are usually species-specific, with asparagine and glutamine as major components in soybeans. PATE and GUNNING (1972) have described the transfer cells of the nodule vascular bundles from which the amino compounds at concentrations several times that of the bacteroid are secreted and exported to the intracellular space of the bundle with water moving osmotically across the bundle endodermis and movement of nitrogen compounds out through the xylem. The export of fixed N compounds is reported to be three times higher during the day than at night (PATE, 1976) and may account for reduced N_2 fixation rates during darkness. It is essential that the newly fixed nitrogen assimilates be rapidly exported to sites of utilization in vegetative and reproductive structures, since accumulation of nitrogen assimilates may inhibit or repress nitrogenase (TUBB and POSTGATE, 1973; GORDON and BRILL, 1974; SHANMUGAM and MORANDI, 1976).

A relatively high amount of photosynthate translocated to the nodule is required for symbiotic N_2 fixation. Recent measurements by MINCHIN and PATE (1974) show that for each 100 units of carbon fixed photosynthetically during a vegetative growth stage, 32 units are translocated to the nodule, of which 5 units are consumed in nodule growth, 12 units lost in respiration and 15 units utilized for N$_2$ fixation and exported back to the shoot as amides and amino acids. These values estimate that a ratio of approximately 4:1 is the minimum amount of photosynthate consumed to the amount of nitrogen fixed. The respiratory efficiency for nitrogen fixation, however, is about the same as for the reduction of nitrate with experimental measurements of 5.9 mg carbon per mg N$_2$-N fixed versus 6.2 mg carbon per mg NO$_3$-N reduced. These values indicate that the plant must pay for N$_2$ fixation and there seems to be the general agreement that it is quite high compared to other biological processes. Therefore, enhancement of biological nitrogen fixation will require considering simultaneously photosynthesis and related biochemical processes regulating carbon metabolism.

References

Bach, M.K., Magee, W.E., Burris, R.H.: Plant Physiol. *33*, 118–124 (1958)
Bergerson, F.J.: Annu. Rev. Plant Physiol. *22*, 121–140 (1971)
Bowes, G., Ogren, W.L., Hageman, R.H.: Biochem. Biophys. Res. Commun. *45*, 716–722 (1971)
Burns, R.C., Hardy, R.W.F.: Nitrogen fixation in bacteria and higher plants. Berlin, Heidelberg, New York: Springer 1975
Burris, R.H., Black, C.C. (eds.): CO$_2$ metabolism and productivity of plants. Baltimore, London, Tokyo: Univ. Park Press 1976
Ching, T.M., Hedthey, S., Russell, S.A., Evans, H.J.: Plant Physiol. *55*, 796–798 (1975)
Chollet, R., Ogren, W.L.: Bot. Rev. *41*, 137–179 (1975)
Criswell, J.G., Havelka, U.D., Quebedeaux, B., Hardy, R.W.F.: Plant Physiol. *58*, 622–625 (1976)
Criswell, J.G., Havelka, U.D., Quebedeaux, B., Hardy, R.W.F.: Crop Sci. *17*, 39–44 (1977)
Denton, M.D., Ginsburg, A.: Biochem. *9*, 617–632 (1970)
Dilworth, M.J.: Annu. Rev. Plant Physiol. *25*, 81–114 (1974)
Dobereiner, J.: Pesq. Agropec. Bras. *1*, 357–365 (1966)
Dobereiner, J., Day, J.M.: In: Symp. Dinitrogen Fixation. Newton, W.E., Nyman, D.J. (eds.), Vol. II, pp. 518–538. Pullman: Wash. State Univ. Press 1975
Dobereiner, J., Day, J.M., Dart, P.J.: Plant Soil *37*, 191–196 (1972)
Eichenbusch, J.D., Scheibe, R., Beck, E.: Z. Pflanzenphysiol. *75*, 375–380 (1975)
Gordon, J.K., Brill, W.J.: Biochem. Biophys. Res. Commun. *59*, 967–971 (1974)
Hardy, R.W.F., Havelka, U.D.: In: Symbiotic nitrogen fixation in plants. Int. Biol. Programme Series. Nutman, P. (ed.), Vol. VII, pp. 421–439. London: Cambridge Univ. Press 1976
Hardy, R.W.F., Holsten, R.D., Jackson, E.K., Burns, R.C.: Planta Physiol. *43*, 1185–1207 (1968)
Hardy, R.W.F., Burns, R.C., Parshall, G.W.: Adv. Chem. Ser. *100*, 219–247 (1971)
Hardy, R.W.F., Burns, R.C., Holsten, R.D.: Soil Biol. Biochem. *5*, 47–81 (1973)
Kelly, G.J., Latzko, E., Gibbs, M.: Annu. Rev. Plant Physiol. *27*, 181–205 (1976)
Kennedy, I.R.: Biochim. Biophys. Acta *130*, 285–294 (1966)
Kock, B., Evans, H.J., Russell, S.: Proc. Natl. Acad. Sci. USA *58*, 1343–1350 (1967)
Lawn, R.J., Brun, W.A.: Crop Sci. *14*, 11–16 (1974a)
Lawn, R.J., Brun, W.A.: Crop Sci. *14*, 22–25 (1974b)
Miller, R.E., Stadtman, E.R.: J. Biol. Chem. *247*, 7407–7419 (1972)
Minchin, F.R., Pate, J.S.: J. Exp. Bot. *24*, 259–271 (1973)
Minchin, F.R., Pate, J.S.: J. Exp. Bot. *25*, 292–308 (1974)
Mortenson, L.E., Valentine, R.C., Carnahan, J.E.: Biochem. Biophys. Res. Commun. *7*, 448–52 (1962)

Nagatani, H., Shimizu, M., Valentine, R.C.: Arch. Mikrobiol. *79*, 164–175 (1971)
O'Gara, F., Shanmugam, K.T.: Biochim. Biophys. Acta *437*, 313–321 (1976)
Oghoghorie, C.G.O., Pate, J.S.: Planta (Berl.) *104*, 35–49 (1972)
Pate, J.S.: Encyclopedia of plant physiology. New Ser. Vol. I, pp. 451–473. Berlin, Heidelberg, New York: Springer 1972
Pate, J.S.: In: A treatise on dinitrogen fixation. Sect. III. Hardy, R.W.F., Silver, W.S. (eds.), pp. 473–517. New York: Wiley-Interscience 1976
Pate, J.S., Gunning, B.E.S.: Annu. Rev. Plant Physiol. *23*, 173–196 (1972)
Quebedeaux, B., Hardy, R.W.F.: Nature (London) *243*, 477–479 (1973)
Quebedeaux, B., Hardy, R.W.F.: Plant Physiol. *53*, S-62 (1974)
Quebedeaux, B., Hardy, R.W.F.: Plant Physiol. *55*, S. 17 (1975a)
Quebedeaux, B., Hardy, R.W.F.: Plant Physiol. *55*, 102–107 (1975b)
Quebedeaux, B., Hardy, R.W.F.: In: CO₂ metabolism and plant productivity. Burris, R.H., Black, C.C. (eds.), pp. 185–202. Baltimore, London, Tokyo: Univ. Park Press 1976
Quebedeaux, B., Havelka, U.D., Livak, K.L., Hardy, R.W.F.: Plant Physiol. *56*, 761–764 (1975)
Radmer, R.J., Kok, B.: Plant Physiol. *58*, 336–340 (1976)
Robertson, J.G., Farnden, K.J.F., Warburton, M.P., Banks, J.M.: Aust. J. Plant Physiol. *2*, 265–272 (1975)
Schubert, K.R., Evans, H.J.: Proc. Natl. Acad. Sci. USA *73*, 1207–1211 (1976)
Scott, D.B., Farnden, K.J.F., Robertson, J.G.: Nature (London) *263*, 703–705 (1976)
Shain, Y., Gibbs, M.: Plant Physiol. *48*, 325–330 (1971)
Shanmugam, K.T., Morandi, C.: Biochim. Biophys. Acta *437*, 322–332 (1976)
Sloger, C., Bezdicek, D., Milberg, R., Boonkerd, N.: In: Nitrogen fixation by free-living microorganisms. Stewart, W.D.P. (ed.), pp. 271–284. London: Cambridge Univ. Press 1975
Sprent, J.I.: New Phytol. *71*, 443 (1972)
Streeter, J.G.: J. Exp. Bot. *25*, 189–198 (1974)
Thomas, J., Wolk, C.P., Shaffer, P.W.: Biochem. Biophys. Res. Commun. *67*, 501–507 (1975)
Tjepkema, J.D., Yocum, C.S.: Planta (Berl.) *115*, 59–72 (1973)
Tubb, R.S., Postgate, J.R.: J. Gen. Microbiol. *79*, 103–117 (1973)
Virtanen, A.I., Moisio, T., Burris, R.H.: Acta Chem. Scand. *9*, 184–186 (1955)
Wheeler, C.T.: New Phytol. *70*, 487–495 (1971)
Whiting, M.J., Dilworth, M.J.: Biochim. Biophys. Acta *371*, 337–351 (1974)
Wittenberg, J.B., Appleby, C.A., Bergersen, F.J., Turner, G.L.: Ann. N. Y. Acad. Sci. *244*, 28–34 (1975)
Woo, K.C., Berry, J.A., Osmond, C.B., Lorimer, G.H.: Proc. Int. Congr. Photosynth.: Abstr. p. 415, 1977
Zelitch, I.: Annu. Rev. Biochem. *44*, 123–145 (1975)
Zelitch, I., Rosenblum, E.D., Burris, R.H., Wilson, P.W.: J. Biol. Chem. *191*, 295–98 (1951a)
Zelitch, I., Rosenblum, E.D., Burris, R.H.: J. Bacteriol. *62*, 747–52 (1951b)

8. Photosynthetic Assimilation of Sulfur Compounds

A. SCHMIDT

A. Introduction

Plants and algae are capable of reducing sulfur for incorporation into amino acids, proteins, coenzymes, and other plant constituents (ROY and TRUDINGER, 1970; SCHIFF and HODSON, 1973; GREENBERG, 1975; SCHWENN and TREBST, 1976). This process, which involves the reduction of oxidized sulfur compounds and subsequent incorporation, will be named assimilatory sulfate reduction. This sulfate assimilation is widespread in nature and can be found in bacteria, fungi, and plants, although not in animals (see MUTH and OLEFELD, 1970), which are dependent on a continuous supply of methionine in their diet. Besides this assimilatory sulfate reduction, a second mechanism of sulfate reduction can be found in certain bacteria, where sulfate is used instead of oxygen as the terminal electron acceptor for respiration (PECK, 1962). This sulfate reduction has been named dissimilatory sulfate reduction (POSTGATE, 1959; PECK, 1961, 1962).

Reduction of sulfate to the level of sulfide requires about 170 kcal/mol (GIBBS and SCHIFF, 1960) and 8 electrons are needed for this process, so that reducing power must be made available. Plants and algae are also capable of oxidizing reduced sulfur to sulfate (MOTHES and SPECHT, 1934; RICHMOND, 1973), however, the extent to which reduced sulfur is oxidized again has not been quantified, and it is not known if part of the energy released during oxidation can be stored in some manner by plant or algal cells.

A summary of possible reactions leading from sulfate to cysteine in a photosynthetic cell is given in Figure 1.

B. Observations with Whole Organisms

Algae and plants are able to accumulate sulfate from the environment, and this sulfate uptake is an energy-dependent process, which has been studied in algae and in plant root and leaf tissue in detail. Since several recent reviews are available on sulfate uptake (NISSEN, 1974; BUOMA, 1975; EPSTEIN, 1976; PITMAN, 1976; RAVEN, 1976) this topic will not be discussed here in detail; however, certain points should be mentioned.

Several authors have noted that sulfate uptake (and further metabolism) was enhanced by light. This stimulation was noticed with algal cells (WEDDING and BLACK, 1960; CLAUSS, 1961; ABRAHAM and BACHAWAT, 1963; DAVIES et al., 1966; KYLIN, 1966; MIYACHI and MIYACHI, 1966; ROBINSON, 1969; RAMUS and GROVES, 1972; DEANE and O'BRIEN, 1975; UTKILEN et al., 1976; MØLLER and EVANS, 1976) and plant leaf tissue (Kylin, 1960 a, b, c; Weigl, 1964; Davies et al., 1965; Penth and Weigl, 1969), suggesting that light energy could be utilized for sulfate accumulation. It should nevertheless be pointed out that plants are also able to meet their requirement for sulfur from other sources besides sulfate.

Although H_2S and SO_2 may have only a small significance for the sulfur nutrition of plants under "normal" conditions, this potential to grow on H_2S and SO_2 may be of great value for detoxification of these substances and it might have some value for studies of regulatory aspects of sulfur metabolism, which will be discussed later.

Higher plants take up the required sulfur as sulfate with the root system. From theoretical considerations, sulfate reduction in the root system should be possible, since isolated roots can grow on sulfate as their sulfur source, indicating that this nongreen tissue is able to reduce sulfate (Ellis, 1963; Pate, 1965; Holobrada, 1971; Ferrari and Renesto, 1972). However, studies with labeled sulfate have revealed that sulfur is reduced mainly in the leaves (Thomas, 1958; Kylin, 1960a–c; Willenbrink, 1964; Yamazoe, 1973), and it was shown that sulfate is transported to the leaves with the transpiration stream and that the only sulfur compound detected was free sulfate when roots were fed with radioactive sulfate (Liverman and Ragland, 1965; Tolbert and Wiebe, 1955).

From these observations it is evident that higher plants transport sulfate to the leaves where it is reduced as needed (Wiebe and Kramer, 1954; Biddulph et al., 1958; Buoma, 1967; Buoma et al., 1972).

Excised tobacco leaves incorporate sulfate into cysteine when fed with labeled sulfate, and the occurrence of sulfite (or a substance exchangeable with sulfite) was strongly stimulated by light (Fromageot and Perez-Milan, 1956, 1959). Similar results were obtained with mung bean leaves (Asahi, 1960; Asahi and Minmikawa, 1960; Kawashima and Asahi, 1961) and during these investigations it was shown that intermediates at the valency state of sulfite and sulfide could be trapped. Stimulation of sulfate reduction by light was observed in tobacco and bean (as discussed above), in wheat, barley, crassula (Kylin, 1960a–c; Torii and Fujiwara, 1967), spinach (Kawashima and Asahi, 1961; Asahi, 1964; Tamura, 1965), tomato (Willenbrink, 1964), and duckweed (Brunold, 1972). In algae, stimulation of sulfate reduction by light has been found in *Chlorella* (Miyachi and Miyachi, 1966), *Euglena* (Davies et al., 1965), *Scenedesmus* (Kylin, 1966), *Porphyridium* (Ramus and Groves, 1972), and *Acetabularia* (Clauss, 1961).

Furthermore it was shown by Sheridan (Sheridan and Castenholz, 1968; Sheridan, 1973) that the blue-green alga *Synechococcus* reduces sulfate and thiosulfate under anaerobic conditions to sulfide, when CO_2 is omitted. This sulfate reduction to sulfide was stimulated about tenfold by light and oxygen production was dependent on sulfate reduction under these conditions. These data suggest that sulfate can be used as a terminal electron acceptor of photosynthesis, and that sulfate reduction can be driven by photosynthesis.

C. Observations with Isolated Organelles

Observations of light-stimulated assimilatory sulfate reduction led to the concept that energy for the reduction should be derived from photosynthesis, and therefore a search for the enzymes in chloroplasts needed for sulfate reduction was initiated. A first report associating sulfate reduction with chloroplasts was published by ASAHI (1964), who showed that isolated chloroplast fragments would reduce sulfate in the light when fortified with the sulfate-activating system from yeast, and formation of APS with these chloroplast fragments was demonstrated. Later it was shown that chloroplast particles from *Euglena* catalyzed the formation of active sulfate (DAVIES et al., 1966) and MAYER (1967) demonstrated that a sulfite reductase activity was associated with spinach chloroplasts. Once suitable preparations of intact chloroplasts became available (JENSEN and BASSHAM, 1966), SCHMIDT was able to demonstrate that isolated chloroplasts reduce sulfate in a light-dependent reaction to cysteine (SCHMIDT, 1968; TREBST and SCHMIDT; 1969; SCHMIDT and TREBST, 1969; SCHWENN, 1970; BALHARRY and NICHOLAS, 1970; SCHMIDT and SCHWENN, 1971; BURNELL and ANDERSON, 1973; SCHWENN and HENNIES, 1973; SCHWENN et al., 1976; SILVIUS, 1976). Even for the phototrophic bacterium *Rhodospirillum rubrum* a light-dependent reduction of sulfate was observed with chromatophores (IBANEZ and LINDSTROM, 1959, 1962).

These data clearly demonstrate that isolated chloroplasts do contain the necessary enzymes to reduce sulfate to cysteine, thus explaining the many observations of light-stimulated sulfate reduction in whole organisms. Whether other cell components are capable of reducing sulfate cannot be answered yet for higher plants. However, BRUNOLD and SCHIFF (1976) reported recently that in *Euglena* sulfate reduction is found at least partly in mitochondria, and it has been reported that an *Aspergillus* mutant with dislocation of cysteine synthase activity outside of mitochondria was unable to grow on sulfate (BAL et al., 1975). Thus other cell compartments could probably be involved in sulfate reduction. It should be emphasized that nongreen tissue like roots or bleached algae will grow on sulfate, showing that they are able to reduce sulfate in the dark. It is not known yet whether proplastids or other cell organelles are involved in assimilatory sulfate reduction in these tissues.

D. Cell-Free Systems

I. Sulfate Activation and Degradation of Active Sulfate

1. ATP-Sulfurylase and Formation of APS

Sulfate activation is achieved by the two-step ATP-dependent system first described in LIPMANN's laboratory (HILZ and LIPMANN, 1955; ROBBINS and LIPMANN, 1957, 1958a, b; LIPMANN, 1958; DEMEIO, 1975). The first enzyme, ATP-sulfurylase (E.C. 2.7.7.4, ATP-sulfate adenylyltransferase) catalyzes the formation of the anhydride between AMP and sulfate with release of pyrophosphate:

$$ATP + H_2SO_4 \xrightarrow{\text{ATP-sulfurylase}} APS + P–P. \tag{1}$$

This reaction has an equilibrium constant determined by ROBBINS and LIPMANN (1958) to be 10^{-8}. From this equilibrium constant the hydrolysis energy for the sulfate-phosphate anhydride has been calculated to be about 19 Kcal (ROBBINS and LIPMANN, 1958; LIPMANN, 1958; ROY and TRUDINGER, 1970; SIEGEL, 1975). ATP-sulfurylase activity has been found in all organisms able to reduce sulfate: phototrophic bacteria, blue-green algae, green algae, and higher plants, as well as in nonphotosynthetic bacteria, and animals (DEMEIO, 1975).

ASAHI reported APS-formation with spinach chloroplast particles (1964) and SCHMIDT (1968) reported the localization of the ATP-sulfurylase within the spinach chloroplast; these data have been confirmed with chloroplasts isolated by the nonaqueous technique (BALHARRY and NICHOLAS, 1970).

2. APS-Kinase and Formation of PAPS

Since the equilibrium constant for APS-formation is low, a second reaction is necessary to remove the product APS, and one possibility is the phosphorylation of APS by a second ATP on the 3'-position of the ribose forming 3'-phosphoadenosine-5'-phosphosulfate (PAPS) according to the following reaction:

$$\text{APS} + \text{ATP} \xrightarrow{\text{APS-kinase}} \text{PAPS} + \text{ADP}. \tag{2}$$

The enzyme catalyzing this reaction has been named APS-kinase (E.C. 2.7.125 ATP-adenylylsulfate-3'-phosphotransferase). This enzyme has been isolated from yeast (ROBBINS and LIPMANN, 1957, 1958; WILSON and BANDURSKI, 1958, 1961). If the formation of PAPS proceeds as suggested, then the demonstration of PAPS formation is evidence for the presence of an APS-kinase. However, we do not know whether other reactions are operative too, for instance phosphate transfer from 3'-AMP and 5'-AMP (BRUNNGRABER and CHARGAFF, 1973 a, b) which has been described for NADP formation from NAD, or transfer of activated sulfate onto PAP forming PAPS. Such a sulfate transfer from p-nitrophenol-sulfate onto PAP forming PAPS has been demonstrated with an enzyme system from liver (GREGORY and LIPMANN, 1957) and by a system in the luciferin-luciferase reaction (STANLEY et al., 1975). Therefore the demonstration of PAPS formation in plant systems does not demonstrate with certainty that APS-kinase is present.

APS-kinase has not been purified from algal or plant sources, although PAPS formation in cell-free systems has been reported (SCHIFF, 1962; HODSON et al., 1968; MERCER and THOMAS, 1969; SCHMIDT, 1968, 1972a; MERCER et al., 1974; SAWHNEY and NICHOLAS, 1976; MOLLER and EVANS, 1976; SCHMIDT, 1977a). From spinach an enzyme has been partially purified which catalyzes PAPS formation, although only in the presence of 3'-AMP (BURNELL and ANDERSON, 1973b). APS-phosphorylation to PAPS can be demonstrated in spinach (SCHMIDT, 1972a), *Chlorella,* and *Synechococcus* via a reaction which is not dependent on 3'-AMP. APS-kinase from *Chlorella* can be separated from the APS-sulfotransferase, demonstrating that the transferase is not in a complex together with the APS-kinase. The coupled system from sulfate to PAPS can be used with high efficiency for PAPS accumulation, and cell-free systems from *Chlorella* have been used for large-scale preparation of PAPS (HODSON and SCHIFF, 1969b).

3. Dephosphorylation of PAPS to APS

During our investigations of sulfate activation and transfer it became evident that plant and algal extracts dephosphorylated PAPS to APS (TSANG and SCHIFF, 1971; SCHMIDT, 1972a, b). Such a phosphatase was purified from *Chlorella* (TSANG and SCHIFF, 1976b). This phosphatase was specific for adenosine-nucleotides containing a phosphate group in the 5'-position and the 3'-position of the ribose. The purified enzyme will not dephosphorylate 3'-AMP or 5'-AMP. However, it was noted that crude extracts from *Chlorella* are able to dephosphorylate PAPS in the presence of Ca^{2+}. Purification of this activity led to a phosphatase able to dephosphorylate PAPS to APS, and in addition 2'-AMP, 3'-AMP, and 5'-AMP to adenosine. It is not known whether either of these enzymes is needed for the sulfate-reducing pathway. Dephosphorylation of PAPS to APS has been used to prepare APS (TSANG et al., 1976).

4. Degradation of Sulfonucleotides

Cell-free extracts of plants and algae have the capacity to degrade sulfonucleotides to sulfate and an adenine containing nucleotide. These reactions in plant and algal sources have not been studied in detail. TSANG and SCHIFF (1976b) described an enzyme from *Chlorella* which splits the sulfate–phosphate anhydride bond and the product of this reaction is either AMP or cyclic-AMP depending on the conditions used; the latter case provides a new pathway for cyclic-AMP formation. In our laboratory we have confirmed cyclic-AMP formation from APS for *Chlamydomonas reinhardii* (KÜHLHORN, 1978). It is not known yet whether this cyclic-AMP has any metabolic function in green algae.

Degradation of sulfonucleotides (APS and PAPS) has been studied in some detail in *Anabaena* (SAWHNEY and NICHOLAS, 1976) and *Chlamydomonas* (KÜHLHORN, 1978) and there is evidence that APS may be an intermediate in PAPS degradation, perhaps indicating that such sulfhydrolases preferentially use APS for degradation. Degradation activities can be inhibited by sodium sulfate at concentrations around 0.5 M (SCHMIDT, 1975a). This is the case for APS degradation in spinach and *Chlorella* extracts, and also for PAPS degradation in cell-free extracts from *Synechococcus* and *Cyanophora*. This effect has been used to optimize conditions of sulfate transfer from sulfonucleotides in cell-free extracts.

II. Transfer of Sulfate from Sulfonucleotides for Further Reduction

1. APS as Sulfate Donor

Investigations of *Chlorella* and spinach recently demonstrated that at least in these species the sulfonucleotide APS was needed for further sulfate reduction (SCHMIDT and SCHWENN, 1971; TSANG et al., 1971; SCHMIDT, 1972a, b; SCHMIDT et al., 1974; GOLDSCHMIDT et al., 1975; SCHMIDT, 1975a, 1976a; TSANG and SCHIFF, 1976c). Differences between these data and earlier reports demonstrating activity with PAPS (HODSON et al., 1968) might have been due to the fact that added PAPS can be dephosphorylated to APS (Sect. D.I.3). A similar sulfotransferase enzyme, specific for APS, has been isolated from the phototrophic bacterium *Rhodospirillum rubrum* (Schmidt, 1977b). The enzyme specific for the sulfate transfer from APS has been named APS-sulfotransferase (Schmidt, 1972b) since it was found to transfer the sulfate group from APS onto suitable acceptors such as glutathione to form the sulphonated thiol according to the following reaction:

$$\text{APS} + \text{GSH} \xrightarrow{\quad\text{APS-sulfotransferase}\quad} \text{G} - \text{S} - \text{SO}_3\text{H} + \text{AMP}. \tag{3}$$

This enzyme is rather unspecific for the thiol acceptors, as it will react with a variety of mono- and dithiols (Levinthal and Schiff, 1968; Schmidt, 1972a, b; Abrams and Schiff, 1973; Schmidt et al., 1974; Tsang and Schiff, 1976c). For instance, it will react with dithioerythritol, forming free sulfite probably due to a displacement of dithioerythritol-bound sulfite. Studies with 2'-3'-dimercapto-propanol have also been reported (Levinthal and Schiff, 1968; Hodson and Schiff, 1971), but it is worth noting that this compound reacts with free sulfite nonenzymically to form thiosulfate.

The monothiol glutathione is probably not the natural acceptor molecule needed for the APS-sulfotransferase in vivo, since it can be demonstrated with cell-free extracts of *Chlorella* that a protein is labeled with sulfate from APS (Schmidt and Schwenn, 1971; Schmidt, 1973; Abrams and Schiff, 1973) and that this bound sulfite can be further reduced (Schmidt, 1973). The nature of the acceptor molecule has not been clarified, nor do we know the exact binding site of the bound sulfite. APS-sulfotransferase activity has been released from isolated spinach chloroplasts by osmotic breakage, and the specific activity found to be ten times that of a normal leaf extract, suggesting that the APS-sulfotransferase in spinach is located within the chloroplast (Schmidt, 1976a). APS-sulfotransferase activity from spinach, corn, and *Chlorella* is inhibited by 5'-AMP (Schmidt, 1975b), one end product of the reaction. The significance of this AMP-inhibition as a possible regulation mechanism (Schmidt, 1976a) is discussed below (Sect. E). APS-sulfo-transferase activity has been detected in higher plants (Schmidt, 1975c), green algae (Tsang and Schiff, 1975), blue-green algae (Schmidt, 1977b) and phototro-phic bacteria (Schmidt, 1977b; Schmidt and Trüper, 1977).

2. PAPS as Sulfate Donor

Although APS-sulfotransferase activity has been found in some plants and algae, the question may be asked whether APS is the sulfate donor for assimilatory sulfate reduction in all photosynthetic organisms. At this point we dare to suggest that higher plants and green algae should be specific for APS. The situation within the blue-green algae is uncertain and has yet to be analyzed carefully. We have demonstrated that *Synechococcus, Synechocystis,* and *Spirulina* have a specificity towards the sulfonucleotide PAPS (Schmidt, 1977a) and the data so far obtained are comparable with PAPS systems from yeast and *Escherichia coli*. The *Synecho-coccus* PAPS-sulfotransferase needs (1) PAPS as sulfate donor (2) has no activity with APS unless ATP is added (3) is stimulated by a heat-stable factor (thioredoxin), and (4) has no activity with the monothiol GSH, although it has good activity with dithioerythritol in the presence of a heat-stable factor identified as thioredoxin (SCHMIDT, 1977b; SCHMIDT and CHRISTEN, 1978). This is in contrast to APS-sulfotransferases so far obtained from photosynthetic organisms; these are (1) specific for APS (2) have no activity with PAPS unless a 3'-phosphatase is present (3) do not need a heat-stable factor (thioredoxin) for activity, and (4) are active with the monothiol GSH as well as the dithiol dithioerythritol (see SCHMIDT, 1977a, b; SCHMIDT and CHRISTEN, 1978). However, it is not suggested that all blue-green algae are specific for PAPS, since it was found that *Plectonema* preferentially uses APS (SCHMIDT, 1977a).

Within the phototrophic bacteria, cell-free extracts from *Rhodospirillum rubrum* have been shown to be active with APS (SCHMIDT, 1977b; SCHMIDT and TRÜPER, 1977) whereas the situation in the Rhodopseudomonaceae is so far uncertain. Chromatiaceae seem to have the possibility to assimilate sulfate via the APS-reductase normally used in dissimilatory systems (THIELE, 1968; TRÜPER and PECK, 1970; TRÜPER and ROGERS, 1971; KIRCHHOFF and TRÜPER, 1974; TRÜPER, 1975).

III. Reduction to the Level of Sulfide

1. Reduction of Sulfite to Sulfide

Systems reducing free sulfite to free sulfide according to the following equation have been isolated from plant material.

$$SO_3^{2-} + 6e^- + 6H^+ \xrightarrow{\text{sulfite reductase}} S^{2-} + 3H_2O. \tag{4}$$

Such an enzyme has been isolated from garlic (TAMURA, 1965). It catalyzed the methylviologen-dependent reduction of sulfite to sulfide. This is a 6-electron step (comparable to the reduction of nitrite to ammonium) and no intermediates have been detected. The natural electron donor for this reductase from garlic has not been described, however it was found that NADPH was not suitable. A similar sulfite reductase was isolated from spinach by Bandurski and coworkers (ASADA et al., 1966; ASADA, 1967; ASADA et al., 1968, 1969). This spinach enzyme was characterized again with methylviologen as electron donor and the statement was made that ferredoxin, which has the same redox potential as the artificial dye methylviologen, could not substitute for it unless supplemented with a heat-stable factor (TAMURA et al., 1967). Thus the natural electron donor for this reductase is unknown. SCHMIDT (1968) and TREBST and SCHMIDT (1969) described the ferredoxin-dependent reduction of sulfite to sulfide with spinach chloroplasts and cell-free preparations of partially purified sulfite reductases. In these experiments photosynthetically reduced ferredoxin was shown to couple to the isolated sulfite reductase (SCHMIDT, 1968; TREBST and SCHMIDT, 1969; SCHMIDT and TREBST, 1969). These results were later confirmed (TAMURA and ITOH, 1974). Methylviologen was not found to couple sulfite reduction to the electron transport chain of chloroplasts under aerobic conditions (SCHMIDT and TREBST, 1969) presumably because methylviologen reacted rapidly with oxygen. Under anaerobic conditions both reductants could be used for reduction of sulfite to sulfide with spinach sulfite reductase, whereas NADPH remained inactive, unless coupled to ferredoxin via ferredoxin-NADP-reductase (HENNIES, 1975). Similar activities were found in wheat (SAWHNEY and NICHOLAS, 1975) and duckweed (BRUNOLD and ERISMANN, 1976). These data clearly demonstrate that plants are capable of reducing free sulfite to free sulfide. Methylviologen-dependent sulfite reductase activities have been detected in a wide variety of algal sources including the blue-green algae *Oscillatoria, Anabaena,* and *Anacystis* (SAITO et al., 1969, 1971) and for the red alga *Porphyra* it has been demonstrated that ferredoxin can be used as natural electron donor (SAITO et al., 1970). Reducing capacity from sulfite to sulfide has been described also for photo-trophic bacteria, but the nature of the electron donor(s) involved is uncertain (PECK et al., 1974).

Most of the sulfite reductases analyzed have been measured with reduced methylviologen as artificial reductant. This may introduce artifact reactions, for instance, methylviologen will reduce the disulfide bond of cystine to cysteine. This implies that bound sulfides also could be reduced by methylviologen to free sulfide, thus altering the products found. In this context the formation of thiosulfate should be mentioned again: it has been shown that some sulfite reductases catalyze the formation of thiosulfate and trithionate (KOBAYASHI et al., 1969; SUH and AKAGI, 1969; LEE and PECK, 1971; KOBAYASHI and ISHIMOTO, 1974; DRAKE and AKAGI, 1976; SIEGEL, 1975), but if bound intermediates are operative, the formation of thiosulfate could easily be explained by sulfitolysis of bound sulfide with free sulfite forming thiosulfate (SCHMIDT, 1973).

2. Reduction of Bound Sulfite to Bound Sulfide

Analysis of sulfate-reducing systems from spinach chloroplasts (SCHMIDT, 1968; TREBST and SCHMIDT, 1969; SCHWENN, 1970; SCHMIDT and SCHWENN, 1971; SCHWENN and HENNIES, 1974; SCHWENN et al., 1976) and *Chlorella* (SCHMIDT, 1973; ABRAMS and SCHIFF, 1973; SCHMIDT et al., 1974) as well as from yeast (HILZ and KITTLER, 1960; TORII and BANDURSKI, 1964, 1967; WILSON and BIERER, 1976) and *Escherichia coli* (TSANG and SCHIFF, 1976a) led to the finding of bound sulfites, which of course opened the possibility that sulfate reduction might proceed with bound intermediates. First suggestions of a bound sulfite as an intermediate in PAPS reduction were made by HILZ and KITTLER (1960) who postulated an organic thiosulfate as an intermediate, and indeed BANDURSKI and coworkers were able to demonstrate that a bound sulfite could be isolated from yeast and reduced further for incorporation into cysteine (TORII and BANDURSKI, 1967). These ideas were followed by Schmidt who isolated S-sulfoglutathione as a product of the reaction between APS and glutathione (SCHMIDT, 1972a) and the enzyme transferring the sulfate group of APS onto glutathione was named APS-sulfotransferase (SCHMIDT, 1972b). The suggestion had already been made in 1968 that the PAPS-reductase from yeast should be a PAPS-sulfotransferase with fraction C as acceptor (SCHMIDT, 1968).

Sulfate from sulfonucleotides can be transferred to an acceptor protein and analysis has shown that this protein-bound sulfite from *Chlorella* can be further reduced to bound sulfide (SCHMIDT, 1973), when the extract was fortified with a ferredoxin-reducing system, according to the following equation:

$$X - SO_3^- + 6 Fd_{red} + 6 H^+ \xrightarrow[\text{reductase}]{\text{thiosulfonate}} X - S^- + 6 Fd_{ox} + 3 H_2O. \tag{5}$$

Sulfate reduction of the same type has been described for spinach chloroplasts (SCHWENN, 1970; SCHMIDT and SCHWENN, 1971; SCHWENN and HENNIES, 1974; SCHWENN et al., 1976). It was demonstrated that in isolated spinach chloroplasts actively reducing sulfate, bound sulfite and bound sulfide were formed, being attached to a protein with a molecular weight of about 5000 daltons (SCHMIDT and SCHWENN, 1971). Kinetic data suggest that these compounds are intermediates in photosynthetic sulfate reduction (SCHWENN, 1970). The enzymes catalyzing this reduction are bound to the chloroplast membrane (SCHWENN et al., 1976).

Heating of protein-bound radioactivity obtained from *Chlorella* extracts released the radioactivity attached to a small substance with a molecular weight of about 1400 daltons. This substance catalyzed chemical reactions typical for bunte salts, suggesting that the sulfate was bound as an organic thiosulfate (ABRAMS and SCHIFF, 1973).

Isolation of an enzyme catalyzing the reduction of bound sulfite using the model substance S-sulfoglutathione led to the isolation of an enzyme named thiosulfonate reductase (SCHMIDT, 1973), which was purified from *Chlorella* and spinach (HENNIES, 1974). This enzyme catalyzed the reduction of bound sulfite to bound sulfide in a ferredoxin-dependent reaction according to the equation given above. Analysis of *Chlorella* mutants blocked for sulfate reduction (HODSON et al., 1971; ABRAMS and SCHIFF, 1973; SCHMIDT et al., 1974) suggest (a) that APS-sulfotransferase activity is needed for sulfate reduction in vivo, and (b) that thiosulfonate reductase is needed for sulfate reduction in vivo. This is based on observations made on the mentioned *Chlorella* mutants which will not grow on sulfate if one of the two enzymes is inactive. Quite recently the formation of bound sulfite with subsequent reduction has been shown in *Escherichia coli* (TSANG and SCHIFF, 1976a) and in yeast (WILSON and BIERER, 1976).

There are some central questions which have to be analyzed for an understanding of the sulfate-reducing pathway involving bound intermediates. (1) What is the chemical nature of the carrier which binds sulfite? (2) What is the actual binding site? (3) What are the relationships between carrier and sulfotransferase and reductase? (4) Is there a relationship between sulfite reductase and thiosulfonate reductase? So far we have no answers to these questions.

IV. Biosynthesis of Cysteine

1. Cysteine Synthase

Cysteine synthase (E.C. 4.2.99.9; O-acetyl-L-serine acetate lyase) catalyzes the following reaction:

$$H_2S + \text{O-acetyl-L-serine} \xrightarrow[\text{synthase}]{\text{cysteine}} \text{cysteine} + \text{acetate}. \tag{6}$$

This synthase has been found in microorganisms (GREENBERG, 1975) and green plants (GIOVANELLI and MUDD, 1967, 1968; THOMPSON and MOORE, 1968; SMITH and THOMPSON, 1971; SMITH, 1972; NGO and SHARGOOL, 1972, 1974; GRANROTH, 1974; GRANROTH and SARNESTO, 1974; MASADA et al., 1975; TAMURA et al., 1976; SCHMIDT, 1977c, d), as well as in nongreen tissue (TAMURA et al., 1976). About 20% of the total cysteine synthase activity of spinach leaves seems to be localized within the chloroplast (FANKHAUSER et al., 1976).

2. L-Serine Hydrolase

A second possibility to form cysteine, in this case from sulfide and serine, was described earlier by LYNEN and coworkers (SCHLOSSMANN and LYNEN, 1957; SCHLOSSMANN et al., 1962; BRÜGGEMANN et al., 1962).

$$H_2S + serine \xleftrightarrow[\text{hydrolase}]{\text{L-serine}} cysteine + H_2O. \tag{7}$$

This enzyme [E.C.4.2.1.22; L-serine hydrolase (adding H_2S)] requires addition of pyridoxal phosphate which is lost during purification. Both enzymes (cysteine synthase and L-serine hydrolase) have been found in spinach (BRÜGGEMANN et al., 1962). It is suggested that L-serine hydrolase is not involved in cysteine biosynthesis because cell-free systems catalyzing the reduction from APS to cysteine will not accept serine (SCHMIDT, 1973; SCHMIDT et al., 1974), indicating that L-serine hydrolase is not operative under these conditions.

3. Isotopic Exchange Reactions Between Cysteine and Sulfide

L-Serine hydrolase catalyzes the reaction between cysteine and sulfide in both directions, resulting in an isotopic exchange reaction between cysteine and sulfide (SCHLOSSMANN and LYNEN, 1957; SCHLOSSMANN et al., 1962). During purification of cysteine synthase from spinach and *Chlorella,* a second activity which catalyzes an isotopic exchange reaction according to the following equation copurifies with the cysteine synthase (SCHMIDT, 1977c, d):

$$Cys - {}^*SH + H_2S \leftrightarrow Cys - SH + {}^*H_2S. \tag{8}$$

This purified enzyme was not active with serine (SCHMIDT, 1977c, d), demonstrating that different enzymes are able to catalyze the exchange reaction.

These exchange reactions may have consequences for tracer studies, since incorporation of sulfide into cysteine might be due simply to the isotopic exchange reaction. Furthermore, addition of free H_2S as a carrier could interact in the same manner.

It was pointed out in Section D.II.1. that free sulfide was not found in chloroplasts actively reducing sulfate, and from cell-free systems of *Chlorella* the sum of bound and free sulfide was found to be about $1-2\,\mu M$. The K_m for sulfide for the cysteine synthase from plants and *Chlorella* is in the order of 2 to 4 mM, which is about 1000-fold higher. There is the possibility that free sulfide is an artificial substrate for the cysteine synthase, which has a relatively unfavorable K_m for free sulfide, but a high turnover number. This could explain the high activity of this enzyme in plant and algal tissue. If bound sulfide is generated

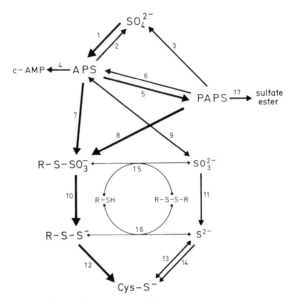

Fig. 1. Scheme of the sulfate-reducing pathway(s) in the phototrophic cell. *1,* ATP-sulfurylase; *2,* APS-sulfhydrolase; *3,* PAPS-sulfhydrolase; *4,* APS-cyclase; *5,* APS-kinase; *6,* PAPS-phosphatase; *7,* APS-sulfotransferase; *8,* PAPS-sulfotransferase; *9,* APS-reductase; *10,* thiosulfonate reductase; *11,* sulfite reductase; *12,* cysteine synthase (if different to 13 and 14 is not known); *13,* isotopic exchange reaction; *14,* cysteine synthase; *15,* sulfitolysis (chemical reaction); *16,* sulfidolysis (chemical reaction); *17,* (O- and N-) PAPS-sulfotransferases. Isotopic exchange reactions between $R-S-SO_3^-$ and free sulfite and probably between $R-S-S^-$ and free sulfide have to be considered, but are not included in this scheme

during assimilatory sulfate reduction, this bound sulfide has to react in some manner with O-acetyl-serine to form cysteine. Two electrons are needed to liberate bound sulfide or bound cysteine. But an electron donor has not yet been found, although it is possible that the fraction C isolated from spinach (SCHWENN et al., 1976) could function in this manner. Another possibility would be that the bound sulfide is transferred to O-acetyl-serine by another protein with the properties of rhodanese. A summary of possible reactions leading from sulfate to cysteine in a photosynthetic cell is given in Figure 1.

E. Regulation of Assimilatory Sulfate Reduction in Photosynthetic Organisms

One approach to study regulatory phenomena of sulfate reduction is by supplying cells with the end product of the reduction i.e., cysteine. These studies led to the finding that sulfate uptake was inhibited (DAVIES et al., 1966; HART and FILNER, 1969; BRUNOLD, 1972; BRUNOLD and ERISMANN, 1974; SMITH, 1975; MAGGIONI and RENESTO, 1977). It was further discovered that sulfate reduction to the level of cysteine was inhibited in *Euglena* under conditions which did not block sulfate flow into sulfolipids (Davies et al., 1966), suggesting that these two pathways can be regulated separately.

Another possibility to study regulatory phenomena is the feeding of plant or algal cells with volatile sulfur compounds such as SO_2 and H_2S. This has been analyzed in detail by Brunold using the duckweed *Lemna minor* (Brunold, 1972; Brunold and Erismann, 1974; Brunold and Schmidt, 1976). These data show that *Lemna minor* can grow on H_2S or SO_2 as the only sulfur source. However, since SO_2 seems to enter the sulfate-reducing pathway as sulfate and not as sulfite (although sulfite reductase activity was detected; Brunold and Erismann, 1976) sulfide was used as the sulfur source to study regulatory phenomena in *Lemna*. Feeding *Lemna* with H_2S decreased the doubling rate from about 40 to 50 h, however, plants grow with H_2S at a new steady rate (Brunold, 1972; Brunold and Erismann, 1974). Analysis of ATP-sulfurylase, APS-sulfotransferase and cysteine synthase showed that only the APS-sulfotransferase was effected significantly. The activity of this enzyme decreased to about 5% of the control grown without H_2S. There was a rapid inactivation of this activity with a half-life of about 5 h and a slower inactivation with a half-life of about 13 h. These inactivation times are fast compared with the doubling times of this organism. Reappearance of the APS-sulfotransferase activity was observed within 24 h after transfer to air without H_2S (Brunold and Schmidt, 1976, 1978). Another method of examining regulatory mechanisms seems to be plant development. This approach led to the finding that APS-sulfotransferase activity in sunflower is low in germinating seeds, cotyledons, and root tissue, but high in the developing leaves. This suggests that root tissue is supplied with reduced sulfur compounds (Schmidt, 1976b).

These data clearly demonstrate that reduced sulfur compounds are a signal to shut off assimilatory sulfate reduction in the organisms studied. This seems to be achieved in *Lemna* and sunflower by a repression or inactivation of the APS-sulfotransferase activity. The signal could be H_2S, cysteine, or a derivative of this amino acid. There are some indications toward cysteine, since this amino acid increases under growth with H_2S (Brunold, 1972; Brunold and Erismann, 1974). Data on algae which show regulation of this pathway are few, but it seems clear that also in algae the sulfate-reducing pathway must be regulated in some manner (Davies et al., 1966). An inhibition of the thiosulfonate reductase activity in crude extracts of *Chlorella* has been reported (Schmidt, 1973).

Sulfate is reduced mainly during the light period. One possibility for such a regulation seems to be the ATP state of the cell. The ATP-sulfurylase from spinach (Schwenn and Depka, 1977) and the APS-sulfotransferases from spinach, corn, *Chlorella* (Schmidt, 1975b), and *Rhodospirillum* (Schmidt, 1977b) are inhibited by 5'-AMP, and it was shown for the spinach APS-sulfotransferase that this inhibition is competitive (Schmidt, 1976a). During the light period the ratio of APS to AMP will change due to photophosphorylation, leading to a decrease in the AMP concentration. Since the APS pool is small, this change in the AMP pool might well have regulatory significance (Schmidt, 1976a). Further, during the light period the pH of the chloroplast stroma changes towards alkaline conditions (Heldt et al., 1973) which favor increased activity of the ATP-sulfurylase and APS-sulfotransferase (with pH optima around 9); this might also be of regulatory significance.

Acknowledgment. The original work reported herein was supported by a grant from the Deutsche Forschungsgemeinschaft.

References

Abraham, A., Bachawat, B.K.: Biochim. Biophys. Acta *70*, 104–107 (1963)
Abrams, W.R., Schiff, J.A.: Arch. Mikrobiol. *94*, 1–10 (1973)
Asada, K.: J. Biol. Chem. *242*, 3646–3654 (1967)
Asada, K., Tamura, G., Bandurski, R.S.: Biochem. Biophys. Res. Commun. *25*, 529–534 (1966)
Asada, K., Tamura, G., Bandurski, R.S.: Biochem. Biophys. Res. Commun. *30*, 554–559 (1968)
Asada, K., Tamura, G., Bandurski, R.S.: J. Biol. Chem. *244*, 4904–4915 (1969)
Asahi, T.: J. Biochem. *48*, 772–773 (1960)
Asahi, T.: Biochim. Biophys. Acta *82*, 58–66 (1964)
Asahi, T., Minmikawa, T.: J. Biochem. *48*, 548–556 (1960)
Bal, J., Maleszka, R., Stephien, P., Cybis, J.: FEBS Lett. *58*, 164–166 (1975)
Balharry, J.E., Nicholas, D.J.D.: Biochim. Biophys. Acta *220*, 513–524 (1970)
Biddulph, O., Biddulph, S., Cory, R., Koontz, H.: Plant Physiol. *33*, 293–300 (1958)
Brüggemann, J., Schlossmann, K., Merkenschlager, M., Waldschmidt, M.: Biochem. Z. *335*, 392–399 (1962)
Brunngraber, E.F., Chargaff, E.: Biochemistry *12*, 3012–3016 (1973a)
Brunngraber, E.F., Chargaff, E.: Biochemistry *12*, 3005–3011 (1973b)
Brunold, Chr.: Thesis, Bern (1972)
Brunold, Chr., Erismann, K.H.: Verh. Schw. Naturforsch. Ges. 1145–1147 (1972)
Brunold, Chr., Erismann, K.H.: Experientia *30*, 465–467 (1974)
Brunold, Chr., Erismann, K.H.: Experientia *32*, 296–297 (1976)
Brunold, Chr., Schiff, J.A.: Plant Physiol. *57*, 430–436 (1976)
Brunold, Chr., Schmidt, A.: Planta (Berl.) *133*, 85–88 (1976)
Brunold, Chr., Schmidt, A.: Plant Physiol. *61*, 342–347 (1978)
Buoma, D.: Aust. J. Biol. Sci. *20*, 613–621 (1967)
Buoma, D.: Sulphur in Australasian agriculture. McLachlan (ed.), pp. 79–86. Sydney Univ. Press 1975
Buoma, D., Titmanis, Z., Greenwood, E.A.N.: Aust. J. Biol. Sci. *25*, 1157–1167 (1972)
Burnell, J.N., Anderson, J.W.: Biochem. J. *133*, 417–428 (1973a)
Burnell, J.N., Anderson, J.W.: Biochem. J. *134*, 565–579 (1973b)
Clauss, H.: Z. Naturforsch. *16b*, 770–771 (1961)
Davies, W.H., Mercer, E.I., Goodwin, T.W.: Phytochemistry *4*, 741–749 (1965)
Davies, W.H., Mercer, E.I., Goodwin, T.W.: Biochem. J. *98*, 369–373 (1966)
Deane, E.M., O'Brien, R.W.: Arch. Microbiol. *105*, 295–312 (1975)
DeMeio, R.H.: Metabolic pathways. Greenberg (ed.), Vol. VII, pp. 287–358. New York: Academic Press 1975
Drake, H.L., Akagi, J.M.: J. Bacteriol. *126*, 733–738 (1976)
Ellis, R.J.: Phytochemistry *2*, 129–136 (1963)
Epstein, E.: Transport in plants. Lüttge, Pitman (eds.), Vol. II, pp. 70–94. Berlin, Heidelberg, New York: Springer 1976
Fankhauser, H., Brunold, Chr., Erismann, K.H.: Experientia *32*, 1494–1497 (1976)
Ferrari, G., Renesto, F.: Plant Physiol. *49*, 114–116 (1972)
Fromageot, P., Perez-Milan, M.: C. R. Acad. Sci. *243*, 1061–1064 (1956)
Fromageot, P., Perez-Milan, M.: Biochim. Biophys. Acta *32*, 457–464 (1959)
Gibbs, M., Schiff, J.A.: Plant physiology: a treatise. Vol. Ib, pp. 279–319. New York: Academic Press 1960
Giovanelli, J., Mudd, S.H.: Biochem. Biophys. Res. Commun. *27*, 150–156 (1967)

Giovanelli, J., Mudd, S.H.: Biochem. Biophys. Res. Commun. *31*, 275–280 (1968)
Goldschmidt, E.E., Tsang, M.L.-S., Schiff, J.A.: Plant Sci. Lett. *4*, 293–300 (1975)
Granroth, B.: Acta Chem. Scand. *28b*, 813–814 (1974)
Granroth, B., Sarnesto, A.: Acta Chem. Scand. *28b*, 814–815 (1974)
Greenberg, D.M.: Metabolic pathways. Greenberg (ed.), Vol. VII, pp. 505–528. New York: Academic Press 1975
Gregory, J.D., Lipmann, F.: J. Biol. Chem. *229*, 1081–1090 (1957)
Hart, J.W., Filner, P.: Plant Physiol. *44*, 1253–1259 (1969)
Heldt, H.W., Werdan, K., Milovanecev, M., Geller, G.: Biochim. Biophys. Acta *314*, 224–241 (1973)
Hennies, H.H.: Thesis, Bochum (1974)
Hennies, H.H.: Z. Naturforsch. *30c*, 359–362 (1975)
Hilz, H., Kittler, M.: Biochem. Biophys. Res. Commun. *3*, 140–142 (1960)
Hilz, H., Lipmann, F.: Proc. Natl. Acad. Sci. USA *41*, 880–890 (1955)
Hodson, R.C., Schiff, J.A.: Plant Physiol. *47*, 296–299 (1969a)
Hodson, R.C., Schiff, J.A.: Arch. Biochem. Biophys. *132*, 151–156 (1969b)
Hodson, R.C., Schiff, J.A.: Plant Physiol. *47*, 296–299 (1971)
Hodson, R.C., Schiff, J.A., Scarsella, A.J., Levinthal, M.: Plant Physiol. *43*, 563–569 (1968)
Hodson, R.C., Schiff, J.A., Mather, J.P.: Plant Physiol. *47*, 306–311 (1971)
Holobrada, M.: Biologia *26*, 27–32 (1971)
Ibanez, L.M., Lindstrom, E.S.: Biochem. Biophys. Res. Commun. *1*, 224–227 (1959)
Ibanez, L.M., Lindstrom, E.S.: J. Bacteriol. *84*, 451–455 (1962)
Jensen, R.G., Bassham, J.A.: Proc. Natl. Acad. Sci. *56*, 1095–1101 (1966)
Kawashima, N., Asahi, T.: J. Biochem. *49*, 52–54 (1961)
Kirchhoff, J., Trüper, H.G.: Arch. Microbiol. *100*, 115–120 (1974)
Kobayashi, K., Tachibana, S., Ishimoto, M.: J. Biochem. *65*, 155–157 (1969)
Kobayashi, K., Seki, Y., Ishimoto, M.: J. Biochem. *75*, 519–530 (1974)
Kühlhorn, F.: Diplomarbeit, München (1978)
Kylin, A.: Physiol. Plant. *13*, 148–154 (1960a)
Kylin, A.: Physiol. Plant. *13*, 366–379 (1960b)
Kylin, A.: Bot. Notiser *113*, 49–54 (1960c)
Kylin, A.: Physiol. Plant. *19*, 883–887 (1966)
Lee, J.P., Peck, H.D., Jr.: Biochem. Biophys. Res. Commun. *45*, 583–589 (1971)
Levinthal, M., Schiff, J.A.: Plant Physiol. *43*, 555–562 (1968)
Lipmann, F.: Science *128*, 575–580 (1958)
Liverman, J.L., Ragland, J.B.: Plant Physiol. *31*, S-7 (1965)
Maggioni, A., Renesto, F.: Physiol. Plant. *39*, 143–147 (1977)
Masada, M., Fukushima, K., Tamura, G.: J. Biochem. *77*, 1107–1115 (1975)
Mayer, A.: Plant Physiol. *42*, 324–326 (1967)
Mercer, E.I., Thomas, G.: Phytochemistry *8*, 2281–2285 (1969)
Mercer, E.I., Thomas, G., Harrison, D.: Phytochemistry *13*, 1299–1302 (1974)
Miyachi, S., Miyachi, S.: Plant Physiol. *41*, 479–486 (1966)
Moller, M.E., Evans, L.V.: Phytochemistry *115*, 1623–1626 (1976)
Mothes, K., Specht, W.: Planta (Berl.) *22*, 800–803 (1934)
Muth, O.H., Olefeld, J.E.: Symposium: Sulfur in nutrition. Westport: A.V.I. Publishing Press 1970
Ngo, T.T., Shargool, P.S.: Biochem. J. *126*, 985–991 (1972)
Ngo, T.T., Shargool, P.S.: Can. J. Biochem. *52*, 435–440 (1974)
Nissen, P.: Annu. Rev. Plant Physiol. *25*, 53–79 (1974)
Pate, J.S.: Science *149*, 547–548 (1965)
Peck, H.D., Jr.: J. Bacteriol. *82*, 933–939 (1961)
Peck, H.D., Jr.: Bacteriol. Rev. *26*, 67–94 (1962)
Peck, H.D., Jr., Tedro, S., Kamen, M.D.: Proc. Natl. Acad. Sci. *71*, 2404–2406 (1974)
Penth, B., Weigl, J.: Z. Naturforsch. *24b*, 342–348 (1969)
Pitman, M.G.: Ion transport in plants. Lüttge, Pitman (eds.), Vol. IIb, pp. 95–128. Berlin, Heidelberg, New York: Springer 1976
Postgate, J.: Annu. Rev. Microbiol. *13*, 505–520 (1959)
Ramus, J., Groves. S.T.: J. Cell. Biol. *54*, 399–407 (1972)

Raven, J.A.: Ion transport in plants. Lüttge, Pitman (eds.), Vol. IIa, pp. 129–188. Berlin, Heidelberg, New York: Springer 1976
Richmond, D.V.: Phytochemistry, Vol. III, pp. 41–73. New York: van Nostrand Reinhold Company 1973
Robbins, P.W., Lipmann, F.: J. Biol. Chem. *229*, 837–851 (1957)
Robbins, P.W., Lipmann, F.: J. Biol. Chem. *233*, 681–685 (1958a)
Robbins, P.W., Lipmann, F.: J. Biol. Chem. *233*, 686–690 (1958b)
Robinson, P.W.: J. Exp. Bot. *20*, 201–211 (1969)
Roy, A., Trudinger, P.A.: The biochemistry of inorganic compounds of sulphur. Cambridge Univ. Press 1970
Saito, E., Tamura, G.: Agric. Biol. Chem. *35*, 491–500 (1971)
Saito, E., Tamura, G., Shinano, S.: Agric. Biol. Chem. *33*, 860–867 (1969)
Saito, E., Wakasa, K., Okuma, M., Tamura, G.: Bull. Assoc. Natl. Sci. Senshu Univ. *3*, 35–50 (1970)
Sawhney, S.K., Nicholas, D.J.D.: Phytochemistry *14*, 1499–1503 (1975)
Sawhney, S.K., Nicholas, D.J.D.: Planta (Berl.) *132*, 189–195 (1976)
Sawhney, S.K., Nicholas, D.J.D.: Plant Sci. Lett. *6*, 103–110 (1976)
Schiff, J.A.: Physiology and biochemistry of algae. pp. 239–246. New York: Academic Press 1962
Schiff, J.A., Hodson, R.C.: Annu. Rev. Plant Physiol. *24*, 381–414 (1973)
Schlossmann, K., Lynen, F.: Biochem. Z. *328*, 591–594 (1957)
Schlossmann, K., Brüggemann, J., Lynen, F.: Biochem. Z. *336*, 258–273 (1962)
Schmidt, A.: Thesis, Göttingen (1968)
Schmidt, A.: Z. Naturforsch. *27b*, 183–192 (1972a)
Schmidt, A.: Arch. Mikrobiol. *84*, 77–86 (1972b)
Schmidt, A.: Arch. Mikrobiol. *93*, 29–52 (1973)
Schmidt, A.: Planta (Berl.) *124*, 267–275 (1975a)
Schmidt, A.: Planta (Berl.) *127*, 93–95 (1975b)
Schmidt, A.: Plant Sci. Lett. *5*, 407–415 (1975c)
Schmidt, A.: Planta (Berl.) *130*, 257–263 (1976a)
Schmidt, A.: Z. Pflanzenphysiol. *78*, 164–168 (1976b)
Schmidt, A.: FEMS Microbiol. Lett. *1*, 137–140 (1977a)
Schmidt, A.: Arch. Microbiol. *112*, 263–270 (1977b)
Schmidt, A.: Z. Naturforsch. *32c*, 219–225 (1977c)
Schmidt, A.: Z. Pflanzenphysiol. *84*, 435–446 (1977d)
Schmidt, A., Christen, U.: Planta (Berl.) *140*, 239–244 (1978)
Schmidt, A., Schwenn, J.D.: II. Int. Congr. Phot. 507–513 (1971)
Schmidt, A., Trebst, A.: Biochim. Biophys. Acta *180*, 529–535 (1969)
Schmidt, A., Trüper, H.G.: Experientia *33*, 1008–1009 (1977)
Schmidt, A., Abrams, W.R., Schiff, J.A.: Eur. J. Biochem. *47*, 423–434 (1974)
Schwenn, J.D.: Thesis, Bochum (1970)
Schwenn, J.D., Depka, B.: Z. Naturforsch. *32c*, 792–797 (1977)
Schwenn, J.D., Hennies, H.H.: III. Int. Congr. Phot. 629–635 (1974)
Schwenn, J.D., Trebst, A.: The intact chloroplast. Barber (ed.), pp. 315–334. Amsterdam, New York, Oxford: Elsevier 1976
Schwenn, J.D., Depka, B., Hennies, H.H.: Plant Cell Physiol. *17*, 165–176 (1976)
Sheridan, R.P.: J. Phycol. *9*, 437–444 (1973)
Sheridan, R.P., Castenholz, W.: Nature (London) *217*, 1064–1065 (1968)
Siegel, L.M.: Metabolic pathways. Greenberg (ed.), Vol. VII, pp. 217–286. New York: Academic Press 1975
Silvius, J.E., Baer, Ch.H., Dodrill, S., Patrick, H.: Plant Physiol. *57*, 799–801 (1976)
Smith, I.K.: Plant Physiol. *50*, 477–479 (1972)
Smith, I.K.: Plant Physiol. *55*, 303–307 (1975)
Smith, I.K., Thompson, J.F.: Biochim. Biophys. Acta *227*, 288–295 (1971)
Stanley, P.E., Kelley, B.C., Tuovinen, O.H., Nicholas, D.J.D.: Anal. Biochem. *67*, 540–551 (1975)
Suh, B., Akagi, J.M.: J. Bacteriol. *99*, 210–215 (1969)
Tamura, G.: J. Biol. Chem. *57*, 207–214 (1965)

Tamura, G., Itoh, S.I.: Agric. Biol. Chem. *38*, 225–226 (1974)
Tamura, G., Asada, K., Bandurski, R.S.: Plant Physiol. *42*, S-36 (1967)
Tamura, G., Iwasawa, T., Masada, M., Fukushima, K.: Agric. Biol. Chem. *40*, 637–638 (1976)
Thiele, H.H.: Anton v. Leuvenhook *34*, 350–356 (1968)
Thomas, M.D.: Handbuch der Pflanzenphysiologie. IX. Ruhland (ed.), Vol. IX, pp. 37–63. Berlin, Heidelberg, New York: Springer 1958
Thompson, J.F., Moore, D.P.: Biochem. Biophys. Res. Commun. *31*, 281–286 (1968)
Tolbert, N.E., Wiebe, H.: Plant Physiol. *30*, 499–504 (1955)
Torii, K., Bandurski, R.S.: Biochim. Biophys. Acta *136*, 286–295 (1967)
Torii, K., Fujiwara, A.: Tohoku J. Agric. Res. *18*, 155–166 (1967)
Trebst, A., Schmidt, A.: Progr. Phot. Res. *III*, 1510–1516 (1969)
Trüper, H.G.: Plant Soil *43*, 29–39 (1975)
Trüper, H.G., Peck, H.D., Jr.: Arch. Mikrobiol. *73*, 125–142 (1970)
Trüper, H.G., Rogers, L.A.: J. Bacteriol. *108*, 1112–1121 (1971)
Tsang, M.L-S., Schiff, J.A.: Plant Sci. Lett. *4*, 301–307 (1975)
Tsang, M.L-S., Schiff, J.A.: J. Bacteriol. *125*, 923–933 (1976a)
Tsang, M.L-S., Schiff, J.A.: Eur. J. Biochem. *65*, 113–121 (1976b)
Tsang, M.L-S., Schiff, J.A.: Plant Cell Physiol. *17*, 1209–1220 (1976c)
Tsang, M.L-S., Goldschmidt, E.E., Schiff, J.A.: Plant Physiol. *47*, S-20 (1971)
Tsang, M.L-S., Lemieux, J., Schiff, J.A., Bojarski, T.B.: Anal. Biochem. *74*, 623–626 (1976)
Utkilen, H.C., Heldal, M., Knutsen, G.: Physiol. Plant. *38*, 217–220 (1976)
Wedding, R.T., Black, M.K.: Plant Physiol. *35*, 72–80 (1960)
Weigl, J.: Z. Naturforsch. *19b*, 845–851 (1964)
Wiebe, H.H., Kramer, P.J.: Plant Physiol. *29*, 341–348 (1954)
Willenbrink, J.: Z. Naturforsch. *19b*, 356–357 (1964)
Wilson, L.G., Bandurski, R.S.: J. Biol. Chem. *233*, 975–981 (1958)
Wilson, L.G., Bandurski, R.S.: J. Biol. Chem. *236*, 1822–1829 (1961)
Wilson, L.G., Bierer, D.: Biochem. J. *158*, 255–270 (1976)
Yamazoe, J.: Jpn. Agric. Res. Q. *7*, 243–247 (1973)

9. Hydrogen Metabolism

A. BEN-AMOTZ

A. Introduction

Chlorophyllous cells under atmospheric conditions photooxidize water, releasing O_2 and simultaneously reducing CO_2 to carbohydrate. This is a reaction unique to chlorophyll a-containing cells and is termed photosynthesis. The photosynthesis of green plants has long been considered a rigid process which can be enhanced or retarded by external conditions, but whose chemical metabolism is unalterable. This is not universally true. Indeed, NAKAMURA (1938) found that certain diatoms and blue-green algae could use H_2S for the reduction of CO_2. Recent reports (STEWART and PEARSON, 1970; COHEN et al., 1975) showed that in contrast to water being the electron donor resulting in the evolution of O_2, in blue-green algae H_2S serves as an electron donor for CO_2 photoassimilation.

The investigation initiated by GAFFRON (1940) was much more detailed. In this study, he observed that certain unicellular green algae, after a period of anaerobic adaptation, become able to utilize molecular hydrogen as a reductant of CO_2 in photosynthesis. Subsequently, GAFFRON and RUBIN (1942) found that algae which were capable of absorbing H_2 liberated hydrogen gas in the absence of CO_2 by fermentative and photochemical pathways. The capacity to include molecular hydrogen in the metabolism of these algae depends on the presence of an adaptable hydrogenase. The presence of hydrogenase introduces several unique metabolic processes. The following list of overall gas exchange reactions occurring in hydrogenase-containing algae under anaerobic conditions may serve as a guideline for various aspects of this review.

Light-dependent reactions

Photosynthesis:

$$CO_2 + 2H_2O \rightarrow (CH_2O) + O_2 + H_2O \tag{1}$$

Photoreduction:

$$CO_2 + 2H_2 \rightarrow (CH_2O) + H_2O \tag{2}$$

H_2 photoproduction:

$$RH_2 \text{ (H}_2\text{O or organic source)} \rightarrow R + H_2 \tag{3}$$

Dark reactions

H_2 production (H_2 fermentation):

$$RH_2 \text{ (organic source)} \to R + H_2 \tag{4}$$

H_2 absorption:

$$H_2 + R \to RH_2 \tag{5}$$

Oxy-hydrogen reaction:

$$O_2 + 2H_2 \to 2H_2O \tag{6a}$$

$$CO_2 + 2H_2 \to (CH_2O) + H_2O \tag{6b}$$

Reactions (2) to (6b), but not (1), require hydrogenase. In these formulations, (CH_2O) represents a subunit of carbohydrate, and R is an acceptor or donor of H_2. A review of the function of hydrogenase and the various aspects of H_2 metabolism associated with an activated hydrogenase is the main concern of this chapter. For a more inclusive review of the subject, the reader is referred to a few articles covering our present knowledge (Spruit, 1962; Bishop, 1966; Kessler, 1974; Lien and San Pietro, 1975).

B. Hydrogenase

I. Occurrence of Hydrogenase in Photosynthetic Cells

Hydrogenase occurs in a few algal strains of the Chlorophyta, i.e., *Scenedesmus*, *Ankistrodesmus* (Gaffron, 1940), *Chlorella* (Kessler, 1967), *Chlamydomonas* (Healey, 1970a, b), *Ulva* (Frenkel and Rieger, 1951; Ward, 1970a, b), in a few representatives of the Cyanophyta (Frenkel et al., 1950; Hattori, 1963; Ward, 1970a, b), in the Euglenophyta (Krasna and Rittenberg, 1954; Hartman and Krasna, 1963), and in the Phaeophyta and Rhodophyta (Frenkel and Rieger, 1951; Ben-Amotz et al., 1975). Hydrogenase has been detected by measuring one or several reactions of (2)–(6). Thus, although hydrogenase has been reported in a variety of algae, whether or not all the reactions listed in Section A are coupled with the action of this enzyme has not been examined in all cases. Only the unconfirmed work of Biochenko (1946) reported a facultative hydrogenase in a higher plant. It has been argued (Bishop, 1966) that because of the aerobic autotrophic nature of higher plants and their conservative metabolism, the occurrence of hydrogenase except in the algae would be of little advantage.

II. Adaptation and Deadaptation

H_2 metabolism is not found in the algae until the organism is put under an anaerobic atmosphere. Adaptation is usually a slow but highly variable process.

The reaction time varies with the reaction and the organism and it ranges from minutes in *Chlamydomonas* (FRENKEL, 1949) to hours in the marine red organisms (BEN-AMOTZ et al., 1975) and about two days in *Chlorella* (DAMASCHKE, 1957). This fact indicates that adaptation in one cell may be merely the modification of a constitutive enzyme, while in another cell it may be the result of de novo synthesis. Evidence favors the former, since chloramphenicol and actinomycin D were reported to prevent adaptation in *Scenedesmus* and *Chlorella* (STILLER and LEE, 1964). The nutritional status of the cell may affect the adaptation period. Thus, heterotrophically grown cells require a shorter anaerobic adaptation time (STILLER, 1966). Apparently, something more than an anaerobic period is required in the slowly adapting cells, which may be a reducing environment within the cell. Under anaerobic-reducing conditions, a slow rate of fermentation may accumulate reduced carbon compounds to a pool source for H_2 production (GIBBS, 1962).

The capacity to include molecular hydrogen in the metabolism of these algae is lost in the presence of O_2 exceeding about 1% of the atmospheric concentration (STUART and GAFFRON, 1972a). Deadaptation might occur in the dark in the presence of O_2 or on return to photosynthesis, resulting in the formation of O_2. The process of deadaptation is less understood. Thus, while adaptation time is variable and lengthy, deadaptation is rapid and irreversible, i.e., H_2 metabolism is not resumed upon return to low partial pressure of O_2. However, the adapted state can be restored much more rapidly immediately after deadaptation than after a prolonged period of anaerobiosis (GAFFRON, 1940; HORWITZ, 1957; GAFFRON, 1960). Inhibition studies on bacterial and algal hydrogenase suggest that oxidation and reduction of ferrous iron is involved in the enzyme activity (HOBERMAN and RITTENBERG, 1943; FISCHER et al., 1954; HARTMAN and KRASNA, 1963). Sensitive measurements of hydrogen-deuterium exchange and hydrogen-tritium exchange for hydrogenase (GOLDSBY, 1961; HARTMAN and KRASNA, 1963) confirmed earlier measurements that hydrogenase activity arises autocatalytically (GAFFRON, 1940). If the reduction of an enzyme is the basis of adaptation, its reoxidation might be the basis of deadaptation. Oxidation may be attributed either to free O_2 or to cellular oxidants. The presence of 3'(3,4-dichlorophenyl)1',1' dimethyl urea (DCMU) (STUART and GAFFRON, 1972a), glucose (STILLER and LEE, 1964), or the absence of photosystem II by mutation (BISHOP, 1962a, 1964) are known to prevent deadaptation by lowering the level of photochemically produced O_2. A tolerance for O_2 may arise after adaptation to 1% or 2%. This is due to the oxy-hydrogen reaction [reaction (6a, b)] which can reduce the O_2 partial pressure to a limited extent provided CO_2 is available (GAFFRON, 1940, 1942; HORWITZ, 1957).

III. Cell-Free Preparations of Hydrogenase

In contrast to bacterial preparations (MORTENSON and CHEN, 1974), a highly purified preparation of algal hydrogenase has not been reported. Crude preparations have been isolated from *Anabaena* (FUJITA et al., 1964; FUJITA and MYERS, 1965), *Chlamydomonas eugametos* (ABELES, 1964), *Chlamydomonas reinhardii* (BEN-AMOTZ and GIBBS, 1975; BEN-AMOTZ et al., 1975), *Chlamydomonas moewusii* (WARD, 1970a, b), *Chlorella pyrenoidosa* (LEE and STILLER, 1967), *Scenedesmus* (WARD,

1970a) and *Chondrus crispus* (BEN-AMOTZ et al., 1975). These preparations usually catalyzed the reduction of a variety of acceptors in the presence of a H_2 atmosphere. The fact that the enzyme is a "true" hydrogenase was demonstrated only with preparations of *Anabaena*, *Chlamydomonas* and *Scenedesmus* by H_2 liberation from reduced substrates of low redox potential such as reduced viologen dyes, reduced ferredoxin, etc. KRASNA and RITTENBERG (1954), HARTMAN and KRASNA (1963) and ABELES (1964) employed the more sensitive hydrogen–deuterium exchange or hydrogen–tritium exchange assay for hydrogenase.

Algal hydrogenase has not been fully characterized, and the information available at present is limited. *Anabaena* hydrogenase, like that of bacteria, seems to be closely linked to ferredoxin (FUJITA and MYERS, 1965), while hydrogenase of *Scenedesmus* requires active SH-groups (HARTMAN and KRASNA, 1964). Iron has been implicated in green algae as a metal constituent of hydrogenase (SASAKI, 1966; KESSLER, 1968). In general, the catalytic activities of hydrogenase from all sources are irreversibly sensitive to O_2 (FISCHER et al., 1954; ABELES, 1964; LEE and STILLER, 1967; BEN-AMOTZ et al., 1975). WARD (1970b), however, showed stability of hydrogenase against O_2 from *Chlamydomonas moewusii*.

C. Evolution of H_2

I. Dark Evolution of H_2

Some algae in the presence of inert gas (nitrogen, helium, or argon) liberate molecular hydrogen in the dark [reaction (4), GAFFRON and RUBIN, 1942; FRENKEL, 1952; KALTWASSER and GAFFRON, 1964]. The release of H_2 is accompanied by the release of CO_2 with a $CO_2:H_2$ ratio of 2.2. Glucose, pyruvate, lactate, and acetate stimulate the rate of the reaction (DAMASCHKE, 1957; DAMASCHKE and LÜBKE, 1958a; SYRETT and WONG, 1963; HEALEY, 1970b) and the uncoupler dinitrophenol inhibits the reaction (GAFFRON and RUBIN, 1942). It has been suggested that oxidative carbon metabolism is the source of reductant for H_2 evolution (GAFFRON and RUBIN, 1942), and that the dark reaction involves energy-dependent reverse electron flow to the redox level equivalent to that of ferredoxin (HEALEY, 1970b).

II. H_2 Photoevolution

In the light, H_2 production by anaerobically adapted algae is greatly stimulated [reaction (3)]. Both light and dark production of H_2 are accelerated by the presence of glucose in the medium. However, uncouplers of phosphorylation that inhibit the dark evolution stimulate the light-dependent evolution of H_2 (GAFFRON and RUBIN, 1942; KALTWASSER et al., 1969; HEALEY, 1970b; WANG, et al., 1971). The light-dependent reaction is therefore independent of demand of phosphorylation and is mediated by an ATP-independent hydrogenase. GREENBAUM (1977) showed that the photosynthetic unit size of H_2 evolution is comparable to the size based on O_2 evolution; however, the principal electron source and the portion of the

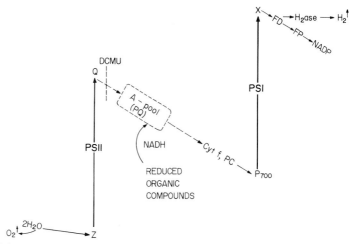

Fig. 1. Scheme to explain H_2 photoproduction by algae and by chloroplast–ferredoxin–hydrogenase systems. H_2ase, hydrogenase; FD, ferredoxin; FP, Ferredoxin–NADP–reductase; PC, plastocyanin; PQ, plastoquinone; X, Z, Q, unknowns

photosynthetic electron transport chain involved in the photo-production of H_2 have been the subject of considerable controversy (Fig. 1). The observed simultaneous release of H_2 and CO_2 (FRENKEL, 1952), the stimulation of H_2 release by glucose (KALTWASSER and GAFFRON, 1964), and the insensitivity toward inhibitors of photosystem II (DAMASCHKE and LÜBKE, 1958c; KALTWASSER et al., 1969; HEALEY, 1970b) suggest that organic compounds rather than water are the source of reducing equivalents only through photosystem I. The prime example here is *Chlamydomonas moewusii* (HEALEY, 1970b). On the other hand, the simultaneous photoproduction of H_2 and O_2 suggested that water may serve as the electron source for H_2 formation (SPRUIT, 1954, 1958; SPRUIT and KOK, 1956). This system is characterized by sensitivity to inhibitors of the Hill reaction but not by respiratory inhibitors such as iodoacetate and monofluoroacetate. The photoevolution of both H_2 and O_2 are inhibited to the same extent by mutations and manganese deficiency, and the reaction exhibits the Emerson enhancement effect. This pathway is characteristic of *Scenedesmus obliquus* (BISHOP, 1962a, b; BISHOP and GAFFRON, 1963; BISHOP et al., 1977). Subsequently, STUART and KALTWASSER (1970), STUART and GAFFRON (1971, 1972a, b, c) and BEN-AMOTZ and GIBBS (1975), through a comprehensive examination of the mechanism of H_2 photoproduction, concluded that the principal source of the H_2 produced by anaerobically adapted algae of both algal systems is NADH or an unknown organic substrate pool which is in equilibrium with the products of glycolysis and fermentation. Whether or not H_2O may be involved in H_2 photoproduction is not clear.

Several attempts have been made to couple the process of water photolysis to H_2 evolution by a chloroplast-ferredoxin-hydrogenase system. BENEMANN et al. (1973), RAO et al. (1976), and FRY et al. (1977) demonstrated in such a constitutive system the evolution of H_2 using water as electron donor. The electrons derived from the photolysis of water are transported to photosystem I where they react with the protons of the medium in the presence of ferredoxin and hydrogenase to evolve H_2. BENEMANN et al. (1973) illustrated that the O_2 produced by the water-splitting reaction had a strong inhibitory effect on the hydrogenase reaction. Low H_2 production rates and absence of net O_2 evolution indicated that even the H_2 evolved is only due to the presence of some O_2 absorbing reactions in the system. The problem of net O_2 production by the chloroplast-ferredoxin-hydrogenase system is central to this biophotolysis reaction (RAO et al., 1976). The inhibition of ferricyanide reduction in chloroplast particles by anaerobicity may explain the cessation of biophotolysis of water in the reconstituted system of chloroplast-ferredoxin-hydrogenase (KING et al., 1977). Other in vitro systems demonstrated the coupling of electron donors such as dithiothreitol, dithioerythritol, NADH, cysteine, and ascorbate with dichlorophenolindophenol to H_2 production by photosystem I in the presence of DCMU (ARNON et al., 1961; MITSUI, 1974; BEN-AMOTZ and GIBBS, 1975; HOFFMAN et al., 1977). Finally, the simultaneous photoevolution of H_2 and O_2 by algae and the constitutive systems [reaction (3)] have been considered as a bio-solar energy conversion mechanism for the generation of H_2 from water (GIBBS et al., 1973; LIEN and SAN PIETRO, 1975).

III. H_2 Evolution by Blue-Green Algae

Blue-green algae are the only organisms which combine green plant photosynthesis and N_2 fixation. Nitrogenase is a very O_2 labile enzyme (FAY and COX, 1967). However, it is generally accepted that nitrogenase activity is localized in heterocysts that lack O_2 evolving photosystem II (THOMAS, 1970; DONZE et al., 1972; TEL-OR and STEWART, 1975), have high respiration rates (FAY and WALSBY, 1966) and are relatively impermeable to O_2. In a proposed model of heterocyst function (WEARE and BENEMANN, 1973), reductant flows into the heterocysts from the vegetative cells, energy is supplied by photophosphorylation and oxidative phosphorylation, and fixed N_2 flows out of the heterocysts to the vegetative cells. N_2 starved cultures of *Anabaena cylindrica* photoevolved H_2 continuously for about two weeks (WEISSMAN and BENEMANN, 1976). The release of H_2 was accompanied by the release of O_2 with a $H_2:O_2$ ratio of 4:1 under conditions of complete N_2 starvation, and 1.7:1 after the addition of ammonium. Maximum level of H_2 photoevolution was 32 µl of H_2 per mg dry weight per hour with a photosynthetic efficiency of converting incident light into H_2 of 0.4%. It has been suggested that H_2 production by filamentous heterocystous blue-green algae could be used for development of a biophotolysis system.

D. Consumption of H_2

I. Dark Absorption of H_2

After anaerobic adaptation, a few algae can utilize H_2 in the dark, presumably to reduce cellular substances [reaction (5); GAFFRON, 1940; GAFFRON and RUBIN, 1942; BEN-AMOTZ et al., 1975].

In the presence of small amounts of O_2 (about 0.1% to 0.2%), a simultaneous absorption of H_2 and O_2 occurs in the so-called oxy-hydrogen reaction coupled to CO_2 fixation [reaction (6a, b); GAFFRON, 1940, 1942]. Low concentrations of dinitrophenol inhibit primarily the coupled reduction of CO_2 and do not impair the oxy-hydrogen reaction. A ratio of 2 O_2 to 1 CO_2 was the best measured by GAFFRON (1942). The observed ratio of 1 H_2 to 1 O_2 in the absence of CO_2 reaches the theoretical value of 2 in the presence of CO_2. HORWITZ (1957) and GIBBS et al. (1970) suggested that the process of oxy-hydrogen reaction has properties of both mitochondrion and chloroplast to account for O_2 consumption and CO_2 assimilation, respectively.

In addition to O_2, inorganic nitrogen compounds and inorganic sulfur compounds such as nitrite, nitrate (KESSLER, 1957b; DAMASCHKE and LÜBKE, 1958b; STILLER, 1966), and sulfite (KESSLER and MAIFARTH, 1960) can be reduced under a H_2 atmosphere with hydrogenase. Unlike the oxy-hydrogen reaction, the reduction of these inorganic compounds is not coupled to CO_2 assimilation.

Crude preparations of hydrogenase and a pure hydrogenase catalyze the reduction with H_2 of different redox dyes and other substances (KESSLER, 1974). Of special interest is the reduction of NAD(P) (ABELES, 1964; FUJITA and MYERS, 1965), ferredoxin (FUJITA and MYERS, 1965), and flavins (LEE and STILLER, 1967).

II. Photoreduction

Algae containing hydrogenase following a period of anaerobic incubation can utilize molecular hydrogen as a reductant in photosynthesis. This reaction of photoreduction [reaction (2)] involves the fixation of CO_2 with H_2 as electron donor (GAFFRON, 1940). CO_2 is apparently reduced to carbohydrate by the Calvin-Benson cycle (BADIN and CALVIN, 1950; GINGRAS et al., 1963; RUSSEL and GIBBS, 1968a). In general, the products are similar to those of photosynthesis and the intramolecular distribution of ^{14}C within the polysaccharide produced in photoreduction or in oxy-hydrogen reaction is identical to photosynthesis. Therefore, the change-over to a H_2 metabolism is limited to the formation of the reductant of CO_2.

Photosynthetic inhibitors that act by preventing O_2 evolution at or close to photosystem II do not affect photoreduction (GAFFRON, 1944; BISHOP, 1958, 1962b). In the presence of DCMU, photoreduction is stabilized against the reversion to photosynthetic O_2 evolution which is usually observed at higher light intensities. From these studies, as well as from manganese deficiency studies (KESSLER, 1957a), it was concluded that photoreduction proceeds independently of the mechanism of water photolysis. Photoreduction is characterized by insensitivity to chloride deficiency and mutation of photosystem II, sensitivity to dibromothymoquinone, an analog of plastoquinone, and by the lack of the Emerson enhancement phenomenon (BISHOP and GAFFRON, 1962, 1963; BISHOP, 1964, 1967; GINGRAS, 1966; PRATT and BISHOP, 1968; GRIMME and KESSLER, 1970; GRIMME, 1972).

GIBBS et al. (1970) suggested two schemes to explain H_2 photoreduction. In the first, H_2 could reduce pyridine nucleotide via ferredoxin and ATP would be synthesized by reoxidation of the ferredoxin in an anaerobic cyclic process. Alternatively, H_2 could reduce plastoquinone or a preceding carrier between the two photosystems to form subsequently ATP and reduced pyridine nucleotide in a noncyclic process.

Both photoreduction and the oxy-hydrogen reaction have been shown to be CO_2 assimilation processes involving the participation of the photosynthetic carbon reduction cycle leading to the net assimilation of the cell. However, the adapted algae have so far not been seen to divide or grow while engaged in photoreduction or the oxy-hydrogen reaction. It seems that the eukaryotic algae are purely aerobic organisms and cannot carry out dismutation reactions involving pyridine nucleotides. In other words, algal fermentation is insufficient for the maintenance of growth (GIBBS, 1962; GIBBS et al., 1970).

References

Abeles, F.B.: Plant Physiol. *39*, 169–176 (1964)
Arnon, D.I., Losada, M., Nozaki, M., Tagawa, K.: Nature (London) *190*, 601–606 (1961)
Badin, E.J., Calvin, M.: J. Am. Chem. Soc. *72*, 5266–5270 (1950)
Ben-Amotz, A., Gibbs, M.: Biochem. Biophys. Res. Commun. *64*, 355–359 (1975)
Ben-Amotz, A., Erbes, D.L., Riederer-Henderson, M.A., Gibbs, M.: Plant Physiol. *56*, 72–77 (1975)
Benemann, J.R., Berenson, J.A., Kaplan, N.O., Kamen, M.D.: Proc. Natl. Acad. Sci. *70*, 2317–2320 (1973)
Biochenko, E.A.: C. R. Acad. Sci. USSR *52*, 521–524 (1946)
Bishop, N.I.: Biochim. Biophys. Acta *27*, 205–206 (1958)
Bishop, N.I.: Nature (London) *195*, 55–57 (1962a)
Bishop, N.I.: Biochim. Biophys. Acta *57*, 186–189 (1962b)
Bishop, N.I.: Rec. Chem. Prog. *25*, 181–195 (1964)
Bishop, N.I.: Annu. Rev. Plant Physiol. *17*, 185–208 (1966)
Bishop, N.I.: Photochem. Photobiol. *6*, 621–628 (1967)
Bishop, N.I., Gaffron, H.: Biochem. Biophys. Res. Commun. *8*, 471–476 (1962)
Bishop, N.I., Gaffron, H.: In: Photosynthetic mechanisms in green plants. Kok, B., Jagendorf, A.T. (eds.), pp. 441–451. Washington, D.C.: NAS-NRC Publication 1145, 1963
Bishop, N.I., Frick, M., Senger, H.: Plant Physiol. (Suppl.) *59*, 130 (1977)
Cohen, Y., Padan, E., Shilo, M.: J. Bacteriol. *123*, 855–861 (1975)
Damaschke, K.: Z. Naturforsch. *12b*, 441–443 (1957)
Damaschke, K., Lübke, M.: Z. Naturforsch. *13b*, 54–55 (1958a)
Damaschke, K., Lübke, M.: Z. Naturforsch. *13b*, 134–135 (1958b)
Damaschke, K., Lübke, M.: Z. Naturforsch. *13b*, 172–182 (1958c)
Donze, M., Haveman, J., Scherick, P.: Biochim. Biophys. Acta *256*, 157–161 (1972)
Fay, P., Cox, R.M.: Biochim. Biophys. Acta *143*, 562–569 (1967)
Fay, P., Walsby, A.E.: Nature (London) *209*, 94–95 (1966)
Fischer, H.F., Krasna, A., Rittenberg, D.: J. Biol. Chem. *209*, 569–578 (1954)
Frenkel, A.W.: Biol. Bull. *97*, 261–262 (1949)
Frenkel, A.W.: Arch. Biochem. Biophys. *38*, 219–230 (1952)
Frenkel, A.W., Rieger, C.: Nature (London) *167*, 1030 (1951)
Frenkel, A.W., Gaffron, H., Battley, E.H.: Biol. Bull. *99*, 157–162 (1950)
Fry, I., Papageorgiou, G., Tel-Or, E., Packer, L.: Z. Naturforsch. *32c*, 110–117 (1977)
Fujita, Y., Myers, J.: Arch. Biochem. Biophys. *111*, 619–625 (1965)
Fujita, Y., Ohama, H., Hattori, A.: Plant Cell Physiol. *5*, 305–314 (1964)
Gaffron, H.: Am. J. Bot. *27*, 273–283 (1940)
Gaffron, H.: J. Gen. Physiol. *26*, 195–217 (1942)
Gaffron, H.: Biol. Rev. *19*, 1–20 (1944)

Gaffron, H.: In: Plant physiology. Stewart, F.C. (ed.), Vol. IB. pp. 3–277. New York, London: Academic Press 1960
Gaffron, H., Rubin, J.: J. Gen. Physiol. *26*, 219–240 (1942)
Gibbs, M.: In: Physiology and biochemistry of algae. Lewin, R.A. (ed.), pp. 91–97. New York, London: Academic Press 1962
Gibbs, M., Hollaender, A., Kok, B., Krampitz, L.O., San Pietro, A.: Proc. Workshop Bio-Solar Conversion. pp. 1–90. NSF/RANN Report 1973
Gibbs, M., Latzko, E., Harvey, M.J., Plaut, Z., Shain, Y.: Ann. N.Y. Acad. Sci. *175*, 541–554 (1970)
Gingras, G.: Physiol. Vég. *4*, 1–65 (1966)
Gingras, G., Goldsby, R.A., Calvin, M.: Arch. Biochem. Biophys. *100*, 178–184 (1963)
Goldsby, R.A.: Lawrence Radiation Lab., Berkeley, Calif., UC-4 Chemistry, 1–71 (1961)
Greenbaum, E.: Science *196*, 879–880 (1977)
Grimme, L.H.: In: Proc. 2nd Int. Congr. Photosynthesis Res. Forti, G., Avron, M., Melandri, A. (eds.), pp. 2011–2019. The Hague: Junk 1972
Grimme, L.H., Kessler, E.: Naturwissenschaften *57*, 133 (1970)
Hartman, H., Krasna, A.I.: J. Biol. Chem. *238*, 749–757 (1963)
Hartman, H., Krasna, A.I.: Biochim. Biophys. Acta *92*, 52–58 (1964)
Hattori, A.: In: Studies on microalgae and photosynthetic bacteria. Japan. Soc. Plant Physiol. (eds.), pp. 485–492. Tokyo: Univ. Tokyo Press 1963
Healey, F.P.: Planta (Berl.) *91*, 220–226 (1970a)
Healey, F.P.: Plant Physiol. *45*, 153–159 (1970b)
Hoberman, H.D., Rittenberg, D.: J. Biol. Chem. *147*, 211 (1943)
Hoffman, D., Thauer, R., Trebst, A.: Z. Naturforsch. *32c*, 257–262 (1977)
Horwitz, L.: Arch. Biochem. Biophys. *66*, 23–44 (1957)
Kaltwasser, H., Gaffron, H.: Plant Physiol. (Suppl) *39*, xiii (1964)
Kaltwasser, H., Stuart, T.S., Gaffron, H.: Planta (Berl.) *89*, 309–322 (1969)
Kessler, E.: Planta (Berl.) *49*, 435–454 (1957a)
Kessler, E.: Arch. Mikrobiol. *27*, 166–181 (1957b)
Kessler, E.: Arch. Mikrobiol. *55*, 346–357 (1967)
Kessler, E.: Arch. Mikrobiol. *61*, 77–80 (1968)
Kessler, E.: In: Algal physiology and biochemistry. Stewart, W.D.P. (ed.), pp. 456–473. Oxford: Blackwell 1974
Kessler, E., Maifarth, H.: Arch. Mikrobiol. *37*, 215–225 (1960)
King, D., Erbes, D.L., Gibbs, M.: Biochem. Biophys. Res. Commun. *78*, 734–738 (1977)
Krasna, A.I., Rittenberg, D.: J. Am. Chem. Soc. *76*, 3015–3020 (1954)
Lee, J.K.H., Stiller, M.: Biochim. Biophys. Acta *132*, 503–505 (1967)
Lien, S., San Pietro, A.: An inquiry into biophotolysis of water to produce hydrogen. pp. 49. NSF/RANN Report 1975
Mitsui, A.: In: Hydrogen energy, Part A. Veziroglu, T.N. (ed.), pp. 309–340. New York: Plenum Press 1974
Mortenson, L.E., Chen, J.S.: In: Microbial iron metabolism. Neilands, J.B. (ed.), pp. 231–269. New York: Academic Press 1974
Nakamura, H.: Acta Phytochim. *10*, 271–281 (1938)
Pratt, L.H., Bishop, N.I.: Biochim. Biophys. Acta *153*, 664–675 (1968)
Rao, K.K., Rosa, L., Hall, D.O.: Biochem. Biophys. Res. Commun. *68*, 21–28 (1976)
Russel, G.K., Gibbs, M.: Plant Physiol. *43*, 649–652 (1968)
Sasaki, H.: Plant Cell Physiol. *7*, 231–241 (1966)
Spruit, C.J.P.: In: Proc. 1st Int. Photobiol. Congr. pp. 323–327. Amsterdam 1954
Spruit, C.J.P.: Landbouwhogesch. Wageningen *58*, 1–17 (1958)
Spruit, C.J.P.: In: Physiology and biochemistry of algae. Lewin, R.A. (ed.), pp. 47–60. New York, London: Academic Press 1962
Spruit, C.J.P., Kok, B.: Biochim. Biophys. Acta *19*, 417–424 (1956)
Stewart, W.D.P., Pearson, H.W.: Proc. R. Soc. B *175*, 293–311 (1970)
Stiller, M.: Plant Physiol. *41*, 348–356 (1966)
Stiller, M., Lee, J.K.H.: Biochim. Biophys. Acta *93*, 174–176 (1964)
Stuart, T.S., Gaffron, H.: Planta (Berl.) *100*, 228–243 (1971)
Stuart, T.S., Gaffron, H.: Plant Physiol. *50*, 136–140 (1972a)

Stuart, T.S., Gaffron, H.: Planta (Berl.) *106*, 91–100 (1972b)
Stuart, T.S., Gaffron, H.: Planta (Berl.) *106*, 101–112 (1972c)
Stuart, T.S., Kaltwasser, H.: Planta (Berl.) *91*, 302–313 (1970)
Syrett, P.J., Wong, H.A.: Biochem. J. *89*, 308–315 (1963)
Tel-Or, E., Stewart, W.D.P.: Nature (London) *258*, 715–716 (1975)
Thomas, J.: Nature (London) *228*, 181–183 (1970)
Wang, R., Healey, F.P., Myers, J.: Plant Physiol. *48*, 108–110 (1971)
Ward, M.A.: Phytochemistry *9*, 259–266 (1970a)
Ward, M.A.: Phytochemistry *9*, 267–274 (1970b)
Weare, N.M., Benemann, J.R.: Arch. Mikrobiol. *93*, 101–112 (1973)
Weissman, J.C., Benemann, J.R.: Appl. Environ. Microbiol. *33*, 123–131 (1976)

Author Index

Page numbers in *italics* refer to the bibliography

Subject Index

Page numbers in *italics* indicate text parts where the corresponding item is given major attention. Page numbers in **bold face** refer to figures or tables where the corresponding item is a principal component.

Designation of a page number implies that the item may appear on the following one or two pages.

A23187 204
abscisic acid 197
Acetabularia 482
acetate 56, 156, 159, 183–186
acetate kinase 184
acetyl CoA 155, 160, 259, 268, 416, 420
acetyl CoA synthetase 184, 420
acetylene reduction 457, 467, 473
acetyl-P 416
acetyl-serine 489
adaptation (see also ecological ...) 82, 126–137, 498
adenylate kinase 108, 265
ADP 41, 46
   as enzyme effector 49, 52, 157, 241
ADPglucose 283, 286, 290, 293
ADPglucose:1,4-α-D-glucan 4-α glucosyltransferase (see starch synthase)
ADPglucose pyrophosphorylase 16, 284,
   285–293, **286–289, 292**
Aeluropus litoralis 196
agronomy 147
alanine 68, **71**, 72, 88, 154, 174
alanine aminotransferase 103, 260
alanine dehydrogenase 448
albino leaves 153
aldolase 14, 60, 133, 245, 264, 314
alfalfa 247, 355
algae (see also blue-green ...) 18, 151, 181, 230,
   232, 245
   light-enhanced dark CO_2 fixation 68–72
   photorespiration 348–351, 358, 373
   respiration *163–171*
allosteric regulation 47, 158, 218, 239, 264, 268,
   285, 289, 291
Amaranthus 89, 91, 153, 258, 379, 389
amino acids 33, 158, 174, 183
   as photosynthetic product 2, 21, 118, 154,
   421
ammonia 23, 154, 197, 343, 435, *445–455*, **451,
   452**, 467, **475**–476
AMP 108
   as enzyme effector 63, 157, 246–248, 486
amylase (α- and β-) 117, 263, 299–300, 305
amylopectin 283, 294, 299
amyloplasts 85
amylose 294, 296, 299
amylosucrase 283

amytal 166
Anabaena 164, 166, 211, 243, 446, 458–466, **460,
   485**, 499, 502
Anacystis nidulans 60, 69, **73**, 95, 164, 166, 169,
   242, 245, 273, 430
Ankistrodesmus 186, 349, 498
antimycin A 48, 109, 164, 187
Aphanocapsa 60, 62
APS 483–486
APS kinase 484
APS sulfotransferase 486, 489, 492
aquatic angiosperms 198, 232, 349, 358, 372
arsenate 47, 221
arsenite 234, 272
ascorbate 43, 106
asparagine 445, 450, 476
aspartate 88, 107, 174, 335, 364
   as product of CO_2 fixation **71**–73, 95, 104,
   106, 133, 176
aspartate aminotransferase 103, 133, 257, 343
aspartate dehydrogenase 448
Astridia 81
Athiorhodaceae 61, **64**, 163
ATP 42, 46, 82, 108, 150, 286, 344, 361, 445,
   453, 462, 466, 474
   as enzyme effector 63, 157, 244–248, 264,
   277
   level in vivo 70, **71**, 166, 462
   translocation across chloroplast envelope 41,
   169
ATPase 202, 406
ATP/ADP ratio 46, 52, 155, 165, 459
ATP regenerating system 47, 460
ATP sulfurylase 483, 492
Atriplex 92, 142, 209, 387, **390**
Atriplex spongiosa 81, 104, 159, 198, 257
autocatalysis of Calvin cycle 50
autotrophy 54
Avena 224, 300
avocado 292
azaserine 446, 450
azide 234, 430
Azotobacter 400, 476

bacteria 400, 405
   photosynthetic 5, 54–58, 163, 181, 212, 246,
   248, 271, 362, 406, *416–423*
bacteroids 231, 473, 477

Planta

An International Journal of Plant Biology

ISSN 0032-0935   Title No. 425

Editorial Board: E. Bünning, Tübingen;
R. Cleland, Seattle; H. Grisebach, Freiburg;
R. Hertel, Freiburg; J. Heslop-Harrison,
Aberystwyth; A. Lang, East Lansing
(Managing Editor); H. F. Linskens,
Nijmegen; H. Mohr, Freiburg (Managing
Editor); A. Sievers, Bonn; P. Sitte, Freiburg;
A. Trebst, Bochum; Y. Vaadia, Bet Dagan;
M. B. Wilkins, Glasgow (Managing Editor);
H. Ziegler, München

Planta publishes original articles in structural and functional botany, covering all
aspects from biochemistry and ultrastructure
to studies with tissues, organs and whole
plants, but excluding evolutionary and popu-
lation botany (taxonomy, floristics, ecology,
etc.). Papers in cytology and genetics, and
papers from applied fields such as phyto-
pathology are accepted only if contributing
to the understanding of specifically botanical
problems.

Subscription information and sample copy
available upon request.

Springer-Verlag
Berlin
Heidelberg
New York

Carlsberg Research
Communications

Continuation of
**Comptes Rendus Des Travaux
Du Laboratoire Carlsberg**

ISSN 0105-1938   Title No. 274

Editorial Board: H. Holter, H. Klenow,
Ebba Lund; M. Ottesen, D. von Wettstein,
Carlsberg Laboratory; E. Zeuthen, Biologi-
cal Institute of the Carlsberg Foundation

Carlsberg Research Communications is the
continuation of **Comptes Rendus des
Travaux du Laboratoire Carlsberg** which has
been published since 1878. The previous
volumes of the journal contain the results
of the work carried out at the scientific
institutions of the Carlsberg Foundation.
Important work such as Kjeldahl's method
for nitrogen determination, S. P. L. Søren-
sen's concept of pH, Schmidt's discovery of
eel migration, Winge's discovery of sexuality
in yeast, and Linderstrøm-Lang's concepts
of protein structure and enzyme function
were made known through this journal.

The **Carlsberg Research Communications**
also report on the scientific work of the
Carlsberg research establishments. These
include the departments of Chemistry and
Physiology of the Carlsberg Foundation and
the departments of Brewing Chemistry, Bio-
technology and Technology of the Carlsberg
Research Laboratory. These laboratories are
involved in fundamental and applied research in the fields of biochemistry, cell and
molecular biology, genetics, microbiology,
and plant and animal physiology.

Language used: Published in English.
Subscription information and sample copy
available upon request.

**Published by Carlsberg Laboratory,
Copenhagen.
Distributed by Springer-Verlag Berlin
Heidelberg New York**

Encyclopedia of Plant Physiology

New Series
Editors: A. Pirson,
M. H. Zimmermann

Springer-Verlag
Berlin
Heidelberg
New York

Volume 1

Transport in Plants I. Phloem Transport

Editors: M. H. Zimmermann, J. A. Milburn
1975. 93 figures. XIX, 535 pages
ISBN 3-540-07314-0

Volume 2A

Transport in Plants IIA Cells

Editors: U. Lüttge, M. G. Pitman
With a Foreword by R. N. Robertson
Contributors: W. J. Cram, J. Dainty, G. P. Findlay,
T. K. Hodges, A. B. Hope, D. H. Jennings, U. Lüttge, C. B. Osmond,
M. G. Pitmann, R. J. Poole, J. A. Raven, F. A. Smith, N. A. Walker
1976. 97 figures, 64 tables, XVI, 419 pages.
ISBN 3-540-07452-X

Volume 2B

Transport in Plants II B Tissues and Organs

Editors: U. Lüttge, M. G. Pitman
With a Foreword by R. A. Robertson
Contributors: W. P. Anderson, E. Epstein, A. E. Hill, B. S. Hill,
T. C. Hsiao, W. D. Jeschke, A. Läuchli, U. Lüttge, J. S. Pate,
M. G. Pitman, E. Schnepf, R. M. Spanswick, R. F. M. Stenveninck,
J. F. van Sutcliffe
1976. 129 figures, 45 tables. XII, 475 pages.
ISBN 3-540-07453-8

Volume 3

Transport in Plants III

Intracellular Interactions and Transport Processes
Editors: C. R. Stocking, U. Heber
With contributions by numerous experts
1976. 123 figures. XXII, 517 pages.
ISBN 3-540-07818-5

Volume 4

Physiological Plant Physiology

Editors: R. Heitefuß, P. H. Williams
1976. 92 figures. XX, 890 pages.
ISBN 3-540-07557-7

Volume 5

Photosynthesis I

Photosynthetic Electron Transport and Photophosphorylation
Editors: A. Trebst, M. Avron
1977. 128 figures. XXIV, 730 pages.
ISBN 3-540-07962-9

Volume 7

Physiology of Movements

Editors: W. Haupt, M. E. Feinleib
1979. Approx. 720 pages.
ISBN 3-540-08776-1